Aircraft Engineering Prin

Roll-Royce RB211-524 Engine fitted to a Boeing 747 Aircraft

Aircraft Engineering Principles

Lloyd Dingle
Mike Tooley

Routledge
Taylor & Francis Group

LONDON AND NEW YORK

First published by Butterworth-Heinemann

This edition published 2011 by Routledge
2 Park Square, Milton Park, Abingdon, Oxon OX14 4RN
711 third Avenue, New York, NY 10017, USA

First edition 2005

Routledge is an imprint of Taylor & Francis Group, an informa business.

British Library Cataloguing in Publication Data
Dingle, Lloyd
    Aircraft engineering principles
    1. Aerospace engineering
    I. Title II. Tooley, Michael H. (Michael Howard), 1946–
    692.1

Library of Congress Cataloging-in-Publication Data
A catalog record for this book is available from the Library of Congress

ISBN: 978-0-7506-5015-1

Contents

Preface

The books in the series have been designed for both independent and tutor assisted studies. For this reason they should prove particularly useful to the "self-starter" and to those wishing to update or upgrade their aircraft maintenance licence. Also, the series should prove a useful source of reference for those taking *ab initio* training programmes in JAR 147 (now EASA, IR, Part-147) and FAR 147 approved organizations and those on related aeronautical engineering programmes in further and higher education establishments.

This book has primarily been written as one in a series of texts, designed to cover the **essential knowledge base** required by aircraft certifying mechanics, technicians and engineers engaged in engineering maintenance activities on commercial aircraft. In addition, this book should appeal to the members of the armed forces, and students attending training and educational establishments engaged in aircraft engineering maintenance and other related aircraft engineering learning programmes.

In this book we cover in detail the underpinning mathematics, physics, electrical and electronic fundamentals, and aerodynamics necessary to understand the function and operation of the complex technology used in modern aircraft. The book is arranged into four major sections:

- Introduction
- Scientific fundamentals
- Electrical and electronic fundamentals
- Fundamentals of aerodynamics

In the *Introductory* section you will find information on the nature of the aircraft maintenance industry, the types of job role that you can expect, the current methods used to train and educate you for such roles and information on the examinations system directly related to civil aviation maintenance engineering. In addition, you will find information on typical career progression routes, professional recognition, and the legislative framework and safety culture that is so much a part of our industry.

In the section on *Scientific fundamentals* we start by studying Module 1 of the JAR 66 (now EASA, IR, Part-66) syllabus (see qualifications and levels) covering the elementary mathematics necessary to practice at the category B technician level. It is felt by *the authors*, that this level of "non-calculator" mathematics is insufficient as a prerequisite to support the study of the physics and the related technology modules, that are to follow. For this reason, and to assist students who wish to pursue other related qualifications, a section has been included on "further mathematics". The coverage of Part 66 Module 2 on physics is sufficiently comprehensive and at a depth, necessary for both category B1 and B2 technicians.

The section on *Electrical and electronic fundamentals* comprehensively covers Part 66 Module 3 and Part-66 Module 4 to a knowledge level suitable for category B2 avionic technicians. Module 5 on *Digital Techniques and Electronic Instrument Systems* will be covered in the fifth book in the series, *Avionic Systems*.

This book concludes with a section on the study of *Aerodynamics*, which has been written to cover Part-66 Module 8.

In view of the international nature of the civil aviation industry, all aircraft engineering maintenance staff need to be fully conversant with the SI system of units *and* be able to demonstrate proficiency in manipulating the "English units" of measurement adopted by international aircraft manufacturers, such as the Boeing Aircraft Company. Where considered important, the English units of measure will be emphasized alongside the universally recognized SI system. The chapter on physics (Chapter 4) provides a thorough introduction to SI units, where you will also find mention of the English system, with conversion tables between each system being provided at the beginning of Chapter 4.

To reinforce the subject matter for each major topic, there are numerous *worked examples* and *test your knowledge* written questions designed to enhance learning. In addition, at the end of each chapter you will find a selection of *multiple-choice questions*, that are graded to simulate the depth and breadth of knowledge required by individuals wishing to practice at the mechanic (category A) or technician (category B) level. These multiple choice question papers should be attempted *after* you have completed your study of the appropriate chapter. In this way, you will obtain a clearer idea of how well you have grasped the subject matter at the module level. Note also that category B knowledge is required by those wishing to practice at the category C or engineer level. Individuals hoping to pursue this route should make sure that they thoroughly understand the relevant information on routes, pathways and examination levels given later.

Further information on matters, such as aerospace operators, aircraft and aircraft component manufacturers, useful web sites, regulatory authorities, training and educational establishments and comprehensive lists of terms, definitions and references, appear as appendices at the end of the book. *References are annotated using superscript numbers at the appropriate point in the text.*

Lloyd Dingle
Mike Tooley

Answers to questions

Answers to the "Test your understanding" questions are given in Appendix F. Solutions to the multiple choice questions and general questions can be accessed by adopting tutors and lecturers. To access this material visit http://books.elsevier.com/manuals and follow the instructions on screen.

Postscript

At the time of going to press JAR 66 and JAR 147 is being superceded by European Aviation Safety Agency (EASA), Implementing Rules (IRs) Parts 66 and 147 and whenever possible references to these and other related aircraft maintenance publications have been amended to reflect the new role and responsibilities of EASA. See Appendix C for more details.

Acknowledgements

The authors would like to express their gratitude to those who have helped in producing this book.

Jeremy Cox and Mike Smith of Britannia Airways, for access to their facilities and advice concerning the administration of civil aircraft maintenance; Peter Collier, chairman of the RAeS non-corporate accreditation committee, for his advice on career progression routes; The Aerospace Engineering lecturing team at Kingston University, in particular, Andrew Self, Steve Barnes, Ian Clark and Steve Wright, for proof reading the script; Jonathan Simpson and all members of the team at Elsevier, for their patience and perseverance. Finally, we would like to say a big 'thank you' to Wendy and Yvonne. Again, but for your support and understanding, this book would never have been produced!

Introduction

Chapter 1 Introduction

1.1 The aircraft engineering industry

The global aircraft industry encompasses a vast network of companies working either as large international conglomerates or as individual national and regional organizations. The two biggest international aircraft manufacturers are the American owned Boeing Aircraft Company and the European conglomerate, European Aeronautic Defence and Space Company (EADS), which incorporates airbus industries. These, together with the American giant Lockheed-Martin, BAE Systems and aerospace propulsion companies, such as Rolls-Royce and Pratt and Whitney, employ many thousands of people and have annual turnovers totalling billions of pounds. For example, the recently won Lockheed-Martin contract for the American Joint Strike Fighter (JSF) is estimated to be worth 200 billion dollars, over the next 10 years! A substantial part of this contract will involve BAE Systems, Rolls-Royce and other UK companies.

The airlines and armed forces of the world who buy-in aircraft and services from aerospace manufacturers are themselves, very often, large organizations. For example British Airways our own national carrier, even after recent downsizing, employs around 50,000 personnel. UK airlines, in the year 2000, employed in total, just over 12,000 aircraft maintenance and overhaul personnel. Even after the events that took place on 11th September 2001, the requirement for maintenance personnel is unlikely to fall. A recent survey by the Boeing Corporation expects to see the demand for aircraft and their associated components and systems rise by 2005, to the level of orders that existed prior to the tragic events of 11th September 2001.

Apart from the airlines, individuals with aircraft maintenance skills may be employed in general aviation (GA), third-party overhaul companies, component manufacturers or airframe and avionic repair organizations. GA companies and spin-off industries employ large numbers of skilled aircraft fitters. The UK armed forces collectively recruit around 1500 young people annually for training in aircraft and associated equipment maintenance activities.

Aircraft maintenance certifying staff are recognized throughout Europe and indeed, throughout many parts of the world, thus opportunities for employment are truly global! In the USA approximately 10,000 airframe and propulsion (A&P) mechanics are trained annually; these are the USA equivalent of our own aircraft maintenance certifying mechanics and technicians.

Recent surveys carried out for the UK suggest that due to demographic trends, increasing demand for air travel and the lack of trained aircraft engineers leaving our armed forces, there exists an *annual shortfall* of around 800 suitably trained aircraft maintenance and overhaul staff. Added to this, the global and diverse nature of the aircraft maintenance industry, it can be seen that aircraft maintenance engineering offers an interesting and rewarding career, full of opportunity.

1.2 Differing job roles for aircraft maintenance certifying staff

Individuals may enter the aircraft maintenance industry in a number of ways and perform a variety of maintenance activities on aircraft or on their associated equipments and components. The nature of the job roles and responsibilities for licensed certifying mechanics, technicians and engineers are detailed below.

The routes and pathways to achieve these job roles, the opportunities for career progression, the certification rights and the nature of the necessary examinations and qualifications are detailed in the sections that follow.

1.2.1 The aircraft maintenance certifying mechanic

Since the aircraft maintenance industry is highly regulated, the opportunities to perform complex maintenance activities are dependent on the amount of time that individuals spend on their initial and aircraft-type training, the knowledge they accrue and their length of experience in post. Since the knowledge and experience requirements are limited for the certifying mechanic (see later), the types of maintenance activity that they may perform, are also limited. Nevertheless, these maintenance activities require people with a sound basic education, who are able to demonstrate maturity and the ability to think logically and quickly when acting under time constraints and other operational limitations.

The activities of the certifying mechanic include the limited rectification of defects and the capability to perform and certify minor scheduled line maintenance inspections, such as daily checks. These rectification activities might include tasks, such as a wheel change, replacement of a worn brake unit, navigation light replacement or a seat belt change. Scheduled maintenance activities might include: replenishment of essential oils and lubricants, lubrication of components and mechanisms, panel and cowling removal and fit, replacement of panel fasteners, etc., in addition to the inspection of components, control runs, fluid systems and aircraft structures for security of attachment, corrosion, damage, leakage, chaffing, obstruction and general wear.

All these maintenance activities require a working knowledge of the systems and structures being rectified or inspected. For example, to replenish the hydraulic oil reservoirs on a modern transport aircraft requires knowledge of the particular system, the type of oil required (Figure 1.1), the replenishment equipment being used, all related safety considerations and knowledge of the correct positioning of the hydraulic services prior to the replenishment.

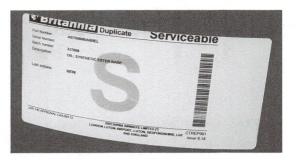

Figure 1.1 Identification label showing the type of oil contained within the drum.

Figure 1.2 Boeing 767 hydraulic reservoir charging point, showing contents gauge, changeover valve and hydraulic hand pump.

In addition, for this task, the mechanic must be able to recognize the symptoms for internal or external hydraulic oil leakage when carrying out these replenishment activities on a particular hydraulic system reservoir.

For example, Figure 1.2 shows the hydraulic reservoir replenishing point for the Boeing 767. The replenishment process requires the changeover valve to be selected and oil sucked into the reservoir, via the replenishment hose (Figure 1.3) which is placed in the oil container. The certifying mechanic then operates the hand pump (see Figure 1.2) to draw the hydraulic fluid up into the reservoir. When the reservoir is full, as indicated by the contents gauge, the hose is withdrawn from the container, blanked and stowed. The changeover valve is put back into the flight position, the panel is secured and the

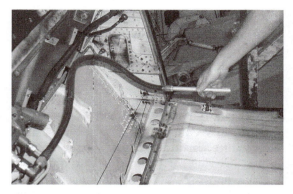

Figure 1.3 Hydraulic reservoir replenishment hose, removed from stowage point.

Figure 1.4 Boeing 767 flap drive motor and associated drive mechanism.

appropriate documentation is completed by the certifying mechanic, who will have a *company approval* to perform this task.

For this job role, like all those that follow, there is a statutory requirement for a particular period of training and experience before a maintenance mechanic is issued with limited certifying privileges.

Within the armed forces a similar job role exists for those who have undergone training as aircraft mechanics, for flight line operations or similar maintenance activities.

1.2.2 The aircraft maintenance category B certifying technician

The role of the category B certifying technician is subdivided into two major sectors: category B1 (mechanical) and category B2 (avionic). B1 maintenance technicians will have an in depth knowledge of airframe, engine and electrical power systems and equipment in addition to a thorough knowledge of aircraft structures and materials. While category B2 maintenance technicians will have an in-depth integrated knowledge of aircraft electrical, instrument, autopilot, radio, radar, communication and navigation systems.

The knowledge and skills gained from their initial training, together with aircraft-type knowledge and a substantial period of practical experience, will enable category B technicians, once granted approvals, to undertake one or

more of the following maintenance operations:

- In-depth scheduled inspection activities.
- Complex rectification activities.
- Fault diagnosis on aircraft systems, propulsion units, plant and equipments.
- Embodiment of modifications and special technical instructions.
- Airframe and other aircraft repairs.
- Strip-down and aircraft re-build activities.
- Major aircraft component removal, fit and replacement tasks.
- Use and interrogate built-in test equipment (BITE) and other diagnostic equipments.
- Functional tests and checks on aircraft systems, propulsion units and sub-systems.
- Trouble-shooting activities on base and away from base.
- Aircraft engine ground running activities.
- Rack and re-rack avionic equipments and carry out operational tests and checks on avionic systems.
- Supervise and certify the work of less experienced technicians and mechanics.

As can be seen from the above list of maintenance operations, the category B maintenance technician can be involved in a very wide and interesting range of possible activities. For example, Figure 1.4 shows a photograph of the Boeing 767 flap drive motor and associated linkage mechanism.

The main source of power is via the hydraulic motor, scheduled servicing may involve the

Figure 1.5 Technicians working at height considering the alignment of the APU prior to fit.

operation and inspection of this complex system, which in turn requires the certifying technician to not only have the appropriate system knowledge, but also the whole aircraft knowledge to ensure that other systems are not operated inadvertently. Figure 1.5 shows two technicians working at height on highway staging, considering the alignment of the aircraft auxiliary power unit (APU), prior to raising it into position in the aircraft.

To perform this kind of maintenance, to the required standards, individuals need to demonstrate maturity, commitment, integrity and an ability to see the job through, often under difficult circumstances.

Similar technician roles exist in the *armed forces*, where the sub-categories are broken down a little more into, mechanical, electrical/instrument and avionic technicians, as well as aircraft weapons specialists known as armament technicians or weaponeers.

In fact, it is planned from January 2004 that the Royal Air Force (RAF) will begin initial training that follows the civil aviation trade categories. That is *mechanical* technicians, who will undertake airframe/engine training and to a lesser extent electrical training and *avionic* technicians, who will eventually cover all avionic systems, in a similar manner to their civil counterparts. Cross-training of existing maintenance personnel is also planned to take place over the next 10 years. The armament technician and weaponeer will still remain as a specialist trade group.

1.2.3 The base maintenance category C certifying engineer

Before detailing the job role of the category C licensed engineer, it is worth clarifying the major differences in the roles performed by *line* maintenance certifying staff and *base* maintenance certifying staff. In the case of the former, the inspections, rectification and other associated maintenance activities take place on the aircraft, on the "live side" of an airfield. Thus the depth of maintenance performed by "line maintenance personnel" is restricted to that accomplishable with the limited tools, equipment and test apparatus available on site. It will include "first-line diagnostic maintenance", as required.

Base maintenance, as its name implies, takes place at a designated base away from the live aircraft movement areas. The nature of the work undertaken on base maintenance sites will be more in-depth than that usually associated with line maintenance and may include: in-depth strip-down and inspection, the embodiment of complex modifications, major rectification activities, off-aircraft component overhaul and repairs. These activities, by necessity, require the aircraft to be on the ground for longer periods of time and will require the maintenance technicians to be conversant with a variety of specialist inspection techniques, appropriate to the aircraft structure, system or components being worked-on.

The category C certifier acts primarily in a maintenance management role, controlling the progress of base maintenance inspections and overhauls. While the actual work detailed for the inspection is carried out by category B technicians and to a limited extent, category A base maintenance mechanics, in accordance with the written procedures and work sheets. These individual activities are directly supervised by category B maintenance certifying technicians, who are responsible for ensuring the adequacy of the work being carried out and the issuing of the appropriate certifications for the individual activities.

The category C certifier will upon completion of all base maintenance activities sign-off the aircraft as serviceable and fit for flight. This is done using a special form known as a certificate of

Figure 1.6 Category C maintenance engineer explaining the complexity of the technical log to the author.

release to service (CRS). Thus the category C certifying engineer has a very responsible job, which requires a sound all-round knowledge of aircraft and their associated systems and major components (Figure 1.6). The CRS is ultimately the sole responsibility of the category C certifying engineer, who confirms by his/her signature that all required inspections, rectification, modifications, component changes, airworthiness directives, special instructions, repairs and aircraft re-build activities have been carried out in accordance with the laid-down procedures and that all documentation have been completed satisfactorily, prior to releasing the aircraft for flight. Thus, the category C certifying engineer will often be the shift maintenance manager, responsible for the technicians and aircraft under his/her control.

The requirements for the issuing of an individual category C licence and the education, training and experience necessary before the issue of such a licence are detailed in the sections that follow.

The military equivalent of the category C licence holder will be an experienced maintenance technician who holds at least senior non-commissioned officer (SNCO) rank and has a significant period of experience on aircraft type. These individuals are able to sign-off the military equivalent of the CRS, for and on behalf of all trade technicians, who have participated in the particular aircraft servicing activities.

1.3 Opportunities for training, education and career progression

Those employed in civil aviation as aircraft certifying staff may work for commercial aircraft companies or work in the field of GA. The legislation surrounding the training and education of those employed in GA is somewhat different (but no less stringent) than those employed by passenger and freight carrying commercial airline companies. The opportunities and career progressions routes detailed below are primarily for those who are likely to be employed with commercial carriers. However, they may in the future, quite easily, be employed by GA organizations.

Commercial air transport activities are well understood. In that companies are licensed to carry fare paying passengers and freight, across national and international regulated airspace. GA, on the other hand, is often misunderstood for what it is and what place it holds in the total aviation scene. Apart from including flying for personal pleasure, it covers medical flights, traffic surveys, pipeline inspections, business aviation, civil search and rescue and other essential activities, including pilot training! With the advent of a significant increase in demand for business aviation, it is likely that those who have been trained to maintain large commercial transport aircraft will find increasing opportunity for employment in the GA field.

In the UK, and indeed in many countries that have adopted our methods for educating and training prospective aircraft maintenance personnel, there have been, historically, a large number of different ways in which these personnel can obtain initial qualifications and improver training. Since the advent of the recent European Aviation Safety Requirements legislation on personnel licensing, the methods for obtaining initial education and training have become somewhat more unified. Although there still exist opportunities for the "self-starter", achievement of the basic license may take longer.

The schematic diagrams that follow are based on those issued by the Civil Aviation Authority (CAA),[1] Safety Regulation Group (SRG) under the auspices of EASA. They show the qualification and experience routes/pathways for the various categories of aircraft maintenance certifying staff, mentioned earlier.

1.3.1 Category A certifying mechanics

EASA Part 147 approved training pathway

The Part 147 approved training organization is able to offer *ab initio* (from the beginning) learning programmes that deliver EASA Part 66 basic knowledge and initial skills training that satisfy the nation aviation (NAA) authority criteria. In the case of the UK our regulatory authority is the CAA.

Note that a list of EASA Part 147 approved training organizations, together with other useful education and training institutions, will be found in Appendix B, at the end of this book.

Ab initio programmes in approved training organizations often encompass the appropriate EASA examinations. If the examinations have been passed successfully, then an individual requires *1 year* of approved maintenance experience before being able to apply for a category A aircraft maintenance license (AML). Note also the minimum age criteria of 21 years, for all certifying staff, irrespective of the category of license being issued (Figure 1.7).

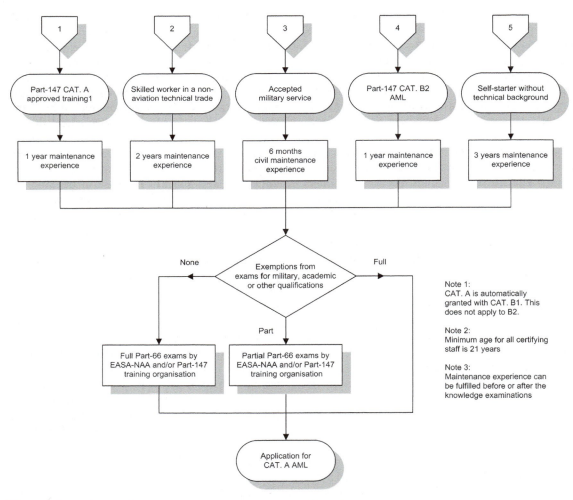

Figure 1.7   Category A qualifications and experience pathways.

Skilled worker pathway

The requirement of practical experience for those entering the profession as non-aviation technical tradesmen is *2 years*. This will enable aviation-orientated skills and knowledge to be acquired from individuals who will already have the necessary basic fitting skills needed for many of the tasks likely to be encountered by the category A certifying mechanic.

Accepted military service pathway

Experienced line mechanics and base maintenance mechanics, with suitable military experience on live aircraft and equipments, will have their practical experience requirement reduced to *6 months*. This may change in the future when armed forces personnel leave after being cross-trained.

Category B2 AML pathway

The skills and knowledge required by category A certifying mechanics is a sub-set of those required by B1 mechanical certifying technicians. Much of this knowledge and many of the skills required for category A maintenance tasks are not relevant by the category B2 avionic certifying technician. Therefore, in order that the category B2 person gains the necessary skills and knowledge required for category A certification, *1 year* of practical maintenance experience is considered necessary.

Self-starter pathway

This route is for individuals who may be taken on by smaller approved maintenance organizations or be employed in GA, where company approvals can be issued on a task-by-task basis, as experience and knowledge are gained. Such individuals may already possess some general aircraft knowledge and basic fitting skills by successfully completing a state funded education programme. For example, the 2-year full-time diploma that leads to an aeronautical engineering qualification (see Section 1.3.4).

However, if these individuals have not practiced as a skill fitter in a related engineering discipline, then it will be necessary to complete

the *3 years* of practical experience applicable to this mode of entry into the profession.

1.3.2 Category B certifying technicians

The qualification and experience pathways for the issue of category B1 and B2 AMLs are shown in Figures 1.8 and 1.9. Having discussed in some detail pathways 1–5 for the category A licence, it will not be necessary to provide the same detail for the category B pathways. Instead you should note the essential differences between the category B1 and B2 pathways as well as the increased experience periods required for both, when compared with the category A license.

Holders of the category A AML require a number of years experience based on their background. This is likely to be less for those wishing to transfer to a category B1 AML, rather than to a B2 AML, because of the similarity in maintenance experience and knowledge that exists between category A and B1 license holders.

Conversion from category B2 to B1 or from B1 to B2 requires *1 year* of practical experience practicing in the new license area. Plus successful completion of the partial Part 66 examinations, as specified by EASA and/or a Part 147 approved training organization.

1.3.3 Category C certifying engineers

The three primary category C qualification pathways are relatively simple to understand and are set out in Figure 1.10.

Qualification is either achieved through practising as a category B1 or B2 certifying technician, for a minimum period of 3 years or entering the profession as an engineering graduate from a recognized degree. Those individuals wishing to gain a category C AML, using the category B route, will already have met the examination criteria in full. However, those entering the profession as engineering graduates will have to take category B1 or B2 knowledge examinations in full or in part, depending on the nature of the degree studied. Examples of

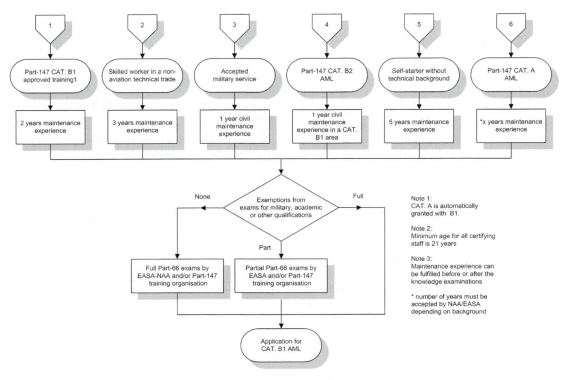

Figure 1.8 Category B1 qualifications and experience pathways.

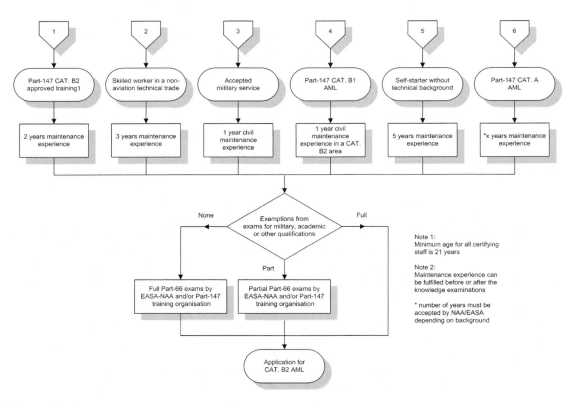

Figure 1.9 Category B2 qualifications and experience pathways.

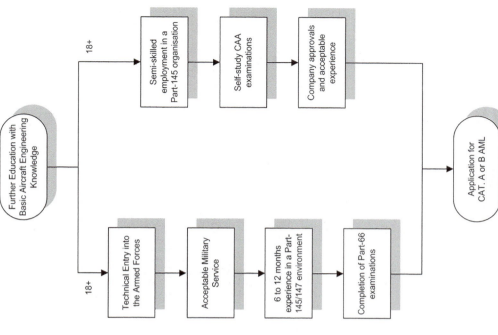

Figure 1.11 Non-standard qualification and experience pathways.

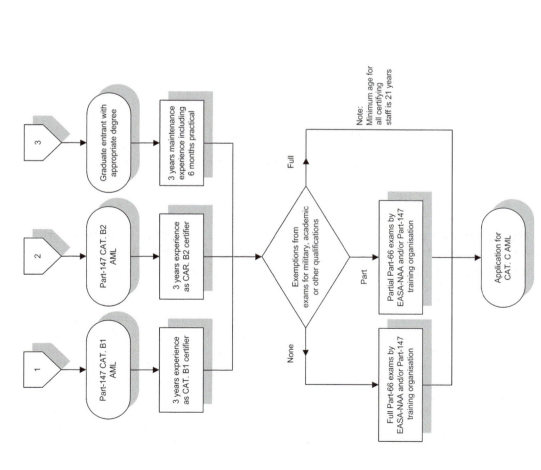

Figure 1.10 Category C qualifications and experience pathways.

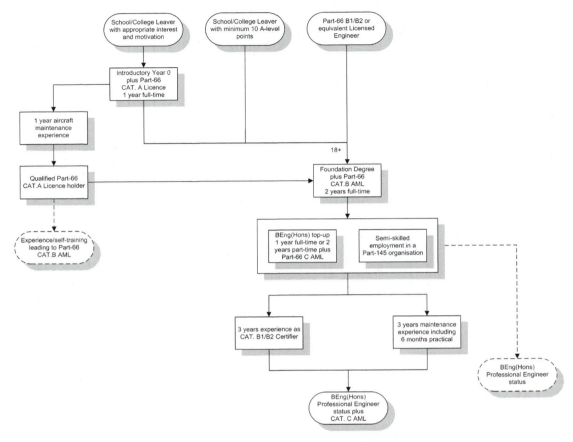

Figure 1.12 Routes to an honours degree and category A, B and C licenses.

non-standard entry methods and graduate entry methods, together with the routes and pathways to professional recognition are given next.

1.3.4 Non-standard qualification and experience pathways

Figure 1.11 illustrates in more detail two possible self-starter routes. The first shows a possible progression route for those wishing to gain the appropriate qualifications and experience by initially serving in the armed forces. The second details a possible model for the 18+ school leaver employed in a semi-skilled role, within a relatively small aircraft maintenance company.

In the case of the semi-skilled self-starter, the experience qualifying times would be dependent on individual progress, competence and motivation. Also note that 18+ is considered to

be an appropriate age to consider entering the aircraft maintenance profession, irrespective of the type of license envisaged.

1.3.5 The Kingston qualification and experience pathway

In this model, provision has been made for qualification and experience progression routes for category A, B and C AML approval and appropriate professional recognition (Figure 1.12).

Figure 1.13 also shows the various stopping-off points, for those individuals wishing to practice as either category A, B or C certifiers.

Figure 1.13 shows two possible fast-track routes for the qualification and award of either a category B or C license. Fast track in this case means that because of the partnership between Kingston University[2] and KLM the

Fast-track Foundation Degree route to Part-66 CAT.B AML with Honours top-up to CAT.C

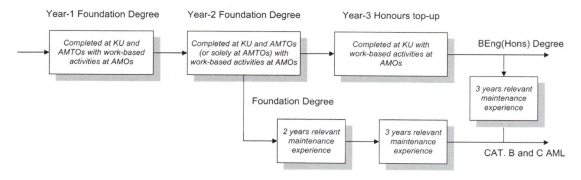

Fast-track Honours Degree route to Part-66 CAT.C AML with Foundation Degree (CAT.B) step-off point

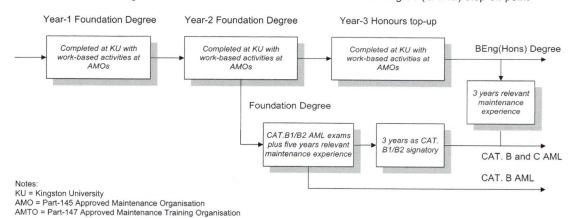

Notes:
KU = Kingston University
AMO = Part-145 Approved Maintenance Organisation
AMTO = Part-147 Approved Maintenance Training Organisation

Figure 1.13 Fast-track routes to category B and C AML.

total programme is recognized by the CAA for *ab initio* approval, which reduces the qualifying times to a minimum, as shown in Figures 1.8–1.10. The appropriate practical experience being delivered at KLM's JAR 147 approved training school at Norwich Airport.

Kingston University also has a partnership with the City of Bristol College, which is a JAR 147 approved organization. With the expansion of Kingston's highly successful programme there will be more opportunities for 18+ school leavers, to undertake *ab initio* training, leading to the CAA examinations and the award of a foundation or full B.Eng.(Hons) degree.

The Royal Aeronautical Society (RAeS) recognizes that full category B Part 66 AML holders, with appropriate experience and responsibilities, meet the criteria for professional recognition as incorporated engineers and may, subject to a professional review, use the initials I.Eng. after their name.

Honours degree holders, who also hold a full category C AML may, with appropriate further learning to masters degree level, apply for recognition as chartered engineers through the RAeS. This is the highest professional accolade for engineers and recognized internationally as the hallmark of engineering ability, competence and professionalism.

Figure 1.14 shows where the full category A, B and C aircraft maintenance certifiers sit, within the professional engineering qualification framework. Thus the category A mechanic, can

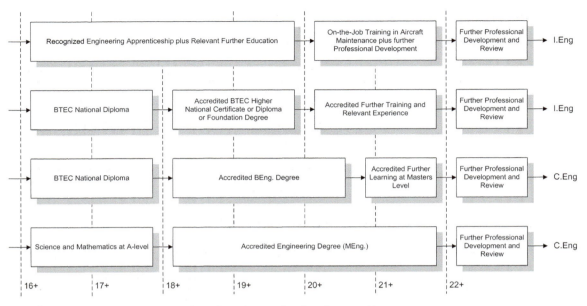

Figure 1.14   Routes to aerospace engineering professional recognition.

Table 1.1

Type of engineering degree	Module exemption
Mechanical engineering bias	Module 1 Mathematics and Module 2 Physics
Aeronautical engineering or Air transport engineering bias	Module 1 Mathematics, Module 2 Physics and Module 8 Basic aerodynamics
Electrical or Electronic engineering bias	Module 1 Mathematics, Module 2 Physics, Module 3 Electrical fundamentals and Module 4 Electronic fundamentals
Avionic engineering bias	Module 1 Mathematics, Module 2 Physics, Module 3 Electrical fundamentals, Module 4 Electronic fundamentals and Module 8 Basic aerodynamics
Kingston University B.Eng.(Hons) aircraft engineering degree (mechanical engineering bias)	Complete exemption from Modules 1 to 10. Approved as fast-track route to "C" licence

with suitable structured training and experience, gain engineering technician status. The full category B technician, again with appropriate structured training and experience, can apply for Incorporated Engineer recognition. The category C engineer, can with an appropriate masters degree or bachelor (Hons) degree and further learning to masters degree level, eventually gain professional recognition as a chartered engineer.

Partial exemptions from Part 66 examinations may be awarded to recognized engineering degrees, dependent on the type of degree being studied. These limited exemptions, by degree type are detailed in Table 1.1. No other exemptions are allowed and all other modules applicable to the licence category need to be passed by EASA approved Part 66 examination.

Note: The one exception, where a large amount of exemption is given for graduates of the Kingston B.Eng.(Hons) aircraft engineering degree, which is directly aimed at preparing aircraft maintenance engineers, for their licence examinations.

1.4 CAA licence – structure, qualifications, examinations and levels

1.4.1 Qualifications structure

The licensing of aircraft maintenance engineers is covered by international standards that are published by the International Civil Aviation Organization (ICAO). In the UK, the Air Navigation Order (ANO) provides the legal framework to support these standards. The purpose of the licence is not to permit the holder to perform maintenance but to enable the issue of certification for maintenance required under the ANO legislation. This is why we refer to licensed maintenance personnel as "certifiers".

At present the CAA issue licences with EASA approval under *two different requirements* depending on the maximum take-off mass of the aircraft.

For aircraft that exceeds 5700 kg, licenses are issued under Part 66. The Part 66 license is common to all European countries who are full members of the Joint Aviation Authority (JAA), regulated by EASA. The ideal being that the issue of a Part 66 licence by any full member country is then recognized as having equal status in all other member countries throughout Europe. There are currently over 25 countries throughout Europe that go to make-up the JAA. In US, the US Federal Aviation Administration (USFAA) is the equivalent of the JAA (now EASA for maintenance personnel licensing). These two organizations have been harmonized to the point where for example, licences issued under Part 66 are equivalent to those licences issued under FAR 66, in countries that adhere to FAA requirements.

Holders of licences issued under Part 66 requirements are considered to have achieved an appropriate level of knowledge and competence that will enable them to undertake maintenance activities on commercial aircraft.

Licences for light aircraft (less than 5700 kg) and for airships, continue to be issued under the UK National Licensing Requirements laid down in British Civil Airworthiness Requirements (BCAR) Section L. The intention is that within a few years, light aircraft will be included within Part 66. At present, this has implications for people who wish to work and obtain licences in GA, where many light aircraft are operated.

Much of the knowledge required for the Part 66 licence, laid down in this series, is also relevant to those wishing to obtain a Section L licence for light aircraft. Although the basic Section L licence is narrower (see Appendix B) and is considered somewhat less demanding than the Part 66 licence it is, nevertheless, highly regarded as a benchmark of achievement and competence within the light aircraft fraternity.

As mentioned earlier, the Part 66 license is divided into categories A, B and C, and for category B license, there are two major career options, either a mechanical or avionic technician. For fear of bombarding you with too much information, what was not mentioned earlier was the *further* subdivisions for the mechanical license. These sub-categories are dependent on aircraft type (fixed or rotary wing) and on engine type (turbine or piston). For clarity, all levels and categories of license that may be issued by the CAA/FAA or member National Aviation Authorities (NAA) under the auspices of EASA, are listed below.

Levels

Category A:	Line maintenance certifying mechanic
Category B1:	Line maintenance certifying technician (mechanical)
Category B2:	Line maintenance certifying technician (avionic)
Category C:	Base maintenance certifying engineer

Note: When introduced, the light AML will be category B3.

Sub-category A

A1:	Aeroplanes turbine
A2:	Aeroplanes piston
A3:	Helicopters turbine
A4:	Helicopters piston

Sub-category B1

B1.1:	Aeroplanes turbine
B1.2:	Aeroplanes piston
B1.3:	Helicopters turbine
B1.4:	Helicopters piston

Note that the experience requirements for all of the above licences are shown in Figures 1.7–1.10.

Aircraft-type endorsements[3]

Holders of an EASA Part 66 aircraft maintenance licences in category B1, B2 and C may apply for inclusion of an *aircraft-type rating* subject to meeting the following requirements.

1. The completion of an EASA Part 147 approved or JAA/NAA approved type training course on the type of aircraft for which approval is being sought and one which covers the subject matter appropriate to the licence category being endorsed.
2. Completion of a minimum period of practical experience on type, prior to application for type rating endorsement.

Type training for category C differs from that required for category B1 or B2, therefore category C type training will not qualify for type endorsement in category B1 or B2. However, type courses at category B1 or B2 level may allow the licence holder to qualify for category C level at the same time, providing they hold a category C basic licence.

Licence holders seeking type rating endorsements from the CAA must hold a basic EASA Part 66 licence granted by the UK CAA.

1.4.2 EASA Part 66 syllabus modules and applicability

The Part 66 syllabus may be taught and examined on a module-by-module basis. The subject matter of individual modules may vary according to the category of licence being studied. The depth of the subject matter may also vary according to the category. Where this is the case, *in this series of books, the greatest depth of knowledge required by category* will always be covered. In all, there are currently 17 modules in the Part 66 syllabus. These modules are tabulated in Table 1.2, together with Table 1.3 indicating their applicability to a particular category and mechanical sub-category.

Table 1.2 Syllabus modules by subject

Module	Content
1	Mathematics
2	Physics
3	Electrical fundamentals
4	Electronic fundamentals
5	Digital techniques and electronic instrument systems
6	Materials and hardware
7	Maintenance practices
8	Basic aerodynamics
9	Human factors
10	Aviation legislation
11	Aeroplane aerodynamics, structures and systems
12	Helicopter aerodynamics, structures and systems
13	Aircraft aerodynamic structures and systems
14	Propulsion
15	Gas turbine engine
16	Piston engine
17	Propeller
18	Airship (to be developed)

1.4.3 Examinations and levels

The Part 66 examinations are modular and designed to reflect the nature of the EASA Part 66 syllabus content. These modular examinations may be taken on CAA/NAA premises, or on the premises of approved Part 147 organizations. The number and type of examination conducted by Part 147 approved organizations will be dependent on the exact nature of their approval. A list of approved organizations and examination venues will be found at the end of this book in Appendix A. For candidates taking the full modular Part 66 examinations, information on the conduct and procedures for these examinations will be found in Chapter 23 of the EASA Administrative and Guidance Material.[4]

The Part 66 module content may vary in terms of the subjects covered within the module and the level of knowledge required according to whether or not a category A, B1 or B2 license is being sought.

Thus, in this book, we will cover in full Part 66 Modules 1, 2, 3, 4 and 8. *Module 1 (Mathematics, Chapter 2 in this book)*, will

Table 1.3 Module applicability to category and mechanical sub-category

Module	A or B1 aeroplanes with:		A or B1 helicopter with:		B2 avionic
	Turbine engine	Piston engine	Turbine engine	Piston engine	
1	✓	✓	✓	✓	✓
2	✓	✓	✓	✓	✓
3	✓	✓	✓	✓	✓
4	✓[a]	✓[a]	✓[a]	✓[a]	✓
5	✓	✓	✓	✓	✓
6	✓	✓	✓	✓	✓
7	✓	✓	✓	✓	✓
8	✓	✓	✓	✓	✓
9	✓	✓	✓	✓	✓
10	✓	✓	✓	✓	✓
11[b]	✓	✓	–	–	–
12	–	–	✓	✓	–
13[c]	–	–	–	–	✓
14[d]	–	–	–	–	✓
15	✓	–	✓	–	–
16	–	✓	–	✓	–
17	✓	✓	–	–	–

[a] This module is not applicable to category A.

[b] Module 11 is applicable only to mechanical certifying staff.

[c] Module 13 is only applicable to B2 avionic certifying technicians.

[d] Module 14 offers a less in depth treatment of propulsion, designed for study by B2 avionic certifying technicians.

be covered to the depth required by the B1 and B2 technician examination. Further mathematics (chapter 3) is also included, which is designed to assist understanding of Module 2, Physics. The further mathematics *is not* subject to Part 66 examination but is still considered by the authors to be very useful foundation knowledge. Those studying for the category A licence should concentrate on fully understanding, the non-calculator mathematics given in Chapter 2 of this book. They should also be able to answer all the test questions at the end of this chapter.

Module 2 (Physics, Chapter 4 in this book) is covered to a depth suitable for category B technicians, no distinction is made between B1 and B2 levels of understanding,[5] the *greatest* depth being covered for both categories, as appropriate. The Module 2 content not required by category A mechanics, is mentioned in the introduction to the chapter and reflected in the physics test questions given at the end.

Module 3 (Electrical fundamentals, Chapter 5 in this book) is covered at the category B technician level, with clear indications given between the levels of knowledge required for the category A and B license requirements. *Module 4 (Electronic fundamentals, Chapter 6 in this book)* is not required by category A mechanics but, as before, the treatment of the differing levels of knowledge for category B1 and B2 will be taken to the greater depth required by B2 technicians. The differences in level again being reflected in the test questions given at the end of the chapter.

Module 8 (Basic aerodynamics, Chapter 7 in this book) will be covered in full to category B level, with no demarcation being made between category A and B levels. For the sake of completeness, this chapter will also include brief coverage of aircraft flight control taken from Module 11.1. The typical examination questions directly related to Module 8 will be clearly identified at the end of the chapter.

Full coverage of the specialist aeroplane aerodynamics, high-speed flight and rotor wing aerodynamics, applicable to Modules 11 and 13 will be covered in the third book in the series, *Aircraft Aerodynamics, Structural Maintenance and Repair*.

Examination papers are mainly multiple-choice type but a written paper must also be passed so that the licence may be issued. Candidates may take one or more papers, at a single examination sitting. The pass mark for each multiple-choice paper is 75%! There is no longer any penalty marking for incorrectly answering individual multiple-choice questions. All multiple-choice questions set by the CAA and by approved organizations have exactly the

Table 1.4 Structure of Part 66 multiple-choice examination papers

Module	Number of questions	Time allowed (min)	Module	Number of questions	Time allowed (min)
1 *Mathematics*			10 *Aviation Legislation*		
Category A	16	20	Category A	40	50
Category B1	30	40	Category B1	40	50
Category B2	30	40	Category B2	40	50
2 *Physics*			11 *Aeroplane aerodynamics, structures and systems*		
Category A	30	40	Category A	100	125
Category B1	50	65	Category B1	130	165
Category B2	50	65	Category B2	–	–
3 *Electrical fundamentals*			12 *Helicopter aerodynamics, structures and systems*		
Category A	20	25	Category A	90	115
Category B1	50	65	Category B1	115	145
Category B2	50	65	Category B2	–	–
4 *Electronic fundamentals*			13 *Aircraft aerodynamics, structures and systems*		
Category A	–	–	Category A	–	–
Category B1	20	25	Category B1	–	–
Category B2	40	50	Category B2	130	165
5 *Digital techniques/electronic instrument systems*			14 *Propulsion*		
Category A	16	20	Category A	–	–
Category B1	40	50	Category B1	–	–
Category B2	70	90	Category B2	25	30
6 *Materials and hardware*			15 *Gas turbine engine*		
Category A	50	65	Category A	60	75
Category B1	70	90	Category B1	90	115
Category B2	60	75	Category B2	–	–
7 *Maintenance practices*			16 *Piston engine*		
Category A	70	90	Category A	50	65
Category B1	80	100	Category B1	70	90
Category B2	60	75	Category B2	–	–
8 *Basic aerodynamics*			17 *Propeller*		
Category A	20	25	Category A	20	25
Category B1	20	25	Category B1	30	40
Category B2	20	25	Category B2	–	–
9 *Human factors*					
Category A	20	25			
Category B1	20	25			
Category B2	20	25			

Note: The time given for examinations may, from time to time, be subject to change. There is currently a review pending of examinations time based on levels. Latest information may be obtained from the CAA website.

same form. That is, each question will contain a stem (the question being asked), two distracters (incorrect answers) and one correct answer. The multiple-choice questions given at the end of each chapter in this book are laid out in this form.

All multiple-choice examination papers are timed, approximately 1 min and 15 s, being allowed for the reading and answering of each question (see Table 1.4). The number of questions asked depends on the module examination being taken and on the category of licence being sought. The structure of the multiple-choice papers for each module together with the structure of the written examination for issue of the license are given in Table 1.4.

More detailed and current information on the nature of the EASA license examinations can be found in the appropriate *CAA documentation*,[6] from which the examination structure detailed in Table 1.4 is extracted.

Written paper

The written paper required for licence issue contains four essay questions. These questions are drawn from the Part 66 syllabus modules as follows:

Module	Paper	Question
7	Maintenance practices	2
9	Human factors	1
10	Aviation legislation	1

1.5 Overview of airworthiness regulation, aircraft maintenance and its safety culture

1.5.1 Introduction

All forms of public transport require legislation and regulation for their operation, in order to ensure that safe and efficient transport operations are maintained. Even with strict regulation, it is an unfortunate fact that incidents and tragic accidents still occur. Indeed, this is only to self-evident with the recent spate of rail accidents where the Potters Bar accident in 2002, may very likely be attributable to poor maintenance!

When accidents occur on any public transport system, whether travelling by sea, rail or air, it is an unfortunate fact, that loss of life or serious injury may involve a substantial number of people. It is also a fact that the accident rate for air travel is extremely low and it is currently one of the safest forms of travel.

The regulation of the aircraft industry can only lay down the framework for the safe and efficient management of aircraft operations, in which aircraft maintenance plays a significant part. It is ultimately the responsibility of the individuals that work within the industry to ensure that standards are maintained. With respect to aircraft maintenance, the introduction of the new harmonized requirements under JAA and more recently ECAR should ensure that high standards of aircraft maintenance and maintenance engineering training are found not only within the UK, but across Europe and indeed throughout many parts of the world.

In order to maintain these high standards, individuals must not only be made aware of the nature of the legislation and regulation surrounding their industry, but also they need to be encouraged to adopt a mature, honest and responsible attitude to all aspects of their job role. Where safety and personal integrity must be placed above all other considerations, when undertaking aircraft maintenance activities.

It is for the above reasons, that a knowledge of the legislative and regulatory framework of the industry and the adoption of aircraft maintenance safety culture, becomes a vital part of the education for all individuals wishing to practice as aircraft maintenance engineers. Set out in this section is a brief introduction to the regulatory and legislative framework, together with maintenance safety culture and the vagaries of human performance. A much fuller coverage of aircraft maintenance legislation and safety procedures will be found in, *Aircraft Engineering Maintenance Practices*, the second book in this series.

1.5.2 The birth of the ICAO

The international nature of current aircraft maintenance engineering has already been mentioned. Thus the need for conformity of standards to ensure the continued airworthiness of aircraft that fly through international airspace is of prime importance.

As long ago as December 1944, a group of forward thinking delegates from 52 countries came together in Chicago, to agree and ratify the convention on international civil aviation. Thus the Provisional International Civil Aviation Organization (PICAO) was established. It ran in this form until March 1947, when final ratification from 26 member countries was received and it became the ICAO.

The primary function of the ICAO, which was agreed in principle at the *Chicago Convention* in 1944, was to develop international air transport in a safe and orderly manner. More formerly, the 52 member countries agreed to undersign:

certain principles and arrangements in order that international civil aviation may be developed in a safe and orderly manner and that international air transport services may be established on the basis of equality of opportunity and operated soundly and economically.

Thus in a spirit of cooperation, designed to foster good international relationships, between member countries, the 52 member states signed up to the agreement. This was a far-sighted decision, which has remained substantially unchanged up to the present. The ICAO Assembly is the sovereign body of the ICAO responsible for reviewing in detail the work of ICAO, including setting the budget and policy for the following 3 years.

The council, elected by the assembly for a 3-year term, is composed of 33 member states. The council is responsible for ensuring that standards and recommended practices are adopted and incorporated as annexes into the convention on international civil aviation. The council is assisted by the Air Navigation Commission to deal with technical matters, the Air Transport Committee to deal with economic matters and the Committee on Joint Support of Air Navigation Services and the Finance Committee.

The ICAO also works closely with other members of the United Nations (UN) and other non-governmental organizations such as the International Air Transport Association (IATA) and the International Federation of Air Line Pilots to name but two.

1.5.3 The UK CAA

The CAA was established by an act of parliament in 1972, as an independent specialist aviation regulator and provider of air traffic services.[7] Under the act it is responsible to the government for ensuring that all aspects of aviation regulation are implemented and regulated in accordance with the ANO formulated as a result of the act.

Following the separation of National Air Traffic Services (NATS) in 2001, the CAA is now responsible for all civil aviation functions, these are: economic regulation, airspace policy, safety regulation and consumer protection.

The *Economic Regulation Group* (ERG) regulates airports, air traffic services and airlines and provides advice on aviation policy from an economic standpoint. Its aim is to secure the best sustainable outcome for users of air transport services.

The *Directorate of Airspace Policy* (DAP) is responsible for the planning and regulation of all UK airspace including the navigation and communication infrastructure to support safe and efficient operations. Both civilian and military experts staff this group.

The *Consumer Protection Group* (CPG) regulates travel organizations, manages the consumer protection organization, air travel organizers' licensing (ATOL) and licenses UK airlines, in addition to other functions.

The *Safety Regulation Group* (SRG) ensures that UK civil aviation standards are set and achieved in a cooperative and cost-effective manner. SRG must satisfy itself that aircraft are properly designed, manufactured, operated and maintained. It is also the responsibility of this group to ensure the competence of flight crews, air traffic controllers and *aircraft maintenance engineers* in the form of personal licensing. All the major functions of this group are shown in Figure 1.15. (Some of the responsibilities of this group have now been transferred to EASA. However this group still "police" and make recommendations to EASA regarding safety standards and regulations, in their

Figure 1.15 CAA-SRG functions and responsibilities.

role as a leading NAA. In particular, EASA are now responsible for the certification of individuals, the approval of organizations and the maintenance of JARs concerned with aircraft maintenance personnel and practice).

1.5.4 Civil aviation requirements

The broad international standards on airworthiness set up by the ICAO were backed up by detailed national standards, overseen in the UK by the National Authority for Airworthiness the CAA. These national standards were known in the UK as BCAR and in the USA as, Federal Airworthiness Regulations (FAR). Many other countries adopted one or the other of these requirements, with their own national variations.

As international collaborative ventures became more wide spread, there was increasing pressure to produce a unified set of standards, particularly in Europe. Thus came into being (under the auspices of the JAA) the European Joint Aviation Requirements or JAR, for short.

Then, with increasing collaborative ventures between Europe, the USA and other major economies around the world, there became a need to harmonize these European requirements (JAR), with those of the USA, FAR. This harmonization process is not without difficulties and with respect to aircraft airworthiness, maintenance and associated safety it has now been taken over by EASA.

It is unnecessary in this brief introduction to go into detail on the exact nature of JAA/EASA in overseeing the European JAR (or in the case of EASA, implementing rules (IR) and certification of specifications (CS) airworthiness requirements and design protocols. Suffice to say[8] that

the Civil Aviation Authorities of certain countries have agreed common comprehensive and detailed aviation requirements (JAR) with a view to minimizing type certification problems on joint aviation ventures, to facilitate the export and import of aviation products, and make it easier for maintenance and operations carried out in one country to be accepted by the CAA in another country.

One or two of the more important requirements applicable to aircraft maintenance organizations and personnel are detailed below:

CS-25 – Certification specifications for large aircraft (over 5700 kg)

CS-E – Certification specifications for aircraft engines

Part 21 – Acceptable means of compliance implementing rules for the airworthiness and environmental certification of aircraft and related products

Part-M – Guidance material for continuing airworthiness

Part-66 – Guidance material for aircraft engineering certifying staff, including the basic knowledge requirements, upon which all the books in this series are based

Part-145 – Guidance material for organizations operating large aircraft

Part-147 – Guidance material for Training Organization requirements to be met by those seeking approval to conduct approved training/ examinations of certifying staff, as specified in JAR 66.

1.5.5 Aircraft maintenance engineering safety culture and human factors

If you have managed to plough your way through this introduction, you cannot have failed to notice that aircraft maintenance engineering is a very highly regulated industry, where **safety** is considered paramount!

Every individual working on or around aircraft and/or their associated equipments, has a personal responsibility for their own safety and the safety of others. Thus, you will need to become familiar with your immediate work area and recognize and avoid, the hazards associated with it. You will also need to be familiar with your local emergency: first aid procedures, fire precautions and communication procedures.

Figure 1.16 Control column, with base cover plate fitted and throttle box assembly clearly visible.

Thorough coverage of workshop, aircraft hangar and ramp safety procedures and precautions will be found in *Aircraft Engineering Maintenance Practices*, the second book in the series.

Coupled with this knowledge on safety, all prospective maintenance engineers must also foster a responsible, honest, mature and professional attitude to all aspects of their work. You perhaps, cannot think of any circumstances where you would not adopt such attitudes? However, due to the nature of aircraft maintenance, you may find yourself working under very stressful circumstances where your professional judgement is tested to the limit!

For example, consider the following scenario.

As an *experienced maintenance technician*, you have been tasked with fitting the cover to the base of the flying control column (Figure 1.16), on an aircraft that is going to leave the maintenance hanger on *engine ground runs*, before the overnight embargo on airfield noise comes into force, in 3 hours time. It is thus important that the aircraft is towed to the ground running area, in time to complete the engine runs before the embargo. This will enable all outstanding maintenance on the aircraft to be carried out over night and so ensure that the aircraft is made ready in good time, for a scheduled flight first thing in the morning.

You start the task and when three quarters of the way through fitment of the cover, you drop a securing bolt, as you stand up. You think that you hear it travelling across the flight deck floor. After a substantial search by torch-light, where you look not only across the floor, but also around the base of the control column and into other possible crevices, in the immediate area, you are unable to find the small bolt. Would you:

(a) *Continue the search for as long as possible and then, if the bolt was not found, complete the fit of the cover plate and look for the bolt, when the aircraft returned from ground runs?*
(b) *Continue the search for as long as possible and then, if the bolt was not found, inform the engineer tasked with carrying out the ground runs, to be aware that a bolt is somewhere in the vicinity of the base of the control column on the flight deck floor. Then continue with the fit of the cover?*
(c) *Raise an entry in the aircraft maintenance log for a "loose article" on the flight deck. Then remove the cover plate, obtain a source of strong light and/or a light probe kit and carry out a thorough search at base of control column and around all other key controls, such as the throttle box. If bolt is not found, allow aircraft to go on ground run and continue search on return?*
(d) *Raise an entry in the aircraft log for a "loose article" on the flight deck. Then immediately seek advice from shift supervisor, as to course of action to be taken?*

Had you not been an experienced technician, you would immediately inform your supervisor (action (d)) and seek advice as to the most appropriate course of action. As an *experienced* technician, what should you do? The course of action to be taken, in this particular case, may not then be quite so obvious, it requires *judgements* to be made.

Quite clearly actions (a) and (b) would be wrong, no matter how much experience the technician had. No matter how long the search continued, it would be essential to remove the cover plate and search the base of the control column to ensure that it was not in the vicinity. Any loose article could dislodge during flight and cause possible catastrophic jamming or fouling of the controls. If the engine run is to proceed, actions (a) and (b) are still not adequate. A search of the throttle box area for the bolt would also need to take place, as suggested by action (c). Action (c) seems plausible, with the addition of a good light source and thorough search of all critical areas, before the fit of the cover plate, seems a reasonable course of action to take, especially after the maintenance log entry has been made, the subsequent search for the bolt, cannot be forgotten, so all is well?

However, if you followed action (c) you would be *making important decisions, on matters of safety, without consultation.* No matter how experienced you may be, you are not necessarily aware of the total picture, whereas your shift supervisor, may well be! *The correct course of action, even for the most experienced engineer would be action (d).*

Suppose action (c) had been taken and on the subsequent engine run the bolt, that had been lodged in the throttle box, caused the throttle to jam in the open position. Then shutting down the engine, without first closing the throttle, could cause serious damage! It might have been the case that if action (d) had been followed, the shift supervisor may have been in a position to prepare *another aircraft* for the scheduled morning flight, thus avoiding the risk of running the engine, before the loose article search had revealed the missing bolt.

In any event, the aircraft would not normally be released for service until the missing bolt had been found, even if this required the use of sophisticated radiographic equipment to find it!

The above scenario illustrates some of the pitfalls, that even experienced aircraft maintenance engineers may encounter, if safety is forgotten or assumptions made. For example, because you thought you heard the bolt travel across the flight deck, you may have assumed that it could not possibly have landed at the base of the control column, or in the throttle box. This, of course, is an *assumption* and one of the golden rules of safety is *never assume, check!*

When the cover was being fitted, did you have adequate lighting for the job? Perhaps with adequate lighting, it might have been possible to track the path of the bolt, as it travelled across

Figure 1.17 Typical aircraft hangar first aid station.

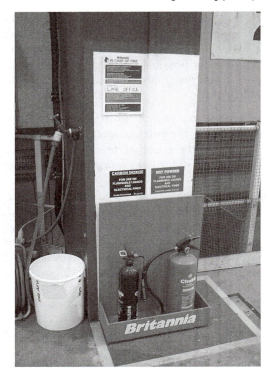

Figure 1.18 Typical aircraft hangar fire point.

the flight deck, thus preventing its loss in the first place.

Familiarity with emergency equipment and procedures, as mentioned previously is an essential part of the education of all aircraft maintenance personnel. Reminders concerning the use of emergency equipment will be found in hangars, workshops, repair bays and in many other areas where aircraft engineering maintenance is practiced. Some typical examples of emergency equipment and warning notices are shown below. Figure 1.17 shows a typical aircraft maintenance hangar first aid station, complete with explanatory notices, first aid box and eye irritation bottles.

Figure 1.18 shows an aircraft maintenance hangar fire point, with clearly identifiable emergency procedures in the event of fire and the appropriate fire appliance to use for electrical or other type of fire.

Figure 1.19 shows a grinding assembly, with associated local lighting and warning signs, for eye and ear protection. Also shown are the

drop-down shields above the grinding wheels to prevent spark burns and other possible injuries to the hands, arms and eyes.

Figure 1.20 shows a warning notice concerning work being carried out on open fuel tanks and warning against the use of electrical power. In addition to this warning notice there is also a *no power* warning at the aircraft power point (Figure 1.21).

You may feel that the module content contained in this book on principles is a long way removed from the working environment illustrated in these photographs. However, consider for a moment the relatively simple task of inflating a ground support trolley wheel (Figure 1.22).

Still it is a common practice to measure tyre pressures in pounds per square inch (psi), as well as in bar (Figure 1.23). Imagine the consequences of attempting to inflate such a tyre to 24 bar, instead of 24 psi, because you mis-read the gauge on the tyre inflation equipment!

The need to understand units, in this particular case is most important. It cannot happen

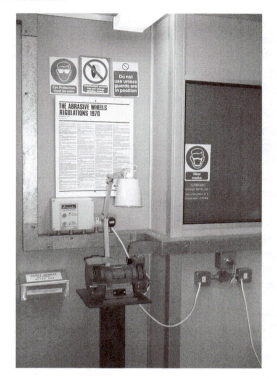

Figure 1.19 Grinding wheel assembly, with associated lighting and warning signs.

Figure 1.20 Open fuel tanks warning notice.

Figure 1.21 Ground power warning.

Figure 1.22 Oxygen bottle trolley, showing trolley wheel.

Figure 1.23 Pressure gauges graduated in bar and in psi.

I hear you say; well unfortunately it can, the above is an account of an actual incident. Fortunately the technician inflating the tyre, followed standard safety procedures, in that he stood behind the tyre, rather than along side it, during the inflation process. The tyre separated from the wheel assembly and shot sideways at high velocity. If the technician had been to the side of the tyre and wheel assembly he would have sustained serious injury! At that time this technician was unaware of the difference in units between the bar and for him, the more familiar imperial units of psi. Thus the need to adopt a mature attitude to your *foundation studies* is just as important as adopting the necessary professional attitude to your on-job practical maintenance activities.

Completing the maintenance documentation

When carrying out any form of maintenance activity on aircraft or aircraft equipment, it is vitally important that the appropriate documentation and procedures are consulted and followed. This is particularly important, if the maintenance technician is unfamiliar with the work, or is new to the equipment being worked on. Even those experienced in carrying out a particular activity should regularly consult the maintenance manual, in order to familiarize themselves with the procedure and to establish the modification state of the aircraft or equipment being worked on.

The modification state of the documentation itself should not only be checked by the scheduling staff, but also by the engineer assigned to the task to ensure currency.

When certifying staff sign-up for a particular maintenance activity, there signature implies that the job has been completed to the best of their ability, in accordance with the appropriate schedule and procedures. Any maintenance engineer, who is subsequently found to have produced work that is deemed to be unsatisfactory, as a result of their negligence, during the execution of such work, may be prosecuted. It should always be remembered by all involved in aircraft maintenance engineering that **mistakes can cost lives**. This is why it is so important that certifying staff always carry out their work to the highest professional standards, strictly adhering to the laid-down safety standards and operational procedures.

Human factors

The above examples concerning the dropped bolt and the mistakes made when attempting to inflate the ground support trolley tyre illustrate the problems that may occur due to *human frailty*.

Human factors[9] impinges on everything an engineer does in the course of their job in one way or another, from communicating effectively with colleagues to ensuring they have adequate lighting to carry out their tasks. Knowledge of this subject has a significant impact on the safety standards expected of the aircraft maintenance engineer.

The above quote is taken from the CAA publication (CAP 715) which provides an introduction to engineering human factors for aircraft maintenance staff, expanding on the human factors syllabus contain in Part-66 Module 9.

A study of human factors, as mentioned earlier, is now considered to be an essential part of the aircraft maintenance engineers education. It is hoped that by educating engineers and ensuring currency of knowledge and techniques, that this will ultimately lead to a reduction in aircraft incidents and accidents which can be attributed to human error during maintenance.

The study of human factors has become so important that for many years the CAA has co-sponsored annual international seminars dedicated to the interchange of information and ideas on the management and practice of eliminating aviation accidents, resulting from necessary human intervention. Numerous learned articles and books have been written on human factors, where the motivation for its study has come from the need to ensure high standards of safety in high risk industries, such as nuclear power and of course air transport!

Aircraft maintenance engineers thus need to understand, how human performance limitations impact on their daily work. For example, if you are the licensed aircraft engineer (LAE) responsible for a team of technicians. It is important that you are aware of any limitations members of your team may have with respect to obvious physical constraints, like their hearing and vision. As well as more subtle limitations, such as their ability to process and interpret information or their fear of enclosed spaces or heights. It is not a good idea to task a technician with a job inside a fuel tank, if they suffer from claustrophobia!

Social factors and other factors that may affect human performance also need to be understood. Issues such as responsibility, motivation, peer pressure, management and supervision need to be addressed. In addition to general fitness, health, domestic and work-related stress, time pressures, nature of the task, repetition, workload and the effects of shift work.

The nature of the physical environment in which maintenance activities are undertaken needs to be considered. Distracting noise, fumes, illumination, climate, temperature, motion, vibration and working at height and in confined spaces, all need to be taken into account.

The importance of good two-way communication needs to be understood and practiced. Communication within and between teams, work logging and recording, keeping up-to-date and the correct and timely dissemination of information must also be understood.

The impact of human factors on performance will be emphasized, wherever and whenever it is thought appropriate, *throughout all the books in this series*. There will also be a section in the second book in this series, on *Aircraft Engineering Maintenance Practices*, devoted to the study of past incidents and occurrences that can be attributed to errors in the *maintenance chain*. This section is called *learning by mistakes*.

However, it is felt by the authors that *human factors* as contained in Part 66 Module 9, is so vast that one section in a textbook, will not do the subject justice. For this reason a list of references are given at the end of this chapter, to which the reader is referred. In particular an excellent introduction to the subject is provided in the CAA publication: CAP 715 – *An Introduction to Aircraft Maintenance Engineering Human Factors for Part 66*.

We have talked so far about the nature of human factors, but how do human factors impact on the integrity of aircraft maintenance activities? By studying previous aircraft incidents and accidents, it is possible to identify the sequence of events which lead to the incident and so implement procedures to try and avoid such a sequence of events, occurring in the future.

1.5.6 The BAC One-Eleven accident

By way of an introduction to this process, we consider an accident that occurred to a BAC One-Eleven, on 10th June 1990 at around 7.30 a.m. At this time the aircraft, which had taken off from Birmingham Airport,

Figure 1.24 A Boeing 767 left front windscreen assembly.

had climbed to a height of around 17,300 ft (5273 m) over the town of Didcot in Oxfordshire, when there was a sudden loud bang. The left windscreen, which had been replaced prior to the flight, was blown out under the effects of cabin pressure when it overcame the retention of the securing bolts, 84 of which, out of a total of 90, were smaller than the specified diameter. The commander narrowly escaped death, when he was sucked halfway out of the windscreen aperture and was restrained by cabin crew whilst the co-pilot flew the aircraft to a safe landing at Southampton Airport.

For the *purposes of illustration*, Figure 1.24 shows a typical front left windscreen assembly of a Boeing 767.

How could this happen? In short, a task deemed to be **safety critical** was carried out by one individual, who also carried total responsibility for the quality of the work achieved. The installation of the windscreen was not tested after fit. Only when the aircraft was at 17,300 ft, was there sufficient *pressure differential* to check the integrity of the work! The shift maintenance manager, who had carried out the work, did not achieve the quality standard during the fitting process, due to inadequate care, poor trade practices, failure to adhere to company standards, use of unsuitable equipment and long-term failure by the maintenance manager to observe the promulgated procedures. The airline's local management product samples and quality audits, had not detected the existence of inadequate standards employed by the shift maintenance manager because they did not monitor directly the work practices of shift maintenance managers.

Engineering factors

There is no room in this brief account of the accident to detail in full all the engineering factors which lead up to the windscreen failure; however, some of the more important factors in *the chain of events* are detailed below:

- Incorrect bolts had been used with the previous installation (A211-7D).
- Insufficient stock of the incorrect A211-7D bolts existed in the controlled spare parts carousel dispenser. Although these bolts were incorrect, they had proved through 4 years of use to be adequate.
- No reference was made to the spare parts catalogue *to check* the required bolts' part number.
- The stores system, available to identify the stock level and location of the required bolts was not used.
- Physical matching of the bolts was attempted and as a consequence, incorrect bolts (A211-8C) were selected from an uncontrolled spare-parts carousel, used by the maintenance manager.
- An uncontrolled torque limiting screwdriver was set up outside the calibration room.
- A bi-hexagonal bit holder was used to wind down the bolts, resulting in the occasional loss of the bit and the covering up of the bolt head. Hence the maintenance manager was unable to see that the countersunk head of the bolts, was further recessed than normal.
- The safety platform was incorrectly positioned leading to inadequate access to the job.
- The warning from the storekeeper that A211-8D bolts were required did not influence the choice of bolts.
- The amount of unfilled countersunk left by the small bolt heads was not recognized as excessive.
- The windscreen was not designated a "vital task" therefore no duplicate (independent) inspection was required.

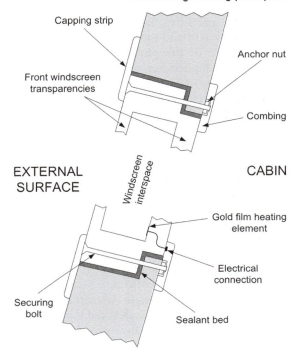

Figure 1.25 Simplified schematic cross-section of a typical windscreen requiring external fit.

- The windscreen was not designed so that internal pressure would hold it in place, but was fitted from the outside (Figure 1.25).
- The shift maintenance manager was the only person whose work on the night shift was not subject to the review of a maintenance manager.
- Poor labelling and segregation of parts in the uncontrolled spare-parts carousel.
- The shift maintenance manager did not wear prescribed glasses when carrying out the windscreen change.

The impact of human factors

The above series of events does not tell the whole story. For example, why was it that the shift maintenance manager was required to perform the windscreen change in the first place? A supervisory aircraft engineer and a further LAE, normally part of the shift, were not available that night. In order to achieve the windscreen change during the night shift and have the aircraft ready for a pre-booked wash

Table 1.5

Part No.	Shank length (in.)	Diameter (in.)	Thread size	Comments
A211-8D	0.8	0.1865–0.1895	10 UNF	Correct bolts
A211-8C	0.8	0.1605–0.1639	8 UNC	84 bolts used
A211-7D	0.7	0.1865–0.1895	10 UNF	Bolts removed

early in the morning, the shift maintenance manager decided to carry out the windscreen change by himself. His supervisory aircraft engineer and other airframe engineer were busy rectifying a fault on another BAC One-Eleven aircraft, which needed to be completed before departure of the aircraft the following morning.

Also in the early hours of the morning when the windscreen change took place, the bodies' circadian rhythms are at a low ebb. This, coupled with a high workload, may have lead to tiredness and a reduced ability to concentrate.

The highway staging platform was incorrectly positioned for easy access to the job, had this been correctly positioned the maintenance manager may have been better able to notice that the bolt heads were recessed in the countersink, significantly more than usual.

The *assumption* that the bolts removed from the aircraft windscreen were correct was made by the maintenance manager. Thus one of the most important dictums was ignored; *never assume, check*!

The non-availability of the bolts (A211-7D) even though incorrect, in the controlled spare parts carousel, lead the manager to search in a non-controlled carousel, where parts were poorly labelled or incorrectly segregated. This in turn lead the manager to select the bolts using visual and touch methods. This resulted in the final error, in the chain, being made. The bolts selected were of the correct length but were crucially 0.026 of an inch, too small in diameter. The illustrated parts catalogue (IPC), which *should have been consulted before replacing the old bolts*, specifies that the attachment bolts should be part number (A211-8D). The specification for these bolts, together with those selected from the carousel (A211-8C) are shown in Table 1.5.

The windscreen change on this aircraft was not considered a *vital point*. The CAA state that the term "vital point" is not intended to refer to multiple fastened parts of the structure, but applies to a single point, usually in an aircraft control system. In September 1985 BCARs introduced a requirement for duplicate inspections of *vital points*, which are defined as: *any point on an aircraft at which a single mal-assembly could lead to a catastrophe, resulting in loss of the aircraft or fatalities*. Had the windscreen been considered a vital maintenance operation, then a *duplicate inspection* would have been performed and the excessive recess of the bolt heads may very well have been noticed.

Also, there are no CAA requirements for a *cabin pressure check* to be called up after the work has been carried out on the pressure hull. Such checks are written into the aircraft maintenance manual at the discretion of the aircraft design team, and were not called up on the BAC One-Eleven. Had they been necessary, then the sub-standard integrity of the incorrectly fitted windscreen would have been apparent.

A *full account of this accident*, the events leading up to it and the subsequent safety recommendations will be found on the *Air Accident Investigation Board website*,[10] from which some of the above account has been taken.

The safety recommendations

As a result of the above accident and subsequent inquiry, eight safety recommendations were given. Briefly, these recommendations are as follows:

- The CAA should examine the applicability of *self-certification* to aircraft engineering safety critical tasks following which the components or systems are cleared for service without functional checks. Such a review should include the interpretation of *single mal-assembly* within the context of *vital points*.

- British Airways should review their quality assurance system and reporting methods, and encourage their engineers to provide feedback from the shop floor.
- British Airways should review the need to introduce job descriptions and terms of reference for engineering grades, including shift maintenance manager and above.
- British Airways should provide the mechanism for an independent assessment of standards and conduct an in depth audit into work practices at Birmingham Airport.
- The CAA should review the purpose and scope of their supervisory visits to airline operators.
- The CAA should consider the need for periodic training and testing of engineers to ensure currency and proficiency.
- The CAA should recognize the need for corrective glasses, if prescribed, in association with the undertaking of aircraft engineering activities.
- The CAA should ensure that, prior to the issue of an air traffic controller (ATC) rating, a candidate undertakes an approved course of training, that includes the theoretical and practical handling of emergency situations.

The above recommendations are far reaching and provide an example of *human factors* involvement, far removed from the direct maintenance activity, but very much impacting on the *chain of events* leading to an accident or serious incident. It is these complex interactions that may often lead to maintenance errors being made, with subsequent catastrophic consequences.

No matter how sophisticated the policies and procedures may be, ultimately due to the influence of human factors, it is the integrity, attitude, education and professionalism of the individual aircraft maintenance engineer, that matters most, in the elimination of maintenance errors.

1.5.7 Concluding remarks

It is hoped that this short introduction into the aircraft maintenance industry has given you an insight into the demanding and yet very rewarding work, offered to aircraft maintenance certifying staff. No matter at what point you wish to enter the industry, you will find routes and pathways that enable you to progress to any level, dependent only, on your own ambitions and aspirations. The training and education to reach the top of any profession is often long and arduous and aircraft maintenance engineering is no exception!

The subject matter that follows may seem a long way removed from the environment portrayed in this introduction and yet, it forms a vital part of your initial educational development. Therefore, you should approach the subjects presented in Parts 2 and 3 of this book, with the same amount of enthusiasm and dedication as you will with the practical activities you find yourself engaged in, when qualified to practice your profession.

The non-calculator mathematics, you are about to meet, may seem deceptively simple. However, do remember that the pass rate is 75%, as it is for all your Part 66 examinations. This is likely to be significantly higher than any other examination pass rate, you may have encountered up till now. It is, therefore, very important that you become familiar with *all* the subject matter contained in the following chapters, if you are to be successful in your future CAA examinations. There are numerous examples, multiple-choice questions and other types of questions provided to assist you in acquiring the necessary standard.

References

1. CAA-SRG Engineer Standards, papers 3–6 (May 2001).
2. Kingston University, Rationale for Aerospace Programmes (May 2001).
3. CAA-SRG, JAR-66 Information for New Applicants Leaflet 2 Issue 16 (October 2001).
4. JAA Administration and Guidance Material (1999).
5. JAR-66 Appendix 2 Section 1 Levels (April 2002).
6. CAA-SRG JAR-66 Syllabus and Examinations No. 6 (issued 16/10/01).
7. CAA Corporate Information, page 1–3. (April 2002).
8. JAR-66 Certifying Staff Maintenance, page F1 (April 2002).
9. CAP715 An Introduction to Aircraft Maintenance Human Factors for JAR-66 (January 2002).
10. UK Air accident investigation branch (AAIB). www.dft. gov.uk/stellent/groups/dft_accidentinvest_page.hcsp

PART 2

Scientific fundamentals

Chapter 2 Mathematics

General introduction

This chapter aims to provide you with a sound foundation in mathematical principles, which will enable you to solve mathematical, scientific and associated aircraft engineering problems at the mechanic and technician level. Mathematics is divided into two major parts: *Non-calculator mathematics*, which covers all of the mathematics laid down in *Joint Aviation Requirements (JAR) 66 Module 1*, up to the level appropriate for aircraft maintenance category B certifying technicians. The other part of mathematics is *Further mathematics* (Chapter 3), which in the opinion of *the authors*, is necessary for a thorough understanding of the physics and electrical principles that follow. A second objective of Further mathematics is to provide the mathematical base necessary for further academic and professional progression, particularly for those individuals wishing to become Incorporated Engineers, after successfully obtaining their category B license.

We start with some elementary arithmetic. In particular, we review the concepts of number and the laws that need to be followed, when carrying out arithmetic operations, e.g. addition, subtraction, multiplication and division. The important concept of arithmetic estimates and estimation techniques involving various forms of number are also covered. While revising the fundamental principles of number, we consider both *explicit numbers* and *literal numbers* (letters), in order to aid our understanding of not only arithmetic operations, but also the algebraic operations that will follow later. Decimal numbers and the powers of 10 are then considered, after which fractional numbers and the manipulation of fractions are covered.

The algebraic content of JAR 66 Module 1 is introduced with the study of powers and exponents (indices) of numbers. This, together with your previous knowledge of fractions and fractional numbers, will provide you with the tools necessary to manipulate algebraic expressions and equations. The essential skill of transposition of formulae is also covered. This will be a particularly useful mathematical tool, when you study your physics and electrical principles. We finish our study of algebra by considering binary and other number systems and their application to simple logic circuits.

In our study of geometry and trigonometry, we start by looking at the methods used for the graphical solution of equations and other functions. This section clearly lays out the idea of graphical axes and scales. We then consider the nature and use of the trigonometric ratios and the solution of right-angled triangles and the circle. The nature and use of rectangular and polar co-ordinate representation systems, for finding bearings and angles of elevation and depression are then considered. We finish our study of non-calculator mathematics with a study of the more important theorems of the circle, together with some geometric constructions, considered particularly useful to solve engineering problems, in particular, as an aid to engineering drawing and marking out.

In our *Further mathematics* (Chapter 3) we build on our initial study of algebra by considering more complex algebraic and logarithmic expressions, functions and formulae. We will use our basic knowledge of graphs to represent more complex algebraic and logarithmic functions and to solve equations and engineering problems, which involve these functions. In addition, we will briefly introduce the concept of complex numbers, which will be found particularly valuable for those wishing to pursue an avionic pathway.

Our further study of trigonometry will include the use of trigonometric ratios to solve engineering problems involving measurement.

Next, we introduce and use a variety of statistical methods to gather, manipulate and display scientific and engineering data. We will then consider the ways in which the elementary rules of calculus arithmetic may be used to solve problems involving simple differentiation and integration of algebraic and trigonometric functions. Finally, we use the calculus to solve some elementary engineering problems, which involve rates of change and the summation of areas and volumes.

In order to aid your understanding of mathematics, you will find numerous fully worked examples and *test your understanding* exercises, spread throughout this chapter. In addition, typical example JAR 66 license questions are given at the end of this chapter.

Important note: Only very familiar units, such as mass, weight, pressure, length, area and volume are used in this part of the mathematics. *The detailed study of units appears in the chapters on physics and electrical principles (Chapters 4 and 5, respectively), where their nature and use is fully explained.* Some of the JAR 66 questions, found at the end of this chapter, require the reader to have some understanding of units, which may be gained by studying other sections of the book (in particular, Chapter 4).

NON-CALCULATOR MATHEMATICS

2.1 Introduction

As mentioned earlier, this part of the mathematics has been written explicitly to cover all of the syllabus content laid down in JAR 66 Module 1. It can thus be studied independently, by those only wishing to gain the knowledge necessary to pass the Civil Aviation Authority (CAA) examination for this module.

However, in order to offer the best chance of success in the JAR 66 physics and electrical and electronic principles modules and as a preparation for further study, *the authors*, strongly recommend that you should also study the further mathematics contained in Chapter 3.

2.2 Arithmetic

2.2.1 Numbers and symbols

It is generally believed that our present number system began with the use of the *natural numbers*, such as $1, 2, 3, 4, \ldots$. These whole numbers, known as the *positive integers*, were used primarily for counting. However, as time went on, it became apparent that whole numbers could not be used for defining certain mathematical quantities. For example, a period in time might be between 3 and 4 days or the area of a field might be between 2 and 3 acres (or whatever unit of measure was used at the time). So the *positive fractions* were introduced, e.g. $\frac{1}{2}$, $\frac{1}{4}$ and $\frac{3}{4}$. These two groups of numbers, the positive integers and the positive fractions, constitute what we call the *positive rational numbers*. Thus, 711 is an integer or whole number, $\frac{1}{4}$ is a positive fraction and $234\frac{3}{5}$ is a rational number. In fact a *rational number is any number that can be expressed as the quotient of two integers*, i.e. any number that can be written in the form a/b where a and b represent any integers. Thus $\frac{2}{5}$, $\frac{8}{9}$ and 1 are all rational numbers. The number 1 can be represented by the quotient $\frac{1}{1} = 1$, in fact *any number divided by itself must always be equal to 1*.

The natural numbers are *positive* integers, but suppose we wish to subtract a larger natural number from a smaller natural number, e.g. 10 subtracted from 7, we obviously obtain a number which is *less than zero*, i.e. $7 - 10 = -3$. So our idea of numbers must be enlarged to include numbers less than zero called *negative numbers*. The number zero (0) is unique, it is not a natural number because all natural numbers represent positive integer values, i.e. numbers above zero and quite clearly from what has been said, it is not a negative number either. It sits uniquely on its own and must be added to our number collection.

Key point

The natural numbers are known as positive integers.

So to the natural numbers (positive integers) we have added negative integers, the concept of zero, positive rational numbers and negative natural numbers. What about numbers like $\sqrt{2}$? This is *not* a rational number because it cannot be represented by the quotient of two integers. So yet another class of number needs to be included, the *irrational* or non-rational numbers. Together all, the above kinds of numbers constitute the broad class of numbers known as *real numbers*.

They include positive and negative terminating and non-terminating decimals (e.g. $\pm\frac{1}{9} = \pm 0.1111\ldots$, 0.48299999, ± 2.5, $1.73205\ldots$). The real numbers are so called to distinguish them from others, such as *imaginary* or *complex* numbers, the latter may be made up of both real and imaginary number parts. Complex numbers will not be considered during our study of mathematics.

Key point

A rational number is any number that can be expressed as the quotient of two integers, i.e. *a/b* where *a* and *b* are any two integers.

Although we have mentioned negative numbers, we have not considered their arithmetic manipulation. All positive and negative numbers are referred to as *signed numbers* and they obey the *arithmetic laws of sign*. Before we consider these laws, let us first consider what we mean by signed numbers.

Conventional representation of signed numbers is shown below, with zero at the midpoint. Positive numbers are conventionally shown to the right of zero and negative numbers to the left:

$$\cdots \; -4 \; -3 \; -2 \; -1 \; 0 \; +1 \; +2 \; +3 \; +4 \; \cdots$$

The number of units a point is from zero, regardless of its direction, is called the *absolute value* of the number corresponding to the point on the above number system when points are drawn to scale. Thus the absolute value of a positive number, or of zero, is the number itself. While the absolute value of a negative number is the number with its sign changed. For example, the absolute value of $+10$ is 10 and the absolute value of -10 is also 10. Now the absolute value of any number n is represented by the symbol $|n|$. Thus $|+24|$ means the absolute value of $+24$. Which is larger, $|+3|$ or $|-14|$?

I hope you said $|-14|$ because its absolute value is 14, while that of $|+3|$ is 3 and of course 14 is larger than 3. We are now ready to consider the laws of signs.

Key point

The absolute value of any number n is always its *positive value* or *modulus* and is represented by $|n|$.

The laws of signs

You are probably already familiar with these laws, here they are:

First law: To add two numbers with like signs, add their absolute values and prefix their common sign to the result.

This law works for ordinary arithmetic numbers and simply defines what we have always done in arithmetic addition.

For example, $3 + 4 = 7$ or in full $(+3) + (+4) = +7$.

After the introduction of the negative numbers, the unsigned arithmetic numbers became the positive numbers, as illustrated above. So now all numbers may be considered either positive or negative, and the laws of signs apply to them all.

Does the above law apply to the addition of two negative numbers? From ordinary arithmetic we know that $(-7) + (-5) = -12$. This again obeys the first law of signs, because we add their *absolute value* and prefix their common sign.

Second law: To add two signed numbers with unlike signs, subtract the smaller absolute value from the larger and prefix the sign of the number with the larger absolute value to the results.

So following this rule, we get for example:

$$5 + (-3) = 2; \quad -12 + 9 = -3; \quad 6 + (-11) = -5$$

and so on.

The numbers written without signs are, of course, positive numbers. Notice that brackets have been removed when not necessary.

Third law: To subtract one signed number from another, change the sign of the number to be subtracted and follow the rules for addition.

For example, if we subtract 5 from −3, we get $-3 - (+5) = -3 + (-5) = -8$.

Now what about the multiplication and division of negative and positive numbers, so as not to labour the point the rules for these operations are combined in our fourth and final law.

Fourth law: To multiply (or divide) one signed number by another, multiply (or divide) their absolute values; then, if the numbers have like signs, prefix the plus sign to the result; if they have unlike signs, prefix the minus sign to the result.

Therefore, applying this rule to the multiplication of two positive numbers, e.g. $3 \times 4 = 12$, $12 \times 8 = 96 \ldots$ and so on, which of course, is simple arithmetic! Now applying the rule to the multiplication of mixed sign numbers we get e.g. $-3 \times 4 = -12$, $12 \times (-8) = -96 \ldots$ and so on. We can show, equally well, that the above rule yields similar results for division.

Example 2.1

Apply the fourth law to the following arithmetic problems and determine the arithmetic result:
(a) $(-4)(-3)(-7) = ?$ (b) $14/-2 = ?$
(c) $5(-6)(-2) = ?$ (d) $-22/-11 = ?$

(a) In this example we apply the fourth law twice, $(-4)(-3) = 12$ (like signs) and so $12(-7) = -84$.
(b) $14/-2$ applying the third law for unlike signs immediately gives -7, the correct result.
(c) Again applying the third law twice. $5(-6) = -30$ (unlike signs) and $(-30)(-2) = 60$.
(d) $-22/-11$ applying the third law for like sign gives 2, the correct result.

The use of symbols

We have introduced earlier the concept of *symbols* to represent numbers when we defined rational numbers where the *letters a and b* were used to represent *any* integer. Look at the symbols below, do they represent the same number?

$$\text{IX}; \quad 9; \quad \text{nine}; \quad +\sqrt{81}$$

I hope you answered *yes*, since each expression is a perfectly valid way of representing the positive integer 9. In *algebra* we use letters to represent Arabic numerals such numbers are called *general numbers* or *literal numbers*, as distinguished from *explicit numbers* like 1, 2, 3, etc. Thus a literal number is simply a number represented by a letter, instead of a numeral. Literal numbers are used to state algebraic rules, laws and formulae; these statements being made in mathematical sentences called *equations*.

If a is a positive integer and b is 1, what is a/b? I hope you were able to see that $a/b = a$. Any number divided by 1 is always itself. Thus, $a/1 = a$, $c/1 = c$, $45.6/1 = 45.6$.

Suppose a is again any positive integer, but b is 0. What is the value of a/b? What we are asking is what is the value of any positive integer divided by zero? Well the answer is that we really do not know! *The value of the quotient a/b, if $b = 0$, is not defined in mathematics.* This is because there is no such quotient that meets the conditions required of quotients. For example, you know that to check the accuracy of a division problem, you can multiply the quotient by the *divisor* to get the *dividend*. For example, if $21/7 = 3$, then 7 is the divisor, 21 is the dividend and 3 is the quotient and so $3 \times 7 = 21$, as expected. So, if $17/0$ were equal to 17, then 17×0 should again equal 17 but it does not! Or, if $17/0$ were equal to zero, then 0×0 should equal 17 but again it does not. *Any number multiplied by zero is always zero.* Therefore, division of any number by zero (as well as zero divided by zero) is excluded from mathematics. *If $b = 0$, or if both a and b are zero, then a/b is meaningless.*

Key point

Division by zero is not defined in mathematics.

When multiplying literal numbers together we try to avoid the multiplication sign ($\times$), this is because it can be easily mistaken for the letter x.

Thus, instead of writing $a \times b$ for the product of two general numbers, we write $a \cdot b$ (the dot notation for multiplication) or more usually just ab to indicate the product of two general numbers a and b.

Example 2.2

If we let the letter n stand for any real number, what does each of the following expressions equal?

(a) $n/n = ?$ (b) $n \times 0 = ?$ (c) $n \times 1 = ?$
(d) $n + 0 = ?$ (e) $n - 0 = ?$ (f) $n - n = ?$
(g) $n/0 = ?$

(a) $n/n = 1$, i.e. any number divided by itself is equal to 1.
(b) $n \times 0 = 0$, any number multiplied by zero is itself zero.
(c) $n \times 1 = n$, any number multiplied *or* divided by 1 is itself.
(d) $n + 0 = n$, the addition of zero to any number will not alter that number.
(e) $n - 0 = n$, the subtraction of zero from any number will not alter that number.
(f) $n - n = 0$, subtraction of any number from itself will always equal zero.
(g) $n/0$, division by zero is *not defined* in mathematics.

The commutative, associative and distributive laws

We all know that $6 \times 5 = 30$ and $5 \times 6 = 30$, so is it true that when multiplying any two numbers together, the result is the same no matter what the order? The answer is yes. The above relationship may be stated as:

The product of two real numbers is the same no matter in what order they are multiplied. That is, $ab = ba$ this is known as the commutative law of multiplication.

If three or more real numbers are multiplied together, the order in which they are multiplied still makes no difference to the product. For example, $3 \times 4 \times 5 = 60$ and $5 \times 3 \times 4 = 60$. This relationship may be stated formally as:

The product of three or more numbers is the same no matter in what manner they are

grouped. That is, $a(bc) = (ab)c$; this is known as the associative law of multiplication.

These laws may seem ridiculously simple, yet they form the basis of many algebraic techniques, which you will be using later!

We also have commutative and associative laws for addition of numbers, which by now will be quite obvious to you, here they are:

The sum of two numbers is the same no matter in what order they are added. That is, $a + b = b + a$. This is known as the commutative law of addition.

The sum of three or more numbers is the same no matter in what manner they are grouped. That is, $(a + b) + c = a + (b + c)$. This is known as the associative law of addition.

You may be wondering where the laws are for subtraction. Well you have already covered these when we considered the laws of signs. In other words, the above laws are valid no matter whether or not the number is positive or negative. So, for example, $-8 + (16 - 5) = 3$ and $(-8 + 16) - 5 = 3$

In order to complete our laws we need to consider the following problem: $4(5 + 6) = ?$ We may solve this problem in one of two ways, firstly by adding the numbers inside the brackets and then multiplying the result by 4, this gives: $4(11) = 44$. Alternatively, we may multiply out the bracket as follows: $(4 \times 5) + (4 \times 6) = 20 + 24 = 44$. Thus, whichever method we choose, the arithmetic result is the same. This result is true in all cases, no matter how many numbers are contained within the brackets!

So in general, using literal numbers we have:

$$a(b + c) = ab + ac$$

This is the *distributive law*. In words, it is rather complicated:

The distributive law states that: the product of a number by the sum of two or more numbers is equal to the sum of the products of the first number by each of the numbers of the sum.

Now, perhaps you can see the power of algebra in representing this law, it is a lot easier to remember than the wordy explanation!

Remember that the distributive law is valid no matter how many numbers are contained in the brackets, and no matter whether the sign connecting them is a plus or minus. As you will see later, this law is one of the most useful and convenient rules for manipulating formulae and solving algebraic expressions and equations.

Key point

The commutative, associative and distributive laws of numbers are valid for both positive and negative numbers.

Example 2.3

If $a = 4$, $b = 3$ and $c = 7$, does

$$a(b - c) = ab - ac$$

The above expression is just the distributive law, with the sign of one number within the bracket, changed. This of course is valid since the sign connecting the numbers within the bracket may be a plus or minus. Nevertheless, we will substitute the arithmetic values in order to check the validity of the expression.

Then:

$$4(3 - 7) = 4(3) - 4(7)$$
$$4(-4) = 12 - 28$$
$$-16 = -16$$

So, our law works irrespective of whether the sign joining the numbers is positive or negative.

Long multiplication

It is assumed that the readers of this book will be familiar with long multiplication and long division. However, with the arrival of the calculator these techniques are seldom used and quickly forgotten. CAA license examinations, for category A and B certifying staff, do not allow the use of calculators; so these techniques will need to be revised. One method of long multiplication is given below. *Long division will be found in Section 2.3*, where the technique is used for both explicit and literal numbers!

Suppose we wish to multiply 35 by 24, i.e. 24×35. You may be able to work this out in your head; we will use a particular method of long multiplication to obtain the result.

The numbers are first set out, one under the other, like this: $\begin{array}{r} 35 \\ \underline{24} \end{array}$ where the right-hand integers 5 and 4 are the units and the left-hand integers are the tens, i.e. 3×10 and 2×10. We multiply the tens on the bottom row by the tens and units on the top row. So to start this process, we place a nought in the units column underneath the bottom row, then multiply the 2 by 5 to get 1×10, carry the 1 into the tens column and add it to the product 2×3; i.e.:

$$\begin{array}{r} 35 \\ \underline{24} \\ 0 \end{array}$$

then multiply the $2 \times 5 = 10$, put in the nought of the ten and carry the one

$$\begin{array}{r} 35 \\ \underline{24} \\ {}^{1}00 \end{array}$$

now multiply $2 \times 3 = 6$ (the tens) and add the carried ten to it, to give 7, then

$$\begin{array}{r} 35 \\ \underline{24} \\ 700 \end{array}$$

We now multiply the 4 units by 35. That is $4 \times 5 = 20$ put down the nought carry 2 into the ten column, then multiply the 4 units by the 3 tens or, $4 \times 3 = 12$ and add to it the 2 we carried to give 140, i.e.:

$$\begin{array}{r} 35 \\ \underline{24} \\ 700 \\ \underline{140} \end{array}$$

All that remains for us to do now is add 700 to 140 to get the result by long multiplication, i.e.:

$$\begin{array}{r} 35 \\ \underline{24} \\ 700 \\ \underline{140} \\ \underline{840} \end{array}$$

So $35 \times 24 = 840$. This may seem a rather long-winded way of finding this product. You should adopt the method you are familiar with.

This process can be applied to the multiplication of numbers involving hundreds, thousands and decimal fractions, it works for them all!

For example, 3.5×2.4 could be set out in the same manner as above, but the columns would be for tenths and units, instead of units and tens. Then we would get:

$$
\begin{array}{r}
3.5 \\
\underline{2.4} \\
7.0 \\
\underline{1.4} \\
8.4
\end{array}
$$

Notice that in this case the decimal place has been shifted two places to the left. If you do not understand why this has occurred you should study carefully the section on *decimals and the powers of 10* that follows.

Example 2.4

Multiply: (1) 350×25 (2) 18.8×1.25

In both the cases, the multiplication is set out as shown before.

1. With these figures, hundreds, tens and units are involved. You will find it easier to multiply it by the smallest or the least complex number.
   $$
   \begin{array}{r}
   350 \\
   \underline{25}
   \end{array}
   $$
   Now we multiply by 25 in a similar manner to the previous example.

Multiply first by the 2×10, which means placing a nought in the units column first. Then multiply 2×0, putting down below the line the result, i.e. zero. Then: $2 \times 5 = 1 \times 10$, again put down the nought and carry the single hundred. So we get:

$$
\begin{array}{r}
350 \\
\underline{25} \\
{}^{1}00
\end{array}
$$

We continue the process by multiplying 2 by the 3 hundreds and adding the single hundred or $2 \times 3 + 1 = 7$ to give 7000 (remembering the nought, we first put down). This part of the process was the equivalent of multiplying $350 \times 20 = 7000$. So we get:

$$
\begin{array}{r}
350 \\
\underline{25} \\
7000
\end{array}
$$

We now multiply the number 350 by 5, where $5 \times 0 = 0$; put it down below the line; $5 \times 5 = 25$ put down the 5 and carry the 2. Finally, $5 \times 3 = 15$, add the 2 you have just carried to give 17. So the total number below the 7000 is $1750 = 350 \times 5$ and we get:

$$
\begin{array}{r}
350 \\
\underline{25} \\
7000 \\
1750
\end{array}
$$

Finally we add the rows below the line to give the result, i.e.:

$$
\begin{array}{r}
350 \\
\underline{25} \\
7000 \\
\underline{1750} \\
8750
\end{array}
$$

Then $350 \times 25 = 8750$.

2. For this example the multiplication is laid out in full, without explanation, just make sure you can follow the steps.

$$
\begin{array}{r}
18.8 \\
\underline{1.25} \\
18800 \\
3760 \\
\underline{940} \\
23.500
\end{array}
$$

Then, $18.8 \times 1.25 = \mathbf{23.5}$.

Note that the decimal point is positioned three places to the left, since there are three integers to the right of the decimal points.

You should now attempt the following exercise, *without* the aid of a calculator!

Test your understanding 2.1

1. 6, 7, 9, 15 are _____ numbers.

2. $\frac{8}{5}$, $\frac{1}{4}$, $\frac{7}{64}$ are _____ numbers.

3. Rewrite the numbers 5, 13, 16 in the form a/b, where $b = 6$.

4. Express the negative integers -4, -7, -12 in the form a/b, where b is the positive integer 4.

5. $+\sqrt{16}$ can be expressed as a positive _____. It is _____.

6. $\sqrt{10}$ cannot be expressed as a _____ number; however, it is a _____.

7. Express as non-terminating decimals: (a) $\frac{1}{3}$, (b) $\frac{1}{7}$, (c) 2.

8. Find the value of:
   (a) $a(b + c - d)$, where $a = 3$, $b = -4$, $c = 6$ and $d = -1$
   (b) $(21 - 6 + 7)3$ (c) $6 \times 4 + 5 \times 3$

9. Which of the following has the largest absolute value: -7, 3, 15, -25, -31?

10. $-16 + (-4) - (-3) + 28 = ?$

11. Find the absolute value of $-4 \times (14 - 38) + (-82) = ?$

12. What is (a) $\frac{15}{-3}$ (b) $3 \times \frac{-12}{2}$ (c) $-1 \times \frac{14}{-2}$?

13. What is (a) $(-3)(-2)(16)$,   (b) $-3 \times -2(15)$.

14. Evaluate $2a(b + 2c + 3d)$, when $a = 4$, $b = 8$, $c = -2$ and $d = 2$.

15. Use long multiplication to find the products of the following:
    (a) 23.4×8.2 (b) 182.4×23.6 (c) 1.25×0.84
    (d) 1.806×1.2 (e) $35 \times 25 \times 32$ (f) $0.014 \times 2.2 \times 4.5$

2.2.2 Decimal numbers, powers of ten and estimation techniques

The powers of ten are sometimes called "*the technicians shorthand*". They enable very large and very small numbers to be expressed in simple terms. You may have wondered why, in our study of numbers, we have not mentioned **decimal numbers**, before now. Well the reason is simple, these are the numbers you are most familiar with, they may be rational, irrational or real numbers. Other numbers, such as the positive and negative integers, are a subset of real numbers. The exception are the complex numbers, these are not a subset of the real numbers and do not form part of our study in this course.

> **Key point**
>
> Decimal numbers may be rational, irrational or real numbers.

Essentially then, decimal numbers may be expressed in index form, using the powers of ten. For example:

$$
\begin{aligned}
1{,}000{,}000 &= 1 \times 10^6 \\
100{,}000 &= 1 \times 10^5 \\
10{,}000 &= 1 \times 10^4 \\
1000 &= 1 \times 10^3 \\
100 &= 1 \times 10^2 \\
10 &= 1 \times 10^1 \\
0 &= 0 \\
1/10 = 0.1 &= 1 \times 10^{-1} \\
1/100 = 0.01 &= 1 \times 10^{-2} \\
1/1000 = 0.001 &= 1 \times 10^{-3} \\
1/10{,}000 = 0.0001 &= 1 \times 10^{-4} \\
1/100{,}000 = 0.00001 &= 1 \times 10^{-5} \\
1/1{,}000{,}000 = 0.000001 &= 1 \times 10^{-6}
\end{aligned}
$$

I am sure you are familiar with the above shorthand way of representing numbers. For example, we show the number one million (1,000,000) as 1×10^6, i.e. 1 multiplied by 10, six times. The *exponent* (*index*) of 10 is 6, thus the number is in exponent or *exponential form*, the *exp* button on your calculator!

Note that we multiply all the numbers, represented in this manner by the number 1. This is because we are representing one million, one hundred thousand, one tenth, etc.

When representing decimal numbers in index (exponent) form, the multiplier *is always a number which is* ≥ 1.0 *or* < 10; i.e. a number greater than or equal to (≥ 1.0) one or less than (< 10) ten.

> **Key point**
>
> A number in exponent or index form, always starts with a multiplier which is ≥ 1.0 and ≤ 10.0.

So, for example, the decimal number is $8762.0 = 8.762 \times 10^3$ in index form. Note that with this *number, greater than 1.0, we displace the decimal point* three (3) places *to the left*; i.e. three powers of ten. Numbers rearranged in this way, using powers of ten, are said to be in *index form* or *exponent form* or *standard form*, dependent on the literature you read.

Key point

When a decimal number is expressed in exponent form, it is often referred to as index form or standard form.

What about the decimal number 0.000245? Well I hope you can see that in order to obtain a multiplier that is greater than or equal to one and less than 10, we need to *displace the decimal point* four (4) places *to the right*. Note that the zero in front of the decimal point is placed there to indicate that a whole number has not been omitted. Therefore, the number in index form now becomes 2.45×10^{-4}. Notice *that for numbers less than 1.0, we use a negative index*. In other words, *all decimal fractions represented in index form have a negative index and all numbers greater than 1.0, represented in this way, have a positive index*.

Every step in our argument up till now has been perfectly logical but how would we deal with a mixed whole number and decimal number such as 8762.87412355? Well again to represent this number exactly, in index form, we proceed in the same manner, as when dealing with just the whole number. So displacing the decimal point three places to the left to obtain our multiplier gives $8.76287412355 \times 10^3$. This is all very well but one of the important reasons for dealing with numbers in index form is that the manipulation should be easier! In the above example we still have 12 numbers to contend with plus the powers of ten.

In most areas of engineering, there is little need to work to so many places of decimals. In the above example for the original number, we have eight decimal place accuracy, this is unlikely to be needed, unless we are dealing with a subject like rocket science or astrophysics! So this leads us into the very important skill of being able to provide *approximations or estimates to a stated degree of accuracy*.

Example 2.5

For the numbers (a) 8762.87412355 and (b) 0.0000000234876

(i) Convert these numbers into *standard form* with three decimal place accuracy.
(ii) Write down these numbers in *decimal form*, correct to two significant figures.

(i)(a) We have already converted this number into standard form, it is: $8.76287412355 \times 10^3$. Now looking at the decimal places for the stated accuracy we must consider the first four places 8.7628 and since the last *significant figure* is 8, in this case (*greater than 5*) *we round up* to give the required answer as $\mathbf{8.763 \times 10^3}$.

(b) $0.0000000234876 = 2.34876 \times 10^{-8}$ and now following the same argument as above this number is to three decimal places $= \mathbf{2.349 \times 10^{-8}}$.

(ii)(a) For the number 8762.87412355, the two required significant figures are to the left of the decimal place. So we are concerned with the whole number 8762 and the first two figures are of primary concern again to find our approximation we need to first consider the three figures 876, again since 6 is above halfway between 1 and 10, then we round up to give the required answer 8800.

Note that we had to add two zeros to the left of the decimal point. This should be obvious when you consider that all we have been asked to do is approximate the number 8762 to within two significant figures.

(b) For the number 0.0000000234876 the significant figures are any integers to the right of the decimal point and the zeros. So, in this case, the number to the required number of significant figures is 0.000000023.

We are now in a position to be able to determine estimates, not just for single numbers but also for expressions involving several numbers. The easiest way of achieving these estimates is to place all numbers involved into standard form and then determine the estimate to the correct degree of accuracy. You may wonder why we do not simply use our calculators and determine values to eight decimal place accuracy. Well, you need only to press one button incorrectly on your calculator to produce an incorrect answer, but how will you know if your answer is incorrect, if you are unable to obtain a rough estimate of what the correct answer should be? Just imagine the consequences if you only put one tenth of the fuel load into the aircraft's fuel tanks, just prior to take-off! This is where the use of estimation techniques proves to be the most useful; these techniques are best illustrated by the following example.

Example 2.6

(a) Determine an estimate for $3.27 \times 10.2 \times 0.124$ correct to one significant figure.
(b) Simplify:
$$\frac{3177.8256 \times 0.000314}{(154025)^2}$$
giving your answer correct to two significant figures.

(a) You might be able to provide an estimate for this calculation, without converting to standard form. For the sake of completeness and to illustrate an important point we will solve this problem, using the complete process.

First we convert all numbers to standard form, this gives:
$$(3.27 \times 10^0)(1.02 \times 10^1)(1.24 \times 10^{-1})$$

Note that $3.27 \times 10^0 = 3.27 \times 1 = 3.27$; in other words, it is already in standard form! Now considering each of the multipliers and rounding to *one significant figure* gives:
$$(3 \times 10^0)(1 \times 10^1)(1 \times 10^{-1})$$
and remembering your first law of indices!
$$(3 \times 1 \times 1)(10^{0+1-1}) = (3)(10^0) = 3(1) = 3.0$$

You may feel that this is a terribly long-winded way to obtain an estimation because the numbers are so simple, but with more complex calculations, the method is very useful indeed.

(b) Following the same procedure as above gives:
$$\frac{(3.1778256 \times 10^3)(3.14 \times 10^{-4})}{(1.54025 \times 10^5)^2}$$
$$= \frac{(3.2 \times 10^3)(3.1 \times 10^{-4})}{(1.5 \times 10^5)^2}$$

Now again applying the laws of indices and the distributive law of arithmetic we get:
$$\frac{(3.2 \times 3.1)(10^{3-4})}{2.25 \times 10^{5 \times 2}} = \frac{(3.2 \times 3.1)10^{-1}}{2.25 \times 10^{10}}$$
$$= \left(\frac{3.2 \times 3.1}{2.25}\right)10^{-11}$$
$$= 4.4 \times 10^{-11}$$

Note that if you were unable to work out the multiplication and division, in your head!, then to one significant figure we would have $3 \times 3/2 = 4.5$, very near our approximation using two significant figures. The calculator answer to 10 significant figures is $4.206077518 \times 10^{-11}$. The error in this very small number (compared with our estimation) is something like two in one thousand million! Of course, the errors for very large numbers, when squared or raised to greater powers, can be significant!

Before leaving the subject of estimation, there is one important convention which you should know. Consider the number 3.7865. If we require an estimate of this number correct to four significant figures, what do we write? In this case, the last significant figure is a 5, so should we write this number as 3.786 or 3.787, correct to four significant figures? The convention states that we *round up when confronted with the number 5*. So the correct answer in this case would be 3.787.

Test your understanding 2.2

1. Express the following numbers in normal decimal notation:
   (a) $3 \times 10^{-1} + 5 \times 10^{-2} + 8 \times 10^{-2}$
   (b) $5 \times 10^3 + 81 - 10^0$
2. Express the following numbers in standard form:
   (a) 318.62   (b) 0.00004702
   (c) 51,292,000,000   (d) -0.00041045

3. Round-off the following numbers correct to three significant figures:
   (a) 2.713 (b) 0.0001267 (c) 5.435×10^4

4. Evaluate: (a) $(81.7251 \times 20.739)^2 - 52,982$
   (b) $\dfrac{(56.739721)^2 \times 0.0997}{(19787 \times 10^3)^2}$
   correct to two significant figures. Show all your working and express your answers in standard form.

2.2.3 Fractions

Before we look at some examples of algebraic manipulation, using the techniques we have just learnt, we need to devote a little time to the study of fractions. In this section, we will only consider fraction using *explicit* numbers. Later, in the main syllabus on algebra we also consider simple fractions using *literal* numbers; i.e. algebraic fractions. A study of the work that follows should enable you to manipulate simple fractions, without the use of a calculator!

I am often asked, why do we need to use fractions at all? Why not use only decimal fractions? Well, one very valid reason is that fractions provide *exact* relationships between numbers. For example, the fraction 1/3 is exact, but the decimal fraction equivalent has to be an approximation, to a given number of decimals 0.3333, is corrected to four decimal places. Thus, $1/3 + 1/3 + 1/3 = 1$ but $0.3333 + 0.3333 + 0.3333 = 0.9999$, not quite 1.

A fraction is a *division* of one number by another. Thus, the fraction 2/3 means two divided by three. The fraction x/y means the literal number x divided by y. The number above the line is called the *numerator*; the number below the line is the *denominator*, as you learnt before. Thus, fractions are represented as:

$$\frac{numerator}{denominator}$$

Fractions written in this form, with integers in the numerator and denominator are often known as *vulgar fractions*, e.g. $\frac{1}{2}$, $3\frac{1}{4}$, $\frac{3}{4}$, etc. Whereas fractions written in decimal form 0.5, 3.25, 0.75, 0.333, etc. are known, as their name implies, as *decimal fractions*.

Having defined the vulgar fraction, let us now look at how we multiply, divide, add and subtract these fractions. We start with multiplication, because unlike arithmetic on ordinary numbers, multiplication of fractions is the easiest operation.

Multiplication of fractions

In order to multiply two or more fractions together all that is necessary is to multiply all the numbers in the numerator together and all the numbers in the denominator together, in order to obtain the desired result.

For example:

$$\frac{1}{3} \times \frac{2}{3} \times \frac{1}{4} = \frac{1 \times 2 \times 1}{3 \times 3 \times 4} = \frac{2}{36}$$

now we are not quite finished, because the fraction 2/36 has numbers in the numerator and denominator which can be further reduced, without affecting the actual value of the fraction. I hope you can see that, if we divide the numerator and denominator by 2, we reduce the fraction to 1/18 without affecting the value. Because we have divided the fraction by $2/2 = 1$ the *whole fraction* is unaltered. You can easily check the validity of the process by dividing 1 by 18 and also 2 by 36 on your calculator, in both cases we get the *recurring decimal fraction* 0.055555. Note that the exact value of this fraction cannot be given in decimal form.

Division of fractions

Suppose we wish to divide 1/3 by 2/3, in other words $\frac{1/3}{2/3}$, the trick is *to turn the devisor (the fraction doing the dividing) upside down and multiply*. In the above example we get $1/3 \times 3/2$ and we proceed as for multiplication, i.e.:

$$\frac{1 \times 3}{3 \times 2} = \frac{3}{6} = \frac{1}{2}$$

Note that again by cancelling numerator and denominator by 3, we get the lowest vulgar fraction. Now again, if you are not convinced that division can be turned into multiplication by using the above method, check on your calculator, or use decimal fractions, to confirm the result.

Addition of fractions

To add fractions, we are required to use some of our previous knowledge concerning *factors*. In particular, we need to determine the *lowest common multiple (LCM)* of two or more numbers. That is, the smallest possible number which is a common multiple of two or more numbers. For example, 10 is a multiple of 5, 30 is a common multiple of 5 and 3, but 15 is the LCM of 5 and 3. Thus, 15 is the smallest possible number which is exactly divisible by both 5 and 3. What is the LCM of 2, 3 and 4? One multiple is found simply as $2 \times 3 \times 4 = 24$, but is this the lowest? Well, of course, it is not. The number 4 is exactly divisible by 2, so is the number 24, to give 12 which is the LCM, well 12 is divisible by the numbers 2, 3 and 4, but the number 6 is not. So, 12 *is the LCM* of the three numbers 2, 3 and 4.

So when adding fractions, why may it be necessary to find the LCM? We will illustrate the process by example.

Example 2.7

Add the following fractions:
(a) $\frac{1}{3} + \frac{1}{4}$ (b) $\frac{2}{5} + \frac{1}{3} + \frac{1}{2}$

(a) We first determine the LCM of the numbers in the denominator. In this case the lowest number divisible by both 3 and 4 is 12. So 12 is the LCM.

Now remembering that the whole idea of adding fractions together is to create one fraction as their sum then we place the LCM below the denominators of all the fractions we wish to add. In this case we get:

$$\frac{\frac{1}{3} + \frac{1}{4}}{12}$$

we now divide 3 into 12 to give 4 and then multiply 4 by the number in the numerator of the fraction 1/3, in this case it is 1; so $4 \times 1 = 4$, which is the result, that will now be placed above the 12. In a similar way, we now consider the fraction 1/4 to be added, where 4 into 12 is 3 and $3 \times 1 = 3$. Thus, we

now have the numbers to be added as:

$$\frac{\frac{1}{3} + \frac{1}{4}}{12} = \frac{4 + 3}{12} = \frac{7}{12}$$

make sure you follow the rather complex logic to obtain the numbers 4 and 3 above the denominator 12, as shown above. Again just to remind you, let us consider the first fraction to be added 1/3. We take the denominator of this fraction 3 and divide it into our LCM, to give the result 4. We then multiply this result (4 in our case) by the numerator of the fraction 1/3, which gives $4 \times 1 = 4$. This process is then repeated on the second fraction to be added, and so on. We then add the numbers in the numerator to give the required result.

(b) We follow the same process, as above to add these three fractions together. The LCM is 30, I hope you can see this. Remember, even if you cannot find the LCM, multiplying all the numbers in the denominator together will always produce a *common multiple*, which can always be used in the denominator of the final fraction.

So we get:

$$\frac{2}{5} + \frac{1}{3} + \frac{1}{2} = \frac{12 + 10 + 15}{30} = \frac{37}{30} = 1\frac{7}{30}$$

Again the number 12 was arrived at by dividing 5 into 30 to give 6 and then multiplying this result by the numerator of the first fraction to give $2 \times 6 = 12$. The numbers 1 and 15 were derived in the same way.

The result of adding the numbers in the numerator of the final fraction gives 37/30, this is known as an *improper fraction*, because it contains a whole integer of 1 or more and a fraction. The final result is found simply, by dividing the denominator (30) into the numerator (37) to give 1 and a remainder of 7/30.

Subtraction of fractions

In the case of subtraction of fractions we follow the same procedure as with addition, until we obtained the numbers above the common denominator. At which point we subtract them, rather than add them. For example, for the fractions given below, we get:

$$\frac{2}{5} + \frac{1}{3} - \frac{1}{2} = \frac{12 + 10 - 15}{30} = \frac{7}{30}$$

Similarly:

$$\frac{3}{8} - \frac{1}{4} + \frac{1}{2} - \frac{1}{8} = \frac{3 - 2 + 4 - 1}{8} = \frac{4}{8} = \frac{1}{2}$$

Note that for these fractions the LCM is not just the product of the factors, but is truly the lowest number, which is divisible by all the numbers in the devisors of these fractions.

Example 2.8

Simplify the following fractions:

(a) $\frac{2}{3} + \frac{3}{5} - \frac{1}{2}$ (b) $\left(\frac{3}{4}\right) \times \left(\frac{3}{8} + \frac{5}{16} - \frac{1}{2}\right)$
(c) $2\frac{5}{8} \div \frac{7}{16} - \frac{3}{8}$

(a) Recognizing that the LCM is 30, which enables us to evaluate this fraction using the rules for addition and subtraction of fractions given before, then:

$$\frac{2}{3} + \frac{3}{5} - \frac{1}{2} = \frac{20 + 18 - 15}{30} = \frac{23}{30}$$

(b) In this example, we need to simplify the right-hand bracket, *before* we multiply. So we get:

$$\left(\frac{3}{4}\right) \times \left(\frac{6 + 5 - 8}{16}\right) = \left(\frac{3}{4}\right) \times \left(\frac{3}{16}\right) = \frac{9}{64}$$

(c) This example involves a whole number fraction to apply the rules, the fraction $2\frac{5}{8}$ is best put into improper form, which is $\frac{21}{8}$. Note, to obtain this form we simply multiply the denominator by the whole number and add the existing numerator, i.e. $(2 \times 8) + 5 = 21$ to obtain the new numerator. We next need to apply the rules of arithmetic, *in the correct order*, to solve the fraction. This follows on from the number laws you learnt earlier. The *arithmetic law of precedence* tells us that we must carry out the operations in the following order: *brackets, of, division,*

multiplication, addition, subtraction (you might have remembered this order using the acronym BODMAS).

This tells us (for our example) that we must carry out division before subtraction, there is no choice!

So following the process discussed above, we get:

$$\left(\frac{21}{8} \div \frac{7}{16} - \frac{3}{8}\right) = \left(\frac{21}{8} \times \frac{16}{7} - \frac{3}{8}\right)$$
$$= \left(\frac{6}{1} - \frac{3}{8}\right) = \left(\frac{48 - 3}{8}\right)$$
$$= \frac{45}{8} = 5\frac{5}{8}$$

Note that the brackets have been included for clarity.

Test your understanding 2.3

1. Simplify the following fractions:
   (a) $\frac{3}{16} \times \frac{8}{15}$ (b) $\frac{3}{5} \div \frac{9}{125}$ (c) $\frac{1}{4}$ of $\frac{18}{5}$
2. Simplify the following fractions:
   (a) $\frac{2}{9} + \frac{15}{9} - \frac{2}{3}$ (b) $3\frac{2}{3} - 2\frac{1}{5} + 1\frac{5}{6}$ (c) $\frac{17}{7} - \frac{3}{14} \times 2$
3. What is the average of $\frac{1}{8}$ and $\frac{1}{16}$?
4. What is $\frac{5}{3} \div 1\frac{2}{3}$?
5. What is the value of $\left(\frac{1}{6} + \frac{4}{5}\right) + \frac{1}{10}$?
6. Simplify the following fraction: $\left(\frac{7}{12} \div \frac{21}{8}\right) \times \left(\frac{4}{5}\right) + \frac{3}{4}$ of $\frac{8}{9}$

2.2.4 Percentages and averages

Percentages

When comparing fractions it is often convenient to express them with a denominator of one hundred. So, for example:

$$\frac{1}{4} = \frac{25}{100} \quad \text{and} \quad \frac{4}{10} = \frac{40}{100}$$

Fractions like these with a denominator of 100 are called *percentages*.

Thus,

$$\frac{7}{10} = \frac{70}{100} = 70 \text{ percent or } 70\%$$

where the percentage sign (%) is used instead of the name in words. To obtain the percentage we have simply multiplied the fraction by 100.

Example 2.9

Convert the following fractions to percentages
(1) $\frac{4}{5}$ (2) $\frac{11}{25}$

(1) Then $\frac{4}{5} \times 100 = \frac{400}{5} = \mathbf{80\%}$
(2) Similarly $\frac{11}{25} \times 100 = \frac{1100}{25} = \mathbf{44\%}$

Decimal numbers can be converted into percentages in a similar way. For example:

$$0.45 = \frac{45}{100} = \frac{45}{100} \times 100 = 45\%$$

We can find the same result, simply by multiplying the decimal number by 100, omitting the intermediate step, so that $0.45 \times 100 = 45\%$.

Key point

To convert a vulgar fraction or decimal fraction into a percentage multiply by 100.

The reverse process, turning a percentage into a fraction, simply requires us to divide the fraction by 100. Thus,

$$52.5\% = \frac{52.5}{100} = 0.525$$

Remembering from your powers of ten, that dividing by 100 requires us to move the decimal place two places to the left.

Key point

To convert a percentage into a fraction divide the fraction by 100.

To find the percentage of a quantity is relatively easy, provided you remember to first express the quantities, as a fraction *using the same units*.

Example 2.10

1. Find 10% of 80
2. What percentage of £6.00 is 90 pence?

3. The total wing area of an aircraft is $120 \, \text{m}^2$. If the two main undercarriage assemblies are to be stored in the wings and each takes up $3.0 \, \text{m}^2$ of the wing area, what percentage of the total wing area is required to store the main undercarriage assemblies?

1. Units are not involved so expressing 10% as a fraction we get $\frac{10}{100}$ and so we require,

$$\frac{10}{100} \text{ of } 80 \quad \text{or} \quad \frac{10}{100} \times 80 = \frac{800}{100} = \mathbf{8\%}$$

2. Here we are involved with units, so converting £6.00 into pence gives 600 and so all that remains for us to do, is express 90 pence as a fraction of 600 pence and multiply by 100. Then:

$$\frac{90}{600} \times 100 = \frac{9000}{600} = \mathbf{15\%}$$

3. We need first to recognize that this problem is none other than finding what percentage of $120 \, \text{m}^2$ is $3.0 \times 2 \, \text{m}^2$ (since there are two main undercarriage assemblies). So following the same procedure as above and expressing the areas as a fraction, we get:

$$\frac{6}{120} \times 100 = \frac{600}{120} = \mathbf{5\%}$$

i.e. the undercarriage assemblies take up 5% of the total wing area.

Another, non-engineering use of percentages is to work out *profit and loss*. You might find this skill particularly useful to work out the effect of any pay rise or deductions on your wages!

Very simply, profit = selling price − cost price and similarly, loss = cost price − selling price. Now both of these can be expressed as a percentage, i.e:

$$\text{Profit } \% = \frac{\text{selling price} - \text{cost price}}{\text{cost price}} \times 100$$

and

$$\text{Loss } \% = \frac{\text{cost price} - \text{selling price}}{\text{cost price}} \times 100$$

Example 2.11

1. An aircraft supplier buys 100 packs of rivets for £60.00 and sells them to the airline operator for 80 pence each. What percentage profit does the supplier make?
2. The same supplier buys an undercarriage door retraction actuator for £1700.00 and because it is reaching the end of its shelf life, he must sell it for £1400.00. What is the suppliers percentage loss?

1. To apply the profit formula to this example we must first find the total selling price, in consistent units. This is 100×80 pence or £100 × 0.8 = £80. Then on application of the formula we get:

$$\text{Profit } \% = \frac{£80 - £60}{£60} = \frac{£20}{£60} \times 100$$
$$= \frac{2000}{60} = 33.3\%$$

2. This is somewhat easier than the previous example and only requires us to apply the percentage loss formula. Then:

$$\text{Loss } \% = \frac{£1700.00 - £1400.00}{£1700}$$
$$= \frac{£300.00}{£1700.00} \times 100$$
$$= \frac{30000}{1700} = 17.65\%$$

Averages

To find the average of a set of values, all we need to do is to add the values together and divide by the number of values in the set. This may be expressed as:

$$\text{Average} = \frac{\text{Sum of the values}}{\text{Total number of values}}$$

Example 2.12

The barometric pressure, measured in mm of mercury (mmHg), was taken everyday for a week. The readings obtained are shown below. What is the average pressure for the week in mmHg?

Day	1	2	3	4	5	6	7
mmHg	75.2	76.1	76.3	75.7	77.1	75.3	76.3

So, average pressure mmHg

$$= \frac{\begin{array}{c} 75.2 + 76.1 + 76.3 + 75.7 \\ + 77.1 + 75.3 + 76.3 \end{array}}{7}$$
$$= 76\,\text{mm}$$

Example 2.13

A light aircraft is loaded with 22 boxes. If nine boxes have a mass of 12 kg, eight boxes have a mass of 14 kg and five boxes have a mass of 14.5 kg. What is the total mass of the boxes *and* the average mass per box?

By finding the total mass of all 22 boxes, we can then find the average mass per box. So we have:

$$9 \times 12 = 108\,\text{kg}$$
$$8 \times 14 = 112\,\text{kg}$$
$$5 \times 15.5 = 77.5\,\text{kg}$$
$$\text{Total mass } = 297.5\,\text{kg}$$

Then average mass of all 22 boxes is $\frac{297.5}{22} = 13.52\,\text{kg}$ (by long division).

The above example illustrates the process we use to find the weighted average. A lot more will be said about averages and mean values when you study the statistics in Chapter 3.

2.2.5 Ratio and proportion

A *ratio* is a comparison between two similar quantities. We use ratios when determining the scale of things. For example, when reading a map we may say that the scale is 1 in 25,000 or 1 to 25,000. We can express ratios mathematically, either as fractions or in the form *1 : 25,000* read as, *one to twenty-five thousand*.

Apart from maps, we as aircraft technicians and engineers, are more likely to meet the idea of ratio when we need to read technical drawings or produce vector drawings to scale. For example, if we have a force of 100 N and we wish to represent its magnitude by a straight line of a specific length, then we may choose a scale say, 1 cm = 10 N, so effectively we are using a scale with a ratio of 1:10. When dealing with ratios it is important to deal with the *same* quantities. If we need to work out the ratio between 20 pence and £2.0, then first, we must put these quantities into the *same units*, i.e. 20 and 200 pence, so the ratio becomes 20:200 and in its simplest terms this is a ratio of 1:10, after division of both quantities by 20.

We may also express ratios as fractions, so in the case of 20 to 200 pence, then this is 1:10 as before or $\frac{1}{10}$ as a fraction.

Key point

A ratio can be presented as a *fraction* or using the *is to* (:) sign.

Example 2.14

Two lengths have a ratio of 13:7. If the second length is 91 m, what is the first length?

The first length $= \frac{13}{7}$ of the second length $= \left(\frac{13}{7}\right)91 = \mathbf{169}$ m.

Suppose now, we wish to split a long length of electrical cable into *three parts* that are *proportional* to the amount of money contributed to the cost of the cable by three people. Then if the overall length of the cable is 240 m and the individuals payed, £30.00, £40.00 and £50.00, respectively. How much cable do they each receive?

This is a problem that involves *proportional parts*. The amount of money paid by each individual is in the ratio 3:4:5, giving a total of $3 + 4 + 5 = 12$ parts. Then the length of each part $= \frac{240}{12}$ or 20 m. So each individual receives, respectively, $20 \times 3 = 60$ m, $20 \times 4 = 80$ m, $20 \times 5 = 100$ m. A quick check will show that our calculations are correct, i.e. $60 + 80 + 100 = 240$ m, the total length of the original electrical cable.

Direct proportion

Two quantities are said to vary directly, or be in *direct proportion*, if they increase or decrease at the same rate. For example, we know that the fraction $\frac{6}{4}$ reduces to $\frac{2}{3}$ so we can write the proportion $\frac{6}{4} = \frac{3}{2}$ we read this as *6 is to 4 as 3 is to 2* or expressed mathematically as $6:4::3:2$, where the double colon :: represents the word *as* in the proportion.

Now in this form, the *first* and *fourth* numbers, in the proportion, 6 and 2 in this case, are called the *extremes* and the *second* and *third* numbers, 4 and 3 in this case, are called the *means*. Now it is also true that from our proportion $\frac{6}{4} = \frac{3}{2}$ then, $6 \times 2 = 4 \times 3$. So that we can say that in any *true* proportion, *the product of the means equals the product of the extremes*.

Example 2.15

A train travels 200 km in 4 h. How long will it take to complete a journey of 350 km, assuming it travels at the same average velocity?

The key is to recognize the *proportion*; 200 km is proportional to 4 h as 350 km is proportional to x h. Then in symbols:

$$200:4::350:x$$

and using our rule for *means* and *extremes*, we get:

$$200x = (4)(350) \quad \text{or} \quad 200x = 1400$$

$$\text{and } x = \frac{1400}{200} \quad \text{or} \quad x = 7 \text{ h}$$

The rule for the products of the means and extremes is very useful and should be remembered!

We can generalize the above rule, using algebra (literal numbers), then:

$$\frac{x}{y} = \frac{a}{b} \quad \text{or} \quad x:y::a:b \qquad \text{then} \quad bx = ay$$

In general, we may also represent a proportion by use of the proportionality sign, $\propto$. For example, $2a \propto 4a$, where $\propto$ is read as '*is proportional to*'.

Inverse proportion

If 30 men are working on a production line and produce 6000 components in 10 working days, we might reasonably assume that if we double the amount of men, we can produce the components in half the time. Similarly, if we employ 20 men it would take longer to produce the same number of components. This situation is an example of *inverse proportion*. So in the above case, the number of men is reduced in the proportion of

$$\frac{20}{30} = \frac{2}{3}$$

therefore, it will take the *inverse proportion* of days to complete the same number of components, i.e.:

$$\left(\frac{3}{2}\right)10 \quad \text{or} \quad 15 \text{ days}$$

Two gear wheels mesh together as shown in Figure 2.1. One has 60 teeth, the other has 45 teeth. If the larger gear rotates at an angular velocity of 150 rpm, what is the angular velocity of the smaller gear wheel, in rpm?

I hope you can see from Figure 2.1 that the larger gear wheel will make less revolutions than the smaller gear wheel, in a given time. Therefore, we are dealing with inverse proportion.

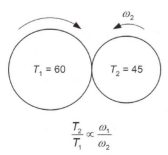

$$\frac{T_2}{T_1} \propto \frac{\omega_1}{\omega_2}$$

Figure 2.1 Two gear wheels in mesh.

The *ratio of teeth* of the smaller gear wheel compared to the larger gear wheel is

$$\frac{45}{60} = \frac{3}{4}$$

Therefore, the ratio of angular velocities must be in the inverse proportion $\frac{4}{3}$.

Then the velocity of the smaller gear wheel is $= \left(\frac{4}{3}\right)150 = 200 \text{ rpm}$.

Constant of proportionality

We can write down the general expression for inverse proportion as: $y \propto \frac{1}{x}$, where y is said to be inversely proportional to x.

Algebraically, using the proportion sign, *direct proportion* between any two quantities may be represented as $y \propto x$.

Now in order to *equate* the above expressions we need to introduce **the constant of proportionality** k. For example, if $2 \propto 4$, then $2 = 4k$ when $k = \frac{1}{2}$; we say that k *is the constant of proportionality*. It allows us to replace the proportionality sign ($\propto$), with the equals ($=$) sign. In our simple example above $k = \frac{2}{4}$ after transposition, or $k = \frac{1}{2}$.

Now if in general, $y \propto x$ then, $y = kx$, or $\frac{y}{x} = k$, where k is the constant of proportionality. Similarly for *inverse proportion*, where $y \propto \frac{1}{x}$, then $y = \frac{k}{x}$ or $xy = k$

The electrical resistance of a wire varies *inversely* as the *square* of its radius.
1. Write down an algebraic expression for this proportionality.
2. Given that the resistance is 0.05 Ω when the radius of the wire is 3 mm. Find the resistance when the wire used has a radius of 4.5 mm.

1. It is not always the case that variables are proportional only to their first powers. In this case, the resistance of the wire varies *inversely* as the *square of the radius*. Now, if R is

the resistance and r the radius, then: $R \propto \frac{1}{r^2}$, or $R = \frac{k}{r^2}$. This is the required algebraic expression.

2. When $R = 0.05$ and $r = 3$, then $0.05 = \frac{k}{3^2}$ and $k = 0.45$.

Therefore, the final connecting equation is $R = \frac{0.45}{r^2}$; when $r = 4.5$, then $R = \frac{0.45}{4.5} = 0.1\ \Omega$.

The above example shows a typical engineering use for proportion. In the example that follows, we can write down some familiar scientific relationships, using the rules for direct and inverse proportion.

Example 2.18

Write down the formulae to express the following:

1. The volume of a gas at constant temperature is inversely proportional to the pressure.
2. The electrical resistance of a wire varies directly as the length and inversely as the square of the radius.
3. The kinetic energy of a body is jointly proportional to its mass and the square of its velocity, when the constant of proportionality $= \frac{1}{2}$.

1. This should be familiar to you as Boyle's law. If we use the symbol V for volume and p for pressure, then $V \propto \frac{1}{p}$ and introducing the constant of proportionality k give the required relationship as $V = \frac{k}{p}$, or $pV = k$ (a constant).
2. This is the same relationship that you met earlier, except the length l of the conductor is involved. So, if we again use R for resistance and r for radius, then $R \propto \frac{l}{r^2}$ and again introducing the constant of proportionality we get $R = \frac{kl}{r^2}$.

Note that in the above case the resistance R is a function of two variables: the length l and the radius r.

3. The kinetic energy (KE) is also dependent on two variables the mass (m) and the square of the velocity (v^2), both variables being in direct proportion. So you may write down

the relationship as $KE \propto mv^2$ and introducing the constant of proportionality, which in this case we are given as $\frac{1}{2}$, then the required relationship is $KE = \frac{1}{2}mv^2$. You will be studying this relationship in your physics.

You will be using the ideas of proportion in the next section on algebra, where we consider the surface area and volume of regular solids.

Test your understanding 2.4

1. What is 15% of 50?

2. An airline engine repair bay has test equipment valued at £1.5 million. Each year 10% of the value of the test equipment is written off as depreciation. What is the value of the equipment after 2 full years?

3. An aircraft flies non-stop for 2.25 h and travels 1620 km. What is the aircraft's average speed?

4. A car travels 50 km at 50 km/h and 70 km at 70 km/h. What is its average speed?

5. A car travels 205 km on 20 L of petrol. How much petrol is needed for a journey of 340 km?

6. Four men are required to produce a certain number of components in 30 h. How many men would be required to produce the same number of components in 6 h?

7. The cost of electroplating a square sheet of metal varies as the square of its length. The cost to electroplate a sheet of metal with sides of 12 cm is £15.00. How much will it cost to electroplate a square piece of metal with sides of 15 cm.

8. If $y - 3$ is directly proportional to x^2 and $y = 5$ when $x = 2$, find y when $x = 8$.

9. Write down the formula to express the height of a cone, when it varies directly as its volume and inversely as the square of the radius.

Before we leave our study of number, we need to consider one or two other number system, other than to the base 10.

2.2.6 Number systems

The decimal system of numbers we have been studying up till now use the integers 0–9. There are in fact 10 integers and for this reason we often refer to the decimal system as the denary (ten) system.

Thus, for example, the denary number 245.5 is equivalent to:

$$(2 \times 10^2) + (4 \times 10^1) + (5 \times 10^0) + (5 \times 10^{-1})$$

This arrangement of the number consists of an integer ≥ 1.0 and ≤ 10.0 multiplied by the *base* raised to the *power*; you have met this idea earlier, when studying decimal numbers, powers of ten and estimation techniques.

In the binary system of numbers, the base is 2 and so, for example, the denary number 43 to the base 10, written as 43_{10} is equivalent to the number:

$$2^5 + 2^3 + 2^1 + 2^0 = 32_{10} + 8_{10} + 2_{10} + 1_{10}$$

Key point

In the binary system of numbers the base is 2.

As a reminder and source of reference the binary and denary equivalents for some important numbers related to computing are detailed below:

Binary$_2$	2^7	2^6	2^5	2^4	2^3	2^2	2^1	2^0
Denary$_{10}$	128	64	32	16	8	4	2	1

We are now in a position to be able to convert from denary to binary and binary to denary number systems.

To convert **denary to binary**, we repeatedly divide by 2 and note the remainder at each stage. Then, for example, to convert the number 25_{10} to binary, we proceed as follows:

$25/2 = 12$	remainder 1	Least significant digit (LSD)
$12/2 = 6$	remainder 0	
$6/2 = 3$	remainder 0	
$3/2 = 1$	remainder 1	
$1/2 = 0$	remainder 1	Most significant digit (MSD)

The binary$_2$ equivalent of 25_{10} is 11001_2.

Note the *order* in which the digits of the binary number are laid out from the MSD to the LSD; i.e. in reverse order to the successive division.

To convert **binary to denary**, we lay out the number in successive powers. For example, to convert binary number 1101_2 into denary, we proceed as follows:

$$1101_2$$

$$= (1 \times 2^3) + (1 \times 2^2) + (0 \times 2^1) + (1 \times 2^0)$$
$$= (1 \times 8) + (1 \times 4) + (0 \times 2) + (1 \times 1)$$
$$= 8 + 4 + 0 + 1 = 13_{10}$$

When numbers are placed in binary form, we can see from above that they consist of a number of ones (1) and noughts (0). If in electronic logic circuits we allow the binary digit "1" to represent "ON" and the binary digit "0" to represent "OFF". We can apply this *binary code* to electronic logic systems. It is this powerful application of binary numbers that makes their study important. In order to get more digital information down computer communication lines, we can use another number system that allows us to send 16 individual pieces of information (bytes) down parallel lines, all at the same time. This type of communication may be coded using **hexadecimal** representation. Thus, for hexadecimal numbers their *base* is 16. However, because in our decimal number counting system we only have 10 digits (0–9), we make up for this in the hexadecimal system by allocating capital letters to the remaining decimal numbers 10–15 (remembering that decimal zero is counted as part of the 16 digit base). Hexadecimal representation, together with their denary and binary equivalents, are shown in the Table 2.1.

Thus in a similar manner to before, the denary number 542_{10} may be represented as:

$$542_{10} = (5 \times 10^2) + (4 \times 10^1) + (2 \times 10^0)$$

which is equivalent to

$$21E_{16} = (2 \times 16^2) + (1 \times 16^1) + (E \times 16^0).$$

To convert **denary to hexadecimal**, we repeatedly divide by 16 in a similar manner to the way in which we converted denary to binary. To convert the denary number 5136_{10} to hexadecimal, we proceed as follows:

$5136/16 = 321$	remainder 0	LSD
$321/16 = 20$	remainder 1	
$20/16 = 1$	remainder 4	
$1/16 = 0$	remainder 1	MSD

Table 2.1 Denary, binary and hexadecimal number systems representation

Denary$_{10}$	Binary$_2$	Hexadecimal$_{16}$
0	0000	0
1	0001	1
2	0010	2
3	0011	3
4	0100	4
5	0101	5
6	0110	6
7	0111	7
8	1000	8
9	1001	9
10	1010	A
11	1011	B
12	1100	C
13	1101	D
14	1110	E
15	1111	F

So the hexadecimal$_{16}$ equivalent of 5136_{10} is 1410_{16}.

Similarly, to convert the number 94_{10} to hexadecimal$_{16}$, we proceed as follows:

$$94/16 = 5 \quad \text{remainder } 14 \ (= E_{16})$$
$$5/16 = 0 \quad \text{remainder } 5$$

So the hexadecimal$_{16}$ equivalent of 94_{10} is $5E_{16}$.

To convert **hexadecimal to denary**, we proceed in a similar manner as for binary to denary. For example, to convert $BA45_{16}$ to denary, we proceed as follows:

$$BA45_{16} = (B \times 16^3) + (A \times 16^2) + (4 \times 16^1)$$
$$+ (5 \times 16^0)$$
$$= (11 \times 4096) + (10 \times 256) + (4 \times 16)$$
$$+ (5 \times 1)$$
$$= (45056) + (2560) + (64) + (5)$$
$$= 47685_{10}$$

The denary$_{10}$ equivalent of hexadecimal number $BA45_{16}$ is 47685_{10}.

To complete our very short study of number systems, it is worth considering how we convert a denary number that has a decimal fraction as part of the number. The process is quite logical

and relatively easy to follow. When dealing with the fraction part of a binary number we apply successive multiplication until we reach unity for the fractional part of the denary number. However, since we have applied multiplication the inverse arithmetic operation to that of division, then the MSD is the *first remainder* in the multiplication process.

Key point

When converting decimal fractions to binary we apply successive multiplication to the fractional part of the denary number.

Example 2.19

Convert denary number 39.625_{10} to binary$_2$.

Then proceeding in the normal way for the non-fractional part of this number we get:

$$39/2 = 19 \quad \text{remainder } 1 \quad \text{LSD}$$
$$19/2 = 9 \quad \text{remainder } 1$$
$$9/2 = 4 \quad \text{remainder } 1$$
$$4/2 = 2 \quad \text{remainder } 0$$
$$2/2 = 1 \quad \text{remainder } 0$$
$$1/2 = 0 \quad \text{remainder } 1 \quad \text{MSD}$$

Then $39_{10} = 100111$

Also for the decimal fraction, applying successive *multiplication*, we get:

$$0.625 \times 2 = 1.[250] \quad \text{MSD}$$
$$0.250 \times 2 = 0.[500]$$
$$0.500 \times 2 = 1.000 \quad \text{LSD}$$

Then $0.625_{10} = 0.101_2$

So, the denary number $39.625_{10} = 100111.101_2$.

Test your understanding 2.5

1. Convert the denary numbers to binary.
   (a) 17 (b) 23 (c) 40

2. Convert the binary numbers to denary.
   (a) 1011 (b) 11111 (c) 1010101

3. Convert the denary numbers to hexadecimal.
   (a) 5890 (b) 16892

4. Convert the hexadecimal numbers to denary.
   (a) 6E (b) CF18

2.3 Algebra

2.3.1 Factors, powers and exponents

Factors

When two or more numbers are multiplied together, each of them, or the product of any number of them (apart from them all), is a factor of the product. This applies to explicit arithmetic numbers and to literal numbers.

So, for example, if we multiply the numbers 2 and 6, we get that $2 \times 6 = 12$, thus 2 and 6 are factors of the number 12. However, the number 12 has more than one set of factors, $3 \times 4 = 12$ so 3 and 4 are also factors of the number 12. We can also multiply $2 \times 2 \times 3$ to get 12. Since the numbers 2, 2 and 3 are yet another set of factors of the number 12. Finally, you will remember that any number n multiplied by 1 is itself, or $n \times 1 = n$. So every number has itself and 1 as factors; 1 and n are considered *trivial factors* and when asked to find the factors of an explicit or literal number, we will exclude the number itself and 1.

Example 2.20

Find the factors of: (a) 8 (b) xy (c) 24 (d) abc (e) $-m$

(a) Apart from the trivial factors 1 and 8, which we agreed to ignore, the number 8 has only the factors 2 and 4, since $2 \times 4 = 8$, remember that these factors can be presented in reverse order, $4 \times 2 = 8$, but **2 and 4** are still the only factors.

(b) Similarly, the literal number xy can only have the factors x and y, if we ignore the trivial factors. Thus, the numbers x **and** y multiplied together to form the product xy are factors of that product.

(c) The number 24 has several sets of factors, with varying numbers in each set. First we find the number of sets with two factors, these are:

$$24 = 6 \times 4$$
$$24 = 8 \times 3$$
$$24 = 12 \times 2$$

More than two factors:

$$24 = 2 \times 2 \times 6$$
$$24 = 4 \times 3 \times 2$$
$$24 = 2 \times 2 \times 2 \times 3$$

However, if we look closely we see that the number 24 has only six *different* factors: **12**, 8, 6, 4, 3 and 2.

(d) So what about the factors in the number abc? Well, I hope you can see that the product of each individual factor a, b and c constitute one set of factors. Also ab and c; a and bc; and b and ac, form a further three sets. So extracting the different factors from these sets we have: a, b, c, ab, ac and bc as the six factors of the number abc.

(e) We have two sets of factors here 1 and $-n$, which is the trivial factor, but also the set n and -1, notice the subtlety with the sign change. When dealing with minus numbers, any two factors must have opposite signs.

Powers and exponents

When a number is the product of the same factor multiplied by itself, this number is called a *power* of the factor. For example, we know that $3 \times 3 = 9$. Therefore, we can say that 9 is a power of 3. To be precise, it is the second power of 3, because two 3s are multiplied together to produce 9. Similarly, 16 is the second power of 4. We may use literal terminology to generalize the relationship between powers and factors.

So the second power of a means $a \times a$ or $(a \cdot a)$, this is written as a^2, where a is known as the *base* (factor) and 2 is the *exponent* (or index). Thus writing the number 9 in exponent form we get $9 = 3^2$ where; *9 is the second power, 3 is the base (factor) and 2 is the exponent (index)*.

The above ideal can be extended to write arithmetic numbers in exponent or index form. For example, $5^2 = 25$, $9^2 = 81$ and $3^3 = 27$. Notice that the second power of 5 gives the number 25 or $5 \times 5 = 25$; similarly 3^3 means the third power of 3, literally $3 \times 3 \times 3 = 27$. The idea of powers and exponents (indices) can be extended to literal numbers. For example: $a \times a \times a \times a \times a$ or a^5 *or in general* a^m *where a is the base (factor) and the exponent m (or index) is any positive integer.* a^m means *a* used as a factor *m* times and is read as the "*m*th power of *a*". *Note* that since any number used as a factor once would simply be the number itself, the index (exponent) is not usually written; in other words *a* means a^1.

Now, providing the base of two or more numbers expressed in index (exponent) form are the same, we can perform multiplication and division on these numbers, by adding or subtracting the *indices* accordingly.

We will from now on refer to the exponent of a number as its index, in order to avoid confusion with particular functions, such as the exponential function, which we study later.

Consider the following literal numbers in index form:

$$x^2 \times x^2 = (x \times x)(x \times x) = x \times x \times x \times x = x^4$$

$$x^2 \times x^4 = (x \times x)(x \times x \times x \times x)$$
$$= x \times x \times x \times x \times x \times x = x^6$$

$$\frac{x^2}{x^2} = \frac{x \times x}{x \times x} = x^0 = 1$$

$$\frac{x^2}{x^4} = \frac{x \times x}{x \times x \times x \times x} = \frac{1}{x \times x} = x^{-2}$$

What you are looking for is a pattern between the first two literal numbers, which involve multiplication and the second two which involve division.

For multiplication of numbers with the same base, we add the indices and for division of numbers with the same base, we subtract the indices in the *denominator* (below the line) from those in the *numerator* (above the line). Remember also that the base number $x = x^1$.

We will now generalize our observations and so formulate the *laws of indices*.

2.3.2 The laws of indices

In the following laws *a* is the common *base*, *m* and *n* are the *indices* (exponents). Each law has an example of its use alongside:

1. $a^m \times a^n = a^{m+n}$ $2^2 \times 2^4 = 2^{2+4} = 2^6 = 64$
2. $\frac{a^m}{a^n} = a^{m-n}$ $\frac{3^4}{3^2} = 3^{4-2} = 3^2 = 9$
3. $(a^m)^n = a^{mn}$ $(2^2)^3 = 2^{2 \times 3} = 2^6 = 64$

4. $a^0 = 1$ Any number raised to the power 0 is always 1
5. $a^{\frac{m}{n}} = \sqrt[n]{a^m}$ $27^{\frac{4}{3}} = \sqrt[3]{27^4} = 3^4 = 81$
6. $a^{-n} = \frac{1}{a^n}$ $6^{-2} = \frac{1}{6^2} = \frac{1}{36}$

We need to study these laws carefully in order to understand the significance of each.

Law 1: As you have already met, it enables us to *multiply numbers* given in index form that have a common base. In the example the common base is 2, the first number raises this base (factor) to the power 2 and the second raises the same base to the power 3. In order to find the result we simply *add* the indices.

Law 2: We have again used when *dividing numbers* with a common base in this case the base is 3. Note that since division is the opposite arithmetic operation to multiplication. It follows that we should perform the opposite arithmetic operation on the indices, that of *subtraction*. Remember we always subtract the index in the denominator from the index in the numerator.

Law 3: It is concerned with raising the powers of numbers. Do not mix this law up with law 1. When *raising powers of numbers* in index form, we *multiple* the indices.

Law 4: As you have also met, this law simply states that *any number raised to the power 0 is always 1*. Knowing that any number divided by itself is also 1, we can use this fact to show that a number raised to the power 0 is also 1. What we need to do is use the second law concerning the division of numbers in index form.

We know that

$$\frac{9}{9} = 1 \text{ or } \frac{3^2}{3^2} = 3^{2-2} = 3^0 = 1$$

which shows that $3^0 = 1$ and in fact because we have used the second law of indices, this must be true in all cases.

Law 5: This, rather complicated looking, law simply enables us to find the decimal equivalent of a number in index form; where the index is a fraction. All that you need to remember is that the index number above the fraction line is raised to that power and the index number below the fraction line has that number root.

So for the number $8^{\frac{2}{3}}$, we raise 8 to the power 2 and then take the cube root of the result.

It does not matter in which order we perform these operations. So we could have just as easily taken the cube root of 8 and then raised it to the power 2.

Law 6: This is a very useful law, when you wish to convert the division of a number to multiplication. In other words, bring a number from underneath the division line to the top of the division line. *As the number crosses the line we change the sign of its index*. This is illustrated in the example, which accompanies this law.

The following examples further illustrate the use of the above laws, when evaluating or simplifying expressions that involve numbers and symbols.

Example 2.21

Evaluate the following expressions:

(a) $\dfrac{3^2 \times 3^3 \times 3}{3^4}$ (b) $(6)(2x^0)$ (c) $36^{-\frac{1}{2}}$ (d) $16^{-\frac{3}{4}}$

(e) $\dfrac{(2^3)^2(3^2)^3}{(3^4)}$

(a) $\dfrac{3^2 \times 3^3 \times 3}{3^4} = \dfrac{3^{2+3+1}}{3^4}$ (law 1)

$= \dfrac{3^6}{3^4} = 3^{6-4}$ (law 2)

$= 3^2 = 9$

(b) $(6)(2x^0) = (6)(2) = 12$ remembering that $x^0 = 1$ (law 4)

(c) $36^{-\frac{1}{2}} = \dfrac{1}{36^{\frac{1}{2}}} = (\text{law 6}) = \dfrac{1}{\sqrt{36}}$ (law 5)

$= \pm\dfrac{1}{6}$ (note $\pm$ square root)

(d) $16^{-\frac{3}{4}} = \dfrac{1}{16^{\frac{3}{4}}}$ (law 6)

$= \dfrac{1}{\sqrt[4]{16^3}}$ (law 5)

$= \dfrac{1}{2^3} = \dfrac{1}{8}$

(e) $\dfrac{(2^3)^2(3^2)3}{3^4} = \dfrac{(2^{3\times 2})(3^{2\times 1})}{3^4}$ (law 3)

$= \dfrac{2^6 \times 3^3}{3^4} = 2^6 \times 3^{3-4}$ (law 2)

$= 2^6 \times 3^{-1} = 64 \times \dfrac{1}{3}$ (law 6) $= \dfrac{64}{3}$

Example 2.22

Simplify the following expressions:

(a) $\dfrac{12x^3y^2}{4x^2y}$ (b) $\left(\dfrac{a^3b^2c^4}{a^4bc}\right)\left(\dfrac{a^2}{c^2}\right)$

(c) $\left[(b^3c^2)(ab^3c^2)(a^0)\right]^2$

(a) $\dfrac{12x^3y^2}{4x^2y} = 3x^{3-2}y^{2-1}$(rule 2 and simple division of integers) $= 3xy$

(b) $\left(\dfrac{a^3b^2c^4}{a^4bc}\right)\left(\dfrac{a^2}{c^2}\right) = a^{3+2-4}b^{2-1}c^{4-1-2}$

(rule 2 and operating on like bases) $= abc$

Note also that in the above problem there was no real need for the second set of brackets, since all numbers were multiplied together.

(c)
$\left[(b^3c^2)(ab^3c^2)(a^0)\right]^2$
$= \left[(b^3c^2)(ab^3c^2)(1)\right]^2$ (rule 4)
$= \left[ab^{3+3}c^{2+2}\right]^2$ (rule 1)
$= \left[ab^6c^4\right]^2 = a^2b^{12}c^8$ (rule 3)

Test your understanding 2.6

1. Find the factors (other than the trivial factors) of:
   (a) 16   (b) n^2 (c) *wxyz*

2. Find the common factors in the expression $ab^2c^2 + a^3b^2c^2 + ab^2c$.

3. Simplify:

   (a) $\dfrac{1}{2^3} \times 2^7 \times \dfrac{1}{2^{-5}} \times 2^{-4}$ (b) $\left(\dfrac{16}{81}\right)^{\frac{3}{4}}$ (c) $\dfrac{b^3b^{-8}b^2}{b^0b^{-5}}$

4. Simplify: (a) $(2^2)^3 - 6 \times 3 + 24$ (b) $\dfrac{1}{2^{-2}} + \dfrac{1}{3^2} - \dfrac{1}{3^{-1}}$

2.3.3 Factorization and products

There are many occasions when we are required to determine the factors and products of algebraic expressions. Literal numbers are used in expressions and formulae to provide a precise, technically accurate way of generalizing laws and statements associated with mathematics, science and engineering, as mentioned previously. When manipulating such expressions, we are often required to multiply them together (determine their *product*) or carry out the reverse process (that of *factorization*). You will see, in your later studies, that these techniques are very useful when it comes to changing the subject of a particular algebraic formula. In other words, when you are required to *transpose a formula* in terms of a particular variable.

We begin by considering the products of some algebraic expressions. Once we are familiar with the way in which these expressions are "built-up", we can then look at the rather more difficult inverse process (that of factorization).

Products

Consider the two factors $(1 + a)$ and $(1 + b)$, noting that each factor consists of a *natural number* and a *literal number*. Suppose we are required to find $(1 + a)(1 + b)$; in other words, their product. Provided we follow a set sequence, obeying the laws of multiplication of arithmetic, then the process is really quite simple!

In order to describe the process accurately, I need to remind you of some basic terminology. In the factor $(1 + a)$ the natural number 1 is considered to be a *constant* because it has no other value; on the other hand, the literal number a can be assigned any number of values. Therefore, it is referred to as a *variable*. Any number or group of numbers, whether natural or literal, separated by a $+$, $-$ or $=$ sign, is referred to as a *term*. For example, the expression $(1 + a)$ has *two* terms.

When multiplying $(1 + a)$ by $(1 + b)$ we start the multiplication process from the left and work to the right, in the same manner as reading a book. We multiply each term in the left-hand bracket by each of the terms in the right-hand bracket as follows:

$$(1 + a)(1 + b)$$
$$= (1 \times 1) + (1 \times b) + (a \times 1) + (a \times b)$$
$$= 1 + b + a + ab = 1 + a + b + ab$$

Note: 1. The "dot" notation $(1 \cdot a)\,(1 \cdot b)$ for multiplication may be used to avoid confusion with the variable x.

2. It does not matter in which order the factors are multiplied, refer back to the commutative law of arithmetic, if you do not understand this fact.

Example 2.23

Determine the product of the following algebraic factors:

(a) $(a + b)(a - b)$ (b) $(2a - 3)(a - 1)$
(c) $(abc^3 d)(a^2 bc^{-1})$

(a) In this example we proceed in the same manner as we did above, i.e.:

$$(a + b)(a - b)$$
$$= (a \times a) + (a)(-b) + (b \times a) + (b)(-b)$$
$$= a^2 + (-ab) + (ba) + (-b^2)$$

which by the laws of signs $= a^2 - ab + ba - b^2$ and by the commutative law this can be written as $a^2 - ab + ab - b^2$ or $(a + b)(a - b) = a^2 - b^2$. I hope you have followed this process and recognize the notation for multiplying two bracketed terms.

The product $a^2 - b^2$ is a special case and is known as *the difference between two squares*. This enables you to write down the product of any two factors that take the form $(x + y)(x - y)$ as equal to $x^2 - y^2$, where x and y are any two variables.

(b) Again for these factors, we follow the process, where we get:

$$(2a - 3)(a - 1)$$
$$= 2a \times a + (2a)(-1) + (-3)(a) + (-3)(-1)$$
$$= 2a^2 - 2a - 3a + 3$$

and so $(2a - 3)(a - 1)$

$$= 2a^2 - 5a + 3$$

(c) In this case we simply multiply together *like variables*, using the *laws of indices*! So we get:

$$(abc^3 d)(a^2 bc^{-1})$$
$$= (a^1 \times a^2)(b^1 \times b^1) \times (c^3 c^{-1})(d^1)$$
$$= (a^{1+2})(b^{1+1})(c^{3-1})(d^1)$$
$$= a^3 b^2 c^2 d$$

Note that the brackets in the above solution have only been included for clarity, they are not required for any other purpose.

I hope you are getting the idea of how to multiply factors to produce products. So far we have restricted ourselves to just two factors. Can we adopt the process for three or more factors? Well, if you did not know already, you will be pleased to know that we can!

Example 2.24

Simplify the following:
(a) $(x + y)(x + y)(x - y)$
(b) $(a + b)(a^2 - ab + b^2)$

(a) This expression may be simplified by multiplying out the brackets and collecting like terms. I hope you recognize the fact that the product of $(x + y)(x - y)$ is $x^2 - y^2$. Then all we need to do is multiply this product by the remaining factor, we get:

$$(x + y)(x^2 - y^2) = x^3 - xy^2 + x^2y - y^3$$

Note that the convention of putting the variables in alphabetical order and the fact that it does not matter in what order we multiply the factors, the result will be the same.

(b) This is a straightforward product, where

$$(a + b)(a^2 - ab + b^2)$$
$$= a^3 - a^2b + ab^2 + a^2b - ab^2 + b^3$$
$$= a^3 + b^3$$

Note that there are six terms resulting from the necessary six multiplications. When we collect like-terms and add we are left with the product known as the *addition of cubes*.

Factorization

Factorizing is the process of finding two or more factors which, when multiplied together, will result in the given expression. Therefore, factorizing is really the opposite of multiplication or finding the product. It was for this reason that we first considered the simpler process of finding the product.

Thus, for example, $x(y + z) = xy + xz$. This product resulted from the multiplication of the two factors x and $(y + z)$. If we now unpick the product you should be able to see that x is a *common factor*, that appears in *both terms* of the product.

What about the expression $x^2 - 16$? I hope you are able to recognize the fact that this expression is an example of the *difference between two squares*. Therefore, we can write down the factors immediately as $(x + 4)$ and $(x - 4)$, look back at Example 2.11, if you are unsure. We can check the validity of our factors by multiplying and checking that the product we get is identical to the original expression, that we were required to factorize, i.e.:

$$(x + 4)(x - 4) = x^2 - 4x + 4x - 16$$
$$= x^2 - 16 \text{ as required.}$$

Suppose you are asked to factorize the expression $a^2 - 6a + 9$, how do we go about it? Well, a good place to start is with the term involving the highest power of the variable, i.e. is a^2. Remember that when *convention* dictates, we layout our expression in descending powers of the unknown, starting with the highest power positioned at the extreme left-hand side of the expression. a can only have factors of itself and 1 or a and a, therefore, ignoring the trivial factors, $a^2 = a \times a$. At the other end of the expression, we have the natural number 9, this has the trivial factors 1 and 9 or the factors 3 and 3 or -3 and -3. Note the importance of considering the *negative* case, where from the laws of signs $(-3)(-3) = 9$. So now, we have several sets of factors we can try, these are:

(i) $(a + 3)(a + 3)$ or   (ii) $(a - 3)(a - 3)$ or
(iii) $(a + 3)(a - 3)$?

Now we could try multiplying up each set of factors until we obtained the required result, i.e. determine the factors by *trial and error*. This does become rather tedious when there are a significant number of possibilities. So, before resorting to this method, we need to see if we can eliminate some combinations of factors by applying one or two simple rules.

I hope you can see why the factors $(a + 3)$ $(a - 3)$ can be immediately excluded. These are the factors for the difference between squares, which is not the original expression we needed to factorize.

What about the factors $(a + 3)(a + 3)$? Both the factors contain only positive terms, therefore, any of their products must also be positive by the laws of signs! In our expression $a^2 - 6a + 9$ there is a *minus sign*, so again this set of factors may be eliminated. This leaves us with the factors $(a-3)(a-3)$ and on multiplication we get:

$$(a-3)(a-3) = a^2 - 3a - 3a + 9 = a^2 - 6a + 9$$

giving us the correct result.

You might have noticed that we have left out the sets of factors $(a-1)(a-9)$, $(a-1)(a+9)$, $(a+1)(a-9)$ and $(a+1)(a+9)$, from our original group of possibles! Well, in the case of $(a+1)(a+9)$, this would be eliminated using the laws of signs, but what about the rest?

There is one more very useful technique we can employ, when considering just two factors. This technique enables us to check the accuracy of our factors by determining the middle term of the expression we are required to factorize. So in our case for the expression $a^2 - 6a + 9$, then $-6a$ is the middle term.

The middle term is derived from our chosen factors by *multiplying the outer terms, multiplying the inner terms and adding.*

So in the case of the correct factors $(a-3)$ $(a-3)$, the outer terms are a and -3, which on multiplication $(a)(-3) = -3a$ and similarly the inner terms $(-3)(a) = -3a$ and so their sum $= -3a + (-3a) = -6a$, as required.

If we try this technique to any of the above factors involving 1 and 9, we will see that they can be quickly eliminated. For example, $(a-1)(a-9)$ has an outer product of $(a)(-9) = -9a$ and an inner product of $(-1)(a) = -a$, which when added $= -9a - a = -10a$, which of course is incorrect.

Example 2.25

Factorize the expressions:
(a) $x^2 + 2x - 8$ and (b) $12x^2 - 10x - 12$

(a) To determine the factors for this expression we follow the same procedure as detailed above.

First we consider the factors for the outer term x^2 (apart from the trivial factors), we have $x^2 = x \times x$ and the factors of -8 are $(2)(4)$ or $(-2)(4)$ or $(4)(-2)$ or $(1)(8)$ or $(-1)(8)$ or $(8)(-1)$. So by considering only outer and inner terms we have the following

possible combination of factors:

$(x+2)(x+4)$, $(x+2)(x-4)$,
$(x-2)(x-4)$ and $(x+1)(x+8)$,
$(x+1)(x-8)$, $(x-1)(x+8)$.

Now we eliminate the sets of factors that only have positive terms (by the law of signs). This leaves $(x+2)(x-4)$, $(x-2)(x+4)$, $(x+1)(x-8)$ and $(x-1)(x+8)$. *The last two sets of factors can be eliminated by applying the outer and inner term rule.* If you apply this rule, neither of these sets give the correct middle term. We are therefore left with the two sets of factors: $(x+2)(x-4)$ or $(x-2)(x+4)$.

So let us try $(x+2)(x-4)$, applying the outer and inner term rules, we get (x) $(-4) = -4x$ and $(2)(x) = 2x$, which on addition give $-2x$, but we require $+2x$, so these are not the factors. Finally, we try $(x-2)$ $(x+4)$, where on application of the rule we get $(x)(4) = 4x$ and $(-2)(x) = -2x$, which on addition give $4x - 2x = 2x$ as required, where the factors of the expression $x^2 + 2x - 8$ are $(x-2)$ $(x+4)$.

(b) For the expression $12x^2 - 10x - 12$, we have the added complication of several possibilities for the term involving the square of the variable x, i.e. $12x^2$. This term could be the product of the factors: $(x)(12x)$ or $(2x)(6x)$ or $(3x)(4x)$ and the right-hand term could be the product of the factors $(-1)(12)$ or $(1)(-12)$ or $(-2)(6)$ or $(2)(-6)$ or $(-3)(4)$ or $(3)(-4)$. By the *rule of signs*, no set of factors can have all positive terms, so these can be eliminated from the possible solutions. This leaves us with:

set 1 $(3x+1)(x-12)$, $(3x-1)(x+12)$
          $(x-1)(3x+12)$ or $(x+1)(3x-12)$
set 2 $(3x+2)(x-6)$, $(3x-2)(x+6)$
          $(x+2)(3x-6)$, $(x-2)(3x+6)$
set 3 $(3x+3)(x-4)$, $(3x-3)(x+4)$
          $(x+3)(3x-4)$, $(x-3)(3x+4)$

The choice of possible solution does seem to be getting complicated! However, if we apply the *multiplication of outer terms, multiplication of inner terms rule* to sets 1 and 3, they are quickly eliminated leaving us with just set 2. Application of the rule, once more, to the factors in set 2 gives us our required solution, where the factors of the expression $12x^2 - 10x - 12$ are $(3x + 2)(4x - 6)$.

Example 2.26

Factorize the expression $3x^3 - 18x^2 + 27x$.

We are now dealing with an unknown variable x raised to the third power! Do not worry, in this particular case the trick is to recognize the common factor. If we first consider the integers that multiply the variable we have: $3x^3 - 18x^2 + 27x$, all these numbers are divisible by 3, therefore 3 is a common factor.

Also, in a similar manner the variable itself has a common factor, since all are divisible by x.

So all we need to do is to remove these common factors to produce the expression: $3x(x^2 - 6x + 9)$. Note that on multiplication you will obtain the original expression, so that $3x$ and $x^2 - 6x + 9$ must be the factors.

This expression now has one factor where the greatest power of the unknown is 2. This factor can itself be broken down into two *linear factors* (i.e. where the unknown is raised to the power 1) using the techniques described before, where, the factors of the expression $3x^3 - 18x^2 + 27x$ are $(3x)(x - 3)(x - 3)$.

Finally, before we leave our study of factorization, some common algebraic expressions are tabulated in Table 2.2 in general form, with their factors.

For example, recognizing that $z^3 + 8 = z^3 + 2^3$, then the factors of the expression $z^3 + 8$ are from expression $5 = (z + 2)(z^2 - 2z + 4)$, where in this case $z = x$ and $y = 2$.

Table 2.2

	Expression	Factors
1	$xy + xz$	$x(y + z)$
2	$x^2 - y^2$	$(x + y)(x - y)$
3	$x^2 + 2xy + y^2$	$(x + y)^2$
4	$x^2 - 2xy + y^2$	$(x - y)^2$
5	$x^3 + y^3$	$(x + y)(x^2 - xy + y^2)$
6	$x^3 - y^3$	$(x - y)(x^2 + xy + y^2)$

Test your understanding 2.7

1. Simplify:
   (a) $(a^2 b^3 c)(a^3 b^{-4} c^2 d)$ (b) $(12x^2 - 2)(2xy^2)$
2. Reduce the following fractions to their lowest terms:
   (a) $\dfrac{21a^3 b^4}{28a^9 b^2}$ (b) $\dfrac{abc}{d} \div \dfrac{abc}{d^2}$
3. Determine the product of the following:
   (a) $(3a - 1)(2a + 2)$ (b) $(2 - x^2)(2 + x^2)$ (c) $ab(3a - 2b)(a + b)$ (d) $(s - t)(s^2 + st + t^2)$
4. Factorize the following expressions:
   (a) $x^2 + 2x - 3$ (b) $a^2 - 3a - 18$ (c) $4p^2 + 14p + 12$ (d) $9z^2 - 6z - 24$
5. Find all factors of the expressions:
   (a) $3x^2 + 27x^2 + 42x$ (b) $27x^3 y^3 + 9x^2 y^2 - 6xy$
6. Evaluate:
   (a) $a^2 + 0.5a + 0.06$, when $a = -0.3$ (b) $(x - y)(x^2 + xy + y^2)$, when $x = 0.7$, $y = 0.4$

2.3.4 Algebraic operations

Having met earlier, the addition and subtraction of literal numbers, together with algebraic factors, products and indices, you are now in a position to simplify, transpose and evaluate algebraic expressions and formulae. After which, you will have all the necessary tools to solve simple algebraic equations.

Simplifying algebraic expressions

As a reminder of some of the techniques and laws you have already covered (with respect to the manipulation of bracketed expressions) some examples are given below. Make sure you are able to work through the following examples, if in any doubt refer back to our earlier work on literal numbers, fractions, factors, powers and exponents.

Example 2.27

Simplify, the following algebraic expressions:

(i) $3ab + 2ac - 3c + 5ab - 2ac - 4ab + 2c - b$
(ii) $3x - 2y \times 4z - 2x$
(iii) $(3a^2 b^2 c^2 + 2abc)(2a^{-1} b^{-1} c^{-1})$
(iv) $(3x + 2y)(2x - 3y + 6z)$

(i) All that is required here, is to add or subtract *like*-terms, so we get:
$$3ab + 5ab - 4ab + 2ac - 2ac - 3c + 2c - b$$
$$= 4ab - b - c$$

(ii) Here you need to be aware of the *law of precedence*, this is derived from the laws of arithmetic you learnt earlier. As an aide memoir we use the acronym: BODMAS, i.e. brackets, of, division, multiplication, addition and finally subtraction. These operations being performed in this order. From this law we carry out multiplication before addition or subtraction. So we get:

$$3x - 8yz - 2x = x - 8yz$$

(iii) With this expression, when multiplying up the brackets, we need to remember *the law of indices for multiplication*. Using this law we get:

$$6a^{2-1}b^{2-1}c^{2-1} + 4a^{1-1}b^{1-1}c^{1-1}$$
$$= 6a^1b^1c^1 + 4a^0b^0c^0 = 6abc + 4$$

(Do not forget the 4! Remember that any number raised to the power zero is 1 and $4 \times 1 \times 1 \times 1 = 4$.)

(iv) This is just the multiplication of brackets, where we multiply all terms in the right-hand bracket by both terms in the left-hand bracket. We perform these multiplications as though we are reading a book from left to right. Starting with $(3x) \times (2x) = 6x^2$, then $(3x) \times (-3y) = -9xy$ and so on. We then repeat the multiplications, using the right-hand term in the first bracket, i.e. $(2y) \times (2x) = 4xy$ and so on. So that before any simplification we should end up with $2 \times 3 = 6$ terms:

$$(3x + 2y)(2x - 3y + 6z)$$
$$= 6x^2 - 9xy + 18xz + 4xy - 6y^2$$
$$+ 12yz \quad \text{(6 terms)}$$

and so after simplification which involves only two like-terms in this case, we get:

$$6x^2 - 5xy + 18xz - 6y^2 - 12yz$$

Key point

Remember the *laws of precedence* by the acronym BODMAS, brackets, of, division, multiplication, addition, subtraction.

Example 2.28

Factorize the following algebraic expressions:
(i) $-x^2 + x + 6$ (ii) $5x^2y^3 - 40z^3x^2$
(iii) $x^2 - 4x - 165$ (iv) $8x^6 + 27y^3$

(i) This is a straightforward example of factorizing a *trinomial* (an algebraic expression of three terms, with ascending powers of the unknown).

We simply follow the rules you studied earlier. I will go through the procedure once more to remind you.

We first consider the left-hand term $-x^2$, which obeying the rules of multiplication must have factors $-x$ and x (ignoring trivial factors). Also for the right-hand term we have 2 and 3 or the trivial factors 1 and 6. Again ignoring the trivial factors, we first try 2 and 3.

So we have the following sets of factors:

$$(-x + 2)(x + 3), \ (x + 2)(-x + 3)$$
$$(-x - 2)(x - 3), \ (x - 2)(-x - 3)$$

Now remembering the rule for determining the middle term, i.e. *addition of outer and inner products*, then by trial and error, we eliminate all sets of factors, except the correct solution, which is $(x + 2)(-x + 3)$.

(ii) Here the trick is to recognize the *common factor/s* and pull them out behind a bracket. In this case I hope you can see that x^2 is common to both terms as is the number 5.

Then we can write the factors as:

$$5x^2(y^3 - 8z^3)$$

Your answer can always be checked by multiplying up the factors. You should, of course, obtain the original expression providing your factors are correct and your subsequent multiplication is also correct!

(iii) With this example, the only difficulty is in recognizing possible factors for the rather large number 165. Well, this is where it is useful to know your 15 times table! With trial and error you should eventually find that apart from the trivial factors, the numbers 15 and 11 are factors of 165. Also recognizing that $15 - 11 = 4$, we know that some combination of these numbers will produce the required result. Then by obeying the rules of signs you should eventually find the correct factors as: $(x - 15)(x + 11)$.

(iv) If you have faithfully completed all the exercises in *test your understanding* 2.6, you would have met this example before! The trick is to recognize that the expression $8x^6 + 27y^3$ may be written as $(2x^2)^3 + (3y)^3$ by application of the laws of indices. Then, all that is needed is to apply rule 5 for the sum of two cubes (found in Table 2.1 at the end of your work on factorization), to obtain the required solution.

$$\text{Thus, } 8x^6 + 27y^3 = (2x^2)^3 + (3y)^3$$
$$= (2x^2 + 3y)(4x^4 - 6x^2y + 9y^2)$$

where, using rule 5, $2x^2$ is equivalent to x and $3y$ is equivalent to y. Make sure you are able to multiply out the factors to obtain the original expression!

In our study of algebraic operations we have not, so far, considered *division* of algebraic expressions. This is in part due to the fact that division is the inverse arithmetic operation of multiplication, so there are ways in which division may be avoided, by turning it into multiplication using the laws of indices. However, there are occasions when division simply cannot be avoided. It is, therefore, useful to master the art of division of both *natural numbers*, as well as *literal numbers*. To aid your understanding of division of algebraic expressions, we first look at the *long division of natural numbers*.

Algebraic division

When dividing the number 5184 by 12, you would use your calculator to obtain the result, which of course is 432. I would like to take you back to the time when you were asked to carryout *long division* to obtain this answer! My reason for doing so is quite logical: once you remember this technique using natural numbers, it will be easy to adapt this same technique to the division of literal numbers or *to algebraic*

expressions. You will also remember that no calculators are permitted when taking the CAA examinations, so long division of natural numbers becomes an essential skill!

So we may set the above division out, as follows: $12\overline{)5184}$ we reason that 12 will not go into 5, so we consider the next number, i.e. 5 and 1 or 51, 12 will go into 51 four (4) times, with 3 left over, so we now have:

$$\begin{array}{r} 4 \\ 12 \overline{)5184} \\ \underline{48} \\ 3 \end{array}$$

We now bring down the 8 because 12 does not go into 3 and get 38, 12 will go into 38 three (3) times ($3 \times 12 = 36$), so we put the 3 on top as we did the 4 then we are left with a remainder of 2. We now have:

$$\begin{array}{r} 43 \\ 12 \overline{)5184} \\ \underline{48} \\ 38 \\ \underline{36} \\ 2 \end{array}$$

We continue this process by bringing down the final figure 4, since again 12 will not go into the remainder 2. We get 24, and 12 goes into 24 two times, leaving no remainder. We place the 2 on top, as before, to finish the division. So the completed long division looks like this:

$$\begin{array}{r} 432 \\ 12 \overline{)5184} \\ \underline{48} \\ 38 \\ \underline{36} \\ 24 \\ \underline{24} \\ 0 \quad \text{leaving a remainder of zero} \end{array}$$

This division is easily checked by carrying out the inverse arithmetic operation, i.e. $(12 \times 432) = 5184$.

I hope this reminded you of the long division process, which I am sure you are familiar with. We are now going to use this process to carry out *long division of algebra*, this is best illustrated by an example.

Example 2.29

Given that $a + b$ is a factor of $a^3 + b^3$, find all remaining factors.

We can approach this problem using long division, since the factors of any expression when multiplied together produce that expression. So we can determine the factors, using the inverse of multiplication, i.e. *division*. Now, we are dividing by two literal numbers a and b, so starting with the unknown a, we see that a divides into a^3. Think of it as 3 into 27, leaving 9 or 3^2, then a into a^3 is a^2. Another approach is simply to apply the laws of indices $a^3/a^1 = a^2$, thus a^1 and a^2 are factors of a^3. This first part of the division is shown below:

$$
\begin{array}{r}
a^2 \qquad\qquad\quad \\
a+b\,\overline{)\,a^3 + b^3} \\
a^3 + a^2b \\
\hline
-a^2b + b^3 \quad\text{(after subtraction)}
\end{array}
$$

Note that the second row underneath the division is obtained by multiplying the *divisor* (the expression doing the dividing, $a + b$ in our case) by the *quotient* (the result above the division line, a^2 in our case). The remainder is obtained after subtraction of the second row from the original expression.

Next we need to find a quotient which when multiplied by the divisor gives us $-a^2b$ (the first term in the bottom line). I hope you can see that $-ab$ when multiplied by the first term in the divisor a, gives us $-a^2b$, then $-ab$ is the next term in our quotient as shown below:

$$
\begin{array}{r}
a^2 - ab \qquad\qquad \\
a+b\,\overline{)\,a^3 + b^3} \\
a^3 + a^2b \\
\hline
-a^2b + b^3 \\
-a^2b\ - ab^2 \\
\hline
+ab^2 + b^3 \quad\text{(again after}\\
\text{subtraction)}
\end{array}
$$

Finally we need the next term in our quotient to yield $+ab^2$, when multiplied by the first term of our divisor a. Again, I hope you can see that this is b^2. This completes the division as shown below:

$$
\begin{array}{r}
a^2 - ab + b^2 \qquad\qquad \\
a+b\,\overline{)\,a^3 + b^3} \\
a^3 + a^2b \\
\hline
-a^2b + b^3 \\
-a^2b - ab^2 \\
\hline
+ab^2 + b^3 \\
+ab^2 + b^3 \\
\hline
0 \quad\text{(after subtraction}\\
\text{the remainder is zero)}
\end{array}
$$

Then the factors of the expression $a^3 + b^3$ are $(a + b)$ and $(a^2 - ab + b^2)$.

We know that these two expressions are *factors*, because there is no remainder after division and if we multiply them together we obtain the original expression. Look back at Table 2.1 where we listed these factors, which you may wish to commit to memory!

> **Key point**
>
> In long division of algebra, always line up terms *in order of powers* leaving gaps where appropriate, *before* carrying out subtraction.

The above process may at first appear rather complicated but I hope you can see the pattern and symmetry that exists in the process.

Below is another completed long division shown without explanation. Study it carefully and make sure you can identify the pattern and sequence of events that go to make up the process:

$$
\begin{array}{r}
a^2 + b^2 \qquad\qquad \\
a^2-b^2\,\overline{)\,a^4 - b^4} \\
a^4 - a^2b^2 \\
\hline
+a^2b^2 - b^4 \\
+a^2b^2 - b^4 \\
\hline
0
\end{array}
$$

You might have been able to write down the factors of $a^4 - b^4$ straight away, recognizing that it is the difference between two squares, where the factors are themselves, literal numbers raised to the power 2.

> **Key point**
>
> The factors of the difference between two squares $x^2 - y^2$ are $(x - y)(x + y)$.

The need for long division of algebra may occur in your future studies, should you be required to deal with *partial fractions*. It is often useful to be able to simplify rather complex algebraic fractions into their simpler components when trying to *differentiate* or *integrate* them. You will meet *calculus arithmetic* later, where you will be asked to carry out *differentiation* and *integration* of simple functions. You will be pleased to know that in this course you are not required to find *partial fractions*!

So far we have concentrated on long division of algebraic expressions, where the division is exact, but what happens if we are left with a *remainder*? Below is shown the division of two expressions which both yield a remainder:

$$x^2 - 1 \overline{) \begin{array}{c} 1 \\ x^2 + 1 \\ \underline{-(x^2 - 1)} \\ 2 \end{array}}$$

Therefore, $\dfrac{x^2+1}{x^2-1} \equiv 1 + \dfrac{2}{x^2-1}$ where $\equiv$ *means* "always equal to".

Similarly:

$$x^3 - x \overline{) \begin{array}{c} 3 \\ 3x^3 - x^2 + 2 \\ \underline{-(3x^3 - 3x)} \\ -x^2 + 3x + 2 \end{array}}$$

therefore, $\dfrac{3x^3 - x^2 + 2}{x^3 - x} \equiv 3 + \left(\dfrac{-x^2 + 3x + 2}{x^3 - x} \right)$

In both the cases, the division has converted an improper fraction into a proper fraction. An *improper algebraic fraction* is one in which the highest power in the numerator *is greater than or equal to* ($\geq$), the highest power in the denominator. Just to make sure you can distinguish between these two types of fraction, let us substitute the natural number 2 for the unknown variable x in the first of the two examples shown above, i.e.:

$$\frac{x^2 + 1}{x^2 - 1} = \frac{(2)^2 + 1}{(2)^2 - 1} = \frac{5}{3} \quad \text{or} \quad 1\frac{2}{3}$$

so $\frac{5}{3}$ is a fraction in *improper* form and $1\frac{2}{3}$ is a *proper* fraction.

Note also that the proper fraction $1\frac{2}{3}$ *is the same as*: $1 + \frac{2}{3}$ or $\frac{3}{3} + \frac{2}{3} = \frac{5}{3}$

With your study of mathematical fundamentals and mastery of the above techniques you should now be ready to tackle problems involving the manipulation and transposition of formulae, which we consider next.

Test your understanding 2.8

1. Simplify the following algebraic expressions:
   (a) $2xy + 4xyz - 2x + 4y + 2z + 2xz - xy + 4y - 2z - 3xyz$
   (b) $2a(3b - c) + abc - ab + 2ac$

2. Multiply out the brackets and simplify the expression:
   $p(2qr - 5ps) + (p - q)(p + q) - 8s(p^2 + 1) - p^2 + q^2$

3. Factorize the following expressions:
   (a) $u^2 - 5u + 6$ (b) $6a^2b^3c - 30abc + 12abc^2$
   (c) $12x^2 - 8x - 10$ (d) $2a^3 + 2b^3$

4. Divide $a^3 - b^3$ by $a - b$ and show that $a - b$ is a factor of $a^3 - b^3$.

5. What is the quotient, when $2x^3 + x^2 - 2x - 1$ is divided by $x^2 - 1$?

2.3.5 Transposition of formulae

As mentioned earlier, formulae provide engineers with a method of writing down some rather complex relationships and ideas in a very precise and elegant way. For example, the formula $v = u + at$ tells us that the final velocity (v) of say, an aircraft, is equal to its initial velocity (u) plus its acceleration (a), multiplied by the time (t) the aircraft is accelerating down the runway. If the aircraft is neither accelerating or decelerating (negative acceleration), then $v = u$ because the acceleration $a = 0$ and $0 \times t = 0$ as you already know. I think you are already beginning to realize that to explain the meaning of one simple formula requires rather a lot of words! It is for this reason that formulae are used rather than just words to convey engineering concepts.

Note also that once the techniques for transposing (rearranging) formulae have been mastered, then solving algebraic equations becomes an easy application of these techniques!

Key point

Formulae enable engineers to write down complex ideas in a very precise way.

Terminology

Before considering the techniques needed to manipulate or transpose formulae, we first need to define some important terms, we will use our equation of motion $v = u + at$, for this purpose.

Term is defined as any variable or combination of variables separated by a $+$, $-$ or $=$ sign. You have already met this definition, in our study on the laws of arithmetic. Therefore in our formula, according to the definition, there are three (3) terms; they are v, u and at.

Variables are represented by literal numbers, which may be assigned various values. In our case v, u, a and t are *all variables.* We say that v is a *dependent variable* because its value is determined by the values given to the *independent variables* u, a and t.

The subject of a formula *sits on its own on one side of the equals sign.* Convention suggests that *the subject is placed to the left of the equals sign.* In our case v is the subject of our formula. However, the position of the subject whether to the left or right of the equals sign makes no difference to the sense of a formula. So, $v = u + at$ is identical to $u + at = v$, the subject is simply pivoted about the equals sign.

Key point

A *term* in an algebraic formula or expression is always separated by a plus ($+$), minus ($-$) or equals ($=$) sign.

Transposition of simple formulae

In the following examples we simply apply the basic arithmetic operations of addition, subtraction, multiplication and division to rearrange the subject of a formula, in other words, to *transpose a formula.*

Example 2.30

Transpose the following formula to make the letter in brackets the subject:

1. $a + b = c$ (b) 2. $y - c = z$ (c)
3. $x = yz$ (y) 4. $y = \frac{a}{b}$ (b)

1. In this formula we are required to make b the subject; therefore, b needs to sit on its own on the left-hand side of the equals sign. To achieve this we need to remove a term from the left-hand side . We ask the question, how is a attached to the left-hand side? It is in fact *added*, so to remove it to the right-hand side of the equals sign we apply the *inverse* arithmetic operation, i.e. we *subtract* it. To maintain the equality in the formula we need in effect to subtract it *from both sides*, i.e.:

   $a - a + b = c - a$, which of course gives
   $$b = c - a$$

   You will remember this operation as: *whatever we do to the left-hand side of a formula or equation, we must do to the other* **or** *when we take any term over the equals sign we change its sign.*

2. Applying the procedure we used in our first example to $y - c = z$, then we subtract y from both sides to give $y - y - c = z - y$, which again gives $-c = z - y$. Now unfortunately, in this case we are left with $-c$ on the left-hand side and we require $+c$ or just c as we normally write it, when on its own. Remembering from your study of fundamentals that a minus multiplied by a minus gives a plus and that any number multiplied by one is itself, then:

   $$(-1)(-c) = (-1)(z) - (y)(-1) \text{ or } c = -z + y$$

   and exchanging the letters on the right-hand side gives $c = y - z$.

   Now all that we have done, in this rather long-winded procedure is to multiply every term in the formula by (-1) or as you may remember it, we have *changed the sign of every term*, in order to eliminate the negative sign from the subject of our formula.

3. Now with the formula, $x = yz$ we have just two terms and our subject z is attached to y by *multiplication*. So all we need to do is *divide* it out. In other words, apply the inverse arithmetic operation, then we get:

   $$\frac{x}{y} = \frac{yz}{y} \quad \text{or} \quad \frac{x}{y} = z$$

and reversing the formula about the equals sign gives: $z = \frac{x}{y}$

4. With the formula $y = \frac{a}{b}$, then b is attached to a by *division*, so we *multiply* it out to give:

$$by = \frac{ab}{b} \quad \text{or} \quad by = a$$

This leaves us with y attached to b by *multiplication* so as to eliminate y we *divide* it out, then:

$$\frac{by}{y} = \frac{a}{y} \quad \text{or} \quad b = \frac{a}{y} \quad \text{as required}$$

In the above examples I have shown every step in full. We often leave out the intermediate steps; e.g. if $p = (q - m)/r$ and we wish to make q the subject of the formula, then multiplying both sides by r gives $pr = q - m$ and adding m to both sides we get $pr + m = q$ and reversing the formula, $q = pr + m$.

Key point

When transposing a formula for a variable, you are making that variable the subject of the formula.

Key point

Always change the sign of a term, variable or number, when you cross the equals ($=$) sign.

Transposition of formulae with common factors

What about transposing simple formula with *common factors*? You have already learnt to factorize, now we can put that knowledge to good use.

Example 2.31

Transpose the following formulae to make c the subject:

1. $a = c + bc$;   2. $2c = pq + cs$;   3. $x = \frac{ab + c}{a + c}$

1. All we need to do here is to take out c as a common factor, then: $a = c(1 + b)$; now

dividing through by the *whole* of the bracketed expression, we get:

$$\frac{a}{1 + b} = c$$

and reversing the formula we get:

$$c = \frac{a}{1 + b}$$

2. Transposition of this formula is essentially the same as in (1), except that we first need to collect all the terms involving the common factor onto one side of the formula so subtracting cs from both sides gives $2c - cs = pq$ and after taking out the common factor we get $c(2 - s) = pq$ and after division by the *whole of the bracketed expression* we get:

$$c = \frac{pq}{(2 - s)} \quad \text{or} \quad c = \frac{pq}{2 - s}$$

since there is no longer any need for the bracket.

3. Now on multiplication of both sides by $a + c$ we get: $x(a + c) = ab + c$. Notice that we have placed $a + c$ in brackets. This is very important because x is multiplied by both a and c. When transferring complicated expressions from one side of the formula to the other, then a convenient way of doing it is to *place the expression in a bracket*, then move it.

Now we can remove the brackets by multiplying out, having transferred the whole expression, we get: $ax + cx = ab + c$ and collecting the terms containing c onto one side gives $cx - c = ab - ax$ and taking out c as a common factor we get: $c(x - 1) = ab - ax$ and again after dividing out the bracketed expression, we get:

$$c = \frac{(ab - ax)}{x - 1} \quad \text{or} \quad c = \frac{a(b - x)}{x - 1}$$

Key point

When transposing for a variable that appears more than once, always collect the terms containing the variable, together, then factorize using a bracket.

Transposition of formulae involving powers and roots

You will remember from your early studies that when we write a number, say 25, in index form we get $5^2 = 25$, where 5 is the base and 2 is the *index* or *power*. Look back at the work we did on indices, in particular, on *powers* and the *laws of indices*. We are going to use this knowledge to transpose formulae that involve terms with powers, they may be *positive, negative or fractional*, e.g. p^2, p^{-3} and $p^{\frac{1}{2}} = \sqrt{p}$, respectively.

If $x^2 = yz$ and we wish to make x, the subject of the formula. All we need to do is to take the *square root of both sides*, i.e.:

$$\sqrt{x^2} = \sqrt{yz} \quad \text{or} \quad x = \sqrt{yz}$$

In index form this is the equivalent to:

$$x^{(2)(\frac{1}{2})} = y^{(1)(\frac{1}{2})}z^{(1)(\frac{1}{2})} \quad \text{or} \quad x^1 = y^{\frac{1}{2}}z^{\frac{1}{2}}$$

Similarly, if we are required to make x the subject of the formula $\sqrt{x} = yz$, then all that we need to do is to square both sides, then:

$$\left(\sqrt{x}\right)^2 = (yz)^2 \quad \text{or} \quad x = y^2z^2$$

Suppose we wish to make p the subject in the formula:

$$\left(\sqrt[3]{p}\right)^2 = abc$$

then writing this formula in index form we have:

$$p^{\frac{2}{3}} = a^1b^1c^1$$

and to get p^1, we need to multiply both sides of the formula by the power $\frac{3}{2}$

so : $p^{(\frac{2}{3})(\frac{3}{2})} = (a^1b^1c^1)^{\frac{3}{2}} \quad \text{or} \quad p = (abc)^{\frac{3}{2}} \quad \text{or}$

$$p = \left(\sqrt{abc}\right)^3$$

What the above working shows is that if we wish to find the subject of a formula, that itself has been raised to a power, we multiply it by its inverse power. It does not matter whether this power is greater than one (>1), or less than one (<1); in other words, whether it is a power or a root, respectively.

Example 2.32

1. If $x = y\sqrt{z}$, make z the subject of the formula.
2. If $Z = \sqrt{R^2 + X^2}$, transpose the formula for X.
3. If $a^{\frac{3}{4}} + b^2 = \frac{c-d}{f}$, make a the subject of the formula.

1. Our subject z is under the square root sign, so our first operation must be to square both sides and release it!
   Squaring both sides:

   $$x^2 = \left(y\sqrt{z}\right)^2 \quad \text{or} \quad x^2 = y^2\left(\sqrt{z}\right)^2,$$

   then $x^2 = y^2z$

   Dividing through by y^2:

   $$\frac{x^2}{y^2} = z \text{ and reversing, } z = \frac{x^2}{y^2}$$

   so that: $z = \left(\frac{x}{y}\right)^2$

2. Again we need to release X from underneath the square root sign.
   Squaring both sides: $Z^2 = R^2 + X^2$
   Subtracting R^2 from both sides: $Z^2 - R^2 = X^2$ and reversing, $X^2 = Z^2 - R^2$
   Then taking the square root of both sides, we get: $X = \sqrt{Z^2 - R^2}$
   Note that we square root the *whole* of both sides!

3. Then isolating the term involving a by subtracting b^2 from both sides, we get:

   $$a^{\frac{3}{4}} = \left[\frac{c-d}{f}\right] - b^2$$

   Now, multiplying *all* of both sides by the inverse power, i.e. by $\left(\frac{4}{3}\right)$ we get: $a^{(\frac{3}{4})(\frac{4}{3})} = \left[\left(\frac{c-d}{f}\right) - b^2\right]^{\frac{4}{3}}$ and so $a = \left[\left(\frac{c-d}{f}\right) - b^2\right]^{\frac{4}{3}}$

Key point 2.9

When carrying out any transposition, remember that the object of the transposition is to, isolate the term involving the subject, then obtain the subject by using multiplication or division.

2.3.6 Evaluation of formulae

So far in our study of formulae, we have concentrated on their transposition or rearrangement. This may be a necessary step, especially in more complex formulae, before we can evaluate them. Evaluation is the process whereby we replace the literal numbers in the formula, with numerical values. A simple example will serve to illustrate the technique.

Example 2.33

The final velocity of an aircraft subject to linear acceleration is given by the formula, $v = u + at$, where u is the initial velocity, a is the linear acceleration and t is the time. Given that $u = 70$ m/s, $a = 4$ m/s^2 and $t = 20$ s, find the final velocity of the aircraft.

I hope you can see that all that is required in this case is to replace the literal letters in the formula with their given numerical values and evaluate the result. So we get: $v = 70 + (4)(20)$ or the final velocity $v = 150$ m/s.

For this simple example no initial transposition of the formula was necessary, before we substituted the numerical values. Suppose we wished to find the initial velocity u? Then using the same values, it would be better to transpose the formula for u, *before* we substitute the numerical values.

Since $v = u + at$ on rearrangement $v - at = u$ or $u = v - at$ and on substituting our values, we get $u = 150 - (4)(20)$, which gives $u = 70$ m/s as we would expect!

In the next example we combine the idea of substitution with that for solving a simple equation, where the power of the unknown is one.

If you are unsure what this means look back at your work on powers and exponents, where you will find numbers written in index form. As a brief reminder 5^2 is the number 5 raised to the power 2; in other words, five squared. If the literal number z is an unknown it is, in

index form, z^1 or z raised to the power one. We normally ignore writing the power of a number when raised to the power one, *unless* we are simplifying expressions where numbers are given in *index form* and we need to use the *laws of indices*, which you have met earlier.

Example 2.34

If $a^2x + bc = ax$, find x, given $a = -3$, $b = -4$, $c = -1$.

In this case we will substitute the numerical values, *before* we simplify the formula. Then:

$$(-3)^2 x + (-4)(-1) = (-3)x$$
$$9x + 4 = -3x$$
$$9x + 3x = -4$$
$$12x = -4$$
$$x = \frac{-4}{12} \quad \text{then } x = -\frac{1}{3}$$

Note that the important use of brackets on the first line; this prevents us from making mistakes with signs!

In the next example, where we use the formula for centripetal force, we will solve for the unknown m (the mass), using both direct substitution and by transposing first and then substituting for the values.

Example 2.35

If $F = \frac{mV^2}{r}$, find m when $F = 2560$, $V = 20$ and $r = 5$; then, by direct substitution:

$$2560 = \frac{m(20)^2}{5} \quad \text{so } (2560)(5) = m(400)$$
$$400m = 12{,}800$$
$$m = \frac{12{,}800}{400} \quad \text{then } m = 32$$

Alternatively, we can transpose the formula for m and then substitute for the given values:

$$F = \frac{mV^2}{r} \quad \text{and} \quad Fr = mV^2 \quad \text{so } \frac{Fr}{V^2} = m$$

then $m = \dfrac{Fr}{V^2}$ and $m = \dfrac{(2560)(5)}{(20)^2}$

$= \dfrac{12,800}{400} = 32$

giving the same result as before.

In our final example on substitution, we use a formula that relates electric charge Q, resistance R, inductance L and capacitance C.

Example 2.36

Find C if $Q = \dfrac{1}{R}\sqrt{\dfrac{L}{C}}$, where $Q = 10, R = 40\,\Omega$, $L = 1.0$.

$$QR = \sqrt{\dfrac{L}{C}}$$

and squaring both sides gives:

$$(QR)^2 = \dfrac{L}{C} \quad \text{or} \quad Q^2R^2 = \dfrac{L}{C}$$

$$C\left(Q^2R^2\right) = L \quad \text{then} \quad C = \dfrac{L}{Q^2R^2}$$

Substituting for the given values, we get:

$$C = \dfrac{1.0}{10^2\,40^2} = 6.25 \times 10^{-6}\,\text{F}$$

Note in the above examples that you are expected to be able to obtain the numerical results, without the use of a calculator!

2.3.7 Using logarithms as an aid to calculation

This short section is not concerned with the laws of logarithms, or the more complex theory, that we leave until you study the further mathematics. Here, we will concentrate only on the use of logarithms and logarithm tables to convert the more complex arithmetic operations, such as long multiplication and long division into addition and subtraction, respectively.

Logarithms and logarithm tables

You are already aware that any positive number can be expressed as a power of 10, from your previous study of indices. Thus, e.g. $1000 = 10^3$; similarly the number $82 = 10^{1.9138}$. These powers of 10 are called *logarithms* to the base 10. That is, any number in index form with base 10 has a logarithm as its power. The logarithm tables in *Appendix D* provide logarithms for numbers between 1 and 10.

Knowing that the logarithm to the base ten of ten equals 1, i.e. $10 = 10^1$ (from the laws of indices that any number raised to the power one is itself) and that the logarithm to the base ten of 1 is equal to zero, i.e. $1 = 10^0$ (any number raised to the power zero is 1). We know that all the logarithms in the table must lie between 0 and 1.

So from the table, e.g. $\log 2.5 = 0.3979$. Now we can find the logarithm of numbers with three decimal place accuracy, using the table in the appendix. We do this by considering the numbers across the top row and across the differences.

For example, make sure that you can work out from the table that the logarithm to the base 10 of 2.556 is $0.4065 + 10 = 0.4075$, i.e. $\log 2.556 = 0.4075$.

To find numbers outside this range we make use of numbers in standard form and use the laws for the multiplication of indices, which by now you should be familiar with!

For example,

$$4567 = 4.567 \times 10^3 \text{ then}$$

$$\log 4.567 = 0.6597$$

$$4567 = 10^{0.6597} \times 10^3$$

$$4567 = 10^{3.6597}$$

$$\log 4567 = 3.6597$$

A logarithm consists of two parts: a whole number part called the *characteristic* and a decimal part called the *mantissa*, which is found directly from the log table in *Appendix D*. In the above case 3 is the characteristic and 0.6567 is the mantissa, found directly from the log tables.

Note that for positive numbers the characteristic is the positive number of powers of 10 required to place that number in standard form. Hence, the characteristic of the number 456,000 is 5, we have to move the decimal place five places to the left to put the number in standard form, i.e. 4.56×10^5.

Negative characteristics will be found in numbers that are less than 1.0. For example, $0.8767 = 8.767 \times 10^{-1}$, when placed in standard form. Then:

$$\log 8.767 = 0.9428$$

$$0.8767 = 10^{0.9428} \times 10^{-1}$$

$$= -1 + 0.9428$$

The characteristic is therefore -1 and the mantissa is 0.9428. However, it is not very convenient to write $-1 + 0.9428$, so we use a short-hand method of representation where the minus sign is placed above the characteristic thus: $\log 0.8767 = \bar{1}.9428$. You should always remember that this representation is the equivalent of $-1 + 0.9428$. Similarly, e.g. the $\log \bar{3}.1657 = -3 + 0.1657$ and $\log \bar{2}.5870 = -2 + 0.5870$.

Antilogarithms

The table of antilogarithms (given in *Appendix D*) contains the numbers which correspond to the given logarithms. When finding the antilogarithm only the decimal part (mantissa) of the logarithm is used.

Example 2.37

Find the number whose logarithm is: (a) 2.7182 and (b) $\bar{3}.5849$

(a) To find the number from this logarithm we first use the mantissa to find the required numerals. Thus from the antilogarithm tables for 0.7182 we find the numerals 5226. Now because the characteristic is 2, then the number must be **522.6**. Therefore, the **log 522.6 = 2.7182**.

(b) Again we use the mantissa 0.5849, these give the significant figures 3845. Now since the characteristic is $\bar{3}$, the number must be **0.003845**, i.e. three decimal places to the left of standard form.

Note that the $\log 0.003845 = \bar{3}.5849 = -3 + 0.5849 = -2.4151$ (this logarithm is the value you would find on your calculator if you inputted the number 0.003845!).

Using logarithms to perform arithmetic operations

Logarithms can be used to simplify, long multiplication and long division as well as finding the roots and powers of awkward or complicated numbers. In order to achieve these arithmetic operations using logarithms, we first need to define a simple set of rules:

1. To carry out multiplication using logarithms, *we find the logarithm of the numbers and add them together, then finding the antilogarithm of the sum* gives the required result.
2. For *division, find the logarithm of each number, then subtract the logarithm of the denominator from the logarithm of the numerator*. Look back at your study of fractions if you cannot remember the numerator and denominator!
3. For *powers, find the logarithm of the number and multiply it by the index denoting the power*.
4. For *roots, find the logarithm of the number and divide it by the number denoting the order of the root*.

Example 2.38

Using logarithm tables:
(a) Find the product $12.78 \times 0.00541 \times 0.886$
(b) Divide $\frac{21.718}{0.08432}$
(c) Find the value of $(0.4781)^3$
(d) Find the value of $\sqrt{0.8444}$

(a) Here we simply find the logarithms of the numbers involved, add them and find the antilogarithm of the result. Remember that when you carry over the figure from the mantissa of the characteristic it is always positive.

Then for $12.78 \times 0.00541 \times 0.886$, we get:

Number	Logarithm
12.78	1.1066
0.00541	$\bar{3}.7332$
0.886	$\bar{1}.9474$
0.006125	$\bar{3}.7872$

Make sure you follow the addition, in the above example.

(b) Then for $\frac{21.718}{0.08432}$, we get:

Number	Logarithm
21.718	1.3369
0.08432	$\bar{2}.9259$
257.5	2.4108

Again make sure you follow the subtraction remembering that subtracting a bar number (negative number) gives a positive number.

(c) Then for $(0.4781)^3$

Obeying the rule we may write:

$$\log(0.4781)^3 = 3 \times \log 0.4781$$
$$= 3 \times \bar{1}.6795 = \bar{1}.0385$$

This gives the answer as 0.1092, after taking the antilogarithm. Again it is important to obtain the correct value of the characteristic after multiplication by 3 (in this case). Note that the product of the mantissa is 2.0385, so carrying positive 2 and adding it to $3 \times \bar{1}$ gives the result $\bar{1}$ for the characteristic, as shown above.

(d) Then for $\sqrt{0.8444}$ we may write: $\log 0.8444 = \bar{1}.9265$ and

$$\frac{\bar{1}.9265}{2} = \frac{\bar{2} + 1.9265}{2} = \bar{1}.9633$$

So after taking antilogarithms, $\sqrt{0.8444} = 0.9189$.

This rather complex arithmetic can be much simplified using logarithms. This is one method of performing arithmetic operations, that is open to you, when attempting answers to non-calculator mathematical questions.

Test your understanding 2.9

1. Transpose the formula $F = \dfrac{mv^2}{r}$ for v and find the value of v, when $F = 14 \times 10^6$, $m = 8 \times 10^4$ and $r = 400$.

2. Transpose the formulae $v = \pi r^2 h$ for r.

3. Rearrange the formula $y = 8\sqrt{x} - 16$ to make x the subject.

4. If the value of the resistance to balance a Wheatstone bridge is given by the formula, $R_1 = \dfrac{R_2 R_3}{R_4}$ Find R_2, if $R_1 = 3$, $R_3 = 8$ and $R_4 = 6$.

5. Using the laws of indices and the rules for transposition, rearrange the formula $y = \dfrac{5}{\sqrt[3]{x^4}} + 20$ to make x the subject.

6. Transpose the formula $s = 18at^2 - 6t^2 - 4$ for t.

7. Make a the subject of the formula $S = \dfrac{n}{2}[2a + (n-1)d]$.

8. Transpose the equation $\dfrac{x-a}{b} + \dfrac{x-b}{c} = 1$, for x.

9. If $X = \dfrac{1}{2\pi f C}$ calculate the value of C, when $X = 405.72$ and $f = 81.144$.

10. Use logarithms to find the values of:
   (a) 192.5×0.714 (b) $\dfrac{0.413}{27.182}$ (c) $\dfrac{792 \times 27.34}{0.9876}$
   (d) $(6.125)^3$ (e) $\sqrt[3]{986.78}$ (f) $\dfrac{3.142 \times (2.718)^3}{0.9154 \times \sqrt{0.6473}}$

2.3.7 Surface area and volume of regular solids

Before considering the surface area and volume of solids we will use some common formulae to find the area of the triangle, circle and parallelogram. The complete solution of triangles using trigonometric ratios and radian measure is left until we deal with these topics in the outcome on trigonometry. The formulae we are going to use are given, without proof, in the table shown below.

Shape	Area
Triangle	Half the base multiplied by the perpendicular height, or $A = \frac{1}{2}bh$
Triangle	$A = \sqrt{s(s-a)(s-b)(s-c)}$ where a, b and c are the lengths of the sides and $s = \frac{1}{2}(a+b+c)$
Parallelogram	$A =$ base multiplied by the perpendicular height between the parallel sides. The base can be any side of the parallelogram
Circle	$A = \pi r^2$ or $A = \frac{\pi d^2}{4}$ where $r =$ radius and $d =$ diameter of circle

Trapezium	Half the sum of the parallel sides (a, b) multiplied by the vertical distance (h) between them, or $A = \left(\frac{a+b}{2}\right)h$

Example 2.39

In the triangle ABC shown in Figure 2.2, side $AB = 3$ cm and side $AC = 4$ cm. Find the area of the triangle, using *both* the formulae, given in the table.

Now we can see from the diagram that this is a *right-angled triangle*; therefore, the area A is found simply by using the formula $A = \frac{1}{2}bh$, where the base can be taken as either side containing the right-angle. Then, $A = \frac{1}{2}(3)(4) = 6$ cm^2. Note that the other side, not used as the base, is at right angles to the base and is therefore the *perpendicular* height. If the triangle was not right angled, we would need to find the perpendicular height or all of the sides in order to find the area.

In our second formula, involving the sides of the triangle, we need to know side AC. Since this is a right-angled triangle we can find the third side (opposite the right angle) by using *Pythagoras theorem*. I am sure you are familiar with this theorem, it states that: *the sum of the square on the hypotenuse is equal to the sum of the squares of the other two sides*. In our case we have that $(AC)^2 = 3^2 + 4^2 = 9 + 16 = 25$ or $AC = \sqrt{25} = 5$.

We now have three sides and $s = \frac{1}{2}(a + b + c) = \frac{1}{2}(3 + 4 + 5) = 6$; therefore, the area of the triangle:

$$A = \sqrt{s(s-a)(s-b)(s-c)}$$

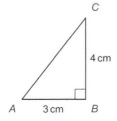

Figure 2.2 Triangle *ABC*.

$$= \sqrt{6(6-3)(6-4)(6-5)}$$
$$= \sqrt{6(3)(2)(1)} = \sqrt{36} = 6 \text{ cm}^2 \text{ as before.}$$

We will now demonstrate the use of the parallelogram formula through another example.

Example 2.40

The cross section of a metal plate is shown in Figure 2.3; find its area correct to four significant figures.

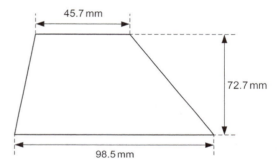

Figure 2.3

Then using the area rule for a trapezium, where in this case the vertical height is 72.7 mm. Then:

$$A = \left(\frac{a+b}{2}\right)h = \left(\frac{45.7 + 98.5}{2}\right)72.7$$
$$= (72.1)(72.7) = 5241.67$$
$$= 5242 \text{ mm}^2$$

The rule for the area of a circle, I am sure you are familiar with, will be used to find the area of an *annulus*.

Example 2.41

Determine the area of the annulus shown in Figure 2.4, which has an inner radius of 5 cm and an outer radius of 8 cm.

The shaded area (similar to a doughnut in shape) is the area of the *annulus* we require. We know both the inner and outer radii, therefore, we can treat this shape as the *difference* between the *outer* and *inner* circles. We know that the area of a circle is πr^2. Now our two circles have two different radii, where $R = 8$ cm and $r = 5$ cm. Then since the area of the *annulus A* is the difference between these two circles we may write:

$$A = \pi R^2 - \pi r^2 \quad \text{or} \quad A = \pi(R^2 - r^2)$$

then, substituting the appropriate values of the radii gives:

$$A = \pi(8^2 - 5^2) = \pi(64 - 25) = (39)\left(\frac{22}{7}\right)$$
$$= 122.6 \, \text{cm}^2$$

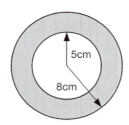

Figure 2.4 The annulus.

Note that with respect to the circle that its *circumference* $C = 2\pi r$ or $C = \pi d$, where again $r = radius$ and $d = diameter$.

> **Key point**
>
> The circumference of a circle $= 2\pi r = \pi d$.

> **Key point**
>
> The area of a circle $= \pi r^2 = \frac{\pi d^2}{4}$.

We are now in a position to tabulate some of the more common formulae we need to calculate the surface area and volume of regular solids (Table 2.3).

Example 2.42

1. Find the volume and total surface area of a right cylinder, with a top and bottom, if the cylinder has a height of 12 cm and a base radius of 3 cm.

Table 2.3 Formulae for regular solids

Solid	Volume	Surface area
Right circular cylinder *without* base and top	$V = \pi r^2 h$	$S = 2\pi r h$
Right circular cylinder *with* base and top	$V = \pi r^2 h$	$S = 2\pi r h + 2\pi r^2$ or $S = 2\pi r(h + r)$
Cone, *without* base	$V = \frac{1}{3}\pi r^2 h$	$S = \pi r l$ where $l =$ the slant height
Cone, *with* base	$V = \frac{1}{3}\pi r^2 h$	$S = \pi r l + \pi r^2$ or $S = \pi r(l + r)$
Sphere	$V = \frac{4}{3}\pi r^3$	$S = 4\pi r^2$
Hollow pipe of uniform circular cross section	$V = \pi(R^2 - r^2)l$	$S = 2\pi(R^2 - r^2) + 2\pi(R + r)$
Spherical shell	$V = \frac{4}{3}\pi(R^3 - r^3)$	$S = 4\pi(R^2 + r^2)$

Notes on table:

1. For the cylinder the height h is the vertical height. There are two formulae for the surface area of a cylinder dependent on whether or not it has a base and top. The area πr^2 is for the addition of the base *or* top, thus $2\pi r^2$ is for both.

2. The formulae for the surface area of the cone also take into consideration the cone, with and without circular base. In the volume formula, the height h is again the vertical height from the base. While the surface area formulae, use the slant height l.

3. The hollow pipe takes into account the surface area at the ends of the pipe, when the cross section is cut at right angles to its length. The volume is given by the cross-sectional area of the annulus, multiplied by the pipe length.

4. The surface area of the spherical shell includes both the inside and outside surface of the shell.

In this example it is simply a question of applying the appropriate formula.

Then for the volume $V = \pi r^2 h = \pi(3)^2 12 = 108\pi = (108)(\frac{22}{7}) = 339.4$ cm^3.

Now the cylinder has a base and a top, therefore, the surface area $S = 2\pi r(h + r)$. Then, $S = 2\pi(3)(12 + 3) = 90\pi = 282.86$ cm^2.

We finish this short section on areas and volumes, with one more example, leaving you to practice the use of these formulae, by completing the exercises in Test your understanding 2.10.

Example 2.43

Water flows through a circular pipe of internal radius 10 cm at 5 m/s. If the pipe is always three-quarters full, find the volume of water discharged in 30 min.

This problem requires us to find the volume of water in the pipe per unit time; in other words, the volume of water in the pipe per second. Note that no length has been given.

The area of the circular cross section $= \pi r^2 = \pi(10)^2 = 100\pi$; therefore, the area of the cross section of water:

$$= (\tfrac{3}{4})100\pi = 75\pi \text{ cm}^3 = (75\pi)10^{-4} \text{ m}^3$$

Now since water flows at 5 m/s, then the volume of water discharged *per second*:

$$= \frac{(5)(75\pi)10^{-4}}{1} = (375\pi)10^{-4} \text{ m}^3/\text{s}$$

Then the number of m^3 discharged in 30 min $= (30)(60)(375\pi)(10^{-4}) = 67.5\pi = 212$ m^3.

Test your understanding 2.10

Use $\pi = 22/7$ to answer the following questions.

1. Find the volume of a circular cone of height 6 cm and base radius 5 cm.

2. Find the area of the curved surface of a cone (not including base) whose base radius is 3 cm and whose *vertical* height is 4 cm.
   Hint: you need first to find the slant height.

3. If the area of a circle is 80 mm^2, find its diameter to two significant figures.

4. A cylinder of base radius 5 cm has a volume of 1 L (1000 cm^3). Find its height.

5. A pipe of thickness 5 mm has an external diameter of 120 mm. Find the volume of 2.4 m of pipe.

2.4 Geometry and trigonometry

In this final section on non-calculator mathematics we look at both the analytical and graphical representation and solution of equations and functions. Although their analytical solution should, more rightly, come under the previous section on algebra, you will find their solution easier to understand if we combine their analytical solution with their graphical representation.

We then consider the basic trigonometric ratios, the use of tables, and the nature and use of rectangular and polar co-ordinate systems. Finally, we briefly consider the methods we adopt to produce simple geometrical constructions, which sometimes involve the use of the trigonometric ratios. We start with a simple example on the analytical solution of linear equations.

2.4.1 Solution of simple equations

Although you may not have realized, you have already solved simple *equations analytically*. However, before we start our study of the *graphical solution* of equations, here is an example, which shows that in order to solve simple equations analytically, all we need to do is apply the techniques you have learnt when transposing and manipulating formula. The important point about equations is that the *equality sign must always be present*!

Example 2.44

Solve the following equations:
1. $3x - 4 = 6 - 2x$
2. $8 + 4(x - 1) - 5(x - 3) = 2(5 + 2x)$
3. $\frac{1}{2x + 3} + \frac{1}{4x + 3} = 0$

1. For this equation, all we need to do is to collect all terms involving the unknown x on to the left-hand side of the equation, simply by using our rules for transposition of formula.

Then:

$$3x + 2x - 4 = 6 \quad \text{so,} \quad 3x + 2x = 6 + 4 \quad \text{or,}$$
$$5x = 10 \quad \text{and so} \quad x = 2$$

2. In this equation first we need to multiply out the brackets, then collect all terms involving the unknown x onto one side of the equation and the numbers onto the other side, then divide out to obtain the solution. So:

$$8 + 4(x - 1) - 5(x - 3) = 2(5 + 2x)$$
$$8 + 4x - 4 - 5x + 15 = 10 + 4x$$
$$4x - 5x - 4x = 10 + 4 - 8 - 15$$
$$-5x = -9$$

and on division by -5

$$x = \frac{-9}{-5} \quad \text{or,} \quad x = \frac{9}{5}$$

Note the care taken with the signs! Also remember from your earlier work that a minus number divided by a minus number leaves us with a plus number. Alternatively, multiply top and bottom of the fraction $\frac{-9}{-5}$ by (-1), then from $(-)(-) = (+)$ we get $\frac{9}{5}$ as required.

3. To solve this equation we need to manipulate fractions, or apply the inverse arithmetic operation to every term! The simplification to obtain x using the rules for transposition is laid out *in full* below:

$$\frac{1}{2x + 3} + \frac{1}{4x + 3} = 0$$

$$\frac{1(2x + 3)}{2x + 3} + \frac{1(2x + 3)}{4x + 3} = 0(2x + 3)$$

$$1 + \frac{2x + 3}{4x + 3} = 0 \quad \text{and}$$

$$1(4x + 3) + \frac{(2x + 3)(4x + 3)}{4x + 3} = 0(4x + 3)$$

$$(4x + 3) + (2x + 3) = 0 \quad \text{or}$$

$$4x + 3 + 2x + 3 = 0$$

$$6x = -6 \quad \text{and so} \quad x = -1$$

We could have carried out the multiplication by the terms in the denominator, in just one operation simply by multiplying every term by the product $(2x + 3)(4x + 3)$. *Notice also that when multiplying any term by zero, the product is always zero.*

Key point

For all linear equations the highest power of the unknown is 1 (one).

2.4.2 Graphical axes, scales and co-ordinates

To plot a graph, you know that we take two lines at right angles to each other (Figure 2.5(a)). These lines being the axes of reference, where their intersection at the point zero is called the origin. When plotting a graph a suitable scale must be chosen, this scale need not be the same for both axes. In order to plot points on a graph, they are identified by their co-ordinates. The points $(2,4)$ and $(5,3)$ are shown in Figure 2.5(b). Note that the *x-ordinate* or *independent variable* is always quoted first. Also remember that when we use the expression plot *s against t*, then all the values of the *dependent* variable *s*, are plotted up the *vertical axis* and the other *independent*

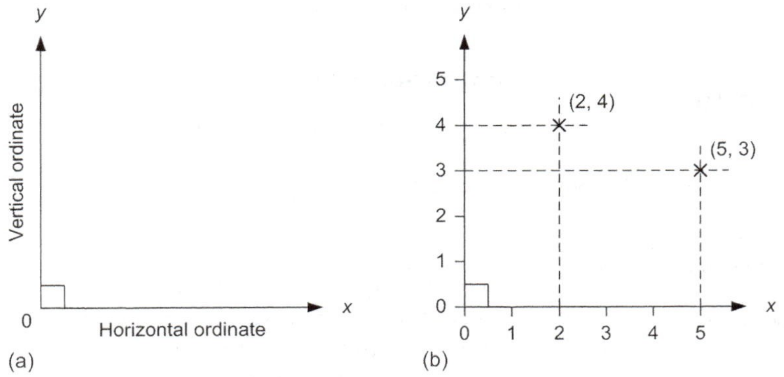

Figure 2.5 Axes and co-ordinates of graphs.

variable (in this case *t*) are plotted along the *horizontal* axis.

You met the concept of dependent and independent variables during your earlier study. Just remember that the values of the *dependent variables* are determined by the values assigned to the *independent variables*. For example, in the simple equation $y = 3x + 2$, if $x = 2$ then $y = 8$ and if $x = -2$ then $y = -4$ and so on. So to plot a graph all we need to do is:

1. Draw the two axis of reference at right angles to each other.
2. Select a suitable scale for the dependent and independent variable, or both.
3. Ensure that values of the dependent variable are plotted up the vertical axis.
4. Produce a table of values, as necessary to aid your plot.

If the graph is either a straight line or a smooth curve, then it is possible to use the graph to determine other values of the variables, apart from those given.

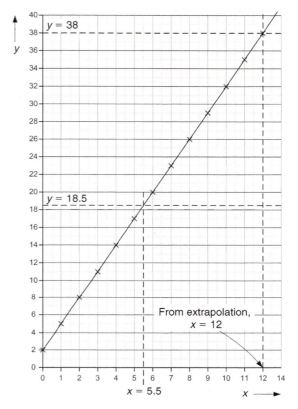

Figure 2.6 A straight line graph.

Example 2.45

Plot the graph of *y* against *x*, given the following co-ordinates

x (m)	0	1	2	3	4	5	6	7	8	9	10
y (m)	2	5	8	11	14	17	20	23	26	29	32

and find the corresponding value of *y* when $x = 5.5$ and the value *x* when $y = 38$.

The graph is plotted in Figure 2.6, note that when we join the co-ordinate points, we get a straight line. The *x*-axis scale is 1 cm = 1 m and the *y*-axis scale is 1 cm = 2 m (Figure 2.6).

To find the value of *y* corresponding to $x = 5.5$, we find 5.5 on the horizontal axis and draw a vertical line up until it meets the graph at point *P*, then draw a horizontal line until it meets the vertical *y*-ordinate and read of the value which is **18.5**.

Should we wish to find a value of *x* given *y*, we reverse this procedure. So to find the value of *x* corresponding to $y = 38$, we first find 38 on

the *y*-axis and draw a horizontal line across to meet the line. However, in this case the line does not extend this far, using the tabulated values. It is, therefore, necessary to *extend or extrapolate* the line. In this particular case it is possible to do this, as shown above, where reading vertically down we see that the intercept is at $x = 12$. This process involved extending the graph, without data being available to verify the accuracy of our extended line. Great care must be taken when using this process to prevent excessive errors. In the case of a straight line graph or linear graph, this is an acceptable practice. This process is commonly known as graphical *extrapolation*.

Key point

When plotting any variable *y against x*, the variable *y* is plotted on the vertical axis.

2.4.3 Graphs of linear equations

In the above example all values of the co-ordinates are positive. This is not always the case and to accommodate negative numbers, we need to extend the axes to form a cross (Figure 2.7), where both positive and negative values can be plotted on both axes.

Figure 2.7 not only shows the positive and negative axes, but also the plot of the equation $y = 2x - 4$. To determine the corresponding y-ordinates shown, for values of x between -2 and 3, we use a table.

x	-2	-1	0	1	2	3
$2x$	-4	-2	0	2	4	6
-4	-4	-4	-4	-4	-4	-4
$y = 2x - 4$	-8	-6	-4	-2	0	2

For example, when $x = -2$, $y = 2(-2) - 4 = -4 - 4 = -8$.

The scale used on the y-axis is 1 cm = 1 unit and on the x-axis 2 cm = 1 unit.

This equation, where the highest power of the variable x, y is 1.0 is known as an *equation of the first degree* or a *linear equation. All linear equations produce graphs that are always straight lines.*

Now every linear equation may be written in *standard form*, i.e.:

$$y = mx + c$$

So for our equation $y = 2x - 4$ which is in the standard form, $m = 2$ and $c = -4$.

Also, every linear equation may be re-arranged so that it is in standard form. For example:

$4y + 2 = 2x - 6$ then re-arranging for y, $4y = 2x - 6 - 2$ or $4y = 2x - 8$ and on division by 4

$$y = \frac{2}{4}x - \frac{8}{4} \quad \text{or} \quad y = \frac{1}{2}x - 2$$

where $m = \frac{1}{2}$ and $c = -2$.

Determining m and c for the equation of a straight line

In Figure 2.8, point A is where the straight line cuts the y-axis and has co-ordinates $x = 0$ and $y = c$. Thus c in the equation $y = mx + c$ is the point where the line meets the y-axis, when the value of $x = 0$ or *the variable c = the y intercept when x = 0.*

Also from Figure 2.8, the value $\frac{BC}{AC}$ is called the *gradient* of the line. Now the length:

$$BC = \left(\frac{BC}{AC}\right)AC = AC \times \text{gradient of the line}$$

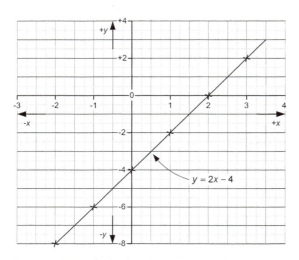

Figure 2.7 Graph of the equation $y = 2x - 4$.

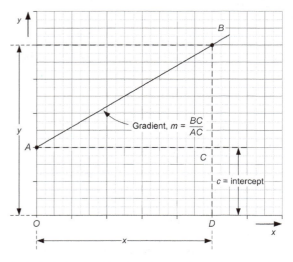

Figure 2.8 Graph showing relationship for variables c and m.

$$y = BC + CD = BC + AO$$

$$= AC \times \text{ the gradient of the line } + AO$$

$$= x \text{ multiplied by the gradient of the line} + c$$

But $y = mx + c$. So it can be seen that:

$$m = \text{ the gradient of the line}$$
$$c = \text{ the intercept on the } y\text{-axis}$$

Example 2.46

1. Find the law of the straight line illustrated in Figure 2.9.
2. If a straight line graph passes through the point $(-1, 3)$ and has a gradient of 4. Find the values of m and c and then write down the equation of the line.

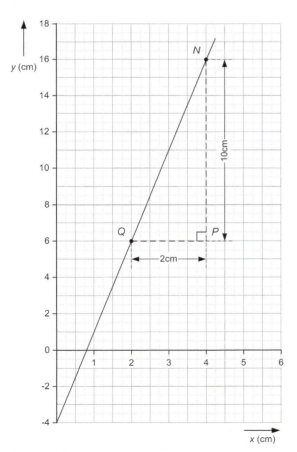

Figure 2.9 Figure for Example 2.46, question 1.

1. Since the intercept c is at the origin, it can be read-off the graph as -4. The value of m, the gradient of the line, is found by taking convenient values of x and y, then the gradient m from the graph $= \frac{NP}{QP} = \frac{10\,\text{cm}}{2\,\text{cm}} = 5$. So the equation of the line $y = mx + c$ is $y = 5x + 4$.
2. We are given the gradient $m = 4$, therefore, $y = 4x + c$ and this line passes through the point $(-1, 3)$. So we know that $y = 3$ when $x = -1$ and substituting these values into the equation of the straight line gives $3 = 4(-1) + c$ and so $c = 7$. Then the equation of the line is $y = 4x + 7$.

Note that in the questions given in Example 2.46 the gradient (or slope) of the straight lines, were both *positive*. Straight line graphs can also have a *negative gradient*, this will occur when the graph of the line slopes downwards to the right of the y-axis. Under these circumstances a negative value of y results, so that $m = \frac{-y}{x}$ and so the gradient m is negative.

We leave our study of linear equations and their straight line graphs, with an example of the application of the law of a straight line, $y = mx + c$, to experimental data.

Example 2.47

During an experiment to verify Ohm's law, the following experimental results were obtained.

E (V)	I (A)
0	0
1.1	0.25
2.3	0.5
3.4	0.75
4.5	1.0
5.65	1.25
6.8	1.5
7.9	1.75
9.1	2.0

Plot voltage against current and so determine the equation connecting E and I.

The resulting plot is shown in Figure 2.10.

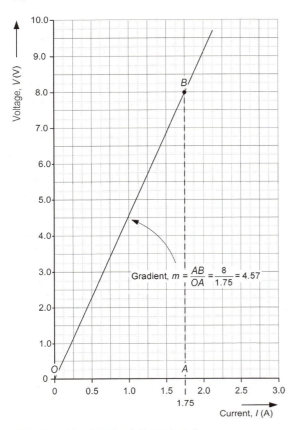

Figure 2.10 Graph of *E* against *I*.

From the plot, it can be seen that the experimental data produces a straight line. Therefore, the equation connecting *E* and *I* is of the form $y = mx + c$. Since the graph goes *directly through the origin*, then the variable $c = 0$. Also from the graph taking suitable values of *E* and *I*, the gradient $m = 4.57$ corrected to three significant figures. So the equation connecting *E* and *I* is $E = 4.57\,I$.

2.4.4 Quadratic equations

A quadratic equation is one in which the unknown variable is raised to the second power. For example, the equation $x^2 = 4$ is perhaps one of the simplest of quadratic equations. We can solve this equation by taking the square root of both sides, something which you are familiar with, when transposing a formula. Then: $\sqrt{x^2} = \sqrt{4}$ or $x = \pm 2$. Note that even

for this simple equation there are two possible solutions, either $x = +2$ or $x = -2$, remembering your laws of signs! When we square a positive number we get a positive number $(+2)(+2) = +4$ or simply 4 also, $(-2)(-2) = 4$, from the laws of signs.

In general, a quadratic equation is of the type $ax^2 + bx + c = 0$, where the constants a, b and c can take *any* numerical value, positive or negative, decimal or fraction. Like linear equations, quadratic equations do not always appear in *standard form*, i.e. they are not always arranged in exactly the same order as their qualifying equation, $ax^2 + bx + c = 0$.

How is our simple equation $x^2 = 4$ related to its qualifying equation? Well the coefficient of x^2 that is the number multiplying the x^2 term $a = 1$. What about the constant b? Well there is no x term in our equation so $b = 0$. What about the constant c? Our equation is not in standard form, because the equation should be equated to zero. Then in standard form our equation becomes $x^2 - 4 = 0$ by simple transposition! So now we know that for our equation the constant term $c = -4$. A quadratic equation may contain only the square of the unknown variable, as in our simple equation, or it may contain the square and the first power of the variable; e.g. $x^2 - 2x + 1 = 0$. Also, the unknown variable may have *up to two* possible *real* solutions. The equations we deal with in this course will always have at least one real solution.

There are several ways in which quadratic equations may be *solved*, that is finding the values of the *unknown* variable. We shall concentrate on just three methods of solution; factorization, using the formula and solving by graphical methods.

Solution of quadratic equations by factorization

Take the equation $x^2 - 2x + 1 = 0$. If we ignore for the moment, the fact that this is an equation and concentrate on the expression $x^2 - 2x + 1$, then you may remember how to find *the factors* of this expression! Look back now at your work on factors to remind yourself.

I hope you were able to identify the factors of this expression as $(x - 1)(x - 1)$. Now all we need to do is equate these factors to *zero*, to solve our equation. Thus, $(x-1)(x-1)=0$ then for the equation to *balance* either the first bracket $(x-1)=0$ or the second bracket (the same in this case), $(x-1)=0$. Thus solving this very simple linear equation gives $x=1$, no matter which bracket is chosen. So, in this case, our equation only has one solution $x=1$. Note that if anyone of the bracketed expressions $(x-1)=0$, then the other bracket is multiplied by zero, i.e. $0(x-1)=0$. This is obviously true, because any quantity multiplied by zero is itself zero.

Key point

For all quadratic equations, the highest power of the independent variable is 2 (two).

Example 2.48

Solve the equation $3x^2-5=-2x-4$ by factorization.

The first thing to note before we attempt a solution is that this equation *is not* in standard form. All we need to do is transpose the equation, to get it into standard form. You should, by now, be able to do the transposition with ease, so make sure you obtain: $3x^2+2x-1=0$. Now using the techniques for factorization, that you learnt earlier, after trial and error, you should find that: $(3x-1)(x+1)=0$ then either:

$$3x-1=0 \quad \text{giving } x=\frac{1}{3}$$
$$\text{or } (x+1)=0 \quad \text{giving } x=-1$$

Note that in this case the equation has two different solutions, both can be checked for accuracy by substituting them into the *original* equation. Then either:

$$3\left(\frac{1}{3}\right)^2-5=-2\left(\frac{1}{3}\right)-4 \quad \text{or} \quad \frac{3}{9}-5=-\frac{2}{3}-4$$

therefore:

$$-4\frac{2}{3}=-4\frac{2}{3}$$

which is correct or $3(-1)^2-5=-4-2(-1)$ or $3-5=-4+2$; therefore, $-2=-2$, which is also correct. Note the need to manipulate fractions and be aware of the laws of signs, skills I hope you have acquired, at this stage in your learning.

Solution of quadratic equations using formula

It is not always possible to solve quadratic equations by factorization. When we cannot factorize a quadratic expression, we may resort to use of the standard formula. Now we know that the standard form of the quadratic equation is $ax^2+bx+c=0$ and it can be shown that the solution of this equation is:

$$x=\frac{-b\pm\sqrt{b^2-4ac}}{2a}$$

Now this equation may look complicated but it is relatively simple to use. The coefficients a, b and c are the same coefficients, as in the standard form of the quadratic. So in finding a solution for the variable x, all we need to do is substitute the coefficients into the above formulae, for the quadratic equation we are considering. All you need to remember is that, *before* using the above formula, *always put the equation to be solved, into standard form*. Also note that in the above formula, the whole of the numerator, including the $-b$, is divided by $2a$.

Example 2.49

Solve the equation $5x(x+1)-2x(2x-1)=20$.

The above equation is not in standard form, in fact until we simplify it, we may not be aware that it is a quadratic equation. So simplifying, by multiplying out the brackets and collecting like-terms gives:

$$5x^2+5x-4x^2+2x=20 \text{ and so } x^2+7x-20=0$$

This equation is now in standard form and may be solved using the formula. You may have attempted to try a solution by factorization first. If you cannot find the factors, reasonably quickly, then you can always resort to the formula, unless told otherwise!

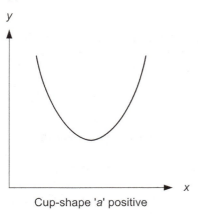

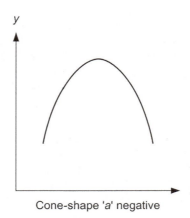

Cup-shape 'a' positive   Cone-shape 'a' negative

Figure 2.11 Curves of quadratic functions.

Then from

$$x = \frac{-b \pm \sqrt{b^2 - 4ac}}{2a}$$

we get

$$x = \frac{-7 \pm \sqrt{7^2 - (4)(1)(-20)}}{2(1)}$$

and simplifying gives

$$x = \frac{-7 \pm \sqrt{129}}{2} \quad \text{or} \quad x = \frac{-7 \pm 11.358}{2}$$

and so

$$x = \frac{-7 + 11.358}{2} \quad \text{or} \quad x = \frac{-7 - 11.358}{2}$$

Giving, the values of the unknown x, corrected to three significant figures as:

$$x = 2.18 \quad \text{or} \quad x = -9.18$$

We now consider our final method of solution of quadratic equations, using a graphical method.

Key point

Quadratic equations may have *up to* two (2) *real* solutions.

Solution of quadratic equations using a graphical method

If we plot a quadratic function of the form $ax^2 + bx + c$ against x, the resulting curve is known as a parabola, and depending on the sign of the coefficient a will determine which way up the curve sits (Figure 2.11).

The plotting of such curves requires a table of values to be set up in terms of the values of the independent and dependent variables. This procedure is best illustrated by the following example.

Example 2.50

Draw the graph of $y = x^2 - 3x + 2$, taking values of the independent variable x, between 0 and 4.

x	0	1	2	3	4
x^2	0	1	4	9	16
$-3x$	0	-3	-6	-9	-12
2	2	2	2	2	2
y	2	0	0	2	6

So, from the equation when $x = 1.5$, $y = 2.25 - 4.5 + 2 = -0.25$.

The resulting plot is shown in Figure 2.12.

Now the points on the curve where it crosses the x-axis are $x = 1$ and $x = 2$. These are the points on the curve for which $y = 0$ or $x^2 - 3x + 2 = 0$. Therefore, $x = 1$ and $x = 2$ are the solutions of the quadratic equation $x^2 - 3x + 2 = 0$.

Now from our graph, we can also solve any equation of the type $x^2 - 3x = k$, where k is a constant. If, for example, we wish to solve $x^2 - 3x + 1 = 0$, then comparing this equation

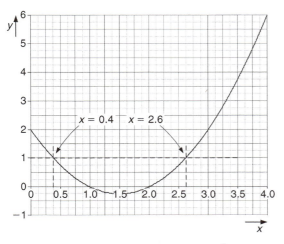

Figure 2.12 Graph of the function $y = x^2 - 3x + 2$.

with the equation of the plot, all we need to do is add 1 to both sides to acquire the equation, $y = x^2 - 3x + 2 = 1$. So that to solve this equation, we need the points on the curve, where $y = 1$. We then draw the line $y = 1$ and read of the corresponding values of x at these points. From the dashed line on the graph we obtain the solution of this modified equation as $x = 2.6$ or $x = 0.4$.

Key point

The graphs of quadratic expressions and equations will always be parabolic in shape.

We finish our study of equations by considering simultaneous equations.

2.4.5 Simultaneous equations

Simultaneous equations, involve more than one variable, or unknown. We can solve a simple linear equation with one unknown, using the laws of algebra, you have already learnt. It is often required to represent an engineering problem that involves more than one unknown. For example, if an engineering problem involves the solution of an equation, such as $3x + 2y = 12$. How do we go about solving it? Well the answer is that a single equation with two unknowns, is unsolvable, unless we know the value of one of

the variables. However, if we have *two equations*, with *two unknowns*, it is possible to solve these equations *simultaneously*, that is at the same time. Three linear equations, with three variables, can also be solved simultaneously; in fact, any number of linear equations, with a corresponding number of unknowns (variables) can be solved simultaneously. However, when the number of variables is greater than three, it is better to solve the system of equations using a computer!

These systems of equations, occur in many aspects of engineering, particularly when we model static and dynamic behaviour of solids and liquids. You will be pleased to know that we will only be considering *two equations* simultaneously, involving *two unknowns*! Even so, the distribution of currents and voltages for example, in electrical networks, sometimes involves the solution of such equations, with just two unknowns.

Analytical solution of simultaneous equations

Consider the pair of equations:

$$3x + 2y = 12 \qquad (1)$$
$$4x - 3y = -1 \qquad (2)$$

Now to solve these equations, all we need to do is to use *elimination* and *substitution* techniques, working on both equations simultaneously.

Let us try to eliminate the variable x, from both the equations. This can be achieved by multiplying each equation by a constant. When we do this, we do not alter the nature of the equations. If we multiply equation (1) by the constant 4, and equation (2) by the constant 3, we get:

$$12x + 8y = 48$$
$$12x - 9y = -3$$

Note that we have multiplied *every term* in the equations, by the constant! Now, how does this help us to eliminate x? Well if we now add, both equations together we end up with the first term being, $24x$, this is not very helpful. However, if we subtract Equation (2) from (1), we get:

$$12x + 8y = 48$$
$$-(12x - 9y = -3)$$
$$\overline{ 0 + 17y = 51}$$

From which we see that $y = 3$. Now having found one of the unknown variables, we can substitute its value into *either one of the original equations*, in order to find the other unknown. Choosing equation (1), then from, $3x + 2y = 12$ we get, $3x + (2)(3) = 12$ or $3x = 6$, and therefore $x = 2$. So the required solution is $y = 3$ and $x = 2$.

When solving any equation, the solutions can always be checked by substituting their values into the original equation, so substituting the values into Equation (2) gives: $4(2) - (3)(3) = -1$ which is correct.

Key point

To solve equations simultaneously, we require the same number of equations, as there are unknowns.

Graphical solution of two simultaneous equations

The method of solution is shown in the next example. For each of the linear equations, we plot their straight line graphs and where the plots intersect is the unique solution for both equations.

Example 2.51

Solve the following simultaneous equations graphically:

$$\frac{x}{2} + \frac{y}{3} = \frac{13}{6}; \quad \frac{2x}{7} - \frac{y}{4} = \frac{5}{14}$$

Now we first need to simplify these equations and rearrange them in terms of the independent variable y. I hope you can remember how to simplify fractions! Make sure that you are able to rearrange the equations and obtain:

$$2y = 13 - 3x$$
$$-7y = 10 - 8x$$

Now transposing in terms of y, we get:

$$y = \frac{13}{2} - \frac{3}{2}x$$
$$y = -\frac{10}{7} + \frac{8}{7}x$$

Now we can find the corresponding values of y, for our chosen values of x. Using just four values of x, say 0, 1, 2 and 3, will enable us to plot the straight lines. Then:

x	0	1	2	3
$y = \dfrac{13}{2} - \dfrac{3}{2}x$	$\dfrac{13}{2}$	5	$\dfrac{7}{2}$	2
$y = -\dfrac{10}{7} + \dfrac{8}{7}x$	$-\dfrac{10}{7}$	$-\dfrac{2}{7}$	$\dfrac{6}{7}$	2

From the plot shown in Figure 2.13, the intersection of the two straight lines, yields the required result, that is $x = 3$ and $y = 2$.

In this particular example, it would have been easier to solve these equations using an algebraic method.

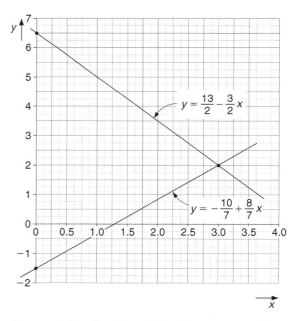

Figure 2.13 Graphs of simultaneous equations $y = \frac{13}{2} - \frac{3}{2}x$ and $y = -\frac{10}{7} + \frac{8}{7}x$.

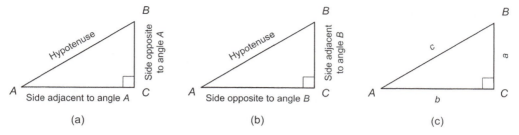

Figure 2.14 The right-angled triangle.

Test your understanding 2.11

1. The values in the table below show how instantaneous current I varies with voltage V. Plot a graph of V against I and so find the value of V when $I = 3.0$.

V	15	25	35	50	70
I	1.1	2.0	2.5	3.2	3.9

2. Solve the following linear equations:
   (a) $5x - 1 = 4$ (b) $3(x - 2) = 2(x - 1)$
   (c) $\dfrac{1}{p} + \dfrac{1}{p+1} = \dfrac{2}{p-1}$

3. Solve the following simultaneous equations:
   (a) $2x + 3y = 8$; $2x - 3y = 2$
   (b) $5x + 4y = 22$; $3x + 5y = 21$
   (c) $\dfrac{a+b}{a-b} = \dfrac{1}{2}$; $\dfrac{a+1}{b+1} = 2$
   (d) $\dfrac{p}{2} + \dfrac{q}{3} = 2$; $2p + 3y = 13$

4. If $y = ax + b$, find the value of y, when $x = 4$, given that $y = 4$ when $x = 1$ and that $y = 7$ when $x = 2$.

5. Solve graphically, the following simultaneous equation:
$$7x - 4y = 37$$
$$6x + 3y = 51$$

6. Solve the following quadratic equations:
   (a) $6x^2 + x - 2 = 0$ (b) $-2x^2 - 20x = 32$
   (c) $f + \dfrac{1}{f} = 3$ (d) $\dfrac{1}{a+1} + \dfrac{1}{a+2} - \dfrac{2}{3} = 0$

7. Solve the equation $\frac{3}{4}x^2 - x = \frac{5}{4}$ graphically.

8. Draw, using the same scale and axes, the graphs of $s = 2u + 3$ and $s = u^2 + u + 1$, from your graphs, solve the equation $u^2 - u = 2$.

2.4.6 The trigonometric ratios, use of tables and the solution of right-angled triangles

We start by identifying the *notation* used for right-angled triangles. We label the points (vertices) of the triangle using capital letters A, B and C as in Figure 2.14.

The side AB lies opposite the right angle ($90°$) and is called the *hypotenuse*. The side BC lies opposite the angle A and is called the side *opposite* to A. Finally in Figure 2.14(a), the side AC is known as the side *adjacent* to A. Another way of distinguishing between the side opposite the angle and the side adjacent to the angle is to imagine you are looking from behind the angle with your eye, then what you see is the side opposite. Figure 2.14(b) shows this when we consider the sides in relationship to angle B. For convenience the sides, opposite their angle, are often distinguished by a lower case letter, as shown in Figure 2.14(c).

When we consider any angle, rather than using capital letters, we use symbols from the Greek alphabet! The most common Greek letter used is theta (θ), but equally the letters α, β, γ and ϕ (alpha, beta, gamma and phi, respectively) may also be used.

Key point

The side opposite the right angle in a right-angled triangle is the hypotenuse.

The trigonometric ratios

Figure 2.15 shows the angle θ, which is bounded by the lines OA and OB. If we take any point P on the line OB and from this point we drop a perpendicular to the line OA to meet it at the point Q.

Then the ratio

$\frac{QP}{OP}$ is called the *sine* of angle AOB

$\frac{OQ}{OP}$ is called the *cosine* of angle AOB and

$\frac{QP}{OQ}$ is called the *tangent* of the angle AOB.

The sine ratio

If we consider the triangle OPQ (Figure 2.16) from the point of view of the angle θ, then the sine (abbreviated to *sin*) of this angle

$$= \frac{opposite}{hypotenuse}, \quad \text{i.e.} \quad \sin\theta = \frac{QP}{OP} \quad \text{or} \quad \sin\theta = \frac{o}{q}$$

similarly, the sine of the angle α

$$= \frac{opposite}{hypotenuse}, \quad \text{i.e.} \quad \sin\alpha = \frac{OQ}{OP} \quad \text{or} \quad \sin\alpha = \frac{p}{q}$$

If we know either the angle θ or α then we can find the value of the sine ratio, for that particular angle. To do this you may, up to now, have simply used your calculator. Since we are considering *non-calculator mathematics* we can only use drawing or tables to find the value of the sine ratio.

> **Key point**
>
> For any angle θ, then $\sin\theta = \frac{side\ opposite}{hypotenuse}$.

Example 2.52

Find by drawing a suitable triangle, the value of sin 30°.

Then using a protractor, or by another method, draw the lines AP and AQ which intersect at A, so that the angle $PAQ = 30°$, as shown in Figure 2.17.

Along AQ measure off to a suitable scale AC (the hypotenuse), say 100 units. From point C, draw line CB perpendicular to AP. Measure CB, which will be found to be 50 units. Then:

$$\sin 30° = \frac{50}{100} = 0.5.$$

This method could be used to find the sine of any angle; however, it is rather tedious and somewhat limited in accuracy. Tables of sine ratios have been compiled that allow us to find the sine of any angle. Table 2.4 shows an extract from the full table of natural sines, which will be found in *Appendix D*.

From Table 2.4 it can be seen that angles are subdivided into degrees (°) and minutes (′) where 1 min is equal to $\frac{1}{60}$ of a degree. The equivalent decimal fraction of each degree is also given at the top of the table. We will demonstrate how to read Table 2.4, using an example.

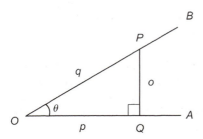

Figure 2.15 The right-angled triangle.

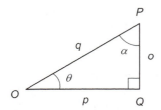

Figure 2.16 The sine of an angle.

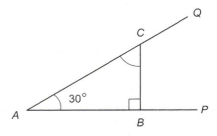

Figure 2.17 Triangle *ABC*.

Table 2.4 Extract from table of natural sines

	0' 0.0°	6' 0.1°	12' 0.2°	18' 0.3°	24' 0.4°	30' 0.5°	36' 0.6°	42' 0.7°	48' 0.8°	54' 0.9°	Mean Differences				
											1'	2'	3'	4'	5'
0°	0.0000	0017	0035	0052	0070	0087	0105	0122	0140	0157	3	6	9	12	15
1	0.0175	0192	0209	0227	0244	0262	0279	0297	0314	0332	3	6	9	12	15
2	0.0349	0366	0384	0401	0419	0436	0454	0471	0488	0506	3	6	9	12	15
3	0.0523	0541	0558	0576	0593	0610	0628	0645	0663	0680	3	6	9	12	15
4	0.0698	0715	0732	0750	0767	0785	0802	0819	0837	0854	3	6	9	12	15
5	0.0872	0889	0906	0924	0941	0958	0976	0993	1011	1028	3	6	9	12	14
6	0.1045	1063	1080	1097	1115	1132	1149	1167	1184	1201	3	6	9	12	14
7	0.1219	1236	1253	1271	1288	1305	1323	1340	1357	1374	3	6	9	12	14
8	0.1392	1409	1426	1444	1461	1478	1495	1513	1530	1547	3	6	9	12	14
9	0.1564	1582	1599	1616	1633	1650	1668	1685	1702	1719	3	6	9	12	14
10°	0.1736	1754	1771	1788	1805	1822	1840	1857	1874	1891	3	6	9	11	14
11	0.1908	1925	1942	1959	1977	1994	2011	2028	2045	2062	3	6	9	11	14
12	0.2079	2096	2113	2130	2147	2164	2181	2198	2215	2233	3	6	9	11	14
13	0.2250	2267	2284	2300	2317	2334	2351	2368	2385	2402	3	6	8	11	14
14	0.2419	2436	2453	2470	2487	2504	2521	2538	2554	2571	3	6	8	11	14
15	0.2588	2605	2622	2639	2656	2672	2689	2706	2723	2740	3	6	8	11	14
16	0.2756	2773	2790	2807	2823	2840	2857	2874	2890	2907	3	6	8	11	14
17	0.2924	2940	2957	2974	2990	3007	3024	3040	3057	3074	3	6	8	11	14
18	0.3090	3107	3123	3140	3156	3173	3190	3206	3223	3239	3	6	8	11	14
19	0.3256	3272	3289	3305	3322	3338	3355	3371	3387	3404	3	5	8	11	14
20°	0.3420	3437	3453	3469	3486	3502	3518	3535	3551	3567	3	5	8	11	14
21	0.3584	3600	3616	3633	3649	3665	3681	3697	3714	3730	3	5	8	11	14
22	0.3746	3762	3778	3795	3811	3827	3843	3859	3875	3891	3	5	8	11	14
23	0.3907	3923	3939	3955	3971	3987	4003	4019	4035	4051	3	5	8	11	14
24	0.4067	4083	4099	4115	4131	4147	4163	4179	4195	4210	3	5	8	11	13
25	0.4226	4242	4258	4274	4289	4305	4321	4337	4352	4368	3	5	8	11	13
26	0.4384	4399	4415	4431	4446	4462	4478	4493	4509	4524	3	5	8	10	13
27	0.4540	4555	4571	4586	4602	4617	4633	4648	4664	4679	3	5	8	10	13
28	0.4695	4710	4726	4741	4756	4772	4787	4802	4818	4833	3	5	8	10	13
29	0.4848	4863	4879	4894	4909	4924	4939	4955	4970	4985	3	5	8	10	13
30°	0.5000	5015	5030	5045	5060	5075	5090	5105	5120	5135	3	5	8	10	13
31	0.5150	5165	5180	5195	5210	5225	5240	5255	5270	5284	2	5	7	10	12
32	0.5299	5314	5329	5344	5358	5373	5388	5402	5417	5432	2	5	7	10	12
33	0.5446	5461	5476	5490	5505	5519	5534	5548	5563	5577	2	5	7	10	12
34	0.5592	5606	5621	5635	5650	5664	5678	5693	5707	5721	2	5	7	10	12
35	0.5736	5750	5764	5779	5793	5807	5821	5835	5850	5864	2	5	7	9	12
36	0.5878	5892	5906	5920	5934	5948	5962	5976	5990	6004	2	5	7	9	12
37	0.6018	6032	6046	6060	6074	6088	6101	6115	6129	6143	2	5	7	9	12
38	0.6157	6170	6184	6198	6211	6225	6239	6252	6266	6280	2	5	7	9	11
39	0.6293	6307	6320	6334	6347	6361	6374	6388	6401	6414	2	4	7	9	11
40°	0.6428	6441	6455	6468	6481	6494	6508	6521	6534	6547	2	4	7	9	11
41	0.6561	6574	6587	6600	6613	6626	6639	6652	6665	6678	2	4	7	9	11
42	0.6691	6704	6717	6730	6743	6756	6769	6782	6794	6807	2	4	6	9	11
43	0.6820	6833	6845	6858	6871	6884	6896	6909	6921	6934	2	4	6	8	11
44	0.6947	6959	6972	6984	6997	7009	7022	7034	7046	7059	2	4	6	8	10

Example 2.53

Find, using Table 2.4:

(a) sin 32°   (b) sin 32°24'   (c) sin 32°28'.

(a) The sine of an angle with an exact number of degrees is shown in the column headed 0'. Thus sin 32° = 0.5299.

(b) To find sin 32°24'. The required value is found underneath the 24' column as 0.5358.

(c) If the number of minutes is not an exact multiple of 6, as in this case, we use the tables of differences, shown on the right of Table 2.4. Thus since sin 32°24' = 0.5358 and 28' is 4' more than 24'. Then, looking in the difference column headed 4' we find the value 10. This is *added* on to the sine of 32°24'. Then the sin 32°28' = 0.5358 + 10 = 0.5368.

Suppose we wish to carry out the inverse operation to that of finding the sine of an angle. In other words, if we are required to find the angle whose sine is 0.3878 (in symbols, $\sin^{-1} 0.3878$) then we proceed as follows. Look in Table 2.4

Table 2.5 Extract from table of natural cosines

	0′ 0.0°	6′ 0.1°	12′ 0.2°	18′ 0.3°	24′ 0.4°	30′ 0.5°	36′ 0.6°	42′ 0.7°	48′ 0.8°	54′ 0.9°	SUBTRACT Mean Differences 1′	2′	3′	4′	5′
0°	1.000	1.000	1.000	1.000	1.000	1.000	.9999	9999	9999	9999	0	0	0	0	0
1	0.9998	9998	9998	9997	9997	9997	9996	9996	9995	9995	0	0	0	0	0
2	0.9994	9993	9993	9992	9991	9990	9990	9989	9988	9987	0	0	0	1	1
3	0.9986	9985	9984	9983	9982	9981	9980	9979	9978	9977	0	0	1	1	1
4	0.9976	9974	9973	9972	9971	9969	9968	9966	9965	9963	0	0	1	1	1
5	0.9962	9960	9959	9957	9956	9954	9952	9951	9949	9947	0	1	1	1	2
6	0.9945	9943	9942	9940	9938	9936	9934	9932	9930	9928	0	1	1	1	2
7	0.9925	9923	9921	9919	9917	9914	9912	9910	9907	9905	0	1	1	2	2
8	0.9903	9900	9898	9895	9893	9890	9888	9885	9882	9880	0	1	1	2	2
9	0.9877	9874	9871	9869	9866	9863	9860	9857	9854	9851	0	1	1	2	2
10°	0.9848	9845	9842	9839	9836	9833	9829	9826	9823	9820	1	1	2	2	3
11	0.9816	9813	9810	9806	9803	9799	9796	9792	9789	9785	1	1	2	2	3
12	0.9781	9778	9774	9770	9767	9763	9759	9755	9751	9748	1	1	2	3	3
13	0.9744	9740	9736	9732	9728	9724	9720	9715	9711	9707	1	1	2	3	3
14	0.9703	9699	9694	9690	9686	9681	9677	9673	9668	9664	1	1	2	3	4
15	0.9659	9655	9650	9646	9641	9636	9632	9627	9622	9617	1	2	2	3	4
16	0.9613	9608	9603	9598	9593	9588	9583	9578	9573	9568	1	2	2	3	4
17	0.9563	9558	9553	9548	9542	9537	9532	9527	9521	9516	1	2	3	3	4
18	0.9511	9505	9500	9494	9489	9483	9478	9472	9466	9461	1	2	3	4	5
19	0.9455	9449	9444	9438	9432	9426	9421	9415	9409	9403	1	2	3	4	5
20°	0.9397	9391	9385	9379	9373	9367	9361	9354	9348	9342	1	2	3	4	5
21	0.9336	9330	9323	9317	9311	9304	9298	9291	9285	9278	1	2	3	4	5
22	0.9272	9265	9259	9252	9245	9239	9232	9225	9219	9212	1	2	3	4	6
23	0.9205	9198	9191	9184	9178	9171	9164	9157	9150	9143	1	2	3	5	6
24	0.9135	9128	9121	9114	9107	9100	9092	9085	9078	9070	1	2	4	5	6
25	0.9063	9056	9048	9041	9033	9026	9018	9011	9003	8996	1	3	4	5	6
26	0.8988	8980	8973	8965	8957	8949	8942	8934	8926	8918	1	3	4	5	6
27	0.8910	8902	8894	8886	8878	8870	8862	8854	8846	8838	1	3	4	5	7
28	0.8829	8821	8813	8805	8796	8788	8780	8771	8763	8755	1	3	4	6	7
29	0.8746	8738	8729	8721	8712	8704	8695	8686	8678	8669	1	3	4	6	7
30°	0.8660	8652	8643	8634	8625	8616	8607	8599	8590	8581	1	3	4	6	7
31	0.8572	8563	8554	8545	8536	8526	8517	8508	8499	8490	2	3	5	6	8
32	0.8480	8471	8462	8453	8443	8434	8425	8415	8406	8396	2	3	5	6	8
33	0.8387	8377	8368	8358	8348	8339	8329	8320	8310	8300	2	3	5	6	8
34	0.8290	8281	8271	8261	8251	8241	8231	8221	8211	8202	2	3	5	7	8
35	0.8192	8181	8171	8161	8151	8141	8131	8121	8111	8100	2	3	5	7	8
36	0.8090	8080	8070	8059	8049	8039	8028	8018	8007	7997	2	3	5	7	9
37	0.7986	7976	7965	7955	7944	7934	7923	7912	7902	7891	2	4	5	7	9
38	0.7880	7869	7859	7848	7837	7826	7815	7804	7793	7782	2	4	5	7	9
39	0.7771	7760	7749	7738	7727	7716	7705	7694	7683	7672	2	4	6	7	9
40°	0.7660	7649	7638	7627	7615	7604	7593	7581	7570	7559	2	4	6	8	9
41	0.7547	7536	7524	7513	7501	7490	7478	7466	7455	7443	2	4	6	8	10
42	0.7431	7420	7408	7396	7385	7373	7361	7349	7337	7325	2	4	6	8	10
43	0.7314	7302	7290	7278	7266	7254	7242	7230	7218	7206	2	4	6	8	10
44	0.7193	7181	7169	7157	7145	7133	7120	7108	7096	7083	2	4	6	8	10

and find the nearest number *lower* than 0.3878. This is 0.3875, corresponding to the angle 22°48′.

Now 0.3875 is 0.0003 less than 0.3878. So we look in the difference table at the column marked 3 and at the top of this column find 1′. So the angle who's sine is 0.3878 or $\sin^{-1} 0.3878 = 22°48′ + 1′ = 22°49′$ or 22.817°.

The cosine ratio

Looking back to Figure 2.15 you can see that the cosine of angle $AOB = \frac{OQ}{OP}$; in other words, the $\cos AOB = \frac{adjacent}{hypotenuse}$.

Before we consider an example using the cosine ratio, we first need to ensure that we can find the cosine of any angle between 0° and 90° using the table of natural cosines, an extract is shown in Table 2.5.

The only difference in the use of this table, when compared to the table of natural sines, is that to find the *cosine* ratio for the angle, we *subtract* the numbers in the difference columns.

Key point

For any angle θ, $\cos \theta = \frac{Side\ adjacent}{Hypotenuse}$

Example 2.54

Find from Table 2.5:
(a) $\cos 27°34'$ and (b) $\cos^{-1} 0.9666$.

(a) We first find $\cos 27°30' = 0.8870$ and looking in the difference column below 4' we find the value 5 which we *subtract* from 0.8870, i.e. the cosine of $\cos 27°34' = 0.8870 - 5 = \mathbf{0.8865}$.

(b) To find the angle who's cosine is 0.9666 we first find the angle with the nearest value *above* that required. In this case 0.9668 which corresponds to the angle 14°48'. Now the difference between 0.9668 and 0.9666 is 0.0002. Then looking across to the column containing 2, we now go to the top of the column which shows 3'. This value is now *added* to 14°48' to give the required result as $\mathbf{14°51'}$. Note that we have performed the *inverse operation* to that of finding the sine of the angle.

We are now in a position to look at a simple example that uses the cosine ratio.

Example 2.55

In the triangle shown in Figure 2.18, find the length of side AC, i.e. side "b".

Then the cosine of angle A, i.e.:

$$\cos 40° = \frac{Adjacent}{Hypotenuse} = \frac{AC}{AB} = \frac{b}{c} = \frac{b}{160}$$

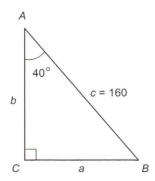

Figure 2.18

Now the cosine of $40° = 0.7660$ (from Table 2.5). So that:

$$0.7660 = \frac{b}{160} \text{ or } (0.7660) \times (160) = b,$$

so (that by long multiplication)

$$b = \mathbf{122.56}$$

The tangent ratio

Again from Figure 2.15 we can see that the tangent of the angle $AOB = \frac{QP}{OQ}$, i.e. the $\tan AOB = \frac{opposite}{adjacent}$. Again we will illustrate the use of this ratio by example. Table 2.6 (see overleaf) is an extract from the table of common tangents which will be found in Appendix D. In this table the common differences are added in the same way as in the table of sines (Table 2.4).

Key point

For any angle θ, then $\tan \theta = \frac{side\ opposite}{side\ adjacent}$

Example 2.56

Find the length of the side "a" shown in Figure 2.19. Then applying the tangent ratio to angle A, we see that, $\tan A = \frac{opposite}{adjacent} = \frac{BC}{AB} = \frac{a}{c} = \frac{a}{80}$.

Now from Table 2.6, it can be seen that the $\tan 28° = 0.5317 = \frac{a}{80}$, from which we see that:

$$\text{side } a = (0.5317)(80) = \mathbf{42.54}$$

Trigonometric ratios for the 45°/45° and 30°/60° triangles

In the special case where the two remaining angles of a right-angled triangle are both 45°,

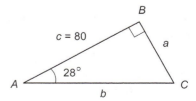

Figure 2.19 Figure for Example 2.56.

Table 2.6 Extract from the table of common tangents

| | 0' | 6' | 12' | 18' | 24' | 30' | 36' | 42' | 48' | 54' | Mean Differences | | | | |
	0.0°	0.1°	0.2°	0.3°	0.4°	0.5°	0.6°	0.7°	0.8°	0.9°	1'	2'	3'	4'	5'
0°	0.0000	0017	0035	0052	0070	0087	0105	0122	0140	0157	3	6	9	12	15
1	0.0175	0192	0209	0227	0244	0262	0279	0297	0314	0332	3	6	9	12	15
2	0.0349	0367	0384	0402	0419	0437	0454	0472	0489	0507	3	6	9	12	15
3	0.0524	9542	0559	0577	0594	0612	0629	0647	0664	0682	3	6	9	12	15
4	0.0699	0717	0734	0752	0769	0787	0805	0822	0840	0857	3	6	9	12	15
5	0.0875	0892	0910	0928	0945	0963	0981	0998	1016	1033	3	6	9	12	15
6	0.1051	1069	1086	1104	1122	1139	1157	1175	1192	1210	3	6	9	12	15
7	0.1228	1246	1263	1281	1299	1317	1334	1352	1370	1388	3	6	9	12	15
8	0.1405	1423	1441	1459	1477	1495	1512	1530	1548	1566	3	6	9	12	15
9	0.1584	1602	1620	1638	1655	1673	1691	1709	1727	1745	3	6	9	12	15
10°	0.1763	1781	1799	1817	1835	1853	1871	1890	1908	1926	3	6	9	12	15
11	0.1944	1962	1980	1998	2016	2035	2053	2071	2089	2107	3	6	9	12	15
12	0.2126	2144	2162	2180	2199	2217	2235	2254	2272	2290	3	6	9	12	15
13	0.2309	2327	2345	2364	2382	2401	2419	2438	2456	2475	3	6	9	12	15
14	0.2493	2512	2530	2549	2568	2586	2605	2623	2642	2661	3	6	9	12	16
15	0.2679	2698	2717	2736	2754	2773	2792	2811	2830	2849	3	6	9	13	16
16	0.2867	2886	2905	2924	2943	2962	2981	3000	3019	3038	3	6	9	13	16
17	0.3057	3076	3096	3115	3134	3153	3172	3191	3211	3230	3	6	10	13	16
18	0.3249	3269	3288	3307	3327	3346	3365	3385	3404	3424	3	6	10	13	16
19	0.3443	3463	3482	3502	3522	3541	3561	3581	3600	3620	3	7	10	13	16
20°	0.3640	3659	3679	3699	3719	3739	3759	3779	3799	3819	3	7	10	13	17
21	0.3839	3859	3879	3899	3919	3939	3959	3979	4000	4020	3	7	10	13	17
22	0.4040	4061	4081	4101	4122	4142	4163	4183	4204	4224	3	7	10	14	17
23	0.4245	4265	4286	4307	4327	4348	4369	4390	4411	4431	3	7	10	14	17
24	0.4452	4473	4494	4515	4536	4557	4578	4599	4621	4642	4	7	11	14	18
25	0.4663	4684	4706	4727	4748	4770	4791	4813	4834	4856	4	7	11	14	18
26	0.4877	4899	4921	4942	4964	4986	5008	5029	5051	5073	4	7	11	15	18
27	0.5095	5117	5139	5161	5184	5206	5228	5250	5272	5295	4	7	11	15	18
28	0.5317	5340	5362	5384	5407	5430	5452	5475	5498	5520	4	8	11	15	19
29	0.5543	5566	5589	5612	5635	5658	5681	5704	5727	5750	4	8	12	15	19
30°	0.5774	5797	5820	5844	5867	5890	5914	5938	5961	5985	4	8	12	16	20
31	0.6009	6032	6056	6080	6104	6128	6152	6176	6200	6224	4	8	12	16	20
32	0.6249	6273	6297	6322	6346	6371	6395	6420	6445	6469	4	8	12	16	20
33	0.6494	6519	6544	6569	6594	6619	6644	6669	6694	6720	4	8	13	17	21
34	0.6745	6771	6796	6822	6847	6873	6899	6924	6950	6976	4	9	13	17	21
35	0.7002	7028	7054	7080	7107	7133	7159	7186	7212	7239	4	9	13	18	22
36	0.7265	7292	7319	7346	7373	7400	7427	7454	7481	7508	5	9	14	18	23
37	0.7536	7563	7590	7618	7646	7673	7701	7729	7757	7785	5	9	14	18	23
38	0.7813	7841	7869	7898	7926	7954	7983	8012	8040	8069	5	9	14	19	24
39	0.8098	8127	8156	8185	8214	8243	8273	8302	8332	8361	5	10	15	20	24
40°	0.8391	8421	8451	8481	8511	8541	8571	8601	8632	8662	5	10	15	20	25
41	0.8693	8724	8754	8785	8816	8847	8878	8910	8941	8972	5	10	16	21	25
42	0.9004	9036	9067	9099	9131	9163	9195	9228	9260	9293	5	11	16	21	27
43	0.9325	9358	9391	9424	9457	9490	9523	9556	9590	9623	6	11	17	22	28
44	0.9657	9691	9725	9759	9793	9827	9861	9896	9930	9965	6	11	17	23	29

the sides opposite to these two angles are also equal.

In Figure 2.20 these two sides have been given the arbitrary value of 1.0 and by Pythagoras (which you have already met) we have: $(AC)^2 = (AB)^2 + (BC)^2 = 1^2 + 1^2 = 2$; therefore, the hypotenuse side $AC = \sqrt{2}$, as shown.

Therefore:

$\sin 45 = \frac{1}{\sqrt{2}} = \frac{\sqrt{2}}{2}$ after multiplying both the top and the bottom by $\sqrt{2}$, similarly

$\cos 45 = \frac{1}{\sqrt{2}} = \frac{\sqrt{2}}{2}$ and

$\tan 45 = \frac{1}{1} = 1$

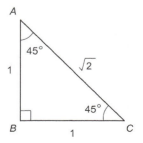

Figure 2.20 The 45°/45° right-angled triangle.

> **Key point**
>
> In a 45°/45° right-angled triangle, the sides have the ratios $1:1:\sqrt{2}$.

The square root of 2 is equal to 1.4142 corrected to four decimal places and is worth committing to memory. Thus, for example, the sine and cosine of $45° = \frac{\sqrt{2}}{2} = \frac{1.4142}{2} = 0.7071$ you might like to check this in Tables 2.4 and 2.5!

Also note an important relationship between the sine, cosine and tangent ratios. From above:

$$\frac{\sin 45}{\cos 45} = \frac{\frac{\sqrt{2}}{2}}{\frac{\sqrt{2}}{2}} = \left(\frac{\sqrt{2}}{2}\right)\left(\frac{2}{\sqrt{2}}\right) = 1 = \tan 45$$

This relationship is not just true for 45°, but is true for *any angle* and may be generalized as:

$$\frac{\sin\theta}{\cos\theta} = \tan\theta$$

We now consider the 30°/60° right-angled triangle, in a similar way to our 45°/45° triangle.

An equilateral triangle is one in which all the sides are equal.

> **Key point**
>
> An equilateral triangle is one in which all three sides are equal in length.

Figure 2.21 shows a triangle *ABC* in which each of the equal sides = 2 units. A perpendicular is drawn from *C* to *D*, which bisects *AB*.

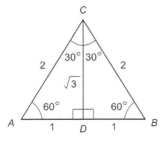

Figure 2.21 Construction for 30°/60° triangle.

Now from Pythagorus for the right-angled triangle *ACD*, we have that:

$$(CD)^2 = (AC)^2 - (AD)^2 = 2^2 - 1^2 = 3,$$

$$\text{therefore, side } CD = \sqrt{3}$$

Now noting that all the angles of the triangle $ABC = 60$ (remembering that there are 180 in a triangle) and that angle $ACD = 30$. Then the trigonometric ratios for these two angles are given as follows:

$$\sin 30 = \frac{1}{2}, \quad \cos 30 = \frac{\sqrt{3}}{2}, \quad \tan 30 = \frac{1}{\sqrt{3}} = \frac{\sqrt{3}}{3}$$

$$\sin 60 = \frac{\sqrt{3}}{2}, \quad \cos 60 = \frac{1}{2}, \quad \tan 60 = \frac{\sqrt{3}}{1} = \sqrt{3}$$

> **Key point**
>
> In a 30°/60° right-angled triangle the sides have the ratio $1:\sqrt{3}:2$.

Rectangular and polar co-ordinates

Before we consider one or two simple applications of the trigonometric ratios, such as angles of elevation and aircraft bearings, first we need to familiarize ourselves with the rectangular and polar co-ordinate systems. You have already used rectangular co-ordinates, in your earlier graphical work. Here we formalize this co-ordinate system and discover how we can convert from rectangular to polar co-ordinates and vice versa.

> **Key point**
>
> Rectangular co-ordinates are also known as Cartesian co-ordinates.

A point on a graph can be defined in several ways. The two most common ways use either rectangular or polar co-ordinates.

Rectangular co-ordinates (Figure 2.22) use two *perpendicular* axis, normally labelled *x* and *y*. Where any point *P* is identified by its horizontal distance along the *x*-axis and its vertical distance up the *y*-axis. Polar co-ordinates give the distance *r*, from the origin *O* and the angle

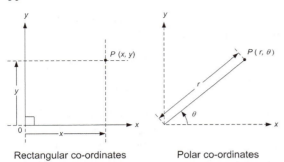

Figure 2.22 Rectangular and polar co-ordinate systems.

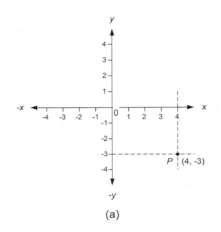

(a)

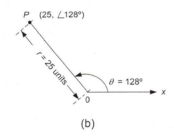

(b)

Figure 2.23 Identification of a point *P* using rectangular and polar co-ordinates.

θ of the line, joining the origin and the point *P* with the *x*-axis.

Thus, for example, the point $(4, -3)$ are the rectangular or Cartesian co-ordinates for the point, i.e. 4 units to the right along the *x*-axis (Figure 2.23(a)) and 3 units in the negative *y* direction, i.e. downwards.

The point $(25\angle128)$ gives the polar co-ordinates for the point *P* (Figure 2.23(b)),

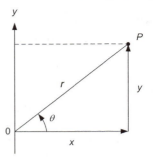

Figure 2.24 Combined rectangular and polar co-ordinates.

that is 25 units in magnitude, from the origin, at an angle of 128°, measured anticlockwise from the horizontal *x*-axis.

Converting rectangular and polar co-ordinates

A useful skill is to be able to convert rectangular to polar co-ordinates and vice versa. This is particularly helpful, when dealing with sinusoidal functions and other oscillatory functions that you may meet in your later studies.

Consider Figure 2.24, which shows a set of rectangular and polar axes combined.

Then to convert *rectangular to polar co-ordinates*, we use Pythagoras theorem and the tangent ratio to give:

$$r = \sqrt{x^2 + y^2} \quad \text{and} \quad \tan\theta = \frac{y}{x}$$

To convert *polar to rectangular co-ordinates*, we use the sine and cosine ratios to give:

$$\sin\theta = \frac{y}{r} \quad \text{therefore } y = r\sin\theta$$

and

$$\cos\theta = \frac{x}{r} \quad \text{therefore } x = r\cos\theta$$

Example 2.57

(a) Convert the rectangular co-ordinates $(-5, -12)$ into polar co-ordinates.
(b) Convert the polar co-ordinates $(150\angle300)$ into rectangular co-ordinates.

(a) Then using Pythagoras and the tangent ratio, we get:

$$r = \sqrt{(-5)^2 + (-12)^2}$$
$$= \sqrt{25 + 144}$$
$$= \sqrt{169} = 13$$

and $\tan\theta = \dfrac{-12}{-5} = 2.4$, therefore $\theta = 67.4°$.

So the polar co-ordinates are $13\angle67.4$

(b) Then using the sine and cosine ratios, to find y and x, respectively, we get:

$$y = r\sin\theta = 150\sin300 = (150)(-0.866)$$
$$= -129.9$$

and

$$x = r\cos\theta = 150\cos300 = (150)(0.5) = 75$$

So the rectangular co-ordinates are $(75, -129.9)$

Angles of elevation and depression

If you look up at a distant object, say a low flying aircraft then the angle formed between the horizontal and your line of sight is known as the angle of elevation. Similarly, if you look down at a distant object say from on top of a hill, the angle formed between the horizontal and your line of sight is called the angle of depression. These two angles are illustrated in Figure 2.25.

Example 2.58

To find the height of an airfield radio mast positioned on top of the control tower, the surveyor sets up his theodolite 200 m from the base of the tower. The surveyor finds that the angle of elevation to the top of the mast is 20°. If the instrument is held 1.6 m from the level ground, what is the height of the tower?

The situation is illustrated in Figure 2.26. Since we have both the opposite and adjacent sides to the angle of elevation, we use the tangent ratio to solve the problem.

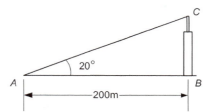

Figure 2.26 Situation for airfield control tower and radio mast.

Then from Figure 2.26, $\tan20 = \frac{BC}{AB}$ so $BC = (\tan 20) \times (AB) = (0.364) \times (200) = 72.8$ m, all we now need to do is add the height of the theodolite viewing piece from the ground. Then the height to the top of the mask is $72.8 + 1.6 = 74.4$ m.

Example 2.59

A aerial erector is positioned 50 m up a radio mask, in line with two landing lights, who's angles of depression are 20° and 22°. Calculate the distance between the landing lights.

The situation is shown in Figure 2.27, where in the triangle ABC, angle $ABC = 90° - 22° = 68°$.

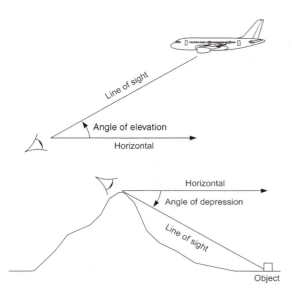

Figure 2.25 Angles of elevation and depression.

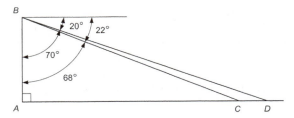

Figure 2.27 Angles of depression to landing lights.

And in the triangle ABD the angle $ABD = 90° − 20° = 70°$.

Then $\tan BAC = \dfrac{AC}{AB}$

so that

$$AC = (\tan BAC) \times (AB)$$
$$= (\tan 68°) \times (50)$$
$$= (2.4751) \times (50)$$

so that length $AC = 123.755$ m. Similarly:

$$\tan ABD = \dfrac{AD}{AB}$$

so that:

$$AD = (\tan ABD) \times (AB)$$
$$= (\tan 70°) \times (50)$$
$$= (2.7475) \times (50)$$

so that length $AD = 137.375$.

Then the distance between the landing lights $= 137.375 − 123.755 = \mathbf{13.62\,m}$.

Bearings

The four primary points of the compass are north (N), south (S), east (E) and west (W). Remembering that there are 360° in a circle, then the eight points of the compass that include NE, SE, SW and NW are each off-set from one another by 45° as shown in Figure 2.28.

A bearing N30°W means an angle of 30° measured from north towards west. A bearing of S20°E means an angle of 20° measured from south towards east. However, bearings

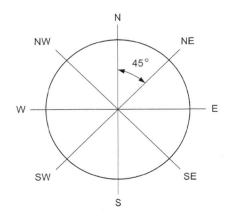

Figure 2.28 Bearings.

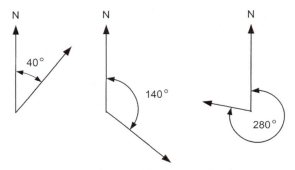

Figure 2.29 Example bearings measured conventionally from north.

are normally measured from north in a clockwise direction, unless stated differently, north is taken as 0°. Three digits are used to indicate the bearing, so that all points of the compass may be considered. Figure 2.29 shows example bearings measured in this way.

Example 2.60

A navigator notes a point B is due east of point A on the coast. Another point C on the coast is noted, 8 km due south of A. The distance BC is 10 km. As the navigator calculate the bearing of C from B.

The most difficult problem with bearings is to picture what is going on, Figure 2.30 illustrates the situation. From the figure we first determine angle B. Then the bearing of position C can

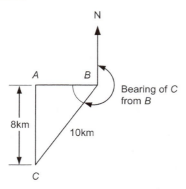

Figure 2.30 Situation diagram.

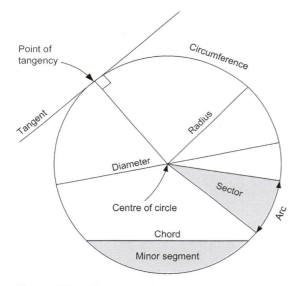

Figure 2.31 Elements of a circle.

be determined, conventionally clockwise from north. Then using the sine ratio:

$$\sin B = \frac{AC}{BC} = \frac{8}{10} = 0.8, \text{ so angle}$$
$$B = 53°8' \quad \text{or} \quad 53.133°$$

Then the bearing of C from $B = 270° - 53.133° = \mathbf{216.867°}$.

Key point

In arc measurement there are 60 min (60′) in 1° and 60 s (60″) in 1 min of arc.

2.4.7 Trigonometry and the circle

In this short section we concentrate on the geometric properties of the circle and the use of trigonometry to solve problems involving the circle.

We have already been introduced into the way in which we find the circumference and area of a circle. Here we extend our knowledge of the circle by identifying and defining certain elements of the circle. This is essentially about the geometry of the circle, which you will find useful when finding particular cross sections, or when considering circular motion.

Elements and properties of the circle

The major elements of the circle are shown in Figure 2.31. You will be familiar with most, if not all of these elements. However, for the sake of completeness, we will formally define them.

A point in a plane whose distance from a fixed point in that plane is constant lies on the *circumference* of a circle. The fixed point is called the center of the circle and the constant distance is called the *radius*.

A circle may be marked out on the ground by placing a peg or spike at its center. Then using a length of chord for the radius, we simply walk round with a pointer at the end of the cord and mark out the circumference of the circle.

A *chord* is a straight line which joins two points on the circumference of a circle. A *diameter* is a chord drawn through the center of a circle.

A *tangent* is a line which just touches the circumference of a circle, at one point (the point of tangency). This tangent line, that lies at right angles to a radius, is drawn from the point of tangency.

A chord line cuts a circle into a *minor segment* and *major segment*. A *sector* of a circle is an area enclosed between two radii, and a length of the circumference, the *arc* length.

Key point

A tangent line touches the circle at only one point and lies at right angles to a radius drawn from this point.

Some important theorems of the circle

These theorems relate to the angles contained in a circle and the tangent to a circle. They are given here without proof, in order to aid the trigonometric solution of problems involving the circle.

Theorem 1

The angle that an arc of a circle subtends at its center, is twice the angle which the arc subtends at the circumference.

Thus in Figure 2.32(a), angle $AOB =$ twice angle ACB. The next two theorems result from Theorem 1.

Theorem 2

Angles in the same segment of a circle are equal. Figure 2.32(b) illustrates this fact, where angle $C =$ angle D.

Theorem 3

The triangle on a semi-circle is always right angled.

Figure 2.33 illustrates this theorem. No matter where the position P is placed on the circumference of the semi-circle, a right angle is always produced opposite the diameter.

Theorem 4

The opposite angles of any cyclic quadrilateral are equal to $180°$.

> **Key point**
>
> A cyclic quadrilateral is one that is inscribed in a circle.

> **Example 2.61**

Find the angles A and B for the cyclic quadrilateral shown (Figure 2.34).

Then by Theorem 4:
$\angle B + \angle D = 180°$ therefore angle $\angle B = 110°$, Similarly $\angle A + \angle C = 180°$ therefore $\angle A = 85°$,

There are many theorems related to the tangent of a circle. In order to understand them, you should be able to define a tangent to a circle, as given above.

Theorem 5

A tangent to a circle is at right angles to a radius drawn from the point of tangency. Figure 2.35(a) illustrates this theorem.

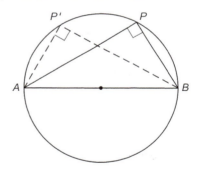

Figure 2.33

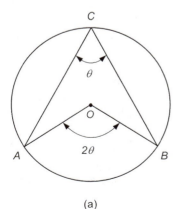

(a)

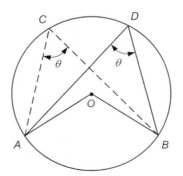

(b)

Figure 2.32

Theorem 6

The angle between a tangent and a chord drawn from the point of tangency equals half of the angle at the center subtended by the chord. Figure 2.35(b) above illustrates this theorem.

Theorem 7

The angle between a tangent and a chord that is drawn from the point of tangency is equal to the angle at the circumference subtended by the chord (Figure 2.36).

Theorem 8

If two circles touch either internally or externally, then the line that passes through their centers also passes through the point of tangency (Figure 2.36).

Example 2.62

The pitch circle diameter of three gear wheels are as illustrated in Figure 2.37.

Given that the gear teeth mesh tangentially to each other. Find width w of the combination.

Then since pitch circles are tangential to each other $PQ = 15 + 7.5 = 22.5$ cm, $QR = 15 + 7.5 = 22.5$ cm and $PR = 15 + 15 = 30$ cm.

The triangle PQR is therefore isosceles. So $PS = (0.5) \times (30) = 15$ cm, from the fact that in an isosceles triangle the perpendicular dropped from the apex, bisects the opposite side.

So using Pythagoras on triangle PQS, $(QS)^2 = (PQ)^2 - (PS)^2 = 22.5^2 - 15^2 = 506.25 - 225 = 281.25$. Then from square root tables $QS = 16.77$ cm and $w = 15 + 16.77 + 7.5 = 39.27$ cm. So width $w = 39.27$ cm.

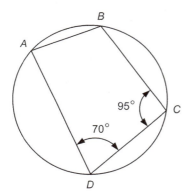

Figure 2.34 A cyclic quadrilateral.

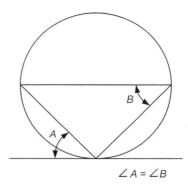

$\angle A = \angle B$

Figure 2.36 Angle between a tangent and a chord.

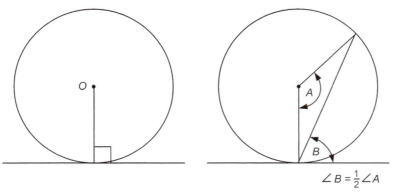

(a)

(b)

$\angle B = \frac{1}{2} \angle A$

Figure 2.35 Tangent theorems.

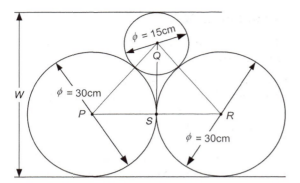

Figure 2.37 Three gear wheels in mesh.

Test your understanding 2.12

1. Find using the appropriate tables:
   (a) $\sin 57°$ (b) $\cos 82°$ (c) $\tan 13°$
   (d) $\sin 12°38'$ (e) $\cos 27°14'$ (f) $\tan 52.56°$

2. Find by drawing the sine of the angles:
   (a) $30°$ (b) $70°$

3. Find by drawing the angles whose sine is:
   (a) $\frac{3}{4}$ (b) $\frac{5}{13}$

4. An isosceles triangle has a base of 5.0 cm and the equal sides are each 6.5 cm long. Find all the internal angles of the triangle and its vertical height.

5. Find the angles marked θ, in the right-angled triangles shown in Figure 2.38.

6. If a point P has the rectangular co-ordinates (6, 7). What are the polar co-ordinates of this point? You should use the table of square roots in *Appendix D*, to help you.

7. Calculate the rectangular co-ordinates of the following points:
   (a) (5, 30°) (b) (8, 150°)

8. A surveyor observes the angle of elevation of a building as 26°. If the eye line of the surveyor is 1.8 m above horizontal ground and he is standing 16 m from the building. What is the height of this building?

9. A man stands on the top of a hill 80 m high in line with two traffic cones in the road below. If the angles of depression of the two traffic cones are 17° and 21°. What is the horizontal distance between them?

10. A cylindrical bar rests in a vee-block as shown in Figure 2.39. Determine the vertical height of the vee-section (h) and the height (x) that the cylindrical bar is raised above the top surface of the vee-block.

2.4.7 Geometric constructions

The inclusion of the following short section on simple geometric constructions, will be found useful, when you commence your study of the technical and engineering drawing section contained in JAR 66 Module 7, on Maintenance practices. This subject matter is best explained through the use of illustrated examples, which will identify the important steps required with each technique. We will limit our techniques to those useful for producing simple engineering drawings, marking out and to those, which help with the identification and solution of triangular and circular shapes.

To bisect the given angle AOB, when the arms of the angle meet

From Figure 2.40, it can be seen that with center O, we set out equal arcs to cut the arms of the angle at A and B. With centers A and B we set out equal length arcs to meet at C.

Then *line* OC bisects the angle.

To bisect a given angle when the arms do not meet

This technique simply involves drawing two lines parallel to the given arms, sufficient to

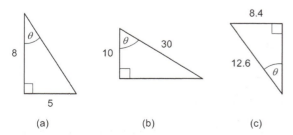

(a) (b) (c)

Figure 2.38

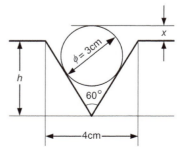

Figure 2.39

make them meet at a point and then using the above technique to bisect the angle formed.

From points on AB and CD draw equal arcs (Figure 2.41(a)). Then using these arcs draw lines parallel to AB and CD that meet at point E (Figure 2.41(b)). Now bisect the angle at point E (Figure 2.41(c)), using the method shown in Figure 2.40.

To set out angles using the trigonometric ratios

This is an extremely accurate method providing the triangles used have a large enough scale. The builders square and the layout of structures, often employ this method. To follow this method you will need to be aware of the basic trigonometric ratios, you have just met, look back to remind yourself. We will use a scale factor of 100, to amplify the ratios found from the tables. Figure 2.42 shows the method.

Figure 2.42(a) shows how to set at an angle using the tangent ratio. In this case the angle is $23°30'$, which from our tables gives a value of 0.4348. Then using a multiplier of 100 units the line $AC = 43.48$ units. Now set out horizontally the line $AB = 100$ units, then set out AC at right angles to AB as shown. Join BC, then the angle ABC will now $= 23°30'$.

Similarly Figure 2.42(b) shows the angle $\theta = 28°36'$, being set out using the sine rule. We first find the sine of $28°36'$ from our tables as 0.4787. Then using our multiplier of 100 we get $R = 47.87$ units, which is our arc length from A. Set out AB as before $= 100$ units. Then from B draw a line that just touches our arc (tangent). Angle ABC will now $= 28°36'$.

To find the center of a given circle

Figure 2.43(a) shows the circle, with three well-spaced points A, B and C marked on its circumference. Bisect the chord between one pair of points, say AB. Figure 2.43(b) shows the circle with the second pair of points BC bisected. The intersect at O is the center of the circle.

To draw a common external tangent to two given circles

Figure 2.44(a) shows two circles with radii R and r. With center O_1 draw a circle of radius $R - r$. Join O_1O_2. Bisect O_1O_2 to obtain center C. With C as center draw a semi-circle of radius CO_1 to cut the inner circle at T. Draw a line

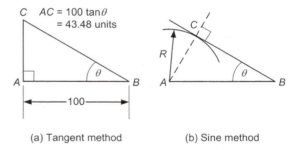

(a) Tangent method   (b) Sine method

Figure 2.42 Setting out angles using trigonometric ratios.

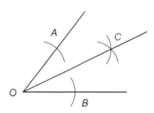

Figure 2.40 Method to bisect the given angle.

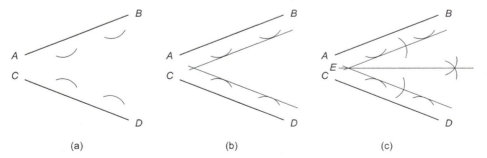

(a)   (b)   (c)

Figure 2.41 Method to bisect given angle when lines do not meet.

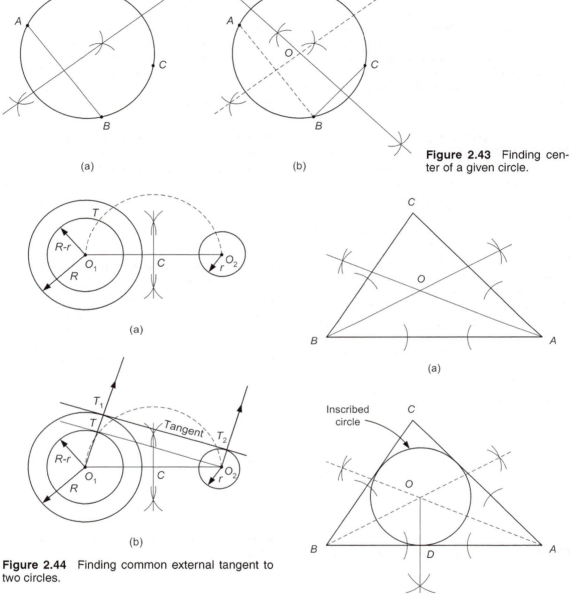

Figure 2.43 Finding center of a given circle.

Figure 2.44 Finding common external tangent to two circles.

Figure 2.45 Finding inscribed circle in given triangle.

from O_1 through T to locate T_1 on the outer circle.

Figure 2.44(b) shows the line O_2 parallel to O_1T_1, drawn to cut the smaller circle at T_2. Now draw a line through T_1 and T_2 to obtain the external tangent to the two circles as shown.

This construction is very useful to accurately portray a belt drive around two pulleys.

To draw the inscribed circle for a given triangle

Figure 2.45(a) shows the given triangle ABC with $\angle A$ and $\angle B$ both having been bisected

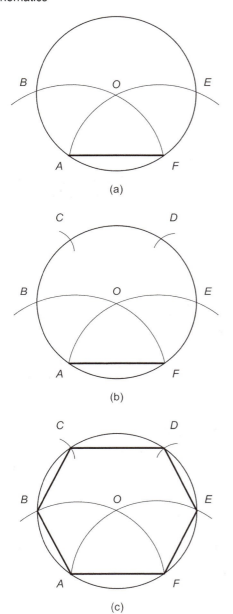

(a)

(b)

(c)

Figure 2.46 Constructing a hexagon given the length of a side.

To draw a hexagon given the length of a side

Draw a straight line AF equal to the given length of the side. With centers A and F, draw the arcs of radius AF to intersect at O. With the center O draw a circle of radius OA to cut the arcs at B and D (Figure 2.46(a)). With centers B and E, draw arcs of radius AF to cut the circle at C and D, respectively (Figure 2.46(b)). Finally join the points on a circle to obtain the required regular hexagon (Figure 2.46(c)).

To blend an arc in a right angle

Set out faint intersecting lines at right angles, for the desired arc. From corner A, set out AB and AD equal to required radius R. From B and D set out arcs of radius R, to intersect at O (Figure 2.47(a)). From O draw an arc radius R to blend with the straight lines (Figure 2.47(b)). Finally erase unwanted construction lines and darken with appropriate grade pencil.

To draw an arc from a point to a circle of radius r

Set out radius R from P and radius $R + r$ from O to meet at C (Figure 2.48(a)). From C draw an arc radius R to touch the circle and point P (Figure 2.48(b)). It is also straightforward to blend an arc from a point to blend with the far side of a circle. In this case set out radius R from P and radius $R - r$ from O. Then from C draw an arc of radius R to touch the circle at P.

This concludes this short section on geometrical construction. There are literally hundreds of techniques that may be used for engineering geometrical drawing, which simply cannot be covered here. The techniques given above are some of the most common and the most useful techniques that you may need for producing engineering and workshop drawings, when you study Module 6.

No test your knowledge questions have been set for this section. However, you are strongly advised to consult any comprehensive text written on engineering and geometrical drawing, to identify and practice the numerous and varied techniques required to enhance your drawing skills.

and the bisectors extended to meet at O. In Figure 2.45(b) a perpendicular is constructed from O to cut AB at D. Then with center O and radius OD draw the inscribed circle of the triangle ABC.

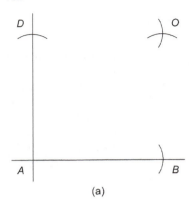
(a)

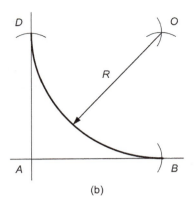
(b)

Figure 2.47 Blending an arc in a right angle.

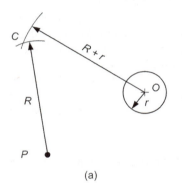
(a)

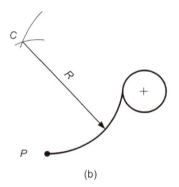
(b)

Figure 2.48 Blending an arc from a point to near side of circle.

The final part of this section on non-calculator mathematics is given over to a number of typical Module 1 examination questions, which you should attempt, once you are sure that you have mastered the mathematics, presented up until now.

2.5 Multiple choice questions

The example mathematics questions set out below follow the sections of Module 1 in the Part 66 syllabus. Note that these questions have been separated by level, where appropriate. Several of the sections (e.g. trigonometry, linear equations, binary numbers, logarithms, etc.) are not required for category A certifying mechanics. Please remember that *ALL of these questions must be attempted without the use of a calculator* and that the pass mark for all Part 66 multiple-choice examination is 75%!

ARITHMETIC

1. The sum of 12,000 and 1200 is:
   [A, B1, B2]
   (a) 12,200
   (b) 13,200
   (c) 23,200

2. The product 230×180 is: [A, B1, B2]
   (a) 4140
   (b) 41,040
   (c) 41,400

3. The number 18,493.4 divided by 18 is:
   [A, B1, B2]
   (a) 0.000973
   (b) 102.74
   (c) 1027.41

4. $0.006432 - 0.0184$ is: [A, B1, B2]
   (a) −0.011968
   (b) −0.177568
   (c) −0.0177568

5. The sum of $329.67 + 1086.14 + 200.2$ is: **[A, B1, B2]**
   (a) 1319.3
   (b) 1616.01
   (c) 1632.31

6. The equivalent of $\frac{326 \times 12.82}{0.62}$, correct to two decimal places is: **[A, B1, B2]**
   (a) 6740.84
   (b) 674.08
   (c) 67.41

7. $21 + 6 \times (8 - 5)$ is equal to: **[A, B1, B2]**
   (a) 39
   (b) 64
   (c) 81

8. p and q are positive integers, $p - q$ *must* be a number that is: **[B1, B2]**
   (a) positive
   (b) natural
   (c) an integer

9. The value of $\sqrt{25 \times 36}$ is: **[A, B1, B2]**
   (a) 30
   (b) 150
   (c) 180

10. $-16 + (-4) - (-4) + 22$ is equal to: **[A, B1, B2]**
    (a) -2
    (b) 6
    (c) 14

11. $3 \times \frac{-12}{2}$ is equal to: **[A, B1, B2]**
    (a) 2
    (b) -2
    (c) -18

12. The value of $5 \times 3 + 4 \times 3$ is: **[A, B1, B2]**
    (a) 57
    (b) 75
    (c) 27

13. The value of $a(b + c - d^2)$ when $a = 2$, $b = -3$, $c = 4$ and $d = -2$ is: **[A, B1, B2]**
    (a) -10
    (b) -6
    (c) 10

14. An estimate for the product $4.28 \times 10.1 \times 0.125$ correct to one significant figure is: **[A, B1, B2]**
    (a) 5.41

(b) 5.4
(c) 5

15. $2\frac{1}{50}$ written in decimal form is: **[A, B1, B2]**
    (a) 2.2
    (b) 2.01
    (c) 2.02

16. The number 0.00009307, expressed in standard form is: **[A, B1, B2]**
    (a) 9.307×10^{-5}
    (b) 9.307×10^{-4}
    (c) 9.307×10^4

17. $\frac{2}{5}$ of a consignment of 600 bolts are distributed to a spares carousel, how many are left? **[A, B1, B2]**
    (a) 240
    (b) 360
    (c) 400

18. An estimate of the value of $(80.125 \times 20.875) - 1600$, correct to three significant figures is: **[A, B1, B2]**
    (a) 74.1
    (b) 80.5
    (c) 85.61

19. The average of $\frac{1}{4}$ and $\frac{1}{12}$ is: **[A, B1, B2]**
    (a) $\frac{1}{3}$
    (b) $\frac{1}{6}$
    (c) $\frac{1}{8}$

20. The value of $\left(\frac{7}{12} \times \frac{3}{14}\right) - \frac{1}{16} + 2\frac{1}{8}$ is: **[A, B1, B2]**
    (a) $2\frac{3}{16}$
    (b) $2\frac{1}{4}$
    (c) $2\frac{5}{16}$

21. The value of $\frac{3}{4}$ of $\frac{1}{3} \div \frac{1}{2} \times \frac{1}{4}$ is: **[A, B1, B2]**
    (a) $\frac{1}{32}$
    (b) $\frac{1}{8}$
    (c) 2

22. $\frac{13}{25}$ as a percentage is: **[A, B1, B2]**
    (a) 5.2%
    (b) 26%
    (c) 52%

23. An aircraft supplier buys 200 packs of rivets for £100.00 and sells them for 70 pence a pack. His percentage profit is: **[A, B1, B2]**
    (a) 30%

(b) 40%
(c) 59%

24. An aircraft is loaded with 20 crates. Eight of the crates each have a mass of 120 kg, the remaining crates each have a mass of 150 kg. The average mass per box is: [A, B1, B2]
(a) 132 kg
(b) 135 kg
(c) 138 kg

25. Two lengths have a ratio of 12 : 5, the second, smaller length is 25 m, the first, larger length is: [A, B1, B2]
(a) 60 m
(b) 72 m
(c) 84 m

26. An aircraft travelling at constant velocity covers the first 800 km of a journey in 1.5 h. How long does it take to complete the total journey of 2800 km, assuming constant velocity? [A, B1, B2]
(a) 3.5 h
(b) 5.25 h
(c) 6.25 h

27. An electrical resistance (R) of a wire varies directly as the length (L) and inversely as the square of the radius (r). This is represented symbolically by: [B1, B2]
(a) $R \propto \frac{r}{L^2}$
(b) $R \propto \frac{L^2}{r}$
(c) $R \propto \frac{L}{r^2}$

28. Given that there are approximately 2.2 lb (pound mass) in 1 kg, then the number of pounds equivalent to 60 kg is: [A, B1, B2]
(a) 132 lb
(b) 60 lb
(c) 27.3 lb

29. One bar pressure is approximately equal to 14.5 psi (pounds per square inch) so the number of bar equivalent to 3625 psi is: [A, B1, B2]
(a) 125 bar
(b) 250 bar
(c) 255 bar

30. There are approximately 4.5 L in a gallon. How many litres will be registered on the fuel gauge if 1600 gallons are dispensed? [A, B1, B2]
(a) 7200 L
(b) 355.6 L
(c) 55.6 L

31. The mass of an electrical part is 23 g, so the total mass, in kg, of 80 such parts is: [A, B1, B2]
(a) 1840
(b) 184
(c) 1.84

32. $2^5 + 2^3 + 1$ may be written as the binary number: [B1, B2]
(a) 10110
(b) 101001
(c) 101010

33. The denary number 37 is the binary number: [B1, B2]
(a) 101001
(b) 10101
(c) 100101

34. The hexadecimal number $6E_{16}$ is equivalent to denary: [B1, B2]
(a) 94
(b) 108
(c) 110

35. The denary number 5138 is equivalent to hexadecimal: [B1, B2]
(a) 412
(b) 214
(c) 321

ALGEBRA

36. The product of $3x$, x, $-2x^2$, is: [A, B1, B2]
(a) $-6x^4$
(b) $-5x^4$
(c) $4x - 2x^2$

37. When simplified the expression $4(a + 3b) - 3(a - 4c) - 5(c - 2b)$ is: [A, B1, B2]
(a) $a + 22b + 17c$
(b) $a + 22b - 17c$
(c) $a + 22b + 7c$

38. When simplified the expression $\frac{(a+b)(a-b)}{a^2-b^2}$ is: **[A, B1, B2]**
    (a) 1
    (b) $a+b$
    (c) $a-b$

39. When simplified the expression $(a-b)^2 - (a^2 - b^2)$ is: **[A, B1, B2]**
    (a) $2a^2 - 2ab$
    (b) $2b^2$
    (c) $2b^2 - 2ab$

40. When simplified the expression $\frac{5a}{4} - \frac{a-1}{3}$ is: **[A, B1, B2]**
    (a) $\frac{11a+1}{12}$
    (b) $\frac{11a+4}{12}$
    (c) $\frac{11a-4}{12}$

41. When simplified $\frac{12x^2 + 16x^4 - 24x^6}{4x^2}$ is equivalent to: **[A, B1, B2]**
    (a) $4x^2 - 9x^4$
    (b) $4x^2 - 2x^4$
    (c) $3 + 4x^2 - 6x^4$

42. When simplified $(x-2)^2 + x - 2$ is equivalent to: **[A, B1, B2]**
    (a) $(x-2)(x-3)$
    (b) $(x-2)(x-1)$
    (c) $(x-2)(x+1)$

43. The expression $\frac{3^3 \times 3^{-2} \times 3}{3^{-2}}$ is equivalent to: **[B1, B2]**
    (a) 81
    (b) 1
    (c) $\frac{1}{27}$

44. The expression $\frac{(2^3)(4^{\frac{1}{2}})^3}{(3^{-3})(2^3)^2}$ simplifies to: **[B1, B2]**
    (a) $\frac{1}{27}$
    (b) $\frac{1}{9}$
    (c) 27

45. The expression $\frac{1}{2^{-3}} + \frac{1}{2^{-4}} - \frac{1}{2^{-2}}$, when simplified is equal to: **[B1, B2]**
    (a) $-\frac{1}{16}$
    (b) 20
    (c) 32

46. The expression simplified $\left[\frac{(a^2 b^3 c)(a^2)(a^2 b)d}{(ab^2 c^2)}\right]$
    $-\left[\frac{(a^6 b^3 c^{-2} d)}{abc^{-1}}\right] + 1$ is: **[B1, B2]**

    (a) -1
    (b) 0
    (c) 1

47. The factors of $3x^2 - 2x - 8$ are: **[A, B1, B2]**
    (a) $(3x-2)(x+4)$
    (b) $(3x+4)(x-2)$
    (c) $(3x+2)(x-4)$

48. Which of the following is a common factor of $x^2 - x - 6$ and $2x^2 - 2x - 12$: **[B1, B2]**
    (a) $x+2$
    (b) $x-3$
    (c) $2x-6$

49. The factors of $a^3 + b^3$ are: **[B1, B2]**
    (a) $(a+b)$ and $(a^2 - ab + b^2)$
    (b) $(a-b)$ and $(a^2 - ab - b^2)$
    (c) $(a+b)$ and $(a^2 - 2ab + b^2)$

50. A correct transposition of the formula, $x = \frac{ab-c}{a+c}$ is: **[A, B1, B2]**
    (a) $c = \frac{a(b-x)}{x+1}$
    (b) $c = \frac{ab-ax}{x-1}$
    (c) $c = \frac{a(b-x)}{2}$

51. The formula, $X = \sqrt{Z^2 - R^2}$, correctly transposed for R is: **[A, B1, B2]**
    (a) $R = \sqrt{Z^2 - X^2}$
    (b) $R = \sqrt{Z^2 + X^2}$
    (c) $R = Z - X$

52. The value of F in the formula $F = \frac{mV^2}{r}$, when $V = 20$, $r = 5$ and $m = 64$ is: **[A, B1, B2]**
    (a) 256
    (b) 512
    (c) 5120

53. The value of L in the formula $Q = \frac{1}{R}\sqrt{\frac{L}{C}}$, when $R = 4$, $C = 0.00625$ and $Q = 1$ is: **[A, B1, B2]**
    (a) 2.5
    (b) 1.0
    (c) 0.1

54. If $\frac{4}{x} = 3 + \frac{3}{x}$, then x is: **[A, B1, B2]**
    (a) $\frac{1}{3}$
    (b) $2\frac{1}{3}$
    (c) 3

55. If the simultaneous equations are $8x + 10y = 35; 2x - 10y = 5$ then x is: **[B1, B2]**
    (a) 5
    (b) 4
    (c) $-\frac{3}{10}$

56. If a is a positive integer and $a^2 + a - 30 = 0$, then the value of a is: **[B1, B2]**
    (a) 6
    (b) 5
    (c) 4

57. An aircraft travels a distance s km in 15 min. It travels at this same average speed for h h. The total distance it travels in km is:
    [A, B1, B2]
    (a) $\frac{sh}{15}$
    (b) $\frac{15h}{s}$
    (c) $4sh$

58. The solution to the equation $(x - 2)^2 + 3 = (x + 1)^2 - 6$ is: **[B1, B2]**
    (a) 2
    (b) -2
    (c) 1

59. The roots of the quadratic equation $x^2 + 10x = 96$ are: **[B1, B2]**
    (a) 6, -16
    (b) -6, 10
    (c) -6, 16

60. From the tables, log 57.68 is: **[B1, B2]**
    (a) $\bar{1}.7610$
    (b) 1.7610
    (c) 1.7598

61. From the tables, the antilogarithm of $\bar{2}.4177$ is: **[B1, B2]**
    (a) 0.02617
    (b) 0.02607
    (c) 0.2607

62. From the tables, the $\sqrt{2587}$ is: **[A, B1, B2]**
    (a) 160.8
    (b) 50.86
    (c) 72.42

63. The product of $(8795.42) \times (76.76)$ correct to six significant figures is: **[B1, B2]**
    (a) 675,136
    (b) 675,000
    (c) 675,100

64. The $(\sqrt{3600}) \times (\sqrt{4900})$ is equal to:
    [A, B1, B2]
    (a) 1764
    (b) 4620
    (c) 4200

65. A circle of diameter $= 10$ cm, will have a circumference of: **[A, B1, B2]**
    (a) 31.4 cm
    (b) 15.7 cm
    (c) 62.8 cm

66. A circle of radius $= 15$ cm, will have an area of: **[A, B1, B2]**
    (a) 707.14 cm^2
    (b) 1414.28 cm^2
    (c) 94.25 cm^2

67. The volume of a right cylinder of height 15 cm and base radius 5 cm is: **[A, B1, B2]**
    (a) 1125π cm^2
    (b) 375π cm^2
    (c) 75π cm^2

68. The surface area of a sphere of radius 10 mm is: **[A, B1, B2]**
    (a) 1333.3π mm^2
    (b) 750π mm^2
    (c) 400π mm^2

69. A hollow fuel pipe is 20 m long and has an internal diameter of 0.15 m and an external diameter of 0.20 m, the volume of the material from which the fuel pipe is made will be:
    [A, B1, B2]
    (a) 0.35π m^3
    (b) 3.5π m^3
    (c) 0.45π m^3

GEOMETRY AND TRIGONOMETRY

70. In the equation of the straight line graph $y = mx + c$, which of the following statements is *true*? **[A, B1, B2]**
    (a) y is the independent variable
    (b) m is the gradient of the line
    (c) c is the dependent variable

71. A straight line passes through the points (3, 1) and (6, 4), the equation of the line is: **[A, B1, B2]**
    (a) $y = x + 2$
    (b) $y = 2x - 2$
    (c) $y = x - 2$

72. The straight line graph shown in Figure 2.49 takes the form $y = mx + c$, the value of m will be approximately: **[A, B1, B2]**
    (a) 40
    (b) −30
    (c) 1.5

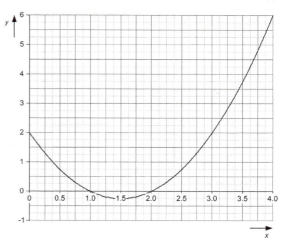

Figure 2.50   Graph of equation $y = x^2 - 3x + 2$.

74. Which one of the graphs shown in Figure 2.51 represents the relationship that $y \propto -x^2$? **[B1, B2]**
    (a) A
    (b) B
    (c) C

75. Which of the following relationships is represented by the graph shown in Figure 2.52? **[A, B1, B2]**
    (a) $y \propto x$
    (b) $y \propto \sqrt{x}$
    (c) $y \propto \frac{1}{x}$

76. Which of the following functions is represented by the graph shown in Figure 2.53? **[B1, B2]**
    (a) $y = \sin \theta$
    (b) $y = 2 \sin \theta$
    (c) $y = \sin 2\theta$

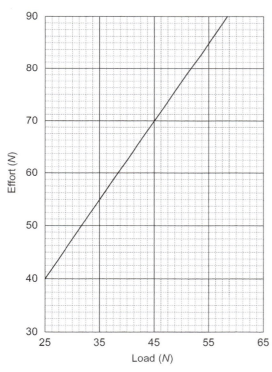

Figure 2.49   Straight line graph of effort against load.

73. The graph of the quadratic equation $y = x^2 - 3x + 2$ is shown in Figure 2.50, from this graph an estimate for the roots of the equation $y = x^2 - 3x + 1$ is: **[A, B1, B2]**
    (a) $x = 2$ and $x = 3$
    (b) $x = 0.5$ and $x = 2.5$
    (c) $x = 0.4$ and $x = 2.6$

77. From tables the cos 57°50′ is: **[B1, B2]**
    (a) 0.5324
    (b) 0.5334
    (c) 0.5319

78. If $\sin A = \frac{3}{5}$, then $\cos A$: **[B1, B2]**
    (a) $\frac{4}{5}$
    (b) $\frac{2}{5}$
    (c) $\frac{3}{4}$

79. From the top of a 40 m high control tower a runway landing light makes an angle of

depression of 30°, how far is the light from the base of the control tower?   **[B1, B2]**

(a) 69.3 m
(b) 56.7 m
(c) 23.1 m

80. In the triangle *ABC* shown in Figure 2.54, the bearing of *B* from *C* is: **[B1, B2]**

(a) 45°
(b) 225°
(c) 245°

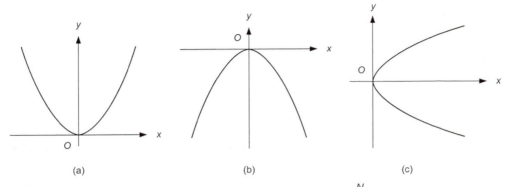

(a)   (b)   (c)

Figure 2.51

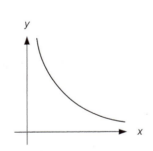

Figure 2.52

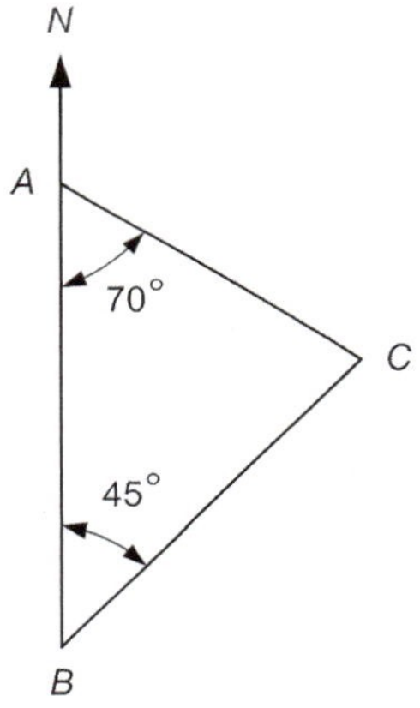

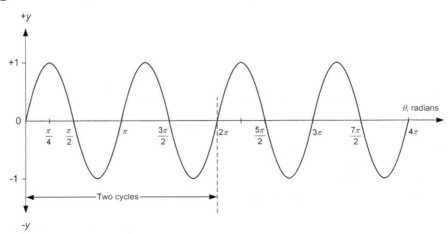

Figure 2.54

Figure 2.53

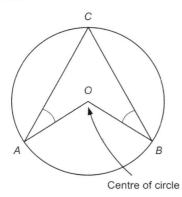

Figure 2.55

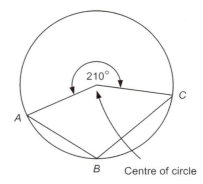

Figure 2.56

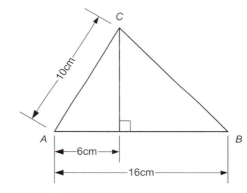

Figure 2.57

81. When converting rectangular to polar co-ordinates, the radius r of the polar co-ordinates is found from: **[B1, B2]**
    (a) $r = x \tan \theta$
    (b) $r = \sqrt{x^2 + y^2}$
    (c) $r = \sqrt{x^2 - y^2}$

82. The rectangular co-ordinates $(5, 12)$ in polar form are: **[B1, B2]**
    (a) $13 \angle 67.4$
    (b) $11.79 \angle 67.4$
    (c) $12 \angle 112.6$

83. The polar co-ordinates $(15 \angle 30)$ in rectangular form are: **[B1, B2]**
    (a) $(12.99, 7.5)$
    (b) $(7.5, 12.99)$
    (c) $(12.99, 8.66)$

84. From Figure 2.55, the $\angle AOB$ is equal to: **[B1, B2]**
    (a) $180 - 2AB$
    (b) $270 - (A + B)$
    (c) $2A + 2B$

85. In Figure 2.56, the $\angle ABC$ is equal to: **[B1, B2]**
    (a) $75°$
    (b) $105°$
    (c) $150°$

86. For the triangle shown in Figure 2.57, what is the value of $\cos B$? **[B1, B2]**
    (a) $\frac{10}{16}$
    (b) $\frac{8}{\sqrt{164}}$
    (c) $\frac{10}{\sqrt{164}}$

Chapter 3

Further mathematics

As mentioned in the introduction of Chapter 2, there is a need to extend the mathematics you have already learnt, so that you are fully prepared for the physics and electrical principles modules (Chapters 4 and 5, respectively) that follow. In addition the study of this chapter will act as a *foundation* for the study of mathematics you are likely to meet in any future higher education programs, such as the Foundation Degree (FD) and/or B.Eng.(Hons) degree in aircraft maintenance engineering or related engineering fields.

This chapter will include further study of some algebraic topics, additional trigonometry and an introduction to statistical methods. Finally, we take a brief look at the nature and use of the calculus. Throughout your study of further mathematics, *the use of a calculator will be assumed*.

3.1 Further algebra

3.1.1 Transposition and evaluation of more complex formulae and equations

So far we have been transposing relatively simple formulae, where the order in which we carried out the operations was reasonably obvious. With more complex formulae and equations, you may have doubts about the order of operations. If you are in any doubt, the following sequence should be followed:

1. remove root signs, fractions and brackets (in an order which suits the particular problem);
2. rearrange the formula for the subject, following the arithmetic operations;
3. collect all terms on one side of the equation that contain the subject;
4. take out the subject as a common factor if necessary;
5. divide through by the coefficient of the subject;
6. take roots, powers, as necessary.

Note that the *coefficient* is a decimal number multiplying a literal number in a formula. For example in the simple formula,

$$3b = cde$$

the number 3 is the *coefficient of b* and on division by 3, we get:

$$b = \frac{cde}{3}$$

The above procedure is best illustrated by the following example.

Example 3.1

1. Given that $\frac{1}{f} = \frac{1}{u} + \frac{1}{v}$, make v the subject of the formula.
2. If $s = ut + \frac{1}{2}at^2$, transpose the formula for a.
3. If $\frac{D}{d} = \sqrt{\frac{f+p}{f-p}}$, transpose the formula for f.

1. Following the procedures, we need first to clear fractions. Remember you *cannot* just turn the fractions upside down! Only when there is a *single fraction* on each side of the equals, we are allowed to invert them. I hope you can remember how to combine fractions! If you are unsure, look back now, and study the method we adopted for combining two or more algebraic fractions. Then:

$$\frac{1}{f} = \frac{1}{u} + \frac{1}{v} \quad \text{or} \quad \frac{1}{f} = \frac{v+u}{uv}$$

and clearing fractions by multiplying both sides by f and uv, we get: $uv = f(v+u)$ and after multiplying out, $uv = fv + fu$ and after gathering all terms containing the subject,

$uv - fv = fu$ then removing the subject as a common factor gives: $v(u - f) = fu$ and after division of both sides by $(u - f)$ we finally get:

$$v = \frac{fu}{u - f}$$

2. Following our procedure, there is really only one fraction which we can eliminate, it is $\frac{1}{2}$, if we multiply every term by the inverse of a $\frac{1}{2}$, i.e. $\frac{2}{1}$ we get:

$$2s = 2ut + at^2$$

Subtracting $2ut$ from both sides gives: $2s - 2ut = at^2$ and dividing both sides by t^2, then:

$$\frac{2s - 2ut}{t^2} = a$$

and reversing the formula and pulling out the common factor gives:

$$a = \frac{2(s - ut)}{t^2}$$

Alternatively, remembering your laws of indices, we can bring up the t^2 term and write the formula for a, as:

$$a = 2t^{-2}(s - ut)$$

3. We again follow the procedure, first clearing roots, then fractions, in the following manner.
Squaring:

$$\left(\frac{D}{d}\right)^2 = \frac{f + p}{f - p} \quad \text{or} \quad \frac{D^2}{d^2} = \frac{f + p}{f - p}$$

and multiplying both sides by the terms in the denominator, or *cross-multiplying* d^2 and $(f - p)$ gives: $D^2(f - p) = d^2(f + p)$ and $D^2 f - D^2 p = d^2 f + d^2 p$; so collecting terms on one side containing the subject, we get: $D^2 f - d^2 f = d^2 p + D^2 p$.
After pulling out common factors, we have: $f(D^2 - d^2) = (d^2 - D^2)p$ and dividing both sides by $(D^2 - d^2)$ yields the result:

$$f = \frac{(d^2 + D^2)p}{D^2 - d^2}$$

Example 3.2

If $F = \frac{mV^2}{r}$, find m when $F = 2560$, $V = 20$ and $r = 5$.
Then, by direct substitution:

$$2560 = \frac{m(20)^2}{5} \quad \text{so} \quad (2560)(5) = m(400)$$

$$400m = 12800$$

$$m = \frac{12{,}800}{400} \quad \text{then} \quad m = 32$$

Alternatively, we can transpose the formula for m and then substitute for the given values:

$$F = \frac{mV^2}{r} \quad \text{and} \quad Fr = mV^2 \quad \text{so} \quad \frac{Fr}{V^2} = m$$

$$\text{then} \quad m = \frac{Fr}{V^2} \quad \text{and}$$

$$m = \frac{(2560)(5)}{(20)^2} = \frac{12{,}800}{400} = 32$$

giving the same result as before.

In our final example on substitution, we use a formula that relates electric charge Q, resistance R, inductance L and capacitance C.

Example 3.3

Find C if $Q = \frac{1}{R}\sqrt{\frac{L}{C}}$ where $Q = 10$, $R = 40\,\Omega$, $L = 0.1$.

$QR = \sqrt{\frac{L}{C}}$ and squaring both sides gives:

$$(QR)^2 = \frac{L}{C} \quad \text{or} \quad Q^2 R^2 = \frac{L}{C}$$

$$C(Q^2 R^2) = L \quad \text{then,} \quad C = \frac{L}{Q^2 R^2}$$

Substituting for the given values, we get:

$$C = \frac{0.1}{10^2\, 40^2} = 6.25 \times 10^{-7}\,\text{F}$$

3.1.2 Logarithms and logarithmic functions

We have already studied the laws and use of indices, and looked at logarithms, as a means of simplifying arithmetic operations. We needed to use logarithm tables and antilogarithm tables in order to do this. As you already know, the logarithm of a number is in fact its index. As a reminder, e.g. $10^3 = 1000$, the left-hand side of this equation 10^3 is the number 1000 written in index form. The index 3 is in fact the *logarithm* of 1000. Check this by pressing your *log* button on your calculator (which is the logarithm to the base 10), then key in the number 1000 and press the $=$ button, you will obtain the number 3!

Manipulation of numbers, expressions and formulae, which are in index form, may be simplified by using logarithms. Another use for logarithms (one which you have already met) is being able to reduce sometimes the more difficult arithmetic operations of multiplication and division to those of addition and subtraction. This is often necessary when manipulating more complex algebraic expressions.

We start by considering the laws of logarithms, in a similar manner to the way in which we dealt with the laws of indices, earlier.

> **Key point**
>
> The power or index of a number, when that number is in index form, is also its logarithm, when taken to the base of the number.

3.1.3 The laws of logarithms

The laws of logarithms are tabulated below, they are followed by simple examples of their use. In all these examples, we use *common logarithms*, i.e. logarithms to the *base 10*. Later we will look at another type, the *Naperian logarithm*, or natural logarithm, where the *base* is the number e($2.71828\ldots$):

Number	Logarithmic law
1	If $a = b^c$, then $c = \log_b a$
2	$\log_a MN = \log_a M + \log_a N$
3	$\log_a \dfrac{M}{N} = \log_a M - \log_a N$
4	$\log_a(M^n) = n \log_a M$
5	$\log_b M = \dfrac{\log_a M}{\log_a b}$

Law 1

All these laws look complicated, but you have already used law 1, when you carried out the

calculator exercise above. So again, we know that $1000 = 10^3$. Now if we wish to put this number into *linear* form (decimal form), then we may do this by *taking logarithms*.

Following *law 1*, where in this case, $a = 1000$, $b = 10$ and $c = 3$ then: $3 = \log_{10} 1000$. You have already proved this fact on your calculator! So you are probably wondering why we need to bother with logarithms? Well in this case we are dealing with *common logarithms*, i.e. numbers in index form where the *base* of the logarithm is 10. We can also consider numbers in index form, that are not to the base 10, as you will see later. We may also be faced with a problem where the index (power) is not known.

Suppose we are confronted with this problem: *find the value of x, where $750 = 10^x$.* The answer is not quite so obvious, but it can easily be solved using our first law of logarithms. So, again following the law, i.e. taking logarithms to the appropriate base we get: $x = \log_{10} 750$ and *now* using our calculator, we get: $x = 2.8751$, correct to four significant figures.

Law 2

One pair of factors for the number 1000 is 10 and 100. Therefore according to the second law: $\log_a(10)(100) = \log_a 10 + \log_a 100$. If we choose logarithms to the base 10, then we already know that the $\log_{10} 1000 = 3$. Then using our calculator again, we see that $\log_{10} 10 = 1$ and $\log_{10} 100 = 2$. What this law enables us to do, is to convert the *multiplication* of numbers in index form into that of *addition*. Compare this law with the *first law of indices* you studied earlier! Also remember that we are at liberty to choose any base we wish, provided we are able to work in this base. Your calculator only gives you logarithms to two bases, 10 and e.

Law 3

This law allows us to convert the *division* of number in index form into that of *subtraction*. When dealing with the *transposition* of more complex formulae, these conversions can be particularly useful and help us with the transposition.

So using the law directly then, e.g.:

$$\log_{10} \frac{1000}{10} = \log_{10} 100 = \log_{10} 1000 - \log_{10} 10$$

or from your calculator $2 = 3 - 1$.

Law 4

This law states that if we take the logarithm of a number in index form M^n this is equal to the logarithm of the base of the number $\log_a M$, multiplied by the index of the number $n\log_a M$. For example, $\log_{10}(100^2) = \log_{10} 10000 = 2\log_{10} 100$. This is easily confirmed on your calculator as $4 = (2)(2)$.

Law 5

This law is rather different from the others, in that it enables us to *change the base* of a logarithm. This of course is very useful, if we have to deal with logarithms, or formulae involving logarithms that have a base, not found on our calculator!

For example, suppose we wish to know the numerical value of $\log_2 64$, then using law 5, we have:

$$\log_2 64 = \frac{\log_{10} 64}{\log_{10} 2} = \frac{1.806179974}{0.301029995}$$

$$= 6, \text{ interesting!}$$

If we *use law 1*, in reverse then $\log_2 64$ is equivalent to the number $64 = 2^6$, which of course is now easily verified by your calculator! This example again demonstrates that given a number in *index form* (the *index* of that number) is also its *logarithm* provided the logarithm has the same base.

We will now consider, through example, one or two engineering uses for the laws of *common* and *natural logarithms*.

Key point

Common logarithms have the base 10.

Example 3.4

An equation connecting the final velocity v of a machine with the machine variables, w, p and z,

is given by the formula $v = 20^{\left(\frac{w}{pz}\right)}$. Transpose the formula for w, and find its numerical value when $v = 15$, $p = 1.24$, and $z = 34.65$.

This formula may be treated as a number in *index form*. Therefore to find w, as the subject of the formula, we need to apply the laws of logarithms. The first step, in this type of problem, is to *take logarithms of both sides*. The base of the logarithm chosen, is not important, provided we are able to find the numerical values of these logarithms, when required. Thus, we generally take logarithms to the base 10 or to the base e. As yet, we have not considered logarithms to the base e, we will take *common logarithms* of both sides. However, if the number or expression is not to a base of logarithms we can manipulate, then we are at liberty to change this base using law 5!

So, $\log_{10} v = \log_{10} 20^{\left(\frac{w}{pz}\right)}$ at this stage, taking logarithms seems to be of little help! However, if we now apply the appropriate logarithmic laws, we will be able to make w the subject of the formula.

Applying law 4 to the right-hand side of the expression we get:

$$\log_{10} v = \left(\frac{w}{pz}\right) \log_{10} 20$$

Then finding the numerical value of $\log_{10} 20 = 1.30103$, we can now continue with the transposition:

$$\log_{10} v = \left(\frac{w}{pz}\right) 1.30103 \quad \text{or} \quad \frac{\log_{10} v}{1.30103} = \frac{w}{pz}$$

and so, $\quad w = \frac{(pz)(\log_{10} v)}{1.30103}$

Having transposed the formula for w, we can substitute the appropriate values for the variables and find the numerical value of w.

Then:

$$w = \frac{(1.24)(34.65)(\log_{10} 15)}{1.30103}$$

$$= \frac{(1.24)(34.65)(1.17609)}{1.30103}$$

$$= 38.84$$

3.1.4 Naperian logarithms and the exponential function

If you look at your calculator you will see the ln or *Naperian logarithm* button. The *inverse* of the Naperian logarithm function is e^x or $\exp x$, the *exponential function*. This logarithm is sometime known as the *natural logarithm*, because it is often used to model naturally occurring phenomena, such as the way things grow or decay. In engineering, e.g. the decay of charge from a capacitor may be modeled using the natural logarithm. It is therefore a very useful function, and both the natural logarithm and its inverse, the exponential function, are very important within engineering.

We will now consider the transposition of a formula that involves the use of natural logarithms and the logarithmic laws.

Example 3.5

Transpose the formula, $b = \log_e t - a \log_e D$ to make t the subject.

First, note that the natural or Naperian logarithm may be expressed as $\log_e$ or ln, as on your calculator. *Do not mix up* the expression $\log_e$, or its inverse e^x or $\exp x$ with the exponential function (EXP) on your calculator, which multiplies a number by powers of 10!

We first use the laws of logarithms as follows:

$$b = \log_e t - \log_e D^a \quad \text{from law 4}$$

$$b = \log_e \left(\frac{t}{D^a} \right) \quad \text{from law 3}$$

Now, for the first time we take the *inverse* of the natural logarithm or *antilogarithm*. Noting that *any* function, multiplied by its inverse is 1 (one). Then multiplying *both sides* of our equation, by e, the *inverse* of $\ln(\log_e)$, we get:

$$e^b = \frac{t}{D^a}$$

(since e is the inverse or antilogarithm of $\log_e = \ln$ or $(e)(\log_e) = 1$), then: $t = e^b D^a$ as required.

As mentioned before, the *exponential function* e^x or $\exp x$ and its inverse ln (natural logarithm) have many uses in aircraft engineering, because they can be used to model growth and decay. So the way solids expand, electrical resistance changes with temperature, a substance cools, pressure changes with altitude or capacitors discharge can *all* be modeled by the *exponential* function.

Here are just two engineering examples of the use of the exponential function.

Key point

Naperian or natural logarithms have the base e, where $e \simeq 2.718281828$ corrected to nine decimal places.

Key point

The *inverse* function of the Naperian logarithm is the *exponential function* which in symbols is expressed as $\exp x$ or e^x.

Example 3.6

If the pressure p at height h (in m) above the ground is given by the relationship:

$$p = p_0 e^{\frac{h}{k}}$$

where p_0 is the sea-level pressure of $101325 \, \text{Nm}^{-2}$. Determine the value of the height h, when the pressure at altitude p is $70129 \, \text{Nm}^{-2}$ and $k = -8152$.

First we need to transpose the formula for h, this will involve taking *natural logarithms*, the inverse function of $e^{\frac{h}{k}}$. Before we do so we will first isolate the exponential term, then:

$$\frac{p}{p_0} = e^{\frac{h}{k}}$$

and taking logarithms gives:

$$\log_e \left(\frac{p}{p_0} \right) = \frac{h}{k} \quad \text{then} \quad k \log_e \left(\frac{p}{p_0} \right) = h.$$

then substituting the given values,

$$h = (-8152) \log_e \left(\frac{70129}{101325} \right)$$

$$= (-8152) \log_e (0.692)$$

$$= (-8152)(-0.368) = 3000 \, \text{m}$$

corrected to four significant figures. Thus the altitude $h = 3000$ m.

Our final example is concerned with the information contained in a radio communications message. It is not necessary to understand the background physics in order to solve the problem, as you will see.

Example 3.7

It can be shown that the information content of a message is given by:

$$I = \log_2\left(\frac{1}{p}\right)$$

Using the laws of logarithms show that the information content may be expressed as $I = -\log_2(p)$ and find the information content of the message if the chances of receiving the code (p) is $\frac{1}{16}$.

So we are being asked to show that:

$$I = \log_2\left(\frac{1}{p}\right) = -\log_2(p)$$

the left-hand side of this expression may be written as $\log_2(p^{-1})$. I hope you remember the laws of indices! Now if we compare this expression with law 4, where $\log_a(M^n) = n\log_a M$ then in this case $M = p$ and $n = -1$ so, $\log_2(p^{-1}) = -1 \log_2 p = -\log_2 p$, as required.

Now to find the information content of the message, we need to substitute the given value of $p = \frac{1}{16}$ into the equation:

$$\log_2(p^{-1}) = \log_2\left(\frac{1}{p}\right) = \log_2(16)$$

Now our problem is that we cannot easily find the value of logarithms, to the base 2. However, if we use logarithmic law 5, then we get:

$$\log_2 16 = \frac{\log_{10} 16}{\log_{10} 2} = 4$$

Then the information content of the message = **4**.

I hope you were able to follow the reasoning, in the above two, quite testing examples. There is just one more application of the laws of logarithms that we need to cover. It is sometimes very useful when considering experimental data to determine if such data can be related to a particular law. If we can relate this data to the law of a straight line $y = mx + c$, then we can easily determine useful results. Unfortunately, the data is not always related in this form. However, a lot of engineering data follows the general form $y = ax^n$, where as before, x is the *independent variable*, y is the *dependent variable* and in this case, a and n, are *constants* for the particular experimental data being considered.

We can use a technique, involving *logarithms* to reduce equations of the form $y = ax^n$ to a linear form, following the law of the straight line, $y = mx + c$. The technique is best illustrated by the following example.

Example 3.8

The pressure p and volume v of a gas, at constant temperature are related by Boyle's law, which can be expressed as $p = cv^{-0.7}$, where c is a constant. Show that the experimental values given in the table follow this law, and from an appropriate graph of the results, determine the value of the constant c:

Volume v (m^3)	1.5	2.0	2.5	3.0	3.5
Pressure p (10^5 Nm^{-2})	7.5	6.2	5.26	4.63	4.16

The law is of the form $p = ax^n$. So taking common logarithms of both side of the law $p = cv^{-0.7}$

we get, $\log_{10} p = \log_{10}(cv^{-0.7})$

and applying laws 2 and 4 to the right-hand side of this equation gives:

$$\log_{10} p = -0.7 \log_{10} v + \log_{10} c;$$

make sure you can see how to get this result. Then comparing this equation with the equation of a straight line $y = mx + c$, we see that:

$$y = \log_{10} p, \quad m = -0.7,$$
$$x = \log_{10} v \quad c = \log_{10} c.$$

So we need to plot $\log_{10} p$ against $\log_{10} v$ (Figure 3.1). A table of values and the resulting plot

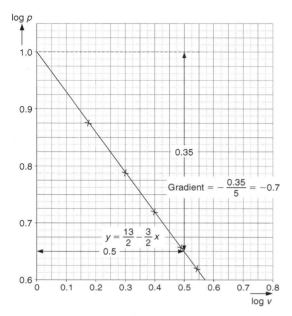

Figure 3.1   Plot of $\log_{10} p$ against $\log_{10} v$.

is shown below:

Volume v (m^3)	1.5	2.0	2.5	3.0	3.5
$\log_{10} v$	0.176	0.301	0.398	0.447	0.544
Pressure p ($10^5\,Nm^{-2}$)	7.5	6.2	5.26	4.63	4.16
$\log_{10} p$	0.875	0.792	0.721	0.666	0.619

Then from the plot it can be seen that the slope of the graph is -0.7, and the y intercept at $\log_{10} v = 0$ is given as 1.0 or $\log_{10} c = 1.0$ and so $c = 10$. Therefore, the plotted results do follow the law, $p = 10v^{-0.7}$.

This use of logarithms to manipulate the experimental data is very useful.

Test your understanding 3.1

1. If $Q = A_2 \sqrt{\dfrac{2gh}{1 - \left(\dfrac{A_2}{A_1}\right)^2}}$ find Q, when $A_1 = 0.0201$, $A_2 = 0.005$, $g = 9.81$ and $h = 0.554$.

2. If $X = \dfrac{1}{2\pi f C}$ calculate the value of C, when $X = 405.72$ and $f = 81.144$.

3. Simplify $\dfrac{1}{x-1} - \dfrac{1}{x+1} - \dfrac{3}{2(x^2-1)}$

4. The Bernoulli equation may be written as:

$$\frac{p_1}{\gamma} + \frac{v_1^2}{2g} + h_1 = \frac{p_2}{\gamma} + \frac{v_2^2}{2g} + h_2$$

given that $(h_1 - h_2) = 2$, $(v_1^2 - v_2^2) = 8.4$, $p_1 = 350$ and $\gamma = 10$; transpose the formula in a suitable way to find the value of the pressure p_2, when $g = 9.81$.

5. Transpose the formula $q = rx^{\frac{s}{t}}$ for (t) and then find its value when $q = 30\pi$, $r = 3\pi$, $x = 7.5$ and $s = 16$.

6. The formula $P = T(1 - e^{-\mu\theta})v$ relates the power (P), belt tension (T), angle of lap (θ), linear velocity (v) and coefficient of friction (μ) for a belt drive system. Transpose the formula for (μ) and find its value when: $P = 2500$, $T = 1200$, $V = 3$ and $\theta = 2.94$.

7. In an experiment, values of current I and resistance R were measured, producing the results tabulated below:

R	0.1	0.3	0.5	0.7	0.9	1.1	1.3
I	0.00017	0.0015	0.0043	0.0083	0.014	0.021	0.029

Show that the law connecting I and R has the form $I = aR^b$, where a and b are constants and determine this law.

3.1.5 Complex numbers

The useful formulae for manipulating and applying complex numbers to engineering problems are given below without proof. Their use will be demonstrated primarily through examples.

Formulae

1. $z = x + iy$ where real $z = x$ and imaginary $z = y$, $i = \sqrt{-1}$ and so $i^2 = -1$.
2. $z = x - iy$ is the *conjugate* of the complex number $z = x + iy$.
3. $z\bar{z} = x^2 + y^2$.
4. Modulus $|z| = \sqrt{x^2 + y^2}$.
5. Distance between two points z_1 and z_2 is $|z_1 - z_2| = |z_2 - z_1|$.
6. Polar form $x + iy = r(\cos\theta + i\sin\theta)$, where $r = |z|$ (modulus) and θ is the *argument* of z, denoted by $\theta = \arg z$. Also $\cos\theta = x/r$ and $\sin\theta = y/r$.
   Thus: $\tan\theta = x/y$ and

$$z_1 z_2 = r_1 r_2 [\cos(\theta_1 + \theta_2)] = r_1 r_2 \angle \theta_1 + \theta_2$$

and

$$\frac{z_1}{z_2} = \frac{r_1[\cos(\theta_1 - \theta_2) + i\sin(\theta_1 - \theta_2)]}{r_2}$$

$$= \frac{r_1}{r_2}\angle(\theta_1 - \theta_2)$$

7. Exponential form: $z = re^{i\theta} = \cos\theta + i\sin\theta$ and $|e^{i\theta}| = 1$.
8. De Moivre's theorem: If n is any integer, then: $(\cos\theta + i\sin\theta)^n = \cos n\theta + i\sin\theta$.

Then from above, the complex number (z) consists of a real (Re) part $= x$ and an imaginary part $= y$, the imaginary unit (i or j) multiplies the imaginary part y. In normal form, complex numbers are written as: $z = x + iy$ or $z = x + jy$ where in all respects $i = j$ and j is often used by engineers for applications.

Let us look at one or two examples to see how we apply these formulae.

Example 3.9

Add, subtract, multiply and divide the following complex numbers:
(a) $(3 + 2j)$ and $(4 + 3j)$; and (b) in general, $(a + bj)$ and $(c + dj)$.

Addition
(a) $(3 + 2j) + (4 + 3j) = 3 + 4 + 2j + 3j = 7 + 5j$
(b) In general, $(a + bj) + (c + dj) = (a + c) + (b + d)j$.

Subtraction
(a) $(3 + 2j) - (4 + 3j) = (-1 - j)$
(b) In general, $(a + bj) - (c + dj) = (a - c) + (b - d)j$.

Multiplication
(a) $(3 + 2j) \times (4 + 3j) = 3(4 + 3j) + 2j(4 + 3j)$

$$= 12 + 9j + 8j + 6j^2$$

Now from the definition $j = \sqrt{-1}$, therefore $j^2 = -1$ and the right-hand side becomes,

$$= 12 + 17j + (6)(-1) \quad \text{or} \quad = 6 + 17j$$

For the general case,

$$(a + bj) \times (c + dj) = ac + adj + bcj + bdj^2$$

and the right-hand side becomes

$$= ac + adj + bcj - bd \quad (\text{where } j^2 = -1)$$
$$= (ac - bd) + (ad + bc)j$$

so the result of multiplication is still a complex number.

Division
(a) $\frac{3 + 2j}{4 + 3j}$ here we use an algebraic trick to assist us. We multiply top and bottom by the *conjugate* of the complex number in the denominator.

So here, $z = 4 + 3j$ then $\bar{z} = 4 - 3j$, where you will observe from the formulae above that $\bar{z}$ is the conjugate of the complex number z. So now, we proceed as follows:

$$\left(\frac{3 + 2j}{4 + 3j}\right)\left(\frac{4 - 3j}{4 - 3j}\right) = \frac{12 - 9j + 8j - 6j^2}{16 + 12j - 12j - 9j^2}$$
$$= \frac{18 - j}{25}$$

(b) Note that the denominator became real and in general:

$$\frac{a + bj}{c + dj} = \left(\frac{a + bj}{c + dj}\right)\left(\frac{c - dj}{c - dj}\right)$$
$$= \frac{ac - adj + bcj - bdj^2}{c^2 + cdj - cdj - dj^2}$$
$$= \frac{(ac + bd) + (-adj + bcj)}{c^2 + d^2}$$

Complex numbers may be transformed from Cartesian (rectangular) to polar form by finding their *modulus* and *argument*, as defined in formulae 4 and 6, above.

Example 3.10

Express the complex numbers: (a) $z = 2 + 3j$; (b) $z = 2 - 5j$ in polar form.

You will remember from your study of co-ordinates, in your graphical work, that polar co-ordinates are represented by an angle θ and a magnitude r. Complex numbers may be represented in the same way as in Figure 3.2.

To express complex numbers in polar form we will first find their modulus and argument. So from formulae 4 and 6, for $z = 2 + 3j$ the modulus $= r = \sqrt{2^2 + 3^2} = \sqrt{13}$. The argument $= \theta$ where: $\tan\theta = y/x = 3/2 = 1.5\theta = 56.3$ Then, $z = 2 + 3j = \sqrt{13}(\cos 56.3 + j\sin 56.3)$ or in the short-hand form, $= \sqrt{13}\angle 56.3$.

Similarly for $z = 2 - 5j$ then modulus $= |z| = r = \sqrt{2^2 + (-5)^2}$ so $r = \sqrt{29}$ and the argument $= \theta$

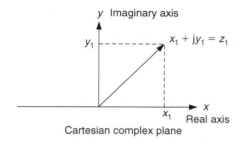
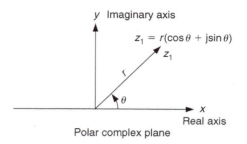

Figure 3.2 Complex number co-ordinate systems.

so that, $\tan\theta = -5/2 = -2.5$ and $\theta = -68.2$. Then, in polar form: $z = 2 - 5j = \sqrt{29}[\cos(-68.2) + j\sin(-68.2)]$ and in short-hand form,

$$z = \sqrt{29}\angle -68.2$$

The argument of a complex number in polar form represents the angle θ in *radians* (see the definition of the radian in the Section 3.2), it can take on an infinite number of values, which are determined up to 2π radians.

When we consider complex number in Cartesian (rectangular) form then each time we multiply the complex numbers by $(i = j)$ the complex *vector*, shifts by 90° or $\pi/2$ rad. This fact is used when complex vectors represent *phasors* (electrical vectors). Then, successive multiplication by j, shift the phase by $\pi/2$, as shown in the next example. Under these circumstances the imaginary unit j is known as the *j-operator*.

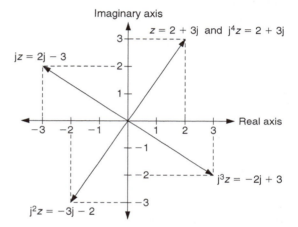

Figure 3.3 j-operator rotation.

Example 3.11

Multiply the complex number, $z = 2 + 3j$ by the j-operator three times in succession. The situation is illustrated in Figure 3.3.

Then successive multiplication gives:

$$jz = j(2 + 3j) = 2j + 3j^2 = 2j - 3$$
$$j^2z = j(2j - 3) = 2j^2 - 3j = -3j - 2$$
$$j^3z = j(-3j - 2) = -3j^2 - 2j = -2j + 3$$
$$j^4z = j(-2j + 3) = -2j^2 + 3j = 2 + 3j$$

Note that $z = j^4z$ we have rotated the vector (phasor) through 2π rad (360°), back to its original position, as shown in the diagram.

We leave this short study of complex numbers by considering the arithmetic operations of multiplication and division of complex numbers in polar form. Addition and subtraction are not considered because we have to convert the complex number from polar form to Cartesian form, before we can perform these operations!

When we multiply a complex number in polar form, we multiply their moduli and add their arguments. Conversely for division, we divide their moduli and subtract their arguments.

Example 3.12

For the complex number given below, find their product (z_1z_2) and their quotient (z_1/z_2):

$$z_1 = 3(\cos 120 + j\sin 120)$$
$$z_2 = 4(\cos(-45) + j\sin(-45))$$

Then

$$z_1z_2 = (3)(4)[\cos(120 - 45) + j\sin(120 - 45)]$$
$$= 12(\cos 75 + j\sin 75)$$

And similarly for division, we get:

$$\frac{z_1}{z_2} = \frac{3}{4}[\cos(120+45) + j\sin(120-45)]$$

$$= 0.75(\cos 165 + j\sin 165)$$

For the abbreviated version of complex numbers in polar form, we can multiply and divide in a similar manner. Once again they need to be converted to Cartesian form to be added and subtracted.

So in abbreviated form $z_1 = 3\angle 120$ and $z_2 = 4\angle 45$. Hence, once again,

$$z_1 z_2 = r_1 r_2 \angle(\theta_1 + \theta_2) = 12\angle 120 - 45 = 12\angle 75°$$

similarly,

$$\frac{z_1}{z_2} = \frac{r_1}{r_2}\angle(\theta_1 - \theta_2) = \frac{3}{4}\angle 120 + 45$$

$$= 0.75\angle 165°$$

as before.

Test your understanding 3.2

1. Perform the required calculation on the following complex numbers and express your results in the form $a + ib$
   (a) $(3-2i) - (4+5i)$; (b) $(7-3i)(3+5i)$; (c) $\frac{1+2i}{3-4i}$

2. Represent the following complex numbers in polar form:
   (a) $6 - 6j$; (b) $3 + 4j$; (c) $(4+5j)^2$

3. Express the following complex numbers in Cartesian form:
   (a) $\sqrt{30}\angle 60°$; (b) $\sqrt{13}\angle\frac{\pi}{4}$ (rad)

4. If $Z_1 = 20 + 10j$, $Z_2 = 15 - 25j$, $Z_3 = 30 + 5j$ find:
   (a) $|Z_1||Z_2|$; (b) $Z_1 Z_2 Z_3$; (c) $\frac{Z_1 Z_2}{Z_3}$; (d) $\frac{Z_1 Z_2}{Z_1 Z_3}$

3.2 Further trigonometry

We start our short study of further trigonometry by introducing the rules necessary to solve any triangle, right angled or otherwise. We then look briefly at the radian and its engineering application. This then leads onto the study of the sine and cosine functions and their graphical analysis. We finish this section by considering the use of trigonometric identities, as an aid to engineering calculations and as a method of simplification of functions, prior to the application of the calculus, which you will meet later.

3.2.1 Angles in any quadrant

So far in our study of angles, we have only considered angles between 0° and 90°. We will now consider angles in any quadrant i.e. all angles between 0° and 360°.

If you key into your calculator cos 150, you get the value -0.866. This is the same numerically, as cos 30 = 0.866, except that there has been a *sign* change. Whether any one trigonometric ratio is positive or negative, depends on whether the projection is on the positive or negative part of the co-ordinate system. Figure 3.4 shows the rectangular co-ordinate system, on which two lines have been placed, at angles of 30° and 150°, respectively, from the positive horizontal *x-ordinate*.

Now if we consider the sine ratio for both angles, then we get:

$$\sin 30 = \frac{+ab}{+ob} \quad \text{and} \quad \sin 150 = \frac{+cd}{+od}$$

thus both these ratios are *positive* and therefore a positive value for sin 30 and sin 150 will result. In fact from your calculator sin 30 = sin 150 = 0.5.

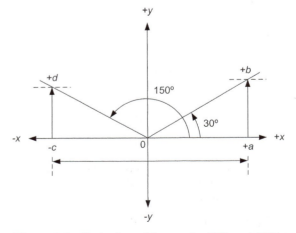

Figure 3.4 Projection of the angles 30° and 150°.

Now, from the diagram we find that:

$$\cos 30 = \frac{+oa}{+ob}$$

which will again yield a positive value, in fact $\cos 30 = 0.866$, but

$$\cos 150 = \frac{-oc}{+od}$$

is a *negative* ratio, that yields the negative value -0.866, which you found earlier.

If we continue to rotate our line in an *anticlockwise* direction, we will find that $\cos 240 = -0.5$ and $\cos 300 = 0.5$. Thus, dependent on which *quadrant* (quarter of a circle, so each 90°) the ratio is placed, depends whether or not the ratio is positive or negative. This is true for all three of the fundamental trigonometric ratios. Figure 3.5 shows the *signs*, for the sine, cosine and tangent functions.

Figure 3.5 also shows a way of remembering when the *sign* of these ratios is *positive* using the word Cosine All Sine Tangent (CAST) positives. Your calculator automatically shows the correct sign for any ratio of any angle, but it is worth knowing what to expect from your calculator!

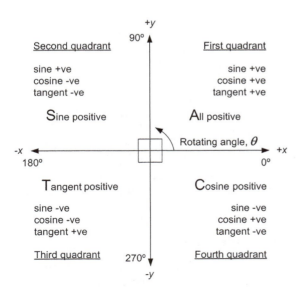

Figure 3.5 Signs of angles of any quadrant.

Example 3.13

Find, on your calculator, the value of the following trigonometric ratios and verify that the *sign* is correct, by consulting Figure 3.5:

(a) sin 57   (b) cos 236   (c) tan 97
(d) sin 320   (e) cos 108   (f) tan 347
(g) sin 137   (h) cos 310   (i) tan 237

The values with their appropriate sign are tabulated below:

(a) 0.8387   (b) −0.5592   (c) −8.144
(d) −0.6428   (e) −0.3090   (f) −0.2309
(g) 0.6819   (h) 0.6428   (i) 1.5397

You can easily verify that all these values are in accordance with Figure 3.5.

We finish solving triangles by considering triangles of any internal angles. This involves the use of the sine and cosine rules, which are given without proof.

3.2.2 General solution of triangles

We now extend our knowledge to the solution of triangles, which are not right angled. In order to do this we need to be armed with just two additional formulae. These are tabulated below for reference:

Sine rule	$\dfrac{a}{\sin a} = \dfrac{b}{\sin b} = \dfrac{c}{\sin c}$
Cosine rule	$a^2 = b^2 + c^2 - 2bc\cos A$
	$b^2 = a^2 + c^2 - 2ac\cos B$
	$c^2 = a^2 + b^2 - 2ab\cos C$

The above rules can only be used in specific circumstances.

For the general triangle ABC shown in Figure 3.6, with sides a, b, c and angles,

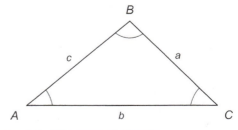

Figure 3.6 The general triangle.

∠A, ∠B, ∠C. Then the *Sine rule may only be used when either*:

One side and any two angles are known or
If two sides and an angle (not the angle between the sides) are known.

The cosine rule may only be used when either:

Three sides are known or
Two sides and the included angle are known.

Note 1

When using the sine rule, the equality signs allow us to use any parts of the rule that may be of help. For example, if we have a triangle to solve, for which we know the angles ∠A and ∠C and side a. We would first use the rule with the terms:

$$\frac{a}{\sin A} = \frac{c}{\sin C}$$

to find side c.

Note 2

When using the cosine rule, the version chosen will also depend on the information given. For example if you are given sides a, b and the included angle C, then the formula: $c^2 = a^2 + b^2 - 2ab \cos C$ would be selected to find the remaining side c.

Only relatively simple examples of these rules are given here, which are sufficient to illustrate their use. You may solve more complex problems using these rules, if you take a higher education program, in your future studies.

Example 3.14

In a triangle ABC, ∠A = 48°, ∠B = 59° and the side a = 14.5 cm. Find the unknown sides and angle.

The triangle ABC is shown in the Figure 3.7.

When the triangle is sketched, it can be seen that we have two angles and one side. So we can use the *sine* rule. Remembering that the sum of the internal angles of a triangle = 180°, then ∠C = 180 − 48 − 59 = 73°. We will use the first two terms of the sine rule, $\frac{a}{\sin A} = \frac{b}{\sin B}$ to find

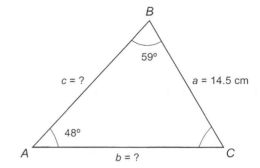

Figure 3.7 Triangle.

the unknown side b. Then:

$$\frac{14.5}{\sin 48} = \frac{b}{\sin 59}$$

or $b = \dfrac{(\sin 59)(14.5)}{\sin 48} = \dfrac{(0.8572)(14.5)}{0.7431}$

$$= 16.72 \text{ cm}$$

similarly, to find side c we use:

$$\frac{a}{\sin A} = \frac{c}{\sin C}$$

which on substitution of the values gives:

$$\frac{14.5}{\sin 48} = \frac{c}{\sin 73}$$

or $c = \dfrac{(\sin 73)(14.5)}{\sin 48} = \dfrac{(0.9563)(14.5)}{0.7431}$

$$= \frac{13.8664}{0.7431} = 18.66 \text{ cm}$$

when using the cosine rule, given three sides. It is necessary to transpose the formula to find the required angles. In the next example, we need to perform this transposition, which you should find, relatively simple. If you have difficulties following the steps, you should refer back to the section on transposition of formula, in outcome 1.

Example 3.15

A flat steel plate is cut with sides of length, 12, 8 and 6 cm. Determine the three angles of the plate.

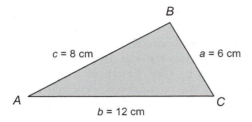

Figure 3.8 Triangle.

A diagram of the plate, suitably labeled, is shown in the Figure 3.8, where side $a = 6$ cm, $b = 12$ cm and $c = 8$ cm.

Now, in this particular case we are free to choose any variant of the formula, to find the corresponding angle. We will use: $b^2 = a^2 + c^2 - 2ac \cos B$. Then transposing for $\cos B$,

$$2ac \cos B = a^2 + c^2 - b^2$$

and

$$\cos B = \frac{a^2 + c^2 - b^2}{2ac}$$

then

$$\cos B = \frac{6^2 + 8^2 - 12^2}{(2)(6)(8)} = \frac{36 + 64 - 144}{96}$$

$$= \frac{-44}{96} = -0.4583$$

Now $\angle B = 117.28°$, using a calculator. Note that $\cos B$ is negative therefore, $\angle B$ must lie outside the first quadrant, i.e. it must be greater than $90°$. However, since it is the angle of a triangle it must also be less than $180°$, thus $\angle B = 117.28°$ is its only possible value.

Now to find another angle, we could again use the cosine rule. However, since we now have an angle and two *non-included* sides, a and b, we are at liberty to use the simpler *sine* rule. Then:

$$\frac{a}{\sin A} = \frac{b}{\sin B} \quad \text{and so,} \quad \frac{6}{\sin A} = \frac{12}{\sin 117.28}$$

or $\quad \sin A = \frac{(6)(\sin 117.28)}{12} = \frac{(6)(0.8887)}{12}$

$$= 0.4444$$

and from calculator, $\angle A = 26.38°$.

Finally $\angle C = 180 - 117.28 - 26.38 = 36.34°$.

Area of any triangle
Now to complete our study of general triangles, we need to be able to calculate their area. Of course, we have already done this during our study of areas and volumes in the non-calculator mathematics. Again, like we did for right-angled triangles, let us use one of the formula we learnt earlier to find the area of any triangle. The formula we will use is:

$$\sqrt{s(s-a)(s-b)(s-c)}$$

where a, b and c were the sides and

$$s = \frac{a+b+c}{2}$$

Then in the case of the triangle we have just been considering in Example 3.15, where: $a = 6$ cm, $b = 12$ cm and $c = 8$ cm

Then, $\quad s = \frac{6 + 12 + 8}{2} = \frac{26}{2} = 13$

and therefore,

the area $= \sqrt{13(13-6)(13-12)(13-8)}$

$$= \sqrt{(13)(7)(1)(5)} = \sqrt{455} = 21.33 \text{ cm}^2$$

Now the area of any triangle ABC, can also be found using any of the following formulae.

Area of any triangle:

$$ABC = \tfrac{1}{2}ab \sin C \quad \text{or}$$

$$= \tfrac{1}{2}ac \sin B \quad \text{or}$$

$$= \tfrac{1}{2}bc \sin A$$

These formulae are quoted here without proof and any variant may be used dependent on the information available. So again, for the triangle in Example 3.15 using the above formulae,

Area of triangle:

$$ABC \tfrac{1}{2}ab \sin C = \tfrac{1}{2}(6)(12)(\sin 36.34)$$

$$= (0.5)(72)(0.5926) = 21.33 \text{ cm}^2$$

as before.

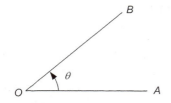

Figure 3.9 The angle as a measure of rotation.

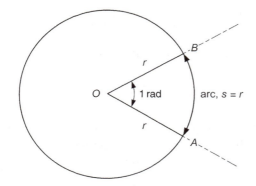

Figure 3.10 Illustration of the radian.

3.2.3 The radian and circular measure

Circular measure using degrees has been with us since the days of the Babylonians, when they divided a circle into 360 equal parts corresponding to what they believed were the days in the year. An angle in degrees is a measure of rotation and an angle is formed when a line rotates with respect to a fixed line (Figure 3.9), when both lines have the same center of rotation.

The degree may be subdivided into minutes and seconds of an arc, where the minute is $\frac{1}{60}$ of a degree and a second is $\frac{1}{60}$ of a minute or $\frac{1}{3600}$ of a degree of an arc. We will restrict ourselves to angular measurement in degrees, and decimal fractions of a degree, as you learnt earlier.

The *degree*, being an arbitrary form of circular measurement, has not always proved an appropriate unit for mathematical manipulation. Another less arbitrary unit of measure has been introduced, known as the **radian** (Figure 3.10), the advantage of this unit is its relationship with an *arc length* of a circle.

A radian is defined as *the angle subtended at the center of a circle by an arc equal in length to the radius of the circle.*

Now we know that the circumference of a circle is given by $C = 2\pi r$ where r is the radius. Therefore, the circumference contain 2π radii. We have just been told that an arc length for 1 rad is $s = r$. Therefore, the whole circle must contain 2π rad, or approximately 6.28 rad. A circle contains 360°, so it follows that: 2π **rad** = 360° or π **rad** = **180°**. We can use this relationship to convert from degrees to radians, and radians to degrees.

Example 3.16

(a) Express 60° in radians
(b) Express $\frac{\pi}{4}$ rad in degrees

(a) Since, $180° = \pi$ rad

then, $1° = \dfrac{\pi \, \text{rad}}{180}$

so, $60° = 60\left(\dfrac{\pi \, \text{rad}}{180}\right)$

$60° = \dfrac{\pi \, \text{rad}}{3}$ or 1.047 rad (three dp)

Note that if we leave radians in terms of π we have an exact value to use for further mathematical manipulation. For this reason, it is more convenient to leave radians expressed in terms of π.

(b) We follow a similar argument, except we apply the reverse operations.

$\pi \, \text{rad} = 180°$ then $1 \, \text{rad} = \dfrac{180°}{\pi}$

so, $\dfrac{\pi}{4} \, \text{rad} = \left(\dfrac{\pi}{4}\right)\dfrac{180°}{\pi}$ and $\dfrac{\pi}{4}$ **rad** = **45°**

To aid your understanding of the relationship between the degree and the radian, Figure 3.11 shows diagrammatically a comparison between some common angles using both forms of measure. Note that, in the figure, all angles in radian measure are shown in terms of π.

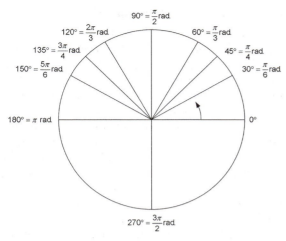

Figure 3.11 Comparison of degree and radian measure.

The area of a sector
It is often useful to be able to find the area of a sector, when considering cross-sectional areas. To determine such areas, we first need to understand the relationship between the *arc length s* and the *angle θ* subtended at the center of a circle by this arc length.

You have seen that the circumference of a circle subtends 2π rad. So if we consider the circumference to be an *arc of length* $2\pi r$, we can say that:

$$2\pi \text{ rad} = \frac{2\pi r}{r} \quad \text{where } r = \text{the radius}$$

or, the angle in radians $= \dfrac{\text{arc length } (s)}{\text{radius } (r)}$

then, $\qquad \theta \text{ rad} = \dfrac{s}{r} \quad \text{or} \quad s = r\theta$

Always remember that when using this formula, the angle θ must be in *radians*.

The area of a sector is now fairly easy to find (Figure 3.12).

We know that the area of a circle $= \pi r^2$. So it follows that when dealing with a portion (sector) of a circle, like that shown in Figure 3.12, the ratio of the angle θ (in rad) of the sector to that of the angle for the whole circle in radians is $\frac{\theta}{2\pi}$, remembering that there are 2π rad in a circle ($360°$). Then the area of any portion of the

circle such as the area of the sector = the area of the circle multiplied by the ratio of the angles, or in symbols.

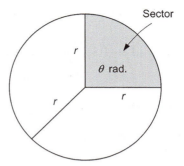

Figure 3.12 Area of sector of a circle.

$$\textbf{Area of sector} = (\pi r^2)\left(\frac{\theta}{2\pi}\right)$$

$$= \frac{r^2\theta}{2} \quad (\boldsymbol{\theta} \textbf{ in rad})$$

Example 3.17

(a) If the angle subtended at the center of a circle by an arc length 4.5 cm is 120°, what is the radius of the circle?
(b) Find the angle of a sector of radius 20 cm and an area 300 cm².

(a) We must first convert 120° into radians. This we can do very easily using the conversion factor, we found earlier, then:

$$120° = \frac{120\pi \text{ rad}}{180} = \frac{2\pi}{3} \text{ rad}$$

we will leave this angle in terms of π.
   Then from $s = r\theta$ we have:

$$r = \frac{s}{\theta} = \frac{4.5}{2\pi/3} = 2.149 \text{ cm}$$

(corrected to three dp)

(b) To find the angle of the sector we use the area of a sector formula, i.e.

$$A = \tfrac{1}{2}r^2\theta \quad \text{or,} \quad \theta = \frac{2A}{r^2}$$

and on substitution of given values, we get:

$$\theta = \frac{(2)(300)}{20^2} = \frac{600}{400} = 1.5 \text{ rad}$$

If we wish to convert this angle to degrees, then:

$$1.5\,\text{rad} = (1.5)\frac{180°}{\pi} = 85.94°$$

(corrected to two dp)

Test your understanding 3.3

1. In a right-angled triangle, the lengths of the shorter sides are 6 and 9 cm. Calculate the length of the hypotenuse.

2. All the sides of a triangle are 8 cm in length. What is the vertical height of the triangle?

3. In the Figure 3.13, the angles of elevation of A and B to D are 32° and 62°, respectively. If $DC = 70$ m, calculate the length of BC.

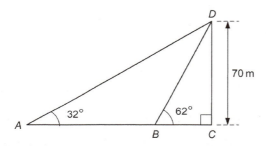

Figure 3.13

4. A vertical radio mast has cable stays of length 64 m, extending from the top of the mast. If each wire makes an angle of 65° with the ground, find:
   (a) the distance each cable is from the base of the mast,
   (b) the vertical height of the mast.

5. State the circumstances under which:
   (a) the sine rule may be used,
   (b) the cosine rule may be used.

6. Use the sine rule to solve the triangle ABC where side $a = 37.2$ cm, side $b = 31.6$ cm and $\angle B = 37°$.

7. Use the cosine rule to solve the triangle ABC, where: $a = 12$ cm, $b = 10$ cm and $c = 6$ cm. Also find the area of this triangle.

8. Define the radian.

9. If an arc of length 8.5 cm subtends an angle of 190.5°, at the center of a circle,
   (a) find its radius,
   (b) determine the area of the sector subtended by the angle 190.5°.

10. An aircraft landing light can spread its illumination over an angle of 40° to a distance of 170 m. Determine the maximum area that is lit up by the landing light in front of the aircraft.

3.2.4 Trigonometric functions

We will limit our study of trigonometric functions to the sine and cosine functions. In particular, we look at the nature of their graphs and the use to which these may be put. The graphs of these functions are very important, as the sine and cosine curves illustrate many kinds of *oscillatory motion*, which you are likely to meet in your future studies. The sine and cosine functions are used to model the oscillatory motion of currents, voltages, springs, vibration dampers, the rise and fall of the tides and many other forms of vibrating system, where the motion is oscillatory.

By *oscillatory*, we mean motion that vibrates back and forth about some mean value, during even *periods* of time. We start by plotting the sine and cosine curves, then consider their use for solving sine and cosine functions, in a similar manner to the graphs of algebraic equations, we considered earlier.

Graphs of sine and cosine functions

The basic sine curve for $y = \sin x$ is a wave which lies between the values $+1$ and -1, it is therefore *bounded*. That is, the value of the dependent variable y reaches a *maximum* value of $+1$ and a *minimum* value of -1 (Figure 3.14). Also the curve is zero at multiples of $180°$ or at multiples of π rad.

The x-axis, in Figure 3.14, is marked out in degrees and radians, which measure angular distance, the maximum and minimum values of y, are also shown. Other things to note about this graph are the fact that it repeats itself every $360°$ or 2π rad. Also this curve reaches it first maximum value at $90°$ or $\frac{\pi}{2}$ rad, it reaches its second maximum $360°$ or 2π rad and later at $450°$ or $\frac{5\pi}{2}$ rad. Similarly, it reaches its first minimum value at $270°$ or $\frac{3\pi}{2}$ rad and again $360°$ or 2π rad and later at $630°$ or $\frac{7\pi}{2}$ rad. These *maximum* and *minimum* values are repeated periodically at $360°$ intervals. We therefore say that the sine wave has periodic motion, where any point on the wave say, p_1, repeats itself, every $360°$ or 2π rad. These repetitions are known as *cycles*, as shown in Figure 3.14, where one complete cycle occurs every $360°$ or every 2π rad.

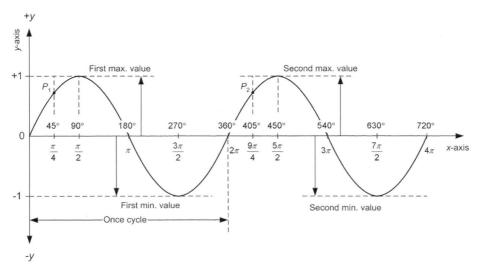

Figure 3.14   Plot of the function $y = \sin x$.

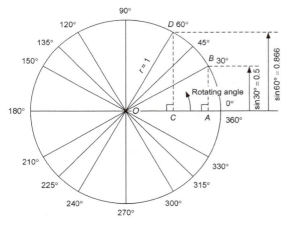

Figure 3.15   The rotating angle and sign function.

Now how do we plot values for sinusoidal functions? Look back at Figure 3.9 and note how we represented angular measure. In Figure 3.15, we represent angular measure on the set of rectangular co-ordinates, *the angle* in degrees or radians, is measured from the *positive x*-axis, and increases as it rotates in an *anticlockwise* direction, reaching a *positive maximum* value at 90° or $\frac{\pi}{2}$ rad. This maximum value is $+1$, when we make the radius of the circle $r = 1$, as in the diagram.

Now, the actual magnitude of this angle (its distance in the *y*-direction) is found using the *sine* function. For example, the height of the line

AB in the triangle OAB can be found by noting that:

$$\sin 30° = \frac{\text{opp}}{\text{hyp}} = \frac{AB}{1} = AB = 0.5$$

Similarly, as the angle increases, say to 60° or $\frac{\pi}{3}$ rad, then $CD = \sin 60° = 0.866$. It reaches its first maximum value when $OE = \sin 90° = 1.0 = \text{radius } r$, compare this value with the value on the curve of the sign function, shown in Figure 3.15! Now as the angle continues to increase, it moves into the second quadrant, where the magnitude of the rotating angle, gradually reduces until it reaches 180° or π rad, when its value becomes zero, once more. As we move into the third quadrant, the magnitude of the rotating angle (vector) once again starts to increase, but in a negative sense, until it reaches it maximum value at 270° or $\frac{3\pi}{2}$ rad where $\sin 270° = -1$. Finally in the fourth quadrant, it reduces from the negative maximum (minimum) value, until it once again reaches zero. The behaviour of this point is plotted as the curve shown in Figure 3.15, where the curve is produced by connecting the magnitude of this point for many values of the angle, between 0° and 360°, after which the pattern repeats itself every 360°.

A table of values for the magnitude of the rotating angle are given below. Check that these values match the plot of the sine curve shown in Figure 3.15.

x = angle θ [values in degrees (rad)]	$y = \sin\theta$	x = angle θ (values in degrees)	$y = \sin\theta$
0	0		
$30(\frac{\pi}{6})$	0.5	210	−0.5
$45(\frac{\pi}{4})$	0.7071	225	−0.7071
$90(\frac{\pi}{2})$	1.0	270	−1.0
$120(\frac{2\pi}{3})$	0.8660	300	−0.866
$135(\frac{3\pi}{4})$	0.7071	315	−0.7071
$150(\frac{5\pi}{6})$	0.5	330	−0.5
$180(\pi)$	0	360	0

The above table is similar to that you would need to produce, when plotting *any sine function* graphically. For example, suppose you were required to plot the curve for the function $y = 2\sin\theta$. What happens to the values of y in the above table? I hope you can see that every value of y is *doubled*. That means the first *maximum* value for this function, will be $y = 2\sin 90° = (2)(1) = 2$, similarly for all other angles, the y, values will be doubled.

I hope you can now appreciate that if $y = 3\sin\theta$, then the magnitude of the y values will all be *trebled*. Then in general, *the magnitude of the plotted y-values is* dependent on the value of the *constant a*, when $y = a\sin\theta$. The magnitude of the y-values are referred to as their *amplitude*. Then the *maximum amplitude a* will occur when $\sin\theta$ *is maximum* i.e. when $\sin\theta = 1.0$. This we know from the table above, to first occur at $\theta = 90°$ and then to occur every 360° or 2π rad, later. The *minimum* value of the *amplitude* will first occur when $\sin\theta = -1.0$, this again can be seen to first occur when $\theta = 270°$ and repeat itself every 360° thereafter.

What do you think will happen if we plot the graph of $y = \sin 2\theta$? Well if $\theta = \frac{\pi}{4}$ rad, then:

$$y = \sin(2)\left(\frac{\pi}{4}\right) = \sin\frac{\pi}{2} = 1.0$$

If we compare this with the plotted values above, then the function, $y = \sin 2\theta$ has reached its first maximum, *twice as fast* as the function, $y = \sin\theta$. The effect of this is to increase the number of oscillations (cycles) in a given angular distance. This is illustrated in Figure 3.16. You should check a few of the plotted values to verify your understanding.

The cosine function

So far we have concentrated our efforts on the sine function. This is because the cosine function is very similar to the sine function, except that it reaches its first maximum and minimum values

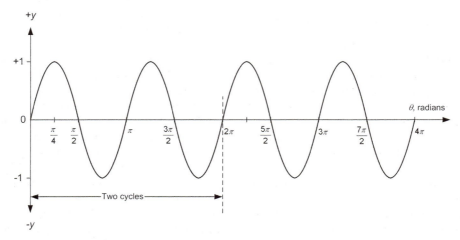

Figure 3.16 Graph of $y = \sin 2\theta$ between 0 and 4π rad.

at different angles to that of the sine function. In all other respects it is identical.

Consider again Figure 3.15 the sine function now in the case of the *cosine function, we start our rotating angle in the vertical position*, i.e. along the line *OE*. This means that what was 90° for the sine function *is now 0° for the cosine function*. This is illustrated in Figure 3.17.

Now, the cosine of the angle 30° is given by the *height of the y-ordinate* in a similar manner to the sine function, then $y = \cos 30° = 0.866$. Similarly, the cosine of 90° is again the height of the *y-ordinate*, which can be seen to be zero, i.e. $\cos 90° = 0$, which can easily be checked on your calculator. The net result is that all the cosine function values, for the given angle, are 90° *in advance* of the sine function. For example, the cosine function starts with its maximum at 0°, which is 90° in advance of the first maximum for the sine function. A plot of the cosine function $y = \cos \theta$ for angles between 0 and 4π rad is shown in Figure 3.18a.

It can be seen from Figure 3.18a that apart from the 90° *advance*, the cosine function follows an identical pattern to that of the sine function.

We finish this short section with a couple of examples of the use of graphical plots of these functions and how they can be used to find solutions to simple trigonometric equations.

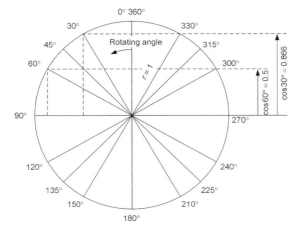

Figure 3.17 Rotating angle to illustrate the cosine function.

Example 3.18

Draw the graph of the function $y = 2\sin\theta + 3\cos\theta$ for values of θ between 0 and 90°. From the graph find:

(a) the maximum amplitude of the function
(b) a value of θ which satisfies the equation $2\sin\theta + 3\cos\theta = 3.5$

(a) Our first task is to set up a table of values and find the corresponding values for θ and y. We will use an interval of 10°:

θ	$2\sin\theta$	$3\cos\theta$	$y = 2\sin\theta + 3\cos\theta$
0	0	3	3
10	0.35	2.95	3.3
20	0.68	2.82	3.5
30	1.0	2.60	3.6
40	1.29	2.30	3.59
50	1.53	1.94	3.47
60	1.73	1.50	3.23
70	1.88	1.03	2.91
80	1.97	0.52	2.49
90	2.0	0	2.0

The table only shows two decimal place accuracy, but when undertaking graphical work, it is difficult to plot values, with any greater accuracy. Note also that we seem to have a maximum value for y when $\theta = 30°$. It is worth plotting a couple of intermediate values on either side of $\theta = 30°$ to see if there is an even higher value of y.

I have chosen $\theta = 27°$ and $\theta = 33°$. Then, when $\theta = 27°$ and $y = 3.58$, and when $\theta = 33°$ and $y = 3.61$, the latter values are very slightly higher, so may be used as the maximum.

The plot is shown in Figure 3.18b where it can be seen that within the accuracy of the plot, the maximum value of the amplitude for the function is $y = 3.5$.

(b) Now the appropriate values for the solution of the equation: $2\sin\theta - 3\cos\theta = 3.5$ are read-off from the graph, where the line $y = 3.5$ intersects with the curve $y = 2\sin\theta + 3\cos\theta$. The solutions are that when $y = 3.5$, **$\theta = 20°$** and **$\theta = 48°$**.

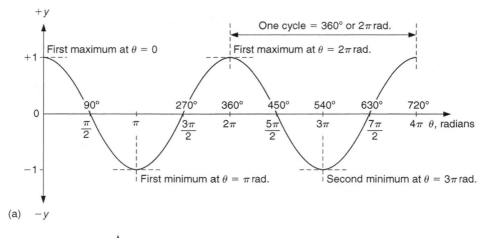

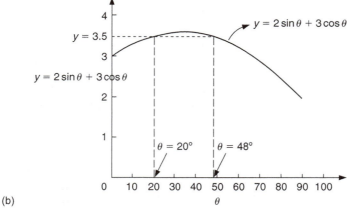

(b)

Figure 3.18 Graph of $y = \cos\theta$.

Example 3.19

For the following trigonometric functions, find the first maximum amplitude and the angular distance it occurs from $\theta = 0°$. Comment on the general form of each function:

1. $y = 4.2\cos\theta$
2. $y = 3\sin 2\theta$
3. $y = \sin\left(\theta - \dfrac{\pi}{2}\right)$

1. The maximum amplitude for all the functions is given when the amplitude a is multiplied by 1.0, in each case.

   We know that for $\cos\theta$ this first occurs when $\theta = 0$, so the maximum amplitude is **4.2** at an angular distance of 0°, from the reference angle.

The graph will follow exactly the form of the graph $y = \cos\theta$, except that every value will be amplified by a factor of 4.2.

2. In this case the maximum amplitude is **3**, and it first occurs when $2\theta = 90°$, i.e. at **+45°** to the reference angle.

   This graph will complete each cycle, in half the angular distance, when compared to $y = \sin\theta$.

3. This function has a maximum amplitude of $a = 1.0$, which first occurs when

$$\theta - \frac{\pi}{2} = \frac{\pi}{2} \text{ rad}$$

   therefore, $\qquad \theta = \dfrac{\pi}{2} + \dfrac{\pi}{2} = \pi \text{ rad}$

   That is, the first maximum which occurs at 180° after the reference angle. When compared to the function $y = \sin\theta$, each value is found to be lagging by $\frac{\pi}{2}$.

If you are finding it difficult to envisage what is happening, sketch these functions, on the same axes and make comparisons.

3.2.5 Trigonometric identities

Below will be found a few of the more common and most useful trigonometric identities. These are given without proof and should be *used as a tool* to simplify expressions or place them in another form for further manipulation. This technique is particularly useful for simplification prior to carrying out *integration*, which you will meet later.

General identities

1. $\operatorname{cosec}\theta = \frac{1}{\sin\theta}$; $\sec\theta = \frac{1}{\cos\theta}$; $\cot\theta = \frac{1}{\tan\theta}$
2. $\tan\theta = \frac{\sin\theta}{\cos\theta}$ (you have met this identity already!)
3. $\sin^2\theta + \cos^2\theta = 1$ where $\sin^2\theta$, is short hand for $(\sin\theta)^2$, etc.
4. $\tan^2\theta + 1 = \sec^2\theta$; $\cot^2\theta + 1 = \operatorname{cosec}^2\theta$
5. $\sin(A \pm B) = \sin A \cos B \pm \cos A \sin B$
6. $\cos(A \pm B) = \cos A \cos B \mp \sin A \sin B$
7. $\tan(A \pm B) = \frac{\tan A \pm \tan B}{1 \pm \tan A \tan B}$

Also from identities 4 to 7 above we get the *Doubles and squares* identities

8. $\sin 2A = 2 \sin A \cos A$
9. $\cos 2A = \cos^2 A - \sin^2 A = 2\cos^2 A - 1$
   $= 1 - 2\sin^2 A$
10. $\tan 2A = \frac{2 \tan A}{1 - \tan^2 A}$

Sums to products

11. $\sin A + \sin B = 2 \sin \frac{A+B}{2} \cos \frac{A-B}{2}$
12. $\sin A - \sin B = 2 \cos \frac{A+B}{2} \sin \frac{A-B}{2}$
13. $\cos A + \cos B = 2 \cos \frac{A+B}{2} \cos \frac{A-B}{2}$
14. $\cos A - \cos B = -2 \sin \frac{A+B}{2} \sin \frac{A-B}{2}$

Products to sums

15. $\sin A \cos B = \frac{1}{2}[\sin(A+B) + \sin(A-B)]$
16. $\cos A \sin B = \frac{1}{2}[\sin(A+B) - \sin(A-B)]$

17. $\cos A \cos B = \frac{1}{2}[\cos(A+B) + \cos(A-B)]$
18. $\sin A \sin B = \frac{1}{2}[\cos(A+B) - \cos(A-B)]$

All of the above identities take some time to become familiar with, which are tabulated above as a source of reference. You will only need to use them when simplification or a change of form of some trigonometric expression is necessary for further manipulation.

There follows one or two examples, illustrating the use of some of the above identities.

Example 3.20

Solve the following trigonometric equations:
(a) $4\sin^2\theta + 5\cos\theta = 5$
(b) $3\tan^2\theta + 5 = 7\sec\theta$

(a) The most difficult problem when manipulating identities is to know where to start! In this equation, we have two unknowns (sine and cosine) so the most logical approach is to try and get the equation in terms of *one* unknown, and this leads us to the use of an appropriate identity. We can in this case use one of the most important identities: $\sin^2\theta + \cos^2\theta = 1$ (identity 3 from above), from which $\sin^2\theta = 1 - \cos^2\theta$ and on substitution into Equation (a) gives:

$$4(1 - \cos^2\theta) + 5\cos\theta = 5 \quad \text{or}$$
$$-4\cos^2\theta + 5\cos\theta - 1 = 0$$

This is now a quadratic equation, which can be solved in a number of ways; the simplest being factorization!
Then,

$$(-4\cos\theta + 1)(\cos\theta - 1) = 0$$
$$\Rightarrow -4\cos\theta = -1 \quad \text{or} \quad \cos\theta = 1$$
$$\Rightarrow \cos\theta = \frac{1}{4} \quad \text{or} \quad \cos\theta = 1$$

so, $\boldsymbol{\theta = 75.5°}$ or $\boldsymbol{0°}$

(b) Proceeding in a similar manner to (a), we need a trigonometric identity which relates to: $\tan\theta$ and $\sec\theta$ (look at identity 4).
Then: $3\tan^2\theta + 5 = 7\sec\theta$ and using $\sec^2\theta = 1 + \tan^2\theta$ or $\tan^2\theta = \sec^2\theta - 1$,

We get:

$$3(\sec^2\theta - 1) + 5 = 7\sec\theta$$

$$\Rightarrow 3\sec^2\theta - 3 + 5 = 7\sec\theta$$

$$\Rightarrow 3\sec^2\theta - 7\sec\theta + 2 = 0$$

(again a quadratic equation) factorizing gives:

$$(3\sec\theta - 1)(\sec\theta - 2) = 0$$

$$\Rightarrow 3\sec\theta = 1 \quad \text{or} \quad \sec\theta = 2$$

$$\left(\text{remembering that } \sec\theta = \frac{1}{\cos\theta}\right)$$

Then: $\sec\theta = \dfrac{1}{3}$ or $\sec\theta = 2$

so $\cos\theta = 3$ or $\cos\theta = \dfrac{1}{2}$

Now, $\cos\theta = 3$ (is not permissible), so there is only one solution, $\cos\theta = 0.5$ so $\theta = 60°$.

The following example shows one or two techniques that may be used to verify trigonometric identities involving, double angle and sums to products.

Example 3.21

Verify the following identities by showing that each side of the equation is equal in all respects:
(a) $(\sin\theta + \cos\theta)^2 \equiv 1 + \sin 2\theta$
(b) $\dfrac{\sin 3\theta - \sin\theta}{\cos\theta - \cos 3\theta} \equiv \cot 2\theta$

(a) Simply requires the left-hand side to be manipulated algebraically to equal the right-hand side. So multiplying out gives:

$$(\sin\theta + \cos\theta)^2 \equiv 1 + \sin 2\theta$$

$$\sin^2\theta + 2\sin\theta\cos\theta + \cos^2\theta \equiv$$

$$\sin^2\theta + \cos^2\theta + 2\sin\theta\cos\theta \equiv$$

(and from $\sin^2\theta + \cos^2\theta = 1$) then,

$$1 + 2\sin\theta\cos\theta \equiv$$

(and from identity 8, above)

$$1 + 2\sin\theta \equiv 1 + \sin 2\theta$$

(as required)

(b) Again considering left-hand side and using sums to products (identities 12 and 14) where:

$$\sin A - \sin B = 2\cos\frac{A+B}{2}\sin\frac{A-B}{2}$$

and $\cos A - \cos B = -2\sin\dfrac{A+B}{2}\sin\dfrac{A-B}{2}$

and $A > B$ (A greater than B) then

$$\sin 3\theta - \sin\theta = 2\cos\left(\frac{3+1}{2}\right)\theta\sin\left(\frac{3-1}{2}\right)\theta$$

$$= 2\cos 2\theta\sin\theta \quad \text{and}$$

$$\cos\theta - \cos 3\theta = -2\sin\left(\frac{1-3}{2}\right)\theta$$

$$\times \sin\left(\frac{1+3}{2}\right)\theta$$

$$\cos\theta - 3\cos\theta = -2\sin(-\theta)(\sin 2\theta)$$

and from the fact that, $\sin(-\theta) = -\sin\theta$ we get

$$\cos\theta - \cos 3\theta = 2\sin 2\theta\sin\theta$$

therefore: $\dfrac{\sin 3\theta - \sin\theta}{\cos\theta - \cos 3\theta} \equiv \dfrac{2\cos 2\theta\sin\theta}{2\sin 2\theta\sin\theta}$

$$\equiv \cot 2\theta$$

In the final example, you will see how trigonometric identities may be used to evaluate trigonometric ratios.

Example 3.22

If A is an acute angle and B is obtuse, where $\sin A = \frac{3}{5}$ and $\cos B = -\frac{5}{13}$, find the values of:
(a) $\sin(A + B)$
(b) $\tan(A + B)$

(a) $\sin(A + B) = \sin A\cos B + \cos A\sin B$ (1)
(from identity 5). In order to use this identity we need to find the values of the ratios for $\sin A$ and $\cos B$. So again we need to choose an identity that allows us to find $\sin\theta$ or $\cos\theta$, in terms of each other.
We know that $\sin^2 B + \cos^2 B = 1$; hence, $\sin^2 B = 1 - \cos^2 B$.   Therefore   inserting

values,

$$\sin^2 B = 1 - (-\tfrac{5}{13})^2 B \quad \text{then}$$
$$\sin B = 1 - \tfrac{25}{169} = \tfrac{144}{169}$$
$$\sin B = \tfrac{12}{13}$$

(Since B is obtuse, $90° < B < 180°$ and sign ratio is positive in second quadrant. Then only positive values of this ratio need to be considered.) Similarly:

$$\sin^2 A + \cos^2 A = 1 \text{ so}, \cos^2 A = 1 - \sin^2 A$$
$$= 1 - \frac{9}{25} \text{ and } \cos A = \frac{4}{5}$$

(Since angle A is $<90°$, i.e. acute, only the positive value is considered.) Now using Equation (1) above:

$$\sin(A + B) = \sin A \cos B + \cos A \sin B$$
$$= \left(\tfrac{3}{5}\right)\left(-\tfrac{5}{13}\right) + \left(\tfrac{4}{5}\right)\left(\tfrac{12}{13}\right)$$
$$= -\tfrac{15}{65} + \tfrac{48}{65} \quad \text{then}$$
$$\sin(A + B) = \frac{33}{56}$$

Note the use of fractions to keep exact ratios!

(b) For this part of the question we simply need to remember that:

$$\frac{\sin A}{\cos A} = \tan A$$

and use inequality (7) in a similar way as before.
Then

$$\tan A = \frac{\sin A}{\cos A} = \frac{\tfrac{3}{5}}{\tfrac{4}{5}} = \left(\tfrac{3}{5}\right)\left(\tfrac{5}{4}\right) = \tfrac{3}{4} \quad \text{and}$$

$$\tan B = \frac{\sin B}{\cos B} = \frac{\tfrac{12}{13}}{-\tfrac{5}{13}} = \left(\tfrac{12}{13}\right)\left(-\tfrac{13}{5}\right) = -\tfrac{12}{5}$$

and using identity 7,

$$\tan(A + B) = \frac{\tan A + \tan B}{1 - \tan A \tan B}$$
$$= \frac{\tfrac{3}{4} - \tfrac{12}{5}}{1 - \left(\tfrac{3}{4}\right)\left(-\tfrac{12}{5}\right)}$$

and after multiplying *every* term in the expression by 20 (the lowest common multiple), we get:

$$\tan(A + B) = \frac{15 - 48}{20 + 36} = -\frac{33}{56}$$

Test your understanding 3.4

1. Given that $\sin(\theta + \phi) = 0.6$ and $\cos(\theta + \phi) = 0.9$, find a value for μ when $\mu = \tan \phi$.

2. Verify the following identities:
   (a) $\tan 3\theta = \dfrac{\sin\theta + \sin 3\theta + \sin 5\theta}{\cos\theta + \cos 3\theta + \cos 5\theta}$
   (b) $\tan 2\theta = \dfrac{1}{1 - \tan\theta} - \dfrac{1}{1 + \tan\theta}$

3. Express the following as ratios of single angles:
   (a) $\sin 5\theta \cos\theta + \cos 5\theta \sin\theta$
   (b) $\cos 9t \cos 2t - \sin 9t \sin 2t$

This ends our short excursion into trigonometric identities and also our further study of trigonometry. In Section 3.3 we will introduce the elementary ideas of statistics.

3.3 Statistical methods

Your view of statistics has probably been formed from what you read in the papers, or what you see on the television. Results of surveys show: which political party is going to win the election, why men grow moustaches, if smoking damages your health, the average cost of housing by area, and all sorts of other interesting data! Well, statistics is used to analyse the results of such surveys and when used correctly, it attempts to eliminate the bias which often appears when collecting data on controversial issues.

Statistics is concerned with collecting, sorting and analysing numerical facts which originate from several observations. These facts are collated and summarized then presented as tables, charts or diagrams, etc.

In this brief introduction to statistics we look at two specific areas. First, we consider the collection and presentation of data in its various forms. Then we look at how we measure such data concentrating on finding *average values* and seeing how these average values may vary.

If you study statistics beyond this course, you will be introduced to the methods used to make predictions based on numerical data and the probability that your predictions are correct. However, at this stage we will only be considering the areas of data manipulation and measurement of central tendency (averages) mentioned above.

3.3.1 Data manipulation

In almost all scientific, engineering and business journals, newspapers and Government reports, statistical information is presented in the form of charts, tables and diagrams as mentioned above. We now look at a small selection of these presentation methods, including the necessary manipulation of the data, to produce them.

> **Key point**
>
> Statistics is concerned with collecting, sorting and analysing numerical facts.

Charts

Suppose, as the result of a survey, we are presented with the following statistical data:

Major category of employment	Number employed
Private business	750
Public business	900
Agriculture	200
Engineering	300
Transport	425
Manufacture	325
Leisure industry	700
Education	775
Health	500
Other	125

Now ignoring, for the moment, the accuracy of this data! let us look at the typical ways of presenting this information in the form of charts, in particular the *bar chart* and *pie chart*.

The bar chart

In its simplest form, the bar chart may be used to represent data by drawing individual bars (Figure 3.19) using the figures from the raw data (the data in the table).

Now the scale for the vertical axis, the number employed, is easily decided by considering the highest and lowest values in the table 900 and 125, respectively. Therefore we use a scale from 0 to 1000 employees. Along the horizontal

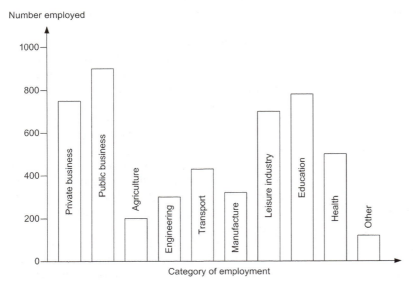

Figure 3.19 Bar chart representing number employed by category.

axis, we represent each category by a bar of even width. We could just as easily have chosen to represent the data using column widths, instead of column heights.

Now the simple bar chart above tells us very little that we could not have determined from the table. So another type of bar chart that enables us to make comparisons, the proportionate bar chart, may be used.

In this type of chart, we use *one bar*, with the same width throughout its height, with horizontal sections marked-off in proportion to the whole. In our example, each section would represent the number of people employed in each category, compared with the total number of people surveyed.

In order to draw a proportionate bar chart for our employment survey, we first need to total the number of people who took part in the survey, this total comes to 5000. Now, even with this type of chart we may represent the data either in proportion by height or in proportion by percentage. If we were to choose height then we need to set our vertical scale at some convenient height say, 10 cm. Then we would need to carry out 10 simple calculations to determine the height of each individual column.

For example, given the height of the total 10 cm represents 5000 people, then the height of the column for those employed in private business: $= \left(\frac{750}{5000}\right)10 = 1.5$ cm.

This type of calculation is then repeated for each category of employment. The resulting bar chart is shown in Figure 3.20.

Example 3.23

Draw a proportionate bar chart for the employment survey shown in the table using the percentage method.

For this method all that is required is to find the appropriate percentage of the total (5000) for each category of employment. Then, choosing a suitable height of column to represent 100% mark on the appropriate percentage for each of the 10 employment categories. To save space only the first five categories of employment have been calculated.

1. Private business $= \left(\frac{750}{5000}\right) \times 100 = 15\%$

2. Public business $= \left(\frac{900}{5000}\right) \times 100 = 18\%$

3. Agriculture $= \left(\frac{200}{5000}\right) \times 100 = 4\%$

4. Engineering $= \left(\frac{300}{5000}\right) \times 100 = 6\%$

5. Transport $= \left(\frac{425}{5000}\right) \times 100 = 8.5\%$

Similarly,

- Manufacture $= 6.5\%$,
- Leisure industry $= 14\%$,
- Education $= 15.5\%$,
- Health $= 10\%$,
- Other category $= 2.5\%$.

Figure 3.21 shows the completed bar chart.

Other categories of bar chart include, *horizontal bar charts*, where for instance Figure 3.19, is turned through 90° in a clockwise direction. One last type may be used to depict data given in chronological (time) order. Thus, e.g. the horizontal *x*-axis is used to represent hours, days, years, etc., while the vertical axis shows the variation of the data with time.

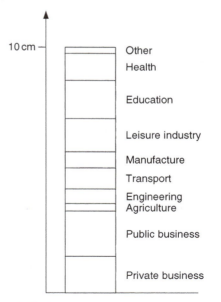

Figure 3.20 Proportionate bar chart graduated by height.

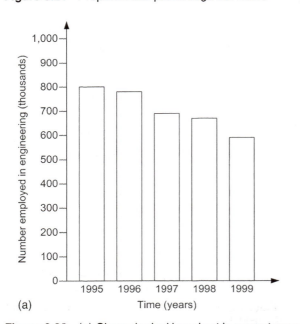

| Other (2.5%) |
| Health (10%) |
| Education (15.5%) |
| Leisure industry (14%) |
| Manufacture (6.5%) |
| Transport (8.5%) |
| Engineering (6%) |
| Agriculture (4%) |
| Public business (18%) |
| Private business (15%) |

Figure 3.21 Proportionate percentage bar chart.

Example 3.24

Represent the following data on a chronological bar chart.

Year	Number employed in general engineering (thousands)
1995	800
1996	785
1997	690
1998	670
1999	590

Since we have not been asked to represent the data on any *specific bar chart* we will use the simplest, involving only the raw data. Then, the only concern is the *scale* we should use for the vertical axis. To present a *true* representation, the scale should start from zero and extend to say, 800 (Figure 3.22(a)). If we wish to emphasize a *trend*, which is the way the variable is rising or falling with time we could use a very much exaggerated scale (Figure 3.22(b)). This immediately emphasizes the *downward trend* since 1995.

Note that this data is *fictitious* (made-up) and used here, merely for emphasis!

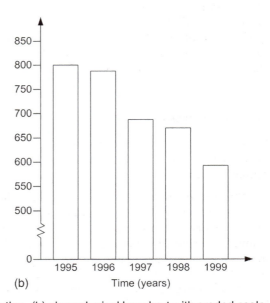

Figure 3.22 (a) Chronological bar chart in correct proportion, (b) chronological bar chart with graded scale.

Pie chart

In this type of chart the data is presented as a proportion of the total using the angle or area of sectors. The method used to draw a pie chart is best illustrated by the following example.

Example 3.25

Represent the data given in Example 3.2.4 on a pie chart.

Then, remembering that there are 360° in a circle and that the total number employed in general engineering (according to our figures) was:

$$800 + 785 + 690 + 670 + 590 = 3535$$

(thousand)

Then, we manipulate the data as follows:

Year	Number employed in general engineering (thousands)	Sector angle
1995	800	$\left(\dfrac{800}{3535}\right) \times 360 = 81.5$
1996	785	$\left(\dfrac{785}{3535}\right) \times 360 = 80$
1997	690	$\left(\dfrac{690}{3535}\right) \times 360 = 70.3$
1998	670	$\left(\dfrac{670}{3535}\right) \times 360 = 68.2$
1999	590	$\left(\dfrac{590}{3535}\right) \times 360 = 60$
Total	3535	360

The resulting pie chart is shown in Figure 3.23.

Other methods of visual presentation include pictograms and ideographs. These are diagrams in pictorial form used to present information to those who have a limited interest in the subject matter or who do not wish to deal with data presented in numerical form. They have

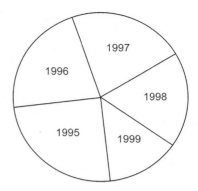

Figure 3.23 Pie chart for Example 3.25 employment in engineering by year.

little or no practical use when interpreting engineering or other scientific data and apart from acknowledging their existence, we will not be pursuing them further.

Key point

Charts and graphs offer an effective visual stimulus for the presentation of statistical data.

Frequency distributions

One of the most common and most important ways of organizing and presenting raw data is through use of *frequency distributions*.

Consider the data given below, which shows the time in hours that it took 50 individual workers to complete a specific assembly line task.

Data for assembly line task

```
1.1  1.0  0.6  1.1  0.9  1.1  0.8  0.9  1.2  0.7
1.0  1.5  0.9  1.4  1.0  0.9  1.1  1.0  1.0  1.1
0.8  0.9  1.2  0.7  0.6  1.2  0.9  0.8  0.7  1.0
1.0  1.2  1.0  1.0  1.1  1.4  0.7  1.1  0.9  0.9
0.8  1.1  1.0  1.0  1.3  0.5  0.8  1.3  1.3  0.8
```

From the data you should be able to see that the shortest time for completion of the task was 0.5 h and the longest time was 1.5 h. The *frequency* of appearance of these values is once. On the other hand, the number of times the job took 1 h appears 11 times, or it has a *frequency of 11*. Trying to sort out the data in this *ad hoc* manner is time consuming and may lead to mistakes.

To assist with the task we use a *tally chart*. This chart simply shows how many times the *event* of completing the task in a specific time takes place. To record the *frequency of events* we use the number 1, in a tally chart and when the frequency of the event reaches 5, we score through the existing four 1 s to show a frequency of 5. The following example illustrates the procedure.

Example 3.26

Use a tally chart to determine the frequency of events, for the data given above on the assembly line task.

Time (hours)	Tally	Frequency
0.5	1	1
0.6	11	2
0.7	1111	4
0.8	1111 1	6
0.9	1111 111	8
1.0	1111 1111 1	11
1.1	1111 111	8
1.2	1111	4
1.3	111	3
1.4	11	2
1.5	1	1
Total		50

We now have a full numerical representation of the *frequency of events*. For example, eight people completed the assembly task in 1.1 h or the time **1.1** h has a frequency of **9**. We will be using the above information later on, when we consider measures of central tendency.

The time in hours given in the above data are simply numbers. When data appears in a form where it can be *individually counted* we say that it is *discrete* data. It goes up or down in *countable* steps. Thus the numbers 1.2, 3.4, 8.6, 9, 11.1, 13.0, are said to be *discrete*. If, however, data is obtained by measurement, e.g.

the heights of a group of people. Then we say that this data is *continuous*. When dealing with continuous data, we tend to quote its limits, i.e. the limit of accuracy with which we take the measurements. For example, a person may be 174 ± 0.5 cm in height. When dealing numerically, with continuous data, or a large amount of discrete data, it is often useful to *group* this data into *classes or categories*. We can then find out the numbers (frequency) of items within each group.

The following Table shows the height of 200 adults grouped into 10 classes.

Table showing height of adults

Height (cm)	Frequency
150–154	4
155–159	9
160–164	15
165–169	21
170–174	32
175–179	45
180–184	41
185–189	22
190–194	9
195–199	2
Total	200

The main advantage of grouping is that it produces a clear overall picture of the frequency distribution. In the table, the first class interval is 150–154. The end number 150 is known as the *lower limit* of the class interval the number 154 is the *upper limit*. The heights have been measured to the nearest centimetre. That means within ± 0.5 cm. Therefore, in effect, the first class interval includes all heights between the range 149.5–154.5 cm, these numbers are known as the lower and upper class *boundaries*, respectively. The *class width* is always taken as the *difference between the lower and upper class boundaries*, not the upper and lower limits of the class interval.

Key point

The grouping of frequency distributions is a means for clearer presentation of the facts.

The histogram

The histogram is a special diagram that is used to represent a frequency distribution, such as that for grouped heights, shown above. It consists of a set of rectangles, whose areas represent the frequencies of the various classes. Often when producing these diagrams, the class width is kept the same. So that the varying frequencies are represented by the height of each rectangle. When drawing histograms for grouped data, the *midpoints* of the rectangles represent the midpoints of the class intervals. Hence for our data, they will be: 152, 157, 162, 167, etc.

An adaptation of the histogram, known as the *frequency polygon*, may also be used to represent a frequency distribution.

Example 3.27

Represent the above data showing the frequency of the height of groups of adults on a histogram and draw in the frequency polygon for this distribution.

All that is required to produce the histogram is to plot frequency against the height intervals, where the intervals are drawn as class widths.

Then, as can been seen from Figure 3.24, the area of each part of the histogram is the *product of frequency × class width*. The frequency polygon is drawn so that it connects the *midpoint* of the class widths.

> **Key point**
>
> The frequencies of a distribution may be added consecutively to produce a graph known as a cumulative frequency distribution or ogive.

Test your understanding 3.5

1. In a particular university, the number of students enrolled by faculty is given in the following table.

Faculty	Number of students
Business and administration	1950
Humanities and social science	2820
Physical and life sciences	1050
Technology	850
Total	6670

   Illustrate this data on both a *bar chart* and *pie chart*.

2. For the group of numbers given below produce a tally chart and determine their frequency of occurrence.

36	41	42	38	39	40	42	41	37	40
42	44	43	41	40	38	39	39	43	39
36	37	42	38	39	42	35	42	38	39
40	41	42	37	38	39	44	45	37	40

3. Given the following frequency distribution produce a histogram, and on it draw the frequency polygon.

Class interval	Frequency (f)
60–64	4
65–69	11
70–74	18
75–79	16
80–84	7
85–90	4

3.3.2 Statistical measurement

When considering statistical data it is often convenient to have one or two values which represent the data as a whole. Average values are often used. For example, we might talk about the average height of females in the UK being 170 cm, or that the average shoe size of British males is size 9. In statistics we may represent these average values using the mean, median or mode of the data we are considering.

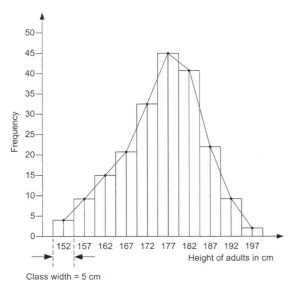

Figure 3.24 Example 3.27 histogram showing frequency distribution.

If we again consider the hypothetical data on the height of females, we may also wish to know how their individual heights vary or deviate from their average value. Thus, we need to consider measures of *dispersion*, in particular, *mean deviation, standard deviation* and *variance* for the data concerned. These *statistical averages* and the way they *vary* are considered next.

The arithmetic mean

The arithmetic mean (AM), or simply the *mean*, is probably the average with which you are most familiar. For example, to find the arithmetic mean of the numbers: 8, 7, 9, 10, 5, 6, 12, 9, 6, 8. All we need to do is to add them all up and divide by how many there are, or more formally:

$$AM = \frac{\text{Arithmetic total of all the individual values}}{\text{Number of values}}$$

$$= \frac{\sum n}{n}$$

where the greek symbol $\sum$ = the sum of the individual values, $x_1 + x_2 + x_3 + x_4 + \cdots + x_n$ and n = the number of these values, in the data.

So, for the *mean* of our 10 numbers, we have:

$$\text{Mean} = \frac{\sum n}{n} = \frac{\begin{array}{c}8 + 7 + 9 + 10 + 5 + 6 \\ + 12 + 9 + 6 + 8\end{array}}{10}$$

$$= \frac{80}{10} = 8$$

Now, no matter how long or complex the data we are dealing with, *provided* that we are only dealing with individual values (discreet data), the above method will always produce the arithmetic mean. The mean of all the *x-values* is given the symbol $\bar{x}$, pronounced, *x-bar*.

Example 3.28

The height of 11 females were measured as follows: 165.6, 171.5, 159.4, 163, 167.5, 181.4, 172.5, 179.6, 162.3, 168.2 and 157.3 cm. Find the mean height of these females.

Then, for $n = 11$:

$$\bar{x} = \frac{\begin{array}{c}165.6 + 171.5 + 159.4 + 163 \\ + 167.5 + 181.4 + 172.5 + 179.6 \\ + 162.3 + 168.2 + 157.3\end{array}}{11}$$

$$\bar{x} = \frac{1848.3}{11} = 168.03 \, \text{cm}$$

Mean for grouped data

What if we are required to find the mean for *grouped data*? Look back at the table on page 136 showing the height of 200 adults, grouped into 10 classes. In this case, the *frequency* of the heights needs to be taken into account.

We select the *class midpoint x* as being the average of that class and then multiply this value by the frequency (f) of the class, so that a value for that particular class is obtained (fx). Then, by adding up all class values in the frequency distribution, the total value for the distribution is obtained $(\sum fx)$. This total is then divided by the *sum of the frequencies* $(\sum f)$ in order to determine the mean. So, for grouped data:

$$\bar{x} = \frac{f_1 x_1 + f_2 x_2 + f_3 x_3 + \cdots + f_n x_n}{f_1 + f_2 + f_3 + \cdots + f_n}$$

$$= \frac{\sum(f \times \text{midpoint})}{\sum f}$$

This rather complicated looking procedure is best illustrated by the following example.

Example 3.29

Determine the mean value for the heights of the 200 adults using the data in Table.

The values for each individual class are best found by producing a table using the class *midpoints* and *frequencies*. Remembering that the *class midpoint* is found by dividing the *sum of the upper and lower class boundaries by 2*. For example, the mean value for the first class interval is $\dfrac{149.5 + 154.5}{2} = 152$. The completed Table is shown below.

Midpoint (x) of height (cm)	Frequency (f)	fx
152	4	608
157	9	1413
162	15	2430
167	21	3507
172	32	5504
177	45	7965
182	41	7462
187	22	4114
192	9	1728
197	2	394
Total	$\sum f = 200$	$\sum fx = 35{,}125$

I hope you can see how each of the values was obtained. When dealing with relatively large numbers, be careful with your arithmetic; especially, when you are keying in variables into your calculator!

Now that we have the required total the mean value of the distribution can be found:

$$\text{Mean value } \bar{x} = \frac{\sum fx}{\sum f} = \frac{35{,}125}{200}$$
$$= 175.625 \pm 0.5\,\text{cm}$$

Notice that our mean value of heights has the same margin of error as the original measurements. The value of the mean cannot be any more accurate than the measured data from which it was found!

Median

When some values within a set of data vary quite widely, the arithmetic mean gives a rather poor representative average of such data. Under these circumstances, another more useful measure of the average is the *median*.

For example, the mean value of the numbers: 3, 2, 6, 5, 4, 93, 7, is 20, which is not representative of any of the numbers given. To find the median value of the same set of numbers, we simply place them in *rank order*, i.e. 2, 3, 4, 5, 6, 7, 93. Then we select the middle (median)

value. Since there are seven numbers (items) we choose the fourth item along, the number 5, as our *median value*.

If the number of items in the set of values is *even* then we add together the value of the *two middle terms* and divide by 2.

<hr/>

Example 3.30

Find the mean *and* median value for the set of numbers: 9, 7, 8, 7, 12, 70, 68, 6, 5, 8.

The arithmetic mean is found as:

$$\text{Mean } \bar{x} = \frac{\begin{array}{c}9 + 7 + 8 + 7 + 12 + 70\\ + 68 + 6 + 5 + 8\end{array}}{10}$$
$$= \frac{200}{10} = 20$$

This value is not really representative of any of the numbers in the set.

To find the *median* value we first put the numbers in *rank order*, i.e:

$$5, 6, 7, 7, 8, 8, 9, 12, 68, 70$$

Then from the 10 numbers, the two middle values, the fifth and sixth values along are 8 and 8. So the *median value* $= \dfrac{8 + 8}{2} = 8$.

Mode

Yet another measure of central tendency for data containing extreme values is the *mode*. Now the *mode* of a set of values containing discreet data is the value that occurs most often. So for the set of values: 4, 4, 4, 5, 5, 5, 5, 6, 6, 6, 7, 7, 7, the *mode* or *modal value* is 5 as this value occurs four times. Now it is possible for a set of data to have more than one mode, e.g. the data used in Example 3.30 above has two modes, 7 and 8, both of these numbers occurring twice and both occurring more than any of the others. A set of data may not have a modal value at all, e.g. the numbers: 2, 3, 4, 5, 6, 7, 8, all occur once and there is no mode.

A set of data that has one mode is called *uni-modal*, data with two modes is *bimodal* and data with more than two modes is known as *multimodal*.

When considering *frequency distributions* for grouped data the *modal class* is that group which occurs most frequently. If we wish to find the actual *modal value* of a frequency distribution, we need to draw a histogram.

Example 3.31

Find the modal class and modal value for the frequency distribution on the height of adults given in the Table on page 136.

Referring back to the table it is easy to see that the class of heights, which occurs most frequently is 175–179 cm, which occurs 45 times.

Now, to find the modal value we need to produce a histogram for the data. We did this for Example 3.27. This histogram is shown again here with the modal shown.

From Figure 3.25 it can be seen that the modal value $= 178.25 \pm 0.5$ cm.

This value is obtained from the intersection of the two construction lines, *AB* and *CD*. The line *AB* is drawn diagonally from the highest value of the preceding class, up to the top right-hand corner of the modal class. The line *CD* is drawn from the top left-hand corner of the modal group to the lowest value of the next class immediately above the modal group. Then, as

can be seen, the *modal value* is read-off where the projection line meets the *x-axis*.

Key point

The mean, median and mode are statistical averages, or measures of central tendency for a statistical distribution.

Mean deviation

We talked earlier of the need not only to consider statistical averages which give us some idea of the position of a distribution, but also the need to consider how the data is *dispersed* or *spread* about this average value. Figure 3.26 illustrates this idea showing how the data taken from two distributions is dispersed about the same mean value.

A measure of dispersion which is often used is the *mean deviation*. To determine the deviation from the statistical average (mean, median or mode) we proceed in the following way.

We first find the statistical average for the distribution, the mean, median or mode ($\bar{x}$). We then find the difference between this average value and each of the individual values in the distribution. We then add up all these difference and divide by the number of individual values in the distribution. This all sounds rather complicated, but the mean deviation may be calculated

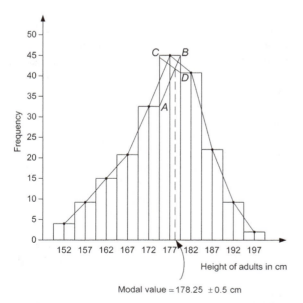

Modal value $\approx 178.25 \pm 0.5$ cm

Figure 3.25 Histogram showing frequency distribution and modal value for height of adults.

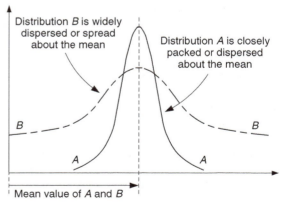

Figure 3.26 Deviation from the mean value for a distribution.

quite easily using the formula:

$$Mean\ deviation = \frac{\sum |x - \bar{x}|}{n}$$

where x = a data value in the distribution, $\bar{x}$ = the statistical average, mean, median or mode, as before and n = the number of individual items in the distribution as before. The || brackets tell us to use the *positive* value of the result contained within the brackets. For example, if $x = 12$ and $\bar{x} = 16$, then $|x - \bar{x}| = |12 - 16| = |-4| = +4$; even though we use the *positive* value in this case, the result was negative.

For *frequency distributions*, using grouped data, we find the deviation from the mean, using a similar formula to that we used to find the arithmetic mean. Where the only addition is to multiply the individual differences from the mean by their frequency. Then, for a frequency distribution:

$$Mean\ deviation = \frac{\sum f |x - \bar{x}|}{\sum f}$$

Example 3.32

Calculate the mean deviation from the arithmetic mean for the data shown in the following table.

Length of rivet (mm)	Frequency
9.8	3
9.9	18
9.95	36
10.0	62
10.05	56
10.1	20
10.2	5

The easiest way to tackle this problem is to set up a *table of values* in a similar manner to the table we produced for Example 3.29. The headings for such a table being taken from the above formula for finding the mean deviation for a frequency distribution.

Table of values

Rivet length (x)	f	fx	$|x - \bar{x}|$	$f |x - \bar{x}|$
9.8	3	29.4	0.208	0.624
9.9	18	178.2	0.108	1.944
9.95	36	358.2	0.058	2.088
10.0	62	620.0	0.008	0.496
10.05	56	562.8	0.042	2.352
10.1	20	202	0.092	1.84
10.2	5	51	0.192	0.96
Total	$\sum f =$ 200	$\sum fx =$ 2001.6		$\sum f|x - \bar{x}| =$ 10.304

Arithmetic mean $\bar{x} = \dfrac{\sum fx}{\sum f} = \dfrac{2001.6}{200} = 10.008$

$\bar{x} = 10.008$ was required to complete the last two columns in the table.

Then the *mean deviation* from the mean of the rivet lengths is:

$$= \frac{\sum f |x - \bar{x}|}{\sum f} = \frac{10.304}{200} = 0.05152\ \text{mm}$$

$$\simeq 0.05\ \text{mm}$$

This small average deviation from the arithmetic mean for rivet length is what we would expect in this case. The deviation being due to very small manufacturing errors. This is, therefore, an example of a frequency distribution tightly packed around the average for the distribution.

Key point

The mean deviation is a measure of the way a distribution deviates from its average value.

Standard deviation

The most important method in determining how a distribution is *dispersed* or *spread* around its average value is known as *standard deviation*. To find this measure of dispersion requires just one or two additional steps from those we used to find the mean deviation.

These additional mathematical steps involve further manipulation of the $|x - \bar{x}|$ or $f |x - \bar{x}|$

values, we needed to find when calculating the mean deviation for discreet or grouped data. The additional steps require us to first *square* these differences, then find *their* mean and finally take their *square root* to reverse the squaring process. This strange way of manipulating these differences is known as the *root mean square deviation or standard deviation*, which is identified using the Greek symbol sigma (σ).

Thus, for frequency distributions with grouped data we can represent these *three further processes* mathematically, as follows:

1. Square the differences and multiply by their frequency $= f\,|x - \bar{x}|^2$.
2. Sum all of these values and find their mean $= \dfrac{\sum f |x - \bar{x}|^2}{\sum f}$ this is a similar step to the way in which we found the mean deviation. The value of the deviation found at this stage is known as the *variance*.
3. Now take the square root of these mean squares to reverse the squaring process $= \sqrt{\dfrac{\sum f |x - \bar{x}|^2}{\sum f}}$

Then the **standard deviation**,

$$\sigma = \sqrt{\frac{\sum f (x - \bar{x})^2}{\sum f}}$$

The || brackets have been replaced by ordinary brackets in this final version of the formula. This is because when we square any quantity whether positive or negative the result is always positive by the law of signs! It is, therefore, no longer necessary to use the special brackets.

This particular value of deviation is more representative than the mean deviation value, we found before, because it takes account of data that may have large differences between items, in a similar way to the use of the mode and median, when finding average values.

When considering discreet ungrouped date, we apply the same steps as above to the differences $|x - \bar{x}|$ and obtain,

$$\sqrt{\frac{\sum (x - \bar{x})^2}{n}}$$

therefore for *ungrouped* data, the standard deviation:

$$\sigma = \sqrt{\frac{\sum (x - \bar{x})^2}{n}}$$

Note that once again we have removed the special brackets for the same reason as given above for grouped data.

> **Key point**
>
> The standard deviation as a measure of deviation from the statistical average takes into account data with extreme values; that is data that it statistically skewed.

Example 3.33

For the set of numbers 8, 12, 11, 9, 16, 14, 12, 13, 10, 9, find the arithmetic mean and the standard deviation.

Like most of the examples concerning central tendency and deviation measure, we will solve this problem by setting up a table of values. We will also need to find the arithmetic mean before we are able to complete the table, where in this case for non-grouped data $n = 10$.

Then,

$$\bar{x} = \frac{\sum x}{n} = \frac{\begin{array}{c} 8 + 12 + 11 + 9 + 16 + 14 \\ + 12 + 13 + 10 + 9 \end{array}}{10}$$

$$= \frac{114}{10} = 11.4$$

Table of values

x	$(x - \bar{x})$	$(x - \bar{x})^2$
8	−3.4	11.56
12	0.6	0.36
11	−0.4	0.16
9	−2.4	5.76
16	4.6	21.16
14	2.6	6.76
12	0.6	0.36
13	1.6	2.56
10	−1.4	1.96
9	−2.4	5.76
$\sum x = 114$		$\sum (x - \bar{x})^2 = 56.4$

Then from the table of values, the *standard deviation*:

$$\sigma = \sqrt{\frac{\sum (x - \bar{x})^2}{n}}$$

$$= \sqrt{\frac{56.4}{10}}$$

$$= \sqrt{5.64} = 2.375$$

Another measure of dispersion the *variance* is simply the value of the standard deviation before taking the square root, so in this example:

$$\text{Variance} = \frac{\sum (x - \bar{x})^2}{n}$$

$$= \frac{56.4}{10} = 5.64$$

So when finding the standard deviation, you can also find the variance.

Finally, make sure you can obtain the values given in the table!

We finish our short study of standard deviation with one more example for grouped data.

Example 3.34

Calculate the standard deviation for the data on rivets given in Example 3.32.

For convenience the data from Example 4.70 is reproduced here.

Length of rivet (mm)	Frequency
9.8	3
9.9	18
9.95	36
10.0	62
10.05	56
10.1	20
10.2	5

Now in Example 3.32 we calculated the arithmetic mean and *mean deviation*, using a table of values, we obtained:

Rivet length (x)	f	fx	$\lvert x - \bar{x} \rvert$	$f \lvert x - \bar{x} \rvert$
9.8	3	29.4	0.208	0.624
9.9	18	178.2	0.108	1.944
9.95	36	358.2	0.058	2.088
10.0	62	620	0.008	0.496
10.05	56	562.8	0.042	2.352
10.1	20	202	0.092	1.84
10.2	5	51	0.192	0.96
Total	$\sum f = 200$	$\sum fx = 2001.6$		$\sum f \lvert x - \bar{x} \rvert = 10.304$

The arithmetic mean we found as

$$\bar{x} = \frac{\sum fx}{\sum f} = \frac{2001.6}{200} = 10.008$$

So, having found the mean, all we need to do now is to find the *standard deviation* modify the table by adding-in the extra steps. We then obtain:

Rivet length (x)	f	fx	$(x - \bar{x})$	$(x - \bar{x})^2$	$f(x - \bar{x})^2$
9.8	3	29.4	−0.208	0.043264	0.129792
9.9	18	178.2	−0.108	0.011664	0.209952
9.95	36	358.2	−0.058	0.003364	0.121104
10.0	62	620	−0.008	0.000064	0.003968
10.05	56	562.8	0.042	0.001764	0.098784
10.1	20	202	0.092	0.008464	0.16928
10.2	5	51	0.192	0.036864	0.18432
Total	200	2001.6			0.9172

Then from the table

$$\sum f = 200 \quad \sum f(x - \bar{x})^2 = 0.9172$$

and standard deviation

$$\sigma = \sqrt{\frac{\sum f (x - \bar{x})^2}{\sum f}} = \sqrt{\frac{0.9172}{200}}$$

$$= 0.067 \, \text{mm}$$

This value is slightly more accurate than the value we found in Example 4.70, for the mean deviation $= 0.05$ mm but as you can see, there is also a lot more arithmetic manipulation! Again, you should make sure that you are able to obtain the additional values shown in the table.

Test your understanding 3.6

1. Calculate the mean of the numbers, 176.5, 98.6, 112.4, 189.8, 95.9 and 88.8.

2. Determine the mean, median and mode for the set of numbers, 9, 8, 7, 27, 16, 3, 1, 9, 4 and 116.

3. Estimates for the length of wood required for a shelf, were as follows:

Length (cm)	35	36	37	38	39	40	41	42
Frequency	1	3	4	8	6	5	3	2

Calculate the arithmetic mean and mean deviation.

4. Calculate the arithmetic mean and the mean deviation for the data shown in the following table.

Length (mm)	167	168	169	170	171
Frequency	2	7	20	8	3

5. Calculate the standard deviation from the median value, for the numbers given in Question 2.

6. Tests were carried out on 50 occasions to determine the percentage of green house gases in the emissions from an internal combustion engine. The results from the tests showing the percentage of greenhouse gas recorded, were as follows:

% Green house gas present	3.2	3.3	3.4	3.5	3.6	3.7
Frequency	2	12	20	8	6	2

Determine the arithmetic mean and the standard deviation for the greenhouse gases present.

3.4 Calculus

3.4.1 Introduction

Meeting the calculus for the first time is often a rather daunting business. In order to appreciate the power of this branch of mathematics we must first attempt to define it. So, what is the calculus and what is its function?

Imagine driving a car or riding a motorcycle starting from rest to over a measured distance, say 1 km. If your time for the run was 25 s, then we can find your average speed over the measured kilometre from the fact that speed $=$ distance/time. Then using consistent units your average speed would be 1000 m/25 s or 40 ms^{-1}. This is fine but suppose you were testing the vehicle and we needed to know its *acceleration* after you had driven 500 m. In order to find this we would need to determine how the vehicle speed *was changing* at this exact point because *the rate at which your vehicle speed changes is its acceleration.* To find such things as rate of change of speed we can use *the calculus.*

The calculus is split into two major areas the *differential calculus* and the *integral calculus.*

The *differential calculus* is a branch of mathematics concerned with finding how things *change with respect to variables, such as time, distance or speed,* especially when these changes are *continually* varying. In engineering, we are interested in the study of motion and the way this motion in machines, mechanisms and vehicles, varies with time and the way in which pressure, density and temperature change with height or time. Also, how electrical quantities vary with time, such as electrical charge, alternating current, electrical power, etc. All these areas may be investigated using the *differential calculus.*

The *integral calculus* has two primary functions. The first function can be used to find the length of arcs, surface areas or volumes enclosed by a surface. Its second function is that of *antidifferentiation.* For example, if we use the differential calculus to find the rate of change of distance of our motorbike, with respect to time. In other words, we have found its instantaneous speed. We can then use the *inverse process,* the integral calculus to determine the original distance covered by the motorbike, from its instantaneous speed.

The mathematical process we use when applying the differential calculus is known as *differentiation* and when using the integral calculus, the mathematical process we apply is known as *integration.*

Before we can apply the calculus to meaningful engineering problems, we need first to understand the notation and ideas that underpin these applications. Thus at this level we spend the majority of our time looking at the basic arithmetic of the calculus, that will enable us to *differentiate* and *integrate* a very small number of mathematical functions. If you study further mathematics, you will gain sufficient knowledge to be able to apply the calculus to realistic engineering problems.

We will start our study with some introductory terminology and notation, which you will need in order to carry out *calculus arithmetic*.

> **Key point**
>
> The differential calculus is concerned with rates of change.

> **Key point**
>
> The integral calculus is antidifferentiation and is concerned with summing things.

Functions

When studying any new topic, you are going to be introduced to a range of new terms and definitions, unfortunately the calculus, is no exception! We have mentioned functions throughout your study of mathematics. It is now time to investigate the concept of *the function* in a little more detail before we consider differentiation and integration of such functions.

A *function is a many–one mapping or a one–one mapping*. An example of a one–one mapping (a function) is a car which has a unique licence plate number. All cars have a licence number, but for each vehicle, that number is unique, so we say that the licence plate is a function of the vehicle. There are many people who may have an intelligence quotient (IQ) of 120, this is an example of a many–one mapping or function. Many people will map to an IQ of 120. What about mathematical functions?

> **Key point**
>
> A function is a many–one or one–one mapping.

Consider the function $y = x^2 + x - 6$ this is a mathematical function because for any one value given to the *independent variable x*, we get a corresponding value for the *dependent variable y*. We say that *y is a function of x*. For example, when $x = 2$ then $y = (2)^2 + (2) - 6 = 0$. When dealing with mathematical functions we often represent them using $f(x)$ as the dependent variable, instead of y, i.e. $f(x) = x^2 + x - 6$, where the letter inside the bracket represents the independent variable. For example, $f(t) = t^2 + t - 6$ is the function f with respect to the independent variable t, which may, for example, represent time.

Now when we assign a value to the *independent variable*, this value is placed inside the bracket and the expression is then evaluated for the chosen value. So if $t = 3$, then we write $f(3) = (3)^2 + (3) - 6 = 6$ similarly, $f(-3) = (-3)^2 + (-3) - 6 = 9 - 3 - 6 = 0$. In fact any value of the independent variable may be substituted in this manner.

Example 3.35

If the distance travelled in metres by a slow moving earth vehicle is given by the function:

$$f(t) = \frac{t^2 + t}{2} + 50$$

Find the distance travelled by this vehicle at $t = 0$, $t = 2.4$ and $t = 5.35$ s. So this is a function that relates distance $f(t)$ and time (t) in seconds. Therefore, to find the dependent variable $f(t)$, all we need to do is substitute the time variable t into the function.

Then for $t = 0$ s,

$$f(0) = \frac{(0)^2 + (0)}{2} + 50 = 50 \, \text{m}$$

and similarly, when $t = 2.4$ s,

$$f(2.4) = \frac{(2.4)^2 + 2.4}{2} + 50 = 54.08 \, \text{m}$$

and similarly, when $t = 5.35$ s,

$$f(5.35) = \frac{(5.35)^2 + 5.35}{2} + 50 = 66.99 \, \text{m}$$

We can extend this idea a little further by considering *how the distance changes with time* for the function:

$$f(t) = \frac{t^2 + t}{2} + 50.$$

We will show graphically how the distance $f(t)$, for this quadratic function, varies with time t, between $t = 0$ and $t = 10$ s.

Example 3.36

1. Draw the graph for the function:

$$f(t) = \frac{t^2 + t}{2} + 50$$

   relating the distance $f(t)$ in metres to the time t in seconds between $t = 0$ and $t = 10$ s using intervals of 1.0 s.
2. From your graph find:
   (i) the distance at time $t = 6.5$ s,
   (ii) the time it takes to reach a distance of 90 m.
3. What does the *slope* of the graph indicates?

1. You have drawn graphs of quadratic functions when you studied your algebra. We will set up a table of values in the normal manner and then use these values to plot the graph:

t	0	1	2	3	4	5	6	7	8	9	10
t^2	0	1	4	9	16	25	36	49	64	81	100
$+t$	0	2	6	12	20	35	42	56	72	90	110
$\div 2$	0	1	3	6	10	15	21	28	36	45	55
$+50$	50	50	50	50	50	50	50	50	50	50	50
$f(t)$	50	51	53	56	60	65	71	78	86	95	105

2. Note that the graph is parabolic in shape, which is to be expected for a quadratic function (Figure 3.27).

   Then from the graph the distance at time 6.5 s is approximately = 74.5 m. The time it takes to reach 75 m approximately = 8.4 s.
3. Unfortunately, the *gradient or slope* of the graph *varies* (it is curved), but an *approximation to the gradient* can be found using a straight line which joins the points (0, 50) and (10, 105) as shown. Then from the graph it can be seen that:

$$\text{The gradient} = \frac{\text{Distance}}{\text{Time}} = \frac{55}{10} = 5.5 \text{ ms}^{-1},$$

which of course is speed.

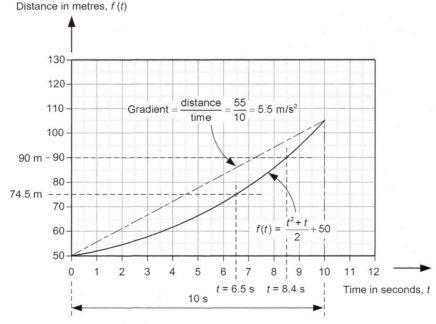

Figure 3.27 Graph of function $f(t) = \frac{t^2 + t}{2} + 50$.

In effect, what we have found is the average speed over the 10 s.

3.4.2 The differential calculus

The gradient of a curve and graphical differentiation

Now, suppose we wish to find the speed of the vehicle identified in Example 3.36, over a slightly shorter period of time, say between 1 and 9 s. Then we know that the speed is given by the *slope or gradient* of the graph at these points (Figure 3.28). This process is continued for time periods of 3–8 s and finally 3–4 s, the resultant speeds can be seen to be, 5.5, 6.2 and 5 ms^{-1}, respectively.

We could continue this process by taking smaller and smaller time periods so that eventually we would be able to find the *gradient or slope of a point* on the graph, In other words, we could find the *gradient at an instant in time*. Now you know from your study of the circle (page 93) that a *tangent line touches a circle (or curve) at just one point. Therefore finding the gradient of the slope of a curve at a point is equivalent to finding the gradient of a tangent line at that same point* (Figure 3.29).

In the case of our vehicle (Figure 3.28) it can be seen that the *gradient is in fact the speed*, so if we were able to find *the gradient of the tangent* at any instant in time we would be finding the *instantaneous* speed.

Now this process of trying to find the gradient at a point (the tangent) is long and tedious. However, it can be achieved very easily using the *differential calculus*, i.e. by *differentiating the function*.

Thus, in the case of our speed example, by finding the slope at a point (the slope of its tangent), we have in effect *graphically differentiated* the function, or found the way that *distance f(x) changes at any instant in time t*, the instantaneous speed!

This may all sound rather complicated, but by *applying certain rules*, we will be able to carry out the *differentiation process* and hence find out how functions change at any instant in time. However, there are a few things to learn before we get there!

> **Key point**
>
> To find the gradient of the tangent at a point of a function we differentiate the function.

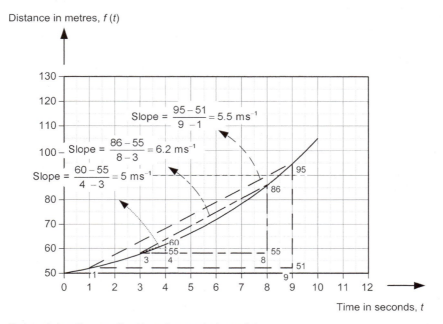

Figure 3.28 Determining the gradient to a tangent at a point.

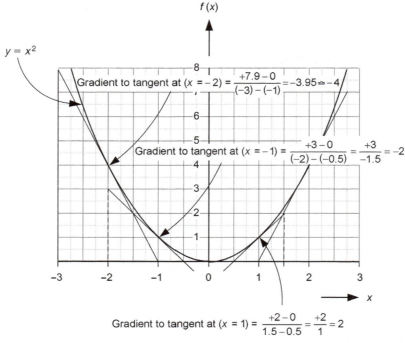

Figure 3.29   Finding the gradient of a curve at a point, for the graph $y = x^2$.

Key point

Finding the slope of a curve at a point is graphical differentiation.

Example 3.37

1. Draw the graph of the function $f(x) = x^2$ for values of x from $x = -3$ to $x = 3$.
2. Find the slope of the tangent lines drawn at $x = -1$, $x = 1$ and $x = -2$ and comment on your results.

1. The graph of the function $f(x) = x^2$ is shown in Figure 3.29. It can be seen that it is symmetrical about zero and is parabolic in shape.
2. From the graph it can be seen that the gradient to the tangent lines at the points $-1, 1$ and -2 are $-2, 2$ and -4, respectively. There seems to be a pattern in that at $x = -1$, the corresponding gradient $= -2$. So the gradient is twice as large as the independent variable x. This is also true for the gradients

at $x = 1$, and $x = -2$, which are again twice as large. This pattern is no coincidence, as you are about to see!

For the function $f(x) = x^2$ we have just shown that on three occasions using three different independent variables that the gradient of the slope of the tangent line is twice the value of the independent variable or more formally:

The gradient of the tangent at $f(x) = 2x$

The process of *finding the gradient of the tangent at a point* is known as *graphical differentiation*. What we have actually done is found the *differential coefficient* of the function $f(x) = x^2$. In other words we have found an algebraic expression of how this function varies, as we increase or decrease the value of the independent variable.

In functional notation the process of *finding the differential coefficient* of a function $f(x)$or *finding the slope to the tangent at a point*, or *finding the derived function*, is given the special symbol, $f'(x)$, read as, "f prime".

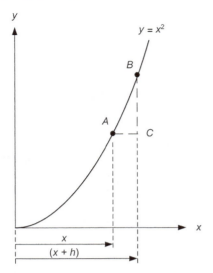

Figure 3.30 Finding the gradient of a curve or finding the derived function.

We can generalize the above procedure for finding the derived function. Consider again part of our function $y = x^2$ (Figure 3.30).

Suppose that A is the point on our curve $y = x^2$ with ordinate x and that B is another point on the curve with ordinate $(x + h)$. The y-ordinate of A is x^2 and the y-ordinate of B is $(x + h)^2$. Then:

$$BC = (x + h)^2 - x^2 = 2hx + h^2 \quad \text{and}$$
$$AC = (x + h) - x = h$$

Then the gradient of $AB = \dfrac{2hx + h^2}{h} = 2x + h$ (if h is not zero).

Now as h gets smaller and smaller (tends to 0) the gradient tends to (approaches) $2x$.

Therefore, as we found graphically, *the gradient of the tangent is* $2x$, or *the derived function is* $2x$.

There are also other ways of representing the *differential coefficient or derived function*, which we now consider.

Example 3.38

Find the derived function (gradient of the curve) for $y = 2x^2 - 2x - 6$ at the point (x, y).

We use the same procedure as before, identifying another point on the curve, say $(x + h, y + k)$.

Then find the slope of the line joining the two points and then we bring the two points closer and closer together until they coincide and become the tangent to the slope of the curve, in other words, the derived function. We proceed as follows:

For the two points on the x-ordinate the y-ordinates are:

$$y + k = 2(x + h)^2 - 2(x + h) - 6 \quad (1)$$
$$y = 2x^2 - 2x - 6 \quad (2)$$

Then expanding Equation (1) and simplifying, *using your algebra*, we get:

$$y + k = 2(x^2 + hx + hx + h^2) - 2x - 2h - 6$$
$$y + k = 2(x^2 + 2hx + h^2) - 2x - 2h - 6$$
$$y + k = 2x^2 + 4hx + 2h^2 - 2x - 2h - 6$$
or $y + k = 2x^2 - 2x - 6 + 4hx + 2h^2 - 2h \quad (1a)$

Now subtracting Equation (2) from (1a) gives,

$$k = 4hx + 2h^2 - 2h$$

therefore on division by h, the gradient of the chord $\frac{k}{h} = 4x + 2h - 2$ and $\frac{k}{h}$ tends to $(4x - 2)$ as h tends to 0.

Thus, $4x - 2$ *is the derived function of* $y = 2x^2 - 2x - 6$, *which is also the gradient of this function at the point* (x, y). For example, at the point $(3, -3)$ the gradient is $[4(3) - 2] = 10$.

You will be pleased to know that we do not need to repeat this rather complicated method of finding the derived function (the tangent to a point of the slope). As you will soon see, *all derived functions* of simple algebraic expressions (polynomials) can be found using a simple rule!

Before we look at this rule we need to consider the different ways in which the derived function can be expressed.

Notation for the derivative

There are several ways in which we can represent and describe *the differential coefficient or derived function*. Below are listed some of the most common methods used to describe the derived function that you will find in text books and literature dealing with the *differential calculus*.

These differing terms for finding *the derived function* include:

- find the derived function of ...
- find the derivative of ...
- find the differential coefficient for ...
- differentiate ...
- find the rate of change of ...
- find the tangent to the function ...
- find the gradient of a function at a point ...

This differing terminology is often confusing to beginners. It is further complicated by the fact that *different symbols* are used for *the differentiation process* (finding the derived function) based on the convention chosen.

We have been dealing with functional notation, where for a function $f(x)$ *the first derivative or first derived function* is given as $f'(x)$. If we were to carry out the *differentiation process again on the first derived function* then we say we have found the *second* derivative $f''(x)$ and so on.

We have used functional notation merely to introduce the idea of the *mathematical function*. We look next at the more common *Leibniz notation*, which we will use from now on throughout the remainder of this book.

In *Leibniz notation*, the mathematical function is represented conventionally as $y(x)$ and its derived function or differential coefficient is represented as $\frac{dy}{dx}$. This expression for the derived function can be thought of as finding the slope to the tangent of a point of a particular function, where we take a smaller and smaller bit of x (dx) and divide it into a smaller and smaller bit of y (dy) until we get the slope of a point $\frac{dy}{dx}$.

So, in **Leibniz notation**, for the function $y = x^2$ the differential coefficient is represented as $\frac{dy}{dx} = 2x$, which we found earlier.

The second derivative in Leibniz notation is represented as $\frac{d^2y}{dx^2}$, the third derivative is $\frac{d^3y}{dx^3}$ and so on.

One other complication arises with all notations, in that *the notation differs according to the variable being used*! For example, if our mathematical function is $s(t)$, then in Leibniz notation, its first derivative would be $\frac{ds}{dt}$, as we are differentiating the variable s with respect to t. In the same way as with $\frac{dy}{dx}$ we are differentiating the variable y with respect to x. So in general, $\frac{dy}{dx}, \frac{ds}{dt}, \frac{du}{dv}$, represent the first derivative of the functions, y, s and u, respectively.

One final type of notation which is often used in mechanics is *dot* notation. For example, where $\dot{v}, \ddot{s}$, etc. means that the function is differentiated once ($\dot{v}$) or twice ($\ddot{s}$) and so on. This notation will not be used in this book, but you may meet it in your further studies.

So much for all the hard theory, we are now going to *use one or two rules* to carry out the differentiation process which, once mastered, is really quite simple!

Key point

In Leibniz notation $\frac{dy}{dx}$ means that we find the first derivative of the function y with respect to x.

Key point

In functional notation $f'(x)$ and $f''(x)$ are the first and second derivatives of the function f, respectively.

Differentiation

As you will be aware by now *the word differentiate* is one of many ways of saying that we wish to *find the derived function*. Again, going back to the simple function $y = x^2$, when we *differentiated* this function, we found that its *derived function* was $\frac{dy}{dx} = 2x$. In a similar manner when we carried out the differentiation process on the function $y = 2x^2 - 2x - 6$ we obtained, $\frac{dy}{dx} = 4x - 2$.

If we were to carry out the rather complex process we used earlier on the following functions, $y = 3x^2$, $y = x^3$ and $y = x^3 + 3x^2 - 2$, we would obtain, $\frac{dy}{dx} = 6x$, $\frac{dy}{dx} = 3x^2$ and $\frac{dy}{dx} = 3x^2 + 6x$, respectively. I wonder if you can see a pattern in these results? They are grouped below for your convenience, can you spot a pattern?

$$y = x^2 \qquad \frac{dy}{dx} = 2x$$

$$y = 3x^2 \qquad \frac{dy}{dx} = 6x$$

$$y = 2x^2 - 2x - 6 \qquad \frac{dy}{dx} = 4x - 2$$

$$y = x^3 \qquad \frac{dy}{dx} = 3x^2$$

$$y = x^3 + 3x^2 - 2 \qquad \frac{dy}{dx} = 3x^2 + 6x$$

I hope you spotted that we seem to multiply by the index (power) of the unknown, then we subtract (1) one from the index of the unknown. For example, with the function $y = 3x^2$ the index is 2 and $(2)(3) = 6$. Also the original index (power) of x was 2 and on subtracting 1 from this index we get: $x^{(2-1)} = x^1 = x$ so we finally get, $\frac{dy}{dx} = 6x$.

This technique can be applied to *any unknown raised to a power*. We can write this rule in general terms:

If $y = x^n$ then, $\frac{dy}{dx} = nx^{n-1}$

Or, in other words, to find the differential coefficient of the function $y = x^n$ we first multiply the unknown variable by the index and then subtract 1 from the index to form the new index.

Again, with this rather wordy explanation, you will appreciate the ease with which we can express this rule, using a formula!

You may be wondering why the constant (number) in the above functions seems to just disappear. If you remember how we performed the differentiation process, graphically, by finding the slope at a point on the function, then for a constant, such as $y = -6$, the graph is simple a straight horizontal line cutting the *y-axis* at -6 therefore, its slope is *zero*, thus its *derived function is zero*. This is true for any constant term no matter what its value.

If the function we are considering has more than one term, e.g. $y = x^3 + 3x^2 - 2$, then we simply *apply the rule in sequence to each and every term*.

Example 3.39

Differentiate the following functions with respect to the variable:

1. $y = 3x^3 - 6x^2 - 3x + 8$
2. $y = \frac{3}{x} - x^3 + 6x^{-3}$
3. $s = 3t^3 - \frac{16}{t^2} + 6t^{-1}$

1. In this example we can simply apply the rule $\frac{dy}{dx} = nx^{n-1}$ to each term in succession, so:

$$\frac{dy}{dx} = (3)(3)x^{3-1} - (2)(6)x^{2-1} - (1)(3)x^{1-1} + 0$$

and remembering that any number raised to the power zero is one, i.e. $x^0 = 1$, then:

$$\frac{dy}{dx} = 9x^{3-1} - 12x^{2-1} - 3x^{1-1} + 0$$

$$\frac{dy}{dx} = 9x^2 - 12x^1 - 3x^0$$

$$\frac{dy}{dx} = 9x^2 - 12x - 3$$

2. In this example, we need to simplify, before we use the rule. The simplification involves clearing fractions. Then remembering that $x = x^1$ and that *from your laws of indices* when you bring a number in index form over the fraction line, we *change its sign*, $y = \frac{3}{x} - x^3 + 6x^{-3}$ becomes:

$$y = \frac{3}{x^1} - x^3 + 6x^{-3} \quad \text{or}$$

$$y = 3x^{-1} - x^3 + 6x^{-3} \text{ and applying the rule,}$$

$$\frac{dy}{dx} = (-1)(3)x^{-1-1} - (3)x^{3-1} - (3)(6)x^{-3-1}$$

$$\frac{dy}{dx} = -3x^{-2} - 3x^2 - 18x^{-4}$$

Notice how we have dealt with *negative* indices. The rule can also be used when *fractional* indices are involved.

3. The only change with this example is that it concerns different variables. In this case we are asked to differentiate the function s with respect to the variable t.

So proceeding as before and *simplifying first*, we get: $s = 3t^3 - 16t^{-2} + 6t^{-1}$ and then differentiating

$$\frac{ds}{dt} = (3)(3)t^{3-1} - (-2)(16)t^{-2-1}$$

$$+ (-1)(6)t^{-1-1}$$

$$\frac{ds}{dt} = 9t^2 + 32t^{-3} - 6t^{-2}$$

Note that you must take care with your signs!

Key point

To find the first derivative of functions of the type $y = ax^n$, we use the rule that $\frac{dy}{dx} = nax^{n-1}$

The second derivative

In the above example we found, in all cases, *the first derivative*. If we wish to find the *second derivative of a function*, all we need to do is differentiate again. So in example 1.100 question 1, for the function $y = 3x^3 - 6x^2 - 3x + 8$, then:

$$\frac{dy}{dx} = 9x^2 - 12x - 3$$

and differentiating this function again, we get,

$$\frac{d^2y}{dx^2} = (2)(9)x^{2-1} + (1)(-12)x^{1-1}$$

$$= 18x - 12x^0 = 18x - 12$$

In the above example, notice the *Leibniz* terminology for the *second differential*.

Similarly for the function, $s = 3t^2 - 16t^{-2} + 6t^{-1}$, then:

$$\frac{ds}{dt} = 9t^2 + 32t^{-3} - 6t^{-2}$$

and on differentiating this function again, we get,

$$\frac{d^2s}{dt^2} = (2)(9)t^{2-1} + (-3)(32)t^{-3-1}$$

$$+ (-6)(-2)t^{-2-1}$$

$$= 18t - 96t^{-4} + 12t^{-3}$$

In this example notice once again the care needed with signs.

Key point

$\frac{d^2y}{dx^2}$, $f''(x)$ and $\ddot{x}$ are all ways of expressing the second derivative.

Rate of change

One application of the differential calculus is to find instantaneous rates of change. The example given at the beginning of this section concerned our ability to find how the speed of a motor vehicle changed at a particular point in time. *In order to find the rate of change of any function*, we simply *differentiate that function* (find its gradient) *at the particular point concerned*.

For example, given that $y = 4x^2$ let us find its rate of change at the points, $x = 2$ and $x = -4$. Then all we need to do is differentiate the function and then substitute in the desired points.

Then, $\frac{dy}{dx} = (2)(4)x = 8x$ and when $x = 2$, $\frac{dy}{dx} = (8)(2) = 16$, thus the slope of the function at $x = 2$ is 16 and this tells us how the function is changing at this point.

Similarly, when $x = -4$, $\frac{dy}{dx} = 8x = (-4)(8) = -32$. In this case the negative sign indicates a negative slope, so the function is changing in the opposite sense compared with what was happening when $x = 2$.

Example 3.40

The distance (s) covered by a missile $s = 4.905t^2 + 10t$. Determine its rate of change of distance with respect to time (its speed), after (i) 4 s and (ii) 12 s have elapsed.

This is a simple rate of change problem hidden in this rather wordy question!

To find *rate of change* of distance with respect to time; we need to find the differential coefficient of the function. Then applying the rule:

$$\frac{ds}{dt} = (2)(4.905)t^{2-1} + 10t^{1-1} = 9.81t + 10$$

Now substituting for the desired times:
when $t = 4$, then

$$\frac{ds}{dt} = (9.81)(4) + 10 = 49.24$$

and when $t = 12$,

$$\frac{ds}{dt} = (9.81)(12) + 10 = 127.72$$

Since, $\frac{ds}{dt} = v$ (speed), then what the above results tell us is that after 4 s the missile has reached a speed of 49.24 ms^{-1} and after 12 s the missile reaches a speed of 127.72 ms^{-1}.

Thus, for very little effort, the differential calculus has enabled us to find instantaneous rates of change, which are of practical use!

Key point

The rate of change of distance with respect to time is velocity.

Turning points

Another useful application of the differential calculus is in finding the turning points of a function. We have already seen differentiation being used to determine rates of change, *turning points* enable us to tell when these rates of change are at a minimum value or maximum value.

Consider Figure 3.31 which shows the graph of the function $y = x^2 - 9$.

If we consider the slope of the function as it approaches the turning point from the left, the slope is negative. The slope of the graph, as it moves away to the right of the turning point, is positive. At *some point* turning points, the slope went from a *negative value to a positive value*, in other words, at the turning point the slope (*gradient*) is *equal to zero*. Now we know that the gradient of the function $y = x^2 - 9$ is found by differentiating the function. Therefore, when $\frac{dy}{dx}$ of $y = x^2 - 9$ is zero, there must be a turning point because at this point the slope of the function is a *horizontal straight* line and its *slope is zero*.

So applying the rule $\frac{dy}{dx} = 2x$ and for a turning point $\frac{dy}{dx} = 2x = 0$, which implies that $x = 0$.

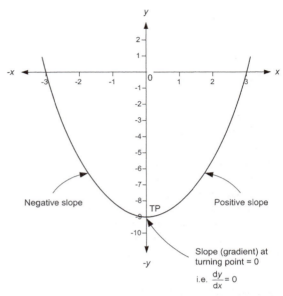

Figure 3.31 Graph of the function $y = x^2 - 9$, showing turning point.

Now if $x = 0$, $y = (0)^2 - 9 = -9$.

So this function turns at the point $(0, -9)$, as can be seen in Figure 3.31.

You should note from Figure 3.31 that at the turning point the function has a *minimum* value. You are not required to find maximum or minimum values at this stage. However, the technique used in finding turning points is the first stage in trying to establish whether such points indicate a maximum or minimum value for a particular function.

Key point

The gradient at a turning point is always zero, i.e. $\frac{dy}{dx} = 0$

Method 1

Determine the rate of change of the gradient function, in other words find the value for the second derivative of the function $\frac{d^2y}{dx^2}$ at the turning point. If this value is *positive* the turning point is a *minimum*, if this value is *negative* then the turning point is a *maximum*.

For example, in the case of the above function $y = x^2 - 9$ the second derivative is $\frac{d^2y}{dx^2} = 2$, which is *positive* and so at the point $(0, -9)$ we have a *minimum*.

Method 2

Consider the **gradient of the curve** close to the turning points, i.e. near either side. Then for a *minimum* the gradient goes from *negative to positive* and for a *maximum* the gradient goes from *positive to negative*.

Quite clearly, for the function $y = x^2 - 9$, we approach the turning point with a negative slope and leave it with a positive slope, therefore, once again, at the point $(0, -9)$ we have a *minimum*.

Key point

A turning point is *minimum* when the gradient goes from *negative to positive*, as we approach and leave the turning point. It is *maximum* when the gradient goes from *positive to negative*.

Differentiation of elementary trigonometric and exponential functions

We have so far concentrated our attention on functions of the type, $ax^n \pm ax^{n-1} \pm ax^{n-2} \pm \cdots \pm ax^3 \pm ax^2 \pm ax \pm a$, this general class of functions are known as *polynomials*.

There are however, other mathematical functions that you have already met. These include *trigonometric functions*, such as the sine and cosine. In addition you have met the *exponential function* e^x and its mathematical inverse the *Naperian logarithm* $\ln x$.

Finding the differential coefficient of these functions can be achieved by graphically differentiating them in a similar manner to the way we originally found the derived function for $y = x^2$. If we were to carry out this exercise, we would be able to establish patterns and subsequent rules, as we did for polynomial functions.

Rather than going through this tedious process you will be pleased to note that these rules have been listed below (without proof) for your convenience!

Table of some standard derivatives

Rule number	y	$\dfrac{dy}{dx}$
1.	x^n	nx^{n-1}
2.	ax^n	nax^{n-1}
3.	$\sin ax$	$a \cos ax$
4.	$\cos ax$	$-a \sin ax$
5.	e^{ax}	ae^{ax}
6.	$\ln ax$	$\dfrac{\frac{dy}{dx}(ax)}{ax}$

Now one or two of the above rules may look a little complex, but in practice they are all fairly straightforward to use. The easiest way to illustrate their use is through the examples that follow.

Example 3.41

Differentiate the following with respect to the variable:

(a) $y = \sin 3x$; (b) $u = \cos 2\theta$; (c) $y = 5 \sin 2\theta - 3 \cos \theta$

(a) In this example we may follow rule 3, in the table directly, noting that $a = 3$ then,

$$\frac{dy}{dx} = a \cos ax = 3 \cos 3x$$

(b) The same approach is needed to solve this little problem, but noting that when we differentiate the cosine function, it has a *sign* change. Also we are differentiating the function u with respect to θ. Then the differential coefficient, using rule 4, is given as:

$$\frac{du}{d\theta} = -2 \sin 2\theta$$

(c) With this final problem, we simply use rule 3 for differentiating sine followed by rule 4 for differentiating cosine. Noting that the numbers 5 and -3 are not the constant a given in the formulae in the table. We simply multiply these numbers by a, when carrying out the differentiation process. So,

$$\frac{dy}{d\theta} = (2)(5)\cos 2\theta(-3)(-1)\sin\theta$$

$$= 10\cos 2\theta + 3\sin\theta$$

Note the effect of the sign change when differentiating the cosine function!

Key point

The sign of the differential of the cosine function is always negative.

Example 3.42

1. Find the differential coefficients of the following function $y = e^{-2x}$
2. Find $\frac{d}{dx}(6\log_e 3x)$
3. Differentiate $v = \frac{e^{3\theta}}{2} - \pi\ln 4\theta$

The above functions involve the use of rules 5 and 6.

1. This is a direct application of rule 5 for the exponential function where $a = -2$. Remember we are differentiating the function y with respect to the variable x. The base e is simply a constant (a number), as mentioned before the value of $e \simeq 2.71828$. It is a number like π, it has a limitless number of decimal places!

   Then $\quad \frac{dy}{dx} = (-2)e^{-2x} = -2e^{-2x}$

2. This is yet another way of being asked to differentiate a function. What it is really saying is find $\frac{dy}{dx}$ of the function $y = 6\log_e 3x$. Remember that when dealing with the Naperian log function $\log_e f(x) = \ln f(x)$, both methods of representing the Naperian log function are in common use; so all we need to do is apply rule 6 where the constant $a = 3$, in this case:

   $$\frac{d}{dx}(6\log_e 3x) = (6)\frac{\frac{dy}{dx}(3x)}{3x} = \frac{(6)(3)}{3x}$$

   $$= \frac{18}{3x} = \frac{6}{x}$$

   Note: when finding this differential we also had to apply rule 1 to the top part of the

fraction. Providing you follow rule 6 exactly laying out all your working you should not make mistakes.

3. For this example, we need to apply rule 5 to the exponential function and then rule 6 to the Naperian log function, noting that $-\pi$ is a constant and does not play any part in the differentiation. We simply multiply the differential by it at the end of the process. Therefore:

   $$\frac{dv}{d\theta} = \frac{3e^{3\theta}}{2} + (-\pi)\frac{\frac{dy}{d\theta}(4\theta)}{4\theta}$$

   and $\quad \frac{dv}{d\theta} = 1.5e^{3\theta} + (-\pi)\frac{4}{4\theta}$

   $$\frac{dv}{d\theta} = 4.5e^{3\theta} - \frac{4\pi}{4\theta}$$

   and $\quad \frac{dv}{d\theta} = 4.5e^{3\theta} - \frac{\pi}{\theta}$

This may look rather complicated but all we have done is followed rule 6, as before.

Now being able to find the differential coefficient of the functions in the above examples, is all very well, but what use is it all?

Well, as was the case with the general rule for differentiating polynomial functions, we can also apply these rules to solving simple *rate of change* problems. In our final example for the *differential calculus*, we apply rules 5 and 6 to rate of change of current in an electrical circuit and rate of discharge from an electrical capacitor. This is not as difficult as it sounds!

Example 3.43

1. An alternating voltage is given by the function $v = \sin 2\theta$, where θ is the angular distance travelled and v is the instantaneous voltage at that angular distance (in radians). Determine the way the voltage is changing with respect to distance, at $\theta = 2$ and $\theta = 4$ rad.
2. Suppose, the charge in a capacitor discharges according to the function $Q = \ln 3t$, where $Q = $ charge (in C) and $t = $ time (ms). Determine the *rate of discharge* at $t = 4$ ms.

1. All we are being asked is to find the rate of change of the voltage after a particular

angular distance has been covered by the alternating (sinusoidal) function. This means we need to find the differential coefficient (the rate of change function) and then simply substitute in the appropriate values.

So $\frac{dv}{d\theta} = 2\cos 2\theta$, which is the rate of change of voltage with respect to distance.

Then at $\theta = 2$ rad, remembering it is *radian* measure! we get:

$$\frac{dv}{d\theta} = 2\cos (2)(2) = 2\cos 4 = (2)(-0.653)$$

$$= -1.3073$$

and the voltage is changing negatively. This value is the slope of the graph of $v = \sin 2\theta$ at the point $\theta = 2$ rad.

Similarly at $\theta = 4$ rad, then

$$\frac{dv}{d\theta} = 2\cos (2)(4) = 2\cos 8 = (2)(-0.1455)$$

$$= -0.291 \text{ to three dp.}$$

Again a negative slope, but with a shallower gradient.

2. The rate of discharge in this case means the rate of change of charge with respect to time. So it is a *rate of change problem* involving the differential coefficient of the function. Then following rule 6 and also using rule 1, we have:

$$\frac{dQ}{dt} = \frac{\frac{dQ}{dt}(3t)}{3t} = \frac{3}{3t} = \frac{1}{t}$$

then when $t = 4$ ms or 4×10^{-3} s,

$$\frac{dQ}{dt} = \frac{1}{t} = \frac{1}{4 \times 10^{-3}} = 250\,\text{C/s}$$

If you were to put in higher values of time you will find that the rate of discharge decreases.

Key point

We always differentiate when finding rates of change.

Test your understanding 3.7

1. When differentiating polynomial functions of the form $y = ax^n$ write down the expression for finding $\frac{dy}{dx}$.

2. For the function $f(x) = 16x^2 - 3x^3 - 12$ find $f(3)$ and $f(-2)$.

3. Differentiate the following functions with respect to the variables given:

   (i) $y = 6x^2 - 3x - 2$ (ii) $s = 3t^2 - 6t^{-1} + \dfrac{t^{-3}}{12}$

   (iii) $p = \dfrac{r^3 - r^2}{r^{-1}} + 12r - 6$ (iv) $y = 3x^{\frac{9}{2}} - 5x^{\frac{3}{2}} + \sqrt{x}$

4. Plot the graph of the function $y = \sin 2\theta$ between $\theta = 0$ and $\theta = 2\pi$ rad, using the techniques you learnt in your trigonometry, making sure that (θ) is in *radians*. Then find the value of the slope at the point where $\theta = 2$ *rad*. Compare your result with the answer to Example 4.81 question 1.

5. If $y = x^2 - 2x + 1$, find the co-ordinates (x, y) at the point where the gradient is 6 (*Hint*: the gradient is $\frac{dy}{dx}$).

6. Determine the rate of change of the function $y = \dfrac{x^4}{2} - 3x^3 + x^2 - 3$, at the point where $x = -2$.

7. What is the rate of change of the function $y = 4e^x$ when $x = 2.32$.

8. Differentiate the functions and comment on your answers
   (i) $y = \ln x$ (ii) $y = 3\ln x$ (iii) $y = \ln 3x$

9. An alternating current is given by the function, $i = \cos 3\theta$, find the rate of change of current when $\theta = 1$ rad.

10. Find the rate at which a capacitor is discharging at $t = 3$ ms and $t = 3.8$ ms, when the amount of charge on the capacitor is given by the function $Q = 2.6 \log_e t$.

3.4.3 The integral calculus

In this short section we are going to look at the *integral calculus* which we mentioned earlier. It has something to do with finding areas and is also the inverse process of finding the derived function. The integral calculus is all about *summing things*, i.e. finding the whole thing from its parts, as you will see shortly.

We start by considering *integration* (the arithmetic of the integral calculus) as the *inverse of differentiation*.

Key point

The integration process is the inverse of the differentiation process.

Integration as the inverse of differentiation

We know that for the function $y = x^2$ the derived function $\frac{dy}{dx} = 2x$. So reversing the process involves finding the function whose derived function is $2x$. One answer will be x^2, but is this the only possibility? The answer is *no* because $2x$ is also the derived function of $y = 2x + 5$, $y = x^2 - 20.51, y = x^2 + 0.345$, etc. In fact $2x$ is the derived function of $y = x^2 + c$ where c is any constant. So when we are finding the inverse of the derived function, in other words, when we are integrating we must always allow for the possibility of a constant being present by putting in this *arbitrary constant c*, which is know as the *constant of integration*. Then in general terms, the inverse of the derived function $2x$ is $x^2 + c$.

Thus, whenever we wish to find the inverse of any derived function, i.e. whenever we *integrate the derived function* we must include the *constant of integration c*.

When carrying out the *antidifferentiation process* or *integration* we can only find a *particular value* for this constant c when we are given some additional information about the original function. For example, if we are told that for $y = x^2 + c$, $y = 2$, when $x = 2$, then by substituting these values into the original function, we find that $2 = 2^2 + c$ from which we find that $c = -2$, so the particular function becomes, $y = x^2 - 2$. This is now one of a *whole family* of functions $y = x^2 + c$ illustrated graphically in Figure 3.32.

Tabulated below are a few familiar *polynomial functions* on which has been carried out this inverse differentiation or *integration* process. When we integrate a derived function, the expression we obtain is often known as the *prime function* (F). See if you can spot a pattern for the derivation of these prime functions.

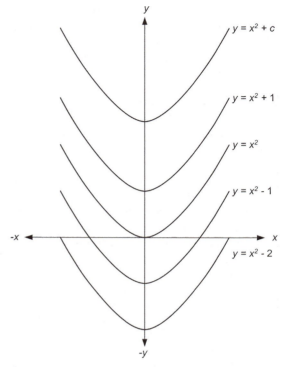

Figure 3.32 Family of curves $y = x^2 + c$.

Apart from the mandatory constant of integration, I hope you can see that the power or index of x increases by 1 over that of the derived function. Then we divide the prime function by this new power or index. Then, in general:

If $\dfrac{dy}{dx} = x^n$,

then the prime function is

$$y = \frac{x^{n+1}}{n+1} + c$$

This rule is valid for all values of n except $n = -1$.

If $n = -1$, then in finding the prime function we would be trying to divide by $n + 1 = -1 + 1 = 0$ and as you are well aware from your earlier study of the laws of arithmetic, division by 0 is not allowed! In this particular case we adopt a special rule, which is given below without proof.

If $\dfrac{dy}{dx} = x^{-1} = \dfrac{1}{x}$,

then the prime function is $y = \ln|x|$.

Derived function	Prime function (F)
$\dfrac{dy}{dx} = 1$	$y = x + c$
$\dfrac{dy}{dx} = x$	$y = \frac{x^2}{2} + c$
$\dfrac{dy}{dx} = x^2$	$y = \frac{x^3}{3} + c$
$\dfrac{dy}{dx} = x^3$	$y = \frac{x^4}{4} + c$

Notice that we have to take the *modulus or positive value of x*, when finding corresponding values of *y*. This is because the *ln* (log$_e$) function *is not valid for all numbers* ≤ 0 (less than or equal to zero).

Notation for the integral

As with differentiation, when we carry out integration, we need to use the appropriate mathematical notation in order to convey our desire to integrate.

If *y* is a function of *x*, then $\int y\,dx$ represents the integral of *y* with respect to the variable *x*. The integral sign $\int$ is the greek letter *S*, and indicates that when carrying out the integration process, we are really carrying out a *summing* process.

Note: that in the same way that the *d* cannot be separated from the *y*, in *dy*, neither can the $\int$ be separated from d*x*, if the integration is with respect to *x*.

For example, if we wish to find the prime function *F*, i.e. if we wish to integrate the function x^2 then this is represented as $\int x^2\,dx$ and using the general rule, we see that

$$\int x^2\,dx = \frac{x^{n+1}}{n+1} + c = \frac{x^{2+1}}{2+1} + c = \frac{x^3}{3} + c,$$

which is in agreement with the prime function or *the integral* shown in the table.

Integration

We have seen from above how to integrate elementary polynomial functions, using the basic rule. In the example given next, we use the rule successively to integrate general polynomial expression with respect to the variable concerned.

Example 3.44

Integrate the following functions with respect to the variables given.

1. $y = 3x^3 + 2x^2 - 6$
2. $s = 5t^{-3} + t^4 - 2t^2$
3. $p = r^{-1} + \frac{r^4}{2}$

1. What we are being asked to do is find the prime function $F(y)$, or using the

conventional notation which we have just learnt then we must find:

$$F(y) = \int 3x^3 + 2x^2 - 6\,dx$$

In this case all we need to do is successively apply the basic rule, i.e.

$$\int 3x^3 + 2x^2 - 6\,dx$$

$$= (3)\frac{x^{3+1}}{3+1} + (2)\frac{x^{2+1}}{2+1} + (-6)x^{0+1} + c$$

$$= \frac{3x^4}{4} + \frac{2x^3}{3} - 6x + c$$

2. In this question we again apply the basic rule but in terms of the different variables, so:

$$\int 5t^{-3} + t^4 - 2t^2\,dt$$

$$= (5)\frac{t^{-3+1}}{-3+1} + \frac{t^{4+1}}{4+1} + (-2)\frac{t^{2+1}}{2+1} + c$$

$$= \frac{5t^{-2}}{-2} + \frac{t^5}{5} - \frac{2t^3}{3} + c$$

With this question I hope you spotted immediately that for part of the function r^{-1}, we cannot apply the general rule but must apply the special case where $n = -1$. So for the integration we proceed as follows:

$$\int r^{-1} + \frac{r^4}{4}\,dr = \ln|r| + \left(\frac{1}{4}\right)\frac{r^{4+1}}{4+1} + c$$

$$= \ln|r| + \frac{r^5}{20} + c$$

Notice also that dividing by 4 is the same as multiplying by a $\frac{1}{4}$ and that we multiply tops by tops and bottoms by bottoms to obtain the final values.

Key point

When finding indefinite integrals, we must always include the constant of integration.

Some common integrals

We have seen now how to integrate polynomial expressions. We can also apply the inverse differentiation process to the *sinusoidal*, *exponential and Naperian logarithm functions.* The table below shows the prime functions (*the integral*) for the basic functions we dealt with during our study of the differential calculus.

Table of some standard integrals

Rule number	Function (y)	Prime function ($\int y \mathrm{d}x$)
1	$x^n (n \neq -1)$	$\frac{x^{n+1}}{n+1} + c$
2	$x^{-1} = \frac{1}{x}$	$\ln|x|$
3	$\sin ax$	$-\frac{1}{a}\cos ax$
4	$\cos ax$	$\frac{1}{a}\sin ax$
5	e^{ax}	$\frac{1}{a}e^{ax}$
6	$\ln x$	$x \ln x - x$

If you compare the integrals of the sine and cosine functions you should be able to recognize that the integral is the inverse of the differential. This is also clearly apparent for the exponential function. The only "strange" integral that seems to have little in common with its inverse is that for the Naperian logarithm function. The mathematical verification of this integral is beyond the level for this unit. However, you will learn the techniques of the calculus necessary for its proof if you study the further mathematics unit.

We will demonstrate the use of these *standard integrals* through the examples that follow.

Example 3.45

1. Find $\int (\sin 3x + 3\cos 2x)\mathrm{d}x$
2. Integrate the function $s = e^{4t} - 6e^{2t} + 2\,\mathrm{d}t$
3. Find $\int 6\log_e t\,\mathrm{d}t$

1. This integral involves using rules 3 and 4, sequentially, the integral may be written as:

$$\int \sin 3x + \int 3\cos 2x$$

$$= -\frac{1}{3}\cos 3x + (3)\frac{1}{2}\sin 2x + c$$

$$= -\frac{1}{3}\cos 3x + \frac{3}{2}\sin 2x + c$$

Any integral involving expressions, separated by $\pm$, may be integrated separately.

Note also that the constant multiplying the function, (3) in this case, *does not play any part in the integration*, it just becomes a multiple of the result.

2. This is just a direct integral involving the successive use of rule 5 and the use of rule 1, for the last term.

Then: $\quad \int e^{4t} - 6e^{2t} + 2\,\mathrm{d}t$

$$= \frac{1}{4}e^{4t} - (6)(\frac{1}{2})e^{2t} + 2t + c$$

$$= \frac{1}{4}e^{4t} - 3e^{2t} + 2t + c$$

3. This integral demonstrates the direct use of rule 6 where the constant is taken behind the integral sign until the process is complete and then brought back in as the multiplier of the integral. Remembering also that, $\log_e t = \ln t$. Then:

$$\int 6\log_e t\,\mathrm{d}t = 6\int \log_e t\,\mathrm{d}t$$

$$= 6(t\log_e t - t) + c \text{ or}$$

$$= 6(t\ln t - t) + c$$

Simple applications of the integral

In the differential calculus we considered *rates of change*. One particular application involved determining *the rate of change of distance with respect to time*. In other words differentiating the function involving distance to find the derived function which gave the *velocity*. Look back to Example 4.78, if you cannot remember this procedure. If we carry out the inverse operation, i.e. we *integrate the velocity function* we will get back to the *distance function*. Taking this idea one step further, if we differentiate the velocity function, we will find the rate of change of velocity with respect to time, in other words we will find the acceleration function (ms^{-2}). So again, if we integrate the acceleration function we get back to the velocity function.

Example 3.46

The acceleration of a missile moving vertically upwards is given by $a = 4t + 4$. Find the formulae for both the velocity and the distance of the missile given that $s = 2$ and $v = 10$ when $t = 0$.

In this application it is important to recognize that *acceleration is rate of change of velocity*, or; $\frac{dv}{dt} = 4t + 4$. This of course is a *derived function* therefore in order to find v, we need to carry out antidifferentiation, i.e. *integration*. When we do this we find the prime function $F(x)$ by integrating both sides of the derived function, as follows:

$$\int \frac{dv}{dt} = \int 4t + 4 \, dt \quad \text{and so}$$

$$F(x) = v = \frac{4t^2}{2} + 4t + c = 2t^2 + 4t + c$$

We now have the original equation for the velocity

$$v = 2t^2 + 4t + c$$

We can now use the given information to find the *particular equation* for the velocity. We know that when the velocity $= 10$, the time $t = 0$. Therefore, substituting into our velocity equation gives: $10 = (2)(0) + (4)(0) + c$, or $10 = c$. So our *particular* equation for velocity is:

$$v = 2t^2 + 4t + 10$$

We are also asked to find the formula for the *distance*. Again, recognizing that *velocity is the rate of change of distance with respect to time* we may write the velocity equation in its derived form as:

$$\frac{ds}{dt} = 2t^2 + 4t + 10$$

then integrating as before to get back to distance, we get:

$$\int \frac{ds}{dt} = \int 2t^2 + 4t + 10 \, dt \quad \text{or,}$$

$$F(x) = s = \frac{2t^3}{3} + \frac{4t^2}{2} + 10t + c$$

$$= \frac{2t^3}{3} + 2t^2 + 10t + c$$

We now have the original equation for distance

$$s = \frac{2t^3}{3} + 2t^2 + 10t + c$$

Again, using the given information, that $s = 2$ and $v = 10$, when $t = 0$, the particular equation for distance can be found. On substitution of time and distance into our distance equation we get that, $2 = 0 + 0 + 0 + c$, or $c = 2$. So our *particular* equation for distance is:

$$s = \frac{2t^3}{3} + 2t^2 + 10t + 2$$

Key point

If we integrate the acceleration function we obtain the velocity function. If we integrate the velocity function we obtain the distance function.

Area under a curve

The above example illustrates the power of the integral calculus in being able to find velocity from acceleration and distance from velocity. We now know:

$$\text{Velocity(speed in a given direction)} = \frac{\text{Distance}}{\text{Time}}$$

that is: $\text{Distance} = \text{Velocity} \times \text{Time}$

So if we set velocity against time on a velocity–time graph the area under the graph (velocity $\times$ time) will be equal to the distance. Therefore if we know the rule governing the motion, we could, in our case, find any distance covered within a particular time period by integrating the velocity–time curve over this period.

Consider Figure 3.33, which shows a velocity–time graph, where the motion is governed by the relationship:

$$v = -t^2 + 3t \quad \text{or} \quad \frac{ds}{dt} = -t^2 + 3t$$

Then to find *the distance* equation, for the motion, all we need do is *integrate* the velocity equation, as in Example 3.46.

The important point to note is that *when we integrate* and *find the distance equation*, this is the same as *finding the area under the graph* because the area under the graph $=$ velocity $\times$ time $=$ distance.

From the graph it can be seen that when time $t = 0$, the velocity $v = 0$, also that when $t = 3$, $v = 0$. So the area of interest is contained between these two time *limits*.

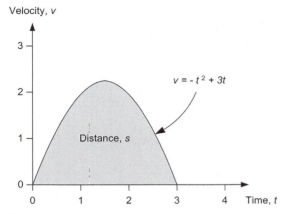

Figure 3.33 Velocity–time graph for the motion $v = -t^2 + 3t$.

Now integrating our velocity equation in the normal manner, gives:

$$\int v = \frac{ds}{dt} = \int -t^2 + 3t\, dt = s = \frac{-t^3}{3} + \frac{3t^2}{2} + c$$

This *distance equation is equivalent to the area under the graph, between time $t = 0$ and $t = 3$*.

At time $t = 0$ the distance travelled $s = 0$ from the graph. The *constant of integration* c can be found by substituting these values of time and distance into our distance equation:

$$s = \frac{-t^3}{3} + \frac{3t^2}{2} + c, \quad \text{so} \quad 0 = 0 + 0 + c$$

therefore $c = 0$ and our particular distance equation is,

$$s = \frac{-t^3}{3} + \frac{3t^2}{2}$$

Now between *our limits* of time $t = 0$ to $t = 3$ s. The *area under the graph indicates the distance travelled*. So at time $t = 0$, the distance travelled $s = 0$. At time $t = 3$, the area under the graph is found by substituting time $t = 3$ into our distance equation, then:

$$s = \frac{-t^3}{3} + \frac{3t^2}{2} = \frac{-(3)^3}{3} + \frac{(3)(3^2)}{2}$$

$$= \frac{-27}{3} + \frac{27}{2} = -9 + 13.5 = 4.5$$

Thus, in the above example, the area under the graph $= 4.5 =$ the distance travelled.

Key point

The area under a velocity time curve is equal to the distance.

The definite integral

When we *integrate between limits*, such as the time limits, given for the motion discussed above. We say that we are finding the *definite integral*. All the integration that we have been doing up till now has involved the constant of integration and we refer to this type of integration as finding the *indefinite integral*, which must contain an arbitrary constant c.

The terminology for indefinite integration is that which we have used so far, for examples:

$$\int -t^2 + 3t\, dt \quad \text{the indefinite integral}$$

When carrying out definite integration we place limits on the integration sign, or summing sign, for examples:

$$\int_0^3 -t^2 + 3t\, dt \quad \text{the definite integral}$$

To evaluate a definite integral, we first integrate the function, then we find the numerical value of the integral at its upper and lower limits and subtract the value of the lower limit from that of the upper limit to obtain the required result.

So following this procedure for the definite integral shown above, which we used to find the distance s, (area under a graph) from the velocity–time graph, we get:

$$s = \int_0^3 -t^2 + 3t\, dt = \left[\frac{-t^3}{3} + \frac{3t^2}{2} + c \right]_0^3$$

$$= \left(\frac{-27}{3} + \frac{27}{2} + c \right) - \left(\frac{0}{3} + \frac{0}{2} + c \right)$$

$$s = (-9 + 13.5 + c) - (0 + c)$$

$$= 4.5 + c - c = 4.5$$

Thus $s = 4.5$, we have found *the area under the graph using definite integration*!

Note that when we *subtract* the upper limit value from the lower limit value, *the constant of integration is eliminated*. This will always be the case when evaluating definite integrals; therefore it need not be shown from now on.

Example 3.47

1. Evaluate $\int_{-1}^{1} \frac{x^5 - 4x^3 + x}{x} \, dx$

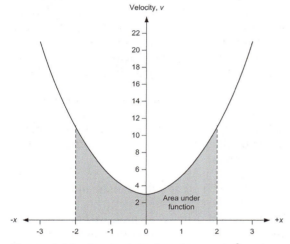

Velocity, *v*

Figure 3.34　Graph of the function $y = 2x^2 + 2$.

2. Determine by integration the area enclosed by the curve $y = 2x^2 + 2$, the x-axis and the ordinates are $x = -2$ and $x = 2$ (Figure 3.34).

1. Before we integrate, it is essential to simplify the function as much as possible. So in this case on division by x, we get:

$$\int_{-1}^{1} x^4 - 4x^2 + 6 \, dx$$

$$= \left[\frac{x^5}{5} - \frac{4x^3}{3} + 6x \right]_{-1}^{1}$$

$$= \left(\frac{1}{5} - \frac{4}{3} + 6 \right) - \left(\frac{-1}{5} - \frac{-4}{3} - 6 \right)$$

$$= 9\frac{11}{15}$$

Note that, in this case, it is easier to manipulate the upper and lower values as fractions!

2. In order to get a picture of the area we are required to find, it is best to draw a sketch of the situation first. The area with the appropriate limits is shown below:

We are required to find the shaded area of the graph between the limits $x = \pm 2$. Then:

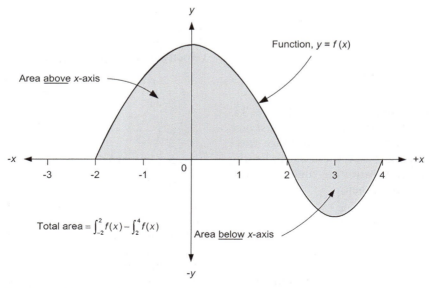

y

Function, $y = f(x)$

Area <u>above</u> *x*-axis

Total area $= \int_{-2}^{2} f(x) - \int_{2}^{4} f(x)$

Area <u>below</u> *x*-axis

-*y*

Figure 3.35　Function with areas above and below the x-axis.

$$\int_{-2}^{2} 2x^2 + 2 \, dx$$

$$= \left[\frac{2x^3}{3} + 2x \right]_{-2}^{2} = \left(\frac{2(2)^3}{3} + (2)(2) \right)$$

$$- \left(\frac{2(-2)^3}{3} + (2)(-2) \right)$$

$$= \left(\frac{16}{3+4} \right) - \left(\frac{-16}{3} - 4 \right)$$

$$= 18\frac{2}{3} \quad \text{square units}$$

A final word of caution when finding areas under curves, using integration. If the area you are trying to evaluate is part above and part below the x-axis. It is necessary to split the limits of integration for the areas concerned.

So for the shaded area shown in Figure 3.35; we find the definite integral with the limits $(2, -2)$ and *subtract* from it the definite integral with limits $(4, 2)$, i.e. the shaded area in Figure $3.35 = \int_{-2}^{2} y \, dx - \int_{2}^{4} y \, dx$. Notice that the higher value always sits at the top of the integral sign. The minus sign is always necessary before the integral of any area that sits *below the x-axis*.

On this important point we finish our study of the integral calculus and indeed your study of Mathematics for this chapter!

Test your understanding 3.8

1. Find the following indefinite integrals using the basic rules:

   (a) $\int 4x^2 + 2x^{-3} dx$ (b) $\int \frac{3x^{\frac{1}{2}}}{6} - \sqrt{x} + x^{\frac{3}{2}} dx$

   (c) $\int -3 \sin 2x \, dx$ (d) $\int \frac{x \cos 3x}{0.5x} dx$

   (e) $\int -0.25 e^{3\theta} d\theta$ (f) $\int -3 \log_e x \, dx$

2. Using your results from Question 1, evaluate the following definite integrals:

   (a) $\int_{0}^{2} 4x^2 + 2x^{-3} dx$ (b) $\int_{0}^{1} \frac{3x^{\frac{1}{2}}}{6} - \sqrt{x} + x^{\frac{3}{2}} dx$

   (c) $\int_{0}^{\frac{\pi}{2}} -3 \sin 2\theta \, d\theta$ (d) $\int_{1}^{2} -0.25 e^{3\theta} d\theta$

   Note, for Questions (c) and (d) θ is in radians.

3. The acceleration of a vehicle is given by the relationship $a = 3t + 4$. Find the formulae for the *velocity* and *distance* of the vehicle given that $s = 0$ and $v = 8$ when $t = 0$. Also find the distance travelled after a time of 25 s, has elapsed.

4. Find the area under the curve $y = x + x^2$ between $x = 1$ and $x = 3$.

5. Sketch the graphs of the line $y = 2x$ and the curve $y = x^2$ on the same axes and determine by integration the area between the line and the curve.

Chapter 4 — Physics

4.1 Summary

This chapter aims to provide you with an understanding of the physical principles that underpin the design and operation of modern aircraft and their associated structures and systems. The study of this chapter will also act as a suitable foundation for those who wish to embark on a higher education qualification, associated with aerospace engineering.

You will be introduced to the nature of matter and elementary mechanics where elements of statics, kinematics, dynamics and fluid dynamics will be considered. You will also study thermodynamics, light and sound.

After introducing units of measure and the fundamental principles of the subjects identified above, their application to aircraft structures and systems will be emphasized. For example a study of statics will enable us to consider, at an elementary level, the nature of the forces imposed on aircraft structures, due to static loading. A study of fluid dynamics will act as a suitable introduction to the study of aerodynamics, which you will meet later. Thermodynamic principles may be applied to cabin conditioning and refrigeration systems as well as to aircraft engine operation. Aircraft engineering applications related to light, optics, wave motion and sound will also be covered.

Each *major section* within this chapter covers the principles of the subject and then provides examples of problems that illustrate the application of this theory, to engineering situations and, wherever possible, actual aircraft engineering problems.

At the end of the chapter you will find a selection of typical *multiple-choice of answer questions* relating to each major section within this chapter. They have been designed, in order to achieve the same academic objectives as those detailed in Chapter 1, Introduction.

In view of the international nature of the civil aviation industry, as aircraft maintenance engineers, you will need to become fully conversant with metric, imperial and United States (US) units and measurements, as mentioned in Chapter 1, Introduction. Therefore, throughout your study of this chapter you will be asked to consider and solve problems using SI units, and also consider a variety of English/US units that you may not be familiar with. Engineering applications using SI units and occasionally, English engineering units, will be emphasized throughout this chapter, as the subject matter is addressed.

4.2 Units of measurement

As mentioned previously, familiarity with international units is an essential tool for all those involved in aircraft engineering. The consequence of making mistakes in the use and conversion of units may prove costly or, in certain circumstances, catastrophic. For example, consider the simple task of inflating a tire on a ground support trolley. If the inflation pressure is $30\,lb/in.^2$, imagine what might happen if the inflation equipment was set to pressurize the tire in bar! Familiarity with international units is not just necessary for the study of physics. You will also be required to consider them, throughout the remainder of this book.

There are in fact, three commonly used "English" systems of measurement, parts of which have been adopted by the US. These are: the English engineering system (force, mass, length, time), the absolute English system (mass, length, time) and the technical English system (force, length, time).

In the past physicists have tended to use the absolute metric or CGS (centimetre–gramme–second) system, whereas engineers used either the English Engineering system or the technical English system. Throughout this book, we shall use both the SI system (metre, kilogram, second) and to a lesser degree when applicable, the English Engineering system.

It should be remembered that all systems apart from SI, are now regarded by the international community as obsolete, therefore, we will concentrate on the use of SI units to develop and illustrate scientific principles. However, in view of the fact that English units are still commonly used by American aircraft manufacturers and airline operators, we will also need to use English units and their conversion factors, when applying scientific principles to aircraft-related problems. Our knowledge of English/US units when referring to aircraft maintenance manuals produced by American manufacturers, will then help us ensure the continued integrity and flight safety of these aircraft when carrying out aircraft maintenance operations.

Key point

The SI system is based on the following units:

- metre (m);
- kilogramme (kg);
- second (s).

Key point

SI units have now replaced all other units for international use.

Key point

Aircraft engineers need to be aware of the use of English/US units and should be able to convert between units when required.

For your convenience and reference seven tables are set out below which contain the SI base units (Table 4.1), supplementary SI units (Table 4.2), SI derived units (Table 4.3), SI prefixes (Table 4.4), some of the more common non-SI metric units (Table 4.5), a table

Table 4.1   SI base units

Basic quantity	SI unit name	SI unit symbol	Other recognized units
Mass (*m*)	kilogram	kg	tonne
Length (*s*)	metre	m	mm, cm, km
Time (*t*)	second	s	ms, min, hour, day
Electric current (*I*)	ampere	A	MA
Temperature (*T*)	kelvin	K	°C
Amount of substance	mole	mol	
Luminous intensity	candela	Cd	

Table 4.2   SI supplementary units

Supplementary unit	SI unit name	SI unit symbol
Plane angle	radian	rad
Solid angle	steradian	srad or sr

of the base units for the English Engineering system (Table 4.6), which was adopted by the USA and is still in frequent use today. Finally Table 4.7 contains multipliers to convert from the SI system to English and other common units of measurement, not directly covered by the SI system.

Definitions of SI base units

What follows are the *true and accurate definitions of the SI base units*, at first these definitions may seem quite strange. They are detailed below for reference, you will meet most of them again during your study of this chapter on physics and when you study Electrical fundamentals (Chapter 5).

Kilogramme

The kilogramme or kilogram is the unit of mass; it is equal to the mass of the international prototype of the kilogram as defined by International Committee for Weights and Measures (CIPM).

Metre

The metre is the length of the path traveled by light in a vacuum during the time interval of 1/299,792,458 s.

Table 4.3 SI derived units

SI name	SI symbol	Quantity	SI unit
Coulomb	C	Quantity of electricity, electric charge	$1\,C = 1\,As$
Farad	F	Electric capacitance	$1\,F = C/V$
Henry	H	Electrical inductance	$1\,H = 1\,kg\,m^2\,s^2/A^2$
Hertz	Hz	Frequency	$1\,Hz = 1\,cycle/s$
Joule	J	Energy, work, heat	$1\,J = 1\,Nm$
Lux	lx	Illuminance	$1\,lx = 1\,cd\,sr/m^2$
Newton	N	Force, weight	$1\,N = 1\,kg\,m/s^2$
Ohm	Ω	Electrical resistance	$1\,\Omega = 1\,kg\,m^2/s^3A^2$
Pascal	Pa	Pressure, stress	$1\,Pa = 1\,N/m^2$
Siemans	S	Electrical conductance	$1\,S = 1\,A/V$
Tesla	T	Induction field, magnetic flux density	$1\,T = 1\,kg/A\,s^2$
Volt	V	Electric potential, electromotive force	$1\,V = 1\,kg\,m^2/s^3A$
Watt	W	Power, radiant flux	$1\,W = 1\,J/s$
Weber	Wb	Induction magnetic flux	$1\,Wb = 1\,kg\,m^2/s^2A$

Table 4.4 SI prefixes

Prefix	Symbol	Multiply by
Peta	P	10^{15}
Tera	T	10^{12}
Giga	G	10^{9}
Mega	M	10^{6}
Kilo	k	10^{3}
Hecto	h	10^{2}
Deca	da	10^{1}
Deci	d	10^{-1}
Centi	c	10^{-2}
Milli	m	10^{-3}
Micro	μ	10^{-6}
Nano	n	10^{-9}
Pico	p	10^{-12}
Femto	f	10^{-15}

Important —

Second

The second is the duration of 9,192,631,770 periods of radiation corresponding to the transition between the two hyperfine levels of the ground state of the cesium 133 atom.

Ampere

The ampere is that constant current which if maintained in two straight parallel conductors of infinite length, of negligible circular cross-section, and placed 1 m apart in a vacuum, would produce between these conductors a force equal to 2×10^{-7} N/m length.

Kelvin

The kelvin, unit of thermodynamic temperature, is the fraction 1/273.16 of the thermodynamic temperature of the triple point of water.

Mole

The mole is the amount of substance of a system which contains as many elementary particles as there are atoms in 0.012 kg of carbon 12. When the mole is used, the elementary entities must be specified and may be atoms, molecules, ions, electrons, or other particles, or specified groups of such particles.

Candela

The candela is the luminous intensity, in a given direction, of a source that emits monochromatic radiation of frequency 540×10^{12} Hz and that has a radiant intensity in that direction of 1/683 W/s rad (see below).

In addition to the seven base units given above, as mentioned before there are two supplementary units, the radian for plane angles (which you will meet later) and the steradian for solid three-dimensional angles. Both of these

Table 4.5 Non-SI units

Name	Symbol	Physical quantity	Equivalent in SI base units
Ampere-hour	Ah	Electric charge	$1\,\text{Ah} = 3600\,\text{C}$
Day	d	Time, period	$1\,\text{d} = 86{,}400\,\text{s}$
Degree	°	Plane angle	$1° = (\pi/180)\,\text{rad}$
Electronvolt	eV	Electric potential	$1\,\text{eV} = (e/C)\,\text{J}$
Kilometre per hour	kph	Velocity	$1\,\text{kph} = (1/3.60)\,\text{ms}^{-1}$
Hour	h	Time, period	$1\,\text{h} = 3600\,\text{s}$
Litre	L, l	Capacity, volume	$1\,\text{L} = 10^{-3}\,\text{m}^3$
Minute	min	Time, period	$1\,\text{min} = 60\,\text{s}$
Metric tonne	t	Mass	$1\,\text{t} = 10^3\,\text{kg}$

Table 4.6 English systems base units

Basic quantity	English engineering name	English engineering symbol	Other recognized units
Mass	slug	Equivalent of: 32.17 lb	pound (lb), ton, hundredweight (cwt)
Length	foot	ft	inch (in.), yard (yd.), mile
Time	second	s	min, hour, day
Electric current	Ampere	A	mA
Temperature	rankine	R	°F (Fahrenheit)
Luminous intensity	foot candle	lm/ft^2	lux, cd/ft^2

Table 4.7 Conversion factors

Quantity	SI unit	Conversion factor $\longrightarrow$	Imperial/other recognized units
Acceleration	metre/second^2 (m/s^2)	3.28084	feet/second^2 (ft/s^2)
Angular measure	radian (rad)	57.296	degrees (°)
	radian/second (rad/s)	9.5493	revolutions per minute (rpm)
Area	metre^2 (m^2)	10.7639	feet^2 (ft^2)
	metre^2 (m^2)	6.4516×10^4	inch^2 (in.^{2})
Density	kilogram/metre^3 (kg/m^3)	0.062428	pound/foot^3 (lb/ft^3)
	kilogram/metre^3 (kg/m^3)	3.6127×10^{-5}	pound/inch^3 (lb/in.^{3})
	kilogram/metre^3 (kg/m^3)	0.010022	pound/gallon (UK)
Energy, work, heat	joule (J)	0.7376	foot pound-force (ft lbf)
	joule (J)	9.4783×10^{-4}	British thermal unit (btu)
	joule (J)	0.2388	calorie (cal)
Flow rate	m^3/s (Q)	35.315	ft^3/s
	m^3/s (Q)	13,200	gal/min (UK)
Force	newton (N)	0.2248	pound-force (lbf)
	newton (N)	7.233	poundal
	kilo-newton	0.1004	ton-force (UK)
Heat transfer	watt (W)	3.412	btu/h
	watt (W)	0.8598	kcal/h
	watt/metre^2 kelvin (W/m^2 K)	0.1761	btu/h ft^2 °F
Illumination	lux (lx)	0.0929	foot candle
	lux (lx)	0.0929	lumen/foot^2 (lm/ft^2)
	candela/metre^2 (cd/m^2)	0.0929	candela/ft^2 (cd/ft^2)

(continued)

Table 4.7 (*continued*)

Quantity	SI unit	Conversion factor $\longrightarrow$	Imperial/other recognized units
Length	metre (m)	1×10^{10}	angstrom
	metre (m)	39.37008	inch (in.)
	metre (m)	3.28084	feet (ft)
	metre (m)	1.09361	yard (yd)
	kilometre (km)	0.621371	mile
	kilometre (km)	0.54	nautical miles
Mass	kilogram (kg)	2.20462	pound (lb)
	kilogram (kg)	35.27392	ounce (oz)
	kilogram (kg)	0.0685218	slug
	tonne (t)	0.984207	ton (UK)
	tonne (t)	1.10231	ton (US)
Moment, torque	newton-metre (Nm)	0.73756	foot pound-force (ft lbf)
	newton-metre (Nm)	8.8507	inch pound-force (in. lbf)
Moment of inertia (mass)	kilogram-metre squared (kgm^2)	0.7376	slug-foot squared $(slug\,ft^2)$
Second moment of area	millimeters to the fourth (mm^4)	2.4×10^{-6}	inch to the fourth $(in.^4)$
Power	watt (W)	3.4121	British thermal unit/hour (btu/h)
	watt (W)	0.73756	foot pound-force/second (ft lbf/s)
	kilowatt (kW)	1.341	horsepower
	horsepower (hp)	550	foot pound-force/second (ft.lbf/s)
Pressure, stress	kilopascal (kPa)	0.009869	atmosphere (atm)
	kilopascal (kPa)	0.145	pound-force/inch^2 (psi)
	kilopascal (kPa)	0.01	bar
	kilopascal (kPa)	0.2953	inches of mercury
	pascal	1.0	newton/metre^2 (N/m^2)
	megapascal (MPa)	145.0	pound-force/inch^2 (psi)
Temperature	kelvin (K)	1.0	celsius (°C)
	kelvin (K)	1.8	rankine (°R)
	kelvin (K)	1.8	fahrenheit (°F)
	kelvin (K)		°C + 273.15
	kelvin (K)		(°F + 459.67)/1.8
	celsius (°C)		(°F − 32)/1.8
Velocity	metre/second (m/s)	3.28084	feet/second (ft/s)
	metre/second (m/s)	196.85	feet/minute (ft/min)
	metre/second (m/s)	2.23694	miles/hour (mph)
	kilometre/hour (kph)	0.621371	miles/hour (mph)
	kilometre/hour (kph)	0.5400	knot (international)
Viscosity (kinematic)	square metre/second (m^2/s)	1×10^6	centi-stoke
	square metre/second (m^2/s)	1×10^4	stoke
	square metre/second (m^2/s)	10.764	square feet/second (ft^2/s)
Viscosity (dynamic)	pascal second (Pa s)	1000	centipoise (cP)
	centipoise (cP)	2.419	pound/feet hour (lb/ft h)
Volume	cubic metre (m^3)	35.315	cubic feet (ft^3)
	cubic metre (m^3)	1.308	cubic yard (yd^3)
	cubic metre (m^3)	1000	litre (l)
	litre (l)	1.76	pint (pt) UK
	litre (l)	0.22	gallon (gal) UK

To convert SI units to Imperial and other recognized units of measurement *multiply* the unit given by the *conversion factor*, i.e. in the direction of the arrow. To *reverse* the *process*, i.e. to convert from non-SI units to SI units divide by the conversion factor.

relationships are ratios and *ratios have no units*, e.g. metres/metres = 1. Again, do not worry too much at this stage, it may become clearer latter, when we look at radian measure in our study of dynamics.

The SI derived units are defined by simple equations relating two or more base units. The names and symbols of some of the derived units may be substituted by special names and symbols. Some of the derived units, which you may be familiar with, are listed in Table 4.3 with their special names as appropriate.

So, for example, 1 millimetre = 1 mm = 10^{-3} m, 1 cm^3 = $(10^{-2}$ m$)^3$ = 10^{-6} m^3 and 1 μm = 10^{-6} m. Note the way in which powers of ten are used. The above examples show us the correct way for representing multiples and sub-multiples of units.

Some of the more commonly used, legally accepted, *non-SI units* are detailed in Table 4.5.

For example, from Table 4.7:

$$14\,\text{kg} = (14)(2.20462) = 30.865\,\text{lb}$$

and

$$70\,\text{bar} = \frac{70}{0.01} = 7000\,\text{kPa} = 7.0\,\text{MPa}$$

Test your understanding 4.1

1. Complete the entries in the table of SI base units shown below:

Base quantity	SI unit name	SI unit symbol
Mass		kg
	metre	m
Time	second	
	ampere	A
Temperature	kelvin	
Amount of substance		mol
	candela	cd

2. What is the SI unit for plane angles?
3. What units are employed in the Absolute Metric (CGS) system?
4. Convert the following quantities using Table 4.7:
   (a) 1.2 UK ton into kg,
   (b) 63 ft^3 into m^3,
   (c) 14 stokes into m^2/s,
   (d) 750 W into horsepower.

5. If we assume that as *an approximation* 14.5 psi = 1 bar then, without the use of a calculator, convert 15 bar into psi.
6. Assuming that *as an approximation* there are 10.75 ft^2 in a m^2. Then estimate, without the use of a calculator, the number of m^2 there are in 215 ft^2.

You will have a lot more practice in the manipulation of units, as your studies progress!

4.3 Fundamentals

Having briefly introduced the idea of units of measurement, we are now going to embark on our study of physics by considering some fundamental quantities, such as mass, force, weight, density, pressure, temperature, the nature of matter and, most importantly, the concept of energy, which plays such a vital role in our understanding of science in general. Knowledge of these fundamental physical parameters will be required when we look at elements of physics in detail.

4.3.1 Mass, weight and gravity

Mass

The **mass** of a body is a measure of the **quantity of matter** in the body. The amount of matter in a body does not change when the position of the body changes. So, **the mass of a body does not change with position.**

As can be seen from Table 4.1, the SI unit of mass is the kilogram (kg). The standard kg is the mass of a block of platinum alloy kept at the Office of Weights and Measures in Sevres near Paris.

Weight

The weight of a body is the gravitational **force** of attraction between the mass of the earth and the mass of a body. The weight of a body decreases as the body is moved away from the earth's center. It obeys the inverse square law, which states that if the distance of the body is doubled, the weight is reduced to a quarter of its previous value. **The SI unit of weight is the newton (N).**

Using mathematical symbols this law may be written as:

$$\text{Weight } (W) \propto \frac{1}{d^2}$$

where $d = $ distance and $\propto$ is the symbol for proportionality.

So, for example, consider a body of weight (W) at an initial distance of 50 m from the gravitational source, then $W \propto 1/50^2 = 4 \times 10^{-4}$.

Now if we double this distance then the weight $(W) \propto 1/100^2 = 1 \times 10^{-4}$, which clearly shows that if the distance is doubled the weight is reduced to a quarter of its original value.

Key point

The mass of a body is unaffected by its position.

Key point

In the SI system, weight is measured in Newton (N).

Gravitational acceleration

When a body is allowed to fall, it moves towards the earth's center with an acceleration caused by the weight of the body. If air resistance is ignored, then at the same altitude all bodies fall with the same gravitational acceleration. Although heavier bodies have more weight, at the same altitude they fall with the same gravitational acceleration because of their greater resistance to acceleration. The concept of resistance to acceleration will be explained more fully, when we deal with Newton's laws of motion.

Like weight, gravitational acceleration depends on distance from the earth's center. At sea level, the **gravitational acceleration** (g) has an accepted standard value of **9.80665 m/s^2**. For the purpose of calculations in this chapter, we will use the approximation $g = 9.81$ **m/s^2**.

Key constant

At sea level, the acceleration due to gravity, g is approximately 9.81 m/s^2.

The mass–weight relationship

From what has already been said, we may define the weight of a body as the product of its mass and the value of gravitational acceleration at the position of the body. This is expressed in symbols as:

$$W = mg$$

Where in the SI system the weight (W) is in N, the mass is in kg and the acceleration due to gravity is taken as 9.81 m/s^2 unless specified differently.

In the "English" systems of units, there is often confusion between mass and weight, because of the inconsistencies with the units. As shown above, *weight = mass × acceleration due to gravity*. This is a special case of Newton's second law: *force = mass × acceleration*, as you will see later. In a coherent system of units, any derived unit must interrelate one-to-one with the system's base units, so that *one force unit equals one mass unit times one acceleration unit*. In the *foot–pound–second* (FPS) system, with the pound as the unit of mass, 1 force unit is required to impart 1 acceleration unit (1 ft/s^2) to a mass of 1 lb. The acceleration due to gravity in the FPS system is approximately 32 ft/s^2, so that the *weight* of 1 lb mass is in fact 32 force units and the force unit must therefore be 1/32 lb. In fact to be accurate, since g is 32.1740486 ft/s^2, it is 1/32.17 or 0.031081 lbf, $= 0.138255$ N, this is termed the **poundal**.

However, because in general use the pound has always been appreciated as a unit of weight, there was a tendency among engineers to continue to use it in this way. In a variant of the FPS system, usually termed *technical, gravitation or engineers' units,* the **pound-force (lbf)** was taken as the base unit, and a unit of mass was derived by reversing the above argument. This unit was named the *slug*, and the mass which when acted upon by 1 lbf experienced an acceleration of 1 ft/s^2, so was equivalent to 32.17 lb. This version of the FPS system was, and to a degree still is, more commonly used in the US than anywhere else.

If you find this confusing, study the conversion factors for mass and force given in Table 4.7

(See also Table E.7 of Appendix E and the examples given at the end of this section). You should then be able to form the connection associating these unfamiliar units for mass and force.

We now know that the mass of a body does not change with changes in altitude but its weight and gravitational acceleration do. However, for bodies that do not move outside the earth's atmosphere, the changes in gravitational acceleration (and therefore weight) are small enough to be ignored for most practical purposes. We may therefore assume our approximation for $g = 9.81$ m/s^2 to be reasonably accurate, unless told otherwise.

To clarify the mass–weight relationship let us consider an example calculation, using standard SI units.

Example 4.1

A missile having a mass of 25,000 kg is launched from sea level on a course for the moon. If the gravitational acceleration of the moon is one sixth that on earth, determine the:
(a) weight of the rocket at launch,
(b) mass of the rocket on reaching the moon,
(c) weight of the rocket on reaching the moon.

(a) Using the relationship $W = mg$, then the weight on earth;

$$W = (25,000 \times 9.81)$$
$$= 245,250 \, \text{N or } \mathbf{245.25 \, kN}$$

(b) We know from our definition of mass, that it does not change with change in position therefore the mass on the moon remains the same as on earth i.e. **25,000 kg**.
(c) We know that the gravitational acceleration on the moon is approximately 1/6 that on earth.
So $g_m = 9.81/6$ m/s$^2 = 1.635$ m/s^2 and again from $W = mg_m$ then weight of rocket on the moon $= (25,000 \times 1.635) = 40,875 \, \text{N} = $ **40.875 kN**.

Note: A much easier method of solution for part (c) would have been to divide the weight on earth by 6.

Test your understanding 4.2

1. What happens to the weight of a body as it is moved away from the center of the earth?
2. What is the SI unit of weight?
3. What is the approximate SI value for the acceleration due to gravity at sea level?
4. If the capacity of a light aircraft's fuel tanks is 800 UK gallons, what is the volume of the fuel in litres?
5. A light aircraft on take-off weighs 42,000 N, what is its mass?
6. Define (a) the poundal and (b) pound-force (lbf).

4.3.2 Density and relative density

Density

The density (ρ) of a body is defined as its mass per unit volume. Combining the SI units for mass and volume gives the unit of density as kg/m^3. Using symbols the formula for density is given as:

$$\rho = \frac{m}{V}$$

where again the mass is in kg and the volume in m^3.

You will see later, when you study the atmosphere, that density is temperature dependent. This is due to the change in volume caused by changing temperature.

Key constant

The density of pure water at 4°C is taken as 1000 kg/m^2.

Relative density

The relative density of a body is *the ratio of the density of the body with that of the density of pure water measured at 4°C*. The density of water under these conditions is 1000 kg/m^3. Since relative density is a ratio it has **no units**. The old name for relative density was *specific gravity* (*SG*) and this is something you need to be aware of in case you meet this terminology in the future.

The density of some of the more common engineering elements and materials is laid out in

Table 4.8 Density of some engineering elements/materials

Element/material	Density (kg/m³)
Acrylic	1200
Aluminum	2700
Boron	2340
Brass	8400–8600
Cadmium	8650
Cast iron	7350
Chromium	7190
Concrete	2400
Copper	8960
Glass	2400–2800
Gold	19,320
Hydrogen	0.09
Iron	7870
Lead	11,340
Magnesium	1740
Manganese	7430
Mercury	13,600
Mild steel	7850
Nickel	8900
Nitrogen	0.125
Nylon	1150
Oxygen	0.143
Platinum	21,450
Polycarbonate	914–960
Polyethylene	1300–1500
Rubber	860–2000
Sodium	971
Stainless steel	7905
Tin	7300
Titanium	4507
Tungsten	1900
UPVC	19,300
Vanadium	6100
Wood (douglas fir)	608
Wood (oak)	690
Zinc	7130

Table 4.8. To find the relative density of any element or material, divide its density by $1000\,\text{kg/m}^3$.

Test your understanding 4.3

1. What is the SI unit of density?
2. Use Tables 4.7 and 4.8 to find the density of aluminum (a) in SI units and (b) in lb/ft³.
3. What is likely to happen to the density of pure water as its temperature increases?
4. Why does relative density have no units?
5. What is the approximate equivalent of 10 lb/gallon (UK) in standard SI units of density?

Example 4.2

A mild steel aircraft component has a mass of 240 g. Using the density of mild steel given in Table 4.8, calculate the volume of the component in cm³.

From Table 4.8 mild steel has a density of $7850\,\text{kg/m}^3$, therefore using our definition for density:

$$\rho = \frac{m}{V},$$

we have

$$V = \frac{m}{\rho} = \frac{240 \times 10^{-3}}{7850}$$

$$= 30.57 \times 10^{-6}\,\text{m}^3$$

Thus, volume of component = $30.57\,\text{cm}^3$.

Note that to obtain the standard unit for mass, the 240 g was converted to kg using the multiplier 10^{-3}, and multiplying m³ by 10^6 converts them into cm³, as required. Be careful with your conversion factors, when dealing with squared or cubic measure!

Example 4.3

An aircraft component made from an aluminum alloy weighs 16 N and has a volume of 600 cm³, determine the relative density of the alloy.

We need to use the mass–weight relationship $m = \dfrac{W}{g}$ to find the mass of the component, i.e.

$$\text{Mass},\ m = \frac{16}{9.81} = 1.631\,\text{kg}$$

Then,

$$\text{Density} = \frac{m}{V} = \frac{1.631}{600 \times 10^{-6}}$$

$$= 2718\,\text{kg/m}^3$$

The relative density (RD) is then given by,

$$\text{RD} = \frac{2718\,\text{kg/m}^3}{1000\,\text{kg/m}^3} = 2.718$$

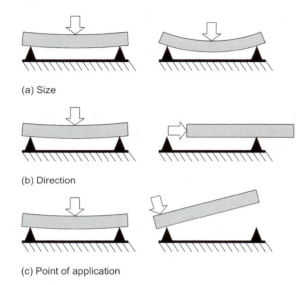

(a) Size

(b) Direction

(c) Point of application

Figure 4.1 Characteristics of a force.

4.3.3 Force

In its simplest sense a force is a push or pull exerted by one object on another. In a member in a static structure, a push causes compression and a pull causes tension. Members subject to compressive and tensile forces have special names. A member of a structure that is in *compression* is known as a *strut* and a member in *tension* is called a *tie*.

Only rigid members of a structure have the capacity to act as both a strut and tie. Flexible members, such as ropes, wires or chains, can only act as ties.

Force cannot exist without opposition, as you will see later when you study Newton's laws. An applied force is called an *action* and the opposing force it produces is called *reaction*.

Key point

The *action* of a force always produces an *opposite reaction*.

The effects of any force depend on its three characteristics, illustrated in Figure 4.1.

In general force (F) = mass (m) × acceleration (a) is used as the measure of force:

$$F = ma$$

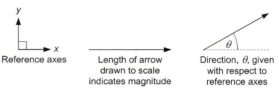

Reference axes | Length of arrow drawn to scale indicates magnitude | Direction, θ, given with respect to reference axes

Figure 4.2 Graphical representation of a force.

The SI unit of force is the Newton. Note that weight force mentioned earlier is a special case where the acceleration acting on the mass is that due to gravity, so *weight force* may be defined as $F = mg$, as mentioned earlier. The Newton is thus defined as follows:

1 N is the force that gives a mass of 1 kg an acceleration of 1 m/s²

It can be seen from Figure 4.1 that a force has size (magnitude), direction and a point of application. A force is thus a *vector quantity*, i.e it has magnitude and direction. A **scalar** quantity has only magnitude, e.g. mass. A force may therefore be represented graphically in two dimensions by drawing an arrow to scale with its length representing the magnitude of a force and the head of the arrow indicating the direction in relation to a set of previously defined axes. Figure 4.2 illustrates the graphical representation of a force.

Note: In the *FPS engineer's system of units*, 1 lbf is the force that gives a mass of 1 slug an acceleration of 1 ft/s². That is *1 lbf = 32.17 lb ft/s²*, where the *slug* is the unit of mass = 32.17 lb.

4.3.4 Pressure

Pressure (as denoted P below) due to the application of force or load, is defined as force per unit area.

$$P = \frac{\text{Force or load applied} \perp \text{to a surface}}{\text{Area over which the force or thrust acts}}$$

The units of pressure in the SI system are normally given as: N/m², N/mm², MN/m² or pascal (Pa) where 1 Pa = 1 N/m². Also pressures in fluid systems is often quoted in bar, where 1 bar = 10^5 Pa or 100,000 N/m².

The bar should not be taken as the value for standard atmospheric pressure at sea level. The value quoted in bar for *standard atmospheric pressure* is 1.0132 bar or 101,320 N/m^2 or 101.32 kPa. Much more will be said about atmospheric pressure, when we study the International Civil Aviation Organization (ICAO) standard atmosphere in the section on atmospheric physics.

Key constant

Standard atmospheric pressure is 1.0132 bar or 101,320 N/m^2.

Example 4.4

The area of ground surface contained by the skirt of a hovercraft is 240 m^2. The unladen weight of the craft is 480 kN and total laden weight is 840 kN. Determine the minimum air pressure needed in the skirt to support the craft when unladen and when fully loaded.

When unladen:

$$\text{Pressure} = \frac{\text{Force}}{\text{Area}} = \frac{480\,\text{kN}}{240\,\text{m}^2} = 2\,\text{kN/m}^2$$

When fully loaded:

$$\text{Pressure} = \frac{840\,\text{kN}}{240\,\text{m}^2} = 3.5\,\text{kN/m}^2$$

In practice the skirt would be inflated to the higher of these two pressures and the craft (when static) would rest in the water at the appropriate level.

Test your understanding 4.4

1. What are the three defining characteristics of any force?
2. Define (a) a scalar quantity and (b) a vector quantity, and give an example of each.
3. Give the general definition of force and explain how weight force varies from this definition.
4. Complete the following statement: "A strut is a member in _____ and a tie is a member in _____".
5. Define pressure and include two possible SI units.
6. Using the appropriate table convert the following to the standard SI unit of pressure: (a) 28 psi (b) 30 in. Hg.

4.3.5 Speed, velocity and acceleration

Speed may be defined as *distance per unit time*. Speed takes no account of direction and is therefore a *scalar* quantity.

The common SI units of speed are: kilometres per hour (kph) or metres per second (m/s).

In the aircraft industry we more commonly talk about aircraft speed in knots (nautical miles per hour) or in miles per hour (mph). Mach number is also used and we will talk more about these units of speed later.

Example 4.5

Convert (a) 450 knots into kph and (b) 120 m/s into mph.

We could simply multiply by the relevant conversion factors in Table 4.7, which for part (a) is the reciprocal of 0.5400 or 1.852. Similarly for part (b) the conversion factor is 2.23694. Let us see if we can derive these conversion factors by addressing the problem in a rather circular manner.

(a) Suppose we know that there are 6080 ft in a knot. Then, since there are 3.28084 ft in 1 m, there are $\frac{6080}{3.28084} = 1853.18$ m in a knot.

Therefore, 450 knots = 450×1853.18 m/h or **833.93 kph**.

Thus to convert knots to kph we need to multiply them by $\frac{833.93}{450} = 1.853$, which to two decimal places agrees with our tabulated value.

(b) A conversion factor to convert m/s into mph may be found in a similar manner. In this case we start with the fact that 1 m = 3.28084 ft and there are 5280 ft in a mile, so:

$$120\,\text{m/s} = 3.28084 \times 120 \text{ feet per second}$$

or

$$\frac{3.28084 \times 120}{5280} \text{ miles per second}$$

We also know that there are 3600 s in an hour therefore:

$$120\,\text{m/s} = \frac{3.28084 \times 120 \times 3600}{5280}$$

$$= 268.4\,\text{mph}$$

Again the multiplying factor is given by the ratio $268.4/120 = 2.2369$ which is in agreement with our tabulated value.

It will aid your understanding of unit conversion, if you attempt to derive your own conversion factors from basic unit conversions.

Velocity is defined as *distance per unit time in a specified direction.* Therefore, velocity is a *vector quantity* and the SI units for the magnitude of velocity are the SI units for speed, i.e. m/s.

The direction of a velocity is not always quoted but it should be understood that the velocity is in some defined direction, even though this direction is unstated.

Key point

Speed is a scalar quantity whereas velocity is a vector quantity.

Acceleration is defined as *change in velocity per unit time or rate of change of velocity,* acceleration is also a *vector quantity* and the SI unit of acceleration is $\frac{\text{m/s}}{\text{s}}$ or m/s^2.

4.3.6 Equilibrium, momentum and inertia

A body is said to be in *equilibrium* when its acceleration continues to be zero, i.e. when it remains at rest or when it continues to move in a straight line with constant velocity (Figure 4.3).

Momentum may be described as the quantity of motion of a body. Momentum is the product of the mass of a body and its velocity. Any change in momentum requires a change in velocity, i.e an acceleration. It may be said that for a fixed quantity of matter to be in equilibrium, it must have constant momentum. A more rigorous definition of momentum is given next, when we consider Newton's second law.

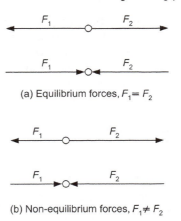

(a) Equilibrium forces, $F_1 = F_2$

(b) Non-equilibrium forces, $F_1 \neq F_2$

Figure 4.3 (a) Equilibrium forces and (b) non-equilibrium forces.

Key point

The momentum of a body is equal to its mass multiplied by its velocity.

All matter resists change. The force resisting change in momentum (i.e. acceleration) is called *inertia*. The inertia of a body depends on its mass, the greater the mass, the greater the inertia.

The inertia of a body is an innate force that only becomes effective when acceleration occurs.

An applied force acts against inertia so as to accelerate (or tend to accelerate) a body.

4.3.7 Newton's laws of motion

Before we consider Newton's laws we need to re-visit the concept of force. We already know that force cannot exist without opposition, i.e. action and reaction. If we apply a 100 N pulling force to a rope, this force cannot exist without opposition.

Force is that which changes, or tends to change, the state of rest or uniform motion of a body. Forces that act on a body may be external (applied from outside the body), such as weight, or internal (such as the internal resistance of a material subject to a compression).

The difference between the forces tending to cause motion and those opposing motion is called the *resultant* or *out-of-balance force*. A body that has no out-of-balance external force acting on it is in equilibrium and will not accelerate. A body that has such an out-of-balance force will accelerate at a rate dependent on the mass of the body and the magnitude of the out-of-balance force. The necessary opposition that permits the existence of the out-of-balance force is provided by the force of inertia.

Newton's first law of motion states that: *a body remains in a state of rest or of uniform motion in a straight line unless it is acted upon by some external resultant force.*

Newton's second law of motion states that: *the rate of change of momentum of a body is directly proportional to the force producing the change and takes place in the direction in which the force acts.*

We defined force earlier as, force = mass × acceleration. We also know that acceleration may be defined as, change in velocity per unit time or rate of change in velocity. If we assume that a body has an *initial velocity u* and a *final velocity v,* then the change in velocity is given by $(v - u)$ and so the rate of change of velocity or acceleration may be written as:

$$\frac{(v - u)}{t} \quad \text{where } t \text{ is unit time}$$

Key point

$F = ma$ is a consequence of Newton's second law of motion.

So since $F = ma$, this may be written as:

$$F = \frac{m(v - u)}{t}$$

and multiplying out the brackets gives:

$$F = \frac{mv - mu}{t}$$

Now we also know that momentum was defined earlier as *mass × velocity*. So the product *mu* gives the initial moment of the body, prior to the application of the force and *mv* gives the final momentum of the body. Thus the expression $(mv - mu)$ is the change in momentum and

so $\frac{(mv - mu)}{t}$ is the rate of change of momentum and therefore Newton's second law may be expressed as:

$$F = \frac{mv - mu}{t} \quad \text{or} \quad F = ma$$

Newton's third law states that: *to every action there is an equal and opposite reaction.*

We will meet Newton's law again when we study aircraft motion and engine thrust.

4.3.8 Temperature

Temperature is a measure of the quantity of energy possessed by a body or substance, it is a measure of the molecular vibrations within the body. The more energetic these vibrations become then the hotter will be the body or substance. For this reason, in its simplest sense, temperature may be regarded as the "degree of hotness of a body". A more scientific definition of temperature will be given when you study thermodynamics.

Test your understanding 4.5

1. Use the appropriate tables to convert the following units:
   (a) 600 kph ⎫ into mph
   (b) 140 m/s ⎬
   (c) 25 m/s² into ft/s²
   (d) 80 ft/s ⎫ into m/s
   (e) 540 mph ⎬
   (f) 240 knot ⎭

2. Determine the acceleration in m/s², when a force of 1000 N is applied to a mass of 500 lb.

3. Define "inertia" and quote its units.

4. What may we write as the equivalent to *the rate of change of momentum* in Newton's second law.

5. What in its simplest sense does temperature measure?

General Questions 4.1

1. A rocket launched into the earth's atmosphere is subject to an acceleration due to gravity of 5.2 m/s². If the rocket has a mass of 120,000 kg. Determine the weight of the rocket: (a) on earth and (b) in orbit.

2. A solid rectangular body measures 1.5 m × 20 cm × 3 cm and has a mass of 54 kg. Calculate: (a) its volume in m³, (b) its density in Pa and (c) its relative density.

3. An aircraft has four fuel tanks, two in each wing. The outer tanks each have a volume of

$20\,\text{m}^3$ and the inner tanks each have a volume of $30\,\text{m}^3$. The fuel used has a specific gravity (RD) of 0.85. Determine the weight of fuel (at sea level) carried when the tanks are full.

4. A body has a weight of $550\,\text{N}$ on the surface of the earth:
   (a) What force is required to give it an acceleration of $6\,\text{m/s}^2$?
   (b) What will be the inertia reaction of the body when given this acceleration?

5. A Cessna 172 and a Boeing 747 are each given an acceleration of $5\,\text{m/s}^2$. To achieve this the thrust force produced by the Cessna's engines is $15\,\text{kN}$ and the thrust force required by the Boeing 747 is $800\,\text{kN}$. Find the mass of each aircraft.

4.4 Matter

4.4.1 Introduction

We have already defined *mass* as the amount of matter in a body but what is the nature of this matter?

All *matter* or *material* is made up from elementary building blocks that we know as atoms and molecules. The *atom* may be further subdivided into *protons*, *neutrons* and *electrons*. Physicists have discovered many more elementary sub-atomic particles that, for the purposes of this discussion, we do not need to consider.

A *molecule* consists of a collection of two or more atoms, which are joined chemically, in a certain way, to give the material its macroscopic properties. The act of joining atoms and/or joining molecules to form parent material is known as *chemical bonding*. The driving force that encourages atoms and molecules to combine in certain ways is energy. Like everything else in nature, matter or material is formed as a consequence of the atoms and/or molecules combining in such a way, that once formed, they attain their lowest energy state. We may define *energy* as the capacity to do work thus, like nature, we measure our efficiency with respect to work, in terms of the least amount of energy we expend. Energy, work and power will be covered more fully later, when we study dynamics.

4.4.2 Chemical bonding

In order to fully understand the mechanisms of bonding you will need to be aware of one or two important facts about the atom and the relationship between the type of bond and the *periodic table* (Table 4.9) *of the elements.*

The nucleus of the atom consists of an association of protons and neutrons; the protons have a minute positive charge and the neutrons, as their name suggests are electrically neutral. Surrounding the nucleus in a series of discreet (measurably separate) *energy bands*, negative charged electrons orbit the nucleus (Figure 4.4). An atom is electrically neutral, in that the number of *positively charged protons* is matched by the equal but opposite, *negatively charged electrons*. Electrons in the energy bands or *shells* closest to the nucleus are held tightly by electrostatic attraction. In the outermost shells they are held less tightly.

Key point

Electrons carry a negative charge and protons carry a positive charge.

An *ion* is formed when an atom gains or loses electrons, which disturbs the electrical neutrality of the original atom. For example, a *positive ion* is formed when an atom loses one or more of its outer electrons.

The *valence* of an atom is related to the ability of the atom to enter into chemical combination with other elements, this is often determined by the number of electrons in the outer most levels, where the binding energy is least. These valence shells are often known as *s* or *p* shells the letters refer to the shell to which the electrons belong. For example, magnesium which has 12 electrons, aluminum which has 13 electrons and germanium which has 32 electrons can be represented as follows:

Mg $1s^2\,2s^2\,2p^6\,3s^2$ valence $= 2$
Al $1s^2\,2s^2\,2p^6\,3s^2\,3p^1$ valence $= 3$
Ge $1s^2\,2s^2\,2p^6\,3s^2$ valence $= 4$
     $3p^6\,3d^{10}\,4s^2\,4p^2$

The numbers 1s, 2s, 2p, etc. relate to the shell level, the superscript numbers relate to the

Table 4.9 Periodic table of the elements

	I A	II A											III A	IV A	V A	VI A	VIIA	0
	s1	s2											s2 p1	s2 p2	s2 p3	s2 p4	s2 p5	s2 p6
1	1 H					*Transition elements*												2 He
2	3 Li	4 Be	IIIB	IVB	VB	VIB	VIIB	←	VIII	→	IB	IIB	5 B	6 C	7 N	8 O	9 F	10 Ne
3	11 Na	12 Mg	d1 s2	d2 s2	d3 s2	d5 s1	d5 s2	d6 s2	d7 s2	d8 s2	d10 s1	d10 s2	13 Al	14 Si	15 P	16 S	17 Cl	18 Ar
4	19 K	20 Ca	21 Sc	22 Ti	23 V	24 Cr	25 Mn	26 Fe	27 Co	28 Ni	29 Cu	30 Zn	31 Ga	32 Ge	33 As	34 Se	35 Br	36 Kr
5	37 Rb	38 Sr	39 Y	40 Zr	41 Nb	42 Mo	43 Tc	44 Ru	45 Rh	46 Pd	47 Ag	48 Cd	49 In	50 Sn	51 Sb	52 Te	53 I	54 Xe
6	55 Cs	56 Ba	57 to 71	72 Hf	73 Ta	74 W	75 Re	76 Os	77 Ir	78 Pt	79 Au	80 Hg	81 Tl	82 Pb	83 Bi	84 Po	85 At	86 Rn
7	87 Fr	88 Ra	89 to 103	104 Ku	105 Ha	106 Sg	107 Bh	108 Hs	109 Mt	110 Uun	111 Uuu	112 Uub						

Inner Transition Elements

Lanthanides	57 La	58 Ce	59 Pr	60 Nd	61 Pm	62 Sm	63 Eu	64 Gd	65 Tb	66 Dy	67 Ho	68 Er	69 Tm	70 Yb	71 Lu
Actinides	89 Ac	90 Th	91 Pa	92 U	93 Np	94 Pu	95 Am	96 Cm	97 Bk	98 Cf	99 Es	100 Fm	101 Md	102 No	103 Lr

Legend

Metals	Metalloids
Non-Metals	Rare Gases

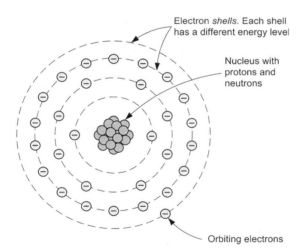

Electron *shells*. Each shell has a different energy level

Nucleus with protons and neutrons

Orbiting electrons

Figure 4.4 A simplified model of the atom.

number of electrons in that shell. Remember the total number of s and p electrons in the outermost shell often account for the valence number. There is an exception to the above rule, the valence may also depend on the nature of the chemical reaction.

> **Key point**
>
> The valence of an element is identified by the column in which it sits within the periodic table.

If an atom has a valence of zero, no electrons enter into chemical reactions and these are all examples of inert or noble elements.

You may be wondering where all this talk of valence is leading us. By studying the periodic table (Table 4.9) you will hopefully be able to see!

The *rows* in the periodic table correspond to the principle energy shells that contain the electrons. The *columns* refer to the number of electrons present in the outermost *sp* energy level and so correspond to the most common valence. Normally the elements in each column have similar properties and behavior.

The transition elements are so named because some of their inner shells are being filled progressively as you move from left to right in the table. For instance scandium (Sc) requires nine electrons to completely fill its *3d* shell, while at the other end copper (Cu) has a filled 3*d* shell which helps to keep the valence electrons tightly held to the inner core, copper as well as silver (Ag) and gold (Au) are consequently very stable and unreactive. Notice that copper, silver and gold all sit in the same column, so they all have similar properties.

In columns I and II the elements have completed inner shells and one or two valence electrons. In column III, e.g. aluminum (Al) has three valence electrons and in column VII chlorine (Cl) has seven valence electrons. The important point to note is that *it is the number of valence electrons in the outermost shells that determines the reactivity of the element* and therefore the way in which that element will combine with others, i.e. *the type of bond* it will form.

All atoms within the elements try to return or sit in their lowest energy levels, this is achieved if they can obtain the *noble gas configuration*, where their outermost *sp* shells are full or empty and they have no spare electrons to combine with other elements. When atoms bond together they try to achieve this noble gas configuration, as you will see next.

Let us now turn our attention to the ways in which atoms and molecules combine or *bond* together.

There are essentially three types of primary bond *ionic*, *covalent* and *metallic* as well as secondary bonds, such as *van der Waals*.

Key point
Ionic bonding involves electron transfer.

When more than one type of atom is present in a material, one atom may donate its valence electrons to a different atom, filling the outer energy shell of the second atom. Both atoms now have completely full or empty outer energy levels but in the process, both have acquired an electrical charge and behave like *ions*. These oppositely charged ions are then attracted to one another and produce an ***ionic bond***. The ionic bond is also sometimes referred to as the *electrovalent bond*. The combination of a sodium atom with that of a chlorine atom illustrates the ionic

bonding process very well, as is shown in Figure 4.5.

Note that in the *transfer* of the electron from the sodium atom to the chlorine atom both the sodium and chlorine ions now have a noble gas configuration, where in the case of sodium the outer valence shell is empty while for chlorine it is full. These two ions in combination are sitting in their lowest energy level and so readily combine. In this classic example of ionic bonding, the metal sodium has combined with the poisonous gas chlorine to form the sodium chloride molecule, common salt!

Key point
In covalent bonding electrons are shared.

In ***covalently bonded*** materials electrons are *shared* among two or more atoms. This sharing between atoms is arranged in such a way that

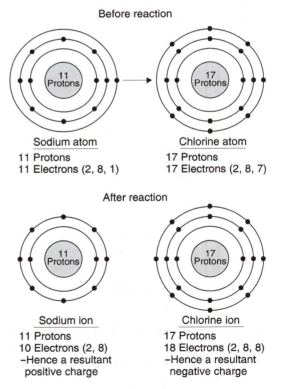

Figure 4.5 Illustration of the ionic bonding process between a sodium and chlorine atom.

each atom has its outer shell filled, so that by forming the molecule each atom again, sits in its lowest energy level and has the noble gas configuration. The covalent bonding of silicon and oxygen to form silica (SiO_2, silicon dioxide) is shown in Figure 4.6.

The metallic elements that have low valence, give up their valence electrons readily to form a "sea of electrons" which surround the nucleus of the atoms. Thus in giving up their electrons the metallic elements form positive ions which are held together by mutual attraction of the surrounding electrons, producing the strong metallic bond. Figure 4.7 illustrates the *metallic* bond.

It is the ease with which the atoms of metals give up their valence electrons (charge carriers) that makes them, in general, very good conductors of electricity.

Key point

van der Waals bonds involve the weak electrostatic attraction of dipoles that sit within the molecules of materials.

---

van der Waals bonds join molecules or groups of atoms by weak electrostatic attraction. Many polymers, ceramics, water and other molecules tend to form electrical dipoles, i.e. some portions of the molecules are positively charged while other portions are negatively charged. The electrostatic attraction between these oppositely charged regions, weakly bond the two regions together (Figure 4.8).

van der Waals bonds are *secondary bonds*, but the atoms within the molecules or groups of molecules are held together by strong covalent or ionic bonds. For example, when water is boiled the secondary van der Waals bonds, which hold the molecules of water together are broken. Much higher temperatures are then required to break the covalent bonds that combine the oxygen and hydrogen atoms. The ductility of polyvinyl chloride (PVC) is attributed to the weak van der Waals bonds that hold the long chain molecules together. These are easily broken allowing these large molecules to slide over one another.

In many materials, bonding between atoms is a mixture of two or more types. For example, iron is formed from a combination of metallic and covalent bonds. Two or more metals may form a metallic compound, by a mixture of metallic and ionic bonds. Many ceramic and semiconducting compounds that are a combination of metallic and non-metallic elements have mixtures of covalent and ionic bonds. The energy necessary to break the bond, *the binding energy* for the bonding mechanisms we have discussed, are shown in Table 4.10.

The electronic structure of an atom may be characterized by the energy levels to which each electron is assigned, in particular to the valence

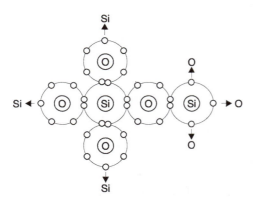

Figure 4.6 Covalent bond formed between silicon and oxygen atoms.

Figure 4.7 Illustration of the metallic bond.

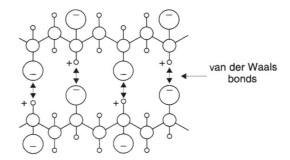

Figure 4.8 van der Waals bonds joining molecules or groups of atoms by weak electrostatic attraction.

Table 4.10 Values of binding energy for primary and secondary bonds

Bond	Binding energy (kJ/mol)
Ionic	625–1550
Covalent	520–1250
Metallic	100–800
van der Waals	<40

of each element. The periodic table of the elements is constructed based on this electronic structure.

The electronic structure plays an important part in determining the bonding between atoms, allowing us to assign general properties to each class of material. Thus metals have good ductility, and electrical and thermal conductivity because of the metallic bond. Ceramics, semiconductors and many polymers are brittle and have poor conductivity because of their covalent and ionic bonds. While van der Waals bonds are responsible for good ductility in certain polymers.

Test your understanding 4.6

1. Define *ion*, stating the condition under which ions are positive or negative.
2. Explain what is meant by the noble gas configuration and state why (when atoms/molecules chemically combine) they try to achieve it.
3. What is the significance of the rows and columns set out in the periodic table of the elements.
4. What is meant when we refer to an element as having a valence of two?
5. Describe the *two stages* of ionic bonding.
6. With reference to the periodic table (Table 4.9), carbon sits in column IV. As a result, what type of bond is carbon likely to form and why?

4.5 The states of matter

During our previous discussion on the way in which matter combines, nothing much was said about the distances over which the binding energy for primary and secondary bonds act. The existence of the *three states of matter* is due to a struggle between the interatomic or intermolecular binding forces and the motion that

these atoms and/or molecules have because of their own *internal energy*.

> **Key point**
>
> Matter is generally considered to exist in solid, liquid and gaseous form.

4.5.1 Solids

When we previously considering interatomic bonding only attraction forces and binding energy were discussed, however, there also exist forces of *repulsion*. Whether or not the force of attraction or repulsion dominates, depends on the atomic distance between the atoms/molecules when combined. It has been shown that at distances greater than 1 atomic diameter the forces of attraction dominate, while at very small separation distances, the reverse is true.

> **Key point**
>
> The atoms within solids tend to combine in such a manner that the interatomic binding forces are balanced by the very short-range repulsion forces.

From what we have said so far, there must be one value of the separation where the resultant interatomic force is zero. This fact is illustrated in Figure 4.9, where the distance at which this interatomic force is zero is identified as r_0. This is the situation that normally exists in a solid. If the atoms are brought closer by compression, they will repel each other and if pulled further apart they attract. Although we have only considered a pair of atoms within a solid, the existence of an equilibrium separation holds good even when we consider the interactions of neighboring atoms.

4.5.2 Liquids

As temperature increases, the *amplitude* of the internal vibration energy of the atoms increases until they are able to partly overcome the interatomic bonding forces of their immediate

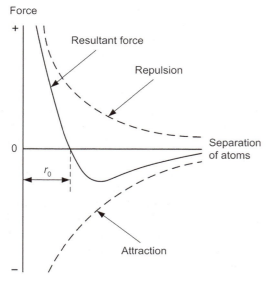

Figure 4.9 Attraction and repulsion forces due to atomic separation.

The idea of a gas filling the vessel in which it is contained has its origins in Newton's first law of motion. Each molecule will, in accordance with this law, travel in a straight line until it collides with another molecule or with the sides of the containing vessel. Therefore, a gas has no particular shape or volume but expands until it fills any vessel into which it is introduced.

The rather scientific discussion laid out above relating to chemical bonding and the states of matter may seen rather far removed from aircraft engineering. However, these important concepts will act as a base when we apply them to the study of engineering materials and thermodynamics. We will be revisiting the behavior of gases and looking closely at the changes that occur between the states of matter, when we study thermodynamics latter in this chapter.

Test your understanding 4.7

1. Explain the essential difference (at the atomic level) between solids and liquids.
2. Over what sort of distances do the atomic repulsion forces act?
3. How is the *internal energy* within matter defined?

neighbors. For short spells they are within range of forces exerted by other atoms which are not quite so near. There is less order and so the solid liquefies. Although the atoms and molecules of a liquid are not much further apart than in a solid, they have greater speeds due to increased temperature and so move randomly in the liquid, while continuing to vibrate. However the primary differences between liquids and solids may be attributed to *differences in structure*, rather than distance between the atoms. It is these differences in the forces between the molecules which give the liquid its flow characteristics while at the same time holding it sufficiently together, to exhibit shape, within a containing vessel.

4.5.3 Gases

In a gas the atoms and molecules move randomly with high speeds and take up all the space in the containing vessel. Gas molecules are therefore relatively far apart when compared with solids and liquids. Because of the relatively large distances involved, molecular interaction only occurs for those brief spells when molecules collide and large repulsive forces operate between them.

4.6 Mechanics

Mechanics is the physical science concerned with the state of rest or motion of bodies under the action of forces. This subject has played a major role in the development of engineering throughout history, and up to the present day. Modern research and development in the fields of vibration analysis, structures, machines, spacecraft, automatic control, engine performance, fluid flow, electrical apparatus, and subatomic, atomic and molecular behavior are all reliant on the basic principles of mechanics.

The subject of mechanics is conveniently divided into two major areas: *statics* which is concerned with the equilibrium of bodies under the action of forces and *dynamics* which is concerned with the motion of bodies. Dynamics may be further subdivided into the motion of rigid bodies and the motion of fluids, the later subject is covered separately under the heading of *fluid dynamics* (Section 4.9.4).

4.7 Statics

4.7.1 Vector representation of forces

You have already met the concept of *force*, when we looked at some important fundamentals. You will remember that the effect of a force was dependent on its magnitude, direction and point of application (Figure 4.1), and that a force may be represented on paper as a *vector* quantity (Figure 4.2).

We will now study the vector representation of a force or combination of forces, in more detail, noting that all vector quantities throughout this book will be identified using emboldened text.

In addition to possessing the properties of magnitude and direction from a given reference (Figure 4.2), vectors must obey the *parallelogram law* of combination. This law requires that two vectors $\mathbf{v}_1$ and $\mathbf{v}_2$ may be replaced by their equivalent vector $\mathbf{v}_T$ which is the diagonal of the parallelogram formed by $\mathbf{v}_1$ and $\mathbf{v}_2$ as shown in Figure 4.10(a). This vector sum is represented by the vector equation:

$$\mathbf{v}_T = \mathbf{v}_1 + \mathbf{v}_2$$

Note that the plus sign in this equation refers to the addition of two vectors and should not be confused with ordinary scalar addition, which is simply the sum of the magnitudes of these two vectors and is written as; $v_T = v_1 + v_2$, in the normal way without emboldening.

Vectors may also be added head-to-tail using the *triangle law* as shown in Figure 4.10(b). It can also be seen from Figure 4.10(c) that the order in which vectors are added does not affect their sum.

Key point

Two vectors may be added using the parallelogram rule or triangle rule.

The vector difference $\mathbf{v}_1 - \mathbf{v}_2$ is obtained by *adding* $-\mathbf{v}_2$ to $\mathbf{v}_1$. The effect of the minus sign is to reverse the direction of the vector $\mathbf{v}_2$ (Figure 4.10(d)). The vectors $\mathbf{v}_1$ and $\mathbf{v}_2$ are known as the components of the vector $\mathbf{v}_T$.

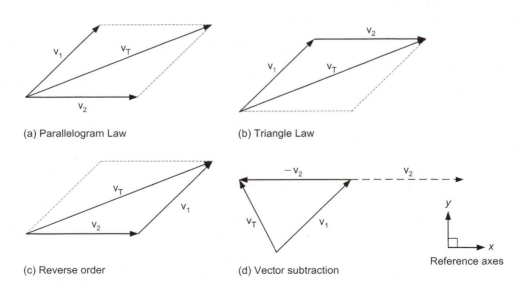

(a) Parallelogram Law

(b) Triangle Law

(c) Reverse order

(d) Vector subtraction

Figure 4.10 Vector addition and subtraction.

Example 4.6

Two forces act at a point as shown in Figure 4.11. Find by vector addition their *resultant* (their single equivalent force).

From the vector diagram the resultant vector **R** is 5 cm in magnitude that (from the scale) is equivalent to 25 N. So the resultant vector **R** has a magnitude of 25 N at an angle of 48°.

Note that a *space diagram* is first drawn to indicate the orientation of the forces with respect to the reference axes, these axes should always

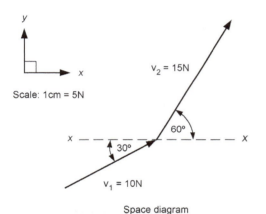

Scale: 1cm = 5N

Space diagram

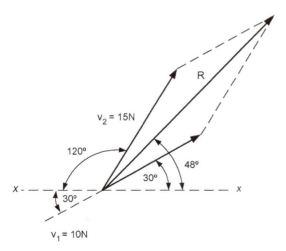

Vector diagram

Figure 4.11 Vector addition using the parallelogram law.

be shown. Also note that the *line of action* of vector v_1 passing through the point 0, is shown in the space diagram and may lie anywhere on this line, as indicated on the vector diagram.

Example 4.7

Find the resultant of the system of forces shown in Figure 4.12, using vector addition.

From the diagram the resultant $= 6.5$ cm $= 6.5 \times 10$ N $= 65$ N. Acting at an angle of 54° from the *x*-reference axis. This result may be written mathematically as, **resultant $= 65$ N$\angle 54°$**

Note that for the force system in Example 4.7, vector addition has produced a polygon. Any number of forces may be added vectorially in any order, providing the *head-to-tail rule* is observed. In this example, if we were to add the vectors in reverse order, the same result will be achieved.

If a force, or system of forces, is acting on a body and is balanced by some other force, or system of forces then the body is said to be in *equilibrium* so, for example, a stationary body is in equilibrium.

The *equilibrant* of a system of forces is that force which, when added to a system, produces equilibrium. It has been shown in Examples 4.6 and 4.7, that the *resultant* is the single force which will replace an existing system of forces and produce the same effect. It therefore follows that if the equilibrant is to produce equilibrium it must be equal in magnitude and direction, but opposite in sense to the resultant, Figure 4.13 illustrates this point.

Bow's notation is a convenient system of labeling the forces for ease of reference, when there are three or more forces to be considered. Capital letters are placed in the space between forces in a clockwise direction, as shown in Figure 4.14.

Any force is then referred to by the letters that lie in the adjacent spaces either side of the vector arrow representing that force. The vectors representing the forces are then given the corresponding *lower case* letters. Thus the forces AB, BC and CA are represented by the vectors **ab**,

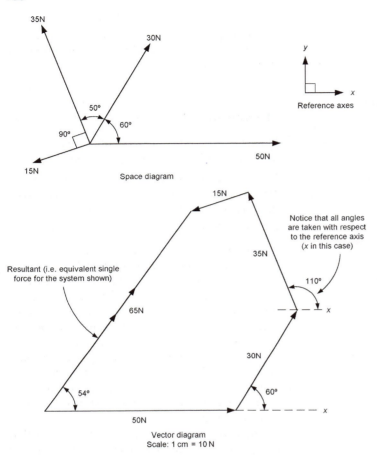

Space diagram

Reference axes

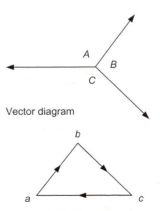

Vector diagram
Scale: 1 cm = 10 N

Figure 4.12 Vector addition using polygon of forces method.

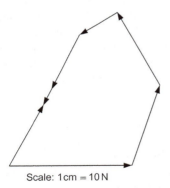

Scale: 1cm = 10 N

Figure 4.13 Equilibrant for Example 4.7.

bc and ca, respectively. This method of labeling applies to any number of forces and their corresponding vectors. Arrowheads need not be used when this notation is adopted, but are shown in Figure 4.14 for clarity.

Vector diagram

Note that arrows are not normally required but are shown here for clarity

Figure 4.14 Bows notation.

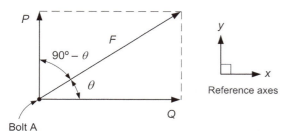

Figure 4.15 Resolving force *F* into its components.

4.7.2 Resolution of forces

Graphical solutions to problems involving forces are sufficiently accurate for many engineering problems and are invaluable for estimating approximate solutions to more complicated force problems. However, it is sometimes necessary to provide more accurate results, in which case a mathematical method will be required. One such mathematical method is known as the *resolution of forces*.

Consider a force *F* acting on a bolt A (Figure 4.15). The force *F* may be replaced by two forces *P* and *Q*, acting at right angles to each other, which together have the same effect on the bolt.

From your knowledge of the trigonometric ratios (Chapter 2) you will know that:

$$\frac{Q}{F} = \cos\theta$$

and so,

$$Q = F\cos\theta$$

Also,

$$\frac{P}{F} = \cos(90 - \theta)$$

and we already know that $\cos(90 - \theta) = \sin\theta$ therefore,

$$P = F\sin\theta$$

So from Figure 4.15,

$$P = F\sin\theta \quad \text{and} \quad Q = F\cos\theta$$

So the single force *F* has been resolved or split into two equivalent forces of magnitude $F\cos\theta$ and $F\sin\theta$, which act at right angles (they are said to be *orthogonal* to each other).

$F\cos\theta$ is known as the *horizontal component of F* and $F\sin\theta$ is known as the *vertical component of F*.

> **Key point**
>
> The resultant of two or more forces is that force which, acting alone, would produce the same effect as the other forces acting together.

Determination of the resultant or equilibrant using the resolution method is best illustrated by the following example.

Example 4.8

Three coplanar forces (forces that act within the same plane) *A*, *B* and *C* are all applied to a pin joint (Figure 4.16(a)). Determine the magnitude and direction of the equilibrant for the system.

Each force needs to be resolved into its two orthogonal (at right-angles) components, which act along the vertical and horizontal axes, respectively. Using the normal algebraic sign convention with our axes, *V* is positive above the origin and negative below it. Similarly, *H* is positive to the right of the origin and negative to the left. Using this convention we need only, consider acute angles for the sine and cosine functions, these are tabulated below:

Mag. (kN)	Horizontal component (kN)	Vertical component (kN)
10	$+10\ (\rightarrow)$	0
14	$+14\cos 60\ (\rightarrow)$	$+14\sin 60\ (\uparrow)$
8	$-8\cos 45\ (\leftarrow)$	$-8\sin 45\ (\downarrow)$

Mag., magnitude.

then total horizontal component $= (10 + 7 - 5.66)\,\text{kN} = 11.34\,\text{kN}\ (\rightarrow)$ and total vertical component $= (0 + 12.22 - 5.66)\,\text{kN} = 6.46\,\text{kN}\ (\uparrow)$

Since both the horizontal and vertical components are positive the resultant force will act upwards to the right of the origin. The three original forces have now been reduced to two which act orthogonally. The magnitude of the

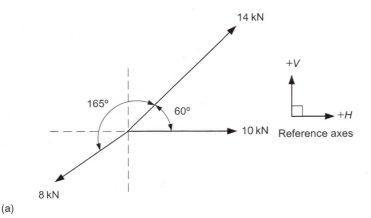

(a)

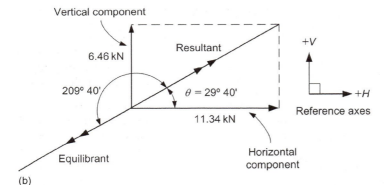

(b)

Figure 4.16 (a) Space diagram for force system. (b) Resolution method.

resultant R or the equilibrant, may now be obtained using the Pythagoras' theorem on the right angle triangle obtained from the orthogonal vectors, as shown in Figure 4.16(b).

From Pythagoras we get,

$$R^2 = 6.46^2 + 11.34^2 = 170.33$$

and so resultant,

$$R = 13.05 \, kN$$

so the magnitude of the *equilibrant* also

$$= 13.05 \, kN.$$

From the right angled triangle shown in Figure 4.16(b), the angle θ that the resultant R makes with the given axes may be calculated using the trigonometric ratios. Then,

$$\tan \theta = \frac{6.46}{11.34} = 0.5697 \quad \text{and} \quad \theta = 29.67°$$

Therefore the resultant R = 13.05 kN∠29.67°.

The *equilibrant* will act in the opposite sense and therefore = **13.05 kN∠209.67°**.

Key point

The equilibrant, is that force which acting alone against the other forces acting on a body in the system, places the body in equilibrium.

To complete our initial study on the resolution of forces, we consider one final example concerned with *equilibrium on a smooth plane*. Smooth in this case implies that the effects of friction may be ignored. When we study dynamics latter on in this chapter, friction and its effects will be covered in some detail.

A body is kept in equilibrium on a plane by the action of three forces as shown in Figure 4.17, these are:

1. The *weight* W of the body acting vertically down.

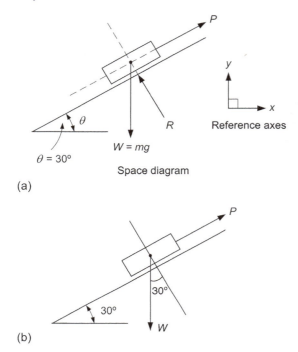

(a)

(b)

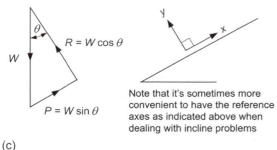

Note that it's sometimes more convenient to have the reference axes as indicated above when dealing with incline problems

(c)

Figure 4.17 Equilibrium on a smooth plane.

2. *Reaction R* of the plane to the weight of the body. R is known as the *normal reaction*, normal in this sense means at right angles to.
3. *Force P* acting in some suitable direction to prevent the body sliding down the plane.

Forces P and R are dependent on the:

- angle of inclination of the plane,
- magnitude of W,
- inclination of the force P to the plane.

It is therefore possible to express the magnitude of both P and R in terms of W and the trigonometric ratios connecting the angles θ and α.

In the example that follows we consider the case when the body remains in equilibrium as a result of the force P being applied parallel to the plane.

Example 4.9

A crate of mass 80 kg is held in equilibrium by a force P acting parallel to the plane as indicated on Figure 4.17(a). Determine, using the resolution method, the magnitude of the force P and the normal reaction R, ignoring the effects of friction.

Figure 4.17(b) shows the space diagram for the problem clearly indicating the nature of the forces acting on the body. W may therefore be resolved into the two forces P and R. Since the force component at right angles to the plane $= W \cos \theta$ and the force component parallel to the plane $= W \sin \theta$ (Figure 4.17(c)).

Equating forces gives,

$$W \cos \theta = R \quad \text{and} \quad W \sin \theta = P$$

So remembering the mass–weight relationship we have:

$$W = mg = (80)(9.81) = 784.8 \, \text{N} \quad \text{then,}$$
$$R = 784.8 \cos 30° = 679.7 \, \text{N} \quad \text{and}$$
$$P = 784.8 \sin 30° = 392.4 \, \text{N}$$

Test your understanding 4.8

1. What is meant by *coplanar* forces?
2. With respect to a system of coplanar forces define: (a) the equilibrant and (b) the resultant.
3. Determine the conditions for static equilibrium of a system of coplanar forces.
4. A body is held in static equilibrium on an inclined plane, ignoring friction, name and show the direction of the forces required to maintain the body in this state.
5. Convert 120 kN into UK tons given that 1 kN = 0.1004 ton.

4.7.3 Moments and couples

A *moment is a turning force*, producing a turning effect. The magnitude of this turning force depends on the size of the *force* applied and the

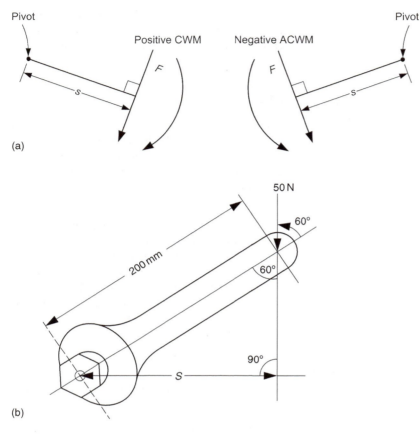

Figure 4.18 Moment of a force.

(a)

(b)

perpendicular distance from the pivot or axis to the line of action of the force (Figure 4.18).

Examples of a turning force are numerous; opening a door, using a spanner, turning the steering wheel of a motor vehicle and an aircraft tailplane creating a pitching moment, are just four examples.

The moment (*M*) of a force is defined as:

the product of the magnitude of force F and its perpendicular distance s from the pivot or axis to the line of action of the force.

This may be written mathematically as:

$$M = Fs$$

The SI unit for a moment is the Nm. You should also note that the *English/American unit* for a moment is the *foot pound-force* (ft/lbf).

From Figure 4.18, you should note that moments can be clockwise, *CWM* or anticlockwise, *ACWM*. Conventionally we

consider CWM to be positive and ACWM to be negative.

Key point

If the line of action passes through the turning point, it has no effect because the effect distance of the moment is zero.

If the line of action of the force passes through the turning point it has no turning effect and so no moment, Figure 4.18a illustrates this point.

Example 4.10

Figure 4.18b shows a spanner being used to tighten a nut, determine the turning effect on the nut.

The turning effect on the nut is equal to the moment of the 50 N force about the nut i.e.

$$M = Fs$$

Remembering that moments are always concerned with perpendicular distances, the distance s is the perpendicular distance or effective length of the spanner. This length is found using trigonometric ratios:

$$s = 200 \sin 60°$$

therefore

$$s = (200)(0.866) = 173.2 \, \text{mm}$$

Then,

Clockwise moment (CWM)

$$= (50)(173.2)$$

$$= 8660 \, \text{Nmm} \quad \text{or} \quad 8.66 \, \text{Nm}$$

So the *turning effect* of the 50 N force acting on a 200 mm spanner at 60° to the center line of the spanner = **8.66 Nm**.

Key point
Moments are always concerned with perpendicular distances.

In engineering problems concerning moments you will meet terminology that is frequently used. You are already familiar with the terms CWM and ACWM. Set out below are three more frequently used terms that you are likely to encounter.

Fulcrum: The *fulcrum* is the point or axis about which rotation takes place. In Example 4.10 above, the geometrical center of the nut is considered to be the fulcrum.

Moment arm: The perpendicular distance from the line of action of the force to the fulcrum is known as the *moment arm*.

Resulting moment: The *resulting moment* is the difference in magnitude between the total CWM and the total ACWM. Note that if the body is in *static equilibrium* this *resultant will be zero*.

Key point
For static equilibrium the algebraic sum of the moments is zero.

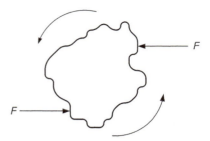

Figure 4.19 Non-equilibrium condition for equal and opposite forces acting on a body.

When a body is in equilibrium there can be no resultant *force* acting on it. However, reference to Figure 4.19 shows that, a body is not necessarily in equilibrium even when there is no resultant force, acting on it. The resultant force on the body is zero but two forces would cause the body to rotate, as indicated. A second condition must be stated to ensure that a body is in equilibrium. This is known as *the principle of moments* which states:

> *When a body is in static equilibrium under the action of a number of forces, the total CWM about any point is equal to the total ACWM about the same point.*

This means that for static equilibrium the *algebraic sum of the moments must be zero*.

Another important fact needs to be remembered about bodies in static equilibrium, consider the uniform beam shown in Figure 4.20. We already know from the principle of moments that the sum of the CWM must equal the sum of ACWM. It is also true that the beam would sink into the ground or rise, if the upward forces did not equal the downward forces. So a further necessary condition for static equilibrium is that:

The upward forces = The downward forces.

We now have sufficient information to readily solve further problems concerning moments.

Example 4.11

A uniform horizontal beam is supported on a fulcrum (Figure 4.21). Calculate the force

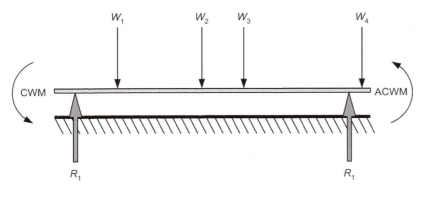

Figure 4.20 Conditions for static equilibrium.

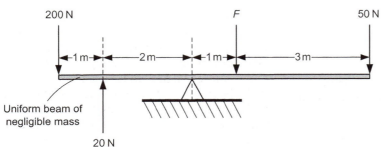

Figure 4.21 Uniform horizontal beam.

F necessary to ensure the beam remains in equilibrium.

We know that the sum of the CWM = the sum of the ACWM, therefore, taking moments about the fulcrum we get:

$$(F \times 1) + (50 \times 4) + (20 \times 2) = (200 \times 3)\,\text{Nm}$$

then,

$$(F \times 1) + 200 + 40 = 600\,\text{Nm} \quad \text{or}$$

$$(F \times 1) = 600 - 200 - 40\,\text{Nm} \quad \text{so}$$

$$F = \frac{360\,\text{Nm}}{1\,\text{m}} = 360\,\text{N}$$

Note:

(a) The 20 N force acting at a distance of 2 m from the fulcrum, tends to turn the beam *clockwise* so is *added* to the sum of the CWM.
(b) The units of F are as required, i.e. they are in N, because the RHS is in Nm and is divided by 1 m.
(c) In this example the weight of the beam has been ignored. If the beam is of uniform

cross-section, then its mass is deemed to act at its geometrical center.

Example 4.12

Figure 4.22 shows an aircraft control system crank lever ABC pivoted at B. AB is 20 cm and BC is 30 cm. Calculate the magnitude of the vertical rod force at C, required to balance the horizontal control rod force of magnitude 10 kN applied at A.

In order to achieve balance of the forces acting on the lever the CWM about B must equal

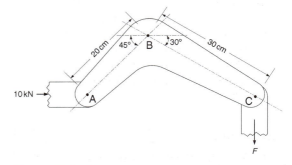

Figure 4.22 Aircraft bell crank control lever.

the ACWM about B. It can also be seen that the 10 kN force produces an ACWM about the fulcrum B. Therefore:

Moment of 10 kN force about B

$$= (10 \times 0.2 \sin 45°) \, \text{kNm}$$

(note the manipulation of units)

$$= (10)(0.2)(0.7071) \, \text{kNm}$$
$$= 1.414 \, \text{kNm}$$

If we now let the vertical force at C, be of magnitude F. Then F produces a CWM about fulcrum B. Therefore:

Moment of force of magnitude F about B

$$= F \times (0.3 \cos 30°) = 0.26 \, F$$

Then applying the principle of moments for equilibrium, we get:

$$1.414 = 0.26 \, F$$

therefore,

$$F = \frac{1.414 \, \text{kNm}}{0.26 \, \text{m}} = \mathbf{5.44 \, kN}$$

Our final example on moments introduces the idea of the *uniformly distributed load (UDL)*. In addition to being subject to point loads, beams can be subjected to loads that are distributed for all, or part, of the beam's length. For UDLs the whole mass of the load is assumed to act as a point load through the center of the distribution.

Example 4.13

For the beam system shown in Figure 4.23, determine the reactions at the supports R_A and R_B, taking into consideration the weight of the beam.

So from what has been said, the UDL acts as a point load of magnitude $(1.5 \, \text{kN} \times 5 = 7.5 \, \text{kN})$ at the center of the distribution, which is 5.5 m from R_A.

In problems involved with reaction it is essential to eliminate one reaction from the calculations because only one equation is formed and only one unknown can be solved at any one time. This is achieved by taking moments about one of the reactions and then, since the distance from that reaction is zero, its moment is zero and, it is eliminated from the calculations.

So taking moments about A (thus eliminating A from the calculations), we get:

$$\left[\begin{array}{l} (2 \times 8) + (5.5 \times 7.5) + (10 \times 5) \\ +(12 \times 12) + (20 \times 20) \end{array} \right] = 16 \, R_B$$

$$\text{or} \quad 651.25 = 16 \, R_B$$

so the *reaction at* B $= \mathbf{40.7 \, kN}$

We could now take moments about B in order to find the reaction at A. However, at this stage it is easier to use the fact that for static equilibrium:

$$\text{Upward forces} = \text{Downward forces}$$

so $\qquad R_A + R_B = 8 + 7.5 + 5 + 12 + 20$

$$R_A + 40.7 = 52.5$$

and so the *reaction at* A $= \mathbf{11.8 \, kN}$.

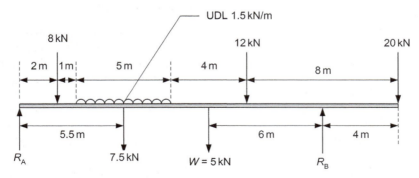

Figure 4.23 Beam system taking account of weight of beam.

4.7.4 Couples

So far we have restricted our problems on moments to the turning effect of forces taken one at a time. A couple occurs when two equal forces acting in opposite directions have their lines of action parallel.

Example 4.14

Figure 4.24 shows the turning effect of a couple, on a beam of regular cross-section.

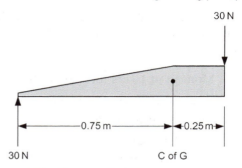

Figure 4.25 Turning effect of a couple with irregular cross-section beam.

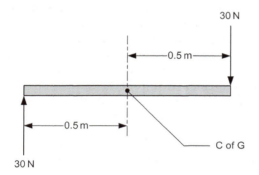

Figure 4.24 Turning effect of a couple.

Taking moments about the *center of gravity* (CG) (the point at which all the weight of the beam is deemed to act), we get:

$$(30 \times 0.5) + (30 \times 0.5) = \text{turning moment}$$

so *turning moment of couple* $= 30\,\text{Nm}$.

Example 4.15

Figure 4.25 shows a beam of irregular cross-section, for this beam the couple will still try to revolve about its CG.

Taking moments about the CG gives:

$$(30 \times 0.75) + (30 \times 0.25) = \text{turning moment}$$

so *the moment of couple* $= 30\,\text{Nm}$.

It can be seen from the above two examples that the moment is the same in both the cases and is independent of the position of the fulcrum. Therefore, if the fulcrum is assumed to

be located at the point of application of one of the forces, *the moment of a couple* is equal to one of the forces multiplied by the perpendicular distance between them. Thus in both cases shown in Examples 4.14 and 4.15 the *moment of the couple* $= 30\,N \times 1\,m = 30\,Nm$, as before.

Another important application of the couple is its *turning moment or torque*. The definition of torque is as follows:

> *Torque is the turning moment of a couple and is measured in Nm:*
> *torque (T) = force (F) × radius (r).*

The *turning moment* of the couple given above in Example 4.15 is $F \times r = (30\,N \times 0.5\,m) = 15\,Nm$.

Key point

The moment of a couple = force × distance between forces and the turning moment = force × radius.

Example 4.16

A nut is to be torque loaded to a maximum of 100 Nm. What is the maximum force that may be applied, perpendicular to the end of the spanner, if the spanner is of length 30 cm?

Since $T = F \times r$, then $F = \dfrac{T}{r} = \dfrac{100\,\text{Nm}}{30\,\text{cm}}$

therefore, $F = 333.3\,\text{N}$

Having studied moments, couples and turning moments, we will now look at how these concepts may be applied to simple aircraft weight and balance calculations.

4.7.5 Aircraft weight and balance calculations

A static aircraft can be represented as a loaded beam with the reactions taken by the under-carriage. So the loads on the undercarriage can be calculated using our previous knowledge of moments. Determining the CG of an aircraft under different loading conditions is an important safety consideration.

Figure 4.26(a) is a pictorial representation of a typical passenger aircraft, showing how the major parts of the aircraft, together with passengers, crew and stores, may be represented as point loads and UDLs.

Figure 4.26(b) shows how, for the purpose of CG calculations, the weights of the

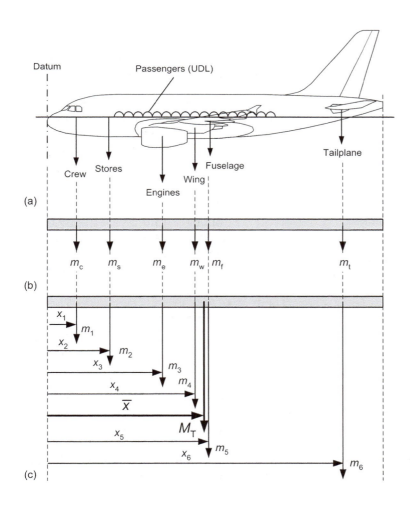

Figure 4.26 Determining aircraft CG.

various parts of the aircraft together with the total weight may be modeled as a simple beam. Figure 4.26(c) show a generalized version of the situation given in Figure 4.26(a) and (b). This generalization enables us to establish a useful formula for determining the *moment arm* (x) for the CG of any aircraft. That is, we can establish how far the CG is from any datum point.

Figure 4.26(c) shows the overall mass of the aircraft (M_T) and the various point and distributed masses labeled as m_1, m_2, m_3, etc. at distances from the datum (which may often be the extreme tip of the nose of the aircraft or at station zero should they differ), labeled x_1, x_2, x_3, etc.

Then the total moment is, in symbols:

$$\bar{x}M_T = m_1 x_1 + m_2 x_2 + m_3 x_3 + \cdots + m_z x_z$$

where $\bar{x}$ = moment arm or distance of the CG from the datum.

If the above equation is divided by M_T we have:

$$\bar{x} = \frac{m_1 x_1 + m_2 x_2 + m_3 x_3 + m_4 x_4 + \cdots + m_z x_z}{M_T}$$

$$= \frac{\sum m_n x_n}{M_T}$$

or in words, the distance of CG from datum

$$\bar{x} = \frac{\text{Sum of the moment of the masses}}{\text{Total mass}}$$

Note that it is not necessary to convert masses to weights for calculation purposes, since each component of the formula would simply be multiplied by a common factor.

Example 4.17

Determine the CG of the aircraft shown in Figure 4.27.

The CG can be determined using the formula,

CG from datum

$$= \frac{\text{Sum of the moments of the masses}}{\text{Total mass}}$$

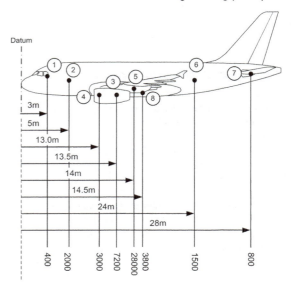

Figure 4.27 Determination of CG position.

It is advisable to display your working in the form of a table as shown below.

Item	Mass (kg)	Distance from datum (m)	Mass-moment (kgm)
1	400	3.0	1200
2	2000	5.0	10,000
3	7200	13.5	97,200
4	3000	13.0	39,000
5	28,000	14.0	392,000
6	1500	24.0	36,000
7	800	28.0	22,400
8	3800	14.5	55,100
Total	46,700		652,900

So position of CG from datum $= \dfrac{652,900 \text{ kgm}}{46,700 \text{ kg}}$

$$= \mathbf{13.98\,m}$$

Having determined the position of the CG it is often necessary to find the change in the CG, which results from either moving a single mass component or altering the magnitude of a mass component. For example, a major modification to say the wing structure may add extra weight, which in turn would alter the mass moment of the wing and thus alter the CG position.

The *change in CG position* as a result of *moving a component* may be determined by multiplying the distance the mass is moved with the ratio of the mass being moved to the total mass.

In symbols, change in CG position is given by,

$$\delta x = \pm \frac{m_1 x_1}{M_T}$$

where δ, the lower case Greek letter *delta* is used to indicate a small change in a variable.

Example 4.18

Find the change in the CG of the aircraft given in Example 4.17, if the CG of the wings is moved forward by 0.2 m.

$$\bar{x} = \pm \frac{(7200)(0.2)}{46,700} = 0.031 \text{ m (forward)}$$

So the new CG position would be $13.98 - 0.031 = 13.95$ m *from nose datum*.

If the *mass* of any single component is changed, the calculation becomes slightly more complicated, this is because the total mass will also be changed. The method of solution is best illustrated by the following example.

Example 4.19

Let us assume that for our previous example (Example 4.17) that 1000 kg of cargo (item 2) is removed from the forward freight bay, at a transit airfield. Our problem is to calculate the new CG position. From our original calculations:

Total mass of aircraft	$= 46,700$ kg
Total moment for aircraft	$= 652,900$ kgm
Cargo removed	$= 1000$ kg
Moment for cargo removed	$= (-1000)(5)$
	$= 5000$ kgm
New total mass for aircraft	$= 46,700 - 1000$
	$= 45,700$ kg
New moment for aircraft	$= 652,900 - 5000$
	$= 647,900$ kgm

So new position of CG from datum

$$= \frac{647,900 \text{ kgm}}{45,700 \text{ kg}} = 14.18 \text{ m}$$

It is important to remember that *if any single mass is altered, this will alter the total mass and total mass moment of the aircraft.*

Alternative method for finding the CG

A standard method of weighing an aircraft is to support the aircraft so that the longitudinal axis and lateral axis are horizontal with the undercarriage resting on weighing units. The readings from the weighing units and the respective distances are used to find the distance of the CG from the relevant datum position of the aircraft. All that is required is for you to remember the criteria for static equilibrium and apply the principle of moments.

4.7.6 Stress and strain

Stress

If a solid, such as a metal bar, is subjected to an external force (or load), a resisting force is set up within the bar and the material is said to be in a state of stress. There are three basic types of stress:

Tensile stress: which is set-up by forces tending to pull the material apart.
Compressive stress: produced by forces tending to crush the material.
Shear stress: resulting from forces tending to cut through the material i.e. tending to make one part of the material slide over the other.

Figure 4.28 illustrates these three types of stress.

Definition of stress

Stress is defined as force per unit area, i.e.

$$\text{Stress}, \sigma = \frac{\text{force}, F}{\text{area}, A}$$

The basic SI unit of stress is N/m^2 other commonly used units include MN/m^2, N/mm^2 and Pa.

Note that the Greek letter σ is pronounced sigma.

In engineering structures, components that are designed to carry tensile loads are known as *ties*, while components design to carry compressive loads are known as *struts*.

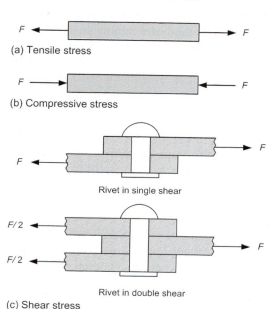

(a) Tensile stress

(b) Compressive stress

Rivet in single shear

Rivet in double shear

(c) Shear stress

Figure 4.28   Basic types of stress.

Strain

A material that is altered in shape due to the action of a force acting on it is said to be *strained*.

This may also mean that a body is strained internally even though there may be little measurable difference in its dimensions, just a stretching of the bonds at the atomic level. Figure 4.29 illustrates three common types of strain resulting from the application of external forces (loads).

Definition of strain

Direct strain may be defined as: *the ratio of change in dimension (deformation) over the original dimension*, i.e.

$$\text{Direct strain}, \, \varepsilon = \frac{\text{deformation}, x}{\text{original length}, l}$$

(both x and l are in metres)

The symbol ε is the Greek lower case letter epsilon. Note also that the deformation for tensile strain will be an extension and for compressive strain it will be a reduction.

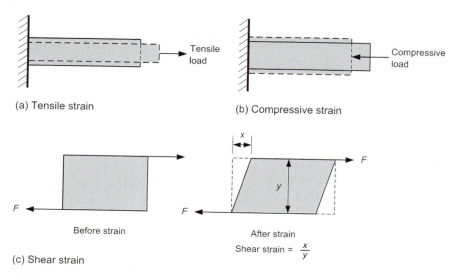

(a) Tensile strain

(b) Compressive strain

(c) Shear strain

Shear strain = $\frac{x}{y}$

Figure 4.29   Common types of strain.

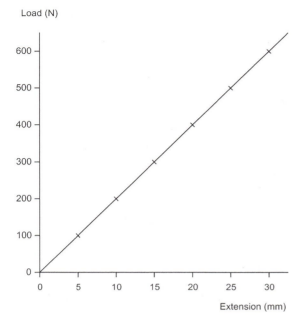

Figure 4.30 Force–extension graph for a spring.

4.7.7 Hooke's law

Hooke's law states that:

> *Within the elastic limit of a material the change in shape is directly proportional to the applied force producing it.*

A good example of the application of Hooke's law is the *spring*. A spring balance is used for measuring weight force, where an increase in weight will cause a corresponding extension (see Figure 4.30).

The stiffness (k) of a spring is the force required to cause a certain (unit deflection):

$$Stiffness\ (k) = \frac{force}{deflection}$$

SI units are N/m or $\mathrm{Nm^{-1}}$.

The concept of *stiffness* will be looked at in a moment, in the mean time here is a question to consider. What does the slope of the graph in Figure 4.30 indicate?

4.7.8 Modulus

Modulus of elasticity

By considering Hooke's law, it follows that stress is directly proportional to strain, while the material remains *elastic*. That is, while the external forces acting on the material are only sufficient to stretch the atomic bonds, without fracture, so that the material may return to its original shape after the external forces have been removed.

Then from Hooke's law and our definition of stress and strain, we know that *stress is directly proportional to strain in the elastic range*, i.e.

$$\text{Stress} \propto \text{strain}$$

or $$\text{Stress} = \text{strain} \times \text{a constant}$$

so $$\frac{\text{Stress}}{\text{Strain}} = \text{a constant } (E)$$

This constant of proportionality, will depend on the material and is given the symbol E. It is known as the *modulus of elasticity* and because strain has no units it has the same units as stress, because the modulus tends to have very high values, $\mathrm{GN/m^2}$ or GPa are the preferred SI units.

Key point

The elastic modulus of a material may be taken as a measure of the stiffness of that material.

Modulus of rigidity

The relationship between the shear stress (τ) and shear strain (γ) is known as the modulus of rigidity (G) i.e.

$$\begin{array}{ll}\text{Modulus} \\ \text{of rigidity} \end{array} = \frac{\text{shear stress } (\tau)}{\text{shear strain } (\gamma)}$$
$$(G) \qquad \mathrm{GPa \ or \ GN/m^2}$$

Note that the symbol τ is the lower case Greek letter *tau* and the symbol γ is the lower case Greek letter *gamma*.

Bulk modulus

If a body of volume v is subject to an increase of external pressure dp which changes its volume by δV, Figure 4.31, the deformation is a change in volume without a change in shape.

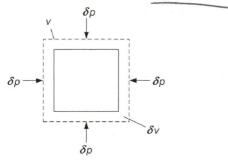

Figure 4.31 Bulk change in volume due to external pressure.

The bulk stress is δp, i.e. an increase in force per unit area, and the bulk strain $\delta v/v$, i.e. change of volume/original volume; the bulk modulus K is then defined by,

$$\text{Bulk modulus} = \frac{\text{bulk stress}}{\text{bulk strain}} = -\frac{\delta p}{\delta v/v}$$

$$= -\frac{v\delta p}{\delta v}$$

The negative sign is introduced to make K positive since the change in volume δv, being a decrease, is negative.

Key point

Solids have all three moduli; liquids and gases only K.

Example 4.20

A rectangular steel bar $10\,\text{mm} \times 16\,\text{mm} \times 200\,\text{mm}$ long extends by $0.12\,\text{mm}$ under a tensile force of $20\,\text{kN}$. Find the:
(a) stress,
(b) strain,
(c) elastic modulus, of the bar material.

(a) Now the

$$\text{Tensile stress} = \frac{\text{Tensile force}}{\text{Cross-sectional area (csa)}}$$

Also tensile force $= 20\,\text{kN} = 20 \times 10^3\,\text{N}$ and csa $= 10 \times 16 = 160\,\text{mm}^2$. Remember tensile loads act against the csa of the material.

Then substituting in above formula we have,

$$\text{Tensile stress } (\sigma) = \frac{20,000\,\text{N}}{160\,\text{mm}^2}$$

$$\sigma = 125\,\text{N/mm}^2$$

(b) Now,

$$\text{Strain, } \varepsilon = \frac{\text{Deformation (extension)}}{\text{Original length}}$$

Also extension $= 0.12\,\text{mm}$ and the original length $= 200\,\text{mm}$, then substituting gives,

$$\varepsilon = \frac{0.12\,\text{mm}}{200\,\text{mm}} = 0.0006\,\text{mm}$$

(c) $\quad E = \dfrac{\text{Stress}}{\text{Strain}} = \dfrac{125\,\text{N/mm}^2}{0.0006}$

$$= 208333.333\,\text{N/mm}^2 \text{ or } \mathbf{208\,GN/m^2}$$

Example 4.21

A $10\,\text{mm}$ diameter rivet holds three sheets of metal together and is loaded as shown in Figure 4.32, find the shear stress in the bar.

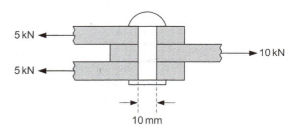

Figure 4.32 Rivet in double shear.

We know that each rivet is in double shear. So the area resisting shear $= 2 \times$ the csa.

$$2\pi r^2 = 2\pi 5^2 = 157\,\text{mm}^2$$

therefore,

$$\text{Shear stress} = \frac{10,000}{157} = 63.7\,\text{N/mm}^2$$

$$= \mathbf{63.7\,MN/m^2}$$

Note that when a rivet is in double shear, the area under shear is multiplied by 2. With respect to the load we know from Newton's laws that to every action there is an equal and opposite reaction, thus *we only use the action or reaction of a force in our calculations, not both.*

Test your understanding 4.10

1. In aircraft weight and balance calculations, write down the formula that enables us to determine the CG from a datum.
2. When determining changes to the CG position. What do we need to remember when the *mass* of an individual component changes?
3. Define: (a) tensile stress, (b) shear stress and (c) compressive stress.
4. State Hooke's law and explain its relationship to the elastic modulus.
5. Define spring stiffness and quote its SI unit.
6. Define in detail the terms: (a) elastic modulus, (b) shear modulus and (c) bulk modulus.
7. Convert the following into N/m^2: (a) $240\,kN/m^2$, (b) $0.228\,GPa$, (c) $600\,N/mm^2$, (d) $0.0033\,N/mm^2$ and (e) $10\,kN/mm^2$.
8. Explain the use of: (a) a strut and (b) a tie.

4.7.9 Some definitions of mechanical properties

The mechanical properties of a material are concerned with its behavior under the action of external forces. This is of particular importance to aeronautical engineers when considering materials for aircraft engineering applications.

The study of aircraft materials, structures and structural maintenance is a major topic in its own right and is considered more fully in a later book in this series. Here, we will concentrate on a few *simple* definitions of the more important mechanical properties of materials that are needed for our study of statics.

These properties include strength, stiffness, specific strength and stiffness, ductility, toughness, malleability and elasticity, in addition to others given below. We have already considered stiffness, which is measured by the elastic modulus. Indirectly, we have also defined strength when we considered the various forms of stress that result from the loads applied on a material. However, a more formal definition of strength follows.

Strength

Strength may be defined simply as the applied force a material can withstand prior to fracture. In fact strength is measured by the *yield stress* σ_y or *proof stress* (see below) of a material. This stress is measured at a known percentage yield, for the material under test. Yielding occurs when the material is subject to loads that cause it to extend by a known fraction of its original length. For metals the measure of strength is often taken at the 0.2% yield or 0.2% proof stress.

Working stress

Following on from the argument given above, we now need to define one or two additional types of stress, since these measure the strength characteristics of materials, under varying circumstances.

Working stress is the stress imposed on the material as a result of the worst possible loads that the material is likely to sustain in service. These loads must be within the elastic range of the material.

Proof stress

Proof stress may be formally defined as: the tensile stress which when applied for a period of 15 s and removed produces a permanent set of a specified amount. For example, 0.2% proof stress will give an elongation of 0.2%, or 0.002 times the original dimension.

Ultimate tensile stress

The ultimate tensile stress (UTS) of a material is given by the relationship, *maximum load/original csa*. Note that the UTS is a measure of the ultimate tensile strength of the material. The point U on the load–extension graph (Figure 4.33), shows maximum load, this must be divided by the original csa not that directly under the point U where the extension may have altered the original csa.

Specific strength

Aircraft materials needs to be as light and strong as possible in order to maximize the payload they may carry, while at the same time meeting the stringent safety requirements laid down

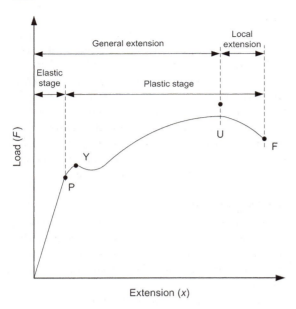

Figure 4.33 Load–extension curve for a mild steel test piece.

for their load bearing structures. Thus to be *structurally efficient* aircraft need to be made of low-density materials, which have the greatest strength. The ratio of the strength of a material (measured by its yield stress) to that of its density is known as *specific strength*, i.e.

$$\text{Specific strength} = \frac{Yield\ stress\ (\sigma_y)}{Density\ (\rho)}$$

SI units are *J/kg.*

Specific stiffness

In a similar manner to the argument given above. The *specific stiffness* of a material is the ratio of its stiffness (measured by its elastic modulus) to that of its density, i.e.

$$\text{Specific stiffness} = \frac{\text{Elastic modulus}\ (E)}{\text{Density}\ (\rho)}$$

SI units are again *J/kg.*

Key point

Specific strength and specific stiffness are measures of the structural efficiency of materials.

Ductility

Ability to be drawn out into threads or wire. Wrought iron, aluminum, copper and low-carbon steels are examples of ductile materials.

Brittleness

Tendency to break easily or suddenly with little or no prior extension. Cast iron, high-carbon steels and glass are examples of brittle materials.

Toughness

Ability to withstand suddenly applied shock loads. Certain alloy steels, some plastics and rubber, are examples of tough materials.

Malleability

Ability to be rolled into sheets or shaped under pressure. Examples of malleable materials include, gold, copper and lead.

Elasticity

Ability of a material to return to its original shape once external forces have been removed. Internal atomic binding forces are stretched but not broken and act like minute springs to return the material to normal, once force has been removed. Rubber, mild- and medium-carbon steels are good examples of elastic materials.

Safety factors

The *safety factor* is used in the design of materials subject to service loads, to give a margin of safety and take account of a *certain factor of ignorance*. Factors of safety vary in aircraft design, dependent on the structural sensitivity of the member under consideration. They are often around 1.5, but can be considerably higher for joints, fittings, castings and primary load bearing structure in general.

4.7.10 Load–extension graphs

These show the results of mechanical tests used to determine certain properties of a material. For instance, as a check to see if heat treatment or processing has been successful, a sample from a batch would be used for such tests.

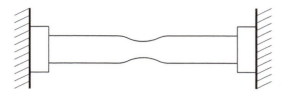

Figure 4.34 Example of waisting where extension is localized.

Load–extension graphs show certain phases, when a material is tested to destruction these include: elastic range, limit of proportionality, yield point, plastic stage and final fracture.

Figure 4.33 shows a typical load–extension curve for a specimen of mild steel which is a ductile material.

The point P at the end of the straight line OP is called the *limit of proportionality*. Between the origin O and P the extension x is directly proportional to the applied force and in this range the material obeys Hooke's law. The *elastic limit* is at or very near the limit of proportionality. When this limit has been passed the extension ceases to be proportional to the load, and at the *yield point* Y the extension suddenly increases and the material enters its plastic phase. At point U (the ultimate tensile strength) the load is greatest. The extension of the test piece has been general up to point U, after which waisting or necking occurs and the subsequent extension is local (Figure 4.34).

Since the area at the waist is considerably reduced then from *stress = force/area*, the stress will increase, resulting in a reduced load for a given stress and so fracture occurs at point F, i.e. at a lower load value than at U.

Remember the elastic limit is at the end of the phase that obeys Hooke's law, after this Hooke's relationship is no longer valid, and full recovery of the material is not possible after removal of the load.

Figure 4.35 shows some typical load–extension curves for some common metals.

The curves in Figure 4.35 shows that annealed copper is very ductile, while hard drawn copper is stronger but less ductile. Hard drawn 70/30 brass is both strong and ductile. Cast iron can clearly be seen as brittle and it is for this reason that cast iron is rarely used under

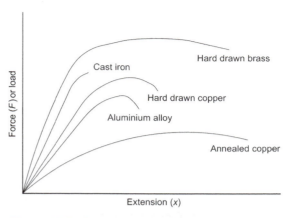

Figure 4.35 Some typical load–extension graphs.

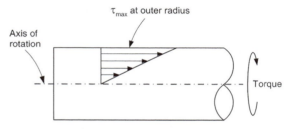

Figure 4.36 Shear stress distribution due to torque.

tensile load. Aluminum alloy can be seen to be fairly strong yet ductile, it has excellent *structural efficiency* and it is for this reason, that it is still used as one of the premier materials for aircraft construction.

4.7.11 Torsion

Drive shafts for aircraft engine driven pumps and motors, propeller shafts, pulley assemblies and drive couplings for machinery, are all subject to torsion or twisting loads. At the same time shear stresses are set up within these shafts (Figure 4.36) resulting from these torsional loads. Aircraft engineers need to be aware of the nature and size of these torsional loads and the subsequent shear stresses in order to design against premature failure and to ensure, through inspection, safe and reliable operation during service.

Drive shafts are, therefore, the engineering components that are used to transmit torsional loads and twisting moments or *torque*. They

Rigid support

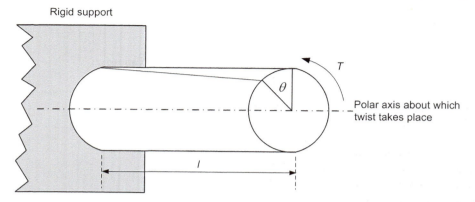

Figure 4.37   Circular shaft subject to torque.

may be of any cross-section but are often circular, since this is the cross-section particularly suited to transmitting torque from pumps, motors and other power supplies used in aircraft engineering systems.

In order to determine the stresses set-up within the drive shaft, we need to use a mathematical relationship often known as *engineers theory of twist* or *the standard equation of torsion of a shaft*. Note from Figure 4.36 that the size of the shear stress increases as we move out from the axis of rotation, in other words as the radius r increases. This axis of rotation is normally called the *polar axis*, because the *angle of twist θ (rad)* of the shaft, which results from the *applied torque or twisting moment T* (Figure 4.37), is measured using *polar co-ordinates*. You will need to refer to page 89, if you are unsure about the use of polar co-ordinates.

One other variable that you have not met, which is used in the engineers theory of twist relationship is known as the *polar second moment of area J*, this variable simply measures the resistance to bending of a shaft, its derivation need not concern us here. It can be shown that the polar second moment of area for a solid circular shaft is given by:

$$J = \frac{\pi D^4}{32}$$

for a hollow shaft (tube):

$$J = \frac{\pi (D^4 - d^4)}{32}$$

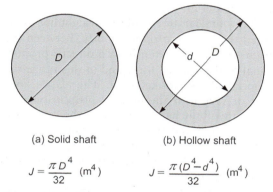

(a) Solid shaft   (b) Hollow shaft

$$J = \frac{\pi D^4}{32} \ (m^4)$$ $$J = \frac{\pi (D^4 - d^4)}{32} \ (m^4)$$

Figure 4.38   Polar second moment of area for solid and hollow shafts.

Figure 4.38 illustrates the polar second moment of area for solid and hollow shafts.

Key point

The polar second moment of area measures the resistance to bending of a shaft.

By combining the above variables with some you have already met, we can produce the standard equation of torsion, which in symbols is:

$$\frac{\tau}{r} = \frac{T}{J} = \frac{G\theta}{l}$$

where:

τ is the shear stress at a distance r from the polar axis of the shaft,
T is the twisting moment on the shaft,

J is the polar second moment of the csa of the shaft,

G is the modulus of rigidity (shear modulus) of the shaft material,

θ is the angle (rad) of twist of a length l of the shaft.

Now the above argument may appear a little complicated but the standard equation of torsion is a very powerful tool which can be used to find any combination of the resulting, torque, angle of twist, or shear stresses acting on the drive shaft.

Example 4.22

A solid circular drive shaft 40 mm in diameter is subjected to a torque of 800 Nm.

(a) Find the maximum stress due to torsion.
(b) Find the angle of twist over a 2 m length of shaft given that the modulus of rigidity of the shaft is 60 GN/m^2.

(a) The maximum stress due to torsion occurs when the radius is a maximum at the outside of the shaft, i.e. when $r = R$. So in this case $R = 20$ mm. Now using the standard relationship,

$$\frac{\tau}{r} = \frac{T}{J}$$

We have the values R and T, so we only need to find the value of J for our solid shaft and then we will be able to find the maximum value of the shear stress τ_{max}.

Then for a solid shaft,

$$J = \frac{\pi D^4}{32}$$

and so

$$\frac{\pi 40^4}{32} = 0.251 \times 10^6 \text{ mm}^4$$

and on substitution into the standard relationship given above we have:

$$\tau = \frac{(20)(800 \times 10^3)}{0.251 \times 10^6} \frac{(\text{mm})(\text{Nmm})}{\text{mm}^4}$$

$$\text{giving} \quad \tau_{max} = 63.7 \text{ N/mm}^2$$

This value is the maximum value of the shear stress, which occurs at the outer surface of the shaft. Notice the manipulation of the units, care must always be taken to ensure consistency of units, especially where powers are concerned!

(b) To find θ we again use engineer's theory of torsion, which after rearrangement gives:

$$\theta = \frac{lT}{GJ}$$

and substituting our known values for l, T, J and G we have:

$$\theta = \frac{(2000)(800 \times 10^3)}{(60 \times 10^3)(0.251 \times 10^6)}$$

$$\frac{(\text{mm})(\text{Nmm})}{(\text{N/mm}^2)(\text{mm}^4)} = 0.106 \text{ rad}$$

So, angle of twist = 6.07°.

Test your understanding 4.11

1. Explain how the strength of solid materials is determined.
2. What is the engineering purpose of the *factor of safety*?
3. What is the difference between ductility and malleability?
4. With respect to tensile testing and the resultant load–extension graph, define: (a) limit of proportionality, (b) UTS, (c) yield point and (d) plastic range.
5. With respect to the theory of torsion define: (a) polar axis, (b) polar second moment of area and (c) torque.
6. Why is the study of torsion important to engineers?

General Questions 4.2

1. For the force system (Figure 4.39) determine *graphically* the magnitude and direction of

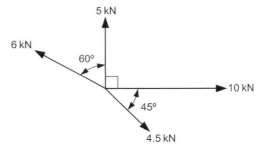

Figure 4.39 Force system.

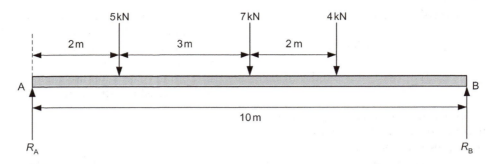

Figure 4.40 Beam system.

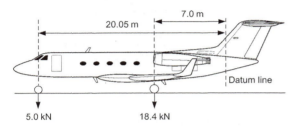

Figure 4.41 Aircraft CG.

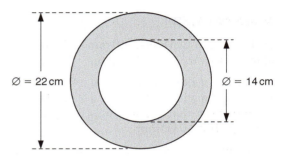

Figure 4.43

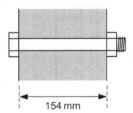

Figure 4.42 Bolt.

the equilibrant. Then use a *mathematical method*, to check the accuracy of your result.

2. Determine the reactions at the supports for the beam system shown in Figure 4.40. Assume the beam has negligible mass.

3. A uniform beam of length 5 m and weight 10 kN has to support a UDL of 1.5 kN/m, along its whole length. It is simply supported at either end. Find the reactions at the supports.

4. Find the distance of the CG from the datum point, for the aircraft shown in Figure 4.41. Note that the weights given are

those at each undercarriage leg and remember that an aircraft has two main undercarriage legs!

5. An aircraft structure contains a steel tie rod that carries a load of 100 kN. If the allowable tensile stress is 75 MN/m^2, find the minimum diameter of the tie rod.

6. The bolt shown in Figure 4.42 has a thread of 1 mm pitch. If the nut is originally tight, and neglecting any compression in the material through which the bolt passes, find the increase in stress in the bolt when the nut is tightened by rotating it through one eighth of a turn. Take the elastic modulus E as 200 GN/m^2.

7. During a test to destruction carried out on a mild steel test specimen, original diameter 24 mm, gauge length 250 mm, the following results were obtained.

8. Calculate the power transmitted by the hollow shaft with the cross-section shown in Figure 4.43, given that the maximum shear stress is 65 MN/m^2.

Load (kN)	Extension (mm)	Load (kN)	Extension (mm)
11.95	0.03	91.8	0.254
19.9	0.056	100	0.274
28.8	0.081	110.6	0.305
40.25	0.118	120	0.355
49.8	0.14	129.5	0.366
61.7	0.173	139.5	0.68 Y.P.
70.7	0.198	198.8	Maximum load
79.7	0.203		

After the test the diameter at fracture was found to be 15 mm and the length was 320 mm.

Draw the load–extension graph and determine the:

(a) elastic stress limit,
(b) ultimate tensile strength,
(c) percentage extension in length,
(d) percentage reduction in area,
(e) 0.1% proof stress.

4.8 Dynamics

4.8.1 Linear equations of motion

You have already been introduced to the concept of force, velocity, acceleration and Newton's laws, these are further exploited through the use of the equations of motion. Look back now, and remind yourself of the relationship between mass, force, acceleration and Newton's laws.

The linear equations of motion rely for their derivation on the one very important fact that the *acceleration is assumed to be constant*. We will now consider the derivation of the four standard equations of motion using a graphical method.

Velocity–time graphs

Even simple linear motion, motion along a straight line, can be difficult to deal with mathematically. However in the case where acceleration is constant it is possible to solve problems of motion by the use of a *velocity–time graph*, without recourse to the calculus. The equations

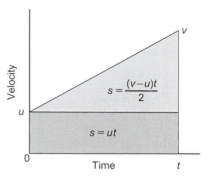

Figure 4.44 Velocity–time graph for uniform acceleration.

of motion use standard symbols to represent the variables, these are shown below:

s = distance (m)
u = initial velocity (m/s)
v = final velocity (m/s)
a = acceleration (m/s^2)
t = time (s)

The *velocity is plotted on the vertical axis and time on the horizontal axis*. Constant velocity is represented by a *horizontal straight line* and acceleration by a *sloping straight line*. Deceleration or *retardation* is also represented by a sloping straight line but with a *negative slope*.

Key point

Velocity is speed in a given direction and is a *vector* quantity.

By considering the velocity–time graph shown in Figure 4.44, we can establish the equation for distance.

The distance traveled in a given time is equal to the velocity m/s multiplied by the time s, this is found from the graph by the *area under the sloping line*. In Figure 4.44, a body is accelerating from a velocity u to a velocity v in time t seconds.

Now the distance traveled,

$$s = \text{area under graph}$$

$$s = ut + \frac{(v - u)}{2} \times t$$

$$s = ut + \frac{vt}{2} - \frac{ut}{2}$$

$$s = \frac{(2u + v - u)t}{2}$$

Thus,

$$s = \frac{(u + v)t}{2}$$

In a similar manner to above, one of the velocity equations can also be obtained from the velocity–time graph. Since the acceleration is the rate of change of velocity with respect to time, the value of the acceleration will be equal to the gradient of a velocity–time graph. Therefore, from Figure 4.44, we have:

$$\text{Gradient} = \frac{Change\ in\ velocity}{Time\ taken} = Acceleration$$

therefore acceleration is given by,

$$a = \frac{v - u}{t} \quad \text{or}$$

$$v = u + at$$

The remaining equations of motion may be derived from the two equations found above. Try now, as an exercise in manipulating formulae, to obtain:

(a) the equation $t = \dfrac{v - u}{a}$,

(b) $s = ut + \frac{1}{2} at^2$,

using the above equations.

Example 4.23

A body starts from rest and accelerates with constant acceleration of 2.0 m/s² up to a speed of 9 m/s. It then travels at 9 m/s for 15 s after which, it is retarded to a speed of 1 m/s. If the complete motion takes 24.5 s, find the:
(a) time taken to reach 9 m/s,
(b) retardation,
(c) total distance traveled.

The solution is made easier if we sketch a graph of the motion, as shown in Figure 4.45.

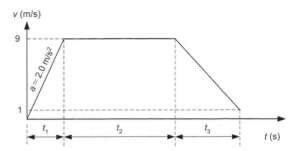

Figure 4.45 Velocity–time graph of the motion.

(a) We first tabulate the known values:

$$u = 0\ \text{m/s (we start from rest)}$$
$$v = 9\ \text{m/s}$$
$$a = 2\ \text{m/s}^2$$
$$t = ?$$

All we need to do now is select an equation which contains all the variables listed above,

i.e. $\qquad\qquad v = u + at$

and on transposing for t and substituting the variables we get,

$$t = \frac{9 - 0}{2}$$

so $\qquad\qquad t = 4.5\ \text{s}$

(b) The retardation is found in a similar manner.

$$u = 9\ \text{m/s}$$
$$v = 2\ \text{m/s}$$
$$t = 5\ \text{s}$$
$$a = ?$$

We again select an equation which contains the variables,

i.e. $\qquad\qquad v = u + at$

and on transposing for a and substituting the variables we get,

$$a = \frac{1 - 9}{5}$$

so $\qquad\qquad a = -1.6\ \text{m/s}^2$

(the −ve sign indicates a retardation)

(c) The total distance traveled requires us to sum the component distances traveled for the times t_1, t_2 and t_3. Again we tabulate the variable for each stage:

$u_1 = 0 \, \text{m/s}$ $u_2 = 9 \, \text{m/s}$ $u_3 = 9 \, \text{m/s}$

$v_1 = 9 \, \text{m/s}$ $v_2 = 9 \, \text{m/s}$ $v_3 = 1 \, \text{m/s}$

$t_1 = 4.5 \, \text{s}$ $t_2 = 15 \, \text{s}$ $t_3 = 5 \, \text{s}$

$s_1 = ?$ $s_2 = ?$ $s_3 = ?$

The appropriate equation is:

$$s = \frac{(u+v)t}{2}$$

and in each case we get,

$$s_1 = \frac{(0+9)4.5}{2} = 20.25$$

$$s_2 = \frac{(9+9)15}{2} = 135$$

$$s_3 = \frac{(9+1)5}{2} = 25$$

Then total distance $S_T = 20.25 + 135 + 25$

$$= 180.25 \, \text{m}$$

4.8.2 Using Newton's laws

You saw earlier that *Newton's second law* may be defined as:

$$F = ma$$

or $$F = \frac{mv - mu}{t}$$

In words, we may say that force is equal to the rate of change of momentum of a body. Look back again and make sure you understand the relationship between force, mass and the momentum of a body. Remembering that *momentum* may be defined as: *the mass of a body multiplied by its velocity*. Also that; *the inertia force is such as to be equal and opposite to the accelerating force that produced it*, this essentially is Newton's third law.

Key point

The inertia force is equal and opposite to the accelerating force.

Example 4.24

A light aircraft of mass 1965 kg accelerates from 160 to 240 kph in 3.5 s. If the air resistance is 2000 N/tonne, find the:
(a) average acceleration,
(b) force required to produce the acceleration,
(c) inertia force on the aircraft,
(d) propulsive effort on the aircraft.

(a) We first need to convert the velocities to standard units:

$$u = 160 \, \text{kph} = \frac{160 \times 1000}{60 \times 60} = 44.4 \, \text{m/s}$$

$$v = 240 \, \text{kph} = \frac{240 \times 1000}{60 \times 60} = 66.6 \, \text{m/s}$$

also $t = 3.5$ s, and we are required to find the acceleration a.

Then using the equation $v = u + at$ and transposing for a we get:

$$a = \frac{v - u}{t} \quad \text{and substituting values}$$

$$a = \frac{66.6 - 44.4}{3.5}$$

$$a = 6.34 \, \text{m/s}^2$$

(b) The accelerating force is readily found using Newton's second law, where:

$$F = ma = 1965 \, \text{kg} \times 6.34 \, \text{m/s}^2$$

$$= 12.46 \, \text{kN}$$

(c) From what has already been said you will be aware that the inertia force = the accelerating force, therefore the **inertia force = 12.46 kN**.

(d) The propulsive force must be sufficient to overcome the inertia force and that of the force due to the air resistance.

$$\text{Force due to air resistance} = \frac{2000 \times 1965}{1000}$$

$$= 3930 \, \text{N}$$

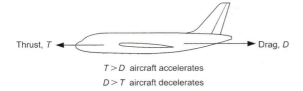

Figure 4.46 Thrust and drag forces.

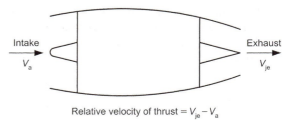

Relative velocity of thrust = $V_{je} - V_a$

Figure 4.47 Jet thrust relative velocities.

Remembering that there are 1000 kg in a metric ton (the tonne). then,

$$\begin{matrix} \text{Propulsive} \\ \text{force} \end{matrix} = \begin{matrix} \text{Inertia} \\ \text{force} \end{matrix} + \begin{matrix} \text{Force due to} \\ \text{air resistance} \end{matrix}$$

$$= 12.46 + 3.93 \text{ kN}$$
$$= \mathbf{16.39 \text{ kN}}$$

Propulsive thrust

When an aircraft is travelling through air in straight and level flight and at constant true airspeed, the engines must produce a total thrust equal to the air resistance (drag force) on the aircraft, as shown in Figure 4.46. This is a consequence of Newton's first law.

If the engine thrust exceeds the drag, the aircraft will accelerate (Newton's laws) and if the drag exceeds the thrust the aircraft will slow down.

Although there are a variety of engine types available for aircraft propulsion, the thrust force must always come from air or gas pressure forces normally acting on the engine or propeller.

A propeller can either be driven by a piston or gas turbine engine, it increases the mass flow rate (kg/s) of the air passing through it and thus produces a net thrust force. One method of calculating this thrust produced by the propeller is provided by Newton's third law,

Force = Mass × Acceleration

$$\text{Thrust} = \begin{matrix} \text{Mass flow rate} \\ \text{of the air through} \\ \text{the propeller} \end{matrix} \times \begin{matrix} \text{Increase in} \\ \text{velocity of} \\ \text{the air} \end{matrix}$$

$$\mathbf{Thrust = \dot{m}(V_{je} - V_a)}$$

where:

$\dot{m}$ = mass flow rate of the air (kg/s);

V_a = true velocity of the aircraft, i.e. true airspeed or TAS, which you will meet later (m/s);

V_{je} = velocity of slipstream (m/s).

Make sure that you understand that mass flow rate multiplied by velocity gives the units of force.

Key point

The mass flow rate of a fluid multiplied by its velocity equals the force produced by the fluid.

If the aircraft uses a jet engine, then a high-velocity exhaust gas is produced. For the air-breathing (turbojet) engine the jet velocity is considerably higher than the TAS of the aircraft. Thrust is again produced according to the equation given above for the propeller engine, except that now V_{je} represents the effective velocity of the gas stream (Figure 4.47) at the exhaust of the jet pipe. Once again the thrust comes from gas pressure forces, but in this case they act on the surface of the engine itself.

Example 4.25

(a) The mass airflow through a propeller is 400 kg/s. If the inlet velocity is 0 m/s and the outlet velocity is 50 m/s. What thrust is developed?
(b) Now assume that the mass airflow through a gas turbine engine is 40 kg/s. If the inlet velocity is 0 m/s and the exhaust jet velocity is 500 m/s. What thrust is developed?

We use the simplified version of the thrust equation to solve both parts (a) and (b).

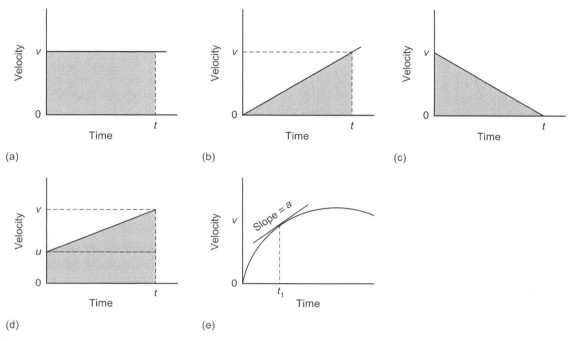

Figure 4.48

(a) Thrust force $= \dot{m}(V_{je} - V_a)$
$$= 400(50 - 0) = 20\,\text{kN}$$
(b) Thrust force $= \dot{m}(V_{je} - V_a)$
$$= 40(500 - 0) = 20\,\text{kN}$$

Make sure you work through the above calculations and understand the units.

This simplified example shows that in order to develop similar amounts of thrust we may accelerate a large mass of air at relatively low speed or accelerate a small mass of air at relatively high speed. If, in the future, you study aircraft propulsion in detail you will see that the former method of developing thrust in a gas turbine engine, is more efficient. This is why these engines are used in most modern commercial airliners.

Engine thrust is often quoted in lb, with the reference to force being ignored. When we use *Imperial units*, the formula for thrust becomes:

$$\text{Thrust force (lb)} = \frac{w}{g}(V_{je} - V_a)$$

where $w =$ flow rate of air (lb/s), $g =$ acceleration due to gravity (32 ft/s²), $V_{je} =$ velocity of slipstream or exhaust (as before) but units are

ft/s, $V_a =$ aircraft velocity (TAS) but units are ft/s.

Using the above formula, with the units stated, will give thrust in lbf. We generally quote thrust in lb and simply ignore the reference to force.

Example 4.26

A twin-engine gas turbine powered aircraft is at rest and preparing for take-off. Each engine mass airflow at take-off is 80 lb/s and the exhaust velocity for each engine is 1400 ft/s. What thrust is being produced by each engine?

Now, $w = 80$ lb/s, $V_a = 0$ and $V_{je} = 1400$,

$$g = 32.2\,\text{ft/s}^2$$

$$\text{Thrust} = \frac{80}{32.2}(1400 - 0) = 3478.3\,\text{lb}$$

Test your understanding 4.12

With reference to the velocity–time graphs shown overleaf (Figure 4.48), answer questions 1 to 8.

Fill in the gaps for questions 1 to 8.

1. The slope of the velocity–time graph measures _____.

2. The area under a velocity–time graph determines _____.

3. Average velocity may be determined by dividing the _____, _____ by _____, _____.

4. Graph (a) is a graph of constant velocity therefore acceleration is given by _____ and the distance traveled is equal to _____.

5. Graph (b) shows uniformly accelerated motion therefore the distance traveled is equal to _____.

6. Graph (c) shows _____, _____, _____.

7. Graph (d) represents uniformly accelerated motion having initial velocity u, final velocity v and acceleration a. So distanced traveled is equal to _____.

8. Graph (e) represents _____.

9. Define the terms: (a) inertia force and (b) momentum.

10. What is the essential difference between speed and velocity?

11. If a rocket is sent to the moon, its mass remains constant but its weight changes, explain this statement.

12. Explain how the expression $F = ma$ is related to the rate of change of momentum with respect to Newton's second law.

13. Define V_{je} for (a) the propeller engine and (b) the jet engine.

14. Under what operating circumstances would the thrust produced by a jet engine be a maximum?

4.8.3 Angular motion

You previously met the equations for linear motion. A similar set of equations exists to solve engineering problems that involve angular motion as experienced, e.g. in the rotation of a drive shaft. The linear equations of motion may be transformed to represent angular motion using a set of equations that we will refer to as the *transformation equations*. These are given below, followed by the equations of angular motion, which are compared with their linear equivalents.

Transformation equations

$$s = \theta r$$
$$v = \omega r$$
$$a = \alpha r$$

where r = radius of body from center of rotation and θ, ω and α are the angular distance, angular velocity and angular acceleration, respectively.

Angular equation of motion	Linear equation of motion
$\theta = (\omega_1 + \omega_2)t/2$	$s = (u + v)t/2$
$\theta = \omega_1 t + \frac{1}{2}\alpha t^2$	$s = ut + \frac{1}{2}at^2$
$\omega_2^2 = \omega_1^2 + 2\alpha\theta$	$v^2 = u^2 + 2as$
$\alpha = (\omega_2 - \omega_1)/t$	$a = (v - u)/t$

Angular velocity

Angular velocity (ω) refers to a body moving in a circular path and may be defined as:

$$\omega = \frac{\text{Angular distance moved (rad)}}{\text{Time taken (s)}}$$

or in symbols $\omega = \theta/s$ (radians per second).

Angular distance is measured in rad, you should refer back to page 122 if you cannot remember the definition of the radian or how to convert radians to degrees and vice versa.

We are often given rotational velocity in rpm. It is therefore useful to be able to convert rpm into rad/s and vice versa.

> **Key point**
>
> 1 rev $= 2\pi$ rad (from the definition of the radian).

> **Key point**
>
> 1 rpm $= 2\pi$ rad/min $= 2\pi/60$ rad/s.

So, e.g. to convert 350 rpm into rad/s we multiply by $2\pi/60$, i.e.

$$350\,\text{rpm} = 350 \times \frac{2\pi}{60} = 36.65\,\text{rad/s}$$

> **Example 4.27**

A 540 mm diameter wheel is rotating at $1500/\pi$ rpm. Determine the angular velocity of wheel in rad/s and the linear velocity of a point on the rim of the wheel.

All we need to do to find the angular velocity is convert rpm to rad/s, i.e.

$$\text{Angular velocity (rad/s)} = \frac{1500}{\pi} \times \frac{2\pi}{60}$$

$$= 50\,\text{rad/s}$$

Now from the transformation equations, linear velocity,

$$v = \text{angular velocity}, w \times \text{radius}, r$$
$$= 50 \, \text{rad/s} \times 0.270 \, \text{m}$$
$$v = \textbf{13.5 m/s}$$

Angular acceleration

Angular acceleration (α) is defined as the rate of change of angular velocity with respect to time, i.e.

$$\alpha = \frac{\text{Change in angular velocity (rad/s)}}{\text{Time (s)}}$$

So, units for angular acceleration are $\alpha = \theta/s^2$.

Example 4.28

A pinion is required to move with an initial angular velocity of 300 rpm and final angular velocity of 600 rpm. If the increase takes place over 15 s, determine the linear acceleration of the rack. Assume a pinion radius of 180 mm.

In order to solve this problem we first need to convert the velocities into rad/s:

$$300 \, \text{rpm} = 300 \times 2\pi/60 = 31.4 \, \text{rad/s}$$
$$600 \, \text{rpm} = 600 \times 2\pi/60 = 62.8 \, \text{rad/s}$$

We can use the equation $\alpha = \frac{(\omega_2 - \omega_1)}{t}$ to find the angular acceleration.

So $\quad \alpha = \dfrac{62.8 - 31.4}{15} = 2.09 \, \text{rad/s}^2$

Now we can use the transformation equation $a = \alpha r$ to find the linear acceleration, i.e.

$$a = (2.09 \, \text{rad/s})(0.18 \, \text{m}) = \textbf{0.377 m/s}^2$$

Torque and angular acceleration

We can apply Newton's third law of motion to angular motion, if it is realized that the distribution of mass relative to the axis of rotation, has some bearing on the calculation. For this reason it is not possible to deal directly with, a

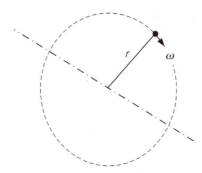

Figure 4.49 A point mass subject to a rotational velocity.

rotating wheel, but rather with a small element of mass whose radius of rotation can be more easily defined.

Figure 4.49 shows a small element of mass δm rotating at a radius, r from the center O, with uniform angular velocity, w (rad/s). We know from the transformation equations that the linear velocity at any instant is given by:

$$v = \omega r$$

and from Newton's third law, to accelerate this mass would require a force such that:

$$F = ma$$

In this case the force would be applied at the radius r and thus would constitute a moment or more correctly a torque T, about the center of rotation thus:

$$T = Fr \quad \text{or} \quad T = mar$$

Since the linear acceleration, $a = \alpha r$,

$$T = m(\alpha r)r \quad \text{or} \quad \textbf{\textit{T} = \textit{m}α\textit{r}}^2$$

The quantity mr^2 is a concentrated mass multiplied by its radius of rotation squared and is known as the, *moment of inertia I*. The quantity I is an important property of a rotating body, in the SI system it has units kgm^2. Therefore, substituting I for mr^2 in our above equation $T = m\alpha r^2$, gives:

$$T = I\alpha$$

The last relationship may be compared with $F = ma$ for linear motion.

Key point

Think of the moment of inertia of a rotating body, as being equivalent to the mass of a body subject to linear motion.

Example 4.29

An aircraft propeller has a moment of inertia of 130 kgm². Its angular velocity drops from 12,000 to 9000 rpm in 6 s, determine the (a) retardation and (b) braking torque.

Now,

$$\omega_1 = 12,000 \times 2\pi/60 = 1256.6\,\text{rad/s}$$
$$\omega_2 = 9000 \times 2\pi/60 = 942.5\,\text{rad/s}$$

and from,

$$\alpha = \frac{\omega_2 - \omega_1}{t}$$

$$\alpha = \frac{942.5 - 1256.6}{6}$$

$$\alpha = -52.35 \quad \text{or}$$

$$\text{retardation} = 52.35\,\text{rad/s}^2$$

Now torque,

$$T = I\alpha$$
$$T = (130)(52.35)$$

so braking torque,

$$T = 6805.5\,\text{Nm}$$

Centripetal acceleration and force

If we consider Figure 4.49 again we can see that that the direction of the mass must be continually changing to produce the circular motion, therefore, it is being subject to an acceleration, which is acting towards the center, this acceleration is known as the *centripetal acceleration* and is equal to $\omega^2 r$. When acting on a mass this acceleration produces a force known as centripetal force, thus:

$$\frac{\text{Centripetal}}{\text{force }(F_c)} = \text{Mass} \times \frac{\text{Centripetal}}{\text{acceleration}}$$

$$F_c = m\omega^2 r$$

and since $v = \omega r$

$$F_c = \frac{mv^2}{r}$$

From Newton's third law, there must be an *equal and opposite force* opposing the centripetal force this is known as the *centrifugal force* and *act outwards* from the center of rotation.

Key point

Centripetal force acts inwards towards the center of rotation, centrifugal force acts in the opposite direction.

Example 4.30

An aircraft with a mass of 80,000 kg is in a steady turn of radius 300 m, flying at 800 kph. Determine the centripetal force required to hold the aircraft in the turn.

Then,

$$\left.\begin{array}{r}\text{the linear velocity} \\ \text{of the aircraft}\end{array}\right\} = \frac{800 \times 1000}{3600}\,\text{m/s}$$
$$= 222.2\,\text{m/s}$$

and from $F_c = mv^2/r$ we get,

$$F_c = \frac{(80,000)(222.2)^2}{300} = 13.17\,\text{MN}$$

4.8.4 Gyroscopes

Gyroscopic motion

Before we leave angular motion we will consider one important aircraft application of the inertia and momentum of a body in circular motion, that of the *gyroscope*. You will remember from our discussion on Newton's laws that we defined *momentum* as: *the product of the mass of a body and its velocity*, it is really a measure of the *quantity of motion* of a body. Also the force that resists a change in momentum (i.e. resists acceleration) is known as **inertia**.

A gyroscope (Figure 4.50(a)) is essentially a rotating mass that has freedom to move at right angles to its plane of rotation. Gyroscopic instruments utilize either or both of two fundamental characteristics of a gyro rotor, that of **rigidity** or gyroscopic inertia and **precession**.

Rigidity is an application of Newton's first law of motion where a body remains in its state of rest or uniform motion unless compelled by some external force to change that state. If a gyro rotor is revolving it will continue to rotate

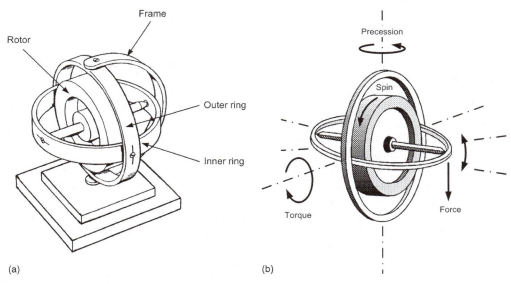

Figure 4.50 (a) A gyroscope; (b) gyroscopic precession.

about that axis unless a force is applied to alter the axis. Now the greater the momentum of the rotor, i.e. the heavier it is and the faster it rotates (*mv*), the greater is the gyro's resistance to change and so it has greater *rigidity* or inertia. The property of rigidity is important since the whole point of a gyroscope is to act as a reference point in space under particular circumstances, no matter what the attitude of the aircraft.

Precession may be defined simply as *the reaction to a force applied to the axis of a rotating assembly.* The actual nature of this reaction is a little more difficult to understand and is illustrated below using *Sperry's rule.*

Laws of gyro-dynamics

The two properties of rigidity and precession provide the visible effects of the laws of gyro-dynamics, which may be stated as follows:

1. If a rotating body is mounted so as to be free to move about any axis through the center

of mass, then its spin axis remains fixed in inertial space no matter how much the frame may be displaced.

2. If a constant torque is applied about an axis, perpendicular to the axis of spin, of an unconstrained, symmetrical, spinning mass, then the spin axis will precess steadily about an axis mutually perpendicular to both spin and torque axis.

Sperry's rule of precession

The direction in which precession takes place is dependent upon the direction of rotation for the mass and the axis about which the torque is applied. *Sperry's rule of precession*, illustrated in Figure 4.50(b), provides a guide as to the direction of precession, knowing the direction of the applied torque and the direction of rotation of the gyro-wheel.

If the applied torque is created by a force acting at the inner gimbol, perpendicular to the spin axis, it can be transferred as a force, to the edge of the rotor, at right angles to the plane of rotation. The point of application of the force should then be carried through 90° in the direction of rotation of the mass and this will be the point at which the force appears to act. It will move that part of the rotor rim, in the direction of the applied disturbing force.

Gyroscopic wander

Movement between the spin axis and its frame of reference may be broken down into two main causes, *real wander*, which is actual misalignment of the spin axis due to mechanical defects in the gyroscope, and *apparent wander*, which is discernable movement of the spin axis due to the reference frame in space, rather than spin axis misalignment. *Wander in a gyroscope is termed drift or topple*, dependent upon the axis about which it takes place. If the spin axis wanders in the azimuth plane it is known as *drift* and in the vertical plane it is referred to as *topple*.

Thus in real wander, the problems of friction in the gimbol bearings and imperfect balancing of the rotor cause torques to be set-up perpendicular to the rotor spin axis, this leads to precession and actual movement, or real wander of the spin axis. There are two main causes of apparent wander, one due to rotation of the earth and the other due to movement over the earth's surface of the aircraft, carrying the gyroscope.

Test your understanding 4.13

1. Define the following, stating their SI units (a) angular velocity and (b) angular acceleration.
2. A body acting at a radius of 175 mm has a tangential (linear) velocity of 25 m/s, find its angular velocity.
3. Convert the following angular velocities into standard SI units: (a) 250 rev/min, (b) 12,500 rev/h (c) 175 rev/s.
4. Define: (a) torque and (b) moment of inertia.
5. Explain why the moment of inertia is used instead of the total mass of the body, when considering objects subject to angular motion?
6. Define the terms: (a) centripetal acceleration and (b) centrifugal force.
7. If an aircraft is in a steady turn, explain the nature of the forces acting on the aircraft during the turn. Which one of these forces holds the aircraft in the turn?
8. Define the terms: (a) momentum and (b) inertia.
9. Define rigidity, explaining the factors upon which the rigidity of a gyro rotor depends.
10. Define precession and explain why the direction of tilt is at right angles to the force producing it.

4.8.5 Vibration and periodic motion

All mechanisms and structures that occur in engineering are capable of vibration or oscillation. This is because they possess both mass and elasticity and are therefore known collectively, as elastic systems.

The result of vibration may be useful as in, for example, a stringed instrument where the string is plucked and made to oscillate to produce a musical sound. The result of vibration may also be harmful, as in an aircraft structure where continuous vibration may lead to premature failure due to metal fatigue.

In all cases oscillations are lessened and may die away completely, due to damping. Damping is the resistance to movement of the system components, due to factors such as air resistance, friction and fluid viscosity (see Section 4.9.4).

Vibrations may be classified as either free or forced. Free vibration refers to an elastic system where having started to vibrate, due to an initial disturbance, it is allowed to continue unhindered. The simply supported spring–mass system shown in Figure 4.51 when subject to an initial push or pull away from its equilibrium position and then allowed to vibrate is a classic example of a *free vibration system*.

In order to examine oscillatory motion, we need first to define some common terms which are used to describe the nature of this type of motion. You have already met these terms, in a slightly different form, when you studied sinusoidal functions in your mathematics

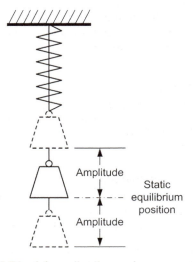

Figure 4.51 A free vibrating spring–mass system.

(Chapter 3). Look back to page 124 and compare the sinusoidal function, with the definitions for general oscillatory motion given below:

Period: This is the time that elapses while the motion repeats itself. Most oscillatory motions repeat themselves in equal intervals of time and are called *periodic*.

Cycle: This is the motion completed in one period.

Frequency: This is the number of cycles completed in unit time. For example, a frequency of 50 Hz, as before, is equal to 50 c/s.

Amplitude: This is the distance of either the highest or lowest point of the motion from the central position.

Forced vibration refers to a vibration that is excited by an external force applied at regular intervals. The system will no longer vibrate at its natural frequency but will oscillate at the frequency of the external exciting force. Thus, e.g. a motor with an out of balance rotor will set up a forced vibration on the supporting structure, on which it rests.

Resonance

The phenomenon known as resonance may be illustrated (Figure 4.52) using an apparatus known as Barton's pendulums.

This consists of a series of paper cone pendulums which are given additional mass by use of plastic rings, or similar. The pendulums progressively vary in length and are all suspended from the same cord. A heavy bob-weight driving pendulum is pulled well aside, so that it oscillates perpendicular to the plane of the paper. The motion settles down after a period of time so that the paper pendulums oscillate at very nearly the same *frequency* as the driver but with different *amplitudes*, thus the pendulums are subject to forced vibration.

The pendulum whose length equals that of the driver has the greatest amplitude and its natural frequency of oscillation is the same as the frequency of the driving pendulum, this is an example of *resonance* (Figure 4.53), where the driving pendulum transfers its energy most easily to the paper cone pendulum having the same length.

The amplitudes of oscillations also depend on system damping. If we remove the plastic rings from the cone pendulums, their mass is reduced and so damping is increased. All amplitudes are reduced, where that of the resonant frequency is less pronounced.

Resonance may be desirable or a source of trouble, dependent on the system. In electronic systems resonance is used in the tuning mechanism, where the frequency of the desired radio signal is matched with the natural frequency of the tuner. In mechanical systems resonance is a problem, e.g. in bridges and other large civil

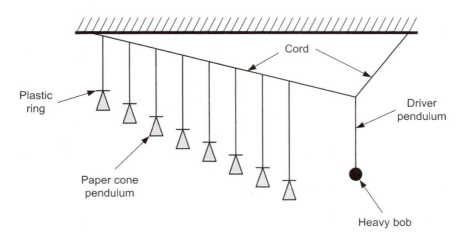

Figure 4.52 Barton's pendulums apparatus.

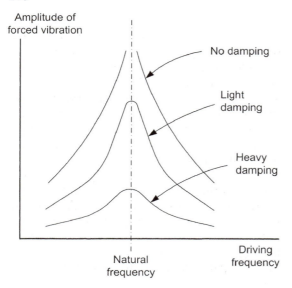

Figure 4.53 Resonance and the effects of damping.

engineering structures, when the wind produces an oscillation that is in harmony with the natural frequency of the structure. The oscillations set-up on the millennium bridge, when it first opened, resulted from the pace of the people walking across it!

Key point

Resonance occurs when a system is forced to vibrate at a frequency equal to its natural frequency.

4.8.6 Simple harmonic motion

Simple harmonic motion (SHM) is defined as the periodic motion of a body where the acceleration is:

(a) *always towards a fixed point in its path,*
(b) *proportional to its displacement from that point.*

Motion closely approximating SHM occurs in a number of natural or free vibration systems. Examples include, springs, spring–mass systems and engineering beams.

In Figure 4.54 point P moves with uniform speed $v = \omega r$ around a circle of radius r. Then the point M projected from P on diameter

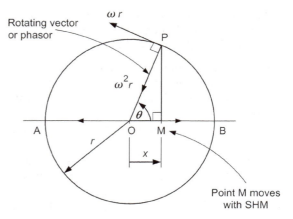

Figure 4.54 Phasor representation of SHM.

AB moves with SHM. The acceleration of P is the centripetal acceleration, $\omega^2 r$. Then the displacement, velocity and acceleration of M are respectively:

Displacement, $x = \text{OM} = r \cos \theta = r \cos \omega t$
where t is the time measured from the instant when P and M are at B and $\theta = 0$.

Velocity, $v = \omega r \sin \theta = -\omega r \sin \omega t$
Acceleration, $a = \omega^2 r \cos \theta$
$$= -\omega^2 r \cos \omega t = -\omega^2 x$$

You should recognize that the expressions for velocity and acceleration can be derived from the expression for displacement by differentiating with respect to time. The negative signs in the expressions for velocity and acceleration show that for the position of M (Figure 4.54), both velocity and acceleration are in the opposite direction from the displacement. Displacement and acceleration are always in opposite directions.

The periodic time, T of the motion is the time taken for one complete oscillation of the point x (see previous definition of period). In this time the phasor OP (rotating vector) makes one complete revolution, therefore:

$T = 2\pi/\omega$ and since $a = \omega^2 x$ or $\omega = \sqrt{a/x}$

then,

$$T = 2\pi \sqrt{\frac{\text{displacement}, x}{\text{acceleration}, a}}$$

hence,

$$T = 2\pi \sqrt{\frac{x}{a}}$$

The frequency f in Hz is given by:

$$f = \frac{\omega}{2\pi} = \frac{1}{T}$$

therefore,

$$\text{frequency } f = \frac{1}{2\pi \sqrt{\frac{x}{a}}}$$

The maximum velocity of x occurs at the midpoint, where it equals the velocity of P, i.e.

$$v_{\max} = \omega r$$

The maximum acceleration of x occurs at the extreme positions A and B where it equals the acceleration of P, i.e.

$$a_{\max} = \omega^2 r$$

The velocity of x is zero at A and B, its acceleration is zero at O. The amplitude of the oscillation is r and the distance AB ($2r$) is sometimes called the stroke or travel of the motion.

I hope you have grasped this rather complicated theory, if you are worried about the derivation of the mathematical expressions, you should revise the work we did on trigonometric functions and the differential calculus, starting on pages 124 and 154, respectively. We have derived several formulae, so lets look at an example that illustrates their use.

Example 4.31

A body moves with SHM with amplitude 50 mm and frequency 2.5 Hz. Find: (a) the maximum velocity and acceleration, stating where they occur and (b) the velocity and acceleration of the motion at a point 25 mm from the mean position.

(a) We first convert the frequency into rad/s, in order to use the expressions for maximum velocity and acceleration. Then,

$$\text{frequency} = 2.5 \, \text{Hz} = \omega/2\pi$$

$$\text{giving } \omega = 5\pi \quad \text{or} \quad 15.71 \, \text{rad/s}$$

so maximum velocity $= \omega r = (15.71)(50)$

$$= 785 \, \text{mm/s}$$

$$= 0.785 \, \text{m/s}$$

$$\text{maximum acceleration} = \omega^2 r = (15.71)^2(50)$$

$$= 12,340 \, \text{mm/s}^2$$

$$= 12.34 \, \text{m/s}^2$$

The velocity is a maximum at the equilibrium position and acceleration occurs at maximum amplitude, the extreme point of the motion.

(b) For a displacement of 25 mm,

$$\cos \theta = 25/50 = 0.5, \quad \text{giving} \quad \theta = 60°$$

Therefore,

$$\text{the velocity} = \omega r \sin \theta$$

$$= (15.71)(50)(\sin 60)$$

$$= 680.3 \, \text{mm/s}$$

$$\text{or} \quad 0.6803 \, \text{m/s}$$

$$\text{the acceleration} = \omega^2 r \cos \theta$$

$$= (15.71)^2(50)(\cos 60)$$

$$= 6.17 \, \text{m/s}^2$$

The spring–mass system

We have derived several equations for SHM, these can be modified to take into account differing systems, that display SHM. Consider the spring–mass system illustrated in Figure 4.55. If, from its position of rest, the mass m is pulled down a distance x and then released, the mass will oscillate vertically.

In the rest position the force in the spring will exactly balance the force of gravity acting on the mass. If s is the *spring stiffness, i.e. the force per unit change of length* (N/m), then for a displacement x from the rest position, the change in force in the spring is sx. This change of force is the unbalanced accelerating force F acting on the mass m. Then:

Force = spring stiffness × the extension
(N)   (N/m)   (m)

or $F = s \times x$

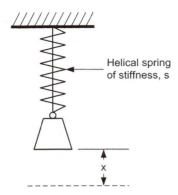

Figure 4.55 Free vibrating spring–mass system.

This demonstrates that the acceleration is directly proportional to the displacement from its rest position. The motion is therefore simple harmonic.

The periodic time is given by:

$$T = 2\pi \sqrt{\frac{x}{a}}$$

and from $F = s \times x$ then acceleration $= F/m = sx/m$ so,

$$T = 2\pi \sqrt{\frac{\frac{x}{sx}}{m}} = 2\pi \sqrt{\frac{xm}{sx}}$$

thus $\boldsymbol{T = 2\pi \sqrt{\dfrac{m}{s}}}$ and frequency $\boldsymbol{f = \dfrac{1}{2\pi} \sqrt{\dfrac{s}{m}}}$

Example 4.32

A helical spring hangs vertically. A load of 10 kg hanging from it causes it to extend 20 mm. The load is pulled down a further distance of 25 mm and then released. Find the frequency of the resulting vibration, the maximum velocity and acceleration of the load and the maximum force in the spring.

The weight of the load $= mg = (10)(9.81)$

$$= 98.1\,\text{N}$$

Spring stiffness, $s = \dfrac{\text{Force}}{\text{Extension}}$

$$= 98.1/20\,\text{mm}$$
$$= 4.905\,\text{N/mm}$$
$$= 4905\,\text{N/m}$$

Now since the frequency of the vibration $f = \frac{1}{T}$, then

$$f = \frac{1}{2\pi} \sqrt{\frac{s}{m}}$$
$$= \frac{1}{2\pi} \sqrt{\frac{4905}{10}}$$
$$= 3.52\,\text{Hz}$$

Now the amplitude x of the vibration is 25 mm. The maximum velocity of the load is ωx where, $\omega = 2\pi f$. You should be able to see that the angular velocity in rad/s is equal to the frequency or cycles per second multiplied by 2π!

So,

$$v_{\max} = \omega x = 2\pi f x = (2\pi)(3.52)(25)$$
$$= 552.64\,\text{mm/s} \quad \text{or} \quad \mathbf{0.553\,m/s}$$

The maximum acceleration of the load is

$$= \omega^2 x = (2\pi \times 3.52)^2(25)$$
$$= 12{,}238.8\,\text{mm/s} \quad \text{or} \quad \mathbf{12.24\,m/s}$$

Finally the maximum force in the spring is the product:

Maximum extension × Spring stiffness

$$= (20\,\text{mm} + 25\,\text{mm})(4.905\,\text{N/mm})$$
$$= \mathbf{220.75\,N}$$

The pendulum

A *simple pendulum* consists of a light inextensible cord that is fixed at one end. The other end is attached to a concentrated mass, which oscillates about the equilibrium position. A *compound pendulum* is one in which the mass is not concentrated, as is the case with most engineering components. We will not consider the compound pendulum, at this stage in your studies.

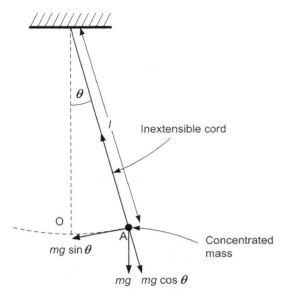

Figure 4.56 The simple pendulum.

From the Figure 4.56, the unbalanced restoring force which acts towards the center O, is given by the tangential component $mg \sin \theta$. If a is the acceleration of the bob along the arc a due to the force $mg \sin \theta$ then the equation of motion of the bob is, $-mg \sin \theta = ma$. The minus sign indicates that the force is towards O, while the displacement x is measured along the arc from O in the opposite direction (remembering that the acceleration always acts in the opposite direction to the displacement).

Now when t is small, $\sin t \cong t$ (rad). Also, from the equation for arc length $s = r\theta$, then $x = l\theta$. You should refer back to radian measure on page 122, if you are unsure of this step!

Now, substituting these values into our equation of motion:

$$-mg \sin \theta = ma$$

gives
$$-mg\theta = -mg\frac{x}{l} = ma$$

where $-gx/l = a$, is the component of g acting along the arc therefore

$$a = \frac{-gx}{l} = -\omega^2 x$$

(from our work before, where $a = \omega^2 x$, then $\omega^2 = g/l$).

The motion of the bob is simple harmonic if the oscillations are of small amplitude, i.e. θ does not exceed 10°. The period T is then given by

$$T = \frac{2\pi}{\omega} = \frac{2\pi}{\sqrt{\frac{g}{l}}}$$

thus,

$$T \propto \sqrt{\frac{l}{g}}$$

T is therefore independent of the amplitude of the oscillations and for constant g, it depends only on the length l of the pendulum.

Example 4.33

A simple pendulum has a period of 4.0 s and an amplitude of swing of 100 mm. Calculate the maximum magnitudes of (a) the velocity of the bob, (b) the acceleration of the bob.

(a) From $T = 2\pi/\omega$, transposing for ω and substituting for T we have $\omega = \pi/2$ per second. The velocity will be a maximum at the equilibrium position where $x = 0$ and using:

$$\omega_{max} = \pm \omega r = \pm (\pi/2)(0.1) = 0.157 \, \text{m/s}$$

since maximum amplitude $r = \pm 100$ mm.

(b) The acceleration is a maximum at the limits of the swing, where $x = r = \pm 100$ mm and using:

$$a = -\omega^2 r$$

then
$$a = -(\pi/2)^2 (0.1) \, \text{m/s}^2$$

$$a = -0.246 \, \text{m/s}^2$$

Test your understanding 4.14

1. Explain the difference between *free* and *forced* vibration.

2. Define the terms: (a) period, (b) cycle, (c) frequency and (d) amplitude with respect to periodic motion.

3. Define resonance *and* give examples of where resonance can be useful *and* where it is considered harmful.

4. Define SHM.

5. In SHM, under what circumstances is (a) the velocity a maximum and (b) the acceleration a maximum?

6. For (a) the mass–spring system and (b) the simple pendulum, explain with the aid of sketches how the amplitude is determined.

7. Define spring stiffness.

8. With respect to radian measurement, explain the expression $s = r\theta$.

4.8.7 Mechanical work, energy and power

Work done

The *energy* possessed by a body is its capacity to do work. So, before we discuss energy, let us first consider the concept of work. Mechanical work is done when a force overcomes a resistance and it moves through a distance.

Mechanical work may be defined as:

Mechanical work done (WD) (J)

$$= \begin{array}{l} \textit{Force required to} \\ \textit{overcome the} \\ \textit{resistance} \text{ (N)} \end{array} \times \begin{array}{l} \textit{Distance moved} \\ \textit{against the} \\ \textit{resistance} \text{ (m)} \end{array}$$

The SI unit of work is Nm or J where $1\,\text{J} = 1\,\text{Nm}$.

Note:

(a) No work is done unless there is both resistance and movement.
(b) The resistance and the force needed to overcome it are equal.
(c) The distance moved must be measured in exactly the opposite direction to that of the resistance being overcome.
(d) The English Engineering unit of work is the ft lbf.

Key point

Mechanical energy may be defined as the capacity to do work.

The more common resistances to be overcome include: *friction*, *gravity* (the weight of the body itself) and *inertia* (the resistance to acceleration of the body) where:

the WD against friction

$= $ Friction force $\times$ Distance moved

WD against gravity

$= $ Weight $\times$ Gain in height

WD against inertia

$= $ Inertia force $\times$ Distance moved

Note:

(a) Inertia force is the out-of-balance force

the Inertia force $=$ mass $\times$ acceleration

(b) Work done in overcoming friction will be discussed in more detail later.

In any problem involving calculation of work done, the first task should be to identify the type of resistance to overcome. If, and only if, there is motion between surfaces in contact, is work done against friction. Similarly, only where there is a gain in height there is work done against gravity and only if a body is accelerated work done is against inertia (look back at our definition of inertia).

Example 4.34

A body of mass 30 kg is raised from the ground at constant velocity through a vertical distance of 15 m. Calculate the work done.

If we ignore air resistance, then the only work done is against gravity.

WD against gravity $=$ Weight $\times$ Gain in height or WD $= mgh$ (and assuming $g = 9.81\,\text{m/s}^2$) then

$$\text{WD} = (30)(9.81)(15)$$
$$\text{WD} = 4414.5\,\text{J} \quad \text{or} \quad \textbf{4.414}\,\textbf{kJ}$$

Work done may be represented graphically and, for linear motion, this is shown in Figure 4.57(a). Where the force needed to overcome the resistance, is plotted against the distance moved. The WD is then given by the area under the graph.

Figure 4.57(b) shows the situation for angular motion, where a varying torque T in Nm is plotted against the angle turned through in rad. Again the work done is given by the area under the graph, where the units are Nm $\times$ rad. Then noting that the radian has no dimensions, the unit for work done remains as Nm or J.

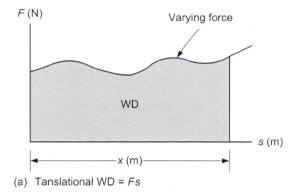

(a) Tanslational WD = Fs

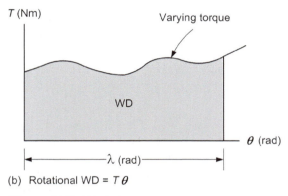

(b) Rotational WD = $T\theta$

Figure 4.57 Work done.

Energy

Energy may exist in many different forms, e.g. mechanical, electrical, nuclear, chemical, heat, light and sound.

The principle of the conservation of energy states that: energy may neither be created nor destroyed, only changed from one form to another.

There are many engineering examples of devices that transform energy, these include the:

- loudspeaker which transforms electrical to sound energy,
- petrol engine which transforms heat to mechanical energy,
- microphone which transforms sound to electrical energy,
- dynamo transforms mechanical to electrical energy,
- battery transforms chemical to electrical energy,
- filament bulb transforms electrical to light energy.

In our study of dynamics we are primarily concerned with mechanical energy and its conservation. Provided no mechanical energy is transferred to or from a body, the total amount of mechanical energy possessed by a body remains constant, unless mechanical work is done. This concept is further explored in the next section.

Mechanical energy

Mechanical energy may be subdivided into three different forms; *potential energy (PE), strain energy and kinetic energy (KE).*

PE is the energy possessed by a body by virtue of its position, relative to some datum. The change in PE is equal to its weight multiplied by the change in height. Since the weight of a body is *mg*, then the change in PE may be written as:

$$\text{Change in PE} = mgh$$

which of course is identical to the work done in overcoming gravity. So, the work done in raising a mass to a height is equal to the PE it possesses at that height, assuming no external losses.

Key point

Strain energy is a particular form of PE.

Strain energy is a particular form of PE possessed by an elastic body that is deformed within its elastic range, e.g. a stretched or compressed spring possesses strain energy.

Consider the spring arrangement shown in Figure 4.58. We know from our previous work that the force required to compress or extend the spring is $F = kx$, where k is the spring constant.

Figure 4.58(a) shows a helical coil spring in the unstrained, compressed and extended positions. The force required to move the spring varies in direct proportion to the distance moved

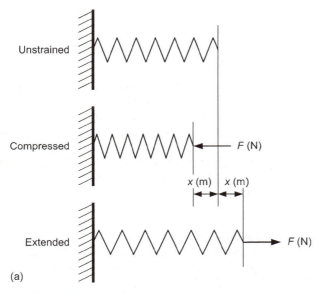

Figure 4.58 Spring system demonstrating strain energy.

(Figure 4.58(b)). Therefore:

Strain energy of spring when compressed or extended	Area under graph = (force × distance moved)
	$= \frac{1}{2}Fx$ J

and since $F = kx$, then substituting for F gives,

Strain energy of spring in tension or compression $= \frac{1}{2}kx^2$ J

A similar argument can be given for a spring which is subject to twisting or torsion about its center (or polar axis). It can be shown that:

Strain energy of a spring when twisted

$= \frac{1}{2}k_{tor}\theta^2$ J   (where $\theta =$ the angle of twist)

KE is energy possessed by a body by virtue of its motion. Translational KE, i.e. the KE of a body travelling in a linear direction (straight line) is:

$$\text{Translational KE (J)} = \frac{\begin{bmatrix} mass\ (kg) \times \\ velocity^2\ (m/s)^2 \end{bmatrix}}{2}$$

Translation KE $= \frac{1}{2}mv^2$

Flywheels are heavy wheel-shaped masses fitted to shafts in order to minimize sudden variations in the rotational speed of the shaft, due to sudden changes in load. A flywheel is therefore a store of rotational KE.

Rotational KE can be defined in a similar manner to translational KE, i.e.

Rotational KE $= \frac{1}{2}I\omega^2$ J

where I = mass moment of inertia (which you met when we studied torsion).

Note: The moment of inertia of a rotating mass I can be defined in general terms by the expression $I = Mk^2$ where $M =$ the *total mass* of the rotating body and $k =$ the *radius of gyration*, i.e. the radius from the center of rotation where all of the mass is deemed to act. When we studied torsion earlier we defined I for *concentrated* or *point masses*, where $I = mr^2$. You should remember that I has different values for different rotating shapes. We will only be considering circular cross-sections, where I is defined as above. One final point, try not to mix-up k for the radius of gyration with k for the spring constant!

Example 4.35

Determine the total KE of a four-wheel drive car which has a mass of 800 kg and is travelling at 50 kph. Each wheel of the car has a mass of 15 kg, a diameter of 0.6 m and a radius of gyration of 0.25 m.

$$\text{Total KE} = \text{Translational (linear) KE}$$
$$+ \text{Angular KE}$$

and

$$\text{Linear KE} = \tfrac{1}{2}mv^2$$

(where $v = 50\,\text{kph} = 13.89\,\text{m/s}$)

$$= \tfrac{1}{2}(800)(13.89)^2$$
$$= 77.16\,\text{kJ}$$

and

$$\text{Angular KE} = \tfrac{1}{2}I\omega^2$$

where $I = Mk^2 = (15)(0.25)^2 = 0.9375\,\text{kgm}^2$ (for each wheel!) and from $v = \omega r$ then $\omega = v/r = 13.89/0.3 = 46.3\,\text{rad/s}$:

$$= \tfrac{1}{2}(4 \times 0.9375)(46.3)^2$$
$$= 4.019\,\text{kJ}$$

Therefore, Total KE of the car

$$= 77.16 + 4.019 = 81.18\,\text{kJ}$$

Conservation of mechanical energy

From the definition of the conservation of energy we can deduce that the total amount of energy within certain defined boundaries, will remain the same. When dealing with mechanical systems, the PE possessed by a body is frequently converted into KE and vice versa. If we *ignore air frictional losses*, then:

$$\text{PE} + \text{KE} = \text{a constant}$$

Thus, if a mass m falls freely from a height h above some datum, then at any height above that datum:

$$\text{Total energy} = \text{PE} + \text{KE}$$

This important relationship is illustrated in Figure 4.59, where at the highest level above the datum the PE is a maximum and is gradually converted into KE, as the mass falls towards the datum, immediately before impact when height $h = 0$, the PE is zero and the KE is equal to the initial PE.

Since the total energy is constant, then:

$$mgh_1 = mgh_2 + \tfrac{1}{2}mv_2^2 = mgh_3 + \tfrac{1}{2}mv_3^2$$
$$= \tfrac{1}{2}mv_4^2$$

Immediately after impact with the datum surface, the mechanical KE is converted into other forms, such as heat, strain and sound.

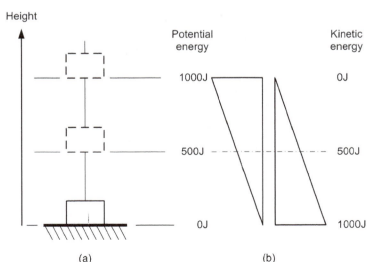

Figure 4.59 PE plus KE equals a constant.

If friction is present then work is done overcoming the resistance due to friction and this is dissipated as heat. Then:

$$\frac{\text{Initial}}{\text{energy}} = \frac{\text{Final}}{\text{energy}} + \frac{\text{WD in overcoming}}{\text{frictional resistance}}$$

Note: KE is not always conserved in collisions. Where KE is conserved in a collision we refer to the collision as *elastic*, when KE is not conserved we refer to the collision as *inelastic*.

Example 4.36

Cargo weighing 2500 kg breaks free from the top of the cargo ramp (Figure 4.60). Ignoring friction, determine the velocity of the cargo the instant it reaches the bottom of the ramp.

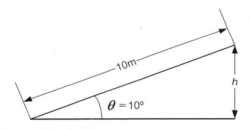

Figure 4.60 Cargo ramp.

The vertical height h is found using the sine ratio, i.e.

$$10 \sin 10 = h \quad \text{so,} \quad h = 1.736 \, \text{m}$$

increase in $\text{PE} = mgh$

$$= (2500)(9.81)(1.736) \, \text{J}$$

$$= 42{,}575.4 \, \text{J}$$

Now, using the relationship $\text{PE} + \text{KE} = \text{Total energy}$. Then immediately prior to the cargo breaking away $\text{KE} = 0$ and so, $\text{PE} = \text{Total energy}$. Also, immediately prior to the cargo striking the base of slope, $\text{PE} = 0$ and $\text{KE} = \text{Total energy}$ (all other energy losses being ignored).

So, at the base of the slope:

$$42{,}575.4 \, \text{J} = \text{KE}$$

and

$$42{,}575.4 = \tfrac{1}{2} mv^2$$

i.e.

$$\frac{(2)(42{,}575.4)}{2500} = v^2$$

and therefore, velocity at bottom of ramp = **5.83 m/s** (check this working for yourself!).

Power

Power is a measure of the rate at which work is done or the rate of change of energy. *Power is therefore defined as the rate of doing work.* The SI unit of power is the watt (W), i.e.

$$\text{\textit{Power} (W)} = \frac{\textit{Work done} \, (\text{J})}{\textit{Time taken} \, (\text{s})}$$

$$= \frac{\textit{Energy change} \, (\text{J})}{\textit{Time taken} \, (\text{s})}$$

or, if the body moves with constant velocity,

$$\textit{Power} \, (\text{W}) = \textit{Force used} \, (\text{N}) \times \textit{Velocity} \, (\text{m/s})$$

Note: Units are $\text{Nm/s} = \text{J/s} = \text{W}$.

> **Key point**
>
> Power is the rate of doing work.

Example 4.37

A packing crate weighing 1000 N is loaded into an aircraft freight bay by being dragged up an incline of 1 in 5 at a steady speed of 2 m/s. The frictional resistance to motion is 240 N. Calculate the:

(a) power needed to overcome friction,
(b) power needed to overcome gravity,
(c) total power needed.

(a)   Power = Friction force
                    × Velocity along surface

$$= 240 \times 2$$

$$= 480 \, \text{W}$$

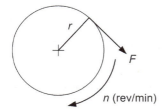

Figure 4.61   Power transmitted by a torque.

(b)   Power = Weight × Vertical component
                     of velocity

$$= 1000 \times 2 \times \frac{1}{5}$$

$$= 400\,\text{W}$$

(c) Since there is no acceleration and therefore,
no work done against inertia,

Total power = Power for friction
                    + Power for gravity
                    = 480 + 400
                    = 880 W

Let us now consider power transmitted by a
torque. You have already met the concept of
torque. Figure 4.61 shows a force F (N) applied
at radius r (m) from the center of a shaft that
rotates at n (rpm).

Since the work done is equal to the force multiplied by the distance, then the work done in
1 rev is given by:

$$\text{WD in 1 rev} = F \times 2\pi r\,\text{J}$$

but Fr is the torque T applied to the shaft,
therefore,

$$\text{WD in 1 rev} = 2\pi T\,\text{J}$$

In 1 min the work done = WD per revolution
                                      × Number of rpm (n)
                                      $= 2\pi n T$

and   WD in 1 s $= 2\pi n T/60$

and since work done per second is equal to
power (1 J/s = 1 W).

Then,

Power (W) transmitted by a torque

$$= 2\pi n T/60$$

Test your understanding 4.15

1. Define work done.
2. Write down the equation for work done against gravity, stating SI units.
3. State the principle of the conservation of energy.
4. Detail the forms of energy input and output for the following devices (a) generator, (b) gas turbine engine, (c) battery and (d) radio.
5. What does the symbol k represent in the formula $F = kx$ and what are its SI units?
6. Write down the formulae for both linear and rotational KE and explain the meaning of each of the symbols within these formulae.
7. Machine A delivers 45,000 J of energy in 30 s, machine B produces 48 kNm of work in 31 s, which machine is more powerful and why?

4.8.8 Friction

We have already met *friction*, in terms of the
*frictional force that tends to oppose relative
motion*, but up till now we have not fully defined
the nature of friction.

When a surface is moved over another surface with which it is in contact, a resistance is
set up opposing this motion. The value of the
resistance will depend on the materials involved,
the condition of the two surfaces, and the force
holding the surfaces in contact; but the opposition to motion will always be present. This
resistance to movement is said to be the result
of *friction* between the surfaces.

We require a slightly greater force to start
moving the surfaces (*static friction*) than we do
to keep them moving (*sliding friction*). As a
result of numerous experiments involving different surfaces in contact under different forces,
a set of rules or laws has been established which,
for all general purposes, materials in contact
under the action of forces, seem to obey. These
rules are detailed below together with one or
two limitations for their use.

Laws of friction

1. The frictional forces always oppose the direction of motion, or the direction in which a
body is tending to move.
2. The sliding friction force F opposing motion,
once motion has started, is proportional to

the normal force N that is pressing the two surfaces together, i.e. $F \propto N$.

3. The sliding frictional force is independent of the area of the surfaces in contact. Thus two pairs of surfaces in contact made of the same materials and in the same condition, with the same forces between them, but having different areas, will experience the same frictional forces opposing motion.

4. The frictional resistance is independent of the relative speed of the surfaces. This is not true for very low speeds or in some cases, for fairly high speeds.

5. The frictional resistance at the start of sliding (*static friction*) is slightly greater than that encountered as motion continues (*sliding friction*).

6. The frictional resistance is dependent on the nature of the surfaces in contact. For example, the type of material, surface geometry, surface chemistry, etc.

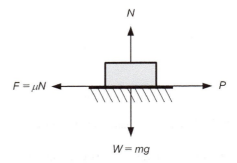

Figure 4.62 Space diagram for arrangement of forces.

It is the coefficient of static friction μ_s, that we use in the examples below, which is considered to be the *limiting friction coefficient*.

You may find the solution of problems involving friction rather difficult. This is because it is often difficult to visualize the nature and direction of all the forces that act on two bodies in contact, as well as resolving these forces into their component parts. Problems involving friction may be solved by calculation or by drawing. The following generalized example involving the simple case of a block in contact with a horizontal surface should help you understand both methods of solution.

Key point

Friction always opposes the motion that produces it.

Solving problems involving friction

From the above laws we have established that the sliding frictional force F is proportional to the normal force N pressing the two surfaces together, i.e. $F \propto N$. You will remember from your mathematical study of proportion, that in order to equate these forces we need to insert a constant, the constant of proportionality, i.e. $F = \mu N$. This constant μ is known as the coefficient of friction and in theory it has a maximum value of 1. Figure 4.62 shows the space diagram for the arrangement of forces on two horizontal surfaces in contact.

Note: The value of the force required to just start to move a body is greater than the force needed to keep the body moving. The difference in these two forces is due to the slightly higher value of the coefficient of static friction (μ_s) between the two surfaces when the body is stationary compared to the coefficient of dynamic friction (μ_d) when the body is rolling.

Example 4.38

(a) Solution by calculation:

Consider again the arrangement of forces shown in Figure 4.62. If the block is in equilibrium, i.e. just on the point of moving, or moving with constant velocity then we can equate the horizontal and vertical forces as follows:

resolving horizontally gives

$$P = F \qquad (1)$$

resolving vertically

$$N = mg \qquad (2)$$

but from the laws of dry friction

$$\mathbf{F = \mu N} \qquad (3)$$

substituting (2) in (3) gives

$$F = \mu mg \qquad (4)$$

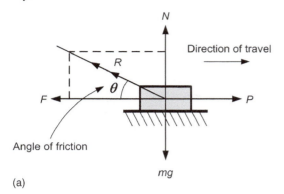

Angle of friction

(a)

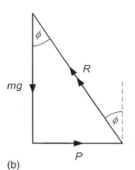

Figure 4.63 (a) Space diagram for horizontal block. (b) Vector diagram.

substituting (4) in (1) gives

$$P = \mu mg$$

(b) Solution by vector drawing:
   You know from your previous work on resolution of coplanar forces (**page 187**) that two forces can be replaced by a single resultant force in a vector diagram. The space diagram for our horizontal block is shown in Figure 4.63(a), where F and N can be replaced by a resultant R at an angle ϕ to the normal force N.
   From the Figure 4.63 it can be seen that:

$$\frac{F}{R} = \sin \phi$$

$$F = R \sin \phi$$

and $$\frac{N}{R} = \cos \phi$$

$$N = R \cos \phi$$

$$\frac{F}{N} = \frac{R \sin \phi}{R \cos \phi} = \tan \phi$$

however, $$\frac{F}{N} = \mu$$

therefore

$$\mu = \tan \phi$$

ϕ is known as the *angle of friction*.

Once F and N have been replaced by R the problem becomes one of three coplanar forces mg, P and R and can therefore be solved using the triangle of forces, you met earlier.
   Then choosing a suitable scale the vector diagram is constructed as shown in the Figure 4.63(b).

Example 4.39

For the situation illustrated in Figure 4.64(a), find the value of the force P to maintain equilibrium.

We can solve this problem by calculation resolving the forces into their horizontal and vertical components or, we can solve by drawing, both methods of solution are detailed below.

(a) Solution by calculation:
   Resolving forces horizontally

$$F = P \cos 30$$

   resolving forces vertically

$$N + P \sin 30 = 80$$

   but

$$F = \mu N$$

   and substituting for N from above gives,

$$F = \mu(80 - P \sin 30)$$

We are told that $\mu = 0.4$ and replacing F in the above equation by $P \cos 30$, in a similar manner to the general example gives:

$$P \cos 30 = 0.4(80 - P \sin 30)$$

and by multiplying out the brackets and rearrangement we get

$$P \cos 30 + 0.4 P \sin 30 = 0.4 \times 80$$

So, $$P(\cos 30 + 0.4 \sin 30) = 32$$

and $$P = 30.02 \, \text{N}$$

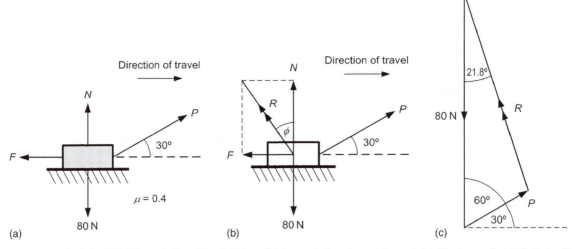

Figure 4.64 (a) Illustration of situation. (b) Magnitude and direction of forces. (c) Diagram showing force *P*.

Make sure you can follow the above trigonometric and algebraic argument, before considering the more difficult example that follows.

(b) Solution by drawing:
The magnitude and direction of all known forces for our block is shown in Figure 4.64(b).
Remembering that $\mu = \tan \phi$ then,

$$\tan \phi = \mu = 0.4 \quad \text{so,}$$

$$\phi = \tan^{-1} 0.4$$
(the angle whose tangent is) and

$$\phi = 21.8°$$

From the resulting vector diagram (Figure 4.64(c)), we find that $P = 30\,\text{N}$.

Key point

The coefficient of friction is given by the tangent of the friction angle.

We finish our short study of friction by considering the forces acting on a body *at rest on a inclined plane* and then the forces that act on a body when *moving on an inclined plane*.

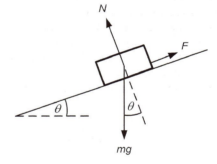

Figure 4.65 Force system for body in equilibrium on an inclined plane.

Forces on a body at rest on an inclined plane

Remembering that the frictional resistance always acts in such a way as to oppose the direction in which the body is tending to move. So in Figure 4.65 where the body is in limiting equilibrium (i.e. on the point of slipping down the plane) the frictional resistance will act up the plane.

It can be seen that there are now three forces acting on this body, the weight *mg* acting vertically downwards, the normal force *N* acting perpendicular to the plane and the frictional resistance *F* acting parallel to the plane. These

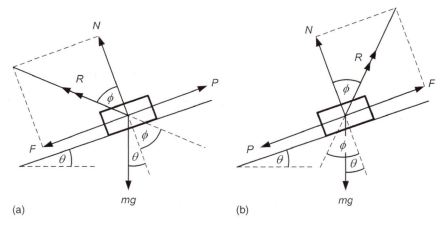

Figure 4.66 Forces acting on a body: (a) when moving up an inclined plane (b) when moving down an inclined plane.

forces are in equilibrium and their values can be found by calculation or drawing.

Again, using simple trigonometry we can resolve the forces parallel and perpendicular to the plane. Resolving parallel to the plane we get $F = mg \sin \theta$ and resolving perpendicular to the plane we get $N = mg \cos \theta$ and from $F = \mu N$ we see that $\mu = \tan \theta$.

Note: *When and only when*, a body on an inclined plane is in limiting equilibrium *and* no external forces act on the body then the angle of slope θ is equal to the angle of friction ϕ, i.e. $\theta = \phi$.

The drawing method would simply require us to produce a triangle of forces vector diagram, from which we could determine $\theta = \phi$ and μ.

Forces on a body moving up and down an inclined plane

Figure 4.66(a) shows the arrangement of forces acting on a body that is moving up an inclined plane and Figure 4.66(b) shows a similar arrangement when a body is moving down an inclined plane.

Study both of these diagrams carefully, noting the arrangement of forces. Also note the clear distinction (in these cases) between the angle of friction ϕ and the angle of slope θ. The weight mg always acts vertically down and the frictional force F always opposes the forces P, tending to cause motion either up or down the slope.

All problems involving bodies moving up or down on an inclined plane, can be solved by calculation or drawing. The resolutions of forces and general vector diagrams for each case are detailed below.

(a) **Forces on body moving up the plane** (Figure 4.65(a)):

resolving forces horizontally

$$P = F + mg \sin \theta$$

resolving vertically

$$N = mg \cos \theta$$
$$F = \mu N = \mu mg \cos \theta$$

therefore

$$P = \mu mg \cos \theta + mg \sin \theta$$

The solution by vector drawing will take the general form shown in Figure 4.67.

(b) **Forces on body moving down the plane** (Figure 4.66(b)):

resolving forces horizontally

$$P + mg \sin \theta = F$$

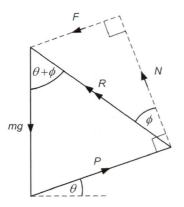

Figure 4.67 Solution by vector drawing when body is moving up the plane.

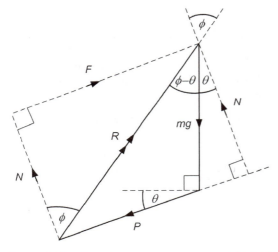

Figure 4.68 Solution by vector drawing when body is moving down the plane.

resolving vertically

$$N = mg \cos \theta$$

$$F = \mu N = \mu mg \cos \theta$$

$$P + mg \sin \theta = \mu mg \cos \theta$$

therefore,

$$\boldsymbol{P = \mu mg \cos \theta - mg \sin \theta}$$

Again the solution by vector drawing will take the general form shown in Figure 4.68.

Example 4.40

(a) A body of mass 400 kg is moved along a horizontal plane by a horizontal force of 850 N, calculate the coefficient of friction.
(b) The body then moves onto a plane made from the same material, inclined at 30° to the horizontal. A force P angle at 15° from the plane is used to pull the body up the plane with constant velocity, determine the value of P.

(a) The space diagram for the arrangement of forces is shown in Figure 4.69(a).
      Then by calculation:
resolving forces horizontally

$$F = 850 \, \text{N}$$

resolving vertically

$$N = (400)(9.81) = 3924 \, \text{N}$$

$$F = \mu N$$

Therefore,

$$\mu = \frac{F}{N} = \frac{850}{3924}$$

$$\mu = 0.217$$

Also from the vector drawing (Figure 4.69(b)) it can be seen that:

$$\phi = 12.2 \quad \text{and so} \quad \mu = 0.217$$

(b) The space diagram for the arrangement of forces is shown in Figure 4.69(c).
      By calculation:
resolving forces horizontally

$$P \cos 15 = (400)(9.81) \sin 30 + F$$

resolving vertically

$$N + P \sin 15 = (400)(9.81) \cos 30$$

$$N = (400)(9.81) \cos 30 - P \sin 15$$

but

$$F = \mu N$$

$$F = 0.217((400)(9.81)(\cos 30) - P \sin 15))$$

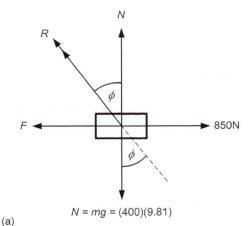

(a) $N = mg = (400)(9.81)$

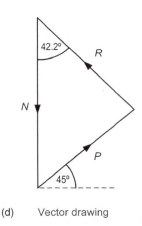

(b)   Scale: 10mm = 500N

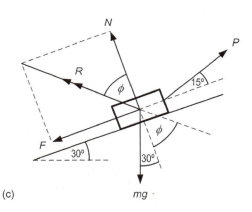

(c)   mg

(d)   Vector drawing

Figure 4.69

therefore,

$$P \cos 15 = (400)(9.81) \sin 30 + 0.217((400) \times (9.81)(\cos 30) - P \sin 15))$$

from which

$$P = 2794.5 \, \text{N}$$

Also if the vector drawing (Figure 4.69d) is drawn to scale, it will be found that $\mathbf{P} \cong 2.8 \, \text{kN}$ at 45° from horizontal.

Test your understanding 4.16

1. On what variables do the value of frictional resistance depend?

2. "The frictional resistance is independent of the relative speed of the surfaces, under all circumstances". Is this statement true or false? You should give reasons for your decision.

3. Define (a) the angle of friction and (b) the coefficient of friction *and* explain how they are related.

4. Sketch a space diagram that shows all the forces that act on a body moving with uniform velocity along a horizontal surface.

5. Explain the relationship between the angles θ and ϕ, (a) when a body on a slope is in static equilibrium and (b) when a body moves down a slope at constant velocity.

6. Sketch diagrams that show all the forces that act on a body when moving with constant velocity (a) up a sloping surface and (b) down a sloping surface.

7. For each case in question 6, resolve the horizontal and vertical components of these forces *and* show that for a body moving up the plane $P = \mu mg \cos \theta + mg \sin \theta$ and that for a body moving down the plane $P = \mu mg \cos \theta - mg \sin \theta$.

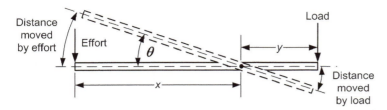

Figure 4.70   The simple lever.

4.8.9 Machines

The maximum force which man can apply unaided is limited. Consequently man has always tried to devise methods by which a load may be moved by a small effort, this can be achieved by use of machines. A *machine* may be defined as *the combination of components that transmit or modify the action of a force or torque to do useful work*. Machines provide us with many examples of the application of the theory associated with work, energy and power.

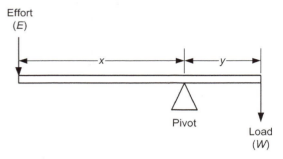

Figure 4.71   Distance moved by effort and load for a simple pivot lever.

Key point

In all practical machines there will be losses, so the mechanical advantage will be less than the velocity ratio.

Key point

For a machine to be of practical value the MA needs to be greater than 1.

Mechanical advantage, velocity ratio and efficiency

One of the most fundamental machines is the simple lever (Figure 4.70) where the pivot or fulcrum point is between the load and the effort.

This machine will only be of use when the *effort* applied is less than the *load* that requires to be moved. The relationship between the ratio of load over effort is known as the *mechanical advantage* of the machine, i.e.

$$Mechanical\ advantage\ (MA) = \frac{load}{effort} = \frac{W}{E} = \frac{x}{y}$$

You may be wondering why the ratio of the pivot arms, x and y, also equals the MA. You will remember from your work on moments that for equilibrium $Ex = Wy$ and so by rearranging this relationship $\frac{W}{E} = \frac{x}{y}$, as above. Also note that since the MA is a ratio it has no units.

Now, as mentioned previously, for the machine to be of practical use the MA will generally need to be greater than 1, but it will not be constant because of the need to overcome losses within the machine, such as friction, windage, backlash, etc. For small loads the MA will be small but as the proportion of the total effort required to overcome losses falls with increases in load, the MA will increase.

It is impossible to obtain a greater work output than work input from any machine. Thus as the effort is smaller than the load, the distance moved by the effort must be greater than the distance moved by the load, this point is illustrated in Figure 4.71. The velocity ratio of a machine is defined as:

$$Velocity\ ratio\ (VR) = \frac{distance\ moved\ by\ effort}{distance\ moved\ by\ load}$$

$$= \frac{x\theta}{y\theta} = \frac{x}{y}$$

Again, because we are dealing with a ratio (in this case of distances) the VR has no units. The distance moved by an effort that always acts at right angles to the lever is given by the arc length $x\theta$ and the distance moved by the vertical load is given by the vertical distance $y\tan\theta$. For a small angle (rad) the distance moved by the load can be approximated to $y\theta$, hence the VR for this machine may be estimated using the ratio, distance x divided by distance y.

The mechanical efficiency η is the ratio of the work output to the work input. Therefore:

$$\text{Efficiency } (\eta) = \frac{\text{Work out}}{\text{Work in}}$$
$$= \frac{\text{Load} \times \text{distance moved by load}}{\text{Effort} \times \text{distance moved by effort}}$$

and since load divided by effort = MA and,

$$\frac{\text{Distance move by load}}{\text{Distance moved by effort}} = \frac{1}{\text{VR}}$$

then

$$\text{Efficiency } (\eta) = \frac{\text{MA}}{\text{VR}}$$

or as a percentage;

$$\text{Efficiency } (\eta) = \frac{\text{MA}}{\text{VR}} \times 100\%$$

For an *ideal machine* (no losses) the efficiency will be 100% and therefore, from above, the $MA = VR$. In all practical machines there will be some losses and so the MA will be less than the VR, in other words the efficiency will always be less than 100%.

Key point

In all practical machines there will be losses so the MA will be less than the VR.

Law of a machine

If an experiment is carried out on a simple lifting machine to determine the effort E required to raise a load W. A graph of E against W is plotted (Figure 4.72) for a range of load values then a straight line graph would be obtained.

The graph shows a straight line with slope a and intercept b. Remembering the law for

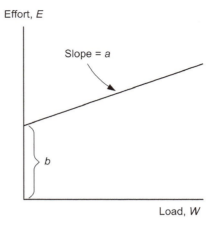

Figure 4.72 Graph illustrating the law of a machine.

a straight line graph, i.e. $y = mx + c$ and comparing this law with the variables in the graph, we obtain the relationship $E = aW + b$, this equation is known as *the law of a machine*.

Example 4.41

The results for a set of measurements of load and corresponding effort carried out on a lifting machine are given below. The effort moved through 1 m while the load was raised 25 mm. By plotting the effort against load find the:

(a) VR of the machine,
(b) law of the machine,
(c) effort needed to raise a load of 1.5 kN,
(d) efficiency of the machine when a load of 800 N is being raised.

Load (N)	100	250	400	550	700	850	1000
Effort (N)	7	11.5	15	19	22.5	26	30

(a) The VR is found quite simply from the given data:

$$\text{VR} = \frac{\text{distance moved by effort}}{\text{distance moved by load}}$$
$$= \frac{1000}{25} = 40$$

(b) By plotting the above effort against load figures, the graph (Figure 4.73) is produced.

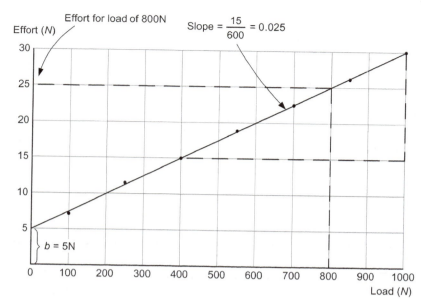

Figure 4.73 Effort against load graph.

From the graph it can be seen that the effort intercept b is 5 N and the slope of the graph a is 0.025.

Therefore the law of the graph is $E = 0.025\,W + 5$

(c) The effort needed to raise 1.5 kN is easily found by substituting the given load value into the law of the machine. This gives,

$$E = (0.025)(1500) + 5 = 42.5\,N$$

(d) We know that the mechanical advantage of the machine varies, as the load varies. When the load is 800 N, the corresponding effort is shown on the graph to be 25 N. Therefore the MA is given by:

$$MA = \frac{Load}{Effort} = \frac{800}{25} = 32 \text{ and the VR} = 40$$

so the efficiency η when the load is 800 N is,

$$\frac{MA}{VR} = \frac{32}{40} = 0.8 = 80\%$$

Key point

The efficiency of a machine is given by the MA/VR.

Pulleys

Pulley systems are widely used with cranes, lifts, hoists and winches, to raise and lower large loads. The cable system within an aircraft engine-hoisting winch, used for engine removal and fit operations, is a good example of their use. For *simple pulleys* the VR of the pulley may be found by counting the number of cable sections supporting the load, Figure 4.74 illustrates this method.

The pulley arrangement shown in Figure 4.75, uses several different cables to hoist the load, under these circumstances the VR cannot be found using the simple counting method.

The load is initially supported equally by the first cable passing from the beam round the pulley P_1 to the shaft of pulley P_2. Thus the tension T is shared and is equal to load/2. At P_2 half of the tension is supported by P_3, therefore the load transferred to $P_3 = load/4$. Again at P_3 half this new load is supported by the tension while the other half passes round P_4 to the effort. Therefore, the ideal effort = load/8. To accommodate an eight-fold decrease in effort in an ideal machine, the distance moved by the effort is eight times the distance moved by the load so the $VR = 8$.

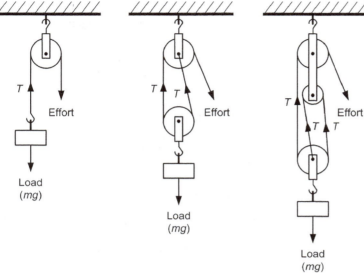

(a) VR =1   (b) VR = 2   (c) VR = 3

Figure 4.74   Determination of VR for a simple pulley system.

(a) $MA = \dfrac{load}{effort} = \dfrac{3000}{30} = 100$

(b) $VR = \dfrac{\text{distance moved by effort}}{\text{distance moved by load}}$

$ = \dfrac{1500\,mm}{10\,mm} = 150$

(c) WD in raising the load by $4\,cm =$ force $\times$ distance $= (3000)(0.04) = 120\,J$

(d) Efficiency, $\eta = MA/VR = 100/150 = 66.6\%$

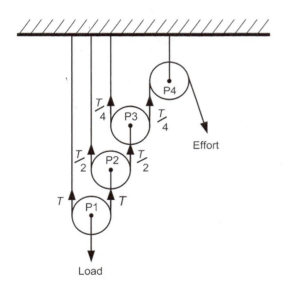

Figure 4.75   Multiple cable–pulley system.

Example 4.42

In a pulley system, an effort of 30 N is required to raise a load of 3 kN. If the effort moves through 1.5 m to raise the load by 1 cm, find the:
(a) MA,
(b) VR,
(c) WD in raising the load by 4 cm,
(d) efficiency of the machine.

The screw jack

The screw jack is a simple machine making use of the screw thread, for raising relatively large loads by means of a small effort. An example of the use of the screw jack may be found in the mechanical trestles used for stabilizing aircraft structures during aircraft jacking operations. In this application normally, a pair of screw jacks are worked in tandem to raise and lower the trestle steadying beam. Figure 4.76(a) shows the general arrangement of a typical screw jack, with the effort being applied at radius r from the center of rotation of the screw thread.

Figure 4.76(b) shows the detail of a typical helical screw thread. The thread *pitch* is the vertical distance from one thread to the next,

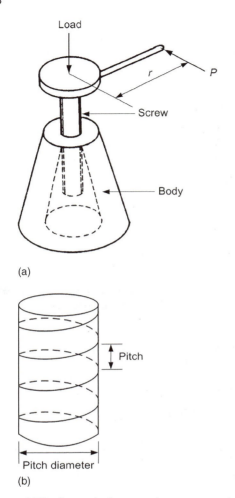

Figure 4.76 Screw jack general arrangement and screw thread detail.

measured along the axis of the screw. The *lead* is the vertical distance traveled by the jack for one complete revolution of the screw thread. For a *single start thread* this will be equivalent to the pitch of the thread. For a *multiple start thread* the lead will be equal to the pitch multiplied by the number of starts. If the effort is applied directly to the screw jack then for one revolution:

$$VR = \frac{\text{distance moved by effort}}{\text{distance moved by load}}$$

$$VR = \frac{\pi \times \text{pitch diameter}}{\text{lead}}$$

If the effort is applied horizontally by a lever, as shown in the general arrangement drawing (Figure 4.76(a)), then:

$$VR = \frac{2\pi r}{\text{lead}}$$

> **Key point**
>
> The lead of a screw thread is equal to the pitch multiplied by the number of starts.

Example 4.43

An effort of 120 N is required to raise a load of 9 kN, applied at a radius of 300 mm on a screw jack. If a double start screw has a pitch of 5 mm, determine the:
(a) VR,
(b) MA,
(c) efficiency of the screw jack.

(a) $VR = \dfrac{2\pi r}{\text{lead}} = \dfrac{(2)(\pi)(0.3)}{(2)(0.005)} \cong \mathbf{189}$

where lead = 2 × pitch, for a two start thread.

(b) $MA = \dfrac{\text{load}}{\text{effort}} = \dfrac{9000}{120} = \mathbf{75}$

(c) Efficiency, $\eta = \dfrac{MA}{VR} = \dfrac{75}{189}$

  $\qquad = 0.397$ or **39.7%**

Gear trains

A simple gear train consists of two meshed gears of different sizes mounted on two separate shafts (Figure 4.77(a)). If gear wheel A is the *driver* then gear wheel B is the *driven*. The driver and driven gears rotate in opposite directions. If rotation in the same direction is required an *idler* gear is added (Figure 4.77(b)).

If the simple gear train without an idler is driven at n rpm and T is the number of teeth on a gear wheel, then assuming no slippage, the number of teeth meshing on each gear wheel must be the same, therefore:

$$N_1 \times T_1 = N_2 \times T_2 \quad \text{and} \quad \frac{T_1}{T_2} = \frac{N_2}{N_1} = \frac{1}{VR}$$

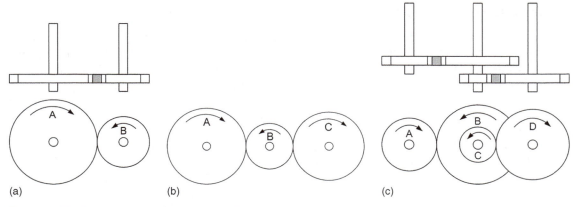

Figure 4.77 Simple gear train with and without idler gear.

Similarly for the gear train with idler:

$$\frac{N_2}{N_1} = \frac{T_1}{T_2} \quad \text{and} \quad \frac{N_3}{N_2} = \frac{T_2}{T_3}$$

so

$$\frac{N_3}{N_2} \times \frac{N_2}{N_1} = \frac{N_3}{N_1} = \frac{T_2}{T_3} \times \frac{T_1}{T_2}$$

therefore

$$\frac{N_3}{N_1} = \frac{T_1}{T_3} = \frac{1}{VR}$$

so for the simple gear train:

$$VR = \frac{\textit{Number of teeth on final gear}}{\textit{Number of teeth on the first gear}}$$

When two or more gears are placed on the same shaft, the gear arrangement is known as a compound train (Figure 4.77(c)). In general, the VR for these systems can be shown to be:

$$VR = \frac{\text{input speed}}{\text{output speed}}$$

$$= \frac{\text{product of no. of teeth on driven wheels}}{\text{product of no. of teeth on drivers}}$$

For example, if in the compound gear system shown above, gear A has 20 teeth, gear B has 80 teeth, gear C has 10 teeth and gear D has 40 teeth, then assuming gear A is a driver:

$$VR = \frac{B \times D}{A \times C} = \frac{(80)(40)}{(20)(10)} = 16$$

The above suggests that this compound gear arrangement will result in a step-down in speed, where the input speed is 16 times faster than the speed at the output.

Key point

An idler gear is used to change the direction of motion of the driven gear; it has no effect on the resulting VR.

Test your understanding 4.17

1. Give a simple definition of a machine.
2. Define: (a) velocity ratio (VR) and (b) mechanical advantage (MA).
3. If, in a machine, the distance moved by the effort is 2.45 m, the distance moved by the load is 10 mm and the efficiency of the machine is 75%, determine the machines MA.
4. Write down the law of a lifting machine and define each of the variables in the law.
5. Detail one method of determining the VR for a simple pulley system.
6. An effort of 150 N is required to raise a load of 10 kN, applied at a radius of 250 mm on a screw jack. If the screw has a lead of 8 mm, determine the VR, MA and efficiency of the screw jack.
7. In a compound gear system the initial independent driver gear has 100 teeth, this drives a second gear with 30 teeth, a third gear connected to the same shaft has 80 teeth and drives a final output gear with 20 teeth. Determine the VR for the system and state whether the system is step-up or step-down.

General Questions 4.3

1. A body starts from rest and constantly accelerates at $1.5\,m/s^2$ up to a speed of 6 m/s. It then travels at 6 m/s for 12 s after which, it is retarded to a speed of 2 m/s. If the complete motion takes 18 s, find the:
   (a) time taken to reach 6 m/s,
   (b) retardation,
   (c) total distance travelled.
2. A light aircraft of mass 2500 kg accelerates from 100 to 150 mph in 3 s. If the air resistance is 1800 N/tonne, find in SI units the:
   (a) average acceleration,
   (b) force required to produce the acceleration,
   (c) inertia force,
   (d) the propulsive force of the aircraft.
3. A twin-engine aircraft is traveling at 450 mph and the exhaust velocity from both engines is identical at 280 m/s. If the mass airflow passing through each engine is 350 lb/s, determine the thrust being produced by each engine, in SI units.
4. An aircraft flap drive motor exerts a torque of 25 Nm at a speed of 3000 rpm, calculate the power being developed.
5. An aircraft of mass 60,000 kg is in a steady horizontal turn of radius 650 m, flying at 600 kph, determine the centrifugal force tending to throw the aircraft out of the turn.
6. A body moves with simple harmonic motion (SHM), with amplitude of 100 mm and frequency of 2 Hz. Find the maximum velocity and acceleration.
7. A locomotive of mass 80 tonne, hauls 11 coaches each having a mass of 20 tonne up an incline of 1 in 80. The frictional resistance to motion is 50 N/tonne. If the train accelerates uniformly from 36 to 72 kph in a distance of 1600 m, determine the:
   (a) change in PE of the train,
   (b) change of KE,
   (c) work done against frictional resistance,
   (d) total mechanical energy required.
8. A vehicle, starting from rest, freewheels down an incline which has a gradient of 1 in 10. Using the conservation of energy principle and neglecting any resistances to motion, find the velocity of the vehicle after it travels a distance of 150 m down the incline.
9. A load of mass 500 kg is positioned at the base of a sloping surface inclined at 30° to the horizontal. A force P, "parallel to the plane," is then used to pull the body up the plane with *constant velocity*. If the coefficient of friction is 0.25, determine the value of the pulling force.
10. A screw jack has a single start thread with a pitch of 5 mm and the effort is applied at a radius of 0.15 m. If a mass of 1000 kg is raised by means of an effort of 250 N, determine the efficiency of the screw jack.

4.9 Fluids

In this section we will study the static and dynamic behaviour of fluids. A fluid may be defined as a *liquid* or a *gas* and both will be considered here.

4.9.1 Pressure

You have already met the concept of pressure, which we defined earlier as, force per unit area. There are, in fact, several types of pressure which were not previously defined; these include **hydrostatic pressure** (the pressure created by stationary bulk liquid), **atmospheric pressure** and **dynamic pressure** due to fluid movement, as well as the pressure applied to solids which we have already considered.

You will meet pressure expressed in many different units, some of the more common units for pressure will be found in Table 4.7, they are repeated here for convenience:

Measurement system	Units
SI	N/m^2, MN/m^2
SI	$1\,Pa = 1\,N/m^2$
SI	$1\,bar = 10^5\,Pa = 10^5\,N/m^2$
SI	millimetres of mercury (mmHg)
Imperial	pounds force per square inch $lbf/in.^2$ (psi)
Imperial	inches of mercury (in. Hg)

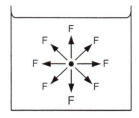

(a) Pressure at a given depth
is equal in all directions

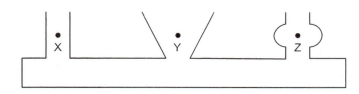

(b) Pressure is independent of the shape
of the containing vessel at a given depth

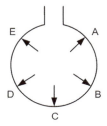

(c) Pressure acts at right angles to the
walls of the containing vessel

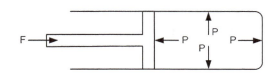

(d) Pressure transmitted through a fluid is equal in all directions

Figure 4.78   Illustration of fluid pressure laws.

The laws of fluid pressure

Four basic factors or laws, govern the pressure within fluids.

With reference to Figure 4.78 these four laws may be defined as follows:

(a) *Pressure at a given depth in a fluid is equal in all directions.*

(b) *Pressure at a given depth in a fluid is independent of the shape of the containing vessel in which it is held.* In Figure 4.78(b) the pressure at point X, Y and Z is the same.

(c) *Pressure acts at right angles to the surfaces of the containing vessel.*

(d) *When a pressure is applied to a fluid it is transmitted equally in all directions.*

Hydrostatic pressure

Pressure at a point in a liquid can be determined by considering the weight force of a fluid above the point. Consider Figure 4.79.

If the density of the liquid is known, then we may express the weight of the liquid in terms of its density and volume, since density is equal to

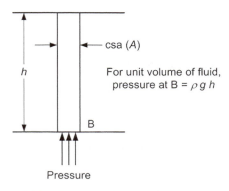

Figure 4.79   Pressure at a point in a liquid.

the mass divided by the volume. The mass of the liquid is given by:

$$m = \rho \times A \times h$$

where:
m = mass of the liquid
ρ = density
A = cross-sectional area
h = height

Now since the weight is equal to the mass multiplied by the acceleration due to gravity then the weight is given by, $W = \rho A g h$ and it follows that the pressure due to the weight of the liquid (hydrostatic pressure) is equal to the weight divided by area A, i.e.

$$\text{Hydrostatic pressure due to weight of liquid} = \rho g h$$

If standard SI units are used for density (kg/m^3), acceleration due to gravity (9.81 m/s^2) and the height (m), then the pressure may be expressed in N/m^2 or Pa.

Note that the *atmospheric pressure* above the liquid was ignored, the above formula refers to *gauge pressure*, this should always be remembered when using this formula. More will be said about the relationship between gauge pressure and atmospheric pressure when we consider atmospheric pressure next.

Key point

$\rho g h$ = gauge pressure.

Example 4.44

Find the head h of mercury corresponding to a pressure of 101.32 kN/m^2, take the density of mercury as $13,600 \text{ kg/m}^3$.

Since pressure $p = \rho g h$ and so $h = p/\rho g$, then using standard SI units

$$h = \frac{1,013,20}{(13,600)(9.81)} = 0.76 \text{ m}$$

or   760 mmHg

Therefore, this is the height of mercury needed to balance standard atmospheric pressure.

Hydraulic press

An application of the use of fluid pressure can be found in the *hydraulic press*, sometimes known as the *Bramah press*. This hydraulic machine may be used for dead-weight testing, hydraulic

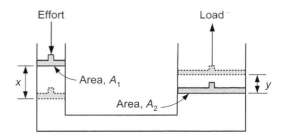

Figure 4.80 The Bramah press.

actuators, lifting loads or for compression and shear testing. Figure 4.80 shows the general arrangement for such a machine. Since, the fluid involved in this machine is a liquid hydraulic oil and therefore virtually incompressible then, the fluid displaced by the effort piston must be equal to the amount of fluid displaced at the load piston.

In other words, the volumes $A_1 x$ and $A_2 y$ must be the same. Therefore the velocity ratio, VR is given by:

$$\text{VR} = \frac{x}{y} = \frac{A_2}{A_1}$$

or in words, the ratio,

$$\frac{\text{Distance moved by effort piston, } x}{\text{Distance moved by load piston, } y}$$

$$= \frac{\text{Area of load piston, } A_2}{\text{Area of effort piston, } A_1}$$

Example 4.45

(a) A force of 500 N is applied to the small cylinder of a hydraulic press, the smaller cylinder has a csa of 10 cm^2. The large cylinder has a csa of 180 cm^2. What load can be lifted by the larger pistons, if the pistons are at the same level?

(b) What load can be lifted by the larger piston if the larger piston is 0.75 m below the smaller?

Take the density of the oil in the press as 850 kg/m^3.

The situation for both cases is shown in Figure 4.81.

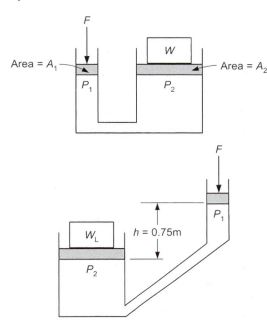

Figure 4.81

(a) We know that $P_1 = P_2$, since pressure is applied equally in all directions.

Therefore,

$$\frac{F}{A_1} = \frac{W}{A_2} \quad \text{or} \quad F = \frac{WA_1}{A_2}$$

then $W = FA_2/A_1$ and substituting values gives,

$$\text{load}, W = \frac{(500)(180 \times 10^{-4})}{1 \times 10^{-3}} = 9000\,\text{N}$$

(b) If the larger piston is 0.75 m below the smaller piston, then pressure P_2 will be greater than P_1 due to the head of liquid.

$$P_2 = P_1 + \rho g h$$

$$P_1 = \frac{F}{A_1} = \frac{500}{1 \times 10^{-3}} = 50 \times 10^4\,\text{N/m}^2$$

then

$$P_2 = (50 \times 10^4) + (850 \times 9.81 \times 0.75)$$

$$P_2 = 50.6254 \times 10^4\,\text{N/m}^2$$

and

$$W_L = P_2 A_2 = (50.6254 \times 10^4)(180 \times 10^{-4})$$

$$= 9112.57\,\text{N}$$

Atmospheric pressure

The air surrounding the earth has mass and is acted upon by the earth's gravity, thus it exerts a force over the earth's surface. This force per unit area is known as *atmospheric pressure*. At the earth's surface at sea level, this pressure is found by measurement to be, *101,320 N/m²* or in *Imperial units 14.7 lbf/in.²*. Thus 1 bar (10^5 N/m²) is approximately 14.5 times larger than 1 lbf/in.², this relationship should be remembered. Imagine the consequences if you inadvertently tried to inflate an aircraft tire to 150 bar, instead of 150 psi!

Outer space is a vacuum and is completely devoid of matter, consequently there is no pressure in a vacuum. Therefore, pressure measurement relative to a vacuum is *absolute*. For most practical purposes it is only necessary to know how pressure varies from the earth's atmospheric pressure. A pressure gauge is designed to read zero, when subject to atmospheric pressure, therefore if a gauge is connected to a pressure vessel it will only read *gauge* pressure. So to convert gauge pressure to absolute pressure, atmospheric pressure must be added to it, i.e.

$$\begin{array}{ccc} \text{Absolute} & \text{Gauge} & \text{atmospheric} \\ \text{pressure} & = \text{pressure} & + \text{pressure} \end{array}$$

Example 4.46

Taking atmospheric pressure as 101,320 N/m², convert the following gauge pressure into absolute pressure, give your answer in kN/m² or kPa: (a) 400 kN/m², (b) 20 MN/m², (c) 5000 Pa and (d) 3000 psi.

We know from above that absolute pressure is equal to gauge pressure plus atmospheric pressure, therefore, the only real problem here is to ensure the correct conversion of units.

Atmospheric pressure = 101.32 kN/m².

(a) $400 + 101.32 = 501.32 \, \text{kN/m}^2$
(b) $20{,}000 + 101.32 = 20{,}101.32 \, \text{kN/m}^2$ (note that MNm^{-2} is the index way of writing MN/m^2)
(c) $5 + 101.32 = 106.32 \, \text{kN/m}^2$ (remember that $1 \, \text{Pa} = 1 \, \text{N/m}^2$)
(d) From Table 4.7, $\frac{3000 \, \text{psi}}{0.145} = 20{,}689.6 \, \text{kN/m}^2$

Buoyancy

It is well known that a piece of metal placed in water will sink and that a piece of cork placed below the surface of the water will rise, also that a steel ship having a large volume of empty space in the hull will float. The study of floating, sinking or rising bodies immersed in a fluid, is known as *buoyancy*. We know from our study of fluid pressure, that there will be an increase in pressure of a fluid with depth, irrespective of the nature of the fluid. This means that eventually, there will be a greater pressure pushing up on the body from underneath than there is pushing down on it from above. So dependent on the relative densities of the fluids and bodies involved, will depend when the upthrust force due to the fluid equals the weight force exerted by the body immersed in it.

Archimedes expresses this relationship very succinctly in his principle: *When a body is immersed in a fluid it experiences an upthrust, or apparent loss of weight, equal to the weight of the fluid displaced by the body.*

This equality relationship is illustrated in Figure 4.82, where it can be seen that the body immersed in the fluid, floats when the upthrust force (equal to the weight of the fluid displaced) equals the weight of the body.

This principle and the concept of buoyancy enable us to determine why and when airships, balloons, ships and submarines, will float. As an example, consider the buoyancy of a helium balloon. The density of the atmosphere reduces with altitude, so when the upthrust force (per unit area) created by the atmospheric air is equal to the weight of the helium and balloon, then the balloon will float at a specified altitude. This assumes of course that the balloon does not burst first!

Measurement of pressure

Devices used to measure pressure will depend on the magnitude (size) of the pressure, the accuracy of the desired readings and whether the pressure is static or dynamic. Here we are concerned with barometers to measure atmospheric pressure and the manometer to measure low pressure changes, such as might be encountered in a laboratory or from variations in flow through a wind tunnel. Further examples of dynamic pressure measurement due to fluid flow will be encountered later, when you study aircraft pitot-static instruments and also, if you study aircraft fluid systems.

The two most common types of barometer used to measure atmospheric pressure are the mercury and aneroid types. The simplest type of *mercury barometer* is illustrated in Figure 4.83. It consists of a mercury-filled tube which is inverted and immersed in a reservoir of mercury.

The atmospheric pressure acting on the mercury reservoir is balanced by the pressure $\rho g h$ created by the mercury column. Thus the atmospheric pressure can be calculated from the height of the column of mercury it can support.

The mechanism of an *aneroid barometer* is shown in Figure 4.84. It consists of an evacuated aneroid capsule, which is prevented from collapsing by a strong spring.

Variations in pressure are felt on the capsule that causes it to act on the spring. These spring movements are transmitted through gearing and amplified, causing a pointer to move over a calibrated scale.

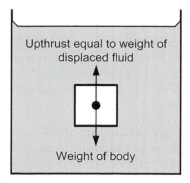

Figure 4.82 Illustration of Archimedes principle.

A common laboratory device used for measuring low pressures is the U-tube manometer (Figure 4.85). A fluid is placed in the tube to a certain level, when both ends of the tube are open to atmosphere the level in the fluid of the two arms is equal. If one of the arms is connected to the source of pressure to be measured it causes the fluid in the manometer to vary in height. This height variation is proportional to the pressure being measured.

The magnitude of the pressure being measured, is the product of the difference in height between the two arms Δh, the density of the liquid in the manometer and the acceleration due to gravity, i.e. pressure being measured *gauge pressure* $= \rho g \Delta h$.

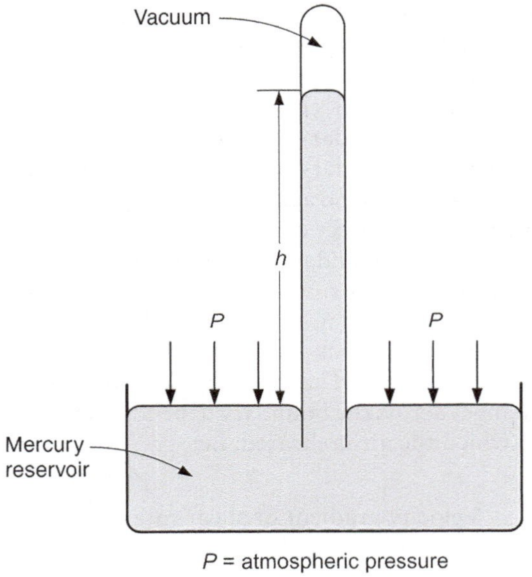

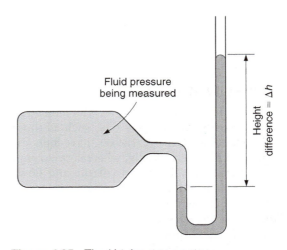

Figure 4.83 Simple mercury barometer.

Figure 4.85 The U-tube manometer.

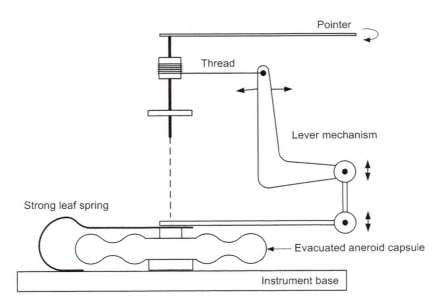

Figure 4.84 Aneroid barometer mechanism.

Example 4.47

A mercury manometer is used to measure the pressure above atmospheric of a water pipe, the water being in contact with the mercury in the left hand arm of the manometer. If the right hand arm of the manometer is 0.4 m above the left hand arm, determine the gauge pressure of the water. Take the density of mercury as $13,600 \, \text{kg/m}^3$.

We know that gauge pressure

$$= \rho g \Delta h = (13,600)(9.81)(0.4)$$
$$= 53,366 \, \text{N/m}^2$$

4.9.2 Fluid viscosity

The ease with which a fluid flows is an indication of its viscosity. Cold heavy oils, such as those used to lubricate large gearboxes, have high viscosity and flow very slowly, whereas petroleum spirit is extremely light and volatile and flows very easily and so has low viscosity.

We thus define **viscosity** as: *the property of a fluid that offers resistance to the relative motion of its molecules.* The energy losses due to friction within a fluid are dependent on its viscosity. As a fluid moves, there is developed a shear stress in it, the magnitude of which depends on the viscosity of the fluid. You have already met the concept of shear stress (τ) and should remember that it can be defined as the force required to slide one unit area of a substance over the other.

Figure 4.86 illustrates the concept of velocity change in a fluid by showing a thin layer of fluid (*a boundary layer*) sandwiched between a fixed and moving boundary.

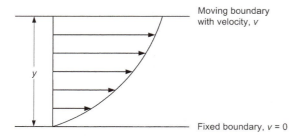

Figure 4.86 Velocity change at boundary layer.

An example of this situation could be an aircraft wing skin traveling through stationary air where the moving boundary is the wing skin and the fixed boundary is the stationary air, a small distance away from the skin.

A fundamental condition exists between a fluid and a boundary, where the velocity of the fluid at the boundary surface is identical to that of the boundary. So again considering our example the air in direct contact with the wing skin (the moving boundary) has the velocity of the wing skin. The air within the boundary further away from the wing skin is gradually reduced in velocity until it has the velocity of the stationary air, i.e. zero. The rate at which this velocity changes across the boundary depends on the rate at which the air is sheared, i.e:

$$\text{Velocity gradient or shear rate} = \frac{\Delta v}{\Delta y}$$

where Δ means "a small change in".

Another way of visualizing this situation is to consider a new pack of playing cards being force to slide over one another, where the card nearest the table has the velocity of the table, and across the whole deck of cards (the fluid) this velocity gradually changes until the outer card has the velocity of the air at this boundary.

Now, from our definition of shear stress we know that *shear stress is directly proportional to the velocity gradient* because the ease with which the fluid shears will dictate the rate at which the velocity of the fluid changes, i.e. its gradient. So using a constant of proportionality μ we have:

$$\text{Shear stress, } \tau = \mu \frac{\Delta v}{\Delta y}$$

The constant of proportionality μ is known as the **dynamic viscosity**. The units of viscosity can be determined by transposing the above formula for μ and then considering the individual units of the terms within the equation, i.e.

$$\mu = \tau \frac{\Delta y}{\Delta v}$$

and substituting the units gives,

$$\frac{N}{m^2} \times \frac{m}{m/s} = Ns/m^2$$

If you have not completely followed the above argument do not worry, it is rather complex. Just remember that viscosity is the resistance to fluid flow and that the units of dynamic viscosity in the SI system are Ns/m². You may be wondering why we keep talking about dynamic viscosity, this is because another form of viscosity exists, which takes into consideration the density of the fluid, this is known as **kinematic viscosity, v**, which is defined as:

Kinematic viscosity, $v = \dfrac{\mu}{\rho}$

and has units m²/s.

For example the dynamic viscosity of air at 20°C is 1.81×10^{-5} Ns/m² and so its kinematic viscosity is given by dividing its dynamic viscosity by the density of air at this temperature, that is $1.81 \times 10^{-5}/1.225$ kg/m³ and thus $v = 1.48 \times 10^{-5}$ m²/s. Dynamic viscosity is frequently quoted in tables in preference to kinematic viscosity because the density of a fluid varies with its temperature.

Key point

Kinematic viscosity is density dependent and therefore varies with temperature.

Test your understanding 4.18

1. Convert 50 mmHg into in.Hg.

2. Convert (a) 200 MN/m², (b) 80 kPa and (c) 72 bar into imperial (psi).

3. State the fluid pressure laws upon which the operation principle of the hydraulic press depends.

4. If a hydraulic press has a VR = 180 and the load piston is raised 10 cm, determine the distance travelled by the effort piston in m.

5. Define: (a) gauge pressure and (b) absolute pressure.

6. State Archimedes principle and explain how this principle relates to buoyancy.

7. Describe the operation of a mercury barometer.

8. If the difference in height of the mercury between the two arms of a U-tube manometer is 12.5 in. determine the: (a) gauge pressure and (b) absolute pressure being measured, in psi.

9. Explain the relationship between velocity gradient, shear stress and dynamic viscosity.

10. Show from the relationship $v = \mu/\rho$ that the SI units for kinematic viscosity are m²/s.

4.9.3 Atmospheric physics

In order to understand the environment in which aircraft fly, you will need to understand the nature of the changes that take place in our atmosphere with respect to temperature, pressure and density.

Gases

In the study of gases we have to consider the above interactions between temperature, pressure and density (remembering that density is mass per unit volume). A change in one of these characteristics always produces a corresponding change in at least one of the other two.

Unlike liquids and solids, gases have the characteristics of being easily compressible and of expanding or contracting readily in response to changes in temperature. Although the characteristics themselves vary in degree for different gases, certain basic laws can be applied to what we call a perfect gas. A *perfect* or *ideal gas is simply one, which has been shown* (through experiment) *to follow or adhere very closely to these gas laws.* In these experiments one factor, e.g. volume is kept constant while the relationship between the other two is investigated. In this way it can be shown that:

1. *The pressure of a fixed mass of gas is directly proportional to its absolute temperature, providing the volume of the gas is kept constant.*

In symbols:

$$\frac{P}{T} = \text{constant}$$

(providing V remains constant)

The above relationship is known as the **pressure law.**

Gas molecules are in a state of perpetual motion, constantly bombarding the sides of the gas-containing vessel. Each molecule produces a minute force as it strikes the walls of the container, since many billion molecules hit the container every second, this produces a steady outward pressure.

Figure 4.87 shows how the pressure of the gas varies with temperature.

If the graph is "extrapolated" downwards, in theory we will reach a temperature where the pressure is zero. This temperature is known as *absolute zero* and is approximately equal to -273 K. Each one degree kelvin (K) is equivalent to one degree celsius (°C). The relationship between the kelvin scale and the celsius scale is shown in Figure 4.88.

Key point

When dealing with the gas equations or any thermodynamic relationship we always use absolute temperature (T) in K.

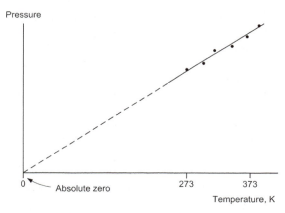

Figure 4.87 Pressure–temperature relationship of a gas.

Figure 4.88 Kelvin–celsius scales.

Returning to the gas laws, it can also be shown experimentally that:

2. *The volume of a fixed mass of gas is directly proportional to its absolute temperature providing the pressure of the gas remains constant.*

So for a fixed mass of gas:

$$\frac{V}{T} = \text{constant}$$

(providing M is fixed and P remains constant)

This relationship is known as **Charles' law.**
A further relationship exists when we keep the temperature of the gas constant, this states that: the volume of a fixed mass of gas is inversely proportional to its pressure providing the temperature of the gas is kept constant. In symbols:

$$P \propto \frac{1}{V}$$

or, for a fixed mass of gas:

$$PV = \text{constant}$$

This relationship is better known as **Boyle's law,** it is illustrated in Figure 4.89.

In dealing with problems associated with the gas laws, remember that we assume that all gases

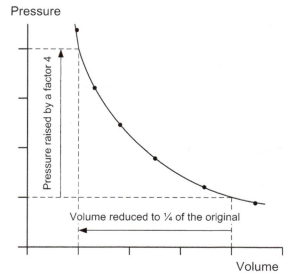

Figure 4.89 Pressure–volume relationships.

are *ideal*, in reality no gas is ideal but at low and medium pressures and temperatures, most gases behave, particularly air, in an ideal way.

The pressure law, Charles' law and Boyle's law can all be expressed in terms of one single equation known as the **combined gas equation**, this is, for a fixed mass of gas:

$$\frac{PV}{T} = \text{constant}$$

If we consider a fixed mass of gas before and after changes have taken place. Then from the combined gas equation, it follows that:

$$\frac{P_1 V_1}{T_1} = \frac{P_2 V_2}{T_2}$$

where the subscript 1 is used for the initial state and subscript 2 for the final state of the gas. The above relationship is very useful when solving problems concerned with the gas laws.

Key point

A perfect gas is one that is assumed to obey the ideal gas laws.

Example 4.48

A quantity of gas occupies a volume of $0.5 \, m^3$. The pressure of the gas is $300 \, kPa$, when its temperature is $30°C$. What will be the pressure of the gas if it is compressed to half its volume and heated to a temperature of $140°C$?

When solving problems involving several variables, always tabulate the information given, in appropriate units.

$P_1 = 300 \, kPa$ $P_2 = ?$
$V_1 = 0.5 \, m^2$ $V_2 = 0.25 \, m^2$
$T_1 = 303 \, K$ $T_2 = 413 \, K$

Remember to convert temperature into K, by adding $273°C$.

Using the combined gas equation and after rearrangement:

$$P_2 = \frac{P_1 V_1 T_2}{T_1 V_2} = \frac{(300)(0.5)(413)}{(303)(0.25)} = 817 \, kPa$$

The atmosphere

The atmosphere is the layer of air which envelopes the earth and its approximate *composition* expressed as a percentage by volume is:

Nitrogen   78
Oxygen   21
Other gases   1

Up to a height of some $8–9 \, km$ water vapor is found in varying quantities. The amount of water vapor in a given mass of air depends on the temperature of the air and whether or not the air has recently passed over large areas of water. The higher the temperature of the air the higher the amount of water vapor it can hold. Thus at altitude where the air temperature is least, the air will be dry.

The earth's atmosphere (Figure 4.90) can be said to consist of five concentric layers. These layers, starting with the layer nearest the surface of the earth, are known as the; *troposphere*, above which are the *stratosphere*, *mesosphere*, *thermosphere* and *exosphere*.

The boundary between the troposphere and stratosphere is known as the *tropopause* and this boundary varies in height above the earth's surface from about $7.5 \, km$ at the poles to $18 \, km$ at the equator. An average value for the tropopause in the "International Standard Atmosphere" (ISA) is around $11 \, km$ or $36,000 \, ft$.

The *thermosphere* and the upper parts of the *mesosphere* are often referred to as the *ionosphere*, since in this region ultraviolet radiation is absorbed in a process known as photo-ionization.

In the above zones, changes in temperature, pressure, density and viscosity take place, but of these aerodynamically at least, only the troposphere and stratosphere are significant. About 75% of the total air mass in the atmosphere is concentrated in the troposphere.

The International Standard Atmosphere

Due to different climatic conditions that exist around the earth, the values of temperature, pressure, density, viscosity and sonic velocity (speed of sound), are not constant for a given height.

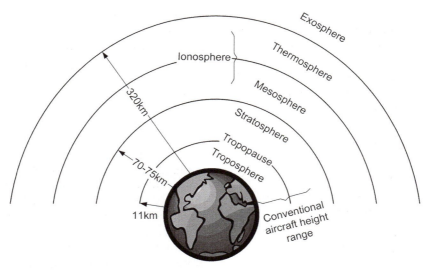

Figure 4.90 Principal zones of the atmosphere.

The **ISA** has, therefore, been set up to provide a standard for:

1. the comparison of aircraft performance; and
2. the calibration of aircraft instruments.

The ISA is a "hypothetical" atmosphere based on world average values. Note that since the performance of aircraft, their engines and their propellers is dependent on the variables quoted in the ISA. It will be apparent that the performance figures quoted by manufacturers in various parts of the world cannot be taken at face value but must be converted to standard values, using the ISA. If the actual performance of an aircraft is measured under certain conditions of temperature, pressure and density, it is possible to deduce, what would have been the performance under the conditions of the ISA, so that it can then be compared with the performance of other aircraft, which have similarly been reduced to standard conditions.

The *sea-level values* of some of the more important *properties of air*, contained in the ISA are tabulated in opposite column.

Property	Symbol	ISA value
Temperature	T_0	288.15 K or 15.15°C
Pressure	P_0	1013.2 mb or 101,320 N/m^2
Density	ρ	1.225 kg/m^3
Speed of sound	a_0	340.3 m/s
Dynamic viscosity	μ_0	1.789×10^{-5} Ns/m^2
Temperature lapse rate	L	6.5 K/km or 6.5°C/km or 1.98°C/1000 ft

Changes in properties of air with altitude

Temperature falls uniformly with height until about 11 km (36,000 ft). This uniform variation in temperature takes place in the troposphere, until a temperature of 216.7 K is reached at the tropopause. This temperature then remains constant in the stratosphere, after which the temperature starts to rise once again.

It is possible to calculate the temperature at a given height h (km) in the *troposphere* from the simple relationship, $T_h = T_0 - Lh$, where T_h is the temperature at height h (km) above sea

---

Key point

The ISA is used to compare aircraft performance and enable the calibration of aircraft instruments.

level and T_0 and L have the meanings given in the table of properties of air at sea level shown above.

The ISA value of *pressure* at sea level is given as 1013.2 mb. As height increases pressure decreases, such that at about 5 km, the pressure has fallen to half its sea-level value and at 15 km it has fallen to approximately one-tenth its sea-level value.

The ISA value of *density* at sea level is 1.225 kg/m^3. As height increases density decreases but not as fast as pressure. Such that, at about 6.6 km the density has fallen to around half its sea-level value and at about 18 km it has fallen to approximately one tenth of its sea-level value.

Humidity levels of around 70% water vapor at sea level drop significantly with altitude. Remember that the amount of water vapor a gas can absorb decreases with decrease in temperature. At an altitude of around 18 km the water vapor in the air is approximately 4%. Thus to ensure passenger comfort during flight it is essential to maintain the correct humidity level, within an aircrafts environmental control system.

Key point

With increase in altitude up to the tropopause; temperature, density, pressure and humidity all decrease.

The relationship between pressure, density and temperature

Having adopted the ISA values at sea level, the conditions at altitude may be calculated based on the temperature lapse rate and the gas laws you met earlier.

We know that,

$$\frac{PV}{T} = \text{constant}$$

It is also true that for a given mass of gas its *volume* is inversely proportional to its *density*, so the above equation may be re-written as:

$$\frac{P}{\rho T} = \text{constant}$$

where $V \propto 1/P$.

So now the combined gas equation may be used to compare values of temperature, density and pressure at two different heights. So we get:

$$\frac{P_1}{\rho_1 T_1} = \frac{P_2}{\rho_2 T_2}$$

Example 4.49

If the density of air at sea level is 1.225 kg/m^3 when the temperature is 288.15 K and the pressure is 101,320 N/m^2. Find the density of air at 10 km, where the temperature is 223 K and the pressure is 26,540 N/m^2.

From the above equation:

$$\rho_h = \frac{\rho_0 T_0 P_h}{P_0 T_h} = \frac{(1.225)(288.15)(26,540)}{(101,320)(223)}$$
$$= 0.414 \text{ kg/m}^3$$

Test your understanding 4.19

1. What is meant by a perfect gas?
2. Convert: (a) 280°C and (b) −170°C into K.
3. What variable is kept constant when formulating Boyle's law?
4. What is the ISA value of the tropopause?
5. Why was the ISA set up?
6. What happens to temperature, pressure, density and humidity with increase in altitude?
7. The ISA value for the speed of sound is 340.3 m/s. Using the appropriate tables and conversion factors find the speed of sound in: (a) mph, (b) knots and (c) ft/s.
8. Given that the sea-level temperature in the ISA is 20°C, what is the temperature in the ISA at 34,000 ft?

4.9.4 Fluids in motion

In order to study aerodynamics a basic understanding of fluids in motion is necessary. The study of fluid in motion or **fluid dynamics** is also important in other areas of engineering, e.g. fluid systems, such as hydraulic, pneumatic, oxygen and fuel systems, all of which provide vital and essential services for safe aircraft operation. We start by considering some important terminology that should also assist you with your study of aerodynamics.

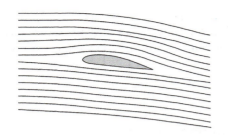

Figure 4.91 Pictorial representation of streamline or laminar flow.

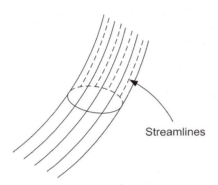

Streamlines

Figure 4.92 Stream tube.

Terminology

Streamline flow sometimes referred to as *laminar flow*, is the flow in which the fluid particles move in an orderly manner and retain the same relative positions in successive cross-sections. In other words, a flow which maintains the shape of the body over which it is flowing (Figure 4.91).

Incompressible flow is flow in which the density does not change from point-to-point. We will base the remainder of our work on fluids, on the assumption that they are incompressible. This is clearly not the case for air, where compressibility effects will need to be considered when we study high-speed flight.

Turbulent flow is flow in which the fluid particles may move perpendicular as well as parallel to the surface of the body and undergo eddying or unsteady motions. This may result in considerable thickening of the airflow and lead to break-up.

A *stream tube or tube of flow* (Figure 4.92) is considered to be an imaginary boundary defined by streamlines drawn so as to enclose a tubular region of fluid. No fluid crosses the boundary of such a tube.

Equation of continuity

This equation simply states that: *fluid mass flow rate is constant*. We will consider this equation only for incompressible fluids, i.e. fluids where the density at successive cross-sections through the stream tube is constant.

Figure 4.92 shows an incompressible fluid flowing through a stream tube, where the density at the inlet *1* is constant and equal to the density at the outlet *2*. v_1 and A_1 are the velocity and csa at cross-section *1* and v_2 and A_2 are the velocity and csa at cross-section *2*.

It is also a fact that the volume of the fluid entering the stream tube per second must be equal to the volume of the fluid leaving the stream tube per second. This follows from the conservation of mass and our stipulation that flow is incompressible. Then from what has just been said:

$$\text{at inlet, volume entering} = \text{area} \times \text{velocity} = A_1 v_1$$
$$\text{at outlet, volume leaving} = \text{area} \times \text{velocity} = A_2 v_2$$

therefore:

$$\dot{Q} = A_1 v_1 = A_2 v_2$$

where $\dot{Q} =$ volumetric flow rate (m^3/s). This equation is known as the *continuity equation for volume flow rate*.

You should ensure that you understand why the units are the same on both sides of this equation. We can also measure mass flow rate as well as volume flow rate, by remembering that density is equal to mass divided by volume so:

$$\text{density} \times \text{volume} = \rho V$$

Therefore to obtain mass flow rate, all we need to do is multiply the volume flow rate by the density. Then:

$$\dot{m} = \rho_1 A_1 v_1 = \rho_2 A_2 v_2$$

where $\dot{m} =$ mass flow rate (kg/s). This equation is known as the *continuity equation for mass flow rate*.

Make sure that you do not mix up the symbols for velocity and volume! For velocity we use lower case (v) and for volume we use upper case (V).

Example 4.50

In the wind tunnel shown in Figure 4.93, the air passes through a converging duct just prior to the working section. The air velocity entering the converging duct is 25 m/s and the duct has a csa of 0.3 m². If the speed of flow in the working section is to be 75 m/s. Calculate the *csa* of the working section. Assume air density is constant at 1.225 kg/m³.

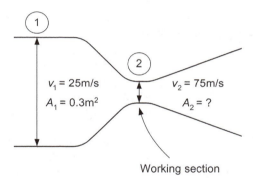

Figure 4.93 Wind tunnel.

We may use our equation for incompressible fluid flow, since $\rho_1 = \rho_2$, therefore:

$$A_1 v_1 = A_2 v_2$$

so

$$A_2 = \frac{A_1 v_1}{v_2}$$

and

$$A_2 = \frac{(0.3)(25)}{75} = \mathbf{0.1\,m}$$

You will note that the continuity equation is far easier to use than to verify!

The Bernoulli equation

The principle of the conservation of energy has already been discussed, earlier in our study of physics. This principle is equally valid for fluids in motion, as it is for solids, except that we now include a *pressure* energy term. The pressure energy of a fluid in motion is defined as:

$$\begin{array}{c}\text{Pressure of}\\\text{fluid}\end{array} \times \begin{array}{c}\text{Volume of the}\\\text{fluid displaced}\end{array} = pV$$

Note that pV gives Nm the correct SI units of energy, since $1\,\text{Nm} = 1\,\text{J}$.

So applying the principle of the conservation of energy to fluids in motion, we know that the total energy is conserved, i.e:

$$\text{PE}_1 + \text{KE}_1 + P_1 = \text{PE}_2 + \text{KE}_2 + P_2$$

where P = fluid static pressure energy and the subscript 1 = inlet, and subscript 2 = outlet. Then in symbols we have the *energy equation*:

$$mgh_1 + \tfrac{1}{2}mv_1^2 + p_1 V_1 = mgh_2 + \tfrac{1}{2}mv_2^2 + p_2 V_2$$

Note that in some texts z is used instead of h in the PE terms, to indicate the height above a datum.

The above formula in terms of energies is not very useful. In fluid dynamics, we wish to compare pressures in terms of an equivalent head of water, i.e. we need each term in our formula to have units of height, this is achieved by a little mathematical manipulation!

Dividing each term in the above energy equation by m gives us energy per unit mass, if at the same time we divide each term by the acceleration due to gravity g, it can be seen immediately from the equation given below, that the PE term now has units of height as required, but what about the other two terms?

$$h_1 + \frac{v_1^2}{2g} + p_1 \frac{V_1}{mg} = h_2 + \frac{v_2^2}{2g} + p_2 \frac{V_2}{mg}$$

The KE term can also be shown to have units of height. Using fundamental units velocity, v is in m/s and so v^2 has units m²/s² and acceleration due to gravity g has units m/s². Then KE term on division by mg has units m²/s² × s²/m giving units of metres, m, as required.

The third term for fluid pressure, can also be shown to have units of height by making the substitution, $\rho = m/V$ or $1/\rho = V/m$ this makes our third term $= p/\rho g$. Then using fundamental units for the newton (kgm/s^2) and thus for pressure ($\text{kgm/m}^2\text{s}^2$) and also for acceleration due to gravity (m/s^2) and density (kg/m^3). Our pressure energy term on division by ρg, also has units of height as required. So our energy equation may be written as the *head equation*:

$$h_1 + \frac{v_1^2}{2g} + \frac{p_1}{\rho g} = h_2 + \frac{v_2^2}{2g} + \frac{p_2}{\rho g}$$

The head equation now shows the total energy at the inlet and total energy at the outlet in terms of the sum:

Head due to PE + head due to KE

+ head due to pressure energy

Thus each of the terms in the head equation is measured in units of equivalent height. Thermodynamicists and aerodynamicists prefer to measure pressures in Pa (N/m^2), rather than in terms of equivalent head. The mathematical manipulation given above to produce the head equation has not been wasted! Since, all we need to do to convert the head equation into an equation involving pressures is multiply each term by density and acceleration due to gravity, to yield the *pressure equation*:

$$\rho g h_1 + \tfrac{1}{2}\rho v_1^2 + p_1 = \rho g h_2 + \tfrac{1}{2}\rho v_2^2 + p_2$$

where, p_1 and p_2 are the static pressures in the fluid flow, $\tfrac{1}{2}\rho v_1^2$ and $\tfrac{1}{2}\rho v_2^2$ are the dynamic pressures in the fluid flow, $\rho g h_1$ and $\rho g h_2$ are the pressures due to change in level of the fluid flow. The units of each term are Pa or N/m^2. You should verify the units of each term by using the fundamental units of N, which are kgm/s^2 these in turn come from the relationship $F = ma$.

The *pressure equation is better known as* **Bernoulli's theorem** *and is only valid for incompressible flow*. If the flow is horizontal then $h_1 = h_2$, then *Bernoulli's theorem* becomes

$$p_1 + \tfrac{1}{2}\rho v_1^2 = p_2 + \tfrac{1}{2}\rho v_2^2 = C$$

This is a most useful equation and yields a wealth of information. The equation tells us that, as the flow progresses from one point to another, an *increase in velocity is accompanied by a decrease in pressure*. This follows because the sum of the static pressure (p) and dynamic pressure ($\tfrac{1}{2}\rho v^2$), is a constant along a streamline. The *constant C represents the total or stagnation pressure*. The total pressure being the sum of the static and dynamic pressures, while the name stagnation arises from the fact that when the velocity is reduced to 0 (stagnation), the stagnation pressure is equal to the total pressure.

Key point

Bernoulli's equation is based on incompressible flow.

Using Bernoulli's equation

We have now found the energy, head and pressure versions of Bernoulli's equation. Before we use the head and pressure version of this equation it should be noted that, from the head version of the equation, collecting like terms:

$(p_2 - p_1)/\rho g$ will give us the pressure energy change

$(v_2^2 - v_1^2)/2g$ will give us the KE change

$(h_2 - h_1)$ will give us the PE change

All these energy changes will be measured in terms of height in m. In problems using Bernoulli's relationships, it will often be necessary to use the *equation of continuity* in order to find all the required information for a solution to the problem.

Example 4.51

A wind tunnel of circular cross-section has a diameter upstream of the contraction of 6 m and a test section diameter of 2 m. The test section pressure is at the ISA value for sea level. If the working section velocity is 270 mph find the:

(a) upstream section velocity,
(b) upstream pressure.

The situation is shown in Figure 4.94.

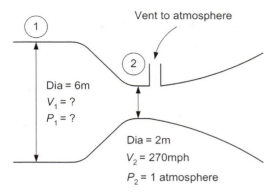

Figure 4.94 Wind tunnel.

(a) We first use the continuity equation to find the upstream velocity v_1. The csa A_1 and A_2 can be found using πr^2, then:

$$A_1 = 9\pi \quad \text{and} \quad A_2 = \pi$$

So from

$$A_1 v_1 = A_2 v_2$$

we have

$$v_1 = A_2 \frac{v_2}{A_1} = \frac{1(270)}{9} = 30 \text{ mph}$$

(b) To find the upstream pressure we will need to use Bernoulli's equation, so we must first convert v_1 and v_2 into m/s.

Using Table 4.7,

$$v_1 = \frac{30}{2.23694} = 13.4 \text{ m/s}$$

and

$$v_2 = \frac{270}{2.23694} = 120.7 \text{ m/s}$$

Then from Bernoulli's equation, assuming the wind tunnel is mounted horizontally,

$$p_1 + \rho \frac{v_1^2}{2} = p_2 + \rho \frac{v_2^2}{2}$$

and on rearrangement and substitution of values then:

$$p_1 = 101{,}320 + \frac{1.225}{2}(120.7^2 - 13.4^2)$$

then,

upstream pressure $p_1 = 110{,}133 \text{ N/m}^2$

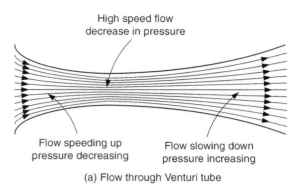

(a) Flow through Venturi tube

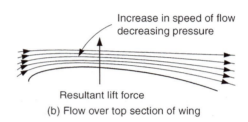

(b) Flow over top section of wing

Figure 4.95 The venturi tube and wing top section.

The venturi tube

An important application of Bernoulli's theorem is provided by the venturi tube (Figure 4.95(a)).

This arrangement shows a tube that gradually narrows to a throat, and then expands even more gradually. If measurements are taken at the throat a decrease in pressure will be observed. Now according to Bernoulli's equation a reduction in static pressure must be accompanied by an increase in dynamic pressure, if the relationship is to remain constant. The increase in dynamic pressure is achieved by an increase in the velocity of the fluid as it reaches the throat. The effectiveness of the venturi tube as a means of causing a decrease in pressure below that of the atmosphere depends very much on its shape.

The venturi tube provides us with the key for the generation of *lift*. Imagine that the bottom cross-section of the tube is the top part of an aircraft wing shown in cross-section (Figure 4.95(b)). Then the increase in velocity of flow over the wing causes a corresponding reduction in pressure, below atmospheric. It is this reduction in pressure which provides the lift force perpendicular to the top surface of the wing, and due to the shape of the lower wing cross-section a slight increase in pressure is achieved, which also provides a component of lift. The nature of lift will be considered in much more detail later, when you study the aerodynamics module.

Compressibility

We finish our short study of fluids with a short note on compressibility and its effects. So far all of our work on fluids has been based on the assumption that fluids are incompressible. This is certainly true for the practical application of fluid theory to liquids such as water but not so for air, which is most definitely compressible!

Our theory based on the incompressible behavior of fluids is still sufficiently valid for air when it flows below speeds of approximately 130–150 m/s. As speed increases compressibility effects become more apparent. The table below shows one or two values of speed against error when we assume that air is incompressible.

Speed of airflow (m/s)	Approximate error when assuming incompressibility (%)
50	0.5
95	2
135	4
225	11
260	15

Therefore when we study high-speed flight where aircraft fly at velocities close to, or in excess of, the *speed of sound* (340 m/s at sea level under standard ISA conditions), then the *compressibility effects of air must be considered*. In reality as seen in the above table, compressibility effects need to be considered at speeds much below the speed of sound. This is particularly true when considering the possible inaccuracies in aircraft pitot–static instruments, where such instruments depend on true static and dynamic air pressures for their correct operation. You will consider the ways in which instruments are calibrated to overcome compressibility effects when you study your specialist systems modules.

General Questions 4.4

1. The head h of mercury corresponding to the ISA value of atmospheric pressure is 0.76 m. What is the corresponding head of water, if it has a density of 1000 kg/m^2?
2. A hydraulic press has a 10 mm diameter small piston and 120 mm diameter large piston. What is the balance load on the large piston if the small piston supports a 5 kN load?
3. Explain the nature of viscosity and differentiate between dynamic viscosity and kinematic viscosity.
4. The pressure of air in an engine air starter vessel is 40 bar and the temperature is 24°C. If a fire in the vicinity causes the temperature of the pressurized air to rise to 65°C, find the new pressure of this air.
5. 70 m^3 of air at an absolute pressure of 7×10^5 Pa expands until the absolute pressure drops to 3.5×10^4 Pa while at the same time the temperature drops from 147°C to 27°C. What is the new volume?
6. What will be the temperature at an altitude, where the pressure of the atmosphere is 44.188 kPa and the density is 0.626 kg/m^3. Assume standard ISA sea-level values of pressure and temperature.
7. Find the "head" change in the KE of air at sea level, if it has an initial velocity of 15 m/s and a final velocity of 25 m/s.
8. Find the change in pressure energy in the form of a "head" if, $p_1 = 2.5$ MPa and $p_2 = 1.8$ MPa, given that the fluid has a *relative* density of 1.2.

4.10 Thermodynamics

Thermodynamics is the science that deals with various forms of energy and their transformation from one form into another. **Applied thermodynamics** is the specialist branch of the subject that deals specifically with heat, mechanical and internal energies and their application to power production, air-conditioning and refrigeration.

We start our study of applied thermodynamics (the area of specific interest to engineers) by considering a number of fundamental thermodynamic properties and relationships.

4.10.1 Fundamentals

Temperature

We have already met temperature on several occasions during our study of physics, but as yet, we have not defined it in thermodynamic terms.

Temperature is a measure of the quantity of energy possessed by a body or substance. It measures the vibration of the molecules, which form the substance. These molecular vibrations only cease when the temperature of the substance reaches *absolute zero*, i.e. −273.15°C.

We have already met the celsius temperature scale and you should now be able to convert degrees centigrade into kelvin and vice versa, for completeness we need to see how the fahrenheit scale relates to these.

Figure 4.96 shows the relationship between these three scales and indicates the common boiling point of pure water and the melting point of pure ice, for each set of units.

> **Key point**
>
> Temperature measures the energy possessed by the vibration of the molecules that go to make up a substance.

Example 4.52

Convert 60°C into (a) K and (b) °F.

(a) You already know that 1°C = 1 K and that to convert °C into K, we simply add 273, therefore 60°C + 273 = 333 K.

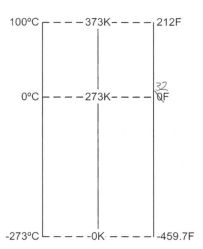

Figure 4.96 Relationship between celsius, kelvin and fahrenheit scales.

Note that to be strictly accurate we should add 273.15, but for all practical purposes the approximate value of 273 is adequate.

(b) Now to convert 60°C into °F we can use the *inverse* of the relationship given in Table E.7 of Appendix E. Then °F = (°C × 1.8) + 32 and substituting value gives,

$$(60°C × 1.8) + 32 = 140°F$$

Alternative versions of the formulae, to convert Fahrenheit to Celsius and vice versa, are:

$$°F = \left(°C × \frac{9}{5}\right) + 32 \text{ and } °C = (°F − 32) × \frac{5}{9}$$

Note: The °F may be converted to absolute temperature (K) by converting to °C and then adding 273. °F may be converted to absolute temperature on the *Rankine scale* by adding 459.67. To convert from Rankine to K simply multiply by 5/9.

So   140°F + 459.67 = 599.67 R = 599.67 (5/9) = 333.15 K. *For all our thermodynamic work we will only use the Kelvin scale for measuring absolute temperature.*

Temperature measurement

The method used to measure temperature depends on the degree of hotness of the body or substance being measured. Measurement apparatus include; liquid-in-glass thermometers, resistance thermometers, thermistor thermometers and thermocouples.

All *thermometers* are based on some property of a material that changes when the material becomes colder or hotter. Liquid-in-glass thermometers use the fact that most liquids expand slightly when they are heated. Two common types of liquid-in-glass thermometer are the mercury thermometer and alcohol thermometer, both have relative advantages and disadvantages.

Alcohol thermometers are suitable for measuring temperatures down to −115°C and have a higher expansion rate than mercury, so a larger containing tube may be used. They have the disadvantage of requiring the addition of a colouring in order to be seen easily. Also, the alcohol tends to cling to the side of the glass tube and may separate.

Mercury thermometers conduct heat well and respond quickly to temperature change. They do not wet the sides of the tube and so flow well in addition to being easily seen. Mercury has the disadvantage of freezing at −39°C and so is not suitable for measuring low temperatures. Mercury is also poisonous and special procedures must be followed in the event of spillage.

Resistance thermometers are based on the principle that current flow becomes increasing more difficult with increase in temperature. They are used where a large temperature range is being measured approximately −200 to 1200°C. *Thermistor thermometers* work along similar lines, except in this case they offer less and less resistance to the flow of electric current as temperature increases.

Thermocouple thermometers are based on the principle that when two different metal wires are joined at two junctions and each junction is subjected to a different temperature, a small current will flow. This current is amplified and used to power an analogue or digital temperature display. Thermocouple temperature sensors are often used to measure aircraft engine and jet pipe temperatures, they can operate over a temperature range from about −200 to 1600°C.

Thermal expansion

We have mentioned in our discussion on thermometers that certain liquids expand with increase in temperature, this is also the case with solids. Thermal expansion is dependent on the nature of the material and the magnitude of the temperature increase. We normally measure the linear expansion of solids, such as the increase in length of a bar of the material, with gases (as you have already seen) we measure volumetric or cubic expansion.

Every solid has a *linear expansivity value, i.e. the amount the material will expand in m/K or m/°C*. This expansivity value is often referred to as the *coefficient of linear expansion* (**α**), some typical values of α are given below:

Material	Linear expansion coefficient α/°C
Invar	1.5×10^{-6}
Glass	9×10^{-6}
Cast iron	10×10^{-6}
Concrete	11×10^{-6}
Steel	12×10^{-6}
Copper	17×10^{-6}
Brass	19×10^{-6}
Aluminium	24×10^{-6}

Given the length of a material (l), its linear expansion coefficient (α) and the temperature rise (Δt), the *increase in its length* can be calculated using:

$$\text{Increase in length} = \alpha l(t_2 - t_1)$$

Note that we are using lower case t to indicate temperature because when we find a *temperature difference* (Δt) we do not need to convert to K.

For solids an estimate of the cubic or volumetric expansion may be found using

$$\text{Change in volume} = 3\alpha V(t_2 - t_1)$$

where V is the original volume.

A similar relationship exists for surface expansion, where a body experiences a change in area. In this case the linear expansion coefficient is multiplied by 2, therefore,

$$\text{Change in area} = 2\alpha A(t_2 - t_1)$$

where A is the original area.

Example 4.53

A steel bar has a length of 4.0 m at 10°C. What will be the length of the bar when it is heated to 350°C? If a sphere of diameter 15 cm is made from the same material what will be the percentage increase in surface area, if the sphere is subject to the same initial and final temperatures?

Using $\alpha = 12 \times 10^{-6}$ from the above table, increase in length of the bar is given by:

$$x = \alpha l(t_2 - t_1) = (12 \times 10^{-6})(4.0)(350 - 10)$$
$$= 0.0163 \, \text{m}$$

This can now be added to the original length to give the final length $= 4.0 + 0.0163 = \mathbf{4.0163\,m}$.

Increase in surface area of the sphere $= 2\alpha A$ $(t_2 - t_1)$. We first need to find the original surface area which is given by:

$$A = 4\pi r^2 = 4\pi \times (0.075)^2 = 0.0707 \, \text{m}^2$$

and from above, the increase in surface area

$$= (2)(12 \times 10^{-6})(0.0707)(340)$$
$$= 5.769 \times 10^{-6} \, \text{m}$$

Therefore the percentage increase in area

$$= \frac{\text{increase in area}}{\text{original area}} \times 100$$
$$= \frac{5.769 \times 10^{-4}}{0.0707} \times 100 = \mathbf{0.82\%}$$

Heat energy

Energy is the most important and fundamental physical property of the universe. We have already defined energy as, *the capacity to do work*, more accurate it may be defined as: *the capacity to produce an effect*. These effects are apparent during the process of energy transfer.

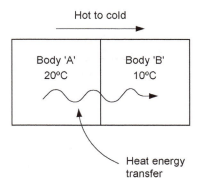

Figure 4.97 Heat energy transfer.

A modern idea of heat is that it is energy in transition and cannot be stored by matter. **Heat** (Q) may be defined as: *transient energy brought about by the interaction of bodies by virtue of their temperature difference when they communicate*. Matter possesses stored energy but not transient (moving) energy, such as heat or work. Heat energy can only travel or *transfer* from a hot body to a cold body, it cannot travel up hill, Figure 4.97 illustrates this fact.

Key point
Heat and work is energy in transit and cannot be stored by matter.

Within matter the amount of molecular vibration determines the amount of *KE* a substance possesses. For incompressible fluids (liquids) the amount of molecular vibration is relatively small and can be neglected. For compressible fluids and gases the degree of vibration is so large that it has to be accounted for in thermodynamics. This KE is classified as *internal energy* (U) and is a form of stored energy.

Heat energy transfer
Literature on heat transfer generally recognizes three distinct modes of heat transmission, the names of which will be familiar to you, i.e. conduction, convection and radiation. Technically only conduction and radiation are true heat transfer processes, because both of these depend totally and utterly on a temperature difference being present. Convection also depends on the transportation of a mechanical mass.

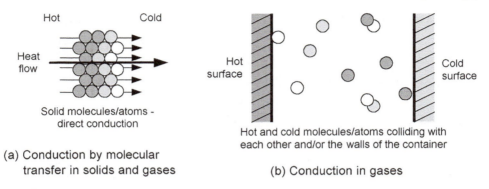

(a) Conduction by molecular
transfer in solids and gases

(b) Conduction in gases

Figure 4.98   Conduction by molecular transfer in solids and liquids.

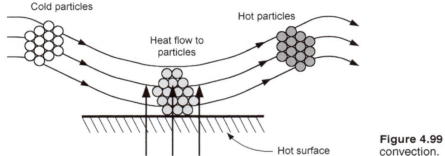

Figure 4.99   Heat transfer by convection.

Nevertheless, since convection also accomplishes transmission of energy from high to low temperature regions, it is conventionally regarded as a heat transfer mechanism.

Thermal conduction in solids and liquids seems to involve two processes the first is concerned with atoms and molecules (Figure 4.98), the second with free electrons.

Atoms at high temperatures vibrate more vigorously about their equilibrium positions than their cooler neighbors. Since atoms and molecules are bonded to one another, they pass on some of their vibrational energy. This energy transfer occurs from atoms of high-vibrational energy to those of low-vibrational energy, without appreciable displacement. This energy transfer has a knock-on effect, since high-vibrational energy atoms increase the energy in adjacent low-vibrational energy atoms, which in turn causes them to vibrate more energetically, causing thermal conduction to occur. In solids (Figure 4.98) the energy transfer is by direct contact between one molecule and another. In gases the conduction process occurs as a result

of collisions between hot and cold molecules and the surface of the containing vessel.

The second process involves material with a ready supply of free electrons. Since electrons are considerable lighter than atoms, then any gain in energy by electrons results in an increase in the electron's velocity and it is able to pass this energy on quickly to cooler parts of the material. This phenomenon is one of the reasons why electrical conductors that have many free electrons are also good thermal conductors. Do remember that metals are not the only good thermal conductors, the first mechanism described above which does not rely on free electrons is a very effective method of thermal conduction, especially at low temperatures.

Heat transfer by *convection* consists of two mechanisms. In addition to energy transfer by random molecular motion (diffusion), there is also energy being transferred by the bulk motion of the fluid.

So in the presence of a temperature difference large numbers of molecules are moving together in bulk (Figure 4.99), at the same time as the individual motion of the molecules takes place. The cumulative effect of both of these energy transfer methods is referred to as heat transfer by convection.

Radiation may be defined as the transfer of energy *not requiring* a medium through which the energy must pass, thus radiation can be transferred through empty space. Thermal radiation is attributed to the electron energy changes within atoms or molecules. As electron energy levels change, energy is released which travels in the form of electromagnetic waves of varying wavelength. You will meet electromagnetic waves again when you study light. When striking a body the emitted radiation is either absorbed by, reflected by, or transmitted through the body.

Specific heat

From what has been said about heat transfer above, it will be apparent that different materials have different capacities for absorbing and transferring thermal energy. The thermal energy needed to produce a temperature rise depends on; the mass of the material, type of material and the temperature rise to which the material is subjected.

Thus the inherent ability of a material to absorb heat for a given mass and temperature rise is dependent on the material itself. This property of the material is known as its *specific heat capacity*. In the SI system, *the specific heat capacity of a material is the same as the thermal energy required to produce a 1 K rise in temperature in a mass of 1 kg*. Therefore knowing the mass of a substance and its specific heat capacity, it is possible to calculate the thermal energy required to produce any given temperature rise, from:

$$\text{Thermal energy, } Q = mc\Delta t$$

where c = specific heat capacity of the material (J/kgK) and Δt is the temperature change.

Example 4.54

How much thermal energy is required to raise the temperature of 5 kg of aluminium from 20 to 40°C? Take the specific heat capacity for aluminium as 900 J/kgK.

All that is required is to substitute the appropriate values directly into the equation:

$$Q = mc\Delta t = (5)(900)(40 - 20)$$
$$= 90,000 \text{ J} = 90 \text{ kJ}$$

Another way of defining the *specific heat capacity* of any substance is: *the amount of heat energy required to raise the temperature of unit mass of the substance through one degree, under specific conditions*.

In *thermodynamics* two specified conditions are used those of constant volume and constant pressure. With *gases* the two specific heats do not have the same value and it is essential that we distinguish between them.

Specific heat at constant volume

If 1 kg of a gas is supplied with an amount of heat energy sufficient to raise the temperature by 1°C or 1 K while the volume of the gas remains constant, then the amount of heat energy supplied is known as the *specific heat capacity at constant volume and is denoted by* c_v. Note that under these circumstances (Figure 4.100(a)) no work is done, but the gas has received an increase in internal energy (U). The specific heat at constant volume for air (c_v air) is 718 J/kgK, this constant is well worth memorizing!

Specific heat at constant pressure

If 1 kg of a gas is supplied with a quantity of heat energy sufficient to raise the temperature of the gas by 1°C or 1 K while the pressure is held constant, then the amount of heat energy supplied is known as the *specific heat capacity at constant pressure and is denoted by* c_p.

This implies that when the gas has been heated it will expand a distance h (Figure 4.100b), so work has been done. Thus for the same amount of heat energy there has been an increase in internal energy (U), plus work. The value of c_p is, therefore, greater than the corresponding value of c_v.

The specific heat capacity at constant pressure for air (c_p air) is 1005 J/kgK, again this is a constant worth remembering.

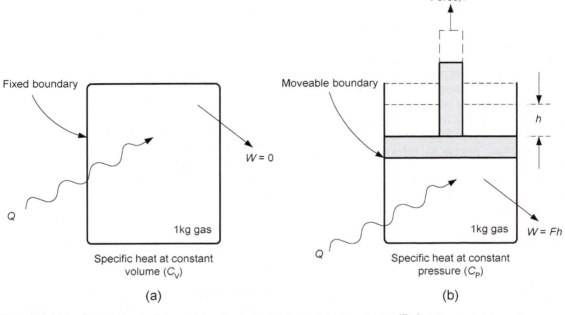

Figure 4.100 Comparison of constant volume and constant pressure specific heats.

The characteristic gas equation

The combined gas law, which you met earlier, stated that for a perfect gas with unit mass:

$$\frac{pV}{T} = \text{a constant}$$

This relationship is of course true for any fixed mass of gas and so we can write that,

$$\frac{pV}{T} = \text{mass} \times \text{a constant}$$

Now for any perfect gas which obeys the ideal gas laws this constant R is specific to that particular gas, i.e. R *is the characteristic gas constant* or specific gas constant for the individual gas concerned. Therefore, the characteristic gas equation may be written as:

$$\frac{pV}{T} = mR$$

or

$$pV = mRT$$

The units for the characteristic gas constant is J/kgK. Note that when the above equation is used both *absolute pressure and absolute temperature must be used.*

The characteristic gas constant for a number of gases is given in the table below:

Gas	Characteristic gas constant (J/kgK)
Hydrogen	4124
Helium	2077
Nitrogen	297
Air	287
Oxygen	260
Argon	208
Carbon dioxide	189

The characteristic gas constant for air, from the above table, is $R = 287$ J/kgK. This is related to the specific heat capacities for air in the following way, i.e. $R = c_p - c_v$, you should check this relationship by noting the above values of R, c_p and c_v for air. This relationship ($R = c_p - c_v$) is not only valid for air, *it is also valid for any perfect gas that follows the ideal laws.*

Example 4.55

0.22 kg of gas at a temperature of 20°C and pressure of 103 kN/m² occupies a volume of 0.18 m³. If the c_v for the gas $= 720$ J/kgK, find the:
(a) characteristic gas constant,
(b) specific heat capacity at constant pressure.

(a) Using $pV = mRT$ then on rearrangement,

$$R = \frac{pV}{mT} = \frac{(103 \times 10^3)(0.18)}{(0.22)(293)}$$

$$= 288 \text{ J/kgK}$$

(b) from $R = c_p - c_v$, then,

$$c_p = R + c_v = 288 + 720 = 1008 \text{ J/kgK}$$

Latent heat

When a substance changes state, i.e. when heat is applied to a solid and it turns into a liquid and with further heating the liquid turns into a gas we say the substance has undergone a *change in state*. The three states of matter are *solid, liquid and gas*. Therefore, the heat energy added to a substance does not necessary give rise to a measurable change in temperature, it may be used to change the state of a substance, under these circumstances we refer to the heat energy as *latent or hidden heat*.

Key point

Latent heat is heat added to a body without change in temperature.

We refer to the thermal energy required to change a solid material into a liquid as, the latent heat of fusion. For water, 334 kJ of thermal

energy are required to change 1 kg of ice at 0°C into water at the same temperature. Thus *the specific latent heat of fusion for water is 334 kJ.* In the case of latent heat, *specific*, refers to unit mass of the material, i.e. per kilogramme. So we define the specific latent heat of fusion of a substance as: the thermal energy required to turn 1 kg of a substance from a liquid into a solid without change in temperature.

If we wish to find the thermal energy required to change any amount of a substance from a solid into a liquid, then we use the relationship:

$$Q = mL$$

where L is the *specific latent heat* of the substance.

In a similar manner to the above argument: *the thermal energy required to change 1 kg of a substance from a liquid into a gas without change in temperature, is known as the specific latent heat of vaporization.* Again, if we wish to find the thermal energy required to change any amount of a substance from a liquid into a gas we use the relationship $Q = mL$, but in this case $L =$ the specific latent heat of vaporization.

The specific latent heat of vaporization for water is 2.26 MJ/kgK.

22. kJ

Example 4.56

(a) How much heat energy is required to change 3 kg of ice at 0°C into water at 30°C?
(b) What thermal energy is required to condense 0.2 kg of steam into water at 100°C?

(a) The thermal energy required to convert ice at 0°C into water at 0°C is calculated using the equation:

$$Q = mL$$

and substituting values we get,

$$Q = (3)(334 \times 10^3) = 1.002 \text{ MJ}$$

The 3 kg of water formed has to be heated from 0 to 30°C. The thermal energy required for this is calculated using the equation $Q = mc\Delta t$, you have already met this equation when we studied specific heat earlier.

So in this case

$$Q = (3)(4200)(30) = 378{,}000 \, \text{J}$$
$$= 0.378 \, \text{MJ}$$

Then total thermal energy required $= 1.002 + 0.378 = \mathbf{1.38 \, MJ}$

(b) In this case we simply use $Q = mL$, since we are converting steam to water at $100°C$ which is the vaporization temperature for water into steam.

Then $Q = (0.2)(2.226 \times 10^6) = \mathbf{445.2 \, kJ}$

Note the large amounts of thermal energy required to change the state of a substance. This energy together with cooling by evaporation is used within aircraft air-conditioning and refrigeration systems.

A liquid does not have to boil in order for it to change state, the nearer the temperature is to the boiling point of the liquid, then the quicker the liquid will turn into a gas. At much lower temperatures the change may take place by a process of *evaporation*. The steam rising from a puddle, when the sun comes out after a rainstorm, is an example of evaporation, where water vapour forms as steam, well below the boiling point of the water.

There are several ways that a liquid can be made to evaporate more readily. These include; an *increase in temperature* that increases the molecular energy of the liquid sufficient for the more energetic molecules to escape from the liquid, *reducing the pressure above the liquid* in order to allow less energetic molecules to escape as a gas, *increase the surface area*, thus providing more opportunity for the more energetic molecules to escape or, *by passing a gas over the surface* of the liquid to assist molecular escape.

An aircraft refrigeration system can be made to work in exactly the same manner as a domestic refrigerator, where we use the fact that *a liquid can be made to vaporize at any temperature by altering the pressure acting on it*. Refrigerators use a fluid that has a very low boiling point, such as *freon*. We know from our laws of thermodynamics that heat can only flow from a point of high temperature to one at a lower temperature. If heat is to be made to flow in the opposite direction some additional energy needs to be supplied. In a refrigeration system

such as that illustrated in the block schematic diagram (Figure 4.101), this additional source of energy is supplied by a compressor or pump. When the gas is compressed, its temperature is raised and when the gas is allowed to expand its temperature is lowered.

A *reverse* flow of heat is achieved by compressing the freon to a pressure high enough so that its temperature is raised above that of the outside air. Heat will now flow from the high-temperature gas to the lower temperature surrounding air, thus lowering the heat energy of the gas. The gas is now allowed to expand to a lower pressure, causing a drop in temperature. This drop in temperature now makes it cooler than the surrounding air, so the air being cooled, acts as the *heat source*. Thus heat will now flow from the heat source, to the freon, which is now compressed again beginning a new cycle.

In a practical sense for the freon refrigeration system shown opposite, the refrigeration cycle operates as follows. Freon as a *liquid* is contained in the receiver under high pressure. It is allowed to flow through a valve into the evaporator at reduced pressure, so now, at reduced pressure, the boiling temperature of the freon is low enough *to cool* the surrounding air through *heat exchange*, this is the purpose of the refrigeration system! In turn, heat now flows from the air to the freon, causing it to boil and vaporize. Cold freon vapor now enters the compressor, where its pressure is raised, thereby raising its boiling point. The refrigerant at high pressure and high temperature flows into the condenser, where heat flows from the freon (refrigerant) to the outside air, condensing the *vapor into a liquid*. The cycle is repeated to maintain the cooling space, through which the air passes, at the desired temperature.

Note that heat flows *into the refrigerant* from the air to be cooled, *via the evaporator* heat exchanger and heat flows *from the refrigerant* to the surrounding air *via the condenser* heat exchanger.

> ### Key point
> A refrigerant is a cryogenic (cold) fluid which has a very low boiling temperature.

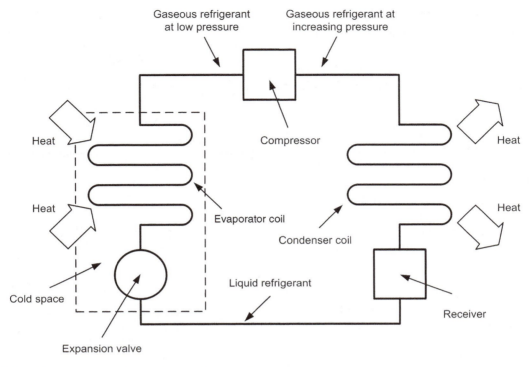

Figure 4.101 A typical aircraft refrigeration system.

Test your understanding 4.21

1. Convert (a) −20°C into K, (b) 120°F into °C and (c) −50°F into K.

2. We are required to measure the jet pipe temperature of an aircraft that, under normal operating conditions, will not exceed 1200°C. Suggest the most suitable temperature-measuring device, giving reasons.

3. Define the linear expansion coefficient for solids *and* explain how it may be used for approximating surface expansion and volumetric expansion.

4. Define heat energy and explain the difference between heat energy and internal energy of a substance.

5. Explain the essential differences between heat transfer by conduction and heat transfer by convection.

6. Why, for a gas, is the specific heat capacity at constant pressure greater than the specific heat capacity at constant volume?

7. State the formula for calculating the thermal energy needed to produce a temperature rise, *and* explain how this formula varies when calculating latent thermal energy (i.e. heat energy input without temperature rise).

8. If the characteristic gas constant for a gas is 260 J/kgK and the specific heat capacity at c_v is 680 kJ/kgK, what is the value of c_p?

9. Detail the three ways in which a liquid can be made to evaporate more readily.

10. What is the purpose of: (a) the evaporator and (b) the compressor, within a typical refrigeration system?

4.10.2 Thermodynamic systems

Thermodynamic systems may be defined as particular amounts of a thermodynamic substance, normally compressible fluids, such as vapors and gases, which are surrounded by an identifiable *boundary*. We are particularly interested in thermodynamic systems which involving *working fluids* (rather than solids) because these fluids enable the system *to do work* or have work done upon it. *Only transient energies in the form of heat (Q) and work (W), can cross the system boundaries*, and as a result there will be a change in the stored energy of the contained substance (working fluid).

Properties of thermodynamic systems

The essential elements that go to make up a thermodynamic system are:

(a) a working fluid, i.e. the matter which may or may not cross the system boundaries, such as water, steam, air, etc.,
(b) a heat source,
(c) a cold body to promote heat flow and enable heat energy transfer,
(d) the system boundaries, which may or may not be fixed.

The *property* of a working fluid is an observable quantity, such as pressure, temperature, etc. The *state* of a working fluid *when it is a gas*, may be defined by any *two* unique properties. For example, Boyle's law defines the state of the fluid by specifying the independent thermodynamic properties of volume and pressure.

When a working fluid is subject to a *process*, then the fluid will have started with one set of properties and ended with another, irrespective of how the process took place or what happened between the start and end states. For example, if a fluid within a system has an initial pressure (p_1) and temperature (T_1) and is then compressed producing an increase in pressure (p_2) and temperature (T_2), then we say that the fluid has undergone a *process* from state 1 to state 2.

We say that **work** is transferred in a thermodynamic system, *if there is movement of the system boundaries*, this concept is further explored when we consider closed systems, next.

The closed system

This type of system has a *closed or fixed boundary containing a fixed amount of vapour or gas*, while an exchange of heat and work may take place. An energy diagram of a typical closed system is shown in Figure 4.102.

> **Key point**
>
> In a closed system there is no mass transfer of system fluid.

The boundary of a closed system is not necessarily rigid, what makes the system closed is the fact that no mass transfer of the system fluid

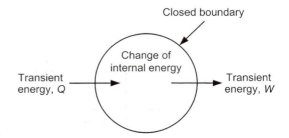

Figure 4.102   Closed system energy exchange.

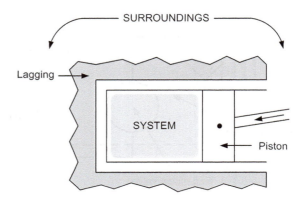

Figure 4.103   Cylinder and piston assembly of the internal combustion engine.

takes place, while an interchange of heat and work take place.

Consider the well known example of a closed system, that of the cylinder and piston assembly of an internal combustion engine (Figure 4.103).

The *closed boundary* is formed by the crown of the piston, the cylinder walls and the cylinder head with the valves closed. The *transient energy* being in the form of combustible fuel that creates a sudden pressure wave which forces the piston down. Therefore as the piston moves, the boundaries of the system move. This movement causes the system to do *work* (force × distance), on its surroundings. In this case the piston connecting rod drives a crank, to provide motive power.

Note that in a *closed system, it requires movement of the system boundary for work to be done by the system or on the system, thus work (like heat) is a transient energy, it is not contained within the system. There is also, no mass transfer of system fluid across the system*

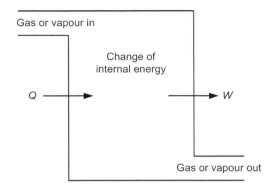

Figure 4.104 Energy exchange for a typical open system.

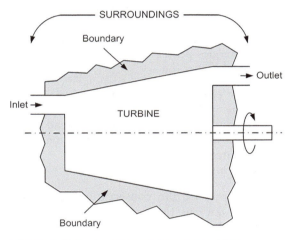

Figure 4.105 An open system gas turbine.

boundary when an interchange of the transient energies of heat (Q) and work (W) is taking place.

The open system

In this type of system there is an opening in the system boundary to allow a mass transfer of fluid to take place while the transient energies of heat (Q) and work (W) are being interchanged. The energy diagram for such a system is shown in Figure 4.104.

A practical example of an open system is the gas turbine engine (Figure 4.105). In this system there is a *transfer of mass across the system boundaries* in the form of airflow, which possesses its own KE, pressure energy and in

some cases its own PE. This energetic air passes through the open system and is subject to an interchange of transient energies in the form of heat and work.

4.10.3 The first law of thermodynamics

In essence this law applies the principle of the conservation of energy to open and closed thermodynamic systems. Formally, it may be stated as follows: *when a system undergoes a thermodynamic cycle then the net heat energy transferred to the system from its surroundings is equal to the net heat energy transferred from the system to its surroundings.* A *thermodynamic cycle* is where the working fluid of the system undergoes a series of processes and final returns to its initial state, more will be said about thermodynamic cycles later. We first consider the application of the first law to closed systems.

First law of thermodynamics applied to a closed system

The principle of the conservation of energy (the first law of thermodynamics) applied to a closed system states that: *given a total amount of energy in a system and its surroundings this total remains the same irrespective of the changes of form that may occur.*

In other words: *the total energy entering a system must be equal to the total energy leaving the system.* This is represented diagramatically (Figure 4.106) where the initial internal energy is U_1, and the final internal energy is U_2 so, *the change in internal energy is shown as $U_2 - U_1$ or ΔU.*

So, in symbol form:

$$U_1 + Q = U_2 + W$$

(i.e. total energy in = total energy out)

In its more normal form:

$$Q - W = \Delta U$$

So the above equation represents the concept of the *first law of thermodynamics applied to*

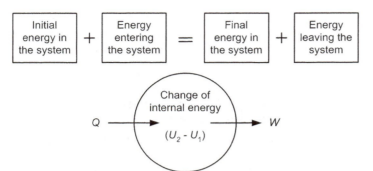

Figure 4.106 First law of thermo-dynamics applied to a closed system.

a closed system. This equation is also known as the *non-flow energy equation (NFEE).*

Heat and work energy transfer are given a sign convention, as shown in Figure 4.106. Heat entering a system is positive, work leaving a system is negative. Another way of expressing the same thing is to say that; *heat supplied to the system, or done on the system, is positive* and *work output or work done by the system is positive.* Naturally the inverse applies, i.e. heat done by the system or leaving the system is negative and work done on the system or entering the system is negative.

> **Key point**
>
> The first law of thermodynamics is a conservation law, where the total energy entering a system is equal to the total energy leaving the system.

Example 4.57

During a non-flow thermodynamic process the internal energy possesed by the working fluid within the system was increased from 10 to 30 kJ, while 40 kJ of work was done by the system. What is the magnitude and direction of the heat energy transfer across the system during the process?

Using $Q - W = U_2 - U_1$, where $U_1 = 10$ kJ, $U_2 = 30$ kJ and $W = 40$ kJ (positive work). Then, $Q - 40 = 30 - 10$ and $Q = 60$ kJ

Since Q is *positive*, it must be *heat supplied to the system*, which may be represented by an arrow pointing into the system, as shown in Figure 4.105.

First law of thermodynamics applied to an open system

Since the fluid is continuously flowing in and out of the system when heat and work transfers are taking place. We need to consider all of the stored energies possessed by the fluid, that we mentioned earlier, i.e.

1. flow or pressure energy = pressure × volume = pV;
2. PE = mgz (notice here we use z instead of h for height, the reason will become clear later!);
3. KE = $\frac{1}{2}mv^2$.

Now applying the conservation of energy (the first law) to the open system shown in Figure 4.106 then:

$$\text{Total energy in} = \text{Total energy out}$$

so,

Transient energy in + Stored energy in
  = Transient energy out + Stored energy out

or,

Heat energy + (IE$_1$ + press E$_1$ + PE$_1$ + KE$_1$)
  = work energy + (IE$_2$ + press E$_2$ + PE$_2$ + KE$_1$)

Now in symbol form we have:

$$Q + U_1 + p_1 V_1 + mgz_1 + \tfrac{1}{2}mv_1^2 = W + U_2$$
$$+ p_2 V_2 + mgz_2 + \tfrac{1}{2}mv_2^2$$

and rearranging gives:

$$Q - W = (U_2 - U_1) + (p_2 V_2 - p_1 V_1)$$
$$+ (mgz_2 - mgz_1) + (\tfrac{1}{2}mv_2^2 - \tfrac{1}{2}mv_1^2)$$

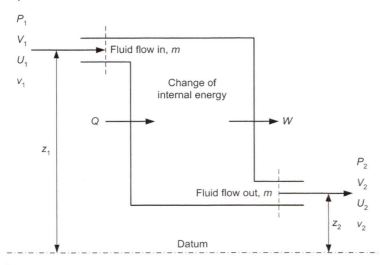

Figure 4.107 First law of thermodynamics applied to an open system.

This is the full equation for the *first law of thermodynamics applied to an open system,* it is called the *steady flow energy equation (SFEE).*

When dealing with flow systems, where there is a mass transfer of fluid. It is convenient to group the internal energy (U) and pressure energy (pV) of the fluid together, when this is done another property of the fluid called *enthalpy* is used for the combination. Then:

Enthalpy (H) = Internal energy (U)
+ Pressure energy (pV)

Now it is also a feature of open systems that the stored energy terms are a function of fluid mass flow rate. It is therefore, convenient to work in specific mass energies, i.e. energy per kilogramme of fluid, i.e in the SI system:

Specific energy of fluid (per kilogramme)

$$= \frac{Energy}{Mass\ in\ kilogrammes\ (m)}$$

The symbols and units for the individual specific energies are:

1. specific internal energy $= u$ (J/kg)
2. specific pressure energy $= (pV/m)$
   $= p/\rho$ (J/kg)
3. specific enthalpy $= h$ (J/kg) where
   $h = u + p/\rho$
4. specific PE $= gz$ (J/kg)
5. specific KE $= 1.2\,v^2$ (J/kg)

Then the steady-flow energy equation in specific terms may be written as:

$$Q - W = (h_2 - h_1) + (gz_2 - gz_1)$$
$$+ \left(\frac{1}{2}v_2^2 - \frac{1}{2}v_1^2\right) \text{(SFEE)}$$

Note that the above equation also implies that the heat and work energy transfers (in addition to all other energies in the equation) are also in specific terms with units in J/kg.

The enthalpy in specific terms has the symbol h, which may have been confused with height in the PE term if we had used it. This is the reason for using z for height when dealing with thermodynamic systems.

Key point

The enthalpy of a system fluid is its internal energy plus its pressure–volume energy.

Example 4.58

At entry to a horizontal steady flow system, the fluid has a specific enthalpy of 2000 kJ/kg and possesses 250 kJ/kg of KE. At the outlet from the system, the specific enthalpy is 1200 kJ/kg and there is negligible KE. If there is no heat energy transfer during the process, determine the magnitude and direction of the work done.

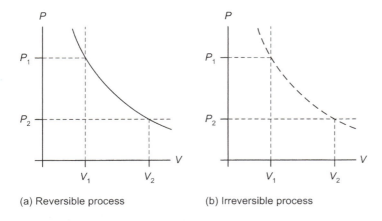

Figure 4.108 Diagrammatic representation of reversible and irreversible processes.

(a) Reversible process

(b) Irreversible process

Using the above SFEE equation, we first note that the PE term $gz_2 - gz_1 = 0$, since there is no change in height between fluid at entry and fluid at exit (horizontal). Also there is negligible fluid KE at exit, in other words $\frac{1}{2}v_2^2 = 0$ and during the process $Q = 0$.

Therefore, substituting appropriate values into the SFEE gives:

$$0 - W = (1200 - 2000) + 0 + (0 - 250)$$
$$-W = -800 - 250$$
$$W = 1050\,\text{kJ/kg}$$

and since work is positive then, work done *by* the system = **1050 kJ/kg**.

4.10.4 Thermodynamic processes

We will now look, very briefly, at one or two processes, which will be of help to us when we discuss the thermodynamic cycles for the internal combustion engine and the gas turbine engine.

Reversible and irreversible processes

Before we consider any specific processes you will need to understand the concepts of *reversibility* and *irreversibility*.

In its simplest sense, a system is said to be *reversible, when it changes from one state to another and at any instant during this process, an intermediate state point can be identified from any two properties that change as a*

result of the process. For reversibility, the fluid undergoing the process passes through a series of *equilibrium* states. Figure 4.108(a) shows a representation of a reversible process where unique equilibrium pressure and volume states can be identified at any time during the process. Reversible processes are represented diagrammatically by solid lines (Figure 4.108(a)).

In practice, because of energy transfers, the fluid undergoing a process cannot be kept in equilibrium in its intermediate states and a continuous path cannot be traced on a diagram of its properties. Such real processes are called *irreversible* and they are usually represented by a dashed line joining the end states (Figure 4.108(b)).

Constant volume process

The constant volume process for a perfect gas is considered to be a reversible process. Although you may not be aware of it, you have already met a constant volume process when we considered specific heat capacities, look back at Figure 4.100(a). This shows the working fluid being contained in a rigid vessel, so the system boundaries are immovable and no work can be done on or by the system. So we make the assumption that a constant volume process, implies that work $W = 0$. Then from the non-flow energy equation (NFEE),

$$Q - W = U_2 - U_1$$

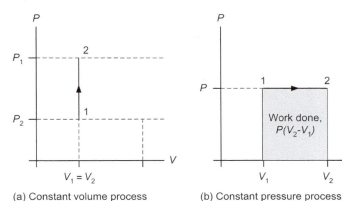

(a) Constant volume process

(b) Constant pressure process

Figure 4.109 Representation of constant volume and constant pressure processes.

where for a constant volume process, $W = 0$, we have

$$Q = U_2 - U_1$$

This implies that for a constant volume process all the heat supplied is used to increase the internal energy of the working fluid. Remember also, that the heat energy $Q = mc\Delta t$.

Constant pressure process

The constant pressure process for a perfect gas is considered to be a reversible process. This process was illustrated in Figure 4.100(b), look back now, to remind yourself. Now consider the pressure–volume diagrams shown in Figure 4.109, it can be seen that when the boundary of the system is rigid as in the constant volume process, then pressure rises when heat is supplied. So for a constant pressure process the boundary must move against an external resistance as heat is supplied and work is done by the fluid on its surroundings.

Now in the SFEE shown above the amount of work energy transferred will be given by $W = p(V_2 - V_1)$, which is simply the change in pressure–volume energy you met when we defined enthalpy as $H = U + pV$.

Isothermal processes

An isothermal process is one in which the temperature remains constant. You may remember that the characteristic gas equation was given as $pV = mRT$. If during the process, the temperature T remains constant (isothermal) then

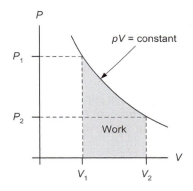

Figure 4.110 Isothermal process.

this equation becomes $pV = $ constant, because the mass is constant and R is a constant.

Figure 4.110 shows the curve for an isothermal process, the area under this curve represents the work energy transfer between state 1 and state 2.

Polytropic process

The most general way of expressing a thermodynamic process is by means of the equation $pV^n = $ constant. This equation represents the general rule for a *polytropic process in which both heat and work energy may be transferred across the system boundary.*

The area under the curve $pV^n = $ constant (Figure 4.111) represents the work energy transfer between state 1 and state 2 of the process.

Reversible adiabatic process

In the special case of, *a reversible process where no heat energy is transferred to or from the*

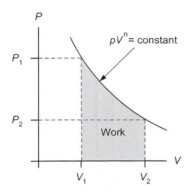

Figure 4.111 Curve for a polytropic process.

working fluid the process will be reversible adiabatic. This special process is often called an *isentropic* process, its importance will be emphasized, when we consider engine thermodynamic cycles. During adiabatic compression and expansion the process follows the curve given by $pV^\gamma = $ constant, where *for the reversible adiabatic case only,* (γ) *replaces* (n) *from the general polytropic case above, and* $\gamma = c_p/c_v$.

Key point

A reversible adiabatic process is also known as an isentropic process, when there is no change in entropy.

4.10.5 The second law of thermodynamics

According to our previous definition for the first law, when a system undergoes a complete cycle, then the net heat energy supplied is equal to the net work done and this definition was based on the principle of the conservation of energy, which is a universal law determined from the observation of natural events. The *second law of thermodynamics* extends this idea. It tells us that although the net heat supplied is equal to the net work done, the *total or gross heat supplied must be greater than the net work done*. This is because some heat must be *rejected* (lost) by the system, during the cycle. Thus in a *heat engine* (Figure 4.112), such as the internal combustion engine, the heat energy supplied by the fuel must be greater than the work done by the crankshaft.

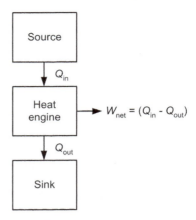

Figure 4.112 The heat engine.

During the cycle, heat energy is rejected or lost to the surroundings of the system through friction, bearing drag and component wear, etc.

A heat engine is a system operating in a complete cycle and developing network from a supply of heat. The second law implies that there is a need for a heat source and a means of rejection or absorption of heat from the system. The heat rejector within the system is often referred to as the heat *sink*. We know from the second law that for a complete cycle, the net heat supplied is equal to the net work done. Then from Figure 4.112, using the symbols:

$$Q_{in} - Q_{out} = W_{net}$$

We also know from the second law that the total heat supplied (heat in) has to be greater than the net work done, i.e.

$$Q_{in} > W$$

Now the thermal efficiency (η) of a heat engine is given by:

Thermal efficiency, $\eta = \dfrac{\text{Net work done } (W_{net})}{\text{Total heat supplied } (Q_{in})}$

or

Thermal efficiency, $\eta = \dfrac{Q_{in} - Q_{out}}{Q_{in}}$

There are many examples of the heat engine, designed to minimize thermal losses, predicted by the second law. These include among others; the steam turbine, refrigeration pack and air-conditioning unit. The internal combustion engine is not strictly a heat engine because the heat source is mixed directly with the working fluid. However, since aircraft propulsion units are based on the internal combustion engine, we will consider it next.

4.10.6 Internal combustion engine cycles

We conclude our study of thermodynamics by considering the theoretical and practical cycles for the internal combustion engine, which may be broadly divided into two types as:

1. Those which make use of a series of non-flow processes to convert heat energy into work energy, e.g. reciprocating piston engines.
2. Those which make use of flow processes to convert heat energy into work energy, e.g. gas turbine engines.

In both types of engine, it is assumed that the working fluid is air. We start by considering the air standard cycle for the constant volume or Otto cycle.

Otto cycle

The *Otto cycle* is the ideal air standard cycle for the spark ignition piston engine. In this cycle it is assumed that the working fluid, air, behaves as a perfect gas and that there is no change in the composition of the air during the complete cycle. Heat transfer occurs at constant volume and there is isentropic (reversible adiabatic) compression and expansion.

This cycle differs from the practical engine cycle in that the same quantity of working fluid is used repeatedly and so an induction and exhaust stroke are unnecessary.

The thermodynamic processes making up a complete Otto cycle (Figure 4.113) are detailed below:

1–2   Adiabatic compression. No heat transfer takes place, temperature and pressure increase and the volume decrease to the clearance volume.

2–3   Reversible constant volume heating, temperature and pressure increase.
3–4   Adiabatic expansion (through swept volume). Air expands and does work on the piston. Pressure and temperature fall. No heat transfer takes place, during the process.
4–1   Reversible constant volume heat rejection (cooling). Pressure and temperature fall to original values.

Note that during the compression and expansion of the working fluid, the ideal Otto cycle assumes that no heat is transferred to or from the working fluid during the process.

The practical four-stroke cycle

The sequence of operations by which the four-stroke spark ignition engine converts heat energy into mechanical energy is known as the four-stroke cycle. A mixture of petrol and air is introduced into the cylinder during the induction stroke and compressed during the compression stroke. At this point the fuel is ignited and the pressure wave produced by the ignited fuel drives the piston down on its power stroke. Finally, the waste products of combustion are ejected during the exhaust stroke.

The cycle of events is illustrated in Figure 4.114 and consists of the following processes:

1–2   Inlet valve is open and piston moves down cylinder sucking in fuel/air mixture (charge).
2–3   With inlet and exhaust valves closed, the piston moves up the cylinder and the charge is compressed. Ignition occurs as cylinder rises and is complete at point 4.
3, 4, 5, 6   The piston moves down the cylinder on the power-stroke, work done by gas on piston.
5   The exhaust valve opens at this point, and pressure decreases to near atmospheric at point 6.
6–1   Spent gases are exhausted as piston rises.

Typical temperatures for key stages in the cycle are given for reference. Temperatures

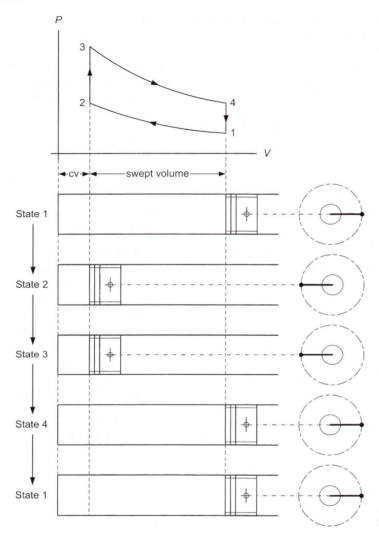

Figure 4.113 The Otto cycle.

cannot be directly superimposed onto a p–V diagram therefore, when temperature and heat need to be considered a *temperature* (*T*) and *entropy* (*S*) diagram is used. Think of entropy as a *measure of the disorder in a process*. If there is no *disorder or change in entropy* during a process, then that process approaches the *ideal*. Thus a T–S diagram may be thought of as comparing temperature with heat. A full explanation of entropy may be found in any text book dedicated to thermodynamics, all that you need to remember at this stage, is that entropy is an abstract way of measuring how a process deviates from the ideal, the larger the change in

entropy displayed on a T–S diagram the larger the degree of disorder within the process or the more inefficient is the process.

Key point

Entropy is a measure of the degree of disorder (or energy loss) in a system, it tells us how the practical system deviates from the ideal.

During the above practical cycle losses will occur. For example, during the expansion and compression processes heat will be transferred from the cylinder walls via the cooling system.

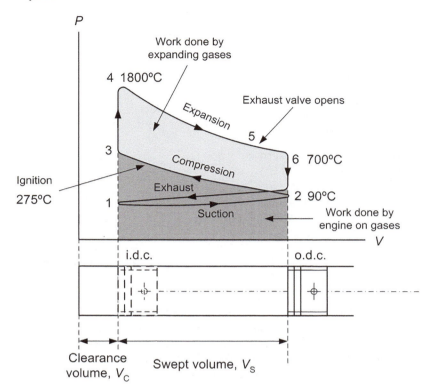

Figure 4.114 Constant volume cycle for four-stroke spark ignition.

The ignition of the charge (heating) takes a finite amount of time and therefore cannot occur at constant volume.

The net work done by the engine is therefore less, than in the ideal case, this can be seen in the diagram by the reduced area of the power loop, when compared with the ideal Otto cycle.

The working cycle of the gas turbine

The working cycle of the gas turbine engine is similar to that of the four-stroke piston engine. In the gas turbine engine combustion occurs at a constant pressure, while in the piston engine it occurs at a constant volume. In both engines there is an induction, compression, combustion and exhaust phase.

As already mentioned in the case of the piston engine we have a non-flow process whereas in the gas turbine we have a continuous flow process. In the gas turbine engine the lack of reciprocating parts gives smooth running and enables more energy to be released for a given engine size.

With the gas turbine engine, combustion occurs at constant pressure with an increase in volume, therefore, the peak pressures which occur in the piston engine are avoided. This allows the use of lightweight, fabricated combustion chambers and lower octane fuels, although the higher flame temperatures require special materials to ensure a long life for combustion chamber components.

The Brayton cycle or constant pressure cycle

The working cycle upon which the gas turbine operates is known as the Brayton cycle. This cycle is illustrated in Figure 4.115, and consists of the following processes:

1–2 Frictionless adiabatic compression where at point 1 atmospheric air is compressed along the line 1–2.

2–3 Frictionless constant pressure heating. Where heat is added from the burnt fuel at constant pressure, thus increasing volume.

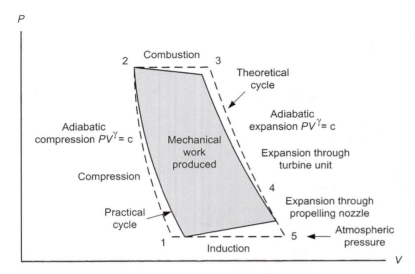

Figure 4.115 The Brayton cycle for a gas turbine.

3–4 Frictioness adiabatic expansion of the gases through the turbine.

4–1 Frictionless constant pressure heat rejection, through the jetpipe nozzle to atmosphere.

To ensure maximum thermal efficiency (see explanation of the second law) we require the highest temperature of combustion (heat in) to give the greatest expansion of the gases. There has to be a limit on the temperature of the combusted gases as they enter the turbine, which is dictated by the turbine materials. Additional cooling within the turbine, helps maximize the gas entry temperature to the turbine.

The practical Brayton cycle

Although it can be seen from Figure 4.115 that the practical cycle follows fairly closely to the ideal Brayton cycle, there are losses, which are detailed as follows:

1. The air is not pure, it contains other gases and water vapour.
2. Heat will be transferred to the materials of the compressor, turbine and exhaust units, so it is not a pure adiabatic process.
3. Due to dynamic problems, such as turbulence and flame stability in the combustion chamber, a constant temperature and hence a constant pressure cannot be maintained.

A further pressure loss occurs as a result of the burnt air causing an increase in volume and hence a decrease in its density. These losses are indicated on the diagram by a drop between points 2 and 3.

4. The Brayton cycle assumes frictionless adiabatic operation and this is not possible in practice.

You will gain a detailed knowledge of the above cycles, related to aircraft engines, should you choose to study the propulsion modules during the course of your career.

Test your understanding 4.22

1. Define: (a) thermodynamic system, (b) heat, (c) work.

2. Under what circumstances is a closed system able to do work on its surroundings?

3. Write down (a) the NFEE, (b) the SFEE and define each term within each equation.

4. What is the essential difference between a closed system and an open system?

5. What is the difference between the enthalpy of a working fluid and the internal energy of a working fluid, *and* under what circumstances is each property used?

6. An irreversible process cannot exist in practice. Explain this statement.

7. Define: (a) an isothermal process, (b) a polytropic process, (c) a reversible adiabatic process.

8. What are the essential elements of a heat engine?

9. What does the second law of thermodynamics tell us about the efficiency of a heat engine?

10. How does the thermodynamic cycle of a practical gas turbine engine differ from the ideal Brayton cycle?

General Questions 4.5

1. A metal bar is heated from 20°C to 120°C and as a result its length increases from 1500 to 1503 mm. Determine the linear expansion coefficient of the metal.

2. (a) Write down the formula for the thermal energy input into a solid and explain the meaning of each term.
   (b) If 3 kg of aluminium requires 54 kJ of energy to raise its temperature from 10 to 30°C, find the specific heat capacity for aluminium.

3. 0.5 kg of a gas at a temperature of 20°C and at standard atmospheric pressure, occupies a volume of 0.4 m^3. If the c_p for the gas = 1000 J/kgK find the:
   (a) characteristic gas constant
   (b) specific heat capacity at constant volume

4. How much heat energy is required to change 2 kg of ice at 0°C into water at 40°C?

5. Describe the operation of a typical refrigeration system, explaining the function of the major components.

6. A fluid enters a steady flow system with an internal energy of 450 kJ/kg, a pressure–volume energy of 1550 kJ/kg and a KE of 500 kJ/kg. At exit from the system the specific enthalpy is 1000 kJ/kg and there is negligible KE. If the difference in PE is 120 kJ/kg and there is no heat transfer during the process, determine the magnitude and direction of the work done.

7. Explain the concept of reversibility and irreversibility.

8. A heat engine is supplied with 150 MJ of heat, if during this time the work done by the heat engine is 65,000 kJ, determine its thermal efficiency.

9. Explain where the losses occur in a practical four-stroke cycle, when compared with the constant volume air standard Otto cycle.

10. What are the essential differences between the air standard cycle for the spark ignition piston engine and the ideal Brayton cycle for the gas turbine engine?

4.11 Light, waves and sound

Communication through the medium of light and sound energy, e.g. through fiber optic cables, sound waves and radio waves, has become an essential part of aircraft design and operation. We start this section by considering the nature of light.

4.11.1 Light

Light is difficult to define, it is a form of energy that travels in straight lines called rays and a collection of rays is a *beam*. The ray treatment of light is termed *geometrical optics*, and is developed from the way light travels in straight lines and the laws of *reflection* and *refraction*.

When light travels through very small objects and apertures it behaves in a similar manner to the waves created by a pebble being dropped into the center of a pond, under these circumstances light travels as a wave. Light waves, which are *electromagnetic*, can travel through empty space and do so at a speed of about 3×10^8 m/s! Light is given out or emitted by very hot objects, such as the Sun and cooler materials when electrons lose energy. In this way light is able to transfer energy from one place to another, e.g. the solar cell converts light energy directly into electrical energy.

Key constant

Light travels in empty space (a vacuum) with a velocity of approximately 3×10^8 m/s.

We first concentrate our attention on the ray nature of light in terms of the laws of reflection and refraction which are an essential part of the study of geometrical optics. These laws enable us to determine the behavior of mirrors and lenses, which are used in optical instruments.

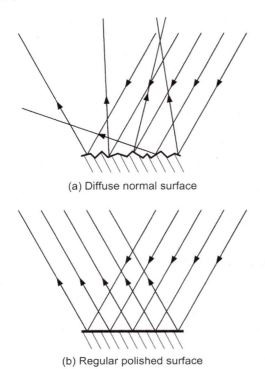

(a) Diffuse normal surface

(b) Regular polished surface

Figure 4.116 Reflection of light.

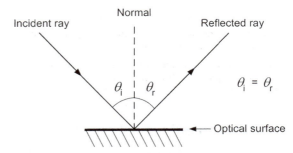

Figure 4.117 Incident and reflected light.

The laws of reflection

Most surfaces are not optically smooth, in other words most surfaces will reflect light in all directions. Figure 4.116(a) shows a normal surface under a microscope which is uneven, under these circumstances light rays will be reflected in all directions, we call this *diffuse* reflection.

Figure 4.116(b) shows light being reflected from a very smooth surface, such as polished metal or glass. Thus reflected light from a mirror, which is essentially metal-coated glass, is *regular* and enables an image to be seen by the human eye. The way in which light is reflected from a surface is governed by the laws of reflection. Figure 4.117 shows an *incident light ray*, which represents the light striking the reflecting surface. A further line leaving the surface represents the *reflected ray*.

The angle that the incident light makes with an imaginary line drawn at right angles to the reflecting surface, the **normal**, is known as the *angle of incidence*. Similarly the angle

that the reflected light makes with the normal is known as the *angle of reflection*. The *angle of reflection equals the angle of incidence* and this relationship, together with the fact that these rays are all in the same *plane*, is laid out in *the laws of reflection*.

1. *The angle of incidence is equal to the angle of reflection.*
2. *The incident ray, the reflected ray, and the normal all lie within the same plane.*

In law 2 above, the word plane means a two-dimensional space, such as a piece of paper, where each of the angles and the normal can be represented as a two-dimensional diagram, similar to coplanar forces you met earlier. Thus a mirror with a flat rather than curved surface is called a *plane* mirror.

For *plane mirrors* the image formed is the same size as the object and the image is as far behind the mirror as the object is in front. The image seen is also *virtual*, in that it cannot be seen on a screen and light rays do not pass through it. Finally the image seen in a plane mirror is *laterally inverted* or back to front. The effect of lateral inversion is easily seen by looking at written text in a mirror.

Key point

Images from plane mirrors are virtual and laterally inverted.

Curved mirrors

Curved mirrors are used as reflectors in car headlamps, aircraft landing lights, searchlights and flash lamps. When a mirror has a curved surface the simple rules for image position and size for plane mirrors no longer apply.

There are two types of spherical mirror, *concave* and *convex* (Figure 4.118).

In a concave mirror the center C of the sphere of which the mirror is a part is in front of the reflecting surface (Figure 4.118(a)) and in a convex mirror (Figure 4.118(b)) it is behind. C is referred to as the *center of curvature* of the mirror, and P which represents the center of the mirror surface is referred to as the *pole*. The line produced by **CP** is called the *principal axis* and **AB** is the *aperture*.

Note also that at the reflecting surface of a curved mirror the angle of incidence is equal to the angle of reflection and the normal is still at right angles to the curved surface of the mirror.

The rays of light reflected from a concave mirror converge at a single point F (Figure 4.119(a)), while the rays reflected from a convex mirror diverge (spread out) from a single point F. In each case *F is the principal focus* of the mirror and the distance from F to P is called the focal length. In both cases, the principal focus is approximately halfway between the center of curvature of the mirror and its pole, in other words:

Focal length = Half the radius of curvature

or in symbols:

$$f = \frac{r}{2}$$

Key point

The light rays from a concave mirror converge at the principal focus and for a convex mirror they diverge from the principal focus.

Images in curved mirrors

It is important when considering the use of curved mirrors to know exactly what type of image will be formed according to the physical characteristics of the mirrors. So we need to be able to determine the position of the image and whether the image is real or imaginary, inverted or upright, magnified or shrunk, etc. This information about the image can be obtained either

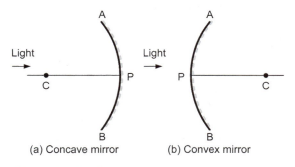

(a) Concave mirror   (b) Convex mirror

Figure 4.118   Curved mirrors.

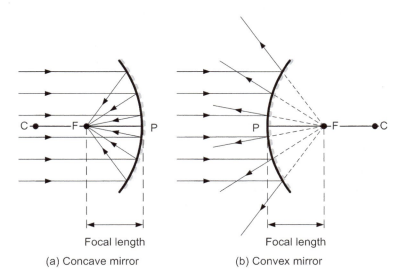

Focal length   Focal length

(a) Concave mirror   (b) Convex mirror

Figure 4.119 Principal focus and focal length.

by drawing a *ray diagram* or by *calculation* using formulae. In order to simplify the construction of a ray diagram we will assume that all *rays are paraxial*, i.e. they are close to the principal axis and therefore the mirror aperture is represented by a straight line.

Ray diagrams

To determine the position and size of the image any two of the following three rays (Figure 4.120) *need to be drawn:*

1. A ray of light parallel to the principal axis, which will be reflected back through the principal focus F.
2. A ray of light through the center of curvature C, which will be reflected back through C.
3. A ray of light through F, which is reflected back parallel.

Note that the rays drawn are for construction purposes and are not necessarily the rays by which the image is seen.

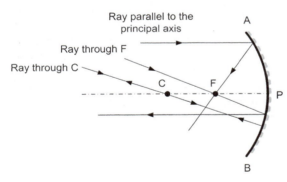

Figure 4.120 Rays used for construction of ray diagrams.

Example 4.59

An object 12 mm high stands on the principal axis of a concave mirror at a distance of 150 mm. If the focal length of the mirror is 50 mm, what is the position, height and nature of the image

We solve this problem by drawing a ray diagram to scale. The object is shown in Figure 4.121, as a thick black triangle. The radius of curvature C is shown at a distance $2f = 100$ mm from the mirror surface.

Since the object stands on the principal axis, then we can use a ray parallel to the principal axis that is reflected through F and a ray through the center of curvature C, to pinpoint the position and height of the image, which in this case is inverted.

Thus from the construction the image is real (see calculations below) and is approximately 74 mm from mirror face and 6 mm high.

Calculation

As mentioned earlier there is an alternative method of working out the position, magnitude and nature of an image formed from a curved mirror, and that is by calculation.

If the object distance from the mirror is u, the image distance v, and the focal length is f, then they may be linked mathematically by the equation:

$$\frac{1}{u} + \frac{1}{v} = \frac{1}{f}$$

Any units may be used for the lengths u, v and f, providing the same type of unit are used in each case.

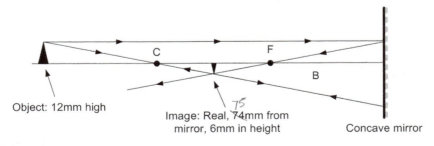

Figure 4.121 Ray diagram.

Note that the above equation can be used for concave and convex mirrors. *If the mirror is concave then the distance f is always treated as positive and if the mirror is convex f is negative. Also if v works out to be positive the image is real and if v works out to be negative the image is imaginary.* You should try to memorize these relationships, so that you place the correct values into the formula and correctly interpret your results.

In Example 4.59, we found the image distance from the concave mirror by constructing a ray diagram, where $u = 150$ mm and $f = 50$ mm. Let us use these values again to find v by calculation. Then:

$$\frac{1}{v} = \frac{1}{f} - \frac{1}{u} = \frac{1}{50} - \frac{1}{150} = \frac{2}{150}$$

we can now invert the fraction,

$$\frac{1}{v} = \frac{2}{150}$$

to give,

$$v = 75 \text{ mm}$$

Now v is positive and therefore *the image is real* and is 75 mm from the mirror face. When we estimated v by scale drawing a larger scale should have been used to obtain a closer estimate, but even so our estimate was fairly close to the calculated value.

We also need to calculate the height of the image. In order to achieve this, we may use the following relationship given here without proof:

$$\frac{h_i}{h_o} = \frac{v}{u}$$

where u and v have their usual meaning and h_i = height of image and h_o = height of object.

So in our example, where $h_o = 12$ mm, $v = 75$ mm and $u = 150$ mm then the height of the object is given by:

$$h_i = \frac{v h_o}{u} = \frac{(75)(12)}{150} = 6 \text{ mm}$$

The image height from calculation = 6 mm, which is the same as that found using the ray diagram.

Key point

Information about the image from curved mirrors may be found by calculation or construction of a ray diagram.

Refraction

When light rays pass from one medium say air to another, say glass, part of the light is reflected back into the first medium and the remainder passes into the second medium with its direction of travel unchanged. The net effect is that the light appears to be *bent or refracted* on entering the second medium and *the angle of refraction is the angle made by the refracted ray and the normal*, as illustrated in Figure 4.122.

For two particular materials (or mediums), the ratio of the sine of the angle of incidence ($\sin \theta_i$) over the sine of the angle of refraction ($\sin \theta_r$) is constant. This relationship is known as *Snell's law* and the constant is known as the refractive index, i.e.:

$$\text{Refractive index } (n) \text{ of a medium} = \frac{\sin \theta_i}{\sin \theta_r}$$

where the *refractive index (n)* is a constant for light passing from one medium to another. This

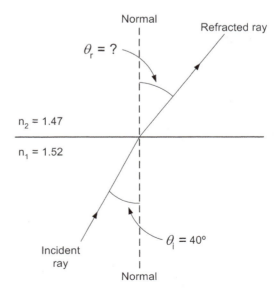

Figure 4.122 Refraction.

index is a measure of the *bending power* of particular materials, when compared with light travelling through a vacuum (or air) and we are able to give these materials a specific refractive index.

> **Key point**
>
> The refractive index of a material is a measure of its bending power or refraction ability as light rays pass through it.

For example, under these circumstances, the refractive index for water $= 1.33$ and refractive index for glass $\cong 1.5$. For all practical purposes we may assume the same values for the refractive index of the medium irrespective of whether the incident light travels through a vacuum or through air.

Snell's law may be written in a different way which relates the refractive indices of *any two materials*, through which light passes. In this form *Snell's law* may be written as:

$$n_1 \sin \theta_i = n_2 \sin \theta_r \qquad (Snell's\ law)$$

where n_1 and n_2 are the refractive indices of the two materials and $\sin \theta_i$ and $\sin \theta_r$ are the angles of incidence and refraction, as previously defined.

Example 4.60

Calculate the angle of refraction (θ_r) shown in Figure 4.123.

From the diagram $n_1 = 1.52$, $n_2 = 1.47$ and $\theta_i = 40°$, then using Snell's law and on substituting values we get:

$1.52 \sin 40 = 1.47 \sin \theta_r$ and on rearrangement

$\dfrac{1.52 \sin 40}{1.47} = \sin \theta_r$ and on simplification

$0.6647 = \sin \theta_r$ and so the required angle

$\theta_r = 41.66°$

Note that the angle of the ray increases as it enters the material having the lower refractive index.

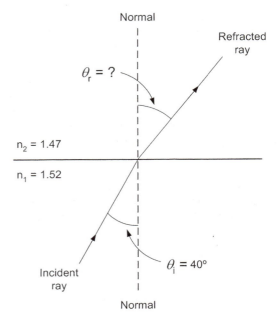

Figure 4.123 Angle of refraction.

An observable example of the effects of refraction is the way objects in water seem nearer than they really are. When you view an object in a swimming pool, the object *appears* to be shallower than it really is, this apparent difference in depth is related to the refractive index of the water. Where *the refractive index (n) = the real depth divided by the apparent depth*. Since, for water, $n = 1.33$ or 4/3, then an object at an apparent depth of 3 m in water will actually be $(3 \times 4/3) = 4$ m down.

Variation in the speed of light

The *speed of light* varies as it travels from medium to medium. The *refractive index* gives us the ratio of this speed change. Thus:

$$\text{Refractive index} = \frac{\text{speed of light in a vacuum}}{\text{speed of light in the medium}}$$

The above relationship implies that the greater the refractive index of the medium or the more the light is bent through the medium then the lower the speed of light.

So, for example light passing from a vacuum through glass with $n = 1.6$, will have an approximate velocity $= 3 \times 10^8 / 1.6 = 1.875 \times 10^8$ m/s.

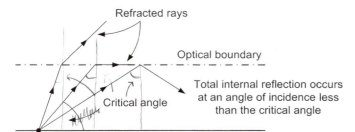

Critical angle

Refracted rays

Optical boundary

Total internal reflection occurs at an angle of incidence less than the critical angle

Light source

Figure 4.124 Total internal reflection.

Key point

The speed of light changes as it travels from medium to medium.

Critical angle and total internal reflection

You have already seen from Example 4.60 that the angle of the ray increases as it enters a material having a lower refractive index. As the angle of the incident ray in the first material is increased, there will come a time when, eventually, the *angle of refraction reaches* 90° and the light ray is refracted along the boundary between the two materials (Figure 4.124). The *angle of incidence* which causes this effect is known as the **critical angle**. We can calculate this critical angle by again considering Snell's law.

We know that $n_1 \sin \theta_1 = n_2 \sin \theta_2$ and that for the critical angle $\sin \theta_2 = \sin 90 = 1$.

Therefore:

$$n_1 \sin \theta_{\text{crit}} = n_2 \quad \text{and} \quad \sin \theta_{\text{crit}} = \frac{n_2}{n_1}$$

so

$$\theta_{\text{crit}} = \arcsin \frac{n_2}{n_1}$$

Consider once again the refractive indices for the materials given in Example 4.60 where $n_1 = 1.52$ and $n_2 = 1.47$ then the critical angle for light passing from material 1 to material 2 is given by:

$\theta_{\text{crit}} = \arcsin(1.47/1.52) = \arcsin 0.9671$ so

$\theta_{\text{crit}} = 75.26°$.

We know that if light approaches at an angle less than the critical angle, the ray is refracted across the boundary between the two materials. If incident light approaches at an angle greater than the critical angle, then the light will be reflected back from the boundary region into the material from which it came. The boundary now acts like a mirror and the effect is known as *total internal reflection*.

Another example of a device that can be used to produce total internal reflection is the *prism*. A typical prism, usually made of glass or perspex, will have a square base with sides at 45°/45° to the base, or be equilateral with each corner angle being 60°. Whichever prism is chosen, total internal reflection occurs because each light ray striking an inside face (Figure 4.125) does so at an angle of incidence of 45° or more, which is greater than the critical angle for glass. Thus prisms may be used to change the direction of light through 90° or 180°.

Fiber optic light propagation

It is the property of total internal reflection which is critical to the operation of *optic fibers*. These effects were illustrated in Figure 4.124 and are the key to the way in which light rays can be made to travel along an optic fiber.

If light rays are initiated at angles *greater* than the *critical angle*, along an optical fiber with parallel sides, then the light rays will travel down the fiber by bouncing from boundary to boundary at the same angle of incidence and reflection, as it goes. In the propagation of light along a fiber optic cable, energy losses occur as a result of dirt at the boundary, impurities in the glass used to propagate the light and the *Fresnel effect*, where light is lost through the boundary as it approaches at angles close to the critical value. To overcome the effects of dirt

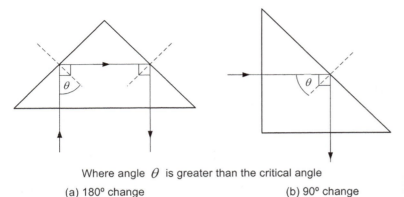

Where angle θ is greater than the critical angle

(a) 180° change   (b) 90° change

Figure 4.125 Internal reflection through a prism.

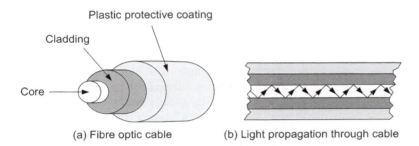

(a) Fibre optic cable   (b) Light propagation through cable

Figure 4.126 A typical fibre optic cable.

at the boundary, fiber optic cables are *clad* with another layer of glass and then protected from *microscopic cracking* by the addition of a plastic coating (Figure 4.126).

It is the fact that fiber optic cables are made, as near as possible, crack free, that we are able to bend and manipulate the fiber bundles in order to route them within an aircraft. Hence the necessity for absolute cleanliness and extreme care, when handling fiber optic cables used, for example, in *fly-by-light* systems.

Key point

Fiber optic cables use the principle of total internal reflection to enable light to travel along the cable.

Lenses

Lenses are of two basic shapes, convex which are thicker in the middle and concave which thin towards the middle (Figure 4.127).

The *principal axis* of a spherical lens is the line joining the center of curvature of its two surfaces. With lenses, like curved mirrors, we will only consider *paraxial* light rays, i.e. rays very close to the principal axis and making very small angles with it. The *principal focus* **F**, in the case of a *convex lens*, is a point on the principal axis towards which all paraxial rays, parallel to the principal axis converge (Figure 4.127). In the case of the *concave lens* these same rays after refraction appear to diverge. Since light can fall on either surface of a lens, it has two principal foci and these are equidistant from its center P. The distance **FP** is the *focal length f* of the lens. A *convex lens* is a converging lens and *has a real focus*, while the *concave lens* is a diverging lens and has an *imaginary focus*.

A parallel beam at a small angle to the axis of a lens (Figure 4.128) is refracted to converge to, or to appear to diverge from, the point in the plane containing F. This plane that is at right angles to the principal axis, is known as the *focal plane*. Thus the focal point for these refracted rays, incident at small angles from the axis, will always lie on this plane.

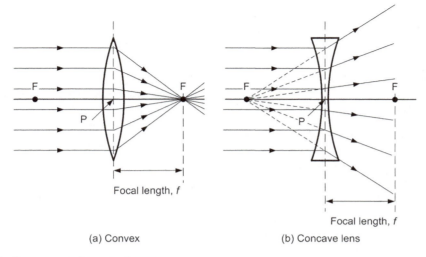

(a) Convex   (b) Concave lens

Figure 4.127 Concave and convex lenses.

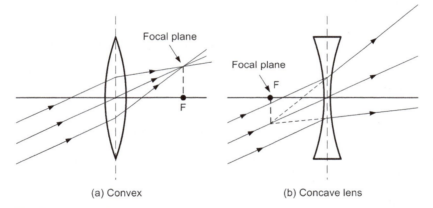

(a) Convex   (b) Concave lens

Figure 4.128 The focal plane of a lens.

Lens ray diagrams

To determine information about the position and nature of the image through a thin lens either a ray diagram or calculations may be used, in a similar manner to those already discussed for concave and convex mirrors. In the case of the lens, as you have seen, the image is produced by refracting light rather than reflecting it. Therefore, to construct the image of an object perpendicular to the axis of the lens (Figure 4.129), *two* of the following rays need to be drawn:

1. A light ray through the center of the lens (the optical center) P. This will pass through the lens in a straight line, undeviated.

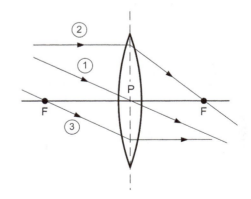

Figure 4.129 Construction lines for a convex lens ray diagram.

2. A light ray parallel to the principal axis, which after refraction passes through the principal focus.

3. A light ray through the principal focus, which is refracted parallel to the principal axis.

Example 4.61

A small object 6 mm high stands on and perpendicular to the principal axis of a convex lens, at a distance of 25 mm from the lens. If the focal length of the lens is 15 mm, what is the position, height and nature of the image?

Any two of the construction lines shown in Figure 4.130 may be used. In our construction we will use the first two lines identified above.

From our ray diagram it can be seen that the distance of the image from the lens is approximately 37 mm, the height of the image is 9 mm and the image is real (convex lens with converging focus) and inverted.

The equation used to solve curved mirror problems may also be used for lenses:

$$\frac{1}{u} + \frac{1}{v} = \frac{1}{f}$$

where u, v and f have the same meaning as for mirrors, make sure you can remember them!

In using this equation the following convention should be used. *If the lens is convex, f is taken as positive, if the lens is concave, then f is taken as negative. When v is positive the image is real, when v is negative the image is virtual.*

Make sure, as you did with mirrors, to follow this convention when using and interpreting the results from the above equation.

> ### Key point
> Convex lenses form real, inverted, small images of distant objects. Concave lenses form upright, smaller, virtual images of objects placed in front of it.

We can now verify our ray diagram results for Example 4.61. The distance of the image from the lens is found using the above equation, where in our case:

$$\frac{1}{v} = \frac{1}{15} - \frac{1}{25} = \frac{4}{150}$$

from which $v = 37.5$ mm and is real since v is positive.

Now to find the height of the image we use the idea of similar triangles to produce the ratios:

$$\frac{\text{Image height}}{\text{Object height}} = \frac{\text{Image distance }(v)}{\text{Object distance }(u)}$$

These ratios are also a measure of the *linear magnification* of the lens. In our case then:

$$\text{Image height} = \frac{\text{Object height} \times v}{u}$$
$$= \frac{(6)(37.5)}{25} = 9 \text{ mm.}$$

Thus, our calculated answers are in good accord with those obtained from the ray diagram.

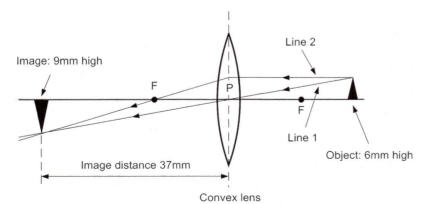

Image: 9mm high

F   P   F

Line 2

Line 1

Image distance 37mm

Object: 6mm high

Convex lens

Figure 4.130 Construction lines for example.

Test your understanding 4.23

1. The speed of light in a vacuum is approximately 3×10^8 m/s. What distance in miles, would light travel through space, in 1 h?

2. Write down the laws of reflection that are applicable to optically smooth surfaces.

3. If the focal length of a curved mirror is 30 cm, what is the radius of curvature of the mirror?

4. What do we mean by the term paraxial rays?

5. Sketch and describe the three principal construction rays used to determine the position, size and nature of an image created by a concave mirror.

6. How does the magnitude of the refractive index affect the angle of rays as they enter materials having different refractive indices?

7. How is the refractive index and the speed of light through a material related?

8. Upon what principle does light propagation through a fiber optic cable depend?

9. Why is fiber optic cable manufactured with a clad layer of glass over the inner glass core?

10. Define the principle focus, with respect to concave and convex lenses.

11. Define the focal plane of a convex and concave lens.

12. How is the image height from a lens determined analytically?

4.11.2 Waves

The study of wave motion is vital to your understanding of the way light energy, electromagnetic radiation and sound energy travels and indeed how we use the properties of waves to explain the principles of radio communication.

We will study two forms of wave motion *transverse waves*, where the vibratory motion is at right angles to the direction of movement of the wave and *longitudinal waves* where particles oscillate (stretch and compress) in the same direction as the wave travels. Light behavior can be modeled by studying transverse wave motion however, even light is a subset of a very much more extensive range of waves known as the *electromagnetic spectrum*. We will first consider transverse waves and their relationship to the behavior of water and light. We then look at the electromagnetic spectrum and in particular radio waves, finally we will look separately at longitudinal waves and sound.

Transverse waves

If a cork is placed into a still pond and then a pebble is dropped into the center of the pond, ripples start to spread out from the source of the disturbance, i.e. where we dropped the pebble, at the same time the cork will bob up and down, these actions are as a result of the energy created by *transverse wave motion*. The cork does not move in the direction of travel of the *wave fronts* (ripples) that travel outwards from the center, but it does *oscillate* about the mid-position of the still water, prior to the disturbance. We know that the waves are *progressive* (moving) because, e.g. sea waves break on the shore, you can see the wave front traveling towards you! However ignoring currents, then in deep water, the effect of hitting the wave front is to cause you to bob up and down, in the same manner as the cork. This oscillatory motion is *transverse motion* because the oscillations are at right angles to the direction of travel of the waves, which are represented diagrammatically by lines known as wave fronts. Figure 4.131 shows the nature of the transverse motion and its relationship to the direction of motion of the wave.

> **Key point**
>
> Transverse waves oscillate at right angles to the direction of travel of the wave motion.

In order to put some scientific precision into the idea of transverse waves, we need to define the properties of this type of wave. In Figure 4.131, it can be seen that the *amplitude* of a transverse wave is the maximum distance a point moves away from its rest position, when the wave passes. The distance occupied by one complete wave is called the *wavelength* and the number of complete waves (oscillations) produced per second is called the *frequency*. When corresponding points on the wave have the same speed and move in the same direction, we say they are in *phase*. The unit of amplitude and wavelength in the SI system is m, the unit of frequency is the cycle per second (c/s) and in the SI system c/s is given the name *hertz* (Hz). Thus, $1\,Hz = 1\,c/s$.

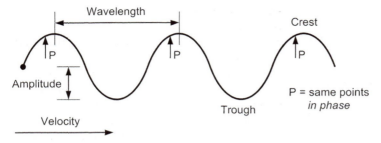

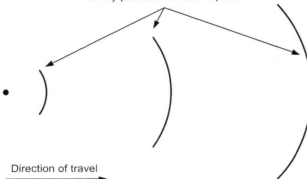

Figure 4.131 Transverse wave motion.

The speed of a wave (wavefront), its frequency and wavelength are linked by a simple formula, which is given without proof below:

Wave speed = frequency × wavelength

or in symbols:

$$v = f\lambda$$

The equation $v = f\lambda$ *applies to any wave*, the wave speed is found in m/s, when the frequency is in Hz and the wavelength is in m. So, for example if waves are produced 10 times a second, i.e. with a frequency of 10 Hz and the speed of wave propagation (wave speed) is 50 m/s, then the wavelength will be 50/10 or, $\lambda = 5$ m.

Wave behavior

The nature of progressive waves can be demonstrated, using a ripple tank, which is a sophisticated water tank where parameters such as wave frequency and amplitude can be varied and the corresponding effects of the wave motion studied. Through these studies, it can be shown that water waves behave in a very similar manner to

light. From observation it has been found that water waves are *reflected* by surfaces in exactly the same way as light, so the same laws of *reflection* apply. It is also true that water waves undergo *refraction* or bending, when they are slowed down, in a similar manner to light. It has been observed, using the ripple tank, that as waves enter shallower water they slow down. This reduction in speed causes a reduction in wavelength and as the waves close up on one another, they change their direction of travel. Two other important properties may be demonstrated using the ripple tank, they are wave *diffraction* and *interference*.

Diffraction

When two plates, with a very narrow gap between them are placed in the path of progressive water waves (Figure 4.132), the waves that pass through them spread out in all directions and produce circular wavefronts.

This effect is known as *diffraction* or bending of waves as they pass through very narrow gaps. If the gap between the plates is made much

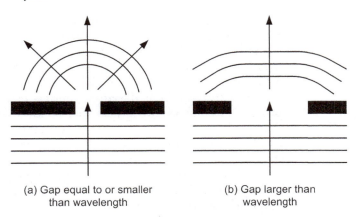

(a) Gap equal to or smaller
than wavelength

(b) Gap larger than
wavelength

Figure 4.132   Diffraction.

wider than the wavelength of the waves passing through it, then the diffraction effect becomes insignificant.

Interference

It can be demonstrated that if two vibration sources with the same frequency produce two identical wave sets. These wave sets can reinforce one another or cancel each other out depending on whether they are *in-phase* or *out-of-phase*.

In Figure 4.133, we see that when the wave sets are *in phase*, reinforcement takes place which is known as **constructive interference**. When the two wave sets are in *anti-phase* (where one wavefront peaks as the other troughs) then cancellation takes place or **destructive interference** occurs. Constructive and destructive reinforcement occur when the wave sets are totally in-phase or totally out-of-phase. There will also be occasions when the wave sets have phase differences between these two extremities, this results in complex wave patterns being formed as the separate wave sets interfere.

Electromagnetic spectrum

As mentioned previously light waves are a subset of a much more extensive range of waves known as the *electromagnetic spectrum*. The electromagnetic waves within the spectrum (Figure 4.133) have differing wavelengths, frequencies and vary tremendously in the amount of energy they are able to transmit.

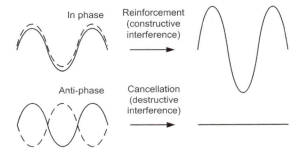

Figure 4.133   Wave interference effects.

You will note from Figure 4.134, that the waves with the smallest wavelength and highest frequency, have the highest energy or intensity. For example, penetrating radioactive gamma rays have wavelengths less than 10^{-10} m and frequencies in the range 10^{19}–10^{21} Hz. While at the other end of the spectrum, we range from microwaves with wavelengths around 1 mm to radio waves with frequencies in the range 10^6–10^5 m and wavelengths around 1–10 km!

Even though the waves in the electromagnetic spectrum may have vastly different frequencies and thus energy levels, they all have the following common characteristics:

1. They all travel in straight lines at the speed of light (3×10^8 m/s) through a vacuum or free space.
2. They are all transverse waves, where the oscillations are produced by changing electrical and magnetic fields.

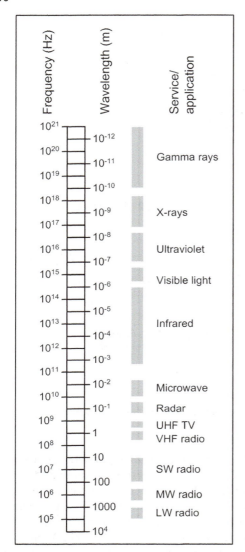

Figure 4.134 The electromagnetic spectrum.

3. They all exhibit, reflection, refraction, interference, diffraction and polarization.
4. The intensity of all waves emitted from a point source in a vacuum, is inversely proportional to the square of the distance from the source, i.e. $I \propto 1/r^2$.
5. They obey the equation $c = f\lambda$ where $c =$ the speed of light.

We have talked about electromagnetic waves having different energy levels but not really explained the source of this energy.

Electromagnetic waves are emitted when electrically charged particles (at the atomic level) change their energy. This occurs when electrons orbiting the nucleus of an atom *jump* to a lower energy level releasing electromagnetic radiation (waves) from the atom, during the process. From our study of heat we also know that the electrons and nuclei of atoms constantly oscillate, their KE is constantly changing, and these atoms release electromagnetic radiation in accord with these changes. The greater the jump or the more rapid the oscillation, the higher the frequency and the more intense is the resulting electromagnetic wave energy.

Radio waves

It should be emphasized right from the outset, that radio waves must not be confused with sound waves, which follow. *Radio waves belong to the series of waves within the electromagnetic spectrum* and have the characteristics identified above. They are transverse progressive waves that are able to travel through free space. Sound waves are longitudinal progressive waves that require a medium, such as air, to pass through.

Figure 4.134 shows that radio waves have the longest wavelengths and the lowest frequencies. They may be produced by making electrons oscillate in an *aerial* or antenna and can be used to transmit sound and picture *information* over long distances.

The electromagnetic information from a transmitting aerial (source) can reach the receiving aerial by three different routes. Via ground waves (Figure 4.135) which travel along the ground following the curvatures of the terrain. Via *sky waves* which leave the transmitting aerial at an angle and are reflected back down to the earths surface via charged particles in the ionosphere. The final possible method of transmission is as *space waves*, which take a straight-line path and effectively use the height of the aerial to hit the earth at a distance related to the curvature of the earth's surface.

Key point
Radio waves travel as ground waves, sky waves or space waves depending on their frequency.

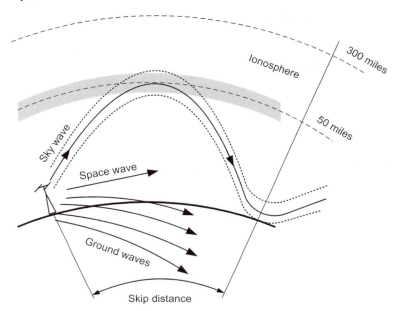

Figure 4.135 Forms of radio wave transmission.

Figure 4.135 also shows the **skip distance**, i.e. the point from the transmitter where the first sky wave can be reached. The area which cannot receive either the ground-wave or first sky-wave reflection is called the **dead space** or **silent zone**. It should be appreciated that the transmitter usually sends out its energy in the form of a wide beam therefore the sky-wave reflection, covers a large area, not just a single point.

By virtue of their wavelength, *long* and *medium* waves will diffract as *ground waves* around hilly terrain, so that a signal can be picked up on these wavelengths, even if hills exist between the transmitter and receiver. *Long* (30–300 kHz) and *medium* (300 kHz–3 MHz) frequency waves may also be transmitted as *sky waves* so that *very long distance* reception is possible. Very high frequency (VHF 30–300 MHz) and ultra high frequency (UHF 300–3000 MHz) waves have shorter wavelengths and are not reflected by the ionosphere and so normally require a straight path between the transmitter and receiver. This is why your television reception and FM (frequency modulated) radio reception is particularly sensitive to the distance from the transmitter, the higher the transmitter, the greater the range of transmission by *space waves*. *Microwaves* with frequencies above 3000 MHz are used for radar, radio astronomy and satellite communications.

The communication process

The essential components that are necessary for radio communication to take place, between two points are shown in Figure 4.135.

The transmitter at station A provides a *radio frequency* (RF) current which, when coupled to the transmitter aerial, produces an *electromagnetic* (EM) wave. This wave is modified by the electrical pulses (caused by the speech) from the microphone, and the wave is then said to be *modulated*. Thus the speech or sound is *carried* on the electromagnetic wave. The modulated wave travels outwards from the aerial in a direction determined by the design of the aerial system.

When the modulated wave is received at station B the wave is *de-modulated* by the radio receiver where the speech being carried is converted back into electrical impulses that act on the speaker of the telephone.

Aircraft radio communications

Since aircraft fly at heights in excess of all ground based-aerials, high-frequency radio

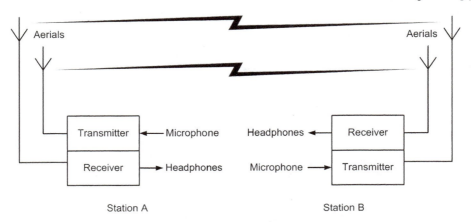

Figure 4.136 Essential components for radio communication.

Table 4.11

Aircraft system	Approximate frequency	Band
Automatic direction finder (ADF)	100 kHz–2 MHz	Long/medium
High frequency communications	2–30 MHz	HF
Instrument landing system (ILS)	108–118 MHz	VHF
VHF communications	118–136 MHz	VHF
Glide path	330 MHz	UHF
Air traffic control transponder	1000 MHz	UHF
Microwave landing system	5000 MHz	Microwave
Weather radar	9375 MHz	Microwave

waves cover a somewhat greater distance as space waves, but are still limited by the curvature of the earth, since at high and very high frequencies they pass through, rather than reflect back from the ionosphere.

Constant HF selections need to be made during flight based on:

- the distance between the aircraft transmitter and other receivers,
- the time of day and year to account for changes in the ionosphere,
- the transmitting power available.

From the operators point of view this does not present a problem because frequencies to be used in given areas are published in the form of tables.

Table 4.11 shows that extremely high frequencies in the form of VHF, UHF and microwave communication are used a lot on aircraft. The reason for this is to reduce the possibility of *static interference*. Static interference (i.e. reception of unwanted crackles and hiss) is worse the lower the frequency but from VHF and above, reception becomes virtually static free.

Unfortunately, as we have already seen VHF and UHF communication has a limited range. Modern communication systems are now able to increase the range of VHF and UHF communications and so eliminate dead space, by employing satellites (Figure 4.137). These receive and transmit radio signals from a great height (normally greater than 20,000 miles) enabling very large areas to be covered. By using a combination of satellite height and speed the satellite can be made *geo-stationary*, in that it will appear to hover, its angular rate around

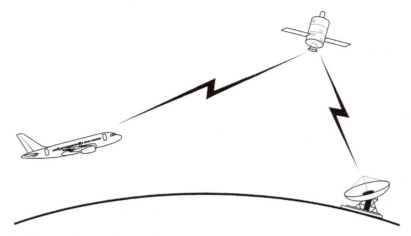

Figure 4.137 Satellite receiver and transmitter.

the world being synchronized to the earth's rotation rate.

Satellite communication can be used to provide airborne telephone systems for passengers and also be used for satellite navigation using geo-stationary or low orbit satellites.

The Doppler effect

When there is relative motion between a wave source and an observer, a change in frequency takes place; this is noticeable with *any wave motion*, light, radio and sound. This change in frequency brought about by the relative motion is known as the *Doppler effect*.

An example with sound waves, commonly quoted, is that of a train using its whistle while passing an observer. As the train approaches, the frequency heard by the observer is higher than that emitted from the source. When the train passes the observer, there is a drop in pitch and as the train moves away from the observer a lower frequency than that generated is heard. The same effect is noted if the observer moves and the sound source is stationary.

With respect to radio transmission, if the relative motion is such that a transmitter and receiver are effectively moving towards each other (e.g. closing aircraft), the received frequency will be higher than that transmitted. If moving away, the received frequency is less than that transmitted. An approximation of the amount of change of frequency, *Doppler shift*, is given by:

$$\text{Doppler shift} = \frac{\text{Transmitter frequency} \times \text{Relative velocity}}{\text{Velocity of radio wave propagation}}$$

So, for example if the transmitter frequency = 100 MHz and the relative velocity between the transmitter and receiver is 3600 km/h (1000 m/s), then:

$$\text{Doppler shift frequency} = \frac{(100 \times 10^6)(1000)}{3 \times 10^8}$$

$$= 333.3 \, \text{Hz}$$

As can be seen this shift is very small, but does have practical applications. For example, the relative motion between a satellite and a survival beacon can give an indication of the location of the beacon. In the case of satellites that have very high velocities, the Doppler shift (change in frequency) could be significant. Thus, if the satellite is traveling with a component of its velocity directed toward the beacon, it receives a higher frequency signal than that transmitted. Traveling away from the beacon, this situation is reversed and the satellite receives a lower frequency signal than that transmitted, therefore there is a *frequency change* at the moment the satellite passes the beacon.

Test your understanding 4.24

1. Infra-red radiation and ultraviolet radiation are two forms of electrostatic waves that sit in the electromagnetic spectrum. Which type of radiation has the highest energy and why?
2. Explain the concept of transverse wave motion.
3. What will happen if a series of linear transverse waves pass through a very narrow slot with an aperture less than the wavelength of the transverse wave passing through it?
4. What is meant by constructive and destructive interference?
5. Detail *three* common characteristics of electromagnetic waves.
6. It is required to transmit by sky wave at a frequency of 32 MHz, is this practical? Explain your answer.
7. What is the approximate range of wavelengths for microwaves?
8. Why are VHF and UHF radio bands, frequently used for aircraft communications?
9. What is (a) skip distance (b) dead space?
10. Describe the nature of the communication process, which enables telephone conversations to take place over large distances.
11. Explain why the pitch of sound of a jet engine changes as the aircraft passes you.

4.11.3 Sound

We begin this very short study of sound by considering the nature of sound waves. You will discover that sound waves are mechanical waves, which unlike radio waves, cannot travel long distances, since their energy is quickly dissipated. This was the reason for carrying them on electromagnetic waves, so that we are able to communicate (speak) over long distances.

Sound waves

We have spent a lot of time talking about light and radio waves, which are both part of the family of transverse waves associated with the electromagnetic spectrum. Sound waves are fundamentally different!

Sound waves are caused by a *source of vibration*, e.g. when a bell rings, it vibrates at a regular rate say 500 times a second (500 Hz) compressing and stretching the air immediately surrounding it. These vibrations set up a series of alternating zones (Figure 4.138) of high pressure (*compression*) and low pressure (*rarefaction*) which travel outwards from the bell in a longitudinal manner. Sound waves, which are mechanical waves, need a medium, such as air, through which to travel. They can travel through all materials: solids, liquids and gases. The sound that we hear generally travels through air, but we are capable of hearing sound

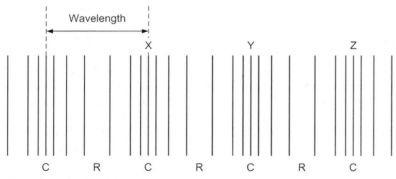

R = rarefaction (low pressure)

C = compression (high pressure)

X, Y and Z = particles in phase

Amplitude, *a*, is the maximum displacement of a particle from its rest position

Figure 4.138 Sound waves.

under water and through solid objects, such as doors, windows and walls.

The amplitude of a sound wave is related to the position of the particles of the material, through which the sound is traveling. We say that: *the amplitude of the sound wave is the maximum displacement of a particle from its rest position* and *the distance between two successive particles in-phase is the wavelength*. Sound waves are **longitudinal progressive waves** where the particles, are compressed and rarefied (oscillate) in the same direction as the *wavefront is traveling*.

Although there are significant differences in the behavior of sound waves, like electromagnetic waves, they are governed by the fundamental equation $v = f\lambda$ except that v replaces c where $v =$ *the speed of sound wave*. You should also remember that sound waves, like other wave forms, can be reflected, refracted, diffracted and display interference effects.

We have already considered the **speed of sound** in some detail when we studied the atmosphere. You should remember that the speed of sound was temperature and density dependent. Thus, *the speed of sound varies according to the nature of the material through which it passes*. For example, the speed of sound in air at $15°C = 340$ m/s or 1120 ft/s, the speed of sound in water at $0°C = 1400$ m/s and the speed of sound through concrete $\cong 5000$ m/s.

Note that the density dependence of the speed of sound is quite apparent from the above examples, the speed of sound increases as it travels through gases, liquids and solids, respectively.

Reflected sound

When we hear an echo, we are hearing a reflected sound a short time after the original sound. The time the echo, or reflected sound, takes to reach us is a measure of how far away the echo source is. This property of reflection can be used with the echo sounder, when we wish to measure the depth of the seabed below the ship. Also by sending out *ultrasound pulses*, we are able to determine the nature of any discontinuities (defects, cavities, cracks, flaws, porosity, etc.) in otherwise sound material.

Ultrasound frequencies above the audible range (in excess of 20 kHz), may be generated using a piezoelectric probe which rests on the surface of the material (Figure 4.139).

Then using the formula $v = f\lambda$ and knowing the frequency and wavelength of the generated ultrasound, the velocity of the ultrasound wave can be determined. If in addition to the pulse transmitter, a receiver measures the reflected pulses from the discontinuity and bottom of the material, then the time difference between the two pulses will enable the depth of the flaw to be established.

Perceiving sound

Through music, speech and other noises, our ears experience a range of different sounds. All these differences in the way we perceive sound, are dependent only on the differences in frequency and amplitude of the sound waves entering our ears. We have already defined *amplitude*, as the maximum displacement of a particle from its rest position. For example, the more distance air particles move from their rest position, when a loud speaker diaphragm oscillates, then the louder the sound we hear. In other words, *the greater the amplitude the louder the sound*.

Key point

The greater the amplitude of the sound wave the louder the sound.

The *intensity* of the sound wave is a measure of the energy passing through unit area every second. More formally: *a sound wave has an intensity of 1 watt per square metre (1 W/m^2) if 1 J of wave energy passes through 1 m^2 every second*. Remember that *1 W is equal to 1 J/s*. Mention has already been made of the *pitch* of sound, when we considered the *Doppler effect* earlier.

Difference in *pitch* is perceived by us when we hear, e.g. different notes in music. Thus, high pitch sound such as that from a whistle, results from high frequencies and low pitch sound such as that from a large drum, results from low frequencies.

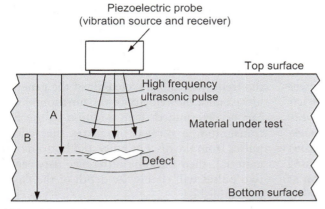

(a) Material on test

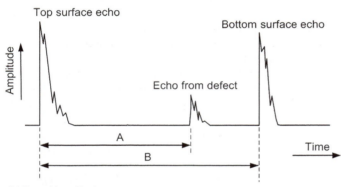

(b) Resulting display

Figure 4.139 Basic principle of ultrasound flaw detection.

Remembering, that the pitch of a train whistle is higher as the train approaches and lower as the train passes should give you a shrewd idea, as to the exact nature of the *Doppler effect*. As the sound waves travel towards you, their relative velocity increases the number of wavefronts present for a given distance which gives an increase in frequency, and a corresponding increase in the pitch of the whistle. As the train reaches you the relative velocity of the whistle sound waves decrease, and the number of wavefronts reaching you is reduced, causing a decrease in frequency and a corresponding decrease in pitch.

Test your understanding 4.25

1. How are sound waves created?
2. Detail the essential differences between sound waves and the waves of the electromagnetic spectrum.

3. Upon what factors, does the speed of sound depend?
4. If the wavelength of ultrasound waves is 6 mm and they have a frequency of 30 kHz. How long would it take the echo to reach you, from a discontinuity 0.5 m deep?
5. With respect to sound waves define: (a) intensity, (b) pitch and (c) amplitude.

General Questions 4.6

1. State the laws of reflection and explain how the images formed by plane and curved mirrors differ.
2. An object 2.5 cm high hangs vertically below the principal axis of a concave mirror at a distance of 1.25 m from the lens. If the focal length of the mirror is 40 cm, determine graphically and confirm by calculating the position, height and nature of the image.
3. A light ray passes from a material with a refractive index of 1.5 and angle of incidence

of 38°, into another material with a refractive index of 1.45, determine the angle of refraction of the light ray.

4. For the situation in question three above, determine the angle that refracts the beam along the boundary between the two materials.

5. Explain, with the aid of a diagram, how light is propagated along a fiber optic cable.

6. Sketch the ray diagram construction lines that can be used to determine the image of an object placed perpendicular to the principal axis of a lens.

7. A wave has a velocity of 400 m/s, its frequency varies between 500 Hz and 5 kHz. What is the variation in wavelength?

8. Describe the *common* properties of *all* electromagnetic waves.

9. Why is it necessary to modulate and de-modulate an electromagnetic carrier wave?

10. A satellite (in empty space) closes on a radio beacon at a relative speed of 18,000 mph, determine the Doppler shift, if the transmitter frequency is 120 MHz.

4.12 Multiple choice questions

The example questions set out below follow the sections of Module 2 in the JAR 66 syllabus. In addition there are questions set on the atmospheric physics contained in Module 8 – Basic aerodynamics. It was felt that the subject matter concerning atmospheric physics was better placed within this chapter.

Also note that the following questions have been separated by level, where appropriate. Much of the thermodynamics and all of the material on light and sound is not required for category A certifying mechanics. The category B questions have all been set at the highest B1 level, for the subject matter in the mechanics and fluid mechanics sections.

Please remember that *ALL these questions must be attempted without the use of a calculator* and that the pass mark for all JAR 66 multiple-choice examinations is 75%!

Units

1. The SI unit of mass is the: [A, B1, B2]
   (a) newton
   (b) kilogram
   (c) pound

2. The SI unit of thermodynamic temperature is the: [A, B1, B2]
   (a) degree celsius
   (b) degree fahrenheit
   (c) kelvin

3. In the English engineering system, the unit of time is: [A, B1, B2]

   (a) second
   (b) minute
   (c) hour

4. In the SI system the radian is a: [A, B1, B2]
   (a) supplementary unit
   (b) base unit
   (c) measure of solid angle

5. In the SI system the unit of luminous intensity is the: [B1, B2]
   (a) lux
   (b) candela
   (c) foot candle

6. 500 mV is the equivalent of: [A, B1, B2]
   (a) 0.05 V
   (b) 0.5 V
   (c) 5.0 V

7. An area 40 cm long and 30 cm wide is acted upon by a load of 120 kN, this will create a pressure of: [A, B1, B2]
   (a) $1 MN/m^2$
   (b) $1 kN/m^2$
   (c) $1200 N/m^2$

8. A light aircraft is filled with 400 imperial gallons of avation gasoline, given that a litre of aviation gasoline equals 0.22 imperial gallons, then the volume of the aircraft fuel tanks is approximately: [A, B1, B2]
   (a) 88 L
   (b) 880 L
   (c) 1818 L

9. If one bar pressure is equivalent to 14.5 psi, then 290 psi is equivalent to: [B1, B2]

(a) 20 kPa
(b) 2.0 MPa
(c) 2000 mbar

10. Given that the conversion factor from mph to m/s is approximately 0.45, then 760 mph is approximately equal to: [A, B1, B2]
    (a) 1680 m/s
    (b) 380 m/s
    (c) 340 m/s

11. If the distance traveled by a satellite from its gravitation source is doubled and the satellite originally weighed 1600 N, then its weight will be reduced to: [B1, B2]
    (a) 1200 N
    (b) 800 N
    (c) 400 N

12. In the engineers version of the FPS system the amount of mass when acted upon by 1 lbf, experiencing an acceleration of 1 ft/s^2 is: [A, B1, B2]
    (a) 1 lb
    (b) 1 lbf
    (c) 32.17 lb

Matter

13. Which of the following statements is true?
                                             [A, B1, B2]
    (a) Protons carry a positive charge neutrons carry a negative charge
    (b) Electrons carry a negative charge protons have no charge
    (c) Protons carry a positive charge, electrons carry a negative charge

14. The valence of an element is identified by the: [B1, B2]
    (a) number of electrons in an atom of the element
    (b) column in which it sits within the periodic table
    (c) number of electrons in all of the p-shells within the atom of the element

15. Ionic bonding involves: [A, B1, B2]
    (a) electron transfer
    (b) the sharing of electrons
    (c) weak electrostatic attraction of dipoles

16. An ion is: [A, B1, B2]
    (a) an atom with loosely bound electrons
    (b) a positively or negatively charged atom
    (c) an atom with a different number of protons and neutrons

17. Matter is generally: [A, B1, B2]
    (a) considered to exist in solid, liquid and gaseous forms
    (b) made up from solid elements
    (c) considered to have an inter-atomic binding force of zero

18. Gases: [A, B1, B2]
    (a) always fill the available space of their containing vessel
    (b) are always made-up from single atoms
    (c) have molecules that always travel in curved paths

Statics

19. A vector quantity: [A, B1, B2]
    (a) is measured only by its sense and direction
    (b) has both magnitude and direction
    (c) is represented by an arrow showing only its magnitude

20. Two vector forces: [A, B1, B2]
    (a) can only be added using the triangle rule
    (b) are always added using the head-to-head rule
    (c) may be added head-to-tail using the triangle law

21. The resultant of two or more forces, is that force which acting alone: [A, B1, B2]
    (a) against the other forces in the system, places the body in equilibrium
    (b) acts normal to all the other forces in the system
    (c) produces the same effect as the other forces acting together in the system

22. Figure 4.139 shows a spring with a pointer attached, hanging next to a scale. Three different weights are hung from it in turn, as shown: [A, B1, B2]

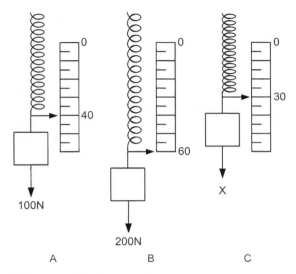

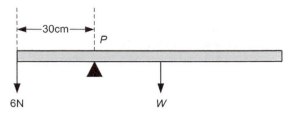

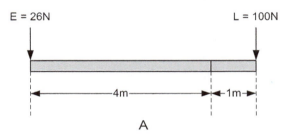

Figure 4.141 A uniform metre rule balanced as shown.

Figure 4.140 Spring with pointer attached.

If all the weight is removed from the spring, which mark on the scale will be indicated by the pointer?
(a) 0
(b) 10
(c) 20

23. With reference to Figure 4.140. What is the weight of X?   [A, B1, B2]
(a) 10 N
(b) 50 N
(c) 0

24. With reference to forces acting on a uniform beam, one of the conditions for static equilibrium is that: [A, B1, B2]
(a) horizontal forces must be equal
(b) vertical forces and horizontal forces must be equal
(c) the algebraic sum of the moments must equal zero

25. A uniform meter rule is balanced as shown in Figure 4.141.   [A, B1, B2]
The weight W of the metre rule is:
(a) 4 N
(b) 5 N
(c) 9 N

26. With respect to Figure 4.141, the force on the rule at point P is: [A, B1, B2]

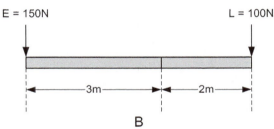

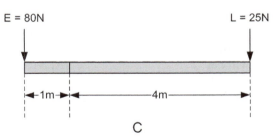

Figure 4.142 Levers, which will rotate clockwise.

(a) 3 N acting vertically down
(b) 15 N acting vertically up
(c) 15 N acting vertically down

27. In Figure 4.142 which lever will rotate clockwise?   [A, B1, B2]
(a) A
(b) B
(c) C

28. Torque may be defined as the: [A, B1, B2]
    (a) turning moment of a couple measured in newton-metres (Nm)
    (b) turning moment of a force measured in newton (N)
    (c) moment of a couple measured in newton (N)

29. When calculating the distance of the center of gravity (CG) of an aircraft from a datum x. This distance is equal to the sum of the: [B1, B2]
    (a) masses multiplied by the total mass
    (b) moments of the masses divided by the total mass
    (c) moments of the masses multiplied by the total mass

30. The stress of a material is defined as: [A, B1, B2]
    (a) $\frac{\text{Force}}{\text{Area}}$, with units in Nm^2
    (b) Force × Area, with units in Nm^2
    (c) $\frac{\text{Force}}{\text{Area}}$, with units in N/m^2

31. The stiffness of a material when subject to tensile loads, is measured by the: [A, B1, B2]
    (a) tensile stress
    (b) modulus of rigidity
    (c) modulus of elasticity

32. When a metal rod 20 cm long is subject to a tensile load, it extends by 0.1 mm, its strain will be: [A, B1, B2]
    (a) 0.0005
    (b) 2.0
    (c) 0.05

33. Ductility may be defined as the: [A, B1, B2]
    (a) tendency to break easily or suddenly with little or no prior extension
    (b) ability to be drawn out into threads of wire
    (c) ability to withstand suddenly applied shock loads

34. Specific strength is a particularly important characteristic for aircraft materials because: [A, B1, B2]
    (a) it is a measure of the energy per unit mass of the material

(b) the density of the material can be ignored
(c) it is a measure of the stiffness of the material

35. You are required to find the shear stress, torque and the polar second moment of area, of a circular section aircraft motor drive shaft, when given the radius of the shaft. Which of the following formulas would be the most useful? [B1, B2]
    (a) $\frac{\tau}{r} = \frac{G\theta}{l}$
    (b) $\frac{\tau}{r} = \frac{T}{J}$
    (c) $\frac{T}{J} = \frac{G\theta}{l}$

36. For an aircraft tubular control rod, subject to torsion, the maximum stress will occur: [B1, B2]
    (a) where the radius is a maximum
    (b) axially through the center of the shaft
    (c) across the shaft diameter

Kinematics and dynamics

37. The linear equations of motion rely for their derivation on one very important fact which is that the: [A, B1, B2]
    (a) velocity remains constant
    (b) velocity is the distance divided by the time taken
    (c) acceleration is assumed to be constant

38. With reference to the graph given in Figure 4.143, [A, B1, B2]
    At the point P the vehicle must be:
    (a) stationary
    (b) accelerating
    (c) traveling at constant velocity

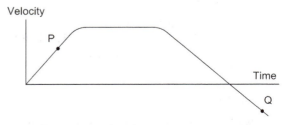

Figure 4.143

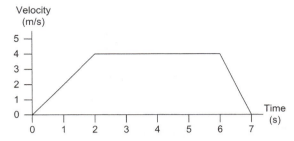

Figure 4.144 Velocity–time graph.

39. With reference to the graph given in Figure 4.143, at the point Q the vehicle must be: **[A, B1, B2]**
    (a) stationary
    (b) traveling down hill
    (c) traveling in the reverse direction

40. Figure 4.144 shows a velocity–time graph for a vehicle, for which the: **[A, B1, B2]**
    (a) initial acceleration is $2 \, m/s^2$
    (b) maximum velocity is $7 \, m/s$
    (c) acceleration between 2 and 6 s is $1 \, m/s^2$

41. Given that an aircraft accelerates from rest at $3 \, m/s^2$, then its final velocity after 36 s will be: **[A, B1, B2]**
    (a) $118 \, m/s$
    (b) $72 \, m/s$
    (c) $12 \, m/s$

42. Newton's third law, essentially states that: **[A, B1, B2]**
    (a) the inertia force is equal and opposite to the accelerating force
    (b) a body stays in a state of rest until acted upon by an external force
    (c) force is equal to mass multiplied by acceleration

43. The force produced by a fluid is the: **[B1, B2]**
    (a) fluid mass flow rate divided by its velocity
    (b) fluid mass flow rate multiplied by its velocity
    (c) mass of the fluid multiplied by its velocity

44. The mass airflow through a propeller is $400 \, kg/s$. If the inlet velocity is $50 \, m/s$ and the outlet velocity is $100 \, m/s$. The thrust developed is: **[B1, B2]**
    (a) $20 \, kN$
    (b) $8 \, kN$
    (c) $2000 \, N$

45. Given that $1 \, rev = 2\pi \, rad$ and assuming $\pi = \frac{22}{7}$, then 14 rev is equivalent to: **[A, B1, B2]**
    (a) $22 \, rad$
    (b) $44 \, rad$
    (c) $88 \, rad$

46. With respect to the torque created by rotating bodies in the formula $T = I\alpha$, the symbol I represents the: **[B1, B2]**
    (a) angular inertial acceleration and has units of m/s^2
    (b) mass moment of inertia and has units of kg/m^2
    (c) mass moment of inertia and has units of kg/m^4

47. Given that the formula for centripetal force is, $F_c = \frac{mv^2}{r}$. Then, the centripetal force required to hold an aircraft with a mass of $90,000 \, kg$ in a steady turn of radius $300 \, m$, when flying at $100 \, m/s$ is: **[B1, B2]**
    (a) $3.0 \, MN$
    (b) $300 \, kN$
    (c) $30 \, kN$

48. Gyroscopes are used within aircraft inertial navigation systems because they possess: **[A, B1, B2]**
    (a) rigidity and precess when their rotor assembly is acted upon by an external force
    (b) agility and process when their rotor assembly is acted upon by an external force
    (c) agility and precess when their rotor assembly is acted upon by an external force

49. With respect to simple harmonic motion, amplitude is defined as the: **[B1, B2]**
    (a) distance completed in one time period
    (b) number of cycles completed in unit time
    (c) distance of the highest or lowest point of the motion from the central position

50. Which of the following devices, has been designed to convert electrical energy into sound energy? **[A, B1, B2]**
    (a) Mains transformer
    (b) Loudspeaker
    (c) Telephone mouthpiece

51. Which of the following expressions defines power? **[A, B1, B2]**
    (a) Work done per unit time
    (b) Force per unit length
    (c) Force per unit time

52. Which of the following quantities has the same units as energy? **[A, B1, B2]**
    (a) Work
    (b) Power
    (c) Velocity

53. Which of the following quantities remains constant for an object falling freely towards the earth? **[A, B1, B2]**
    (a) Potential energy
    (b) Acceleration
    (c) Kinetic energy

54. The force acting on a 10 kg mass is 25 N. The acceleration is: **[A, B1, B2]**
    (a) 0.4 m/s^2
    (b) 25 m/s^2
    (c) 2.5 m/s^2

55. Given that the strain energy of a spring in tension or compression is $= \frac{1}{2}kx^2$. Then the strain energy contained by a spring with a spring constant of 2000 N/m, stretched 10 cm is: **[B1, B2]**
    (a) 10 J
    (b) 100 J
    (c) 100 kJ

56. Figure 4.145 shows a vehicle of mass 4000 kg, sitting on a hill 100 m high, having a potential energy of 50 kJ: **[B1, B2]**
    If all this potential energy is converted into kinetic energy as the vehicle rolls down the hill. Then its velocity at the bottom of the hill will be:
    (a) 5 m/s
    (b) 25 m/s
    (c) 40 m/s

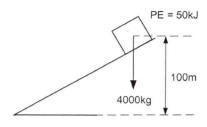

Figure 4.145 Diagram showing vehicle.

57. Which of the following statements concerning friction is true? **[A, B1, B2]**
    (a) Static friction is equal to sliding friction
    (b) The frictional resistance is dependent on the type of surfaces in contact
    (c) The coefficient of friction is equal to the product of the sliding friction force and the normal force

58. A body weighing 3000 N is moved along a horizontal plane by a horizontal force of 600 N, the coefficient of friction will be: **[A, B1, B2]**
    (a) 0.2
    (b) 2.0
    (c) 5.0

59. The mechanical advantage (MA) of a machine is equal to: **[A, B1, B2]**
    (a) $\dfrac{\text{distance moved by load}}{\text{distance moved by effort}}$
    (b) $\dfrac{\text{load}}{\text{effort}}$
    (c) $\dfrac{\text{distance moved by effort}}{\text{distance moved by load}}$

60. The efficiency of a machine is given by the mechanical advantage (MA) divided by the velocity ratio (VR). If a machine is 50% efficient and has a VR = 150, then its MA will be: **[A, B1, B2]**
    (a) 75
    (b) 300
    (c) 7500

Fluid dynamics

61. With reference to the laws of fluid pressure, which one of the statements given below is true? **[A, B1, B2]**

(a) Pressure acts vertically upwards from all surfaces

(b) Pressure at a given depth depends on the shape of the containing vessel

(c) Pressure at a given depth in a fluid is equal in all directions

62. If the gauge pressure of a fluid is 200 kPa and atmospheric pressure is 100 kPa. Then the absolute pressure will be: **[A, B1, B2]**
    (a) 2 kPa
    (b) 100 kPa
    (c) 300 kPa

63. If the density of mercury is 13,600 kg/m^3 and we assume that the acceleration due to gravity is 10 m/s^2. Then a 10 cm column of mercury will be the equivalent to a gauge pressure of: **[A, B1, B2]**
    (a) 1360 Pa
    (b) 13,600 Pa
    (c) 1360 kPa

64. A man weighing 800 N is wearing snow-shoes. The area of each of his snow shoes is ¼ m^2. The pressure exerted on the ground by each of his snow shoes is: **[A, B1, B2]**
    (a) 100 N/m^2
    (b) 400 N/m^2
    (c) 3200 N/m^2

65. An object completely immersed in still water, will remain at a fixed depth, when the: **[A, B1, B2]**
    (a) weight of the fluid displaced equals the weight of the object
    (b) up thrust force reaches a uniform velocity
    (c) apparent loss of weight remains stable

66. The boundary layer: **[B1, B2]**
    (a) remains stable at constant velocity
    (b) is the thin layer of fluid between the fixed and moving boundary
    (c) has an exponential velocity gradient between the fixed and moving boundary

67. Kinematic viscosity is: **[A, B1, B2]**
    (a) equal to the dynamic viscosity multiplied by the velocity
    (b) density dependent and varies with temperature

(c) pressure dependent and varies with weight

68. Streamline flow may be defined as: **[A, B1, B2]**
    (a) flow in which fluid particles move perpendicular and parallel to the surface of the body
    (b) flow where the density does not change from point to point
    (c) flow in which fluid particles move in an orderly manner and retain the shape of the body over which they are flowing

69. Given that a stream tube at a point, has a cross-sectional area of 1.5 m^2 and an incompressible fluid flows steadily past this point at 6 m/s. Then the volume flow rate will be: **[A, B1, B2]**
    (a) 9 m^3/s
    (b) 4 m^3/s
    (c) 0.25 m^3/s

70. A wind tunnel is subject to an incompressible steady flow of air at 40 m/s, upstream of the working section. If the cross-sectional area (csa) in the upstream part of the wind tunnel is twice the csa of the working section then: **[A, B1, B2]**
    (a) the working section velocity will be 1600 m/s
    (b) the working section velocity will be twice that of the upstream velocity
    (c) the working section velocity will be half that of the upstream velocity

71. The Bernoulli's equation, which applies the conservation of energy to fluids in motion, is represented in its energy form by: **[B1, B2]**
    (a) $\rho g h_1 + \frac{1}{2} m v_1^2 + p_1 V_1 = \rho g h_2 + \frac{1}{2} m v_2^2 + p_1 V_1$
    (b) $m g h_1 + \frac{1}{2} m v_1^2 + p_1 V_1 = m g h^2 + \frac{1}{2} m v_2^2 + p_2 V_2$
    (c) $\rho g h_1 + \frac{1}{2} \rho v_1^2 + p_1 = \rho g h_2 + \frac{1}{2} \rho v_2^2 + p_2$

72. As subsonic fluid flow passes through a Venturi tube, at the throat, the fluid pressure: **[A, B1, B2]**
    (a) increases, the fluid velocity decreases

(b) decreases and the fluid velocity decreases

(c) decreases and the fluid velocity increases

Atmospheric physics

73. Starting at sea level the atmosphere is divided into the following regions: [A, B1, B2]
    (a) troposphere, stratosphere and ionosphere
    (b) exosphere, troposphere and stratosphere
    (c) troposphere, ionosphere and stratosphere

74. Boyle's law states that, the volume of a fixed mass of gas is inversely proportional to its: [A, B1, B2]
    (a) temperature providing the pressure of the gas remains constant
    (b) pressure providing the temperature of the gas remains constant
    (c) pressure providing the density of the gas remains constant

76. The equation, $\frac{PV}{T} = \text{constant}$, for an ideal gas, is known as: [A, B1, B2]
    (a) Charles' law
    (b) Combined gas equation
    (c) Boyle's law

77. In the characteristic gas equation given by: $PV = mRT$ the symbol R is the: [B1, B2]
    (a) universal gas constant with a value of 8314.4 J/kmol K
    (b) characteristic gas constant, that has units of J/kg K
    (c) special gas constant, that has units of kg/kmol K

78. If the temperature of the air in the atmosphere increases but the pressure remains constant, the density will: [A, B1, B2]
    (a) decrease
    (b) remain the same
    (c) increase

79. The temperature at the tropopause in the International Standard Atmosphere (ISA) is approximately: [A, B1, B2]

(a) −56 K
(b) −56°F
(c) −56°C

80. The ISA sea-level pressure is expressed as: [A, B1, B2]
    (a) 29.92 mbar
    (b) 1 bar
    (c) 101,320 Pa

81. With increase in altitude, the speed of sound will: [A, B1, B2]
    (a) increase
    (b) decrease
    (c) remain the same

82. Temperature falls uniformly, with altitude, in the:
    (a) ionosphere
    (b) stratosphere
    (c) troposphere

83. The simple relationship $T_h = T_0 - Lh$ may be used to determine the temperature at a given height h in km. Where the symbol L in this equation represents the: [B1, B2]
    (a) linear distance in meters between the two altitudes
    (b) log-linear temperature drop measured in Kelvin
    (c) the temperature lapse rate measured in °C/1000 m

84. A gas occupies a volume of 4 m³ at a pressure of 400 kPa. At constant temperature, the pressure is increased to 500 kPa. The new volume occupied by the gas is: [A, B1, B2]
    (a) 5 m³
    (b) 3.2 m³
    (c) 0.3 m³

Thermodynamics

85. The temperature of a substance is: [A, B1, B2]
    (a) a measure of the energy possessed by the vibrating molecules of the substance
    (b) a direct measure of the pressure energy contained within a substance

(c) directly dependent on the volume of the substance

86. The equivalent of $60\,°C$ in Kelvin is approximately: **[A, B1, B2]**
    (a) $213\,K$
    (b) $273\,K$
    (c) $333\,K$

87. Alcohol thermometers are most suitable for measuring: **[A, B1, B2]**
    (a) jet pipe temperatures
    (b) cyrogenic substances
    (c) temperatures down to $-115°C$

88. The increase in length of a solid bar 5 m in length is $= \alpha l(t_2 - t_1)$. If the linear expansion coefficient for a solid is 2×10^{-6} and the solid is subject to a temperature rise of $100°C$. Then the increase in length will be: **[A, B1, B2]**
    (a) $1 \times 10^{-3}\,m$
    (b) $1 \times 10^{-4}\,m$
    (c) $1 \times 10^{-5}\,m$

89. The temperature of the melting point of ice and the boiling point of water are: **[A, B1, B2]**

	Melting point	Boiling point
(a)	$0\,K$	$373\,K$
(b)	$273\,K$	$373\,K$
(c)	$173\,K$	$273\,K$

90. Heat energy: **[A, B1, B2]**
    (a) is the internal energy stored within a body
    (b) travels from a cold body to a hot body
    (c) is transient energy

91. Heat transfer by conduction: **[B1, B2]**
    (a) is where a large number of molecules travel in bulk in a gas
    (b) involves energy transfer from atoms with high vibration energy to those with low vibration energy
    (c) involves changes in electron energy levels which emits energy in the form of electromagnetic waves

92. How much thermal energy is required to raise the temperature of 2 kg of aluminium by $50°C$, if the specific heat capacity of aluminium is $900\,J/kgK$

(a) $90\,kJ$
(b) $22,500\,J$
(c) $9000\,J$

93. The specific heat capacity at constant pressure, c_p is: **[B1, B2]**
    (a) less than the specific heat capacity at constant volume, c_v for the same substance
    (b) based on constant volume heat transfer
    (c) always greater than c_v

94. The specific latent heat of fusion of a substance is the heat energy required to: **[B1, B2]**
    (a) change any amount of a substance from a solid into a liquid
    (b) turn any amount of a substance from a liquid into a solid
    (c) turn unit mass of a substance from a liquid into a solid

95. A closed thermal system is one: **[B1, B2]**
    (a) that always has fixed system boundaries
    (b) that always allows the mass transfer of system fluid
    (c) in which there is no mass transfer of system fluid

96. The first law of thermodynamics applied to a closed system, may be represented symbolically by: **[B1, B2]**
    (a) $U_1 + Q = U_2 + W$
    (b) $Q + W = \Delta U$
    (c) $U_1 - Q = U_2 + W$

97. The enthalpy of a fluid is the combination of: **[B1, B2]**
    (a) kinetic energy + pressure energy
    (b) internal energy + pressure energy
    (c) potential energy + kinetic energy

98. An isentropic process is one in which: **[B1, B2]**
    (a) the enthalpy remains constant
    (b) no heat energy is transferred to or from the working fluid
    (c) both heat and work may be transferred to or from the working fluid

99. From the second law of thermodynamics the thermal efficiency (η) of a heat engine may be defined as: [B1, B2]

(a) $\eta = \dfrac{\text{total heat suppled}}{\text{work done}}$

(b) $\eta = \dfrac{Q_{out} + Q_{in}}{Q_{out}}$

(c) $\eta = \dfrac{\text{net work done}}{\text{total heat supplied}}$

100. The ideal air standard Otto cycle is: [B1, B2]
(a) based on constant pressure heat rejection
(b) used as the basis for the aircraft gas turbine engine cycle
(c) based on constant volume heat rejection

101. Entropy is a measure of: [B1, B2]
(a) the degree of disorder in a system
(b) is the product of internal energy and pressure–volume energy
(c) the adiabatic index of the system fluid

102. A polytropic process is: [B1, B2]
(a) one that obeys the law $pv^{\gamma} = c$
(b) one in which heat and work transfer may take place
(c) has constant entropy

Light and sound

103. Light: [B1, B2]
(a) is a longitudinal wave that travel through air at 340 m/s
(b) is an electromagnetic wave that travels at 3×10^8 m/s
(c) cannot transfer energy from one place to another

104. With respect to the laws of reflection: [B1, B2]
(a) the angle of incidence is equal to the angle of refraction
(b) the incident ray and the normal lie within the same plane
(c) images from plane mirrors are real and laterally converted

105. The light rays from a concave mirror: [B1, B2]
(a) converge at the principle focus
(b) diverge at the principle focus
(c) diverge at the pole which is approximately twice the radius of curvature

106. Given that $\frac{1}{u} = \frac{1}{v} + \frac{1}{f}$, the object distance = 50 mm and the focal length = 150 mm, then the distance of the object from the mirror is: [B1, B2]
(a) 37.5 mm
(b) 75 mm
(c) 150 mm

107. As light travels from one medium to another medium with a greater refractive index, its speed: [B1, B2]
(a) is increased
(b) remains the same
(c) is decreased

108. Fibre optic cables use the principle of: [B1, B2]
(a) total external reflection to enable light to travel along the cable
(b) internal refraction to enable light to travel along the cable
(c) total internal reflection to enable light to travel along the cable

109. Convex lenses: [B1, B2]
(a) form real, inverted, small images of distant objects
(b) create virtual, inverted, small images of near objects
(c) produce images where the focal length is always negative

110. Sound waves [B1, B2]
(a) are transverse waves that are able to travel through a vacuum
(b) form part of the electromagnetic spectrum, with low or high frequencies
(c) are longitudinal waves that need a medium through which to travel

111. The speed of a wavefront is linked by the relationship $v = f\lambda$. If given that the wave frequency = 1 kHz and the speed of propagation is 100 m/s, then the wavelength is: [B1, B2]

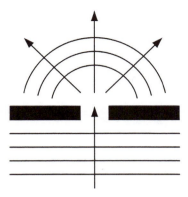

Figure 4.146

(a) 0.1 m
(b) 10 m
(c) 1×10^5 m

112. With respect to the behaviour of waves, Figure 4.146 illustrates: **[B1, B2]**
    (a) diffraction
    (b) reinforcement
    (c) destructive interference

113. Radio waves travel as: **[B1, B2]**
    (a) sound waves, carrier waves, longitudinal waves
    (b) ground waves, sky waves, space waves
    (c) aerial waves, longitudinal waves, Doppler waves

114. An aircrafts microwave landing system is likely to operate at an approximate frequency of: **[B1, B2]**
    (a) 500 kHz
    (b) 5000 kHz
    (c) 5000 MHz

115. The phenomenon where a change in wave frequency is brought about by relative motion is known as: **[B1, B2]**
    (a) radio wave travel effect
    (b) the Doppler effect
    (c) transmitter effect

PART 3

Electrical and electronic fundamentals

Chapter 5

Electrical fundamentals

5.1 Introduction

In today's world, electricity is something that we all take for granted. So, before we get started, it is worth thinking about what electricity means to you and, more importantly, how it affects *your* life.

Think, for a moment, about where and how electricity is used in your home, car, workplace or college. You will quickly conclude that electricity is a means of providing heat, light, motion and sound. You should also conclude that electricity is invisible – we only know that it is there by looking at what it does!

Now let us turn to the world of aircraft and flight. Although it may not be obvious at first sight, it is fair to say that an aircraft just could not fly without electricity. Not only is electricity used to provide a means of ignition for the engines, but it also supplies the lighting and instruments within an aircraft as well as the navigational aids and radio equipment essential for safe flight in a modern aircraft. Electricity is used to heat windows, pump fuel, operate brakes, open and shut valves, and to control numerous other systems within the aircraft. In fact, aircraft that use modern "fly-by-wire" controls could not even get off the ground without

the electrical systems and supplies that make them work!

In this chapter we will explain electricity in terms of electric charge, current, voltage and resistance. We will begin by introducing you to some important concepts, including the Bohr model of the atom and the fundamental nature of electric charge and conduction in solids, liquids and gases. Next we will look briefly at static electricity before moving on to explain some of the terminology that we use with electric circuits and measurements. We also describe some of the most common types of electrical and electronic component including resistors, capacitors, inductors, transformers, generators and motors.

5.1.1 Electrical units and symbols

You will find that a number of units and symbols are commonly encountered in electrical circuits so let us get started by introducing some of them. In fact, it is important to get to know these units and also to be able to recognize their abbreviations and symbols before you actually need to use them. Later we will explain how these units work in much greater detail but for now we will simply list them (Table 5.1) so that at least you can begin to get to know something about them.

Table 5.1

Unit	Abbreviation	Symbol	Notes
Ampere	A	I	Unit of electric current (a current of 1 A flows in a conductor when a charge of 1 C is transported in a time interval of 1 s)
Coulomb	C	Q	Unit of electric charge or quantity of electricity (a *fundamental unit*)
Farad	F	C	Unit of capacitance (a capacitor has a capacitance of 1 F when a charge of 1 C results in a potential difference (*p.d.*) of 1 V across its plates)
Henry	H	L	Unit of inductance (an inductor has an inductance of 1 H when an applied current changing uniformly at a rate of 1 A/s produces a p.d. of 1 V across its terminals)
Hertz	Hz	f	Unit of frequency (a signal has a frequency of 1 Hz if one complete cycle occurs in a time interval of 1 s)

(*continued*)

Table 5.1 (*continued*)

Unit	Abbreviation	Symbol	Notes
Joule	J	W, J	Unit of energy (a *fundamental unit*)
Ohm	Ω	R	Unit of resistance (a *fundamental unit*)
Second	s	t	Unit of time (a *fundamental unit*)
Siemen	S	G	Unit of conductance (the reciprocal of resistance)
Tesla	T	B	Unit of magnetic flux density (a flux density of 1 T is produced when a flux of 1 Wb is present over an area of 1 m^2)
Volt	V	V, E	Unit of electric potential (we sometimes refer to this as *electromotive force* (*e.m.f.*) or *p.d.*
Watt	W	P	Unit of power (equal to 1 J of energy consumed in a time of 1 s)
Weber	Wb	Φ	Unit of magnetic flux (a *fundamental unit*)

Key point

Symbols used for electrical and other quantities are normally shown in italic font whilst units are shown in normal (non-italic) font. Thus *V* and *I* are symbols whilst V and A are units.

5.1.2 Multiples and sub-multiples

Unfortunately, because the numbers can be very large or very small, many of the electrical units can be cumbersome for everyday use. For example, the voltage present at the antenna input of a very high frequency (VHF) radio could be as little as 0.0000015 V. At the same time, the resistance present in an amplifier stage could be as high as 10,000,000 Ω! Clearly we need to make life a little easier. We can do this by using a standard range of multiples and sub-multiples. These use a *prefix* letter in order to add a *multiplier* to the quoted value, as follows:

Prefix	Abbrev.	Multiplier	
Tera	T	10^{12}	(=1,000,000,000,000)
Giga	G	10^9	(=1,000,000,000)
Mega	M	10^6	(=1,000,000)
Kilo	k	10^3	(=1000)
(None)	(None)	10^0	(=1)
Centi	c	10^{-2}	(=0.01)
Milli	m	10^{-3}	(=0.001)
Micro	μ	10^{-6}	(=0.000,001)
Nano	n	10^{-9}	(=0.000,000,001)
Pico	p	10^{-12}	(=0.000,000,000,001)

Example 5.1

An indicator lamp requires a current of 0.15 A. Express this in mA.

Solution

To convert A to mA, we apply a multiplier of 10^3 or 1000. Thus to convert 0.15 A to mA we multiply 0.15 by 1000 as follows:

$$0.15\,\text{A} = 0.15 \times 1000 = 150\,\text{mA}$$

Key point

Multiplying by 1000 is equivalent to moving the decimal point *three* places to the *right* whilst dividing by 1000 is equivalent to moving the decimal point *three* places to the *left*. Similarly, multiplying by 1,000,000 is equivalent to moving the decimal point *six* places to the *right* whilst dividing by 1,000,000 is equivalent to moving the decimal point *six* places to the *left*.

Example 5.2

An insulation tester produces a voltage of 2750 V. Express this in kV.

Solution

To convert V to kV we apply a multiplier of 10^{-3} or 0.001. Thus we can convert 2750 V to kV as follows:

$$2750\,\text{V} = 2750 \times 0.001 = 2.75\,\text{kV}$$

Here, multiplying by 0.001 is equivalent to moving the decimal point three places to the *left*.

Example 5.3

A capacitor has a value of 27,000 pF. Express this in μF.

Solution

There are 1,000,000 pF in 1 μF. Thus, to express the value in 27,000 pF in μF we need to multiply by 0.000,001. The easiest way of doing this is simply to move the decimal point six places to the left. Hence 27,000 pF is equivalent to 0.027 μF (note that we have had to introduce an extra zero before 2 and after the decimal point).

Test your understanding 5.1

1. State the units for electric current.
2. State the units for frequency.
3. State the symbol used for capacitance.
4. State the symbol used for conductance.
5. A pulse has a duration of 0.0075 s. Express this time in ms.
6. A generator produces a voltage of 440 V. Express this in kV.
7. A signal has a frequency of 15.62 MHz. Express this in kHz.
8. A current of 570 μA flows in a resistor. Express this current in mA.
9. A capacitor has a value of 0.22 μF. Express this capacitance in nF.
10. A resistor has a value of 470 kΩ. Express this resistance in MΩ.

5.2 Electron theory

Syllabus
Structure and distribution of electrical charges within atoms, molecules, ions and compounds; Molecular structure of conductors, semiconductors and insulators.

Knowledge level key

A	B1	B2
1	1	1

To understand what electricity is we need to take a look inside the atoms that make up all forms of matter. Since we cannot actually do this with a real atom we will have to use a model. Fortunately, understanding how this model works is not too difficult – just remember that what we are talking about is very, very small!

5.2.1 Atomic structure

As you already know, all matter is made up of atoms or groups of atoms (*molecules*) bonded together in a particular way. In order to understand something about the nature of electrical

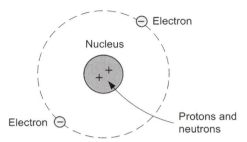

Figure 5.1   The Bohr model of the atom.

charge we need to consider a simple model of the atom. This model known as the Bohr model (see Figure 5.1) shows a single atom consisting of a central nucleus with orbiting electrons.

Within the nucleus there are *protons* which are *positively charged* and *neutrons* which, as their name implies, are electrical neutral and *have no charge*. Orbiting the nucleus are *electrons that have a negative charge, equal in magnitude (size) to the charge on the proton*. These electrons are approximately 2000 times lighter than the protons and neutrons in the nucleus.

In a stable atom the number of protons and electrons are equal, so that overall, the atom is neutral and has no charge. However, if we rub two particular materials together, electrons may be transferred from one to another. This alters the stability of the atom, leaving it with a net positive or negative charge. When an atom within a material *looses electrons* it becomes positively charged and is known as a *positive ion*, when an atom *gains an electron* it has a surplus negative charge and so is known as a *negative ion*. These differences in charge can cause *electrostatic* effects. For example, combing your hair with a nylon comb may result in a difference in charge between your hair and the rest of your body, resulting in your hair standing on end when your hand or some other differently charged body is brought close to it.

The number of electrons occupying a given orbit within an atom is predictable and is based on the position of the element within the periodic table. The electrons in all atoms sit in a particular position (shell) dependent on their energy level. Each of these shells within the atom is filled by electrons from the nucleus outwards, as shown in Figure 5.2. The first, inner most, of these shells can have up to two electrons; the second shell can have up to eight and the third up to 18.

5.2.2 Conductors and insulators

A material which has many free electrons available to act as charge carriers and thus allows current to flow freely is known as a *conductor*. Examples of good conductors include aluminium, copper, gold and iron. Figure 5.2 shows a material with one outer electron that can become easily detached from the parent atom. It requires a small amount of external energy to overcome the attraction of the nucleus. Sources of such energy may include heat, light or electrostatic fields. The atom once detached

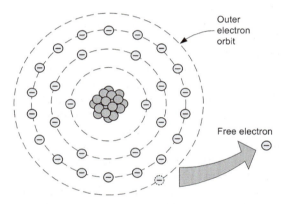

Figure 5.2 A material with a loosely bound electron in its outer shell.

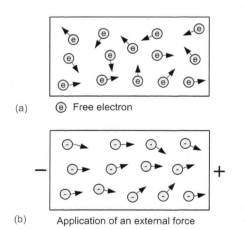

Figure 5.3 Free electrons and the application of an external force: (a) electrons in random motion and (b) current flow.

from the atom is able to move freely around the structure of the material and is called a *free electron*. It is these free electrons that become the *charge carriers*, within a material. Materials that have large numbers of free electrons make good *conductors* of electrical energy and heat.

In a material containing free electrons their direction of motion is random, as shown in Figure 5.3(a), but if an external force is applied that causes the free electrons to move in a uniform manner (Figure 5.3(b)) an electric *current* is said to flow.

Metals are the best conductors, since they have a very large number of free electrons available to act as charge carriers. Materials that do not conduct charge are called *insulators*, their electrons are tightly bound to the nuclei of their atoms. Examples of insulators include plastics, glass, rubber and ceramic materials.

The effects of electric current flow can be detected by the presence of one or more of the following effects: light, heat, magnetism, chemical, pressure and friction. For example, if a piezoelectric crystal is subject to an electrical current it can change its shape and exert pressure. Heat is another, more obvious effect from electric heating elements.

Key point

Metals, like copper and silver, are good conductors of electricity and they readily support the flow of current. Plastics, rubber and ceramic materials are insulators and do not support the flow of current.

5.2.3 Semiconductors

Some materials combine some of the electrical characteristics of conductors with those of insulators. They are known as *semiconductors*. In these materials there may be a number of free electrons sufficient to allow a small current to flow. It is possible to add foreign atoms (called *impurity atoms*) to the semiconductor material that modify the properties of the semiconductor. Varying combinations of these additional atoms are used to produce various electrical devices, such as diodes and transistors. Common types of semiconductor materials are silicon, germanium, selenium and gallium.

> **Key point**
>
> Semiconductors are pure insulating materials with a small amount of an impurity element present. Typical examples are silicon and germanium.

5.2.4 Temperature effects

As stated earlier, all materials offer some resistance to current flow. In *conductors* the free electrons, rather than passing unobstructed through the material, collide with the relatively large and solid nuclei of the atoms. As the temperature increases, the nuclei vibrate more energetically further obstructing the path of the free electrons, causing more frequent collisions. The result is that the *resistance of conductors increases with temperature*.

Due to the nature of the bonding in *insulators*, there are no free electrons, except that when thermal energy increases as a result of a temperature increase, a few outer electrons manage to break free from their fixed positions and act as charge carriers. The result is that the *resistance of insulators decreases as temperature increases*.

Semiconductors behave in a similar manner to insulators. At absolute zero ($-273°C$) both the types of material act as perfect insulators. However, unlike the insulator, as temperature increases in a semiconductor *large numbers* of electrons break free to act as charge carriers.

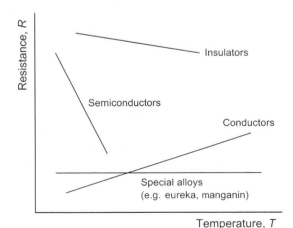

Figure 5.4 Variation of resistance with temperature for various materials.

Therefore, as temperature increases, the resistance of a semiconductor decreases rapidly.

By producing special alloys, such as eureka and manganin that combine the effects of insulators and conductors, it is possible to produce a material where the resistance remains constant with increase in temperature. Figure 5.4 shows how the resistance of insulators, semiconductors and conductors change with temperature.

Test your understanding 5.2

1. In a stable neutral atom the number of _____ and _____ are equal and there is no overall charge.
2. When an atom within a material loses electrons it becomes _____ charged and is known as a _____.
3. When an atom gains an electron it has a surplus _____ charge and so known as a _____.
4. The electrical properties of a material are determined by the number of _____ present.
5. Materials that do not conduct electric charge are called _____.
6. Name two materials that act as good electrical conductors.
7. Name two materials that act as good electrical insulators.
8. Name two semiconductor materials.
9. Explain briefly how the resistance of a metallic conductor varies with temperature.
10. Explain briefly how the resistance of an insulator varies with temperature.

5.3 Static electricity and conduction

Syllabus
Static electricity and distribution of electrostatic charges; Electrostatic laws of attraction and repulsion; Units of charge, Coulomb's law; Conduction of electricity in solids, liquids, gases and a vacuum.

Knowledge level key

A	B1	B2
1	2	2

Electric charge is all around us. Indeed, many of the everyday items that we use in the home and at work rely for their operation on the existence of electric charge and the ability to make that charge do something useful. Electric charge is also present in the natural world and anyone who has experienced an electric storm cannot fail to have been awed by its effects. In this section we begin by explaining what electric charge is and how it can be used to produce conduction in solids, liquids and gases.

5.3.1 Static electricity

We have already found that, if a conductor has a deficit of electrons, it will exhibit a net positive charge. On the other hand, if it has a surplus of electrons, it will exhibit a net negative charge. An imbalance in charge can be produced by friction (removing or depositing electrons using materials, such as silk and fur, respectively) or induction (by attracting or repelling electrons using a second body which is, respectively, positively or negatively charged).

5.3.2 Force between charges

Consider two small charged bodies of negligible weight are suspended as shown in Figure 5.5. If the two bodies have charges with the same polarity (i.e. either both positively or both negatively charged) the two bodies will move apart, indicating that a force of repulsion exists between them. On the other hand, if the charges

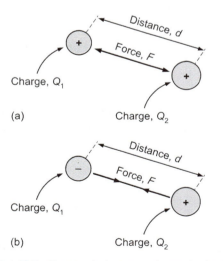

Figure 5.5 Force between charged bodies: (a) charges with same polarity and (b) charges with opposite polarity.

on the two bodies are unlike (i.e. one positively charged and one negatively charged), the two bodies will move together indicating that a force of attraction exists between them. From this we can conclude that *like charges repel* and *unlike charges attract*.

Key point

Charges with the same polarity repel one another whilst charges with opposite polarity will attract one another.

5.3.3 Coulomb's law

Coulomb's law states that if charged bodies exist at two points, the force of attraction (if the charges are of opposite charge) or repulsion (if of like charge) will be proportional to the product of the magnitude of the charges divided by the square of their distance apart. Thus:

$$F = \frac{kQ_1Q_2}{d^2}$$

where Q_1 and Q_2 are the charges present at the two points (in C), d the distance separating the two points (in m), F the force (in N) and k is a constant depending upon the medium in which the charges exist.

In vacuum or *free space*

$$k = \frac{1}{4\pi\varepsilon_0}$$

where ε_0 is the permittivity of free space (8.854×10^{-12} C/Nm^2).

Combining the two previous equations gives:

$$F = \frac{Q_1Q_2}{4\pi\varepsilon_0 d^2}$$

or

$$F = \frac{Q_1Q_2}{4\pi \times 8.854 \times 10^{-12} \times d^2} \text{ N}$$

If this formula looks complex there are only a couple of things that you need to remember. The denominator simply consists of a constant ($4\pi \times 8.854 \times 10^{-12}$) multiplied by the square of the distance, d. Thus we can re-write the formula as:

$$F \propto \frac{Q_1Q_2}{d^2}$$

where the symbol $\propto$ denotes proportionality.

5.3.4 Electric fields

The force exerted on a charged particle is a manifestation of the existence of an electric field. The electric field defines the direction and magnitude of a force on a charged object. The field itself is invisible to the human eye but can be drawn by constructing lines which indicate the motion of a free positive charge within the field; the number of field lines in a particular region being used to indicate the relative strength of the field at the point in question.

Figures 5.6 and 5.7 show the electric fields between isolated unlike and like charges whilst

Figure 5.8 shows the field which exists between the two charged parallel metal plates (note the *fringing* which occurs at the edges of the plates).

5.3.5 Electric field strength

The strength of an electric field (E) is proportional to the applied p.d. and inversely proportional to the distance between the two conductors (see Figure 5.9). The electric field strength is given by:

$$E = \frac{V}{d}$$

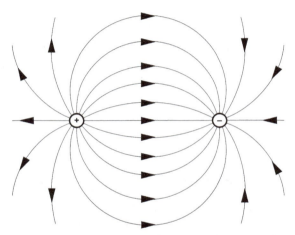

Figure 5.6 Electric field between two isolated unlike charges.

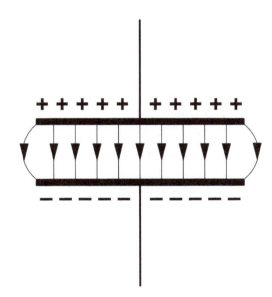

Figure 5.8 Electric field between two charged parallel metal plates.

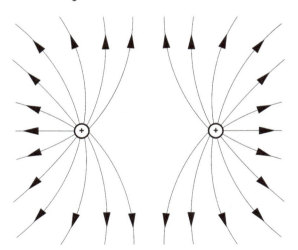

Figure 5.7 Electric field between two isolated like charges.

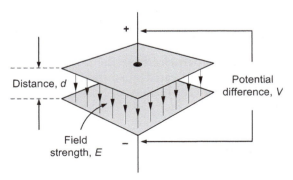

Figure 5.9 Electric field strength.

where E is the electric field strength (in V/m), V is the applied p.d. (in V) and d is the distance (in m).

Example 5.4

Two charged particles are separated by a distance of 25 mm. Calculate the force between the two charges if one has a positive charge of $0.25\,\mu\text{C}$ and the other has a negative charge of $0.4\,\mu\text{C}$. What will the relative direction of the force be?

Solution

Now $\quad F = \dfrac{Q_1 Q_2}{4\pi \times 8.854 \times 10^{-12} \times d^2}$

where $\quad Q_1 = 0.25\,\mu\text{C} = 0.25 \times 10^{-6}$ C, $\quad Q_2 = 0.4\,\mu\text{C} = 0.4 \times 10^{-6}$ C, and $\quad d = 2.5\,\text{mm} = 2.5 \times 10^{-3}$ m, thus:

$$F = \frac{0.25 \times 10^{-6} \times 0.4 \times 10^{-6}}{4\pi \times 8.854 \times 10^{-12} \times (2.5 \times 10^{-3})^2}$$

$$= \frac{0.1 \times 10^{-12}}{4\pi \times 8.854 \times 10^{-12} \times 6.25 \times 10^{-6}}$$

or

$$F = \frac{0.1}{4\pi \times 8.854 \times 6.25 \times 10^{-6}}$$

$$= \frac{0.1}{695.39 \times 10^{-6}} = 1.438 \times 10^2$$

Hence $\quad F = 1.438 \times 10^2\,\text{N} = 143.8\,\text{N}$

Example 5.5

Two charged particles have the same positive charge and are separated by a distance of 10 mm. If the force between them is 0.1 N, determine the charge present.

Solution

Now $F = \dfrac{Q_1 Q_2}{4\pi \times 8.854 \times 10^{-12} \times d^2}$

where $F = 0.1$ N, $d = 0.01$ m and $Q_1 = Q_2 = Q$, thus:

$$0.1 = \frac{QQ}{4\pi \times 8.854 \times 10^{-12} \times (0.01)^2}$$

Re-arranging the formula to make Q the subject gives:

$$Q^2 = 0.1 \times 4\pi \times 8.854 \times 10^{-12} \times (0.01)^2$$

or

$$Q = \sqrt{0.1 \times 4\pi \times 8.854 \times 10^{-12} \times (0.01)^2}$$

$$= \sqrt{4\pi \times 8.854 \times 10^{-17}}$$

$$= \sqrt{111.263 \times 10^{-17}} = \sqrt{11.1263 \times 10^{-16}}$$

thus

$$Q = \sqrt{11.1263} \times \sqrt{10^{-16}} = 3.336 \times 10^{-8}\,\text{C}$$

$$= 0.03336\,\mu\text{C}$$

Example 5.6

Two parallel conductors are separated by a distance of 25 mm. Determine the electric field strength if they are fed from a 600 V direct current (DC) supply.

Solution

The electric field strength will be given by:

$$E = \frac{V}{d}$$

where $\quad V = 600$ V and $\quad d = 25\,\text{mm} = 0.025$ m, thus:

$$E = \frac{600}{0.025} = 24{,}000\,\text{V/m} = 24\,\text{kV/m}$$

Example 5.7

The field strength between the two parallel plates in a cathode ray tube is 18 kV/m. If the plates are separated by a distance of 21 mm, determine the p.d. that exists between the plates.

Solution

The electric field strength will be given by:

$$E = \frac{V}{d}$$

Re-arranging this formula to make V the subject gives:

$$V = E \times d$$

Now $E = 18\,\text{kV/m} = 18,000\,\text{V/m}$ and $d = 21\,\text{mm} = 0.021\,\text{m}$, thus:

$$V = 18,000 \times 0.021 = 378\,\text{V}$$

5.3.6 Conduction of electricity in solids, liquids, gases and a vacuum

In order to conduct an electric current a material must contain charged particles. In solids (such as copper, lead, aluminium and carbon) it is the negatively charged electrons that are in motion. In liquids and gases, the current is carried by the part of a molecule that has acquired an electric charge. These are called *ions* and they can possess either a positive or a negative charge. Examples include *hydrogen ions* (H^+), *copper ions* (Cu^{++}) and *hydroxyl ions* (OH^-). It is worth noting that pure distilled water contains no ions and is thus a poor conductor of electricity whereas salt water contains ions and is therefore a relatively good conductor of electricity.

Finally, you might be surprised to learn that an electric current can pass through a vacuum. It does this in the form of a stream of electrons liberated from a hot metal surface that can be made to travel from a point that has a negative potential (known as a *cathode*) towards another point which has a high positive potential (known as an *anode*). This is the principle of the cathode ray tube that you find in your television set or computer display!

Key point

Current flow in liquids and gases is made possible by means of positively or negatively charged molecules called ions. In a vacuum, current flow is made possible by means of a moving stream of negatively charged electrons, as in the cathode ray tube.

Test your understanding 5.3

1. If a body has a shortage of electrons it will exhibit a _____ charge.
2. Isolated charges having the same polarity will _____ one another.
3. List the factors that determine the force that exists between two charges.
4. Two charges are separated by a distance of 1 mm. If the distance increases to 2 mm whilst the charges remain unchanged, by how much will the force between them change?
5. Two plates are separated by a distance of 100 mm. If the p.d. between the plates is 200 V where what will the electric field strength be?
6. The electric field between two parallel plates is 2 kV/m. If the plates are separated by a distance of 4 mm, determine the p.d. between the plates.
7. Two charged particles have the same positive charge and are separated by a distance of 2 mm. If the force between them is 0.4 N, determine the charge present.
8. In liquids and gases electric current is carried by _____.
9. An electric current can be made to pass through a vacuum by means of a stream of _____ charged _____.
10. Explain why salt water conducts electricity whilst pure distilled water does not.

5.4 Electrical terminology

Syllabus
The following terms, their units and factors affecting them; p.d., e.m.f., voltage, current, resistance, conductance, charge, conventional current flow and electron flow.

Knowledge level key

A	B1	B2
1	2	2

This section will introduce you to some of the terminology that we use in electric circuits. In addition to the syllabus topics listed above we have also included two other important terms, power and energy.

5.4.1 Charge

All electrons and protons have an electro-static *charge*, its value is so small that a more convenient unit of charge is needed for practical use, which we call the *coulomb*. One coulomb C is the total charge Q of 6.21×10^{18} electrons. Thus a single electron has a charge of 1.61×10^{-19} C.

5.4.2 Current

Current, I, is defined as the rate of flow of charge and its unit is the ampere, A. One ampere is equal to one coulomb per second, or:

$$\text{One ampere of current, } I = \frac{Q}{t}$$

where t is time in seconds.

So, for example, if a steady current of 3 A flows for 2 min, then the amount of charge transferred will be:

$$Q = I \times t = 3\,\text{A} \times 120\,\text{s} = 360\,\text{C}$$

Key point

Current is the rate of flow of charge. Thus, if more charge moves in a given time, more current will be flowing. If no charge moves then no current is flowing.

5.4.3 Conventional current and electron flow

In Section 5.2.2 we described electric current in terms of the organized movement of electrons in a metal conductor. Owing to their negative charge, electrons will flow from a negative potential to a more positive potential (recall that like-charges attract and unlike-charges repel). However, when we indicate the direction of current in a circuit we show it as moving from a point that has the greatest positive potential to a point that has the most negative potential. We call this *conventional current* and, although it may seem to be odd, you just need to remember that it flows in the *opposite* direction to that of the motion of electrons!

Key point

Electrons move from negative to positive whilst conventional current is assumed to flow from positive to negative.

5.4.4 Potential difference (voltage)

The force that creates the flow of current (or rate of flow of charge carriers) in a circuit is known as the *e.m.f.* and it is measured in volts (V). The *p.d.* is the voltage difference or voltage drop between two points.

One volt is the p.d. between two points if one joule of energy is required to move one coulomb of charge between them. Hence:

$$V = \frac{W}{Q}$$

where W is the energy and Q is the charge, as before. Energy is defined later in Section 5.4.8.

5.4.5 Resistance

All materials at normal temperatures oppose the movement of electric charge through them, this opposition to the flow of the charge carriers is known as the *resistance R* of the material. This resistance is due to collisions between the charge carriers (electrons) and the atoms of the material. The unit of resistance is the *ohm*, with symbol Ω.

Note that 1 V is the e.m.f. required to move 6.21×10^{18} electrons (1 C) through a resistance of $1\,\Omega$ in 1 s. Hence:

$$V = \left(\frac{Q}{t}\right) \times R$$

where Q is the charge, t is the time and R is the resistance.

Re-arranging this equation to make R the subject gives:

$$R = \frac{V \times t}{Q}\,\Omega$$

We shall be looking at the important relationship between voltage, V, current, I and resistance, R, later on in Sections 5.7.1 and 5.7.2.

5.4.6 Conductance

Conductance is the inverse of resistance. In other words, as the resistance of a conductor increases its conductance reduces, and vice versa. A material that has a low value of conductance will not conduct electricity as well as a material that has a high conductance, and vice versa. You can thus think of conductance as the lack of opposition to the passage of charge carriers. The symbol used for conductance is G and its unit is the *Siemen* (S).

The following table shows the relative conductance of some common metals:

Metal	Relative conductance (copper = 1)
Silver	1.06
Copper (annealed)	1.00
Copper (hard drawn)	0.97
Aluminium	0.61
Mild steel	0.12
Lead	0.08

Key point

Metals, like copper and silver, are good conductors of electricity. Good conductors have low resistance whilst poor conductors have high resistance.

Example 5.8

A current of 45 mA flows from one point in a circuit to another. What charge is transferred between the two points in 10 min?

Solution

Here we will use $Q = I \times t$
where $I = 45\,mA = 0.045\,A$ and $t = 10\,min = 10 \times 60 = 600\,s$, thus:

$$Q = 0.045 \times 600 = 27\,C$$

Example 5.9

A 28 V DC aircraft supply delivers a charge of 5 C to a window heater every second. What is the resistance of the heater?

Solution

Here we will use $R = \dfrac{V \times t}{Q}$
where $V = 28\,V$, $Q = 5\,C$ and $t = 1\,s$, thus:

$$R = \frac{V \times t}{Q} = \frac{28\,V \times 1\,s}{5\,C} = 5.6\,\Omega$$

5.4.7 Power

Power, P, is the rate at which energy is converted from one form to another and it is measured in *Watts*. The larger the amount of power the greater the amount of energy that is converted in a given period of time.

1 watt = 1 joule per second or

$$\text{Power, } P = \frac{\text{Energy, } J}{\text{Time, } t}$$

thus:

$$P = \frac{J}{t}\,W$$

5.4.8 Energy

Like all other forms of energy, electrical energy is the capacity to do work. Energy can be converted from one form to another. For example an electric fire converts electrical energy into heat. A filament lamp converts electrical energy into light, and so on. Energy can only be transferred when a difference in energy levels exists.

The unit of energy is the *joule*. Then, from the definition of power:

$$1\,\text{Joule} = 1\,\text{Watt} \times 1\,\text{second}$$

hence:

$$\text{Energy, } \quad J = (\text{Power, } P) \times (\text{Time, } t)$$
$$\text{with units of (Watts} \times \text{seconds)}$$

thus: $J = P \times t\,W$

Thus joules are measured in *watt-seconds* (Ws). If the power was to be measured in kilowatts and the time in hours, then the unit of electrical energy would be the *kilowatt-hour* (kWh) (commonly knows as a *unit of electricity*). The electricity meter in your home records the number of kilowatt-hours. In other words, it indicates the *amount of energy* that you have used.

Example 5.10

An auxiliary power unit (APU) provides an output of 1.5 kW for 20 min. How much energy has it supplied to the aircraft?

Solution

Here we will use $J = P \times t$

where $P = 1.5\,\text{kW} = 1500\,\text{W}$ and $t = 20$ min $= 20 \times 60 = 1200\,\text{s}$, thus:

$$J = 1500 \times 1200 = 1,800,000\,\text{J} = 1.8\,\text{MJ}$$

Note here that we have converted from J to MJ by moving the decimal point six places to the *left*.

Example 5.11

A smoothing capacitor is required to store 20 J of energy. How much power is required to store this energy in a time interval of 0.5 s?

Solution

Re-arranging $J = P \times t$ to make P the subject gives:

$$P = \frac{J}{t}$$

we can now find P when $J = 20\,\text{J}$ and $t = 0.5\,\text{s}$, thus:

$$P = \frac{J}{t} = \frac{20\,\text{J}}{0.5\,\text{s}} = 40\,\text{W}$$

Example 5.12

A main aircraft battery is used to start an engine. If the starter demands a current of 1000 A for 30 s and the battery voltage remains at 12 V during this period, determine the amount of electrical energy required to start the engine.

Solution

Here we will use $Q = I \times t$

where $I = 1000\,\text{A}$ and $t = 30\,\text{s}$, thus:

$$Q = 1000 \times 30 = 30,000\,\text{C}$$

But

$$V = \frac{W}{Q}$$

where W is the energy and Q is the charge.

$$\therefore W = V \times Q$$

$$= 12 \times 30,000 = 360,000 = 360\,\text{kJ}$$

Test your understanding 5.4

1. Current is defined as the rate of flow of _____ and its unit is the _____.
2. Conventional current flows from _____ to _____.
3. Electron flow is from _____ to _____.
4. The unit of resistance is the _____ and its symbol is _____.
5. Which of the following materials: aluminium, copper, gold and silver is (a) the best and (b) the worst conductor of electricity?
6. A current of 1.5 A flows for 10 min. What charge is transferred?
7. The e.m.f. required to move 6.21×10^{18} electrons through a resistance of $1\,\Omega$ is _____.
8. The energy transferred by an electric circuit is the product of _____ and _____.
9. Explain briefly what is meant by the term, resistance.
10. Explain briefly the relationship between resistance and conductance.

5.5 Generation of electricity

Syllabus

Production of electricity by the following methods: light, heat, friction, pressure, chemical action, magnetism and motion.

Knowledge level key

A	B1	B2
1	1	1

There are electrons and protons in the atoms of all materials but, to do useful work, charges must be separated in order to produce a p.d. that we can use to make current flow and do work. Since the generation of electric current is a fundamental requirement in every aircraft we shall be looking at this topic in much greater detail later on. For now, we will briefly describe some of the available methods for separating charges and creating a flow of current.

5.5.1 Friction

Static electricity can be produced by friction. In this method, electrons and protons in an insulator can be separated by rubbing two materials together in order to produce opposite charges. These charges will remain separated for some time until they eventually leak away due to

Figure 5.10 Static discharging devices.

losses in the insulating material (the *dielectric*) or in the air surrounding the materials. Note that more charge will be lost in a given time if the air is damp.

Static electricity is something that can cause particular problems in an aircraft and special measures are taken to ensure that excessive charges do not build up on the aircraft's structure. The aim is that of equalizing the potential of all points on the aircraft's external surfaces. The static charge that builds up during normal flight can be dissipated into the atmosphere surrounding the aircraft by means of small conductive rods connected to the aircraft's trailing surfaces. These are known as *static dischargers* or *static wicks* (see Figure 5.10).

5.5.2 Chemical action

Another way of producing electricity is a cell or battery in which a chemical reaction produces opposite charges on two dissimilar metals which serve as the negative and positive terminals of the cell. In the common zinc–carbon dry cell, the zinc container is the negative electrode and the carbon electrode in the center is the positive electrode. In the lead–acid wet cell, sulphuric acid diluted with water is the liquid electrolyte, while the negative terminal is lead and the positive terminal is lead peroxide. We shall explore these two types of cell in Section 5.6.

Chemical action can also be responsible for a highly undesirable effect known as *corrosion*. Corrosion is a chemical process in which metals are converted back to salts and other

oxides from which they were first formed. The two basic mechanisms associated with corrosion are direct chemical attack and electrochemical attack. In the latter, the process of corrosion is associated with the presence of dissimilar metals and an electric current. Such corrosion can often be observed at electric contacts and battery terminals.

5.5.3 Magnetism and motion

When a conductor (such as a copper wire) moves through a magnetic field, an e.m.f. will be induced across its ends. In a similar fashion, an e.m.f. will appear across the ends of a conductor if it remains stationary whilst the field moves. In either case, the action of cutting through the lines of magnetic flux results in a generated e.m.f. The amount of e.m.f., e, *induced* in the conductor will be directly proportional to:

- the density of the magnetic flux, B, measured in tesla (T);
- the effective length of the conductor, l, within the magnetic flux;
- the speed, v, at which the lines of flux cut through the conductor measured in metres per second (m/s);
- the sine of the angle, θ, between the conductor and the lines of flux.

The induced e.m.f. is given by the formula:

$$e = B \times l \times v \sin\theta$$

Electricity and magnetism often work together to produce motion. In an electric *motor*, current flowing in a conductor placed inside a magnetic field produces motion. On the other hand, a *generator* produces a voltage when a conductor is moved inside a magnetic field. These two effects are, as you might suspect, closely related to one another and they are vitally important in the context of aircraft electrical systems!

Example 5.13

A copper wire of length 2.5 m moves at right angles to a magnetic field with a flux density of 0.5 T. If the relative speed between the wire and the field is 4 m/s what e.m.f. will be generated across the ends of the conductor?

Solution

Now $e = B \times l \times v \sin \theta$

Since the wire is moving at right angles to the field, the value of θ is 90°.

Hence:

$$e = 0.5 \times 2.5 \times 4 \sin 90° = 0.5 \times 2.5 \times 4 \times 1 = 5 \, V$$

Example 5.14

A copper wire of length 50 cm is suspended at 45° to a magnetic field which is moving at 50 m/s. If an e.m.f. of 2 V is generated across the ends of the conductor determine the magnetic flux density.

Solution

Now $e = B \times l \times v \sin \theta$, thus:

$$B = \frac{e}{l \times v \sin \theta} = \frac{2}{0.5 \times 50 \sin 45°}$$

$$= \frac{2}{25 \times 0.707} = \frac{2}{17.7} = 0.11$$

Hence the magnetic flux density will be 0.11 T.

5.5.4 Light

A photocell uses photovoltaic conversion to convert light into electricity. By doping pure silicon with a small amount of different impurity elements it can be made into either N- or P-type material. A photocell consists of two interacting layers of silicon: the N-layer at the top with an upper conductor and the P-layer with a bottom conductor. Where the two layers meet an internal electrical field exists. When light hits the solar cell, a negatively charged electron is released and a positive hole remains. When such an electron-hole pair is created near the internal electrical field, the two become separated and the P-layer becomes positively charged whilst the N-layer becomes negatively charged. Thus a small voltage is produced and a current will flow when the photocell is connected to an external circuit. As more light hits the photocell, more electrons will be released and thus more voltage and current will be produced. The photovoltaic process continues as long as light hits the cell.

Figure 5.11   Photoconductive resistor (LDR).

Other devices are photoconductive rather than photovoltaic (see Figure 5.11). In other words, whilst they do not generate electric current by themselves, their ability to conduct an electric current (i.e. their *conductivity*) depends upon the amount of incident light present. Both photovoltaic and photoconductive devices are used in aircraft. A particular application of worth noting is that of smoke detection in which a light beam and a photoelectric device are mounted in a light-proof chamber through which any smoke that may be present can pass (see Figure 5.12).

5.5.5 Thermoelectric cells

When two lengths of dissimilar metal wires (such as iron and constantan) are connected at both ends to form a complete electric circuit as shown in Figure 5.13, a small e.m.f. is generated whenever a temperature difference exists between the two junctions. This is now known as the *thermoelectric effect*. Because the two junctions are at different temperatures, one is referred to as the *hot junction*, whilst the other is referred to as the *cold junction*. The whole device is called a *thermocouple* and the small voltage that it generates increases as the difference in temperature between the hot junction and the cold junction increases. We will return to this topic in Section 5.6.8.

5.5.6 Pressure (piezoelectric) cells

Some crystalline materials, such as quartz, suffer mechanical deformation when an electric charge is applied across opposite faces of a crystal of the material. Conversely, a charge will be developed across the faces of a quartz crystal when it

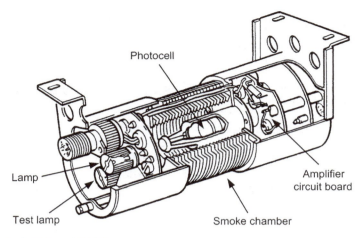

Figure 5.12 An aircraft smoke detector.

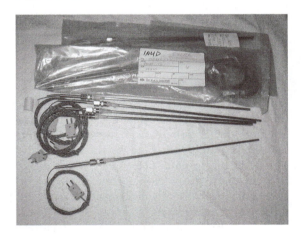

Figure 5.13 A thermocouple.

is mechanically deformed. This phenomenon is known as the "piezoelectric effect" and it has a number of important applications in the field of electronics, including the basis of a device that will convert variations in pressure to a variation in voltage. Such a device is thus able to sense the amount of strain present in a mechanical component such as a beam or strut.

Quartz is a crystalline material that is based on both silicon and oxygen (silicon dioxide). The quartz crystals used in pressure sensors usually consist of one or more thin slices of quartz onto the opposite faces of which film electrodes of gold or silver are deposited. The entire assembly is then placed in a hermetically sealed enclosure with a diaphragm at one end which is mechanically connected to the structural member.

Whilst quartz crystals occur quite naturally, they can also be manufactured to ensure consistency both in terms of physical properties and supply. The growing of quartz crystals simply involves dissolving quartz from small chips and allowing the quartz to grow on prepared seeds. These encompass a batch process that requires about 21 days to produce crystals of the required crystals.

The quartz chips are dissolved in sodium hydroxide solution during which temperatures are maintained above the critical temperature of the solution. The growth process of the quartz is controlled by a two-zone temperature system such that the higher temperature exists in the dissolving zone and the lower in the growth zone. In the actual manufacturing process, the quartz chips (or "nutrients") are placed in the bottom of a long vertical steel autoclave that is specifically designed to withstand very high temperatures and pressures (much like the barrel of a large gun).

Key point

Although there are so many different applications, remember that all electrons are the same, with identical charge and mass. Whether the electron flow results from a battery, rotary generator or photoelectric device the end result is the same, a movement of electrons in a conductor.

Test your understanding 5.5

1. Static electricity can be produced by rubbing two materials together in order to separate _____ and _____ charges.

2. Static electricity that builds up on the aircraft's external structure can be dissipated by means of a _____.

3. The two materials used in a conventional dry cell are _____ and _____.

4. In a lead–acid cell the electrolyte is dilute _____.

5. When a conductor moves in a magnetic field an _____ will be _____ in it.

6. A photocell uses _____ conversion to produce electric current from light.

7. When light hits the surface of a photocell, a _____ charged electron is released and a _____ charged hole remains.

8. A typical application of photoelectricity in an aircraft is a _____.

9. A junction of dissimilar metal wires that generates a small voltage when heated is known as a _____.

10. When a quartz crystal is deformed a small _____ will appear across opposite faces of the crystal. This is often referred to as the _____ effect.

5.6 DC sources of electricity

Syllabus

Construction and basic chemical action of primary cells, secondary cells, lead–acid cells, nickel–cadmium (Ni-Cd) cells, other alkaline cells; Cells connected in series and parallel; Internal resistance and its effect on a battery; Construction, materials and operation of thermocouples; Operation of photocells.

Knowledge level key

A	B1	B2
1	2	2

DC is the current that flows in one direction only (recall from Section 5.4.3 that conventional current flows from positive to negative whilst electrons travel in the opposite direction, from negative to positive). The most commonly used method of generating DC is the electrochemical cell. In this section we shall describe the basic principles of cells and batteries. We shall also be looking at two other important devices that generate electric current, thermocouples and photocells.

5.6.1 Cells and batteries

A *cell* is a device that produces a charge when a chemical reaction takes place. When several cells are connected together they form a *battery*. Most aircraft have several batteries, the most important of which are the *main aircraft batteries*. The two principal functions of the main aircraft batteries are:

- to emergency electrical power in case of electrical generation system failure in flight;
- to provide an autonomous source of electrical power for starting engines or APU on the ground or in flight.

Engine or APU starting requires high initial peak current (sometimes over 1000 A) to overcome mechanical inertia followed by high current discharge (hundreds of amperes) during a time interval of typically 30 s. Several successive attempts, which progressively deplete capacity, may be required, but because the duration is short, starting usually determines what *power* capacity the battery should have. Conversely, emergency loads usually determine what *energy* the battery should have.

The precise configuration of aircraft batteries depends both on aircraft complexity and airworthiness requirements. For example, one or more batteries may be dedicated to supporting essential systems (such as avionics) for 30–60 min without the voltage falling below a minimum level (typically 18 V). Furthermore, one battery may be dedicated to starting whilst the other supports essential equipment during engine start-up. When required, both batteries can then be connected in parallel to support emergency loads. When alternative emergency power generation is available, such as a *ram air turbine* (RAT), the battery may only be needed for a few minutes during RAT deployment or during a final (low-speed) approach to a runway.

Having briefly set the scene, we will now look briefly at the nature of cells and batteries

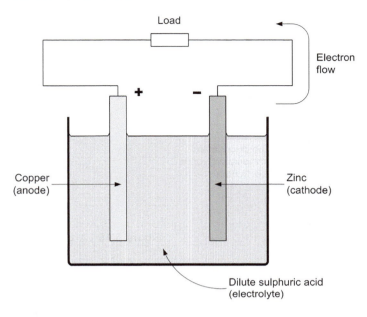

Figure 5.14   A simple primary cell.

but, before we do, we need to introduce you to the concept of *primary* and *secondary* cells. Primary cells produce electrical energy at the expense of the chemicals from which they are made and once these chemicals are used up, no more electricity can be obtained from the cell. In secondary cells, the chemical action is reversible. This means that the chemical energy is converted into electrical energy when the cell is *discharged* whereas electrical energy is converted into chemical energy when the cell is being *charged*.

Key point

In a primary cell the conversion of chemical energy to electrical energy is irreversible and so these cells cannot be recharged. In secondary cells, the conversion of chemical energy to electrical energy is reversible. Thus these cells can be recharged and reused many times.

5.6.2 Primary cells

All cells consist of two electrodes which are dissimilar metals, or carbon and a metal, which are placed into an electrolyte. One of the simplest examples of a primary cell is the voltaic type. This cell (Figure 5.14) consists of a plate of zinc forming the negative electrode, a plate of copper forming the positive electrode and

dilute sulphuric acid as the electrolyte. The negative electrode is known as the cathode and the positive electrode is known as the anode.

When the electrodes are connected outside the cell so that a circuit is completed, a current flows from the copper electrode, through the external circuit to the zinc and from the zinc to the copper, through the electrolyte in the cell.

One of the problems with the voltaic cell is that it only works for a short time before a layer of hydrogen bubbles builds up on the positive copper electrode, drastically reducing the e.m.f. of the cell and increasing its internal resistance. This effect is called *polarization*. The removal of this hydrogen layer from the copper electrode may be achieved by mechanical brushing or adding a depolarizer such as potassium dichromate to the acid solution. The removal of this hydrogen layer is known as *depolarization*.

If the zinc electrode is not 100% pure, which for cost reasons is often the case, then the impurities react with the zinc and the sulphuric acid to produce miniature cells on the surface of the zinc electrode. This reaction takes place in the voltaic cell, irrespective of whether a current is being taken from the cell or not. This *local action*, as it is known, is wasteful and may be eliminated by coating the zinc plate with mercury, or by using

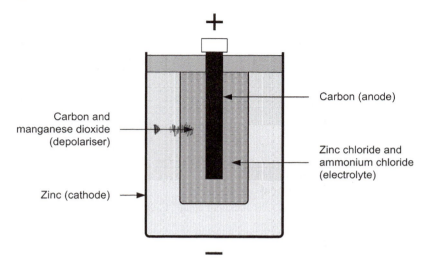

Figure 5.15 A zinc–carbon cell.

the more expensive pure zinc. The e.m.f. of a cell of this type is approximately equal to 1.0 V.

A second type of primary cell is the dry cell. In this type of cell instead of using a dilute acid electrolyte we use ammonium chloride in thick paste form. In one variant of this cell the positive electrode is a centrally positioned carbon rod (Figure 5.15) while the negative electrode is the zinc outer casing the cell. Carbon and manganese dioxide act as the depolarizing agent that surrounds the carbon electrode. This type of cell is often used to power torches and other portable equipment and each cell has an e.m.f. of approximately 1.5 V.

5.6.3 Lead–acid cells

The lead–acid cell is one of the most common secondary cells. In this type of cell, the electrical energy is initially supplied from an external source and converted and stored in the cell as chemical energy. This conversion of energy is reversible and when required this stored chemical energy can be released as a direct electric current. This process of storage leads to the alternative name for this type of cell, the lead–acid *accumulator*.

The manufacture of this cell is quite complex. The positive plate consists of a grid of lead and antimony filled with lead peroxide (Figure 5.16).

The negative plate uses a similar grid, but its open spaces are filled with spongy lead. Thus the cells are made up of a group of positive plates, joined together and interlaced between a stack of negative plates. Porous separators keep the plates apart and hold a supply of electrolyte in contact with the active materials. The electrolyte consists of a mixture of sulphuric acid and water (i.e. dilute sulphuric acid) which covers the plates and takes an active part in the charging and discharging of the cell.

A fully charged lead–acid cell has an e.m.f. of approximately 2.2 V, but when in use this value falls rapidly to about 2.0 V. In the fully charged condition the negative plate is spongy lead and the positive plate is lead peroxide. In the discharged condition, where the e.m.f. is about 1.8 V, the chemical action of the cell converts both positive and negative plates into a lead sulphate mix. When discharged, the cell may then be recharged from an external source and made ready for further use. The condition of this type of cell may be checked by measuring the *relative density* of the electrolyte. In the fully charged condition this will be around 1.26, while in the discharged condition it drops to around 1.15. This type of cell, when joined together as a battery, has many commercial uses, the most familiar of which is as a motor vehicle battery.

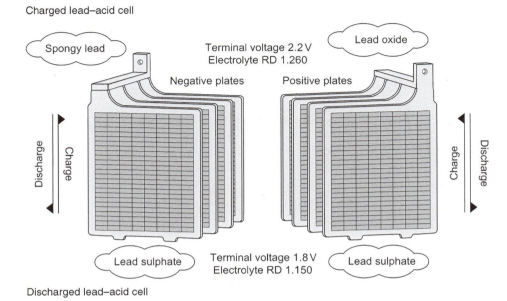

Charged lead–acid cell

Spongy lead

Lead oxide

Terminal voltage 2.2 V
Electrolyte RD 1.260

Negative plates   Positive plates

Discharge   Charge

Charge   Discharge

Lead sulphate

Lead sulphate

Terminal voltage 1.8 V
Electrolyte RD 1.150

Discharged lead–acid cell

Figure 5.16 A lead–acid cell.

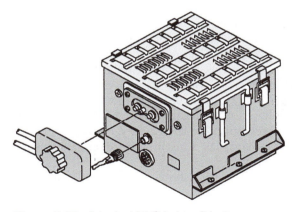

Figure 5.17 A typical Ni-Cd aircraft battery.

5.6.4 Ni-Cd cells

Ni-Cd batteries are now increasingly used in aircraft because they offer a long service life coupled with excellent performance and reliability (Figure 5.17). Like their lead–acid counterparts, Ni-Cd batteries consist of a number of series-connected cells each comprising a set of positive and negative plates, separators, electrolyte, cell vent and a cell container.

The positive plates of a Ni-Cd battery comprise a porous plate on which nickel hydroxide has been deposited. The negative plates are made from similar plates on which cadmium hydroxide has been deposited. A continuous strip of porous plastic separates these two sets of plates from each other. The electrolyte used in a Ni-Cd is a 30% solution (by weight) of potassium hydroxide (KOH) in distilled water. The specific gravity (relative density) of the electrolyte remains between 1.24 and 1.30 at room temperature and (unlike the lead–acid battery) no appreciable change occurs in the specific gravity of the electrolyte during charge and discharge. For this reason it is not possible to infer much about the state of a Ni-Cd battery from a measurement of the specific gravity check of the electrolyte. However, as with a lead–acid battery, the electrolyte level should be maintained just above the tops of the plates.

When a charging current is applied to a Ni-Cd battery, the negative plates lose oxygen and begin to form metallic cadmium. At the same time, the active material of the positive plates, nickel hydroxide, becomes more highly oxidized. This process continues while the charging current is applied or until all the oxygen is removed from the negative plates and only cadmium remains. Toward the end of the charging

Table 5.2

Cell type	Primary or secondary	Wet or dry	Positive electrode	Negative electrode	Electrolyte	Output voltage (nominal)(V)	Notes
Zinc–carbon (Leclanché)	Primary	Dry	Zinc	Carbon	Ammonium chloride	1.5	Used for conventional AA, A, B and C type cells
Alkaline dry cells	Primary	Dry	Manganese dioxide	Zinc	KOH	1.5	
	Secondary	Dry	Manganese dioxide	Zinc	KOH	1.5	Can be recharged about 50 times
Lead–acid	Secondary	Wet	Lead peroxide	Lead	Sulphuric acid	2.2	For general purpose 6, 12 and 24 V batteries
Ni-Fe	Secondary	Wet	Nickel	Iron	Potassium and lithium hydroxides	1.4	Rugged construction for industrial use
Ni-Cd	Secondary	Dry	Nickel	Cadmium hydroxide	KOH	1.2	Can be recharged about 400 times

cycle (and when the cells are overcharged) the water in the electrolyte decomposes into hydrogen (at the negative plates) and oxygen (at the positive plates).

The time taken to charge a battery will depend partly on the charging voltage and partly on the temperature. It is important to note that to *completely* charge a Ni-Cd battery, some gassing, however slight, must take place. Furthermore, when the battery is fully charged the volume of the electrolyte will be at its greatest thus water should only be added in order to bring the electrolyte to its correct level when the battery is fully charged and allowed to rest for a period of several hours. During subsequent discharge, the plates will absorb a quantity of the electrolyte and the level of the electrolyte will fall as a result. The useful life of a Ni-Cd battery depends largely on how well it is maintained and whether or not it is charged and discharged regularly.

5.6.5 Other alkaline cells

Other types of alkaline cells include the nickel–iron (Ni-Fe) cell that is sometimes also referred to as the "Edison battery" after its inventor, Thomas Edison. The positive plates of the Ni-Fe

cell are made from nickel whilst the negative plates are made from iron. As with the Ni-Cd cell, the electrolyte is a solution of KOH having a specific gravity (relative density) of about 1.25. The Ni-Fe cell produces hydrogen gas during charging and has a terminal voltage of approximately 1.15 V. The battery is well suited to heavy-duty industry applications and has a useful life of approximately 10 years.

The Table 5.2 shows the principal characteristics of various common types of cell.

5.6.6 Cells connected in series and parallel

In order to produce a battery, individual cells are usually connected in series with one another, as shown in Figure 5.18(a). Cells can also be connected in parallel (see Figure 5.18(b)).

In the series case, the voltage produced by a battery with n cells will be n times the voltage of one individual cell (assuming that all of the cells are identical). Furthermore, each cell in the battery will supply the same current.

In the parallel case, the current produced by a battery of n cells will be n times the current

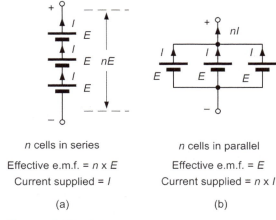

Figure 5.18 Cells connected in series and in parallel: (a) cells connected in series and (b) cells connected in parallel.

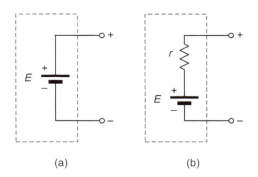

Figure 5.19 Sources of e.m.f.: (a) a perfect source of e.m.f. and (b) a practical source of e.m.f.

produced by an individual cell (assuming that all of the cells are identical). Furthermore, the voltage produced by the battery will be the same as the voltage produced by an individual cell.

Example 5.15

How many individual series-connected zinc–carbon cells are there in a battery that produces a nominal output of 9 V?

Solution

If you refer to the Table 5.2 you will find that the nominal output voltage of a zinc–carbon cell is 1.5 V. Dividing 9 V by 1.5 V will give you the number of cells required (i.e. the value of n).

$$n = \frac{9 \, \text{V}}{1.5 \, \text{V}} = 6$$

Hence the battery will require six zinc–carbon cells connected in series.

5.6.7 Internal resistance of a cell

Every practical source of e.m.f. (e.g. a cell, battery or power supply) has some internal resistance. This value of resistance is usually extremely small but, even so, it has the effect of limiting the amount of current that the source can supply and also reducing the e.m.f. produced by the source when it is connected to a

load (i.e. whenever we extract a current from it). The idea of an "invisible" internal resistance can be a bit confusing so, when we need to take it into account we show it as a fixed resistor connected in series with a "perfect" voltage source. To clarify this point, Figure 5.19(a) shows a "perfect" source of e.m.f. whilst Figure 5.19(b) shows a practical source of e.m.f. It is important to note that the *internal resistance, r*, is actually *inside* the cell (or battery) and is not actually something that we can measure with an ohmmeter!

Key point

Every practical source of e.m.f. (e.g. a cell, battery or power supply) has some internal resistance which limits the amount of current that it can supply. When we need to take the internal resistance of a source into account (e.g. in circuit calculations) we show the source as a perfect voltage source connected in series with its internal resistance.

5.6.8 Thermocouples

We have already briefly mentioned the thermocouple in Section 5.5.5. The output of a thermocouple depends on two factors:

• the difference in temperature between the hot junction and the cold junction (note that any change in either junction temperature will affect the e.m.f. produced by the thermocouple);
• the metals chosen for the two wires that make up the thermocouple.

It is also worth noting that a thermocouple is often pictured as two wires joined at one end, with the other ends not connected; it is important to remember that it is not a true thermocouple unless the other end is also connected! In many practical applications the cold junction is formed by the load to which a hot junction is connected. In measuring applications this load can be a measuring instrument, such as a sensitive voltmeter.

The polarity of the e.m.f. generated is determined by (a) the particular metal or alloy pair that is used (such as iron and constantan) and (b) the relationship of the temperatures at the two junctions.

If the temperature of the cold junction is maintained constant, or variations in that temperature are compensated for, then the net e.m.f. is a function of the hot junction temperature. In most installations, it is not practical to maintain the cold junction at a constant temperature. The usual standard temperature for the junction (referred to as the *reference junction*) is 32°F (0°C). This is the basis for published tables of e.m.f. versus temperature for the various types of thermocouples.

Note that, where additional metals are present in the thermocouple circuit, they will have no effect on the e.m.f. produced provided they are all maintained at the same temperature.

Junction materials	Output voltage (μV/°C)	Temperature range (°C)
Iron and constantan	41	−40 to +750
Chromel and alumel	41	−200 to +1200
Chromel and constantan	68	−270 to +790
Platinum and rhodium	10	+100 to +1800

Example 5.16

The temperature difference between the hot and cold junctions of an iron–constantan thermocouple is 250°C. What voltage will be produced by the thermocouple?

Solution

The voltage produced by the thermocouple will be given by:

$$41\,\mu V \times 250°C = 10,250\,\mu V = 10.25\,mV$$

Example 5.17

The hot junction of a platinum–rhodium thermocouple is suspended in a gas turbine exhaust chamber. If the cold (reference) junction is maintained at a temperature of 30°C and the thermocouple produces an output of 9.8 mV, determine the temperature inside the exhaust chamber.

Solution

The temperature difference between the hot and cold (reference) junctions will be given by $(t - 30°C)$ where t is the temperature (in °C) inside the exhaust chamber.

Now $9.8\,mV = 10\,\mu V \times (t - 30°C)$
From which:

$$t = \frac{9.8\,mV}{10\,\mu V/°C} + 30°C = 980°C + 30°C$$

$$= 1010°C$$

5.6.9 Photocells

We first met photocells in Section 5.5.4. The output of a photocell depends on the amount of light that falls on the surface of the cell. As more light hits the photocell, more electrons will be released and thus more voltage will be produced.

In order to generate a useful voltage and current, photocells are usually connected in large series and parallel arrays. However, they are still rather inefficient in terms of energy conversion (typically only 10–15% of the incident light energy is converted to useful electrical energy). Cells constructed from indium phosphide and gallium arsenide are, in principle, more efficient but conventional silicon-based cells are generally less costly.

Solar photovoltaic cells have long been used to provide electric power for spacecraft and other devices that have no access to a power source. However, recent developments have driven costs down to the point where the silicon photocells are being used more and more as a replacement for conventional energy sources (such as dry cells and lead–acid batteries). Photocells also make it possible for us to "top-up" the charge in a secondary cell battery that can later be used to maintain an electrical supply during the hours of darkness.

Test your understanding 5.6

1. A cell is a device that produces a _____ when a _____ _____ takes place.
2. _____ cells produce electrical energy at the expense of the chemical from which they are made.
3. The negative electrode of a cell is known as the _____.
4. Name the material used for the positive electrode of a typical dry cell.
5. The electrolyte of a lead–acid cell consists of dilute _____ _____.
6. The e.m.f. of a fully charged lead–acid cell is approximately _____V.
7. The e.m.f. of a fully charged Ni-Cd cell is approximately _____V.
8. The relative density of the electrolyte in a fully charged lead–acid cell is approximately _____.
9. The relative density of the electrolyte in a fully discharged lead–acid cell is approximately _____.
10. Explain briefly how a thermocouple operates.

5.7 DC circuits

Syllabus
Ohm's law, Kirchhoff's voltage and current laws; Calculations using the above laws to find resistance, voltage and current; Significance of the internal resistance of a supply.

Knowledge level key

A	B1	B2
1	2	2

DC circuits are found in every aircraft. An understanding of how and why these circuits

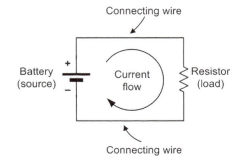

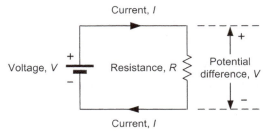

Figure 5.20 A simple DC circuit consisting of a battery (source) and resistor (load).

work is an essential prerequisite to understand more complex circuits. The most basic DC circuit uses only two components: a cell (or battery) acting as a source of e.m.f. and a resistor (or *load*) through which a current is passing. These two components are connected together with wire conductors in order to form a completely closed circuit as shown in Figure 5.20.

5.7.1 Current, voltage and resistance

We have already said that electric current is the name given to the flow of electrons (or negative charge carriers). The ability of an energy source (e.g. a battery) to produce a current within a conductor may be expressed in terms of e.m.f. Whenever an e.m.f. is applied to a circuit a p.d. exists. Both e.m.f. and p.d. are measured in volts (V). In many practical circuits there is *only one* e.m.f. present (the battery or supply) whereas a p.d. will be developed across *each* component present in the circuit.

The *conventional flow* of current in a circuit is from the point of more positive potential to the point of greatest negative potential (note that electrons move in the opposite direction!). DC results from the application of a direct e.m.f. (derived from batteries or a DC supply, such as a generator or a "power pack"). An essential characteristic of such supplies is that the applied e.m.f. does not change its polarity (even though its value might be subject to some fluctuation).

For any conductor, the current flowing is directly proportional to the e.m.f. applied. The current flowing will also be dependent on the physical dimensions (length and cross-sectional area) and material of which the conductor is composed. The amount of current that will flow in a conductor when a given e.m.f. is applied is inversely proportional to its resistance. Resistance, therefore, may be thought of as an "opposition to current flow"; the higher the resistance the lower the current that will flow (assuming that the applied e.m.f. remains constant).

5.7.2 Ohm's law

Provided that temperature does not vary, the ratio of p.d. across the ends of a conductor to the current flowing in the conductor is a constant. This relationship is known as Ohm's law and it leads to the relationship:

$$\frac{V}{I} = \text{a constant} = R$$

where V is the p.d. (or voltage drop) in volts (V), I is the current in amperes (A) and R is the resistance in ohms (Ω) (see Figure 5.21).

The formula may be arranged to make V, I or R the subject, as follows:

$$V = I \times R \quad I = \frac{V}{R} \quad \text{and} \quad R = \frac{V}{I}$$

The triangle shown in Figure 5.22 should help you remember these three important relationships. It is important to note that, when performing calculations of currents, voltages and resistances in practical circuits, it is seldom necessary to work with an accuracy of better than $\pm 1\%$ simply because component tolerances are invariably somewhat greater than this. Furthermore, in calculations involving Ohm's law, it is sometimes convenient to work in units of $k\Omega$ and mA (or $M\Omega$ and μA) in which case p.d. will be expressed directly in volts.

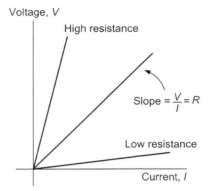

Figure 5.21 Relationship between voltage, V, current, I, and resistance, R.

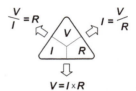

Figure 5.22 The Ohm's law triangle.

Example 5.18

A current of 100 mA flows in a 56 Ω resistor. What voltage drop (p.d.) will be developed across the resistor?

Solution

Here we must use $V = I \times R$ and ensure that we work in units of volts (V), amperes (A) and ohms (Ω).

$$V = I \times R = 0.1\,\text{A} \times 56\,\Omega = 5.6\,\text{V}$$

(note that 100 mA is the same as 0.1 A)

Hence a p.d. of 5.6 V will be developed across the resistor.

Example 5.19

A 18 Ω resistor is connected to a 9 V battery. What current will flow in the resistor?

Solution

Here we must use $I = \dfrac{V}{R}$

where $V = 9$ V and $R = 18\,\Omega$:

$$I = \frac{V}{R} = \frac{9\,\text{V}}{18\,\Omega} = \frac{1}{2}\,\text{A} = 0.5\,\text{A} = 500\,\text{mA}$$

Hence a current of 500 mA will flow in the resistor.

Example 5.20

A voltage drop of 15 V appears across a resistor in which a current of 1 mA flows. What is the value of the resistance?

Solution

Here we must use $R = \dfrac{V}{I}$
where $V = 15$ V and $I = 1$ mA $= 0.001$ A

$$R = \frac{V}{I} = \frac{15\,\text{V}}{0.001\,\text{A}} = 15,000\,\Omega = 15\,\text{k}\Omega$$

Note that it is often more convenient to work in units of mA and V which will produce an answer directly in kΩ, i.e.:

$$R = \frac{V}{I} = \frac{15\,\text{V}}{1\,\text{mA}} = 15\,\text{k}\Omega$$

5.7.3 Kirchhoff's current law

Used on its own, Ohm's law is insufficient to determine the magnitude of the voltages and currents present in complex circuits. For these circuits we need to make use of two further laws: Kirchhoff's current law and Kirchhoff's voltage law.

Kirchhoff's current law states that the algebraic sum of the currents present at a junction (or *node*) in a circuit is zero (see Figure 5.23).

Example 5.21

Determine the value of the missing current, I, shown in Figure 5.24.

Solution

By applying Kirchhoff's current law in Figure 5.24, and adopting the convention that currents flowing towards the junction are positive, we can say that:

$$+2\,\text{A} + 1.5\,\text{A} - 0.5\,\text{A} + I = 0$$

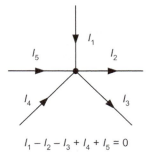

$$I_1 - I_2 - I_3 + I_4 + I_5 = 0$$

Convention:
Current flowing towards the junction is positive (+)
Current flowing away from the junction is negative (−)

Figure 5.23   Kirchhoff's current law.

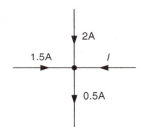

Figure 5.24

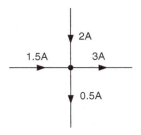

Figure 5.25   In Example 5.21 the unknown current is flowing *away* from the junction.

Note that we have shown I as positive. In other words we have assumed that it is flowing towards the junction.

Re-arranging gives:

$$+3\,\text{A} + I = 0$$

Thus

$$I = -3\,\text{A}$$

The negative answer tells us that I is actually flowing in the other direction, i.e. away from the junction (see Figure 5.25).

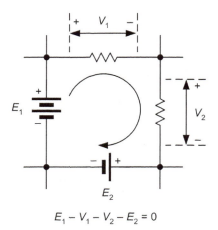

$$E_1 - V_1 - V_2 - E_2 = 0$$

Convention:
Move clockwise around the circuit starting with the positive terminal of the leargest e.m.f.
Voltages acting in the same sense are positive (+)
Voltages acting in the opposite sense are negative (−)

Figure 5.26   Kirchhoff's voltage law.

> ### Key point
>
> If the Kirchhoff's current law equation is a little puzzling, just remember that the sum of the current flowing towards a junction must always be equal to the sum of the current flowing away from it!

5.7.4 Kirchhoff's voltage law

Kirchhoff's second voltage law states that the algebraic sum of the potential drops present in a closed network (or *mesh*) is zero (see Figure 5.26).

Example 5.22

Determine the value of the missing voltage, V, shown in Figure 5.27.

Solution

By applying Kirchhoff's voltage law in Figure 5.27, starting at the positive terminal of the largest e.m.f. and moving clockwise around the closed network, we can say that:

$$+24\,V + 6\,V - 12\,V - V = 0$$

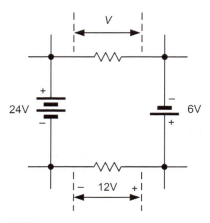

Figure 5.27

Note that we have shown V as positive. In other words we have assumed that the more positive terminal of the resistor is the one on the left.

Re-arranging gives:

$$+24\,V - V + 6\,V - 12\,V = 0$$

From which:

$$+18\,V - V = 0$$

Thus:

$$V = +18\,V$$

The positive answer tells us that we have made a correct assumption concerning the polarity of the voltage drop, V, i.e. the more positive terminal is on the left.

> ### Key point
>
> If Kirchhoff's voltage law equation is a little puzzling, just remember that, in a closed circuit, the sum of the voltage drops must be equal to the sum of the e.m.f. present. Note, also, that it is important to take into account the polarity of each voltage drop and e.m.f. as you work your way around the circuit.

5.7.5 Series and parallel circuit calculations

Ohm's law and Kirchhoff's law can be combined to solve more complex series and parallel circuits. Before we show you how this is done, however, it is important to understand what we mean by "series" and "parallel" circuit!

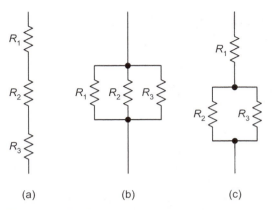

(a)   (b)   (c)

Figure 5.28 Series and parallel circuits: three resistors connected in (a) series, (b) parallel, and (c)series and parallel circuit.

Figure 5.28 shows three circuits, each containing three resistors, R_1, R_2 and R_3.

In Figure 5.28(a), the three resistors are connected one after another. We refer to this as a *series circuit*. In other words the resistors are said to be connected *in series*. It is important to note that, in this arrangement, *the same current flows through each resistor*.

In Figure 5.28(b), the three resistors are all connected across one another. We refer to this as a *parallel circuit*. In other words the resistors are said to be connected *in parallel*. It is important to note that, in this arrangement, *the same voltage appears each resistor*.

In Figure 5.28(c), we have shown a mixture of these two types of connection. Here we can say that R_1 is connected in series with the parallel combination of R_2 and R_3. In other words, R_2 and R_3 are connected *in parallel* and R_2 is connected *in series* with the parallel combination.

We shall look again at the series and parallel connection of resistors in Section 5.8 but, before we do that, we shall put our new knowledge to good use by solving some more complicated circuits!

Example 5.23

Figure 5.29 shows a simple battery test circuit which is designed to draw a current of 2 A from

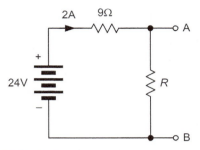

Figure 5.29 Battery test circuit.

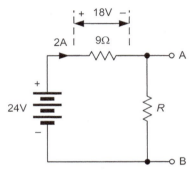

Figure 5.30 Using Ohm's law to find the voltage dropped across the 9 Ω resistor.

a 24 V DC supply. The two test points, A and B, are designed for connecting a meter. Determine:

(a) the voltage that appears between terminals A and B (without the meter connected);
(b) the value of resistor, R.

Solution

(a) We need to solve this problem in several small stages. Since we know that the circuit draws 2 A from the 24 V supply we know that this current must flow both through the 9 Ω resistor and through R (we hope that you have spotted that these two components are connected *in series*!).

We can determine the voltage drop across the 9 Ω resistor by applying Ohm's law (Figure 5.30):

$$V = I \times R = 2\,\text{A} \times 9\,\Omega = 18\,\text{V}$$

Next we can apply Kirchhoff's voltage law in order to determine the voltage drop, V, that appears across R (i.e. the potential drop between terminals A and B) (Figure 5.31).

$$+24\,V - 18\,V - V = 0$$

From which:

$$V = +6\,V$$

(b) Finally, since we now know the voltage, V, and current, I, that flows in R, we can apply Ohm's law again in order to determine the value of R (Figure 5.32):

$$R = \frac{V}{I} = \frac{6\,V}{2\,A} = 3\,\Omega$$

Hence the voltage that appears between A and B will be 6 V and the value of R is 3 Ω.

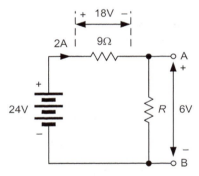

Figure 5.31 Using Kirchhoff's voltage law to find the voltage that appears between terminals A and B.

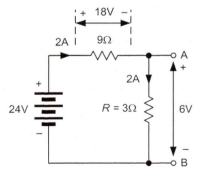

Figure 5.32 Using Ohm's law to find the value of R.

Example 5.24

For the circuit shown in Figure 5.33, determine:

(a) the voltage dropped across each resistor,
(b) the current drawn from the supply,
(c) the supply voltage.

Solution

(a) Once again, we need to solve this problem in several small stages. Since we know the current flowing in the 6 Ω resistor, we will start by finding the voltage dropped across it using Ohm's law (Figure 5.34):

$$V = I \times R = 0.75\,A \times 6\,\Omega = 4.5\,V$$

(b) Now the 4 Ω resistor is connected in parallel with the 6 Ω resistor. Hence the voltage drop across the 4 Ω resistor is also 4.5 V. We can now determine the current flowing in the 4 Ω resistor using Ohm's law (Figure 5.35):

$$I = \frac{V}{R} = \frac{4.5\,V}{4\,\Omega} = 1.125\,A$$

Now, since we know the current in both the 4 and 6 Ω resistors, we can use Kirchhoff's

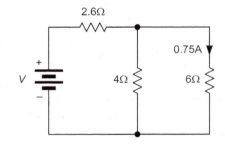

Figure 5.33

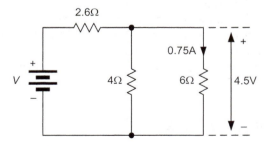

Figure 5.34 Using Ohm's law to find the voltage dropped across the 6 Ω resistor.

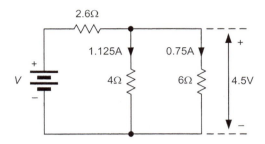

Figure 5.35 Using Ohm's law to find the current flowing in the 4 Ω resistor.

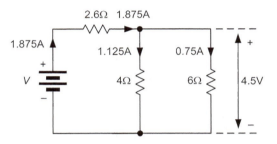

Figure 5.36 Using Kirchhoff's current law to find the current in the 2.6 Ω resistor.

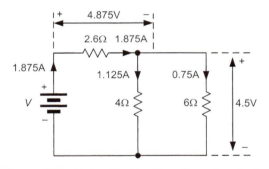

Figure 5.37 Using Kirchhoff's voltage law to find the voltage drop across the 2.6 Ω resistor.

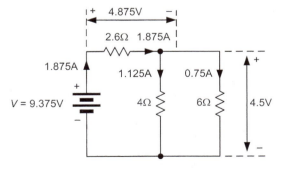

Figure 5.38 Using Kirchhoff's voltage law to find the supply voltage.

law to find the current, I, in the 2.6 Ω resistor (Figure 5.36):

$$+I - 0.75\,\text{A} - 1.125\,\text{A} = 0$$

From which:

$$I = 1.875\,\text{A}$$

Since this current flows through the 2.6 Ω resistor it will also be equal to the current taken from the supply.

Next we can find the voltage drop across the 2.6 Ω resistor by applying Ohm's law (Figure 5.37):

$$V = I \times R = 1.875\,\text{A} \times 2.6\,\Omega = 4.875\,\text{V}$$

(c) Finally, we can apply Kirchhoff's voltage law in order to determine the supply voltage, V (Figure 5.38):

$$+V - 4.875\,\text{V} - 4.5\,\text{V} = 0$$

From which:

$$V = +9.375\,\text{V}$$

Hence the supply voltage is 9.375 V.

5.7.6 Internal resistance

We first met internal resistance in Section 5.6.7. Since we now know how to solve problems involving voltage, current and resistance it is worth illustrating the effect of internal resistance with a simple example. Figure 5.39 shows what happens as the internal resistance of a battery increases. In Figure 5.39(a) a "perfect" 10 V battery supplies a current of 1 A to a 10 Ω load. The output voltage of the battery (when *on-load*) is, as you would expect, simply 10 V. In Figure 5.39(b) the battery has a relatively small value of internal resistance (0.1 Ω) and this causes the output current to fall to 0.99 A and the output voltage to be reduced (as a consequence) to 9.9 V. Figure 5.39(c) shows the effect of the internal resistance rising to 1 Ω. Here the output current has fallen to 0.91 A and the output voltage to 9.1 V. Finally, taking a more extreme case, Figure 5.39(d) shows the effect of

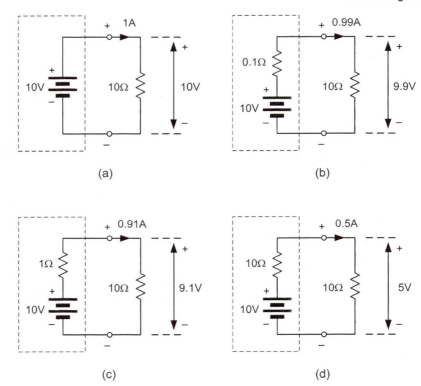

Figure 5.39   Effect of internal resistance.

the internal resistance rising to 10 Ω. In this situation, the output current is only 0.5 A and the voltage applied to the load is a mere 5 V!

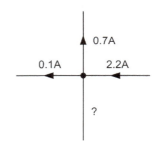

Figure 5.40

Key point

Internal resistance is quite important in a number of applications. When a battery goes "flat" it is simply that its internal resistance has increased to a value that begins to limit the output voltage when current is drawn from the battery.

Test your understanding 5.7

1. Kirchhoff's current law states that the _____ _____ of the currents present at a junction in a circuit is _____.
2. Determine the unknown current in Figure 5.40.
3. Determine the unknown current in Figure 5.41.
4. Kirchhoff's voltage law states that the _____ _____ of the potential drops present in a closed network is _____.

5. Determine the unknown voltage in Figure 5.42.
6. In Figure 5.43 which two resistors are connected in parallel?
7. In Figure 5.44 which two resistors are connected in series?
8. Determine the voltage drop across the 10 Ω resistor shown in Figure 5.45.
9. The _____ resistance of a battery _____ as it becomes exhausted.
10. Determine the value of r in Figure 5.46.

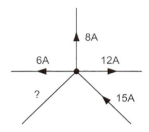

Figure 5.41

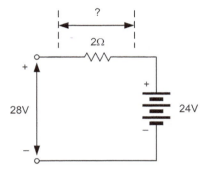

Figure 5.42

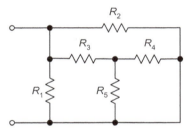

Figure 5.43

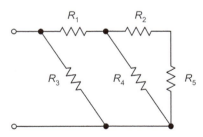

Figure 5.44

5.8 Resistance and resistors

Syllabus
(a) Resistance and affecting factors; Specific resistance; Resistor colour code, values and

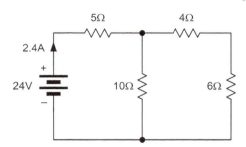

Figure 5.45

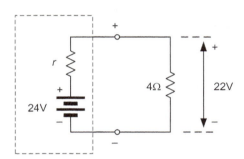

Figure 5.46

tolerances, preferred values, wattage ratings; Resistors in series and parallel; Calculation of total resistance using series, parallel, and series and parallel combinations; Operation and use of potentiometers and rheostats; Operation of Wheatstone bridge.

Knowledge level key

A	B1	B2
–	2	2

Syllabus
(b) Positive and negative temperature coefficient (PTC and NTC, respectively) conductance; Fixed resistors, stability, tolerance and limitations, methods of construction; Variable resistors, thermistors, voltage dependent resistors; Construction of potentiometers and rheostats; Construction of Wheatstone bridge.

Knowledge level key

A	B1	B2
–	1	1

The notion of resistance as opposition to current was discussed in Section 5.7. Resistors provide us with a means of controlling the currents and voltages present in electronic circuits. We also use resistors, as the *loads* simulate the presence of a circuit during testing and as a mean of converting current into a corresponding voltage drop and vice versa.

5.8.1 Specific resistance

The resistance of a metallic conductor is directly proportional to its length and inversely proportional to its area. The resistance is also directly proportional to its specific resistance (or *resistivity*). Specific resistance is defined as the resistance measured between the opposite faces of a cube having sides of 1 m.

The resistance, R, of a conductor is thus given by the formula:

$$R = \frac{\rho l}{A}$$

where R is the resistance (in Ω), ρ is the specific resistance (in Ωm), l is the length (in m) and A is the area (in m^2).

The following table shows the specific resistance of various common metals:

Metal	Specific resistance (Ωm, at 20°C)
Silver	1.626×10^{-8}
Copper (annealed)	1.724×10^{-8}
Copper (hard drawn)	1.777×10^{-8}
Aluminium	2.803×10^{-8}
Mild steel	1.38×10^{-7}
Lead	2.14×10^{-7}

Example 5.25

A coil consists of an 8 m length of annealed copper wire having a cross-sectional area of 1 mm^2. Determine the resistance of the coil.

Solution

Here we will use $R = \dfrac{\rho l}{A}$

where $l = 8$ m and $A = 1$ mm$^2 = 1 \times 10^{-6}$ m^2

From the table shown earlier we find that the value of specific resistance, ρ, for annealed copper is 1.724×10^{-8} Ωm.

Hence:

$$R = \frac{\rho l}{A} = \frac{1.724 \times 10^{-8} \times 8}{1 \times 10^{-6}}$$

$$= 13.792 \times 10^{-2} = 0.13792 \,\Omega$$

Hence the resistance of the wire will be approximately 0.14 Ω.

Example 5.26

A wire having a specific resistance of 1.6×10^{-8} Ωm, length 20 m and cross-sectional area 1 mm^2 carries a current of 5 A. Determine the voltage drop between the ends of the wire.

Solution

First we must find the resistance of the wire and then we can find the voltage drop.

To find the resistance we use:

$$R = \frac{\rho l}{A} = \frac{1.6 \times 10^{-8} \times 20}{1 \times 10^{-6}}$$

$$= 32 \times 10^{-2} = 0.32 \,\Omega$$

To find the voltage drop we can apply Ohm's law:

$$V = I \times R = 5\,\text{A} \times 0.32\,\Omega = 1.6\,\text{V}$$

Hence a potential of 1.6 V will be dropped between the ends of the wire.

5.8.2 Temperature coefficient of resistance

The resistance of a resistor depends on the temperature. For most metallic conductors resistance increases with temperature and we say that these materials have a PTC. For non-metallic conductors, such as carbon or semiconductor materials such as silicon or germanium, resistance falls with temperature and we say that these materials have a NTC.

The variation of resistance with temperature for various materials is shown in Figure 5.47.

Resistance, R

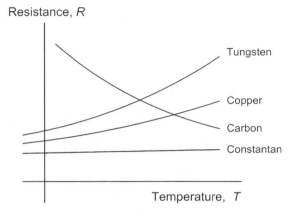

Figure 5.47 Variation of resistance with temperature for various electrical conductors.

The following table shows the temperature coefficient of resistance of various common metals:

Metal	Temperature coefficient of resistance ($°C^{-1}$)
Silver	0.0041
Copper (annealed)	0.0039
Copper (hard drawn)	0.0039
Aluminium	0.0040
Mild steel	0.0045
Lead	0.0040

The resistance of a conductor, R_t, at a temperature, t, can be determined from the relationship:

$$R_t = R_0(1 + \alpha t + \beta t^2 + \gamma t^3 + \cdots)$$

where R_0 is the resistance of the conductor at $0°C$ and α, β and γ are constants. In practice β and γ can usually be ignored and so we can approximate the relationship to:

$$R_t = R_0(1 + \alpha t)$$

where α is the temperature coefficient of resistance (in $°C^{-1}$).

Example 5.27

A copper wire has a resistance of $12.5\,\Omega$ at $0°C$. Determine the resistance of the wire at $125°C$.

Determine the A wire having a specific resistance of $1.6 \times 10^{-8}\,\Omega m$, length $20\,m$ and cross-sectional area $1\,mm^2$ carries a current of $5\,A$. Determine the voltage drop between the ends of the wire.

Solution

To find the resistance at $125°C$ we use:

$$R_t = R_0(1 + \alpha t)$$

where $R_0 = 12.5\,\Omega$, $\alpha = 0.0039°C^{-1}$ (from the table) and $t = 125°C$.
Hence:

$$R_t = R_0(1 + \alpha t) = 12.5 \times (1 + (0.0039 \times 125))$$
$$= 12.5 \times (1 + 0.4875) = 18.6\,\Omega$$

5.8.3 Resistor types, values and tolerances

The value marked on the body of a resistor is not its exact resistance. Some minor variation in resistance value is inevitable due to production tolerance. For example, a resistor marked $100\,\Omega$ and produced within a tolerance of $\pm 10\%$ will have a value which falls within the range $90–110\,\Omega$. If a particular circuit requires a resistance of, for example, $105\,\Omega$, a $\pm 10\%$ tolerance resistor of $100\,\Omega$ will be perfectly adequate. If, however, we need a component with a value of $101\,\Omega$, then it would be necessary to obtain a $100\,\Omega$ resistor with a tolerance of $\pm 1\%$.

Resistors are available in several series of fixed decade values, the number of values provided with each series being governed by the tolerance involved. In order to cover the full range of resistance values using resistors having a $\pm 20\%$ tolerance it will be necessary to provide six basic values (known as the E6 series). More values will be required in the series which offers a tolerance of $\pm 10\%$ and consequently the E12 series provides 12 basic values. The E24 series for resistors of $\pm 5\%$ tolerance provides no less than 24 basic values and, as with the E6 and E12 series, decade multiples (i.e. $\times 1$, $\times 10$, $\times 100$, $\times 1\,k$, $\times 10\,k$, $\times 100\,k$ and $\times 1\,M$) of the basic series.

Other practical considerations when selecting resistors for use in a particular application include temperature coefficient, noise performance, stability and ambient temperature range. The following table summarizes the properties of several of the most common types of resistor.

Example 5.28

A resistor has a marked value of $220\,\Omega$. Determine the tolerance of the resistor if it has a measured value of $207\,\Omega$.

Solution

The difference between the marked and measured values of resistance (in other words the *error*) is $(220\,\Omega - 207\,\Omega) = 13\,\Omega$. The tolerance is given by:

$$\text{Tolerance} = \frac{\text{Error}}{\text{Marked value}} \times 100\%$$
$$= \frac{13\,\Omega}{220\,\Omega} \times 100\% = 5.9\%$$

Example 5.29

A 9 V power supply is to be tested with a $39\,\Omega$ load resistor. If the resistor has a tolerance of 10%, determine:

(a) the nominal current taken from the supply,
(b) the maximum and minimum values of supply current at either end of the tolerance range for the resistor.

Solution

(a) If a resistor of exactly $39\,\Omega$ is used the current, I, will be given by:

$$I = \frac{V}{R} = \frac{9\,V}{39\,\Omega} = 0.231\,A = 231\,mA$$

(b) The lowest value of resistance would be $(39\,\Omega - 3.9\,\Omega) = 35.1\,\Omega$. In which case the current would be:

$$I = \frac{V}{R} = \frac{9\,V}{35.1\,\Omega} = 0.256\,A = 256\,mA$$

At the other extreme, the highest value of resistance would be $(39\,\Omega + 3.9\,\Omega) = 42.9\,\Omega$.

In this case the current would be:

$$I = \frac{V}{R} = \frac{9\,V}{42.9\,\Omega} = 0.210\,A = 210\,mA$$

5.8.4 Power ratings

We have already mentioned that power dissipated by a resistor is determined by the product of the current flowing in the resistor and the voltage (p.d.) dropped across it. The power rating (or "wattage rating") of a resistor, on the other hand, is the maximum power that the resistor can safely dissipate. Power ratings are related to operating temperatures and resistors should be *derated* at high temperatures. For this reason, in all situations where reliability is important resistors should be operated at well below their nominal maximum power rating.

Example 5.30

A resistor is rated at 5 W. If the resistor carries a current of 30 mA and has a voltage of 150 V dropped across it determine the power dissipated and whether or not this exceeds the maximum power rating.

Solution

We can determine the actual power dissipated by applying the formula:

$$P = I \times V$$

where $I = 30\,mA = 0.03\,A$ and $V = 150\,V$, thus

$$P = I \times V = 0.03\,A \times 150\,V = 4.5\,W$$

This is just less than the power (wattage) rating of the resistor (5 W).

Example 5.31

A current of 100 mA ($\pm 20\%$) is to be drawn from a 28 V DC supply. What value and type of resistor should be used in this application?

Solution

The value of resistance required must first be calculated using Ohm's law:

$$R = \frac{V}{I} = \frac{28\,\text{V}}{100\,\text{mA}} = \frac{28\,\text{V}}{0.1\,\text{A}} = 280\,\Omega$$

The nearest preferred value from the E12 series is $270\,\Omega$, which will actually produce a current of 103.7 mA (i.e. within $\pm4\%$ of the desired value). If a resistor of $\pm10\%$ tolerance is used, the current will be within the range 94–115 mA (well within the $\pm20\%$ accuracy specified).

The power dissipated in the resistor can now be calculated:

$$P = \frac{V^2}{R} = \frac{(28\,\text{V} \times 28\,\text{V})}{270\,\Omega}$$
$$= \frac{784}{270} = 2.9\,\text{W}$$

Hence a component rated at 3 W (or more) will be required. This would normally be a vitreous enamel coated wire-wound resistor.

Table 5.3 gives typical characteristics of common types of resistor (Figure 5.48).

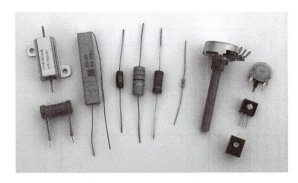

Figure 5.48 Various resistors.

Key point

The specifications for a resistor usually include the value of resistance (expressed in Ω, $k\Omega$ or $M\Omega$), the accuracy or tolerance of the marked value (quoted as the maximum permissible percentage deviation from the marked value) and the power rating (which must be equal to, or greater than, the maximum expected power dissipation). Temperature coefficient and stability are also important considerations in certain applications.

Table 5.3

Characteristic	Resistor type				
	Carbon film	Metal film	Metal oxide	Ceramic wire wound	Vitreous wire wound
Resistance range	$10\,\Omega$ to $10\,M\Omega$	$1\,\Omega$ to $10\,M\Omega$	$10\,\Omega$ to $1\,M\Omega$	$0.47\,\Omega$ to $22\,k\Omega$	$0.1\,\Omega$ to $22\,k\Omega$
Typical tolerance	$\pm5\%$	$\pm1\%$	$\pm2\%$	$\pm5\%$	$\pm5\%$
Power rating	0.25 W to 2 W	0.125 W to 0.5 W	0.25 W to 0.5 W	4 W to 17 W	2 W to 4 W
Temperature coefficient	$+250\,\text{ppm/}^\circ\text{C}$	$+50$ to $+100\,\text{ppm/}^\circ\text{C}$	$+250\,\text{ppm/}^\circ\text{C}$	$+250\,\text{ppm/}^\circ\text{C}$	$+75\,\text{ppm/}^\circ\text{C}$
Stability	Fair	Excellent	Excellent	Good	Good
Temperature range	-45°C to $+125^\circ\text{C}$	-55°C to $+125^\circ\text{C}$	-55°C to $+155^\circ\text{C}$	-55°C to $+200^\circ\text{C}$	-55°C to $+200^\circ\text{C}$
Typical applications	General purpose	Low-noise amplifiers and oscillators	General purpose	Power supplies and loads	Power supplies and loads

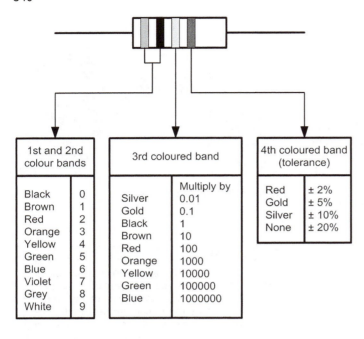

Figure 5.49 The four-band resistor colour code.

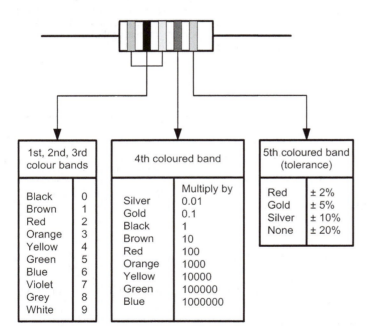

Figure 5.50 The five-band resistor colour code.

5.8.5 Resistor colour codes

Carbon and metal oxide resistors are normally marked with colour codes that indicate their value and tolerance. Two methods of colour coding are in common use: one involves four-coloured bands (see Figure 5.49) whilst the other uses five-coloured bands (see Figure 5.50).

Example 5.32

A resistor is marked with the following coloured stripes: brown, black, red, gold. What is its value and tolerance?

Solution

First digit: brown $= 1$
Second digit: black $= 0$
Multiplier: red $= 2$ (i.e. $\times 100$)
Value: $10 \times 100 = 1000 = 1\,\text{k}\Omega$
Tolerance: gold $= \pm 5\%$
   Hence the resistor value is $1\,\text{k}\Omega, \pm 5\%$

Example 5.33

A resistor is marked with the following coloured stripes: blue, grey, orange, silver. What is its value and tolerance?

Solution

First digit: blue $= 6$
Second digit: grey $= 8$
Multiplier: orange $= 3$ (i.e. $\times 1000$)
Value: $68 \times 1000 = 68,000 = 68\,\text{k}\Omega$
Tolerance: silver $= \pm 10\%$
   Hence the resistor value is $68\,\text{k}\Omega, \pm 10\%$

Example 5.34

A resistor is marked with the following coloured stripes: orange, orange, silver, silver. What is its value and tolerance?

Solution

First digit: orange $= 3$
Second digit: orange $= 3$
Multiplier: silver $= \div 100$
Value: $33/100 = 0.33\,\Omega$
Tolerance: silver $= \pm 10\%$
   Hence the resistor value is $0.33\,\Omega, \pm 10\%$

5.8.6 Series and parallel combinations of resistors

In order to obtain a particular value of resistance, fixed resistors may be arranged in either series or parallel as shown in Figures 5.51 and 5.52.

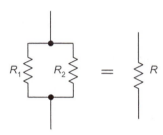

(a) Two resistors in series

(b) Three resistors in series

Figure 5.51 Series combination of resistors.

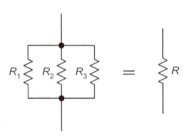

(a) Two resistors in parallel

(b) Three resistors in parallel

Figure 5.52 Parallel combination of resistors.

The equivalent resistance of each of the series circuits shown in Figure 5.51 is simply equal to the sum of the individual resistances.
   Hence, for Figure 5.51(a):

$$R = R_1 + R_2$$

Whilst for Figure 5.51(b):

$$R = R_1 + R_2 + R_3$$

Key point

The equivalent resistance of a number of resistors connected in series can be found by simply adding together the individual values of resistance.

Turning to the parallel resistors shown in Figure 5.52, the reciprocal of the equivalent resistance of each circuit is equal to the sum of the reciprocals of the individual resistances. Hence, for Figure 5.52(a):

$$\frac{1}{R} = \frac{1}{R_1} + \frac{1}{R_2}$$

whilst for Figure 5.52(b):

$$\frac{1}{R} = \frac{1}{R_1} + \frac{1}{R_2} + \frac{1}{R_3}$$

Key point

The *reciprocal* of the equivalent resistance of a number of resistors connected in parallel can be found by simply adding together the *reciprocals* of the individual values of resistance.

In the former case (*for just two resistors connected in parallel*) the equation can be more conveniently re-arranged as follows:

$$R = \frac{R_1 \times R_2}{R_1 + R_2}$$

Key point

The equivalent resistance of two resistors connected in parallel can be found by taking the *product* of the two resistance values and *dividing* it by the *sum* of the two resistance values (in other words, *product over sum*).

Example 5.35

Resistors of 22, 47 and 33 Ω are connected (a) in series and (b) in parallel. Determine the effective resistance in each case.

Solution

(a) In the series circuit:

$$R = R_1 + R_2 + R_3$$

thus

$$R = 22\,\Omega + 47\,\Omega + 33\,\Omega = 102\,\Omega$$

(b) In the parallel circuit:

$$\frac{1}{R} = \frac{1}{R_1} + \frac{1}{R_2} + \frac{1}{R_3}$$

thus

$$\frac{1}{R} = \frac{1}{22} + \frac{1}{47} + \frac{1}{33}$$

or $\frac{1}{R} = 0.045 + 0.021 + 0.03 = 0.096$

thus $R = 10.42\,\Omega$

Example 5.36

Determine the effective resistance of the circuit shown in Figure 5.53.

Solution

The circuit can be progressively simplified as shown in Figure 5.54.

The stages in this simplification are:

(a) R_3 and R_4 are in series and they can be replaced by a single resistance (R_A) of $12 + 27 = 39\,\Omega$.
(b) R_A appears in parallel with R_2. These two resistors can be replaced by a single resistance (R_B) of:

$$\frac{39 \times 47}{39 + 47} = 21.3\,\Omega$$

(c) R_B appears in series with R_1. These two resistors can be replaced by a single resistance, R, of $21.3\,\Omega + 4.7\,\Omega = 26\,\Omega$.

5.8.7 The potential divider

The potential divider circuit (see Figure 5.55) is commonly used to reduce voltage levels in a circuit. The output voltage produced by the circuit is given by:

$$V_{\text{out}} = V_{\text{in}} \times \frac{R_2}{R_1 + R_2}$$

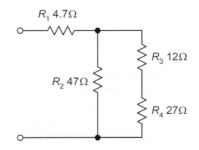

Figure 5.53

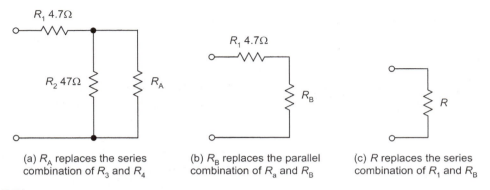

(a) R_A replaces the series combination of R_3 and R_4

(b) R_B replaces the parallel combination of R_a and R_B

(c) R replaces the series combination of R_1 and R_B

Figure 5.54

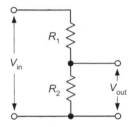

Figure 5.55 The potential divider.

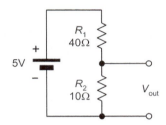

Figure 5.56

It is, however, important to note that the output voltage (V_{out}) will fall when current is drawn away from the arrangement.

Example 5.37

Determine the output voltage of the circuit shown in Figure 5.56.

Solution

Here we can use the potential divider formula:

$$V_{out} = V_{in} \times \frac{R_2}{R_1 + R_2}$$

where $V_{in} = 5\,\text{V}$, $R_1 = 40\,\Omega$

and $R_2 = 10\,\Omega$,

thus: $V_{out} = V_{in} \times \dfrac{R_2}{R_1 + R_2} = 5 \times \dfrac{10}{40 + 10}$

$$= 5 \times \frac{1}{5} = 1\,\text{V}$$

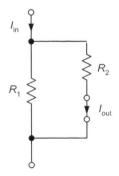

Figure 5.57 The current divider.

5.8.8 The current divider

The current divider circuit (see Figure 5.57) is used to divert current from one branch of a circuit to another. The output current produced by the circuit is given by:

$$I_{out} = I_{in} \times \frac{R_1}{R_1 + R_2}$$

It is important to note that the output current (I_{out}) will fall when the load connected to the output terminals has any appreciable resistance.

Example 5.38

A moving coil meter requires a current of 1 mA to provide full-scale deflection. If the meter coil has a resistance of $100\,\Omega$ and is to be used as a milliammeter reading 5 mA full scale, determine the value of parallel "shunt" resistor required.

Solution

This problem may sound a little complicated so it is worth taking a look at the equivalent circuit of the meter (Figure 5.58) and comparing it with the current divider shown in Figure 5.57.

We can apply the current divider formula, replacing I_{out} with I_m (the meter full-scale deflection current) and R_2 with R_m (the meter resistance). R_1 is the required value of shunt resistor, R_s.

From
$$I_{out} = I_{in} \times \frac{R_1}{R_1 + R_2}$$

we can say that

$$I_m = I_{in} \times \frac{R_s}{R_s + R_m}$$

where $I_m = 1$ mA, $I_{in} = 5$ mA and $R_2 = 100\,\Omega$.
Re-arranging the formula gives:

$$I_m \times (R_s + R_m) = I_{in} \times R_s$$
thus $\qquad I_m R_s + I_m R_m = I_{in} \times R_s$
or $\qquad I_{in} \times R_s - I_m R_s = I_m R_m$
hence $\qquad R_s(I_{in} - I_m) = I_m R_m$

and $$R_s = \frac{I_m R_m}{I_{in} - I_m}$$

Now $I_m = 1$ mA, $R_m = 100\,\Omega$ and $I_{in} = 5$ mA

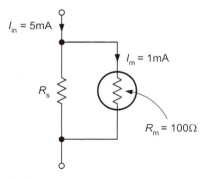

Figure 5.58 Meter circuit.

thus

$$R_s = \frac{I_m R_m}{I_{in} - I_m} = \frac{1\,\text{mA} \times 100\,\Omega}{5\,\text{mA} - 1\,\text{mA}} = 25\,\Omega$$

5.8.9 Variable resistors

Variable resistors are available in two basic forms: those which use carbon tracks and those which use a wirewound resistance element. In either case, a moving slider makes contact with the resistance element. Most variable resistors have three (rather than two) terminals and as such are more correctly known as potentiometers (see Figure 5.59).

Carbon potentiometers are available with linear or semi-logarithmic law tracks (see Figure 5.60) and in rotary or slider formats. Ganged controls, in which several potentiometers are linked together by a common control shaft, may also be encountered.

Preset resistors are used to make occasional adjustments and for calibration purposes. As such, they are not usually accessible without dismantling the equipment to gain access to the circuitry. Variable resistors, on the other

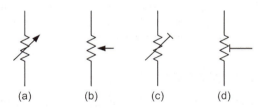

(a)   (b)   (c)   (d)

Figure 5.59 Symbols for variable resistors: (a) variable resistor (rheostat); (b) variable potentiometer; (c) preset resistor and (d) preset potentiometer.

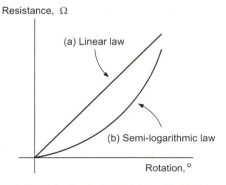

Figure 5.60 Linear and semi-logarithmic laws: (a) linear and (b) semi-logarithmic law potentiometer.

hands, are usually adjustable from the equipment's exterior. Various forms of preset resistor may be commonly encountered including open carbon track skeleton presets (for both horizontal and vertical printed circuit board (PCB) mounting) and fully encapsulated carbon and multi-turn cermet types.

5.8.10 The Wheatstone bridge

The Wheatstone bridge forms the basis of a number of electronic circuits including several that are used in instrumentation and measurement. The basic form of Wheatstone bridge is shown in Figure 5.61. The voltage developed between A and B will be zero when the voltage between A and the junction of R_2 and R_4 is the same as that between B and the junction of R_2 and R_4. In effect, R_1 and R_2 constitute a potential divider (see Section 5.8.7) as do R_3 and R_4. The bridge will be balanced (and $V_{AB} = 0$) when the ratio of $R_1 : R_2$ is the same as ratio $R_3 : R_4$. Hence at balance:

$$\frac{R_1}{R_2} = \frac{R_3}{R_4}$$

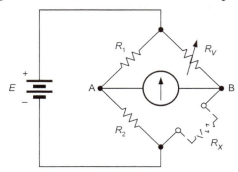

Figure 5.61 Basic form of Wheatstone bridge.

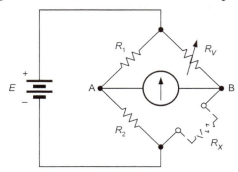

Figure 5.62 Practical form of Wheatstone bridge.

A practical form of Wheatstone bridge that can be used for measuring unknown resistances is shown in Figure 5.62. R_1 and R_2 are known as ratio arms while one arm (occupied by R_3 in Figure 5.62) is replaced by a calibrated variable resistor. The unknown resistor, R_X, is connected in the fourth arm.

At balance:

$$\frac{R_1}{R_2} = \frac{R_V}{R_X}$$

thus

$$R_X = \frac{R_2}{R_1} \times R_V$$

> **Key point**
>
> A Wheatstone bridge is balanced when no current flows in the central link. In this condition, the same voltage appears across adjacent arms.

Example 5.39

Figure 5.63 shows a balanced bridge. Determine the value of the unknown resistor.

Solution

Using the Wheatstone bridge equation:

$$R_X = \frac{R_2}{R_1} \times R_V$$

where $R_1 = 100\,\Omega$, $R_2 = 10\,\text{k}\Omega = 10,000\,\Omega$ and $R_V = 551\,\Omega$ gives:

$$R_X = \frac{10,000}{100} \times 551$$

$$= 100 \times 551 = 55,100 = 55.1\,\text{k}\Omega$$

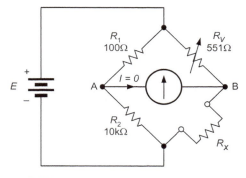

Figure 5.63

5.8.11 Thermistors

Unlike conventional resistors, the resistance of a thermistor is intended to change considerably with temperature. Thermistors are thus employed in temperature sensing and temperature compensating applications. Two basic types of thermistor are available: NTC and PTC.

Typical NTC thermistors have resistances which vary from a few hundred (or thousand) ohms at 25°C to a few tens (or hundreds) of ohms at 100°C (see Figure 5.64). PTC thermistors, on the other hand, usually have a resistance–temperature characteristic which remains substantially flat (typically at around 100 Ω) over the range 0°C to around 75°C. Above this, and at a critical temperature (usually in the range 80–120°C) their resistance rises very rapidly to values of up to, and beyond, 10 kΩ (see Figure 5.65).

A typical application of PTC thermistors is over-current protection. Provided the current passing through the thermistor remains below the threshold current, the effects of self-heating will remain negligible and the resistance of the thermistor will remain low (i.e. approximately the same as the resistance quoted at 25°C). Under fault conditions, the current exceeds the threshold value and the thermistor starts to self-heat. The resistance then increases rapidly and the current falls to the rest value. Typical values of threshold and rest currents are 200 and 8 mA, respectively, for a device which exhibits a nominal resistance of 25 Ω at 25°C.

> **Key point**
>
> Thermistors are available as both NTC and PTC types. The resistance of a PTC thermistor <u>increases</u> with temperature whilst that for an NTC thermistor <u>falls</u> with temperatures.

Test your understanding 5.8

1. A wirewound resistor consists of a 2 m length of annealed copper wire having a cross-sectional area of 0.5 mm². Determine the resistance of this component.

2. A batch of resistors is marked "560 Ω, ±10%". If a resistor is taken from this batch, within what limits will its value be?

3. A resistor is marked with the following coloured stripes: brown, green, red, gold. What is its value and tolerance?

4. Resistors of 10, 15 and 22 Ω are connected (a) in series and (b) in parallel. Determine the effective resistance in each case.

5. Use Ohm's law and Kirchhoff's laws to show that the equivalent resistance, R, of three resistors, R_1, R_2 and R_3, connected in parallel is given by the equation:

$$\frac{1}{R} = \frac{1}{R_1} + \frac{1}{R_2} + \frac{1}{R_3}$$

6. Determine the output voltage of the circuit shown in Figure 5.66.

7. Determine the unknown current in Figure 5.67.

8. When no current flows across between X and Y in Figure 5.68, the _____ _____ is said to be _____.

9. In Figure 5.68, determine the value of R.

10. The resistance of a PTC thermistor _____ as the temperature increases.

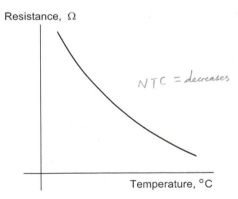

Resistance, Ω

NTC = decreases

Temperature, °C

Figure 5.64 NTC thermistor characteristics.

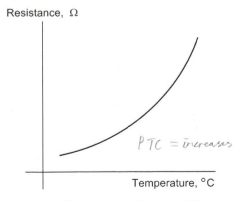

Resistance, Ω

PTC = increases

Temperature, °C

Figure 5.65 PTC thermistor characteristics.

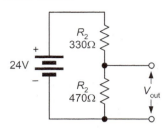

Figure 5.66

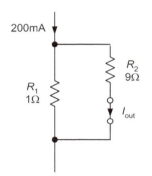

Figure 5.67

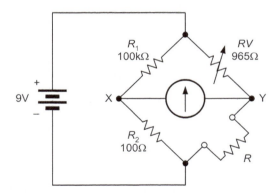

Figure 5.68

5.9 Power

Syllabus

Power, work and energy (kinetic and potential); Dissipation of power by a resistor; Power formula; Calculations involving power, work and energy.

Knowledge level key

A	B1	B2
–	2	2

Earlier in Section 5.4 we briefly mentioned power and energy, and the relationship between them. In this section we will delve a little deeper into these important topics and derive some formulae that will allow us to determine the amount of power dissipated in a circuit as well as the energy that is supplied to it.

5.9.1 Power, work and energy

From your study of physics you will recall that energy can exist in many forms including kinetic, potential, heat, light energy, etc. Kinetic energy is concerned with the movement of a body whilst potential energy is the energy that a body possesses due to its position. Energy can be defined as "the ability to do work" whilst power can be defined as "the rate at which work is done".

In electrical circuits, energy is supplied by batteries or generators. It may also be stored in components such as capacitors and inductors. Electrical energy is converted into various other forms of energy by components such as resistors (producing heat), loudspeakers (producing sound energy) and light emitting diodes (producing light).

The unit of energy is the joule (J). Power is the rate of use of energy and it is measured in watts (W). A power of 1 W results from energy being used at the rate of 1 J/s. Thus:

$$P = \frac{E}{t}$$

where P is the power in watts (W), E is the energy in joules (J) and t is the time in seconds (s).

We can re-arrange the previous formula to make E the subject, as follows:

$$E = P \times t$$

The power in a circuit is equivalent to the product of voltage and current. Hence:

$$P = I \times V$$

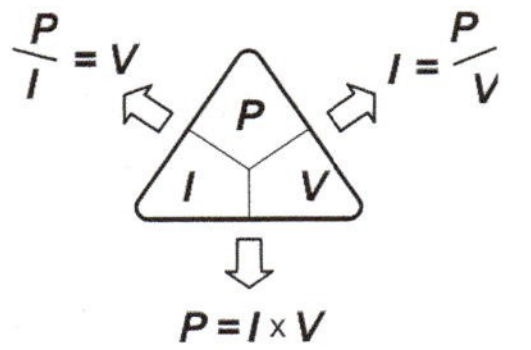

Figure 5.69 Relationship between *P*, *I* and *V*.

where P is the power in watts (W), I is the current in amperes (A) and V is the voltage in volts (V).

The formula may be arranged to make P, I or V the subject, as follows:

$$P = I \times V \quad I = \frac{P}{V} \quad \text{and} \quad V = \frac{P}{I}$$

The triangle shown in Figure 5.69 should helps you remember these three important relationships. It is important to note that, when performing calculations of power, current and voltages in practical circuits it is seldom necessary to work with an accuracy of better than ±1% simply because component tolerances are invariably somewhat greater than this.

Key point

Power is the rate of using energy and a power of 1 W results from energy being used at the rate of 1 J/s.

5.9.2 Dissipation of power by a resistor

When a resistor gets hot it is dissipating power. In effect, a resistor is a device that converts electrical energy into heat energy. The amount of power dissipated in a resistor depends on the current flowing in the resistor. The more current flowing in the resistor the more power will be dissipated and the more electrical energy will be converted into heat.

It is important to note that the relationship between the current applied and the power dissipated is not linear – in fact it obeys a *square law*. In other words, the power dissipated in a resistor is proportional to the square of the applied current. To prove this, we will combine the formula

for power which we have just met with Ohm's law that we introduced in Section 5.7.

Key point

When a resistor gets hot it is converting electrical energy to heat energy and dissipating power. The power dissipated by a resistor is proportional to the *square* of the current flowing in the resistor.

5.9.3 Power formulae

The relationship $P = I \times V$ may be combined with that which results from Ohm's law (i.e. $V = I \times R$), to produce two further relationships. Firstly, substituting for V gives:

$$P = I \times (I \times R) = I^2 \times R$$

Secondly, substituting for I gives:

$$P = \left(\frac{V}{R}\right) \times V = \frac{V^2}{R}$$

Example 5.40

A current of 1.5 A is drawn from a 3 V battery. What power is supplied?

Solution

Here we must use $P = I \times V$
where $I = 1.5$ A and $V = 3$ V:

$$P = I \times V = 1.5\,\text{A} \times 3\,\text{V} = 4.5\,\text{W}$$

Hence a power of 4.5 W is supplied.

Example 5.41

A voltage drop of 4 V appears across a resistor of 100 Ω. What power is dissipated in the resistor?

Solution

Here we must use $P = \dfrac{V^2}{R}$
where $V = 4$ V and $R = 100\,\Omega$:

$$P = \frac{V^2}{R} = \frac{(4\,\text{V} \times 4\,\text{V})}{100\,\Omega} = \frac{16}{100} = 0.16\,\text{W}$$

Hence the resistor dissipates a power of 0.16 W (or 160 mW).

Example 5.42

A current of 20 mA flows in a 1 kΩ resistor. What power is dissipated in the resistor and what energy is used if the current flows for 10 min?

Solution

Here we must use $P = I^2 R$ where $I = 200$ mA and $R = 1000\,\Omega$:

$$P = I^2 \times R = (0.2\,\text{A} \times 0.2\,\text{A}) \times 1000\,\Omega$$
$$= 0.04 \times 1000 = 40\,\text{W}$$

Hence the resistor dissipates a power of 40 W.

To find the energy we need to use $E = P \times t$ where $P = 40$ W and $t = 10$ min:

$$E = P \times t = 40\,\text{W} \times (10 \times 60)\,\text{s}$$
$$= 24{,}000\,\text{J}$$
$$= 24\,\text{kJ}$$

Test your understanding 5.9

1. Power can be defined as the _____ at which _____ is done.
2. A power of 1 W results from _____ being used at the rate of 1 _____ per _____.
3. Give three examples of different forms of energy produced by electrical/electronic components. In each case name the component.
4. A resistor converts 15 J of energy to heat in a time of 3 s. What power is used?
5. A load consumes a power of 50 W. How much energy will be delivered to the load in 1 min?
6. A 24 V battery delivers a current of 27 A to a load. What energy is delivered to the load if the load is connected for 10 min?

7. A 28 V supply is connected to a 3.5 Ω load. What power is supplied to the load?
8. A resistor is rated at "11 Ω, 2 W". What is the maximum current that should be allowed to flow in it?
9. A 28 V DC power supply is to be tested at its rated power of 250 W. What value of load resistance should be used and what current will flow in it?
10. A current of 2.5 A flows in a 10 Ω resistor. What power is dissipated in the resistor and what energy is used if the current flows for 20 min?

5.10 Capacitance and capacitors

Syllabus
Operation and function of a capacitor; Factors affecting capacitance area of plates, distance between plates, number of plates, dielectric and dielectric constant, working voltage, voltage rating; Capacitor types, construction and function; Capacitor colour coding; Calculations of capacitance and voltage in series and parallel circuits; Exponential charge and discharge of a capacitor, time constants; Testing of capacitors.

Knowledge level key

A	B1	B2
–	2	2

5.10.1 Operation and function of a capacitor

The capacitance of a conductor is a measure of its ability to store an electric charge when a p.d. is applied. Thus a large capacitance will store a larger charge for a given applied voltage. Consider, for a moment, the arrangement shown in Figure 5.70. Here three metal conductors,

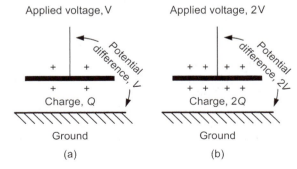

Applied voltage, V — Charge, Q — Ground — (a)

Applied voltage, 2V — Charge, 2Q — Ground — (b)

Applied voltage, 3V — Charge, 3Q — Ground — (c)

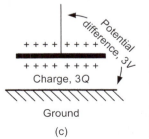

Figure 5.70 Relationship between charge, Q, and voltage, V, for a conductor suspended above a perfect ground: (a) p.d. $= V$, charge $= Q$; (b) p.d. $= 2V$, charge $= 2Q$ and (c) p.d. $= 3V$, charge $= 3Q$.

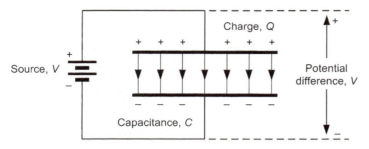

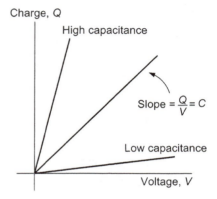

Figure 5.71 A simple parallel-plate capacitor.

of identical size and area, are suspended above a perfectly conducting zero potential plane (or *ground*). In Figure 5.70(a) the p.d., V, between the conductor and ground produces a charge, Q. If the p.d. is doubled to $2V$, as shown in Figure 5.70(b), the charge increases to $2V$. Similarly if the p.d. increases to $3V$, as shown in Figure 5.70(c), the charge increases to $3Q$. This shows that, for a given conductor, there is a linear relationship between the charge present, Q, and the p.d., V.

In practice we can alter the shape of a conductor so that it covers a relatively large area over which the charge is distributed. This allows us to produce a relatively large value of charge for only a relatively modest value of p.d. The resulting component is used for storing charge and is known as a *capacitor* (Figure 5.71).

If we plot charge, Q, against p.d., V, for a capacitor we arrive at a straight line law (as mentioned earlier in relation to Figure 5.70). The slope of this graph is an indication of the capacitance of the capacitor, as shown in Figure 5.72.

From Figure 5.72, the capacitance of a capacitor is defined as follows:

$$\text{Capacitance} = \frac{\text{charge on the capacitor's plates}}{\text{potential difference between the plates}}$$

or   in symbols

$$C = \frac{Q}{V}$$

where the charge, Q, is measured in coulombs (C) and the p.d., V, is measured in volts (V). The unit of capacitance is the *Farad* (F) where one farad of capacitance produces a charge of one

Figure 5.72 Charge, Q, plotted against p.d., V, for three different values of capacitance.

coulomb when a p.d. of one volt is applied. Note that, in practice, the farad is a very large unit and we therefore often deal with sub-multiples of the basic unit such as μF (1×10^{-6} F), nF (1×10^{-9} F) and pF (1×10^{-12} F).

For example if a potential of 200 V is required to create a charge of $400\,\mu C$ on a capacitor then the capacitance of the capacitor (expressed in F) is given by:

$$C = \frac{Q}{V} = \frac{400 \times 10^{-6}\,\text{C}}{200\,\text{V}} = 2 \times 10^{-6}\,\text{F} = 2\,\mu\text{F}$$

We have already said that a capacitor is a device for storing electric charge. In effect, it is a *reservoir* for charge. Typical applications for capacitors include reservoir and smoothing capacitors for use in power supplies, coupling alternating current (AC) signals between the stages of amplifiers and decoupling supply rails (i.e. effectively grounding the supply rails to residual AC signals and noise). You will learn more about these particular applications when you study Chapter 6 but, for now, we will concentrate our efforts on explaining how a capacitor works and what it does!

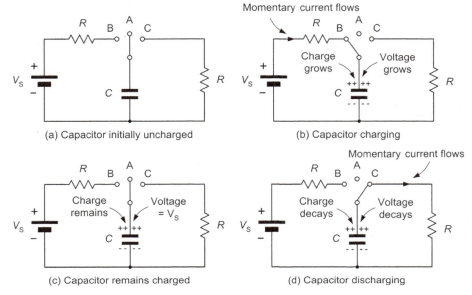

Figure 5.73 Charging and discharging a capacitor.

Consider the arrangement shown in Figure 5.73. If the switch is left open (position A) (Figure 5.73(a)), no charge will appear on the plates and in this condition there will be no electric field in the space between the plates nor will there be any charge stored in the capacitor.

When the switch is moved to position B (Figure 5.73(b)), electrons will be attracted from the positive plate to the positive terminal of the battery. At the same time, a similar number of electrons will move from the negative terminal of the battery to the negative plate. This sudden movement of electrons will manifest itself in a momentary surge of current (conventional current will flow from the positive terminal of the battery towards the positive terminal of the capacitor).

Eventually, enough electrons will have moved to make the e.m.f. between the plates the same as that of the battery. In this state, the capacitor is said to be *charged* and an electric field will be present in the space between the two plates.

If, at some later time the switch is moved back to position A (Figure 5.73(c)), the positive plate will be left with a deficiency of electrons whilst the negative plate will be left with a surplus of electrons. Furthermore, since there is no path for current to flow between the two plates the capacitor will remain charged and a p.d. will be maintained between the plates.

Now assume that the switch is moved to position C (Figure 5.73(d)). The excess electrons on the negative plate will flow through the resistor to the positive plate until a neutral state once again exists (i.e. until there is no excess charge on either plate). In this state the capacitor is said to be *discharged* and the electric field between the plates will rapidly collapse. The movement of electrons during the discharging of the capacitor will again result in a momentary surge of current (current will flow from the positive terminal of the capacitor into the resistor).

Figure 5.74(a) and (b), respectively, shows the direction of current flow in the circuit of Figure 5.73 during charging (switch in position B) and discharging (switch in position C). It should be noted that current flows momentarily in both circuits even though the circuit is apparently broken by the gap between the capacitor plates!

5.10.2 Capacitance, charge and voltage

We have already seen how the charge (or quantity of electricity) that can be stored in the

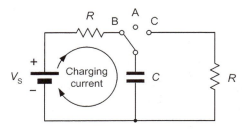

(a) Capacitor charging

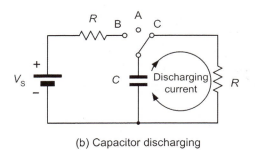

(b) Capacitor discharging

Figure 5.74   Current flow during charging and discharging.

electric field between the capacitor plates is proportional to the applied voltage and the capacitance of the capacitor. Thus, re-arranging the formula that we met earlier, $C = Q/V$, gives:

$$Q = C \times V \quad \text{and} \quad V = \frac{Q}{C}$$

where Q is the charge (in C), C is the capacitance (in F) and V is the p.d. (in V).

Example 5.43

A $10\,\mu\text{F}$ capacitor is charged to a potential of 250 V. Determine the charge stored.

Solution

The charge stored will be given by:

$$Q = C \times V = 10 \times 10^{-6} \times 250$$

$$= 2500 \times 10^{-6}$$

$$= 2.5 \times 10^{-3} = 2.5\,\text{mC}$$

Example 5.44

A charge of $11\,\mu\text{C}$ is held in a 220 nF capacitor. What voltage appears across the plates of the capacitor?

Solution

To find the voltage across the plates of the capacitor we need to re-arrange the equation to make V the subject, as follows:

$$V = \frac{Q}{C} = \frac{11 \times 10^{-6}\,\text{C}}{220 \times 10^{-9}\,\text{F}} = 50\,\text{V}$$

5.10.3 Energy storage

The area under the linear relationship between Q and V that we met earlier (Figure 5.72) gives the energy stored in the capacitor. The area shown shaded in Figure 5.75 is $\frac{1}{2}QV$, thus:

$$\text{energy stored}, W = \tfrac{1}{2}QV$$

Combining this with the earlier relationship, $Q = CV$, gives:

$$W = \tfrac{1}{2}(CV)V = \tfrac{1}{2}CV^2$$

where W is the energy (in J), C is the capacitance (in F) and V is the p.d. (in V).

This shows us that the energy stored in a capacitor is proportional to the product of the capacitance and the square of the p.d. between its plates.

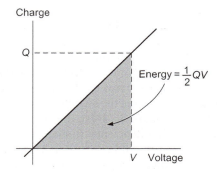

Figure 5.75   Energy stored in a capacitor.

Example 5.45

A $100\,\mu\text{F}$ capacitor is charged from a charge of $20\,\text{V}$ supply. How much energy is stored in the capacitor?

Solution

The energy stored in the capacitor will be given by:

$$W = \tfrac{1}{2}(C \times V^2) = \tfrac{1}{2} \times 100 \times 10^{-6} \times (20)^2$$

$$= 50 \times 400 \times 10^{-6} = 20,000 \times 10^{-6}$$

$$= 2 \times 10^{-2}\,\text{J}$$

Example 5.46

A capacitor of $47\,\mu\text{F}$ is required to store energy of $40\,\text{J}$. Determine the p.d. required to do this.

Solution

To find the p.d. (voltage) across the plates of the capacitor we need to re-arrange the equation to make V the subject, as follows:

$$V = \sqrt{\frac{2E}{C}} = \sqrt{\frac{2 \times 40}{47 \times 10^{-6}}} = \sqrt{\frac{80}{47} \times 10^6}$$

$$= \sqrt{1.702 \times 10^6} = 1.3 \times 10^3 = 1.3\,\text{kV}$$

5.10.4 Factors affecting capacitance

The capacitance of a capacitor depends upon the physical dimensions of the capacitor (i.e. the size of the plates and the separation between them) and the dielectric material between the plates. The capacitance of a conventional parallel plate capacitor is given by:

$$C = \frac{\varepsilon_0 \varepsilon_r A}{d}$$

where C is the capacitance (in F), ε_0 is the permittivity of free space, ε_r is the relative permittivity (or *dielectric constant*) of the dielectric medium between the plates, A is the area of the plates (in m^2) and d is the separation between the plates (in m). The permittivity of free space, ε_0, is $8.854 \times 10^{-12}\,\text{F/m}$.

Some typical capacitor dielectric materials and relative permittivity are given in the table below:

Dielectric material	Relative permittivity (free space = 1)
Vacuum	1
Air	1.0006 (i.e. 1!)
Polythene	2.2
Paper	2–2.5
Epoxy resin	4.0
Mica	3–7
Glass	5–10
Porcelain	6–7
Aluminium oxide	7
Ceramic materials	15–500

Example 5.47

Two parallel metal plates each of area $0.2\,\text{m}^2$ are separated by an air gap of $1\,\text{mm}$. Determine the capacitance of this arrangement.

Solution

Here we must use the formula:

$$C = \frac{\varepsilon_0 \varepsilon_r A}{d}$$

where $A = 0.2\,\text{m}^2$, $d = 1 \times 10^{-3}\,\text{m}$, $\varepsilon_r = 1$ and $\varepsilon_0 = 8.854 \times 10^{-12}\,\text{F/m}$. Hence:

$$C = \frac{8.854 \times 10^{-12} \times 1 \times 0.2}{1 \times 10^{-3}}$$

$$= \frac{1.7708 \times 10^{-12}}{1 \times 10^{-3}}$$

$$= 1.7708 \times 10^{-9}\,\text{F}$$

$$= 1.7708\,\text{nF}$$

Example 5.48

A capacitor of $1\,\text{nF}$ is required. If a dielectric material of thickness $0.5\,\text{mm}$ and relative permittivity 5.4 is available, determine the required plate area.

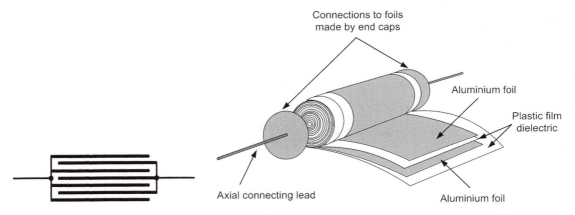

(a) Basic interleaved plate arrangement   (b) Typical construction of tubular capacitor

Figure 5.76   A multiple-plate capacitor.

Solution

Re-arranging the formula $C = \varepsilon_0 \varepsilon_r A/d$ to make A the subject gives:

$$A = \frac{Cd}{\varepsilon_0 \varepsilon_r} = \frac{1 \times 10^{-9} \times 0.5 \times 10^{-3}}{8.854 \times 10^{-12} \times 5.4}$$

$$= \frac{0.5 \times 10^{-12}}{47.811 \times 10^{-12}} = 0.0105 \, \text{m}^2$$

thus $A = 0.0105 \, \text{m}^2$ or $105 \, \text{cm}^2$.

In order to increase the capacitance of a capacitor, many practical components employ multiple plates (see Figure 5.76) in which case the capacitance is then given by:

$$C = \frac{\varepsilon_0 \varepsilon_r (n - 1)A}{d}$$

where C is the capacitance (in F), ε_0 is the permittivity of free space, ε_r is the relative permittivity of the dielectric medium between the plates, n is the number of plates, A is the area of the plates (in m^2) and d is the separation between the plates (in m).

Example 5.49

A capacitor consists of six plates each of area $20 \, \text{cm}^2$ separated by a dielectric of relative permittivity 4.5 and thickness 0.2 mm. Determine the capacitance of the capacitor.

Solution

Using $C = \dfrac{\varepsilon_0 \varepsilon_r (n - 1)A}{d}$

gives:

$$C = \frac{8.854 \times 10^{-12} \times 4.5 \times (6 - 1) \times (20 \times 10^{-4})}{0.2 \times 10^{-3}}$$

$$= \frac{3984.3 \times 10^{-16}}{2 \times 10^{-4}}$$

$$= 1992.15 \times 10^{-12} \, \text{F}$$

$$= 1.992 \, \text{nF}$$

5.10.5 Capacitor types, values and tolerances

The specifications for a capacitor usually include the value of capacitance (expressed in μF, nF or pF), the voltage rating (i.e. the maximum voltage which can be continuously applied to the capacitor under a given set of conditions) and the accuracy or tolerance (quoted as the maximum permissible percentage deviation from the marked value).

Other practical considerations when selecting capacitors for use in a particular application include temperature coefficient, leakage current, stability and ambient temperature range. Electrolytic capacitors require the application of a DC polarizing voltage in order to work

Table 5.4

Characteristic	Capacitor type				
	Ceramic dielectric	Electrolytic	Metalised film	Mica dielectric	Polyester dielectric
Capacitance range	2.2 pF to 100 nF	100 nF to 68 mF	1 µF to 16 µF	2.2 pF to 10 nF	10 nF to 2.2 µF
Typical tolerance	±10% and ±20%	−10% to +50%	±20%	±1%	±20%
Voltage rating	50 V to 250 V	6.3 V to 400 V	250 V to 600 V	350 V	250 V
Temperature coefficient	+100 to −4700 ppm/°C	+1000 ppm/°C	+100 to +200 ppm/°C	+50 ppm/°C	+250 ppm/°C
Stability	Fair	Poor	Fair	Excellent	Good
Temperature range	−85°C to +85°C	−40°C to +85°C	−25°C to +85°C	−40°C to +85°C	−40°C to +100°C
Typical applications	General purpose	Power supplies	High-voltage power supplies	Oscillators, tuned circuits	General purpose

properly. This voltage must be applied with the correct polarity (invariably this is clearly marked on the case of the capacitor) with a positive (+) sign or negative (−) sign or a coloured stripe or other marking. Failure to observe the correct polarity can result in over-heating, leakage and even a risk of explosion!

The typical specifications for some common types of capacitor are shown in Table 5.4.

Key point

The specifications for a capacitor usually include the value of capacitance (expressed in µF, nF or pF), the accuracy or tolerance of the marked value (quoted as the maximum permissible percentage deviation from the marked value) and the voltage rating (which must be equal to, or greater than, the maximum expected voltage applied to the capacitor). The temperature coefficient and stability are also important considerations for certain applications.

Some typical capacitors are shown in Figures 5.77 and 5.78.

5.10.6 Working voltages

Working voltages are related to operating temperatures and capacitors must be de-rated at

Figure 5.77 Various capacitors.

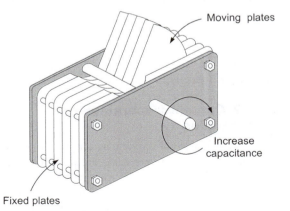

Figure 5.78 Variable air-spaced capacitor.

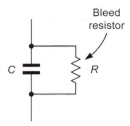

Figure 5.79 A high-voltage capacitor with a bleed resistor.

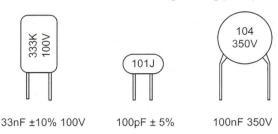

33nF ±10% 100V   100pF ± 5%   100nF 350V

Figure 5.80 A typical capacitor marking.

high temperatures. Where reliability is important capacitors should be operated at well below their nominal maximum working voltages.

Where the voltage rating is expressed in terms of a direct voltage (e.g. 250 V DC) unless otherwise stated, this is related to the maximum working temperature. It is, however, always wise to operate capacitors with a considerable margin for safety which also helps to ensure long-term reliability. As a rule of thumb, the working DC voltage should be limited to no more than 50–60% of the rated DC voltage.

Where an AC voltage rating is specified this is normally for sinusoidal operation. Performance will not be significantly affected at low frequencies (up to 100 kHz, or so) but, above this, or when non-sinusoidal (e.g. pulse) waveforms are involved the capacitor must be de-rated in order to minimize dielectric losses that can produce internal heating and lack of stability.

Special care must be exercised when dealing with high-voltage circuits as large value electrolytic and metalised film capacitors can retain an appreciable charge for some considerable time. In the case of components operating at high voltages, a carbon film *bleed resistor* (of typically 1 MΩ 0.5 W) should be connected in parallel with the capacitor to provide a discharge path (see Figure 5.79).

5.10.7 Capacitor markings and colour codes

The vast majority of capacitors employ written markings which indicate their values, working voltages and tolerance. The most usual method of marking resin dipped polyester (and other) types of capacitor involves quoting the value (in μF, nF or pF), the tolerance (often either 10%

or 20%) and the working voltage (using _ and ~ to indicate DC and AC, respectively). Several manufacturers use two separate lines for their capacitor markings and these have the following meanings:

- First line: capacitance (in pF or μF) and tolerance ($K = 10\%$, $M = 20\%$).
- Second line: rated DC voltage and code for the dielectric material.

A three-digit code is commonly used to mark monolithic ceramic capacitors. The first two digits correspond to the first two digits of the value whilst the third digit is a multiplier that gives the number of zeroes to be added to give the value in pF (Figure 5.80).

The colour code shown in Figure 5.81 is used for some small ceramic and polyester types of capacitor. Note, however, that this colour code is not as universal as that used for resistors and that the values are marked in pF (not F).

Example 5.50

A monolithic ceramic dielectric capacitor is marked with the legend "103". What is its value?

Solution

The value (in pF) will be given by the first two digits (10) followed by the number of zeros indicated by the third digit (3). The value of the capacitor is thus 10,000 pF or 10 nF.

Example 5.51

A polyester capacitor is marked with the legend "0.22/20 250_". What is its value, tolerance and working voltage?

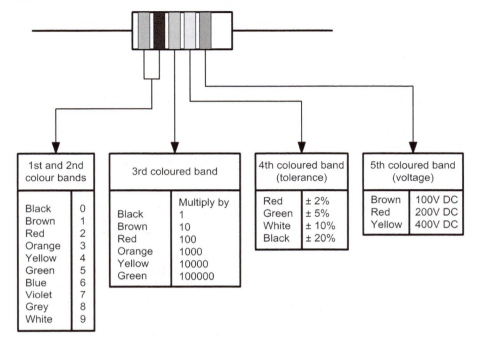

Figure 5.81 Capacitor colour code.

Solution

The value (0.22) is stated in μF and the tolerance (±20%) appears after the "/" character. The voltage rating (250) precedes the "_" character which indicates that the rating is for DC rather than AC. Hence the capacitor has a value, tolerance and working voltage of 0.22 μF, ±20%, 250 V DC, respectively.

Example 5.52

A tubular ceramic capacitor is marked with the following coloured stripes: brown, green, brown, red, brown. What is its value, tolerance and working voltage?

Solution

First digit: brown = 1
Second digit: green = 5
Multiplier: brown = ×10
Value: $15 \times 10 = 150$ pF
Tolerance: red = ±2%
Voltage: brown = 100 V

Hence the capacitor is 150 pF, ±2% rated at 100 V.

Figure 5.82 Two capacitors in series.

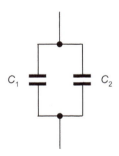

Figure 5.83 Two capacitors in parallel.

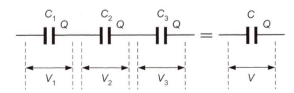

Figure 5.84 Three capacitors in series.

5.10.8 Capacitors in series and parallel

In order to obtain a particular value of capacitance, fixed capacitors may be arranged in either series or parallel (Figures 5.82 and 5.83).

Now consider Figure 5.84 where C is the equivalent capacitance of three capacitors (C_1, C_2 and C_3) connected in series.

The applied voltage, V, will be the sum of the voltages that appear across each capacitor. Thus:

$$V = V_1 + V_2 + V_3$$

Now, for each capacitor, the p.d., V, across its plates will be given by the ratio of charge, Q, to capacitance, C. Hence:

$$V = \frac{Q}{C}, \quad V_1 = \frac{Q_1}{C_1}, \quad V_2 = \frac{Q_2}{C_2} \quad \text{and}$$

$$V_3 = \frac{Q_3}{C_3}$$

Combining these equations gives: In the series circuit the same charge, Q, appears across each capacitor, thus:

$$Q = Q_1 = Q_2 = Q_3$$

Hence:

$$\frac{Q}{C} = \frac{Q}{C_1} + \frac{Q}{C_2} + \frac{Q}{C_3}$$

From which:

$$\frac{1}{C} = \frac{1}{C_1} + \frac{1}{C_2} + \frac{1}{C_3}$$

> **Key point**
>
> The *reciprocal* of the equivalent capacitance of a number of capacitors connected in series can be found by simply adding together the *reciprocals* of the individual values of capacitance.

When two capacitors are connected in series the equation becomes:

$$\frac{1}{C} = \frac{1}{C_1} + \frac{1}{C_2}$$

This can be arranged to give the slightly more convenient expression:

$$C = \frac{C_1 \times C_2}{C_1 + C_2}$$

Note that the foregoing expression is *only correct for two capacitors*. It cannot be extended for three or more!

> **Key point**
>
> The equivalent capacitance of two capacitors connected in series can be found by taking the *product* of the two capacitance values and *dividing* it by the *sum* of the two capacitance values (in other words, *product over sum*).

Now consider Figure 5.85 where C is the equivalent capacitance of three capacitors (C_1, C_2 and C_3) connected in series.

The total charge present, Q, will be the sum of the charges that appear in each capacitor. Thus:

$$Q = Q_1 + Q_2 + Q_3$$

Now, for each capacitor, the charge present, Q, will be given by the product of the capacitance, C, and p.d., V. Hence:

$$Q = CV, \quad Q_1 = C_1 V_1, \quad Q_2 = C_2 V_2$$
$$\text{and} \quad Q_3 = C_3 V_3$$

Combining these equations gives:

$$CV = C_1 V_1 + C_2 V_2 + C_3 V_3$$

In the series circuit the same voltage, V, appears across each capacitor, thus:

$$V = V_1 = V_2 = V_3$$

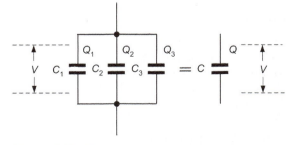

Figure 5.85 Three capacitors in parallel.

Hence:

$$CV = C_1 V + C_2 V + C_3 V$$

From which:

$$C = C_1 + C_2 + C_3$$

When two capacitors are connected in series the equation becomes:

$$C = C_1 + C_2$$

Key point

The equivalent capacitance of a number of capacitors connected in parallel can be found by simply adding together the individual values of capacitance.

Example 5.53

Capacitors of 2.2 and 6.8 μF are connected (a) in series and (b) in parallel. Determine the equivalent value of capacitance in each case.

Solution

(a) Here we can use the simplified equation for *just two* capacitors connected in series:

$$C = \frac{C_1 \times C_2}{C_1 + C_2} = \frac{2.2 \times 6.8}{2.2 + 6.8} = \frac{14.96}{9}$$
$$= 1.66 \,\mu F$$

(b) Here we use the formula for two capacitors connected in parallel:

$$C = C_1 + C_2 = 2.2 + 6.8 = 9 \,\mu F$$

Example 5.54

Capacitors of 2 and 5 μF are connected in series across a 100 V DC supply. Determine:

(a) the charge on each capacitor and (b) the voltage dropped across each capacitor.

Solution

(a) First we need to find the equivalent value of capacitance, C, using the simplified equation for two capacitors in series:

$$C = \frac{C_1 \times C_2}{C_1 + C_2} = \frac{2 \times 5}{2 + 5} = \frac{10}{7} = 1.428 \,\mu F$$

Next we can determine the charge (note that, since the capacitors are connected in series, the *same* charge will appear in each capacitor):

$$Q = C \times V = 1.428 \times 100 = 142.8 \,\mu C$$

(b) In order to determine the voltage dropped across each capacitor we can use:

$$V = \frac{Q}{C}$$

Hence, for the 2 μF capacitor:

$$V_1 = \frac{Q}{C_1} = \frac{142.8 \times 10^{-6}}{2 \times 10^{-6}} = 71.4 \,V$$

Similarly, for the 5 μF capacitor:

$$V_2 = \frac{Q}{C_2} = \frac{142.8 \times 10^{-6}}{5 \times 10^{-6}} = 28.6 \,V$$

We should now find that the total voltage (100 V) applied to the series circuit is the sum of the two capacitor voltages, i.e.:

$$V = V_1 + V_2 = 71.4 + 28.6 = 100 \,V$$

5.10.9 Capacitors charging and discharging through a resistor

Networks of capacitors and resistors (known as *C–R networks*) form the basis of simple timing and delay circuits.

In many electrical/electronic circuits the variation of voltage and current with time is important. In order to satisfy this requirement, simple *C–R* networks have useful properties that we can exploit. A simple *C–R* circuit is shown in Figure 5.86.

When the *C–R* network is connected to a constant voltage source (V_S), as shown in Figure 5.87, the voltage (v_C) across the (initially

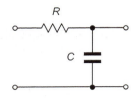

Figure 5.86   Simple *C–R* circuit.

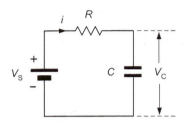

Figure 5.87   *C–R* circuit with *C* charging through *R*.

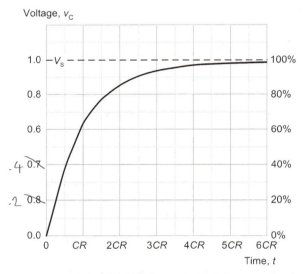

Figure 5.88 Exponential growth of capacitor voltage (*v_c*) in Figure 5.87.

uncharged) capacitor will rise exponentially as shown in Figure 5.88. At the same time, the current in the circuit (*i*) will fall, as shown in Figure 5.89.

The rate of growth of voltage with time and decay of current with time will be dependent upon the product of capacitance and resistance. This value is known as the *time constant* of the circuit. Hence:

$$\text{Time constant, } t = C \times R$$

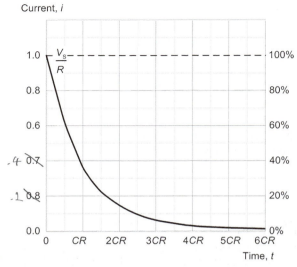

Figure 5.89 Exponential decay of current (*i*) in Figure 5.87.

where *C* is the value of capacitance (in F), *R* is the resistance (in Ω) and *t* is the time constant (in s).

The voltage developed across the charging capacitor (v_C) varies with time (*t*) according to the relationship:

$$v_C = V_S \left(1 - e^{\frac{-t}{CR}} \right)$$

where v_C is the capacitor voltage (in V), V_S is the DC supply voltage (in V), *t* is the time (in s) and *CR* is the time constant of the circuit (equal to the product of capacitance, *C*, and resistance, *R*, in s).

The capacitor voltage will rise to approximately 63% of the supply voltage in a time interval equal to the time constant. At the end of the next interval of time equal to the time constant (i.e. after an elapsed time equal to 2*CR*), the voltage would have risen by 63% of the remainder, and so on. In theory, the capacitor will never quite become fully charged. However, after a period of time equal to 5*CR*, the capacitor voltage will (to all intents and purposes) be equal to the supply voltage. At this point the capacitor voltage will have risen to 99.3% of its final value and we can consider it to be fully charged.

During charging, the current in the capacitor (i) varies with time (t) according to the relationship:

$$i = V_S e^{\frac{-t}{CR}}$$

where i is the current (in A), V_S is the DC supply voltage (in V), t is the time and CR is the time constant of the circuit (equal to the product of capacitance, C, and resistance, R, in s).

The current will fall to approximately 37% of the initial current in a time equal to the time constant. At the end of the next interval of time equal to the time constant (i.e. after a total time of $2CR$ has elapsed) the current will have fallen by a further 37% of the remainder, and so on.

Example 5.55

An initially uncharged capacitor of $1\,\mu F$ is charged from a 9 V DC supply via a $3.3\,M\Omega$ resistor. Determine the capacitor voltage 1 s after connecting the supply.

Solution

The formula for exponential growth of voltage in the capacitor is:

$$v_C = V_S\left(1 - e^{\frac{-t}{CR}}\right)$$

where
$V_S = 9\,V$, $t = 1\,s$ and $CR = 1\,\mu F \times 3.3\,\Omega = 3.3\,s$

$$v_C = 9\left(1 - e^{\frac{-1}{3.3}}\right) = 9(1 - 0.738) = 2.358\,V$$

A charged capacitor contains a reservoir of energy stored in the form of an electric field. When the fully charged capacitor from Figure 5.87 is connected as shown in Figure 5.90, the capacitor will discharge through the resistor, and the capacitor voltage (v_C) will fall exponentially with time, as shown in Figure 5.91. The current in the circuit (i) will also fall, as shown in Figure 5.92. The rate of discharge (i.e. the rate of decay of voltage with time) will once again be governed by the time constant of the circuit (CR).

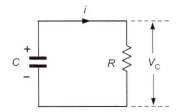

Figure 5.90 *C–R* circuit with *C* discharging through *R*.

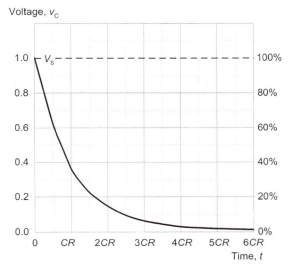

Figure 5.91 Exponential decay of capacitor voltage (v_C) in Figure 5.90.

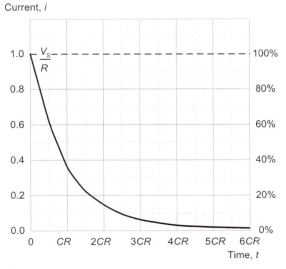

Figure 5.92 Exponential decay of current (i) in Figure 5.90.

The voltage developed across the discharging capacitor (v_C) varies with time (t) according to the relationship:

$$v_C = V_S e^{\frac{-t}{CR}}$$

where v_C is the capacitor voltage (in V), V_S is the DC supply voltage (in V), t is the time (in s) and CR is the time constant of the circuit (equal to the product of capacitance, C, and resistance, R, in s).

The capacitor voltage will fall to approximately 37% of the initial voltage in a time equal to the time constant. At the end of the next interval of time equal to the time constant (i.e. after an elapsed time equal to $2CR$) the voltage will have fallen by 37% of the remainder, and so on. In theory, the capacitor will never quite become fully discharged. After a period of time equal to $5CR$, however, the capacitor voltage will to all intents and purposes be zero. At this point the capacitor voltage will have fallen below 1% of its initial value. At this point we can consider it to be fully discharged.

As with charging, the current in the capacitor (i) varies with time (t) according to the relationship:

$$i = V_S e^{\frac{-t}{CR}}$$

where i is the current (in A), V_S is the DC supply voltage (in V), t is the time and CR is the time constant of the circuit (equal to the product of capacitance, C, and resistance, R, in s).

The current will fall to approximately 37% of the initial current in a time equal to the time constant. At the end of the next interval of time equal to the time constant (i.e. after a total time of $2CR$ has elapsed) the current will have fallen by a further 37% of the remainder, and so on.

Example 5.56

A 10 µF capacitor is charged to a potential of 20 V and then discharged through a 47 kΩ resistor. Determine the time taken for the capacitor voltage to fall below 10 V.

Solution

The formula for exponential decay of voltage in the capacitor is:

$$v_C = V_S e^{\frac{-t}{CR}}$$

In this case, $V_S = 20$ V and $CR = 10\,\mu\text{F} \times 47\,\text{k}\Omega = 0.47$ s and we need to find t when $v_C = 10$ V.

Re-arranging the formula to make t the subject gives:

$$t = -CR \times \ln\left(\frac{v_C}{V_S}\right)$$

thus:

$$t = -0.47 \times \ln\left(\frac{10}{20}\right) = -0.47 \times (-0.693)$$

$$= 0.325\,\text{s}$$

In order to simplify the mathematics of exponential growth and decay, the table below provides an alternative tabular method that may be used to determine the voltage and current in a C–R circuit:

$\dfrac{t}{CR}$	k (ratio of instantaneous value to final value)	
	Exponential growth	Exponential decay
0.0	0.0000	1.0000
0.1	0.0951	0.9048
0.2	0.1812	0.8187 (see Example 5.57)
0.3	0.2591	0.7408
0.4	0.3296	0.6703
0.5	0.3935	0.6065
0.6	0.4511	0.5488
0.7	0.5034	0.4965
0.8	0.5506	0.4493
0.9	0.5934	0.4065
1.0	0.6321	0.3679
1.5	0.7769	0.2231
2.0	0.8647	0.1353
2.5	0.9179	0.0821
3.0	0.9502	0.0498
3.5	0.9698	0.0302
4.0	0.9817	0.0183
4.5	0.9889	0.0111
5.0	0.9933	0.0067

Example 5.57

A $150\,\mu\text{F}$ capacitor is charged to a potential of $150\,\text{V}$. The capacitor is then removed from the charging source and connected to a $2\,\text{M}\Omega$ resistor. Determine the capacitor voltage 1 min later.

Solution

We will solve this problem using the tabular method rather than using the exponential formula. First we need to find the time constant:

$$C \times R = 150\,\mu\text{F} \times 2\,\text{M}\Omega = 300\,\text{s}$$

Next we find the ratio of t to CR. After 1 min, $t = 60\,\text{s}$ therefore the ratio of t to CR is:

$$\frac{t}{CR} = \frac{60}{300} = 0.2$$

Referring to the table (page 366) we find that when $t/CR = 0.2$, the ratio of instantaneous value to final value (k) for decay is 0.8187.
   Thus:

$$\frac{v_C}{V_S} = 0.8187$$

or

$$v_C = 0.8187 \times 150 = 122.8\,\text{V}$$

Key point

The time constant of a C–R circuit is the product of the capacitance, C, and resistance, R.

Key point

The voltage across the plates of a charging capacitor grows exponentially at a rate determined by the time constant of the circuit. Similarly, the voltage across the plates of a discharging capacitor decays exponentially at a rate determined by the time constant of the circuit.

Test your understanding 5.10

1. Capacitance is defined as the ratio of _____ to _____.

2. A capacitor of $220\,\mu\text{F}$ is charged from a $200\,\text{V}$ supply. What charge will be present on the plates of the capacitor?

3. A charge of $25\,\mu\text{C}$ appears on the plates of a $500\,\mu\text{F}$ capacitor. What p.d. appears across the capacitor plates?

4. A capacitor is said to be fully discharged when there is no _____ _____ between the plates.

5. When a capacitor is charged an _____ _____ will be present in the space between the plates.

6. A $10\,\mu\text{F}$ capacitor is charged to a potential of $20\,\text{V}$. How much energy is stored in the capacitor?

7. Which one of the following has the lowest value of dielectric constant (permittivity): air, glass, paper, polystyrene, vacuum?

8. Capacitors of 4 and $2\,\mu\text{F}$ are connected (a) in series and (b) in parallel. Determine the equivalent capacitance in each case.

9. A capacitor consists of plates having an area of $0.002\,\text{m}^2$ separated by a ceramic material having a thickness of $0.2\,\text{mm}$ and a permittivity of 450. Determine the value of capacitance.

10. A $100\,\mu\text{F}$ capacitor is to be charged from a $50\,\text{V}$ supply via a resistor of $1\,\text{M}\Omega$ resistor. If the capacitor is initially uncharged, determine the capacitor voltage at (a) $50\,\text{s}$ and (b) $200\,\text{s}$.

5.11 Magnetism

Syllabus
(a) Theory of magnetism; Properties of a magnet; Action of a magnet suspended in the earth's magnetic field; Magnetization and demagnetization; Magnetic shielding; Various types of magnetic material; Electromagnets construction and principles of operation; Hand clasp rules to determine magnetic field around current-carrying conductor.

Knowledge level key

A	B1	B2
–	2	2

Syllabus
(b) Magnetomotive force (m.m.f.), field strength, magnetic flux density, permeability, hysteresis loop, retentivity, coercive force reluctance, saturation point, eddy currents; Precautions for care and storage of magnets.

Knowledge level key

A	B1	B2
–	2	2

5.11.1 Magnetism and magnetic materials

Magnetism is an effect created by moving the elementary atomic particles in certain materials such as iron, nickel and cobalt. Iron has outstanding magnetic properties and materials that behave magnetically, in a similar manner to iron, are known as *ferromagnetic* materials. These materials experience forces that act on them when placed near a magnet.

The atoms within these materials group in such a way that they produce tiny individual magnets with their own north and south poles. When subject to the influence of a magnet or when an electric current is passed through a coil surrounding them, these individual tiny magnets line up and the material as a whole exhibits magnetic properties.

Figure 5.93(a) shows a ferromagnetic material that has not been influenced by the forces generated from another magnet. In this case, the miniature magnets are oriented in a random manner. Once the material is subject to the influence of another magnet, then these miniature magnets line up (Figure 5.93(b)) and the material itself becomes magnetic with its own north and south poles.

Miniature magnets randomly
orientated. No observable
magnetic effect

(a)

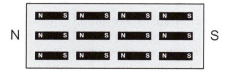

Miniature magnets aligned
(material magnetised). Lines of
magnetic force produced.

(b)

Figure 5.93 The behaviour of ferromagnetic materials.

5.11.2 Magnetic fields around permanent magnets

A magnetic field of *flux* is the region in which the forces created by the magnet have influence. This field surrounds a magnet in all directions, being strongest at the end extremities of the magnet, known as the *poles*. Magnetic fields are mapped by an arrangement of lines that give an indication of strength and direction of the flux as illustrated in Figure 5.94. When freely suspended horizontally a magnet aligns itself north and south parallel with the earth's magnetic field. Now because unlike poles attract, the north of the magnet aligns itself with the south magnetic pole of the earth and the south pole of the magnet aligns itself with the earth's north pole. This is why the extremities of the magnet are known as poles.

Permanent magnets should be carefully stored away from other magnetic components and any systems that might be affected by stray permanent fields. Furthermore, in order to ensure that a permanent magnet retains its magnetism it is usually advisable to store magnets in pairs using soft-iron keepers to link adjacent north and south poles. This arrangement ensures that there is a completely closed path for the magnetic flux produced by the magnets.

Key point

A magnetic field of flux is the region in which the forces created by the magnet have influence. This field surrounds a magnet in all directions and is concentrated at the north and south poles of the magnet.

5.11.3 Electromagnetism

Whenever an electric current flows in a conductor a magnetic field is set up around the conductor in the form of concentric circles. The field is present along the whole length of the conductor and is strongest nearest to the conductor. Now like permanent magnets, this field also has direction. The direction of the magnetic field is dependent on the direction of the current passing through the conductor and may

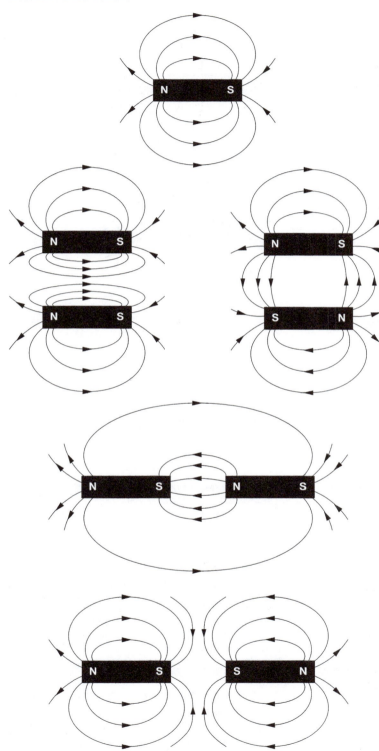

Figure 5.94 Field and flux directions for various bar magnet arrangements.

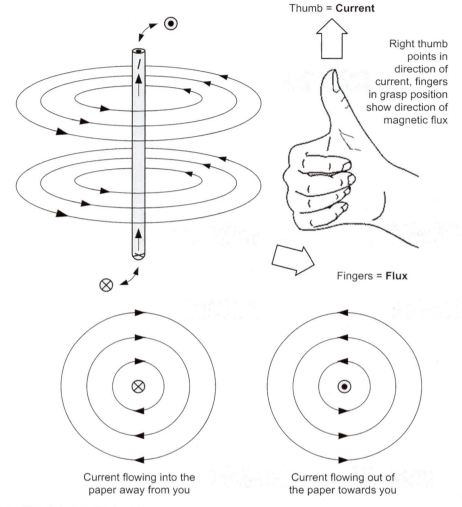

Figure 5.95 The right-hand grip rule.

be established using the *right-hand grip rule*, as shown in Figure 5.95.

If the right-hand thumb is pointing in the direction of current flow in the conductor, then when gripping the conductor in the right hand, the fingers indicate the direction of the magnetic field. In a cross-sectional view of the conductor a point or dot (•) indicates that the current is flowing towards you (i.e. out of the page!) and a cross (×) shows that the current is flowing away from you (i.e. into the page!). This convention mirrors arrow flight, where the dot is the tip of the arrow and the cross is the feathers at the tail of the arrow.

Key point

Whenever an electric current flows in a conductor a magnetic field is set up in the space surrounding the conductor. The field spreads out around the conductor in concentric circles with the greatest density of magnetic flux nearest to the conductor.

5.11.4 Force on a current-carrying conductor

If we place a current-carrying conductor in a magnetic field, the conductor has a force exerted on it. Consider the arrangement shown in

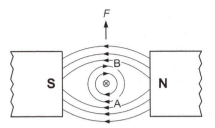

Figure 5.96 A current-carrying conductor in a magnetic field.

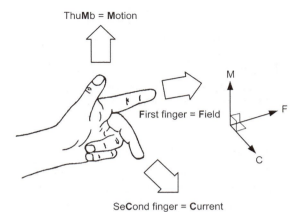

Figure 5.97 Fleming's left-hand rule.

Figure 5.96, in which a current-carrying conductor is placed between the north and south poles of two permanent magnets. The direction of the current passing through it is into the page going away from us. Then by the right-hand screw rule, the direction of the magnetic field, created by the current in the conductor, is clockwise, as shown. We also know that the flux lines from the permanent magnet exit at a north pole and enter at a south pole, in other words, they travel from north to south, as indicated by the direction arrows. The net effect of the coming together of these two magnetic force fields is that at position A, they both travel in the same direction and reinforce one another. While at position B, they travel in the opposite direction and tend to cancel one another. So with a stronger force field at position A and a weaker force at position B the conductor is forced upwards out of the magnetic field.

If the direction of the current was reversed, i.e. if it was to travel towards us out of the page, then the direction of the magnetic field in the current-carrying conductor would be reversed and therefore so would the direction of motion of the conductor.

A convenient way of establishing the direction of motion of the current-carrying conductor is to use *Fleming's left-hand (motor) rule*. This rule is illustrated in Figure 5.97, where the left hand is extended with the thumb, first finger and second finger pointing at right angles to one another. From the figure it can be seen that the first finger represents the magnetic field, the second finger represents the direction of the current in the conductor and the thumb represents the motion of the conductor due to the forces acting on it. The following will help you to remember this:

First finger = Field
SeCond finger = Current
ThuMb = Motion

> **Key point**
>
> If we place a current-carrying conductor in a magnetic field, the conductor has a force exerted on it. If the conductor is free to move this force will produce motion.

The magnitude of the force acting on the conductor depends on the current flowing in the conductor, the length of the conductor in the field and the strength of the magnetic flux (expressed in terms of its *flux density*). The size of the force will be given by the expression:

$$F = BIl$$

where F is the force in newton (N), B is the flux density in tesla (T), I is the current in ampere (A) and l is the length in metre (m).

Flux density is a term that merits a little more explanation. The total flux present in a magnetic field is a measure of the total magnetic intensity present in the field and it is measured in webers (Wb) and represented by the Greek symbol, Φ. The flux density, B, is simply the total flux, Φ, divided by the area over which the flux acts, A. Hence:

$$B = \frac{\Phi}{A}$$

where B is the flux density (T), Φ is the total flux present (Wb) and A is the area (m^2).

Example 5.58

In Figure 5.98, a straight current-carrying conductor lies at right angles to a magnetic field of flux density 1.2 T such that 250 mm of its length lies within the field. If the current passing through the conductor is 15 A, determine the force on the conductor and the direction of its motion.

Solution

In order to find the magnitude of the force we use the relationship $F = BIl$, hence:

$$F = BIl = 1.2 \times 15 \times 250 \times 10^{-3} = 4.5\,\text{N}$$

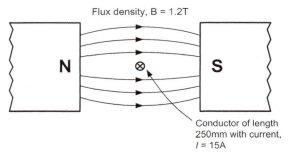

Flux density, B = 1.2T

N $\otimes$ S

Conductor of length
250mm with current,
I = 15A

Figure 5.98

Now the direction of motion is easily found using Fleming's left-hand rule, where we know that the first finger points in the direction of the magnetic field north and south, the second finger points inwards into the page in the direction of the current, which leaves your thumb pointing *downwards* in the direction of motion.

5.11.5 Magnetic field strength and flux density

The strength of a magnetic field is a measure of the density of the flux at any particular point. In the case of Figure 5.99, the field strength, B, will be proportional to the applied current and inversely proportional to the distance from the conductor.

Thus:

$$B = \frac{kI}{d}$$

where B is the magnetic flux density (in T), I is the current (in A), d is the distance from the conductor (in m) and k is a constant. Assuming that the medium is vacuum or *free space*, the density of the magnetic flux will be given by:

$$B = \frac{\mu_0 I}{2\pi d}$$

where B is the flux density (in T), μ_0 is the permeability of free space ($4\pi \times 10^{-7}$ or 12.57×10^{-7} H/m), I is the current (in A) and d is the distance from the center of the conductor (in m).

The flux density is also equal to the total flux, Φ, divided by the area, A, over which the field

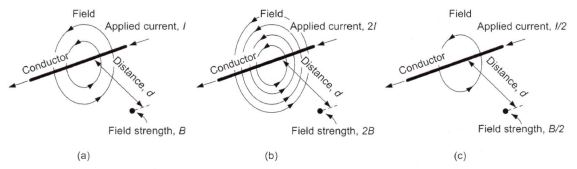

(a)   (b)   (c)

Figure 5.99   Field strength at a point.

acts. Thus:

$$B = \frac{\Phi}{A}$$

where Φ is the flux (in Wb) and A is the area of the field (in m^2).

In order to increase the strength of the field, a conductor may be shaped into a loop (Figure 5.100) or coiled to form a *solenoid* (Figure 5.101).

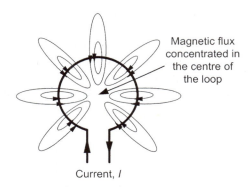

Magnetic flux concentrated in the centre of the loop

Current, I

Figure 5.100 Magnetic field around a single turn loop.

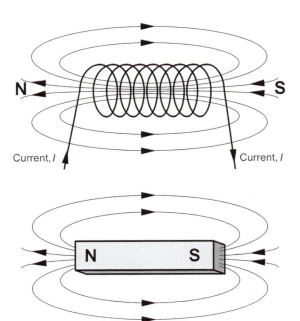

Current, I Current, I

N   S

N   S

Figure 5.101 Magnetic field around a coil or solenoid.

Example 5.59

Determine the flux density produced at a distance of 50 mm from a straight wire carrying a current of 20 A.

Solution

Applying the formula:

$$B = \frac{\mu_0 I}{2\pi d}$$

gives:

$$B = \frac{12.57 \times 10^{-7} \times 20}{6.28 \times 5 \times 10^{-3}} = \frac{251.4}{31.4} \times 10^{-4}$$
$$= 8 \times 10^{-4}\,\text{T}$$

or:

$$B = 0.8\,\text{mT}$$

Example 5.60

A flux density of 2.5 mT is developed in free space over an area of 20 cm^2. Determine the total flux.

Solution

Re-arranging the formula $B = \dfrac{\Phi}{A}$ to make Φ the subject gives:

$$\Phi = BA$$

thus:

$$\Phi = 2.5 \times 10^{-3} \times 20 \times 10^{-4}$$
$$= 50 \times 10^{-7}\,\text{Wb} \quad \text{or} \quad 5\,\mu\text{Wb}.$$

5.11.6 Magnetic circuits

Materials such as iron and steel possess considerably enhanced magnetic properties. Hence they are employed in applications where it is necessary to increase the flux density produced by an electric current. In effect, they allow us to channel the electric flux into a *magnetic circuit*, as shown in Figure 5.102(b).

In the circuit of Figure 5.102(b) the reluctance of the magnetic core is analogous to the resistance present in the electric circuit shown in Figure 5.102(a). We can make the following comparisons between the two types of circuit:

Electric circuit	Magnetic circuit
e.m.f. $= V$	m.m.f. $= N \times I$
Resistance $= R$	Reluctance $= S$
Current $= I$	Flux $= \Phi$
e.m.f. $=$ current $\times$ resistance	m.m.f. $=$ flux $\times$ reluctance
$V = IR$	$NI = S\Phi$

In practice, not all of the magnetic flux produced in a magnetic circuit will be concentrated within the core and some *leakage flux* will appear in the surrounding free space (as shown in Figure 5.103). Similarly, if a gap appears within the magnetic circuit, the flux will tend to spread out as shown in Figure 5.104. This effect is known as *fringing*.

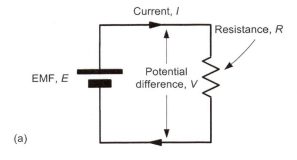

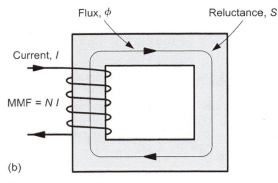

Figure 5.102 Comparison of (a) electric and (b) magnetic circuits.

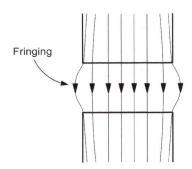

Figure 5.104 Fringing.

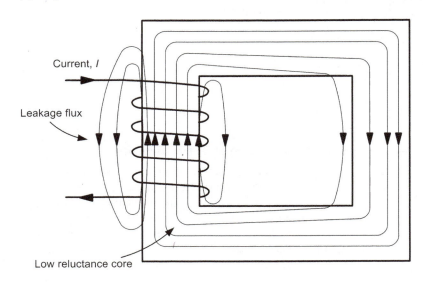

Figure 5.103 Leakage flux.

5.11.7 Reluctance and permeability

The reluctance of a magnetic path is directly proportional to its length and inversely proportional to its cross-sectional area. The reluctance is also inversely proportional to the *absolute permeability* of the magnetic material.

Thus:

$$S = \frac{l}{\mu A}$$

where S is the reluctance of the magnetic path, l is the length of the path (in m), A is the cross-sectional area of the path (in m²) and μ is the absolute permeability of the magnetic material.

Now the absolute permeability of a magnetic material is the product of the *permeability of free space* (μ_0) and the *relative permeability* of the magnetic medium (μ_r).

Thus:

$$\mu = \mu_0 \times \mu_r$$

and:

$$S = \frac{l}{\mu_0 \mu_r A}$$

One way of thinking about permeability is that it is a measure of a magnetic medium's ability to support magnetic flux when subjected to a *magnetizing force*. Thus absolute permeability, μ, is given by:

$$\mu = \frac{B}{H}$$

where B is the flux density (in T) and H is the magnetizing force (in A/m).

The term "magnetizing force" needs a little explanation. We have already said that, in order to generate a magnetic flux we need to have a current flowing in a conductor and that we can increase the field produced by winding the conductor into a coil which has a number of turns of wire.

The product of the number of turns, N, and the current flowing, I, is known as m.m.f. (look back at the comparison table of electric and magnetic circuits if this is still difficult to understand). The magnetizing force, H, is the m.m.f. (i.e. $N \times I$) divided by the length of the magnetic path, l. Thus:

$$H = \frac{\text{m.m.f.}}{l} = \frac{NI}{l}$$

where H is the magnetizing force (in A/m), N is the number of turns, I is the current (in A) and l is the length of the magnetic path (in m).

> **Key point**
>
> The m.m.f. produced in a coil can be determined from the product of the number of turns, N, and the current flowing, I. The units of m.m.f. are "ampere-turns" or simply A (as "turns" strictly has no units). Magnetizing force, on the other hand, is determined from the m.m.f. divided by the length of the magnetic circuit, l, and its units are "ampere-turns per metre" or simply A/m.

5.11.8 *B–H* curves

Figure 5.105 shows three typical curves showing flux density, B, plotted against magnetizing force, H, for some common magnetic materials. It should be noted that each of these B–H curves eventually flattens off due to magnetic saturation and that the slope of the curve (indicating the value of μ corresponding to a particular value of H) falls as the magnetizing force increases. This is important since it dictates the acceptable working range for a particular magnetic material when used in a magnetic circuit.

It is also important to note that, once exposed to a magnetizing force, some magnetic materials (such as soft iron) will retain some of their magnetism even when the magnetizing force is removed. This property of a material to retain some residual magnetism is known as remanance (or retentivity). It is important that the materials used for the magnetic cores of inductors and transformers have extremely low values of remanance.

> **Key point**
>
> *B–H* curves provide us with very useful information concerning the magnetic properties of the material that is used for the magnetic core of an inductor or transformer. The slope of the *B–H* curve gives us an indication of how good the material is at supporting a magnetic flux whilst a flattening-off of the curve shows us when saturation has been reached (and no further increase in flux can be accommodated within the core).

Flux Density, B (T)

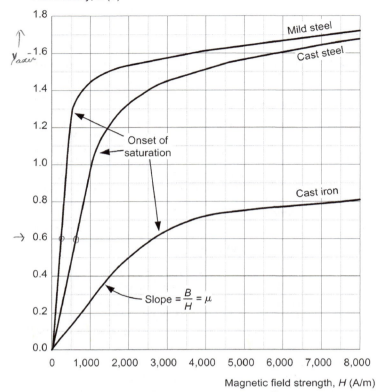

Figure 5.105 Some typical B–H curves for common magnetic materials.

Example 5.61

Estimate the relative permeability of cast steel at (a) a flux density of 0.6 T and (b) a flux density of 1.6 T.

Solution

From Figure 5.105, the slope of the graph at any point gives the value of μ at that point. The slope can be found by constructing a tangent at the point in question and finding the ratio of vertical change to horizontal change.

(a) The slope of the graph, μ, at 0.6 T is

$$\frac{0.5}{500} = 1 \times 10^{-3}$$

Now since $\mu = \mu_0 \times \mu_r$:

$$\mu_r = \frac{\mu}{\mu_0} = \frac{1 \times 10^{-3}}{12.57 \times 10^{-7}} = 795$$

(b) The slope of the graph, μ, at 1.6 T is

$$\frac{0.06}{1500} = 0.04 \times 10^{-3}$$

Now since $\mu = \mu_0 \times \mu_r$:

$$\mu_r = \frac{\mu}{\mu_0} = \frac{0.04 \times 10^{-3}}{12.57 \times 10^{-7}} = 31.8$$

Note: This example very clearly shows the effect of saturation on the permeability of a magnetic material!

Example 5.62

A coil of 800 turns is wound on a closed mild steel core having a length 600 mm and cross-sectional area 500 mm². Determine the current required to establish a flux of 0.8 mWb in the core.

Solution

Now:

$$B = \frac{\Phi}{A} = \frac{0.8 \times 10^{-3}}{500 \times 10^{-6}} = 1.6\,\text{T}$$

From Figure 5.105, a flux density of 1.6 T will occur in mild steel when $H = 3500\,\text{A/m}$.

Recall that:

$$H = \frac{NI}{l}$$

from which:

$$I = \frac{Hl}{N} = \frac{3500 \times 0.6}{800} = 2.625\,\text{A}$$

5.11.9 Magnetic shielding

As we have seen, magnetic fields permeate the space surrounding all current-carrying conductors. The leakage of magnetic flux from one circuit into another can sometimes cause problems, particularly where sensitive electronic equipment is present (such as instruments, navigational aids and radio equipment).

The magnetic field around a conductor or a magnetic component (such as an inductor or transformer) can be contained by surrounding the component in question with a magnetic shield made up of a high permeability alloy (such as *mumetal*). The shield not only helps to prevent the leakage of flux from the component placed inside it but it can also prevent the penetration of stray external fields. In effect, the shield acts as a "magnetic bypass" which offers a much lower reluctance path than the air or free space surrounding it. *like a sponge*

Test your understanding 5.11

1. Whenever an electric _____ flows in a conductor a _____ _____ is set up in the space surrounding the conductor.

2. The unit of magnetic flux is the _____ and its symbol is ____.

3. The unit of flux density is the _____ and its symbol is ____.

4. A straight conductor carrying a current of 12 A is placed at right angles to a magnetic field having a flux density of 0.16 T. Determine the force acting on the conductor if it has a length of 40 cm.

5. State the relationship between flux density, B, total flux, Φ, and area, A.

6. A flux density of 80 mT is developed in free space over an area of 100 cm². Determine the total flux present.

7. The reluctance, S, of a magnetic circuit is _____ proportional to its length and _____ proportional to its cross-sectional area.

8. Sketch a graph showing how flux density, B, varies with magnetizing force, H, for a typical ferromagnetic material.

9. In the linear portion of a B–H curve the flux density increases from 0.1 to 0.3 T when the magnetizing force increases from 35 to 105 A/m. Determine the relative permeability of the material.

10. Briefly explain the need for magnetic shielding and give an example of a material that is commonly used for the construction of a magnetic shield.

5.12 Inductance and inductors

Syllabus

Faraday's law; Action of inducing a voltage in a conductor moving in a magnetic field; Induction principles; Effects of the following on the magnitude of an induced voltage: magnetic field strength, rate of change of flux, number of conductor turns; Mutual induction; The effect the rate of change of primary current and mutual inductance has on induced voltage; Factors affecting mutual inductance: number of turns in coil, physical size of coil, permeability of coil, position of coils with respect to each other; Lenz's law and polarity determining rules; Back e.m.f., self induction; Saturation point; Principal uses of inductors.

Knowledge level key

A	B1	B2
–	2	2

5.12.1 Induction principles

The way in which electricity is generated in a conductor may be viewed as being the exact opposite to that which produces the motor force. In order to generate electricity we require movement in to get electricity out. In fact we need the same components to generate electricity as those needed for the electric motor, namely a closed conductor, a magnetic field and movement.

Whenever relative motion occurs between a magnetic field and a conductor acting at right

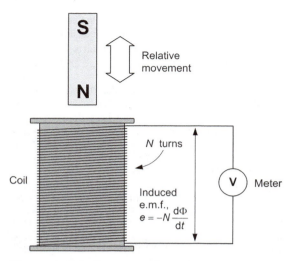

Figure 5.106 Demonstration of electromagnetic induction.

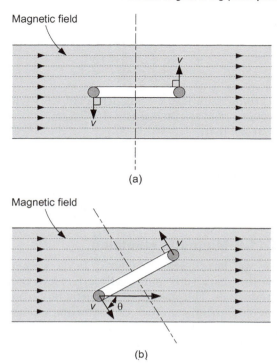

Figure 5.107 Cutting lines of flux and the e.m.f. generated: (a) at 90°, $e = Blv$ and (b) at θ, $e = Blv \sin \theta$.

angles to the field, an e.m.f. is induced or generated in the conductor. The manner in which this e.m.f. is generated is based on the principle of electromagnetic induction.

Consider Figure 5.106, which shows relative movement between a magnet and a closed coil of wire.

An e.m.f. will be induced in the coil whenever the magnet is moved in or out of the coil (or the magnet is held stationary and the coil moved). The magnitude of the induced e.m.f., e, depends on the number of turns, N, and the rate at which the flux changes in the coil, $d\Phi/dt$. Note that this last expression is simply a mathematical way of expressing the *rate of change of flux with respect to time*.

The e.m.f., e, is given by the relationship:

$$e = -N\frac{d\Phi}{dt}$$

where N is the number of turns and $d\Phi/dt$ is the rate of change of flux. The minus sign indicates that the polarity of the generated e.m.f. *opposes* the change.

5.12.2 Induced e.m.f.

Now the number of turns, N, is directly related to the length of the conductor, l, moving through a magnetic field with flux density, B.

Also, the velocity with which the conductor moves through the field determines the rate at which the flux changes in the coil as it cuts the flux field. Thus the magnitude of the induced (generated) e.m.f., e, is proportional to the flux density, length of conductor and relative velocity between the field and the conductor, or in symbols:

$$e \propto Blv$$

where B is the strength of the magnetic field (in T), l is the length of the conductor in the field (in m) and v is the velocity of the conductor (in m/s).

Now you are probably wondering why the above relationship has the proportionality sign. In order to generator an e.m.f. the conductor must cut the lines of magnetic flux. If the conductor cuts the lines of flux at right angles (Figure 5.107(a)) then the maximum e.m.f. is generated; cutting them at any other angle θ (Figure 5.107(b)), reduces this value until $\theta = 0°$,

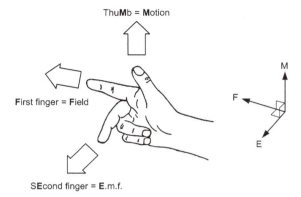

ThuMb = Motion

First finger = Field

SEcond finger = E.m.f.

Figure 5.108 Fleming's right-hand rule.

at which point the lines of flux are not being cut at all and no e.m.f. is induced or generated in the conductor. So the magnitude of the induced e.m.f. is also dependent on $\sin\theta$. So we may write:

$$e = Blv \sin\theta$$

So much for the magnitude of the generated e.m.f., what about its direction in the conductor? Since the conductor offers some resistance, the generated e.m.f. will initiate current flow as a result of the p.d. and the direction of this current can be found using Fleming's right-hand rule. Note that for generators we use the right-hand rule (Figure 5.108), for motors we used the left-hand rule. The first finger, second finger and thumb represent the field, e.m.f. and motion, respectively, as they did when we looked at the motor rule earlier in Section 5.11.4.

5.12.3 Faraday's law

When a magnetic flux through a coil is made to vary, an e.m.f. is induced. The magnitude of this e.m.f. is proportional to the rate of change of magnetic flux.

What this law is saying in effect is that relative movement between the magnetic flux and the conductor is essential to generate an e.m.f. The voltmeter shown in Figure 5.106 indicates the induced (generated) e.m.f. and if the direction of motion changes the polarity of the induced e.m.f. in the conductor changes. Faraday's law also tells us that the magnitude

of the induced e.m.f. is dependent on the relative velocity with which the conductor cuts the lines of magnetic flux.

5.12.4 Lenz's law

Lenz's law states that the current induced in a conductor opposes the changing field that produces it. It is therefore important to remember that the induced current *always* acts in such a direction so as to oppose the change in flux. This is the reason for the minus sign in the formula that we met earlier.

$$e = -N\frac{d\Phi}{dt}$$

Key point

The induced e.m.f. tends to oppose any change of current and because of this we often refer to it as a *back e.m.f.*

Example 5.63

A closed conductor of length 15 cm cuts the magnetic flux field of 1.25 T with a velocity of 25 m/s. Determine the induced e.m.f. when:

(a) the angle between the conductor and field lines is 60°;
(b) the angle between the conductor and field lines is 90°.

Solution

(a) The induced e.m.f. is found using

$$e = Blv \sin\theta,$$

hence:

$$e = 1.25 \times 0.15 \times 25 \times \sin 60°$$
$$= 4.688 \times 0.866 = 4.06\,\text{V}$$

(b) The maximum induced e.m.f. occurs when the lines of flux are cut at 90°. In this case

$$e = Blv \sin\theta = Blv$$

(recall that $\sin 90° = 1$), hence:

$$e = 1.25 \times 0.15 \times 25 = 4.688\,\text{V}$$

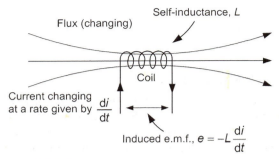

NB: Induced e.m.f. opposes current change

Figure 5.109 Self-inductance.

5.12.5 Self and mutual inductance

We have already shown how an induced e.m.f. (i.e. a back e.m.f.) is produced by a flux change in an inductor. The back e.m.f. is proportional to the rate of change of current (from Lenz's law), as illustrated in Figure 5.109.

This effect is called *self-inductance* (or just *inductance*) which has the symbol, L. Self-inductance is measured in henries (H) and is calculated from:

$$e = -L\frac{di}{dt}$$

where L is the self-inductance, di/dt is the rate of change of current and the minus sign indicates that the polarity of the generated e.m.f. *opposes* the change (you might like to compare this relationship with the one shown earlier for electromagnetic induction).

The unit of inductance is the henry (H) and a coil is said to have an inductance of 1 H if a voltage of 1 V is induced across it when a current changing at the rate of 1 A/s is flowing in it.

Example 5.64

A coil has a self-inductance of 15 mH and is subject to a current that changes at a rate of 450 A/s. What e.m.f. is produced?

Solution

Now $e = -L\dfrac{di}{dt}$

and hence:

$$e = -15 \times 10^{-3} \times 450 = -6.75\,\text{V}$$

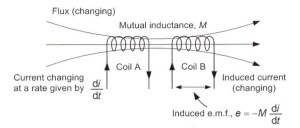

Figure 5.110 Mutual inductance.

Note the minus sign! In other words, a *back e.m.f.* of 6.75 V is induced.

Example 5.65

A current increases at a uniform rate from 2 to 6 A in a time of 250 ms. If this current is applied to an inductor determine the value of inductance if a back e.m.f. of 15 V is produced across its terminals.

Solution

Now $e = -L\dfrac{di}{dt}$

and hence $\quad L = -e\dfrac{dt}{di}$

$$\begin{aligned}
\text{Thus } L &= -(-15) \times \frac{250 \times 10^{-3}}{(6-2)} \\
&= 15 \times 62.5 \times 10^{-3} = 937.5 \times 10^{-3} \\
&= 0.94\,\text{H}
\end{aligned}$$

Finally, when two inductors are placed close to one another, the flux generated when a changing current flows in the first inductor will cut through the other inductor (see Figure 5.110). This changing flux will, in turn, induce a current in the second inductor. This effect is known as *mutual inductance* and it occurs whenever two inductors are *inductively coupled*. This is the principle of a very useful component, the *transformer*, which we shall meet later in Section 5.16.

The value of mutual inductance, M, is given by:

$$M = k\sqrt{L_1 \times L_2}$$

where k is the coupling factor and L_1 and L_2 are the values of individual inductance.

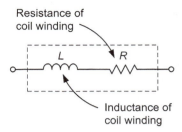

Figure 5.111   A real inductor has resistance as well as inductance.

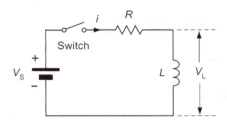

Figure 5.112   Circuit in which a current is applied to an inductor.

5.12.6 Inductors

Inductors provide us with a means of storing electrical energy in the form of a magnetic field. Typical applications include chokes, filters and frequency selective circuits. The electrical characteristics of an inductor are determined by a number of factors including the material of the core (if any), the number of turns and the physical dimensions of the coil.

In practice every coil comprises both inductance and resistance and the circuit of Figure 5.111 shows these as two discrete components. In reality the inductance, L, and resistance, R, are both distributed throughout the component but it is convenient to treat the inductance and resistance as separate components in the analysis of the circuit.

Now let us consider what happens when a current is first applied to an inductor. If the switch in Figure 5.112 is left open, no current will flow and no magnetic flux will be produced by the inductor. If the switch is now closed, current will begin to flow as energy is taken from the supply in order to establish the magnetic field. However, the change in magnetic flux resulting from the appearance of current creates a voltage (an induced e.m.f.) across the coil which opposes the applied e.m.f. from the battery.

The induced e.m.f. results from the changing flux and it effectively prevents an instantaneous rise in current in the circuit. Instead, the current increases slowly to a maximum at a rate which depends upon the ratio of inductance, L, to resistance, R, present in the circuit.

After a while, a steady-state condition will be reached in which the voltage across the inductor will have decayed to zero and the current will

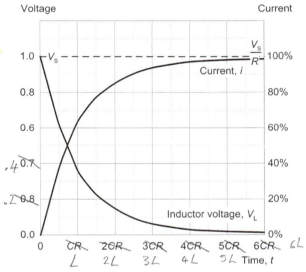

Figure 5.113   Voltage and current in the circuit of Figure 5.112.

have reached a maximum value (determined by the ratio of V to R, i.e. using Ohm's law!).

If, after this steady-state condition has been achieved, the switch is opened again, the magnetic field will suddenly collapse and the energy will be returned to the circuit in the form of a *back e.m.f.* which will appear across the coil as the field collapses (Figure 5.113).

5.12.7 Energy storage

The energy stored in an inductor is proportional to the product of the inductance and the square of the current flowing in it. Thus:

$$W = 0.5LI^2$$

where W is the energy (in J), L is the inductance (in H) and I is the current (in A).

Example 5.66

A current of 1.5 A flows in an inductor of 5 H. Determine the energy stored.

Solution

Now $W = 0.5LI^2$

$$= 0.5 \times 5 \times 1.5^2 = 5.625\,\text{J}$$

Example 5.67

An inductor of 20 mH is required to store an energy of 2.5 J. Determine the current that must be applied to the inductor.

Solution

Now $W = 0.5LI^2$ and hence

$$I = \sqrt{\frac{W}{0.5 \times L}} = \sqrt{\frac{2.5}{0.5 \times 20 \times 10^{-3}}}$$

$$= \sqrt{2.5 \times 10^2} = 15.8\,\text{A}$$

5.12.8 Inductance and physical characteristics

The inductance of an inductor depends upon the physical dimensions of the inductor (e.g. the length and diameter of the winding), the number of turns and the permeability of the material of the core. The inductance of an inductor is given by:

$$L = \frac{\mu_0 \mu_r n^2 A}{l}$$

where L is the inductance (in H), μ_0 is the permeability of free space (12.57×10^{-7} H/m), μ_r is the relative permeability of the magnetic core, l is the length of the core (in m) and A is the cross-sectional area of the core (in m^2).

Example 5.68

An inductor of 100 mH is required. If a closed magnetic core of length 20 cm, cross-sectional area 15 cm^2 and relative permeability 500 is available, determine the number of turns required.

Solution

Now $L = \dfrac{\mu_0 \mu_r n^2 A}{l}$

and hence $n = \sqrt{\dfrac{Ll}{\mu_0 \mu_r A}}$

Thus $n = \sqrt{\dfrac{Ll}{\mu_0 \mu_r A}}$

$$= \sqrt{\frac{100 \times 10^{-3} \times 20 \times 10^{-2}}{12.57 \times 10^{-7} \times 500 \times 15 \times 10^{-4}}}$$

$$= \sqrt{\frac{2 \times 10^{-2}}{94{,}275 \times 10^{-11}}} = \sqrt{21{,}215} = 146$$

Hence the inductor requires 146 turns of wire.

5.12.9 Inductor types, values and tolerances

Inductor specifications normally include the value of inductance (expressed in H, mH, μH or nF), the current rating (i.e. the maximum current which can be continuously applied to the inductor under a given set of conditions) and the accuracy or tolerance (quoted as the maximum permissible percentage deviation from the marked value). Other considerations may include the temperature coefficient of the inductance (usually expressed in parts per million, ppm, per-unit temperature change), the stability of the inductor, the DC resistance of the coil windings (ideally zero), the quality factor (Q-factor) of the coil and the recommended working frequency range.

Table 5.5 summarizes the properties of commonly available types of inductor.

Several manufacturers supply fixed and variable inductors for operation at high and radio frequencies. Fixed components are generally available in the E6 series between 1 μH and 10 mH. Variable components have ferrite dust cores which can be adjusted in order to obtain a precise value of inductance as required, for example, in a tuned circuit. The higher inductance values generally exhibit a larger DC resistance due to the greater number of turns and

Table 5.5

Characteristic	Inductor type					
	Single-layer open		Multi-layer open		Multi-layer pot cored	Multi-layer iron cored
Core material	Air	Ferrite	Air	Ferrite	Ferrite	Iron
Inductance range	50 nH to 10 µH	1 µH to 100 µH	5 µH to 500 µH	10 µH to 1 mH	1 mH to 100 mH	20 mH to 20 H
Typical tolerance	±10%	±10%	±10%	±10%	±10%	±10%
Typical current rating	0.1 A	0.1 A	0.2 A	0.5 A	0.5 A	0.2 A
Typical DC resistance	0.05 Ω to 1 Ω	0.1 Ω to 10 Ω	1 Ω to 20 Ω	2 Ω to 100 Ω	2 Ω to 100 Ω	10 Ω to 400 Ω
Typical Q-factor	60	80	100	80	40	20
Typical frequency range	5 MHz to 500 MHz	1 MHz to 500 MHz	200 kHz to 20 MHz	100 kHz to 10 MHz	1 kHz to 1 MHz	50 Hz to 1 kHz
Typical applications	Tuned circuits	Tuned circuits	Filters and HF transformers	Filters and HF transformers	LF and MF chokes and transformers	LF chokes and transformers

relatively small diameter of wire used in their construction.

At medium and low frequencies, inductors are often manufactured using one of a range of ferrite pot cores. The core material of these inductors is commonly available in several and the complete pot core assembly comprises a matched pair of core halves, a single-section bobbin, a pair of retaining clips and a core adjuster. Effectively, the coil winding is totally enclosed in a high permeability ferrite pot. Typical values of inductance for these components range between 100 µH and 100 mH with a typical saturation flux density of 250 mT.

Inductance values of iron cored inductors are very much dependent upon the applied DC and tend to fall rapidly as the value of applied DC increases and saturation is approached. Maximum current ratings for larger inductors are related to operating temperatures and should be de-rated when high ambient temperatures are expected. Where reliability is important, inductors should be operated at well below their nominal maximum current ratings.

Finally, *ferrite* (a high permeability non-conductive magnetic material) is often used as the core material for inductors used in high-frequency filters and as broadband transformers

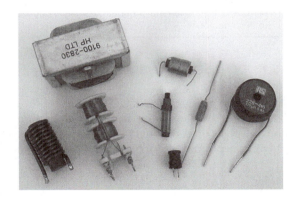

Figure 5.114 Various inductors.

at frequencies of up to 30 MHz. At these frequencies, inductors can be realized very easily using these cores with just a few turns of wire (Figure 5.114)!

Key point

The specifications for an inductor resistor usually include the value of inductance (expressed in H, mH or µH), the current rating (quoted as the maximum permissible percentage deviation from the marked value) and the DC resistance (this is the resistance of the coil windings measured in Ω). The Q-factor and frequency range are also important considerations for certain types of inductor.

Test your understanding 5.12

1. Whenever relative motion occurs between a _____ _____ field and a _____ an e.m.f. is _____ in the _____.

2. Sketch a simple diagram showing how you could demonstrate electromagnetic induction.

3. State Faraday's law.

4. State Lenz's law.

5. Explain what is meant by a "back e.m.f."

6. A closed conductor of length 50 cm cuts a magnetic flux of 0.75 T at an angle of 45°. Determine the induced e.m.f. if the conductor is moving at a velocity of 5 m/s.

7. A current increases at a uniform rate from 1.5 to 4.5 A in a time of 50 ms. If this current is applied to a 2 H inductor determine the value of induced e.m.f.

8. Explain, with the aid of a sketch, what is meant by (a) self-inductance and (b) mutual inductance.

9. An inductor has a closed magnetic core of length 40 cm, cross-sectional area 10 cm² and relative permeability 450. Determine the value of inductance if the inductor has 250 turns of wire.

10. An inductor of 600 mH is required to store an energy of 400 mJ. Determine the current that must be applied to the inductor.

5.13 DC motor/generator theory

Syllabus

Basic motor and generator theory; Construction and purpose of components in DC generator; Operation of, and factors affecting output and direction of current flow in DC generators; Operation of, and factors affecting output power torque, speed and direction of rotation of DC motors; Series wound, shunt wound and compound motors; Starter-Generator construction.

Knowledge level key

A	B1	B2
–	2	2

5.13.1 Basic generator theory

When a conductor is moved through a magnetic field, an e.m.f. will be induced across its ends. An induced e.m.f. will also be generated if the conductor remains stationary whilst the field moves. In either case, cutting at right angles through the lines of magnetic flux (see Figure 5.115) results in a generated e.m.f. and the magnitude of which will be given by:

$$E = Blv$$

where B is the magnetic flux density (in T), l is the length of the conductor (in m), and v is the velocity of the field (in m/s).

If the field is cut at an angle, θ (rather than at right angles), the generated e.m.f. will be given by:

$$E = Blv \sin \theta$$

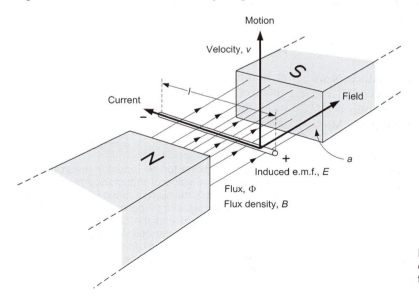

Figure 5.115 Generating an e.m.f. by moving a conductor through a magnetic field.

where θ is the angle between the direction of motion of the conductor and the field lines.

Example 5.69

A conductor of length 20 cm moves at 0.5 m/s through a uniform perpendicular field of 0.6 T. Determine the e.m.f. generated.

Solution

Since the field is perpendicular to the conductor, the angle is 90° ("perpendicular" means the same as "at right angles") we can use the basic equation:

$$E = Blv$$

where $B = 0.6$ T, $l = 20$ cm $= 0.02$ m and $v = 0.5$ m/s. Thus:

$$E = Blv = 0.6 \times 0.02 \times 0.5$$
$$= 0.006 \text{ V}$$
$$= 6 \text{ mV}$$

Key point

An e.m.f. will be induced across the ends of a conductor when there is relative motion between it and a magnetic field. The induced voltage will take its greatest value when moving at right angle to the magnetic field lines and its least value (i.e. zero!) when moving along the direction of the field lines.

5.13.2 A simple AC generator

Being able to generate a voltage by moving a conductor through a magnetic field is extremely useful as it provides us with an easy way of generating electricity. Unfortunately, moving a wire at a constant linear velocity through a uniform magnetic field presents us with a practical problem simply because the mechanical power that can be derived from an aircraft engine is available in rotary (rather than linear) form!

The solution to this problem is that of using the rotary power available from the engine (via a suitable gearbox and transmission) to rotate a conductor shaped into the form of loop as shown in Figure 5.116. The loop is made to rotate inside a permanent magnetic field with opposite poles (north and south) on either side of the loop.

There now remains the problem of making contact with the loop as it rotates inside the magnetic field but this can be overcome by means of a pair of carbon brushes and copper slip rings. The brushes are spring loaded and held against the rotating slip rings so that, at any time, there is a path for current to flow from the loop to the load to which it is connected (Figure 5.117).

The opposite sides of the loop consist of conductors that move through the field. At 0° (with the loop vertical as shown in Figure 5.118) the opposite sides of the loop will be moving in the same direction as the lines of flux. At that instant, the angle, θ, at which the field is cut is 0°

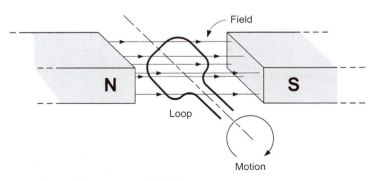

Figure 5.116   A loop rotating within a magnetic field.

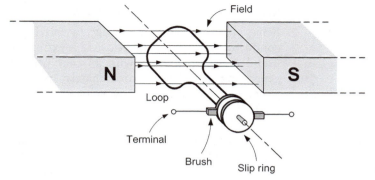

Figure 5.117 Brush arrangement.

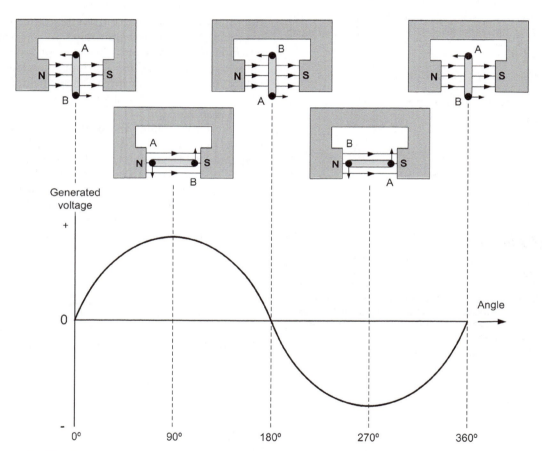

Figure 5.118 The e.m.f. generated at various angles.

and since the sine of 0° is 0 the generated voltage (from $E = Blv\sin\theta$) will consequently also be zero.

If the loop has rotated to a position which is 90° from that shown in Figure 5.118, the two conductors will effectively be moving at right angles to the field. At that instant, the generated e.m.f. will take a maximum value (since the sine of 90° is 1).

At 180° from the starting position the generated e.m.f. will have fallen back to zero since, once again, the conductors are moving along

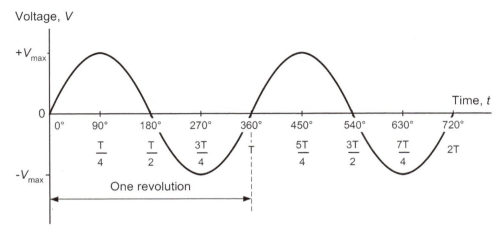

Figure 5.119   Sinusoidal voltage produced by the rotating loop.

the flux lines (but in the direction opposite to that at 0°).

At 270° the conductors will once again be moving in a direction which is perpendicular to the flux lines (but in the direction opposite to that at 90°). At this point, a maximum generated e.m.f. will once again be produced. It is, however, important to note that the e.m.f. generated at this instant will be of opposite polarity to that which was generated at 90°. The reason for this is simply that the relative direction of motion (between the conductors and flux lines) has effectively been reversed.

Since $E = Blv \sin \theta$, the e.m.f. generated by the arrangement shown in Figure 5.118 will take a sinusoidal form, as shown in Figure 5.119. Note that the maximum values of e.m.f. occur at 90° and 270°, and that the generated voltage is zero at 0°, 180° and 360°.

In practice, the single loop shown in Figure 5.118 would comprise a coil of wire wound on a suitable non-magnetic former. This coil of wire effectively increases the length of the conductor within the magnetic field and the generated e.m.f. will then be directly proportional to the number of turns on the coil.

Key point

In a simple AC generator a loop of wire rotates inside the magnetic field produced by two opposite magnetic poles. Contact is made to the loop as it rotates by means of slip rings and brushes.

5.13.3 DC generators

When connected to a load, the simple generator shown in Figure 5.118 produces a sinusoidal AC output. In many applications a steady DC output may be preferred. This can be achieved by modifying the arrangement shown in Figure 5.118, replacing the brushes and slip rings with a *commutator* arrangement, as shown in Figure 5.120. The commutator arrangement functions as a rotating reversing switch which ensures that the e.m.f. generated by the loop is reversed after rotating through 180°. The generated e.m.f. for this arrangement is shown in Figure 5.121. It is worth comparing this waveform with that shown in Figure 5.118.

The generated e.m.f. shown in Figure 5.121, whilst unipolar (i.e. all positive or all negative), is clearly far from ideal since a DC power source should provide a constant voltage output rather than a series of pulses. One way of overcoming this problem is with the use of a second loop (or coil) at right angles to the first, as shown in Figure 5.122. The commutator is then divided into four (rather than two) *segments* and the generated e.m.f. produced by this arrangement is shown in Figure 5.123.

In real generators, a coil comprising a large number of turns of conducting wire replaces the single-turn rotating loop. This arrangement effectively increases the total length of the conductor within the magnetic field and, as a result, also increases the generated output voltage. The

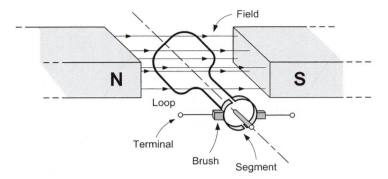

Figure 5.120   Commutator arrangement.

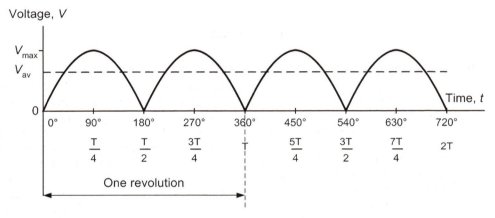

Figure 5.121   The e.m.f. generated (compare with Figure 5.119).

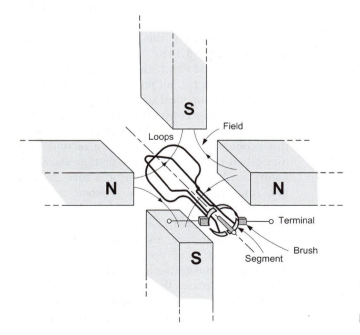

Figure 5.122   An improved DC generator.

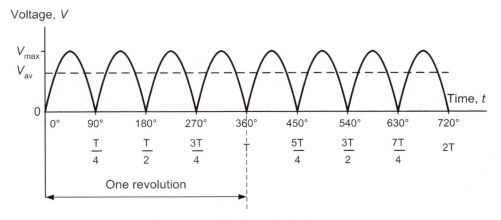

Figure 5.123 The e.m.f. generated (compare with Figure 5.121).

output voltage also depends on the density of the magnetic flux through which the current-carrying conductor passes. The denser the field the greater the output voltage will be.

> **Key point**
>
> A simple DC generator uses an arrangement similar to that used for an AC generator but with the slip rings and brushes replaced by a commutator that reverses the current produced by the generator every 180°.

5.13.4 DC motors

A simple DC motor consists of a very similar arrangement to that of the DC generator that we met earlier. A loop of wire that is free to rotate is placed inside a permanent magnetic field (see Figure 5.124). When a DC current is applied to the loop of wire, two equal and opposite forces are set up which act on the conductor in the directions indicated in Figure 5.124.

The direction of the forces acting on each arm of the conductor can be established by again using the right-hand grip rule and Fleming's left-hand rule. Now because the conductors are equidistant from their pivot point and the forces acting on them are *equal and opposite*, they form a *couple*. The *moment* of this couple is equal to the magnitude of a single force multiplied by the distance between them and this

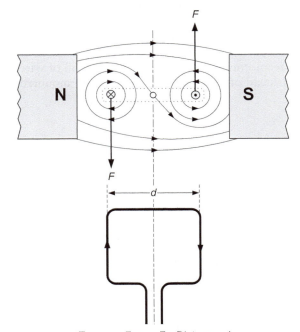

Torque = Force, F x Distance, d

Figure 5.124 Torque on a current-carrying loop suspended within a magnetic field.

moment is known as *torque*, T. Now,

$$T = Fd$$

where T is the torque (in Newton-metres, Nm), F is the force (in N) and d is the distance (in m).

We already know that the magnitude of the force F is given by $F = BIl$; therefore, the torque produced by the current carrying thus the torque expression can be written:

$$T = BIld$$

where T is the torque (in Nm), B is the flux density (in T), I is the current (in A), l is the length of conductor (in m) and d is the distance (in m).

In a practical situation the conductor would be wound to form a coil. If the coil has N turns and each loop of the coil has a length, l, then the torque produced will be given by:

$$T = BlINd$$

(You can more easily remember this as "BLIND"!)

The torque produces a *turning moment* such that the coil or loop rotates within the magnetic field. This rotation continues for as long as a current is applied. A more practical form of DC motor consists of a rectangular coil of wire (instead of a single turn loop of wire) mounted on a former and free to rotate about a shaft in a permanent magnetic field, as shown in Figure 5.125.

In real motors, this rotating coil is known as the *armature* and consists of many hundreds of turns of conducting wire. This arrangement is needed in order to maximize the force imposed on the conductor by introducing the *longest possible* conductor into the magnetic field. Also from the relationship $F = BIl$ it can be seen that the force used to provide the torque in a motor is directly proportional to the size of the magnetic flux, B. Instead of using a permanent magnet to produce this flux, in a real motor, an electromagnet is used. Here an electromagnetic field is set up using the *solenoid* principle (Figure 5.126). A long length of conductor is wound into a coil consisting of many turns and a current passed through it. This arrangement constitutes a *field winding* and each of the turns in the field winding assists each of the other turns in order to produce a strong magnetic field, as shown in Figure 5.126.

As in the case of the DC generator, this field may be intensified by inserting a ferromagnetic core inside the coil. Once the current is applied to the conducting coil, the core is magnetized and all the time the current is on it acts in combination with the coil to produce a

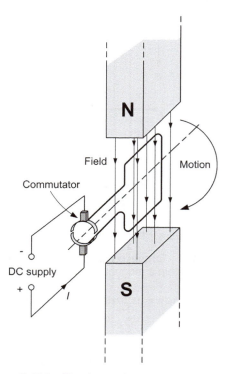

Figure 5.125 Simple electric motor with commutator.

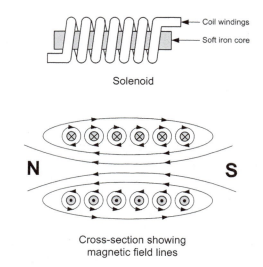

Figure 5.126 Magnetic field produced by a solenoid.

permanent magnet, having its own north–south poles.

Now returning to the simple motor illustrated in Figure 5.125, we know that when current is supplied to the armature (*rotor*) a torque is produced. In order to produce continuous rotary motion, this torque (turning moment) must always act in the same direction.

Therefore, the current in each of the armature conductors must be reversed as the conductor passes between the north and south magnetic field poles. The *commutator* acts like a rotating switch, reversing the current in each armature conductor at the appropriate time to achieve this continuous rotary motion. Without the presence of a commutator in a DC motor, only a half-turn of movement is possible!

In Figure 5.127(a) the rotation of the armature conductor is given by Fleming's left-hand rule. When the coil reaches a position mid-way between the poles (Figure 5.127(b)), no rotational torque is produced in the coil. At this stage the commutator reverses the current in the coil. Finally (Figure 5.127(c)) with the current reversed, the motor torque now continues to rotate the coil in its original direction.

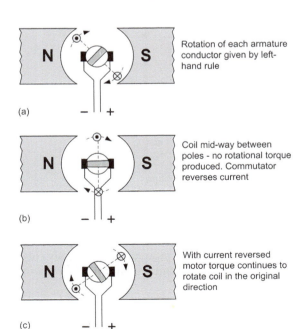

(a) Rotation of each armature conductor given by left-hand rule

(b) Coil mid-way between poles - no rotational torque produced. Commutator reverses current

(c) With current reversed motor torque continues to rotate coil in the original direction

Figure 5.127   Action of the commutator.

Key point

The torque produced by a DC motor is directly proportional to the product of the current flowing in the rotating armature winding.

Example 5.70

The rectangular armature shown in Figure 5.128 is wound with 500 turns of wire. When situated in a uniform magnetic field of flux density 300 mT, the current in the coil is 20 mA. Calculate the force acting on the side of the coil and the maximum torque acting on the armature.

Solution

With this arrangement the ends of the conductor are not within the influence of the magnetic field and therefore have no force exerted on them. Therefore, the force acting on one length of conductor can be found from $F = BIl$, thus:

$$F = BIl$$
$$= (300 \times 10^{-3})(20 \times 10^{-3})(30 \times 10^{-3})$$
$$= 1.8 \times 10^{-4} \, \text{N}$$

Then the force on one side of the coil is 500 times this value, thus:

$$F = 500 \times 1.8 \times 10^{-4} = 9 \times 10^{-2} \, \text{N}$$

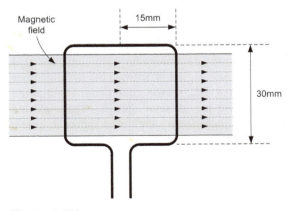

Figure 5.128

Now from our definition of torque $T = Fd$, the torque acting on the armature winding is:

$$T = (9 \times 10^{-2})(30 \times 10^{-3}) = 2.7 \times 10^{-3} \, \text{N}$$

This is a relatively small amount of torque. Practical motors can be made to produce output torques with very small values as demonstrated here, up to several hundred Nm!

One other application of the motor principle is used in simple analogue measuring instruments. Some meters, including multimeters used to measure current, voltage and resistance, operate on the principle of a coil rotating in a magnetic field. The basic construction is shown in Figure 5.129, where the current, I, passes through a pivoted coil and the resultant motor force (the *deflecting torque*) is directly proportional to the current flowing in the coil windings which of course is the current being measured. The magnetic flux is concentrated within the coil by a solid cylindrical ferromagnetic core, in exactly the same manner as the flux is concentrated within a solenoid.

5.13.5 Series wound, shunt wound and compound motors

The field winding of a DC motor can be connected in various different ways according to the application envisaged for the motor in question. The following configurations are possible:

- series wound;
- shunt wound;
- compound wound (where both series and shunt windings are present).

In the series-wound DC motor the field winding is connected in series with the armature and the full armature current flows through the field winding (see Figure 5.130). This arrangement results in a DC motor that produces a large starting torque at slow speeds. This type of motor is ideal for applications where a heavy load is applied from rest. The disadvantage of this type of motor is that on light loads the motor speed may become excessively high. For this reason this type of motor should not be used in situations where the load may be accidentally removed. A typical set of torque and

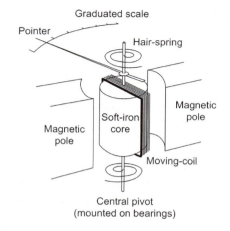

Figure 5.129   The moving-coil meter.

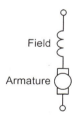

Figure 5.130   Series-wound DC motor.

speed characteristics (plotted against supply current) for a series-wound DC motor is shown in Figure 5.131.

In the shunt-wound DC motor the field winding is connected in parallel with the armature and thus the supply current is divided between the armature and the field winding (see

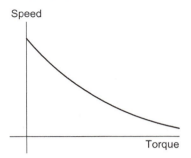

Figure 5.131 Typical torque and speed characteristics for a series-wound DC motor.

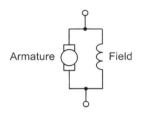

Figure 5.132 Shunt-wound DC motor.

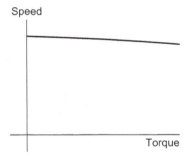

Figure 5.133 Typical torque and speed characteristics for a shunt-wound DC motor.

Figure 5.132). This arrangement results in a DC motor that runs at a reasonably constant speed over a wide variation of load but does not perform well when heavily loaded. A typical set of torque and speed characteristics (plotted against supply current) for a shunt-wound DC motor is shown in Figure 5.133.

The compound-wound DC motor has both series and shunt field windings (see Figure 5.134) and is therefore able to combine some of the advantages of each type of motor. A typical set of torque and speed characteristics (plotted against supply current) for a compound-wound DC motor is shown in Figure 5.135.

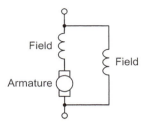

Figure 5.134 Compound-wound DC motor.

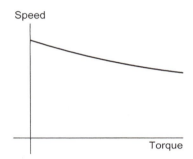

Figure 5.135 Typical torque and speed characteristics for a compound-wound DC motor.

> **Key point**
>
> In order to avoid the need for a large permanent magnet, a separate field winding can be used in a DC machine (i.e. a motor or generator). This field winding is energized with DC. In the case of a DC generator, this current can be derived from the output of the generator (in which case it is referred to as *self-excited*) or it can be energized from a separate DC supply.

5.13.6 Starter-generator

Starter-generators eliminate the need for separate starter motors and DC generators. They usually have separate field windings (one for the starter motor and one for the generator) together with a common armature winding. When used for starting, the starter-generator is connected as a series-wound DC motor capable of producing a very high starting torque. However, when used as a generator the connections are changed so that the unit operates as shunt-wound generator producing reasonably constant current over a wide range of speed.

In the start condition, the low-resistance starter field and common armature windings of the starter-generator are connected in series

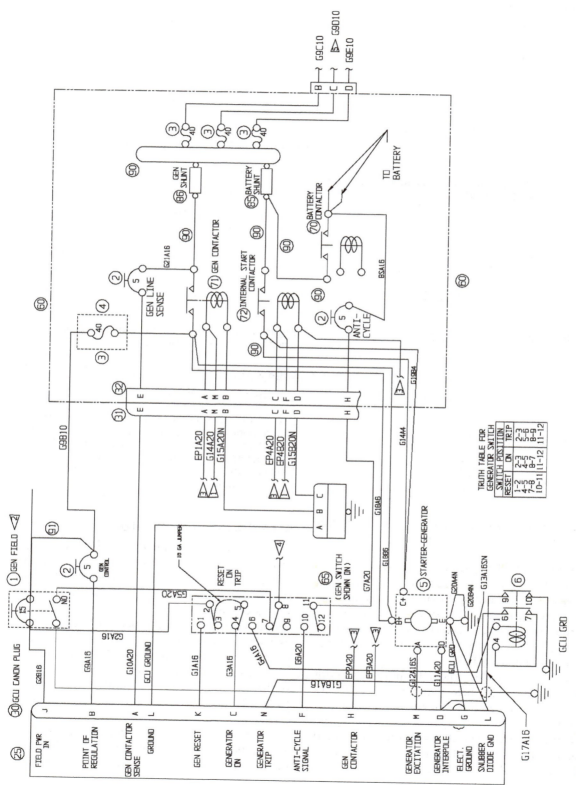

Figure 5.136 Starter-generator circuit showing contactors.

across the DC supply via a set of contactors. This arrangement ensures that a torque is produced that is sufficient to start an aircraft's turbine engine.

When the engine reaches self-sustaining speed, the current is broken through the first set of contactors and a second set of contactors operate, removing the external DC power supply from the starter-generator and reconnecting the arrangement so that the armature voltage generated is fed to the higher-resistance shunt field and the aircraft's main voltage regulator.

The advantage of this arrangement is not only that the starter-generator replaces two individual machines (i.e. a starter and a generator) with consequent savings in size and weight, but additionally that only a single mechanical drive is required between the engine and the starter-generator unit. The disadvantage of this arrangement is that the generator output is difficult to maintain at low engine revolutions per minute (rpm) and therefore starter-generators are mainly found on turbine powered aircraft that maintain a relatively high engine rpm (Figure 5.136).

Test your understanding 5.13

1. When a _____ is moved through a magnetic field an _____ will be _____ across its ends.

2. A wire of length 80 cm moves at right angles to a magnetic field of 0.5 T. Determine the e.m.f. generated if the conductor has a velocity of 15 m/s.

3. In a simple AC generator the problem of making contact with a rotating loop of wire can be solved using _____ and _____.

4. In a simple DC generator the current is reversed every _____ by means of a _____.

5. Maximum e.m.f. will be generated when a conductor moves at _____ with respect to a magnetic field.

6. An e.m.f. of 50 mV appears across the ends of a conductor when it is moved at 1.5 m/s through a magnetic field. What e.m.f. will be produced if the speed is increased to 6.0 m/s?

7. A rectangular loop is suspended in a magnetic field having a flux density of 0.4 T. If the total length of the loop is 0.2 m and the current flowing is 3 A, determine the torque produced.

8. Sketch the field and armature arrangement of each of the following types of DC motor: (a) series wound, (b) shunt wound and (c) compound wound.

9. Explain the advantages and disadvantages of series-wound DC motors.

10. Sketch a graph showing the typical variation of torque and speed for a series-wound DC motor.

5.14 AC theory

Syllabus

Sinusoidal waveform: phase, period, frequency, cycle; Instantaneous, average, root mean square (r.m.s), peak, peak-to-peak current values and calculations of these values, in relation to voltage, current and power; Triangular/square waves; Single- or three-phase principles.

Knowledge level key

A	B1	B2
1	2	2

5.14.1 Alternating current

DC are currents which, even though their magnitude may vary, essentially flow only in one direction. In other words, DC are unidirectional. AC, on the other hand, are bi-directional and continuously reversing their direction of flow. The polarity of the e.m.f. which produces an AC must also consequently be changing from positive to negative, and vice versa, as shown in Figure 5.137.

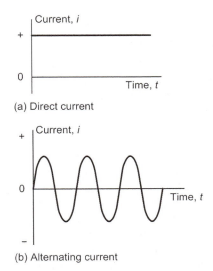

(a) Direct current

(b) Alternating current

Figure 5.137 Direct and alternating voltages.

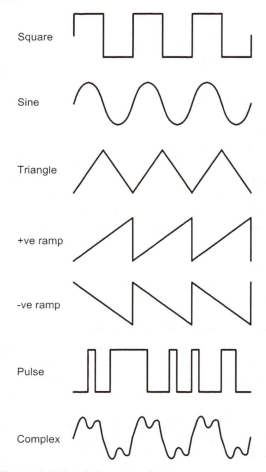

Figure 5.138 Various waveforms.

5.14.2 Waveforms

A graph showing the variation of voltage or current present in a circuit is known as a waveform. There are many common types of waveforms encountered in electrical circuits including sine (or sinusoidal), square, triangle, ramp or sawtooth (which may be either positive or negative going) and pulse. Complex waveforms like speech or music usually comprise many components at different frequencies. Pulse waveforms are often categorized as either repetitive or nonrepetitive (the former comprises a pattern of pulses which regularly repeats whilst the latter comprises pulses which constitute a unique event). Several of the most common waveform types are shown in Figure 5.138.

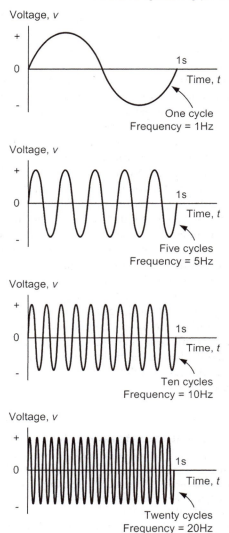

Figure 5.139 Waveforms with different frequencies shown to a common time scale.

5.14.3 Frequency and periodic time

The *frequency* of a repetitive waveform is the number of cycles of the waveform that occur in unit time. Frequency is expressed in hertz (Hz). A frequency of 1 Hz is equivalent to one cycle per second. Hence, if a voltage has a frequency of 400 Hz, then 400 cycles will occur in every second (Figure 5.139).

The *periodic time* (or *period*) of a waveform is the time taken for one complete cycle of the wave

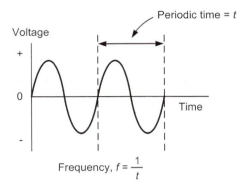

Figure 5.140 Periodic time.

(see Figure 5.140). The relationship between periodic time and frequency is thus:

$$t = \frac{1}{f} \quad \text{or} \quad f = \frac{1}{t}$$

where t is the periodic time (in s) and f is the frequency (in Hz).

Example 5.71

A waveform has a frequency of 400 Hz. What is the periodic time of the waveform?

Solution

Now $t = \dfrac{1}{f} = \dfrac{1}{400\text{Hz}} = 0.0025\,\text{s} = 2.5\text{ms}$

Hence the waveform has a periodic time of 2.5 ms.

Example 5.72

A waveform has a periodic time of 20 ms. What is its frequency?

Solution

Now $f = \dfrac{1}{t} = \dfrac{1}{20\,\text{ms}} = \dfrac{1}{0.02\,\text{s}} = 50\,\text{Hz}$

Hence the waveform has a frequency of 50 Hz.

5.14.4 Average, peak, peak-to-peak and r.m.s. values

The *average* value of an AC which swings symmetrically above and below zero will obviously be zero when measured over a long period of time. Hence average values of currents and voltages are invariably taken over one complete half-cycle (either positive or negative) rather than over one complete full-cycle (which would result in an average value of zero).

The *peak* value (or *maximum* value or *amplitude*) of a waveform is the measure of an extent of its voltage or current excursion from the resting value (usually zero). The *peak-to-peak* value for a wave which is symmetrical about its resting value is twice its peak value.

The *r.m.s.* or *effective* value of an alternating voltage or current is the value which would produce the same heat energy in a resistor as a direct voltage or current of the same magnitude. Since the r.m.s. value of a waveform is very much dependent upon its shape, values are only meaningful when dealing with a waveform of known shape. Where the shape of a waveform is not specified, r.m.s. values are normally assumed to refer to sinusoidal conditions.

For a given waveform, a set of fixed relationships exist between average, peak, peak-to-peak and r.m.s. values. The required multiplying factors are summarized below for sinusoidal voltages and currents (see Figure 5.141).

Given quantity	Wanted quantity			
	Avg.	Peak	Peak-to-peak	r.m.s.
Average	1	1.57	3.14	1.11
peak	0.636	1	2	0.707
Peak-to-peak	0.318	0.5	1	0.353
r.m.s.	0.9	1.414	2.828	1

From the table we can conclude that, e.g.:

$$V_{\text{av}} = 0.636 \times V_{\text{pk}}$$

$$V_{\text{pk-pk}} = 2 \times V_{\text{pk}}$$

$$V_{\text{r.m.s.}} = 0.707 \times V_{\text{pk}}$$

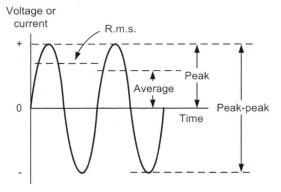

Figure 5.141 Average, r.m.s., peak and peak-to-peak values of a sine wave.

Similar relationships apply to the corresponding AC, thus:

$$I_{av} = 0.636 \times I_{pk}$$

$$I_{pk-pk} = 2 \times I_{pk}$$

$$I_{r.m.s.} = 0.707 \times I_{pk}$$

Example 5.73

A sinusoidal voltage has an r.m.s. value of 220 V. What is the peak value of the voltage?

Solution

Now $V_{pk} = 1.414 \times V_{r.m.s.}$

$$= 1.414 \times 220\,V = 311\,V$$

Hence the sinusoidal voltage has a peak value of 311 V.

Example 5.74

A sinusoidal current has a peak-to-peak value of 4 mA. What is its r.m.s. value?

Solution

Now $I_{r.m.s} = 0.353 \times I_{pk-pk}$

$$= 0.353 \times 40\,mA = 14.12\,mA$$

Hence the sinusoidal current has an r.m.s. value of 14.12 mA.

5.14.5 Expression for a sine wave voltage

We can derive an expression for the instantaneous voltage, v, of a sine wave in terms of its peak voltage and the sine of an angle, θ. Thus:

$$v = V_{pk} \sin \theta$$

The angle, θ, will in turn depend on the exact moment in time, t, and how fast the sine wave is changing (in other words, its angular velocity, ω).

Hence:

$$v = V_{pk} \sin(\omega t) \qquad (1)$$

Since there are 2π radians in one complete revolution or cycle of voltage or current, a frequency of one cycle per second (i.e. 1 Hz) must be the same as 2π radians per second. Hence, a frequency, f, is equivalent to:

$$f = \frac{\omega}{2\pi}\,Hz$$

Making ω the subject of the equation gives:

$$\omega = 2\pi f \qquad (2)$$

By combining equations (1) and (2) we can obtain a useful expression that will allow us to determine the voltage (or current) at any instant of time provided that we know the peak value of the sine wave and its frequency:

$$v = V_{pk} \sin(2\pi f t)$$

Example 5.75

A sine wave voltage has a maximum value of 100 V and a frequency of 50 Hz. Determine the instantaneous voltage present at (a) 2.5 ms and (b) 15 ms from the start of the cycle.

Solution

We can determine the voltage at any instant of time using:

$$v = V_{max} \sin(2\pi ft)$$

where $V_{max} = 100$ V and $f = 50$ Hz.

In (a), $t = 2.5$ ms hence:

$$v = 100 \sin(2\pi \times 50 \times 0.0025)$$

$$= 100 \sin(0.785) = 100 \times 0.707 = 70.7 \text{ V}$$

In (b), $t = 15$ ms hence:

$$v = 100 \sin(2\pi \times 50 \times 0.015) = 100 \sin(4.71)$$

$$= 100 \times (-1) = -100 \text{ V}$$

5.14.6 Three-phase supplies

The most simple method of distributing an AC supply is a system that uses two wires. In fact, this is how AC is distributed in your home (the third wire present is simply an earth connection for any appliances that may require it for safety reasons). In many practical applications, including aircraft, it can be advantageous to use a *multi-phase* supply rather than a *single-phase* supply (here the word *phase* simply refers to an AC voltage source). The most common system uses three separate voltage sources (and three wires) and is known as *three phase*. The voltages produced by the three sources are spaced equally in time such that the *phase angle* between them is 120° (or 360°/3). The waveforms for a three-phase supply are shown in Figure 5.142 (note that each is a sine wave and all three sine waves have the same frequency and periodic time). We will look at this again in much greater detail in Section 5.18.

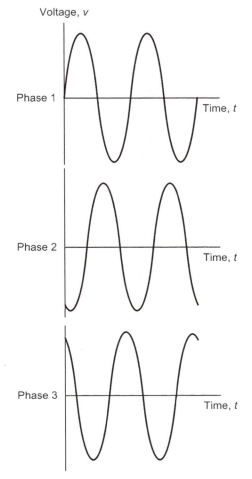

Figure 5.142 Waveforms for a three-phase AC supply.

Test your understanding 5.14

1. The average value of a sine wave *over one complete cycle* is _____.

2. The average value of a sine wave *over one half-cycle* is _____ of its peak value.

3. To convert a sinusoidal r.m.s. value to a peak value you need to multiply by _____.

4. To convert a sinusoidal peak value to an r.m.s. value you need to multiply by _____.

5. A waveform having a periodic time of 40 ms will have a frequency of _____ Hz.

6. A waveform having a frequency of 500 Hz will have a periodic time of _____ ms.

7. Another name for the r.m.s. value of a waveform is its _____ value.

8. Amplitude is another name for the _____ value of a waveform.

9. The r.m.s. value of an AC is the value which would produce the same amount of _____ in a resistor as the same value of DC.

10. Sketch each of the following waveforms: a sine wave, a square wave and a triangle wave. Label your waveforms with axes of time and voltage.

5.15 Resistive, capacitive and inductive circuits

Syllabus
Phase relationship of voltage and current in L, C and R circuits, parallel, series and series and parallel; Power dissipation in L, C and R circuits; Impedance, phase angle, power factor and current calculations; True power, apparent power and reactive power calculations.

Knowledge level key

A	B1	B2
–	2	2

5.15.1 AC flowing through pure resistance

Ohm's law is obeyed in an AC circuit just as it is in a DC circuit. Thus, when a sinusoidal voltage, V, is applied to a resistor, R (as shown in Figure 5.143), the current flowing in the resistor will be given by:

$$I = \frac{V}{R}$$

This relationship must also hold true for the instantaneous values of current, i, and voltage, v, thus:

$$i = \frac{v}{R}$$

and since $v = V_{\max} \sin \omega t$

$$i = \frac{V_{\max} \sin(\omega t)}{R}$$

The current and voltage in Figure 5.143 both have a sinusoidal shape and since they rise and fall together, they are said to be *in-phase* with one another. We can represent this relationship by means of the *phasor diagram* shown in Figure 5.144. This diagram shows two rotating phasors (of magnitude I and V) rotating at an angular velocity, ω. The applied voltage (V) is referred to as the *reference phasor* and this is aligned with the horizontal axis (i.e. it has a phase angle of $0°$).

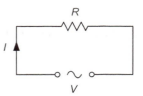

Figure 5.143   AC flowing in a resistor.

Figure 5.144   Phasor diagram showing current and voltage in a resistor.

Key point

Phasor diagrams provide us with a quick way of illustrating the relationships that exist between sinusoidal voltages and currents in AC circuits without having to draw lots of time-related waveforms. Figure 5.145 will help you to understand how the previous phasor diagram relates to the time-related waveforms for the voltage and current in a resistor.

Example 5.76

A sinusoidal voltage $20\,V_{pk-pk}$ is applied to a resistor of $1\,k\Omega$. What value of r.m.s. current will flow in the resistor?

Solution

This problem must be solved in several stages. First we will determine the peak-to-peak current in the resistor and then we shall convert this value into a corresponding r.m.s. quantity.
Since $I = \frac{V}{R}$, we can conclude that

$$I_{pk-pk} = \frac{V_{pk-pk}}{R}$$

Thus, $I_{pk-pk} = \dfrac{20\,V_{pk-pk}}{1\,k\Omega} = 20\,mA_{pk-pk}$

Next, $I_{pk} = \dfrac{I_{pk-pk}}{2} = \dfrac{20}{2} = 10\,mA_{pk}$

Finally,

$$I_{r.m.s.} = 0.707 \times I_{pk-pk} = 0.353 \times 10\,mA$$
$$= 3.53\,mA$$

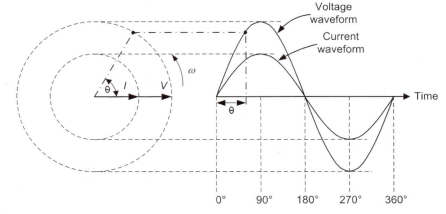

Figure 5.145 A rotating phasor.

5.15.2 Reactance

Reactance, like resistance, is simply the ratio of applied voltage to the current flowing. Thus:

$$X = \frac{V}{I}$$

where X is the reactance in ohms (Ω), V is the alternating p.d. in volts (V) and I is the AC in amperes (A).

In the case of *capacitive reactance* (i.e. the reactance of a capacitor) we use the suffix, C, so that the reactance equation becomes:

$$X_C = \frac{V_C}{I_C}$$

Similarly, in the case of *inductive reactance* (i.e. the reactance of an inductor) we use the suffix, L, so that the reactance equation becomes:

$$X_L = \frac{V_L}{I_L}$$

The voltage and current in a circuit containing pure reactance (either capacitive or inductive) will be out of phase by 90°. In the case of a circuit containing pure capacitance the current will lead the voltage by 90° (alternatively we can say that the voltage lags the current by 90°). This relationship is illustrated by the waveforms shown in Figure 5.146 and the phasor diagram shown in Figure 5.147.

In the case of a circuit containing pure inductance the voltage will lead the current by 90°

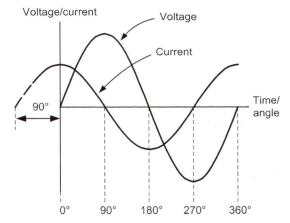

Figure 5.146 Voltage and current waveforms for a pure capacitor (the current leads the voltage by 90°).

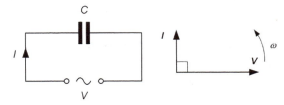

Figure 5.147 Circuit and phasor diagram for the voltage and current in a pure capacitor.

(alternatively we can also say that the current lags the voltage by 90°). This relationship is illustrated by the waveforms shown in Figure 5.148 and the phasor diagram shown in Figure 5.149.

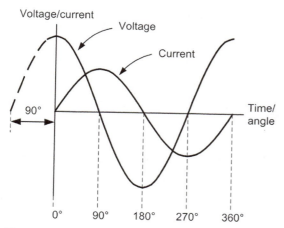

Figure 5.148 Voltage and current waveforms for a pure inductor (the voltage leads the current by 90°).

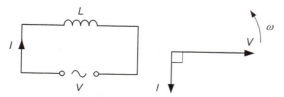

Figure 5.149 Circuit and phasor diagram for the voltage and current in a pure inductor.

Key point

A good way of remembering leading and lagging phase relationships is to recall the word *CIVIL*, as shown in Figure 5.150. Note that, in the case of a circuit containing pure capacitance (C) the current (I) will lead the voltage (V) by 90° whilst in the case of a circuit containing pure inductance (L) the voltage (V) will lead the current (I) by 90°.

5.15.3 Inductive reactance

Inductive reactance is directly proportional to the frequency of the applied AC and can be determined from the following formula:

$$X_L = 2\pi f L$$

where X_L is the reactance (in Ω), f is the frequency (in Hz) and L is the inductance (in H).

Since inductive reactance is directly proportional to frequency ($X_L \propto f$), the graph of inductive reactance plotted against frequency takes the form of a straight line (see Figure 5.151).

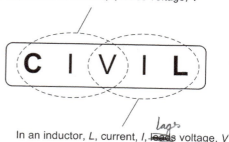

Figure 5.150 Relationship between current and voltage in a circuit with reactance (CIVIL).

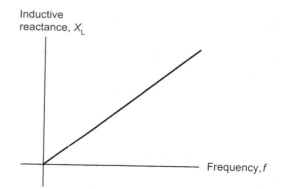

Figure 5.151 Variation of inductive reactance, X_L, with frequency, f.

Example 5.77

Determine the reactance of a 10 mH inductor at (a) 100 Hz and (b) 10 kHz.

Solution

(a) At 100 Hz, $X_L = 2\pi \times 100 \times 10 \times 10^{-3} = 6.28\,\Omega$

(b) At 10 kHz, $X_L = 2\pi \times 10,000 \times 10 \times 10^{-3} = 628\,\Omega$

5.15.4 Capacitive reactance

Capacitive reactance is inversely proportional to the frequency of the applied AC and can be determined from the following formula:

$$X_C = \frac{1}{2\pi f C}$$

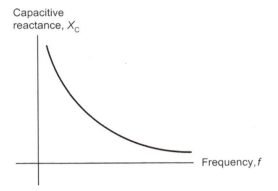

Figure 5.152 Variation of capacitive reactance, X_C, with frequency, f.

where X_C is the reactance (in Ω), f is the frequency (in Hz) and C is the capacitance (in F).

Since capacitive reactance is inversely proportional to frequency ($X_L \propto 1/f$), the graph of inductive reactance plotted against frequency takes the form of a rectangular hyperbola (see Figure 5.152).

Example 5.78

Determine the reactance of a $1\,\mu F$ capacitor at (a) 100 Hz and (b) 10 kHz.

Solution

(a) At 100 Hz,

$$X_C = \frac{1}{2\pi f C} = \frac{1}{2\pi \times 100 \times 1 \times 10^{-6}}$$

$$= \frac{0.159}{10^{-4}} = 1.59\,k\Omega$$

(b) At 10 kHz,

$$X_C = \frac{1}{2\pi f C} = \frac{1}{\begin{array}{c}2\pi \times 10 \times 10^3 \\ \times 1 \times 10^{-6}\end{array}}$$

$$= 0.159 \times 10^2 = 15.9\,\Omega$$

Key point

When alternating voltages are applied to capacitors or inductors the magnitude of the current flowing will depend upon the value of capacitance or inductance and on the frequency of the voltage. In effect, capacitors and inductors oppose the flow of current in much the same way as a resistor. The important difference being that the effective resistance (or *reactance*) of the component varies with frequency (unlike the case of a conventional resistor where the magnitude of the current does not change with frequency).

5.15.5 Impedance

Circuits that contain a mixture of both resistance and reactance (either capacitive reactance or inductive reactance or both) are said to exhibit *impedance*. Impedance, like resistance and reactance, is simply the ratio of applied voltage to the current flowing. Thus:

$$Z = \frac{V}{I}$$

where Z is the impedance in ohms (Ω), V is the alternating p.d. in volts (V) and I is the AC in amperes (A).

Because the voltage and current in a pure reactance are at 90° to one another (we say that they are in *quadrature*) we can not simply add up the resistance and reactance present in a circuit in order to find its impedance. Instead, we can use the *impedance triangle* shown in Figure 5.153. The impedance triangle takes into account the 90° phase angle and from it we can infer that the impedance of a series circuit (R in series with X) is given by:

$$Z = \sqrt{R^2 + X^2}$$

where Z is the impedance (in Ω), X is the reactance, either capacitive or inductive (expressed in Ω) and R is the resistance (also in Ω).

We shall be explaining the significance of the *phase angle*, ϕ, later on. For now you simply need to be aware that ϕ is the angle between the impedance, Z, and the resistance, R. Later on

Figure 5.153 The impedance triangle.

we shall obtain some useful information from the fact that:

$$\sin \phi = \frac{\text{opposite}}{\text{hypotenuse}} = \frac{X}{Z}$$

from which $\phi = \arcsin\left(\dfrac{X}{Z}\right)$

$$\cos \phi = \frac{\text{adjacent}}{\text{hypotenuse}} = \frac{R}{Z}$$

from which $\phi = \arccos\left(\dfrac{R}{Z}\right)$

and

$$\tan \phi = \frac{\text{opposite}}{\text{adjacent}} = \frac{X}{R}$$

from which $\phi = \arctan\left(\dfrac{X}{R}\right)$

Key point

Resistance and reactance combine together to make *impedance*. In other words, impedance is the *resultant* of combining resistance and reactance in the impedance triangle. Because of the *quadrature* relationship between voltage and current in a pure capacitor or inductor, the angle between resistance and reactance in the impedance triangle is always 90°.

Example 5.79

A resistor of 30 Ω is connected in series with a capacitive reactance of 40 Ω. Determine the impedance of the circuit and the current flowing when the circuit is connected to a 115 V supply.

Solution

First we must find the impedance of the C–R series circuit:

$$Z = \sqrt{R^2 + X^2} = \sqrt{30^2 + 40^2} = \sqrt{2500} = 50 \, \Omega$$

The current taken from the supply can now be found:

$$I = \frac{V}{Z} = \frac{115}{50} = 2.3 \, \text{A}$$

Example 5.80

A coil is connected to a 50 V AC supply at 400 Hz. If the current supplied to the coil is 200 mA and the coil has a resistance of 60 Ω, determine the value of inductance.

Solution

Like most practical forms of inductor, the coil in this example has both resistance *and* reactance (see Figure 5.154). We can find the impedance of the coil from:

$$Z = \frac{V}{I} = \frac{50}{0.2} = 250 \, \Omega$$

Since $Z = \sqrt{R^2 + X^2}$, $Z^2 = R^2 + X^2$ and $X^2 = Z^2 - R^2$
from which:

$$X^2 = Z^2 - R^2 = 250^2 - 60^2$$
$$= 62{,}500 - 3600 = 58{,}900$$

Thus $X = \sqrt{58{,}900} = 243 \, \Omega$
  Now since $X_L = 2\pi f L$

$$L = \frac{X_L}{2\pi f} = \frac{243}{6.28 \times 400} = \frac{243}{2512} = 0.097 \, \text{H}$$

Hence $L = 97$ mH.

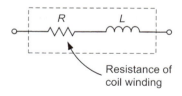

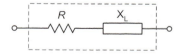

Figure 5.154   A coil with resistance and inductive reactance (see Example 5.80).

5.15.6 Resistance and inductance in series

When a sinusoidal voltage, V, is applied to a series circuit comprising resistance, R, and

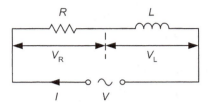

Figure 5.155 A series $R–L$ circuit.

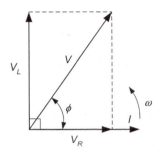

Figure 5.156 Phasor diagram for the series $R–L$ circuit.

inductance, L (as shown in Figure 5.155), the current flowing in the circuit will produce separate voltage drops across the resistor and inductor (V_R and V_L, respectively). These two voltage drops will be 90° apart with V_L *leading* V_R.

We can illustrate this relationship using the phasor diagram shown in Figure 5.156. Note that we have used current as the reference phasor in this series circuit for the simple reason that the same current flows through each component (recall that earlier we used the applied voltage as the reference).

From Figure 5.156 you should note that the supply voltage (V) is simply the result of adding the two voltage phasors, V_R and V_L. Furthermore, the angle between the supply voltage (V) and supply current (I) is the phase angle, ϕ.

Now $\sin\phi = V_L/V$, $\cos\phi = V_R/V$ and $\tan\phi = V_L/V_R$.

Since $X_L = V_L/I$, $R = V_R/I$ and $Z = V/I$ (where Z is the impedance of the circuit), we can illustrate the relationship between X_L, R and Z using the impedance triangle shown in Figure 5.157.

Note that $Z = \sqrt{R^2 + X_L^2}$ and $\phi = \arctan(X_L/R)$.

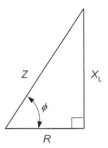

Figure 5.157 Impedance triangle for the series $R–L$ circuit.

Example 5.81

An inductor of 80 mH is connected in series with a 100 Ω resistor. If a sinusoidal current of 20 mA at 50 Hz flows in the circuit, determine:

(a) the voltage dropped across the inductor,
(b) the voltage dropped across the resistor,
(c) the impedance of the circuit,
(d) the supply voltage,
(e) the phase angle.

Solution

(a) $V_L = IX_L = I \times 2\pi fL$
$\quad = 0.02 \times 25.12 = 0.5$ V
(b) $V_R = IR = 0.02 \times 100 = 2$ V
(c) $Z = \sqrt{(R^2 + X_L^2)} = \sqrt{(100^2 + 25.12^2)}$
$\quad = \sqrt{10,631} = 103.1$ Ω
(d) $V = I \times Z = 0.02 \times 103.1 = 2.06$ V
(e) $\phi = \arctan(X_L/R) = \arctan(25.12/100)$
$\quad = \arctan(0.2512) = 14.1°$

5.15.7 Resistance and capacitance in series

When a sinusoidal voltage, V, is applied to a series circuit comprising resistance, R, and inductance, L (as shown in Figure 5.158) the current flowing in the circuit will produce separate voltage drops across the resistor and capacitor (V_R and V_C, respectively). These two voltage drops will be 90° apart – with V_C lagging V_R.

We can illustrate this relationship using the phasor diagram shown in Figure 5.159. Note that once again we have used current as the reference phasor in this series circuit.

From Figure 5.159 you should note that the supply voltage (V) is simply the result of adding the two voltage phasors, V_R and V_C. Furthermore, the angle between the supply voltage (V) and supply current (I), ϕ, is the phase angle.

Now $\sin\phi = V_C/V$, $\cos\phi = V_R/V$ and $\tan\phi = V_C/V_R$.

Since $X_C = V_C/I$, $R = V_R/I$ and $Z = V/I$ (where Z is the impedance of the circuit), we can illustrate the relationship between X_C, R and Z using the impedance triangle shown in Figure 5.160.

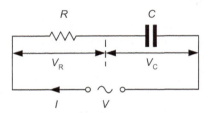

Figure 5.158 A series R–C circuit.

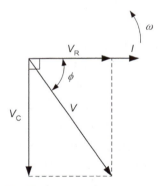

Figure 5.159 Phasor diagram for the series R–C circuit.

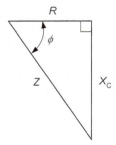

Figure 5.160 Impedance triangle for the series R–C circuit.

Note that $Z = \sqrt{(R^2 + X_C^2)}$ and $\phi = \arctan(X_C/R)$.

Example 5.82

An capacitor of 22 μF is connected in series with a 470 Ω resistor. If a sinusoidal current of 10 mA at 50 Hz flows in the circuit, determine:

(a) the voltage dropped across the capacitor,
(b) the voltage dropped across the resistor,
(c) the impedance of the circuit,
(d) the supply voltage,
(e) the phase angle.

Solution

(a) $V_C = IX_C = I \times 1/(2\pi fC)$
    $= 0.01 \times 144.5 = 1.4$ V
(b) $V_R = IR = 0.01 \times 470 = 4.7$ V
(c) $Z = \sqrt{(R^2 + X_C^2)} = \sqrt{(470^2 + 144.5^2)}$
    $= \sqrt{241,780} = 491.7\ \Omega$
(d) $V = I \times Z = 0.01 \times 491.7 = 4.91$ V
(e) $\phi = \arctan(X_C/R) = \arctan(144.5/470)$
    $= \arctan(0.3074) = 17.1°$

5.15.8 Resistance, inductance and capacitance in series

When a sinusoidal voltage, V, is applied to a series circuit comprising resistance, R, inductance, L and capacitance, C (as shown in Figure 5.161) the current flowing in the circuit will produce separate voltage drops across the resistor, inductor and capacitor (V_R, V_L and V_C, respectively). The voltage drop across the inductor will lead the applied current (and voltage dropped across V_R) by 90° whilst the voltage drop across the capacitor will lag the applied current (and voltage dropped across V_R) by 90°.

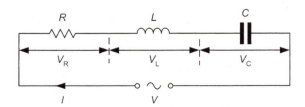

Figure 5.161 Series R–L–C circuit.

When the inductive reactance (X_L) is greater than the capacitive reactance (X_C), V_L will be greater than V_C and the resulting phasor diagram is shown in Figure 5.162. Conversely, when the capacitive reactance (X_C) is greater than the inductive reactance (X_L), V_C will be greater than V_L and the resulting phasor diagram will be shown in Figure 5.163. Note that once again we have used current as the reference phasor in this series circuit.

From Figures 5.162 and 5.163, you should note that the supply voltage (V) is simply the result of adding the three voltage phasors, V_L, V_C and V_R, and that the first stage in simplifying the diagram is that of resolving V_L and V_C into a single voltage ($V_L - V_C$ or $V_C - V_L$ depending upon whichever is the greater). Once again, the phase angle, ϕ, is the angle between the supply voltage and current.

Figures 5.164 and 5.165 show the impedance triangle for the circuit for the cases when $X_L > X_C$ and $X_C > X_L$, respectively.

Note that, when $X_L > X_C$, $Z = \sqrt{[R^2 + (X_L - X_C)^2]}$ and $\phi = \arctan(X_L - X_C)/R$, similarly, when $X_C > X_L$, $Z = \sqrt{[R^2 + (X_C - X_L)^2]}$ and $\phi = \arctan(X_C - X_L)/R$.

It is important to note that a special case occurs when $X_C = X_L$ in which case the two equal but opposite reactances effectively cancel each other out. The result of this is that the circuit behaves as if only resistance, R, is present (in other words, the impedance of the circuit, $Z = R$). In this condition the circuit is said to be *resonant*. The frequency at which resonance occurs is given by:

$$X_C = X_L$$

thus

$$\frac{1}{2\pi fC} = 2\pi fL$$

from which

$$f^2 = \frac{1}{4\pi^2 LC}$$

and thus

$$f = \frac{1}{2\pi\sqrt{LC}}$$

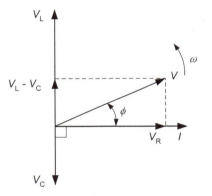

Figure 5.162 Phasor diagram for the series R–C circuit when $X_L > X_C$.

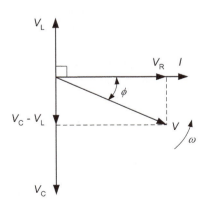

Figure 5.163 Phasor diagram for the series R–C circuit when $X_C > X_L$.

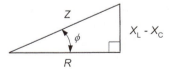

Figure 5.164 Impedance triangle for the series R–C circuit when $X_L > X_C$.

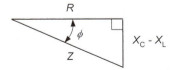

Figure 5.165 Impedance triangle for the series R–C circuit when $X_C > X_L$.

where f is the resonant frequency (in Hz), L is the inductance (in H) and C is the capacitance (in F).

Example 5.83

A series circuit comprises an inductor of 80 mH, a resistor of 200 Ω and a capacitor of 22 μF. If a sinusoidal current of 40 mA at 50 Hz flows in this circuit, determine:

(a) the voltage developed across the inductor,
(b) the voltage dropped across the capacitor,
(c) the voltage dropped across the resistor,
(d) the impedance of the circuit,
(e) the supply voltage,
(f) the phase angle.

Solution

(a) $V_L = IX_L = I \times 2\pi fL = 0.04 \times 25.12$
     $= 1\,V$
(b) $V_C = IX_C = I \times 1/(2\pi fC) = 0.04 \times$
     $144.5 = 5.8\,V$
(c) $V_R = IR = 0.04 \times 200 = 8\,V$
(d) $Z = \sqrt{R^2 + (X_C - X_L)^2}$
     $= \sqrt{200^2 + (144.5 - 25.12)^2}$
     $= \sqrt{54{,}252} = 232.9\,\Omega$
(e) $V = I \times Z = 0.04 \times 232.9 = 9.32\,V$
(f) $\phi = \arctan(X_C - X_L)/R = \arctan(119.38/$
     $200) = \arctan(0.597) = 30.8°$

Example 5.84

A series circuit comprises an inductor of 10 mH, a resistor of 50 Ω and a capacitor of 40 nF. Determine the frequency at which this circuit is resonant and the current that will flow in it when it is connected to a 20 V AC supply at the resonant frequency.

Solution

Using:

$$f = \frac{1}{2\pi\sqrt{LC}}$$

where $L = 10 \times 10^{-3}\,H$ and $C = 40 \times 10^{-9}\,F$ gives:

$$f = \frac{1}{6.28\sqrt{10 \times 10^{-3} \times 40 \times 10^{-9}}}$$

$$= \frac{0.159}{\sqrt{4 \times 10^{-10}}} = \frac{0.159}{2 \times 10^{-5}} = \frac{0.159}{2 \times 10^{-5}}$$
$$= 7950 = 7.95\,kHz$$

At the resonant frequency the circuit will behave as a pure resistance (recall that the two reactances will be equal but opposite) and thus the supply current at resonance can be determined from:

$$I = \frac{V}{Z} = \frac{V}{R} = \frac{20}{50} = 0.4\,A$$

5.15.9 Parallel and series–parallel AC circuits

As we have seen, in a series AC circuit the *same current* flows through each component and the supply voltage is found from the phasor sum of the voltage that appears across each of the components present. In a parallel AC circuit, by contrast, the *same voltage* appears across each branch of the circuit and the total current taken from the supply is the phasor sum of the currents in each branch. For this reason we normally use *voltage* as the reference quantity for a *parallel* AC circuit rather than current. Rather than simply quote the formulae, we shall illustrate the techniques for solving parallel, and series–parallel AC circuits by taking some simple examples.

Example 5.85

A parallel AC circuit comprises a resistor, R, of 30 Ω connected in parallel with a capacitor, C, of 80 μF. If the circuit is connected to a 240 V, 50 Hz supply determine:

(a) the current in the resistor,
(b) the current in the capacitor,
(c) the supply current,
(d) the phase angle.

Solution

Figure 5.166 shows the parallel circuit arrangement showing the three currents present; I_1 (the current in the resistor), I_2 (the current in the capacitor) and I_S (the supply current). Figure 5.167 shows the phasor diagram for the parallel

Electrical fundamentals

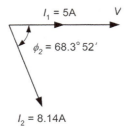

Figure 5.169 Phasor diagram for the circuit shown in Figure 5.168.

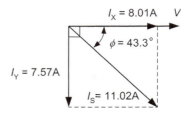

Figure 5.170 Phasor diagram showing total in-phase and total quadrature components in Example 5.86.

(b) The current flowing in the capacitor can be determined from:

$$I_2 = \frac{V}{Z} = \frac{V}{\sqrt{R^2 + X_L^2}} = \frac{V}{\sqrt{R^2 + (2\pi f L)^2}}$$

from which:

$$I_2 = \frac{110}{\sqrt{5^2 + (6.28 \times 400 \times 5 \times 10^{-3})^2}}$$

$$= \frac{110}{\sqrt{5^2 + (12.56)^2}} = \frac{110}{\sqrt{182.75}} = \frac{110}{13.52}$$

$$= 8.14\,\text{A}$$

Thus $I_2 = 8.14\,\text{A}$ (lagging the supply voltage by ϕ_2).

The phase angle for the inductive branch, ϕ_2, can be determined from:

$$\cos\phi_2 = \frac{R_2}{Z} = \frac{5}{13.52} = 0.37$$

or

$$\sin\phi_2 = \frac{X_L}{Z} = \frac{12.56}{13.52} = 0.93$$

from which $\phi_2 = 68.3°$

Hence the current in the inductive branch is 8.14 A lagging the supply voltage by 68.3°.

(c) In order to determine the supply current we need to find the total in-phase current and the total quadrature current (i.e. the total current at 90°) as shown in Figure 5.170.

The total in-phase current, I_x, is given by:

$$I_x = I_1 + I_2 \cos\phi_2 = 5 + (8.14 \times 0.37)$$

$$= 5 + 3.01 = 8.01\,\text{A}$$

The total quadrature current, I_y, is given by:

$$I_y = I_2 \sin\phi_2 = 8.14 \times 0.93 = 7.57\,\text{A}$$

The supply current, I_S, can now be determined from:

$$I_S = \sqrt{8.01^2 + 7.57^2} = \sqrt{64.16 + 57.3}$$

$$= \sqrt{121.46} = 11.02\,\text{A}$$

(d) The phase angle, ϕ, can be determined from:

$$\cos\phi = \frac{\text{in-phase current}}{\text{supply current}} = \frac{8.01}{11.02} = 0.73$$

from which:

$$\phi = 43.4° \text{ (lagging)}$$

Key point

We use current as the reference phasor in series AC circuit because the same current flows through each component. Conversely, we use voltage as the reference phasor in a parallel AC circuit because the same voltage appears across each component.

5.15.10 Power factor

The power factor in an AC circuit containing resistance and reactance is simply the ratio of true power to apparent power. Hence:

$$\text{Power factor} = \frac{\text{true power}}{\text{apparent power}}$$

The true power in an AC circuit is the power that is actually dissipated as heat in the resistive component. Thus:

$$\text{True power} = I^2 R$$

where I is r.m.s. current and R is the resistance. True power is measured in watts (W).

The apparent power in an AC circuit is the power that is apparently consumed by the circuit and is the product of the supply current and supply voltage (which may not be in phase). Note that, unless the voltage and current are in phase (i.e. $\phi = 0°$) the apparent power *will not* be the same as the power which is actually dissipated as heat. Hence:

$$\text{Apparent power} = IV$$

where I is r.m.s. current and V is the supply voltage. To distinguish apparent power from true power, apparent power is measured in volt-amperes (VA).

Now since $V = IZ$ we can re-arrange the apparent power equation as follows:

$$\text{Apparent power} = IV = I \times IZ = I^2Z$$

Now returning to our original equation:

$$\text{Power factor} = \frac{\text{true power}}{\text{apparent power}} = \frac{I^2R}{IV}$$

$$= \frac{I^2R}{I \times IZ} = \frac{I^2R}{I^2Z} = \frac{R}{Z}$$

From the *impedance triangle* shown earlier in Figure 5.153, we can infer that:

$$\text{Power factor} = \frac{R}{Z} = \cos\phi$$

Example 5.87

An AC load has a power factor of 0.8. Determine the true power dissipated in the load if it consumes a current of 2 A at 110 V.

Solution

Now since:

$$\text{Power factor} = \cos\phi = \frac{\text{true power}}{\text{apparent power}}$$

$$\text{True power} = \text{power factor} \times \text{apparent power}$$
$$= \text{power factor} \times VI$$

Thus:

$$\text{True power} = 0.8 \times 2 \times 110 = 176\,\text{W}$$

Example 5.88

A coil having an inductance of 150 mH and resistance of 250 Ω is connected to a 115 V, 400 Hz AC supply. Determine:

(a) the power factor of the coil,
(b) the current taken from the supply,
(c) the power dissipated as heat in the coil.

Solution

(a) First we must find the reactance of the inductor, X_L, and the impedance, Z, of the coil at 400 Hz.

$$X_L = 2\pi \times 400 \times 150 \times 10^{-3} = 376\,\Omega$$

and

$$Z = \sqrt{R^2 + X_L^2} = \sqrt{250^2 + 376^2} = 452\,\Omega$$

We can now determine the power factor from:

$$\text{Power factor} = \frac{R}{Z} = \frac{250}{452} = 0.553$$

(b) The current taken from the supply can be determined from:

$$I = \frac{V}{Z} = \frac{115}{452} = 0.254\,\text{A}$$

(c) The power dissipated as heat can be found from:

$$\text{True power} = \text{power factor} \times VI$$
$$= 0.553 \times 115 \times 0.254$$
$$= 16.15\,\text{W}$$

Key point

In an AC circuit the power factor is the ratio of true power to apparent power. The power factor also the cosine of the phase angle between the supply current and supply voltage.

Test your understanding 5.15

1. In a circuit containing pure capacitance the _____ _____ will lead the _____ by an angle of _____.

2. Determine the reactance of a 220 nF capacitor at (a) 400 Hz and (b) 20 kHz.

3. Determine the reactance of a 60 mH inductor at (a) 20 Hz and (b) 4 kHz.

4. A 0.5 μF capacitor is connected to a 110 V, 400 Hz supply. Determine the current flowing in the capacitor.

5. A resistor of 120 Ω is connected in series with a capacitive reactance of 160 Ω. Determine the impedance of the circuit and the current flowing when the circuit is connected to a 200 V AC supply.

6. A capacitor or 2 μF is connected in series with a 100 Ω resistor across a 24 V, 400 Hz AC supply. Determine the current that will be supplied to the circuit and the voltage that will appear across each component.

7. An 80 mH coil has a resistance of 10 Ω. Calculate the current flowing when the coil is connected to a 250 V, 50 Hz supply.

8. Determine the phase angle and power factor for Question 7 (*supra*).

9. An AC load has a power factor of 0.6. If the current supplied to the load is 5 A and the supply voltage is 110 V determine the true power dissipated by the load.

10. An AC load comprises a 110 Ω resistor connected in parallel with a 20 μF capacitor. If the load is connected to a 220 V, 50 Hz supply, determine the apparent power supplied to the load and its power factor.

5.16 Transformers

Syllabus
Transformer construction principles and operation; Transformer losses and methods for overcoming them; Transformer action under load and no-load conditions; Power transfer, efficiency, polarity markings; Primary and secondary current, voltage, turns ratio, power, efficiency; Auto transformers.

Knowledge level key

A	B1	B2
–	2	2

5.16.1 Transformer principles

The principle of the transformer is illustrated in Figure 5.171. The primary and secondary windings are wound on a common low-reluctance magnetic core consisting of a number of steel laminations. All of the alternating flux generated by the primary winding is therefore coupled into the secondary winding (very little flux escapes

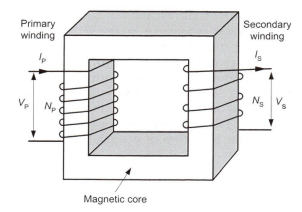

Figure 5.171 The principle of the transformer.

due to leakage). A sinusoidal current flowing in the primary winding produces a sinusoidal flux within the transformer core.

At any instant the flux, Φ, in the transformer core is given by the equation:

$$\Phi = \Phi_{max} \sin(\omega t)$$

where Φ_{max} is the maximum value of flux (in Wb) and t is the time in seconds. You might like to compare this equation with the one that you met earlier for a sine wave voltage in Section 5.14.5.

The r.m.s. value of the primary voltage (V_P) is given by:

$$V_P = 4.44 f N_P \Phi_{max}$$

Similarly, the r.m.s. value of the secondary voltage (V_S) is given by:

$$V_S = 4.44 f N_S \Phi_{max}$$

From these two relationships (and since the same magnetic flux appears in both the primary and secondary windings) we can infer that (Figure 5.172):

$$\frac{V_P}{V_S} = \frac{N_P}{N_S}$$

If the transformer is loss-free the primary and secondary powers will be equal.
Thus:

$$P_P = P_S$$

Now $P_P = I_P \times V_P$ and $P_S = I_S \times V_S$

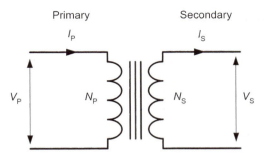

Figure 5.172 Transformer turns and voltages.

So $I_P \times V_P = I_S \times V_S$

From which $\frac{I_P}{I_S} = \frac{V_S}{V_P}$ and thus $\frac{I_P}{I_S} = \frac{N_S}{N_P}$

Furthermore, assuming that no power is lost in the transformer (i.e. as long as the primary and secondary powers are the same) we can conclude that:

$$\frac{I_P}{I_S} = \frac{N_S}{N_P}$$

The ratio of primary turns to secondary turns (N_P/N_S) is known as the *turns ratio*.

Furthermore, since ratio of primary voltage to primary turns is the same as the ratio of secondary turns to secondary voltage, we can conclude that, for a particular transformer:

$$\text{Turns-per-volt (t.p.v.)} = \frac{V_P}{N_P} = \frac{V_S}{N_S}$$

The t.p.v. rating can be quite useful when it comes to designing transformers with multiple secondary windings.

Example 5.89

A transformer has 2000 primary turns and 120 secondary turns. If the primary is connected to a 220 V AC mains supply, determine the secondary voltage.

Solution

Since $\frac{V_P}{V_S} = \frac{N_P}{N_S}$ we can conclude that:

$$V_S = \frac{V_P N_S}{N_P} = \frac{220 \times 120}{2000} = 13.2\text{V}$$

Example 5.90

A transformer has 1200 primary turns and is designed to operated with a 110 V AC supply. If the transformer is required to produce an output of 10 V, determine the number of secondary turns required.

Solution

Since $\frac{V_P}{V_S} = \frac{N_P}{N_S}$ we can conclude that:

$$N_S = \frac{N_P V_S}{V_P} = \frac{1200 \times 10}{110} = 109.1$$

Example 5.91

A transformer has a t.p.v. rating of 1.2. How many turns are required to produce secondary outputs of (a) 50 V and (b) 350 V?

Solution

Here we will use $N_S = \text{t.p.v.} \times V_S$

(a) In the case of a 50 V secondary winding:
$$N_S = 1.2 \times 50 = 60 \text{ turns}$$

(b) In the case of a 350 V secondary winding:
$$N_S = 1.2 \times 350 = 420 \text{ turns}$$

Example 5.92

A transformer has 1200 primary turns and 60 secondary turns. Assuming that the transformer is loss-free, determine the primary current when a load current of 20 A is taken from the secondary.

Solution

Since $\frac{I_S}{I_P} = \frac{N_P}{N_S}$ we can conclude that:

$$I_P = \frac{I_S N_S}{N_P} = \frac{20 \times 60}{1200} = 1\text{ A}$$

5.16.2 Transformer applications

Transformers provide us with a means of coupling AC power from one circuit to another without a direct connection between the two.

Table 5.6

	Core material			
	Air	Ferrite	Laminated steel (low volume)	Laminated steel (high volume)
Typical power rating	Less than 100 mW	Less than 10 W	100 mW to 50 W	3 VA to 500 VA
Typical frequency range	10 MHz to 1 GHz	1 kHz to 10 MHz	50 Hz to 20 kHz	45 Hz to 500 Hz
Typical efficiency	See note	95% to 98%	95% typical	90% to 98%
Typical applications	Radio receivers and transmitters	Pulse circuits, switched mode power supplies	Audio and low-frequency amplifiers	Power supplies

A further advantage of transformers is that voltage may be stepped-up (secondary voltage greater than primary voltage) or stepped-down (secondary voltage less than primary voltage). Since no increase in power is possible (like resistors, capacitors and inductors, transformers are *passive* components) an increase in secondary voltage can only be achieved at the expense of a corresponding reduction in secondary current, and vice versa (in fact, the secondary power will be very slightly less than the primary power due to losses within the transformer).

Typical applications for transformers include stepping-up or stepping-down voltages in power supplies, coupling signals in audio frequency amplifiers to achieve impedance matching and to isolate the DC potentials that may be present in certain types of circuit. The electrical characteristics of a transformer are determined by a number of factors including the core material and physical dimensions of the component.

The specifications for a transformer usually include the rated primary and secondary voltages and currents the required power rating (i.e. the rated power, usually expressed in VA), which can be continuously delivered by the transformer under a given set of conditions, the frequency range for the component (usually stated as upper and lower working frequency limits) and the per-unit regulation of a transformer. As we shall see, this last specification is a measure of the ability of a transformer to maintain its rated output voltage under load.

Table 5.8 summarizes the properties of some common types of transformer (note how the choice of core material is largely responsible

Figure 5.173 Various transformers.

for determining the characteristics of the transformer) (Figure 5.173).

5.16.3 Transformer regulation

The output voltage produced at the secondary of a real transformer falls progressively, as the load imposed on the transformer increases (i.e. as the secondary current increases from its no-load value). The *voltage regulation* of a transformer is a measure of its ability to keep the secondary output voltage constant over the full range of output load currents (i.e. from *no-load* to *full-load*) at the same power factor. This change, when divided by the no-load output voltage, is referred to as the *per-unit regulation* for the transformer. This can be best illustrated by the use of an example.

Example 5.93

A transformer produces an output voltage of 110 V under no-load conditions and an output voltage of 101 V when a full-load is applied. Determine the per-unit regulation.

Solution

The per-unit regulation can be determined for:

$$\text{Per-unit regulation} = \frac{V_{S(\text{no-load})} - V_{S(\text{full-load})}}{V_{S(\text{no-load})}}$$

$$= \frac{110 - 101}{110}$$

$$= 0.081 \text{ (or } 8.1\%)$$

5.16.4 Transformer efficiency and losses

As we saw earlier, most transformers operate with very high values of efficiency. Despite this, in high power applications the losses in a transformer cannot be completely neglected. Transformer losses can be divided into two types of loss:

- Losses in the magnetic core (often referred to as *iron loss*).
- Losses due to the resistance of the coil windings (often referred to as *copper loss*).

Iron loss can be further divided into *hysteresis loss* (energy lost in repeatedly cycling the magnet flux in the core backwards and forwards) and *eddy current loss* (energy lost due to current circulating in the steel core).

Hysteresis loss can be reduced by using material for the magnetic core that is easily magnetized and has a very high permeability (see Figure 5.174 – note that energy loss is proportional to the area inside the B–H curve). Eddy current loss can be reduced by laminating the core (e.g. using E- and I-laminations) and also ensuring that a small gap is present. These laminations and gaps in the core help to ensure that there is no closed path for current to flow.

Copper loss results from the resistance of the coil windings and it can be reduced by using wire of large diameter and low resistivity.

It is important to note that, since the flux within a transformer varies only slightly between the no-load and full-load conditions, iron loss is substantially constant regardless of

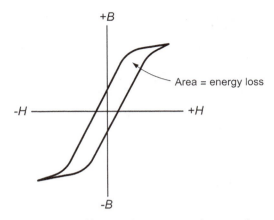

Figure 5.174 Hysteresis curves and energy loss.

the load actually imposed on a transformer. On the other hand, copper loss is zero when a transformer is under no-load conditions and rises to a maximum at full-load.

The efficiency of a transformer is given by:

$$\text{Efficiency} = \frac{\text{output power}}{\text{input power}} \times 100\%$$

from which

$$\text{Efficiency} = \frac{\text{input power} - \text{losses}}{\text{input power}} \times 100\%$$

and

$$\text{Efficiency} = 1 - \frac{\text{losses}}{\text{input power}} \times 100\%$$

As we have said, the losses present are attributable to iron and copper loss but the copper loss appears in both the primary and the secondary windings. Hence:

$$\text{Efficiency} = 1 - \frac{\left[\begin{array}{l}\text{iron loss} \\ + \text{primary copper loss} \\ + \text{secondary copper loss}\end{array}\right]}{\text{input power}}$$
$$\times 100\%$$

Once again, we shall explain this with the aid of some examples.

Example 5.94

A transformer rated at 500 VA has an iron loss of 3 W and a full-load copper loss (primary plus secondary) of 7 W. Calculate the efficiency of the transformer at 0.8 power factor.

Solution

The input power to the transformer will be given by the product of the apparent power (i.e. the transformer's *VA* rating) and the power factor. Hence:

$$\text{Input power} = 0.8 \times 500 = 400\,\text{W}$$

Now

$$\text{Efficiency} = 1 - \frac{(7+3)}{400} \times 100\% = 97.5\%$$

Test your understanding 5.16

1. Sketch a diagram to illustrate the principle of the transformer. Label your diagram.

2. The core of a power transformer is _____ in order to reduce _____ _____ current loss.

3. Sketch a *B–H* curve for the core material of a transformer and explain how this relates to the energy loss in the transformer core.

4. A transformer has 480 primary turns and 120 secondary turns. If the primary is connected to a 110 V AC supply determine the secondary voltage.

5. A step-down transformer has a 220 V primary and a 24 V secondary. If the secondary winding has 60 turns, how many turns are there on the primary?

6. A transformer has 440 primary turns and 1800 secondary turns. If the secondary supplies a current of 250 mA, determine the primary current (assume that the transformer is loss-free).

7. Show that, for a loss-free transformer, $\frac{I_P}{I_S} = \frac{N_S}{N_P}$.

8. Explain how copper loss occurs in a transformer. How can this loss be minimized?

9. A transformer produces an output voltage of 220 V under no-load conditions and an output voltage of 208 V when full-load is applied. Determine the per-unit regulation.

10. A 1 kVA transformer has an iron loss of 15 W and a full-load copper loss (primary plus secondary) of 20 W. Determine the efficiency of the transformer at 0.9 power factor.

5.17 Filters

Syllabus
Operation, application and uses of the following filters: low pass, high pass, band pass and band stop.

Knowledge level key

A	B1	B2
–	1	1

5.17.1 Types of filter

Filters provide us with a means of passing or rejecting AC signals within a specified frequency range. Filters are used in a variety of applications including amplifiers, radio transmitters and receivers. They also provide us with a means of reducing noise and unwanted signals that might otherwise be passed along power lines.

Filters are usually categorized in terms of the frequency range that they are designed to accept or reject. Simple filters can be based around circuit (or *networks*) of passive components (i.e. resistors, capacitors and inductors) whilst those used for signal (rather than power) applications can be based on active components (i.e. transistors and integrated circuits).

Most filters are networks having four terminals; two of these terminals are used for the *input* and two are used for the *output*. Note that, in the case of an *unbalanced network*, one of the input terminals may be linked directly to one of the output terminals (in which case this connection is referred to as *common*). This arrangement is shown in Figure 5.175.

The following types of filter are available:

- low-pass filter,
- high-pass filter,

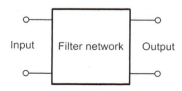

Figure 5.175   A four-terminal network.

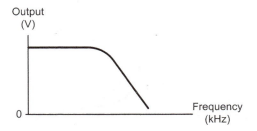

Figure 5.176 Frequency response for a low-pass filter.

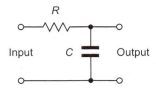

Figure 5.177 A simple C–R low-pass filter.

- band-pass filter,
- band-stop filter.

> **Key point**
>
> Filters are circuits that pass or reject AC signals within a specified frequency range. Simple passive filters are based on networks of resistors, capacitors and inductors.

5.17.2 Low-pass filters

Low-pass filters exhibit very low attenuation of signals below their specified *cut-off frequency*. Beyond the cut-off frequency they exhibit increasing amounts of attenuation, as shown in Figure 5.176.

A simple C–R low-pass filter is shown in Figure 5.177. The cut-off frequency for the filter occurs when the output voltage has fallen to 0.707 of the input value. This occurs when the reactance of the capacitor, X_C, is equal to the value of resistance, R. Using this information we can determine the value of cut-off frequency, f, for given values of C and R.

Since

$$R = X_C$$

or

$$R = \frac{1}{2\pi f C}$$

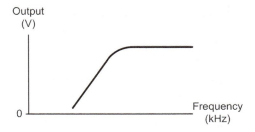

Figure 5.178 Frequency response for a high-pass filter.

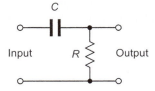

Figure 5.179 A simple C–R high-pass filter.

from which:

$$f = \frac{1}{2\pi C R}$$

where f is the cut-off frequency (in Hz), C is the capacitance (in F) and R is the resistance (in Ω).

5.17.3 High-pass filters

High-pass filters exhibit very low attenuation of signals above their specified *cut-off frequency*. Below the cut-off frequency they exhibit increasing amounts of attenuation, as shown in Figure 5.178.

A simple C–R high-pass filter is shown in Figure 5.179. Once again, the cut-off frequency for the filter occurs when the output voltage has fallen to 0.707 of the input value. This occurs when the reactance of the capacitor, X_C, is equal to the value of resistance, R. Using this information we can determine the value of cut-off frequency, f, for given values of C and R.

Since

$$R = X_C$$

or

$$R = \frac{1}{2\pi f C}$$

and once again:

$$f = \frac{1}{2\pi C R}$$

where f is the cut-off frequency (in Hz), C is the capacitance (in F) and R is the resistance (in Ω).

Example 5.95

A simple C–R low-pass filter has $C = 100\,\text{nF}$ and $= 10\,\text{k}\Omega$. Determine the cut-off frequency of the filter.

Solution

Now $f = \dfrac{1}{2\pi CR}$

$$= \frac{1}{6.28 \times 100 \times 10^{-9} \times 10 \times 10^4}$$

$$= \frac{100}{6.28}$$

$$= 15.9\,\text{Hz}$$

Example 5.96

A simple C–R low-pass filter is to have a cut-off frequency of $1\,\text{kHz}$. If the value of capacitance used in the filter is to be $47\,\text{nF}$, determine the value of resistance.

Solution

Now

$$f = \frac{1}{2\pi CR}$$

from which

$$R = \frac{1}{2\pi fC} = \frac{1}{6.28 \times 1 \times 10^3 \times 47 \times 10^{-9}}$$

$$= \frac{10^6}{295.16} = 3.39\,\text{k}\Omega$$

5.17.4 Band-pass filters

Band-pass filters exhibit very low attenuation of signals within a specified range of frequencies (known as the *pass band*) and increasing

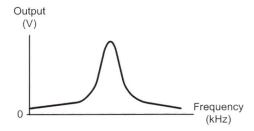

Figure 5.180 Frequency response for a band-pass filter.

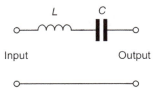

Figure 5.181 A simple L–C band-pass filter (or *acceptor*).

attenuation outside this range. This type of filter has two cut-off frequencies: a *lower cut-off frequency* (f_1) and an *upper cut-off frequency* (f_2). The difference between these frequencies ($f_2 - f_1$) is known as the *bandwidth* of the filter. The response of a band-pass filter is shown in Figure 5.180.

A simple L–C band-pass filter is shown in Figure 5.181. This circuit uses an L–C resonant circuit (see Section 5.15.8) and is referred to as an *acceptor circuit*.

The frequency at which the band-pass filter in Figure 5.181 exhibits minimum attenuation occurs when the circuit is resonant, i.e. when the reactance of the capacitor, X_C, is equal to the value of resistance, R. This information allows us to determine the value of frequency at the center of the pass band, f_0:

$$X_C = X_L$$

thus

$$\frac{1}{2\pi f_0 C} = 2\pi f_0 L$$

from which

$$f_0^2 = \frac{1}{4\pi^2 LC}$$

and thus

$$f_0 = \frac{1}{2\pi\sqrt{LC}}$$

where f_0 is the resonant frequency (in Hz), L is the inductance (in H) and C is the capacitance (in F).

The bandwidth of the band-pass filter is determined by its *Q-factor*. This, in turn, is largely determined by the loss resistance, R, of the inductor (recall that a practical coil has some resistance as well as inductance). The bandwidth is given by:

$$\text{Bandwidth} = f_2 - f_1 = \frac{f_0}{Q} = \frac{2\pi f_0 L}{R}$$

where f_0 is the resonant frequency (in Hz), L is the inductance (in H) and R is the loss resistance of the inductor (in Ω).

5.17.5 Band-stop filters

Band-stop filters exhibit very high attenuation of signals within a specified range of frequencies (know as the *stop-band*) and negligible attenuation outside this range. Once again, this type of filter has two cut-off frequencies: a *lower cut-off frequency* (f_1) and an *upper cut-off frequency* (f_2). The difference between these frequencies ($f_2 - f_1$) is known as the *bandwidth* of the filter. The response of a band-stop filter is shown in Figure 5.182.

A simple *L–C* band-stop filter is shown in Figure 5.183. This circuit uses an *L–C* resonant circuit (see Section 5.15.8) and is referred to as a *rejector circuit*.

The frequency at which the band-stop filter in Figure 5.183 exhibits maximum attenuation occurs when the circuit is resonant, i.e. when the reactance of the capacitor, X_C, is equal to the value of resistance, R. This information allows us to determine the value of frequency at the center of the pass band, f_0:

$$X_C = X_L$$

thus

$$\frac{1}{2\pi f_0 C} = 2\pi f_0 L$$

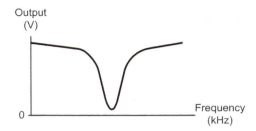

Figure 5.182 Frequency response for a band-stop filter.

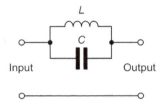

Figure 5.183 A simple *L–C* band-stop filter (or *rejector*).

from which

$$f_0^2 = \frac{1}{4\pi^2 LC}$$

and thus

$$f_0 = \frac{1}{2\pi\sqrt{LC}}$$

where f_0 is the resonant frequency (in Hz), L is the inductance (in H) and C is the capacitance (in F).

As with the band-pass filter, the bandwidth of the band-pass filter is determined by its *Q-factor*. This, in turn, is largely determined by the loss resistance, R, of the inductor (recall that a practical coil has some resistance as well as inductance). Once again, the bandwidth is given by:

$$\text{Bandwidth} = f_2 - f_1 = \frac{f_0}{Q} = \frac{2\pi f_0 L}{R}$$

where f_0 is the resonant frequency (in Hz), L is the inductance (in H) and R is the loss resistance of the inductor (in Ω).

Example 5.97

A simple acceptor circuit uses $L = 2\,\text{mH}$ and $C = 1\,\text{nF}$. Determine the frequency at which minimum attenuation will occur.

Solution

Now

$$f_0 = \frac{1}{2\pi\sqrt{LC}} = \frac{1}{2\pi\sqrt{2 \times 10^{-3} \times 1 \times 10^{-9}}}$$

$$= \frac{10^6}{8.88} = 112.6\,\text{kHz}$$

Example 5.98

A $15\,\text{kHz}$ rejector circuit has a Q-factor of 40. Determine the bandwidth of the circuit.

Solution

Now

$$\text{Bandwidth} = \frac{f_0}{Q} = \frac{15 \times 10^3}{40} = 375\,\text{Hz}$$

5.17.6 More complex filters

The simple C–R and L–C filters that we have described in earlier sections have far from ideal characteristics. In practice, more complex circuits are used and a selection of these (based on T- and π-section networks) are shown in Figure 5.184. The design equations for these circuits are as follows:

Characteristic impedance: $\quad Z_0 = \sqrt{\dfrac{L}{C}}$

Cut-off frequency: $\quad f_C = \dfrac{1}{2\pi\sqrt{LC}}$

Inductance: $\quad L = \dfrac{Z_0}{2\pi f_C}$

Capacitance: $\quad C = \dfrac{1}{2\pi f_C Z_0}$

where Z_0 is the characteristic impedance (in Ω), f_C is the cut-off frequency (in Hz), L is the inductance (in H) and C is the capacitance

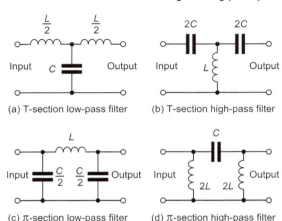

(a) T-section low-pass filter

(b) T-section high-pass filter

(c) π-section low-pass filter

(d) π-section high-pass filter

Figure 5.184 Improved T-section and π-section filters.

(in F). Note that the *characteristic impedance* of a network is the impedance seen looking into an infinite series of identical networks. This can be a difficult concept to grasp but, for now, it is sufficient to know that single section networks (like the T- and π-section filters shown in Figure 5.184) are normally terminated in their characteristic impedance at both the source (input) and load (output).

Example 5.99

Determine the cut-off frequency and characteristic impedance for the filter network shown in Figure 5.185.

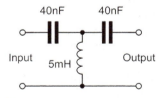

Figure 5.185

Solution

Comparing the circuit shown in Figure 5.185 with that shown in Figure 5.184 shows that the filter is a high-pass type with $L = 5\,\text{mH}$ and $C = 20\,\text{nF}$.

Now

$$f_C = \frac{1}{2\pi\sqrt{LC}} = \frac{1}{6.28\sqrt{5 \times 10^{-3} \times 20 \times 10^{-9}}}$$

$$= \frac{10^5}{6.28} = 15.9 \,\text{kHz}$$

and

$$Z_0 = \sqrt{\frac{L}{C}} = \sqrt{\frac{5 \times 10^{-3}}{20 \times 10^{-9}}} = \sqrt{\frac{5}{20}} \times 10^3$$

$$= 0.5 \times 10^3 = 500 \,\Omega$$

Test your understanding 5.17

1. Sketch the typical circuit for a simple C–R low-pass filter.

2. Sketch the circuit of (a) a simple C–R lowpass filter (b) a simple C–R high-pass filter.

3. A simple C–R high-pass filter has $R = 5\,\text{k}\Omega$ and $C = 15\,\text{nF}$. Determine the cut-off frequency of the filter.

4. Signals at 115, 150, 170 and 185 kHz are present at the input of a band-stop filter with a center frequency of 160 kHz and a bandwidth of 30 kHz. Which frequencies will be present at the output?

5. Identify the type of filter shown in Figure 5.186.

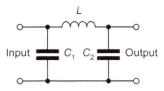

Figure 5.186 See Question 5 of Test your knowledge 5.17.

6. The cut-off frequency of a filter is the frequency at which the _____ voltage has fallen to _____ of its _____ voltage.

7. The output of a low-pass filter is 2 V at 100 Hz. If the filter has a cut-off frequency of 1 kHz what will the approximate output voltage be at this frequency?

8. An L–C tuned circuit is to be used to reject signals at 15 kHz. If the value of capacitance used is 22 nF determine the required value of inductance.

9. Sketch the frequency response for (a) a simple L–C acceptor circuit and (b) a simple L–C rejector circuit.

10. A T-section filter has $L = 10\,\text{mH}$ and $C = 47\,\text{nF}$. Determine the characteristic impedance of the filter.

5.18 AC generators

Syllabus
Rotation of loop in a magnetic field and waveform produced; Operation and construction of revolving armature and revolving field type AC generators; Single-, two- and three-phase alternators; Three phase star and delta connections advantages and uses; Calculation of line and phase voltages and currents; Calculation of power in a three phase system; Permanent magnet generators (PMG).

Knowledge level key

A	B1	B2
–	2	2

5.18.1 AC generators

AC generators, or *alternators*, are based on the principles that relate to the simple AC generator that we met earlier in Section 5.13.2. However, in a practical AC generator the magnetic field is rotated rather than the conductors from which the output is taken. Furthermore, the magnetic field is usually produced by a rotating electromagnet (the *rotor*) rather than a permanent magnet. There are a number of reasons for this including:

(a) The conductors are generally lighter in weight than the magnetic field system and are thus more easily rotated.

(b) Thicker insulation can be used for the conductors because there is more space and the conductors are not subject to centrifugal force.

(c) Thicker conductors can be used to carry the large output currents. It is important to note that the heat generated in the output windings limits the output current that the generator can provide. By having the output windings on the outside of the machine they are much easier to cool!

Figure 5.187 shows the simplified construction of a *single-phase AC generator*. The *stator* consists of five coils of insulated heavy gauge wire located in slots in the high-permeability laminated core. These coils are connected in series

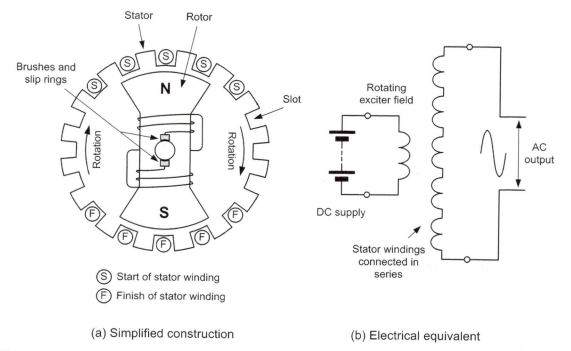

Start of stator winding

Finish of stator winding

(a) Simplified construction

(b) Electrical equivalent

Figure 5.187   Simplified construction of a single-phase AC generator.

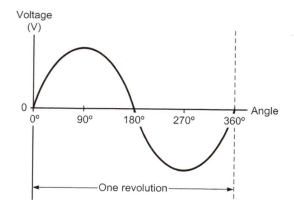

Figure 5.188 Output voltage produced by the single-phase AC generator shown in Figure 5.187.

to make a single stator winding from which the output voltage is derived.

The two-pole rotor comprises a field winding that is connected to a DC field supply via a set of slip rings and brushes. As the rotor moves through one complete revolution the output voltage will complete one full cycle of a sine wave, as shown in Figure 5.188.

By adding more pairs of poles to the arrangement shown in Figure 5.187, it is possible to produce several cycles of output voltage for one single revolution of the rotor. The frequency of the output voltage produced by an AC generator is given by:

$$f = \frac{pN}{60}$$

where f is the frequency of the induced e.m.f. (in Hz), p is the number of pole pairs and N is the rotational speed (in rpm).

Example 5.100

An alternator is to produce an output at a frequency of 60 Hz. If it uses a four-pole rotor, determine the shaft speed at which it must be driven.

Solution

Re-arranging $f = \frac{pN}{60}$ to make N the subject gives:

$$N = \frac{60f}{p}$$

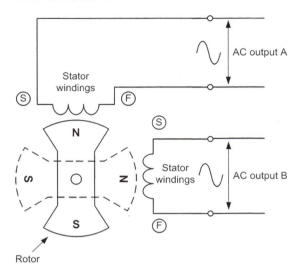

Figure 5.189 Simplified construction of a two-phase AC generator.

A four-pole machine has two pairs of poles thus $p = 2$ and:

$$N = \frac{60 \times 60}{2} = 1800\,\text{rpm}$$

Key point

In a practical AC generator, the magnetic field excitation is produced by the moving *rotor* whilst the conductors from which the output is taken are stationary and form part of the *stator*.

5.18.2 Two-phase AC generators

By adding a second stator winding to the single-phase AC generator shown in Figure 5.187, we can produce an alternator that produces two separate output voltages which will differ in phase by 90°. This arrangement is known as a *two-phase AC generator* (Figures 5.189 and 5.190).

When compared with a single-phase AC generator of similar size, a two-phase AC generator can produce more power. The reason for this is attributable to the fact that the two-phase AC generator will produce two positive and two negative pulses per cycle whereas the single-phase generator will only produce one positive and one negative pulse. Thus, over a period

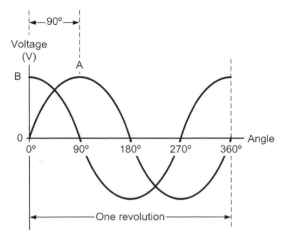

Figure 5.190 Output voltage produced by the two-phase AC generator shown in Figure 5.189.

of time, a multi-phase supply will transmit a more evenly distributed power and this, in turn, results in a higher overall efficiency.

Key point

Three-phase AC generators are more efficient and produce more constant output than comparable single-phase AC generators.

5.18.3 Three-phase AC generators

The three-phase AC generator has three individual stator windings, as shown in Figure 5.191. The output voltages produced by the three-phase AC generator are spaced by 120° as shown in Figure 5.192. Each phase can be used independently to supply a different load or the generator outputs can be used with a three-phase distribution system like those described in Section 5.18.4. In a practical three-phase system the three output voltages are identified by the colours red, yellow and blue or by letters, A, B and C, respectively.

5.18.4 Three-phase distribution

When three-phase supplies are distributed there are two basic methods of connection:

- star (as shown in Figure 5.193);
- delta (as shown in Figure 5.194).

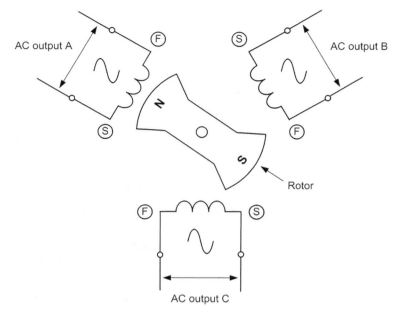

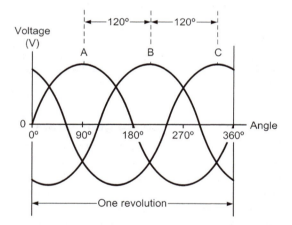

Figure 5.192 Output voltage produced by the three-phase AC generator shown in Figure 5.191.

Figure 5.191 Simplified construction of a three-phase AC generator.

Figure 5.193 Star connection.

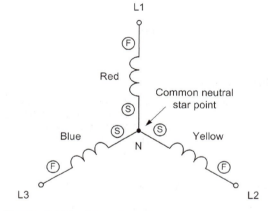

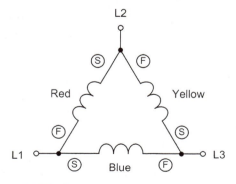

Figure 5.194 Delta connection.

A complete star-connected three-phase distribution system is shown in Figure 5.195. This shows a three-phase AC generator connected a three-phase load. Ideally, the load will be balanced in which case all three-load resistances (or impedances) will be identical.

The relationship between the line and phase voltages shown in Figure 5.195 can be determined from the phasor diagram shown in Figure 5.196. From this diagram it is important to

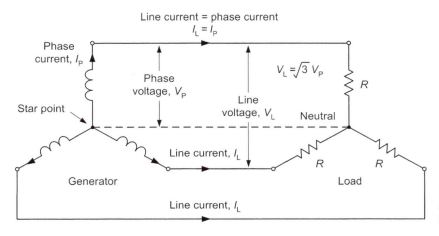

Figure 5.195 A complete star-connected three-phase distribution system.

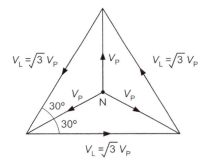

Figure 5.196 Phasor diagram for the three-phase system shown in Figure 5.195.

note that three line voltages are 120° apart and that the line voltages lead the phase voltages by 30°. In order to obtain the relationship between the line voltage, V_L, and the phase voltage, V_P, we need to resolve any one of the triangles, from which we find that:

$$V_L = 2(V_P \times \cos 30°)$$

Now $\cos 30° = \frac{\sqrt{3}}{2}$
and hence:

$$V_L = 2\left(V_P \times \frac{\sqrt{3}}{2}\right)$$

from which:

$$V_L = \sqrt{3}V_P$$

Note also that the phase current is the same as the line current, hence:

$$I_P = I_L$$

An alternative, delta-connected three-phase distribution system is shown in Figure 5.197. Once again this shows a three-phase AC generator connected a three-phase load. Here again, the load will ideally be balanced in which case all three-load resistances (or impedances) will be identical.

In this arrangement the three line currents are 120° apart and that the line currents lag the phase currents by 30°. We can also show that:

$$I_L = \sqrt{3}I_P$$

It should also be obvious that:

$$V_P = V_L$$

Example 5.101

In a star-connected three-phase system the phase voltage is 240 V. Determine the line voltage.

Solution

$$V_L = \sqrt{3}V_P = \sqrt{3} \times 240 = 415.68 \text{ V}$$

Example 5.102

In a delta-connected three-phase system the line current is 6 A. Determine the phase current.

Solution

$$I_L = \sqrt{3}I_P$$

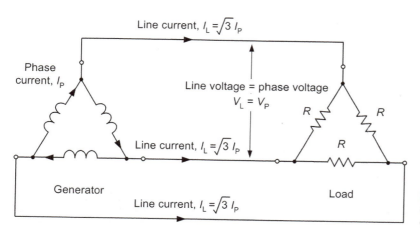

Figure 5.197 A complete delta-connected three-phase distribution system.

from which:

$$I_P = \frac{I_L}{\sqrt{3}} = \frac{6}{1.732} = 3.46 \, A$$

5.18.5 Power in a three-phase system

In an unbalanced three-phase system the total power will be the sum of the individual phase powers. Hence:

$$P = P_1 + P_2 + P_3$$

or

$$P = (V_1 I_1)\cos\phi_1 + (V_2 I_2)\cos\phi_2 + (V_3 I_3)\cos\phi_3$$

However, in the balanced condition the power is simply:

$$P = 3 \, V_P I_P \cos\phi$$

where V_P and I_P are the phase voltage and phase current, respectively, and ϕ is the phase angle.

Using the relationships that we derived earlier, we can show that, for both the star and delta-connected systems the total power is given by:

$$P = \sqrt{3} V_L I_L \cos\phi$$

Example 5.103

In a three-phase system the line voltage is 110 V and the line current is 12 A. If the power factor is 0.8 determine the total power supplied.

Solution

Here it is important to remember that:

$$\text{Power factor} = \cos\phi$$

and hence:

$$P = \sqrt{3} V_L I_L \times \text{power factor}$$
$$= \sqrt{3} \times 110 \times 12 \times 0.8 = 1829 = 1.829 \, \text{kW}$$

Key point

The total power in a three-phase system is the sum of the power present in each of the three phases.

5.18.6 A practical three-phase AC generator

Finally, Figure 5.198 shows a practical AC generator which uses a "brushless" arrangement based on a rotating rectifier and PMG. The generator is driven from the engine at 8000 rpm and the PMG produces an output of 120 V at 800 Hz which is fed to the PMG rectifier unit. The output of the PMG rectifier is fed to the voltage regulator which provides current for the primary exciter field winding.

The primary exciter field induces current into a three-phase rotor winding. The output of this winding is fed to three shaft-mounted rectifier diodes which produce a pulsating DC output which is fed to the rotating field winding.

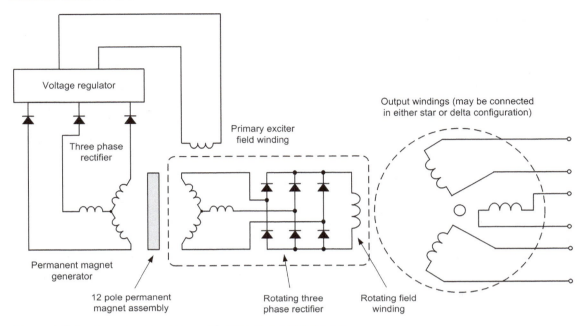

Figure 5.198 A practical brushless AC generator arrangement.

The main exciter winding is wound so as to form six poles in order to produce an output at 400 Hz. The output voltage from the stator windings is typically 115 V phase, 200 V line at 20 kVA, or more. Finally, it is important to note that the excitation system is an integral part of the rotor and that there is no direct electrical connection between the stator and rotor.

Key point

A three-phase AC generator can be made "brushless" by incorporating an integral excitation system in which the field current is derived from a rotor-mounted rectifier arrangement. In this type of generator the coupling is entirely magnetic and no brushes and slip rings are required.

Test your understanding 5.18

1. Sketch the arrangement of a simple two-pole single-phase AC generator.
2. An alternator with a four-pole rotor is to produce an output at a frequency of 400 Hz. Determine the shaft speed at which it must be driven.
3. Sketch (a) a star-connected and (b) a delta-connected three-phase load.
4. Explain the advantage of two- and three-phase AC generators compared with single-phase AC generators.

5. In a star-connected three-phase system the phase voltage is 220 V. Determine the line voltage.
6. In a star-connected three-phase system the line voltage is 120 V. Determine the phase voltage.
7. In a delta-connected three-phase system the line current is 12 A. Determine the phase current.
8. A three-phase system delivers power to a load consisting of three 8 Ω resistors. Determine the total power supplied if a current of 13 A is supplied to each load.
9. In a three-phase system the line voltage is 220 V and the line current is 8 A. If the power factor is 0.75 determine the total power supplied.
10. Explain, with a simple diagram, how a brushless AC generator works.

5.19 AC motors

Syllabus

Construction, principles of operation and characteristics of AC synchronous and induction motors both single and poly-phase; Methods of speed control and direction of rotation; Methods of producing a rotating field: capacitor, inductor, shaded or split pole.

Knowledge level key

A	B1	B2
–	2	2

5.19.1 Principle of AC motors

AC motors offer significant advantages over their DC counterparts. AC motors can, in most cases, duplicate the operation of DC motors and they are significantly more reliable. The main reason for this is that the commutator arrangements (i.e. brushes and slip rings) fitted to DC motors are inherently troublesome. As the speed of an AC motor is determined by the frequency of the AC supply that is applied, AC motors are well suited to constant-speed applications.

The principle of all AC motors is based on the generation of a rotating magnetic field. It is this rotating field that causes the motor's rotor to turn.

AC motors are generally classified into two types:

• synchronous motors,
• induction motors.

The synchronous motor is effectively an AC generator (i.e. an alternator) operated as a motor. In this machine, AC is applied to the stator and DC is applied to the rotor. The induction motor is different in that no source of AC or DC power is connected to the rotor. Of these two types of AC motor, the induction motor is by far the most commonly used.

5.19.2 Producing a rotating magnetic field

Before we go any further it is important to understand how a rotating magnetic field is produced. Take a look at Figure 5.199 which shows a three-phase stator to which three-phase AC is applied. The windings are connected in delta configuration, as shown in Figure 5.200. It is important to note that the two windings for each phase (diametrically opposite to one another) are wound in the *same* direction.

At any instant the magnetic field generated by one particular phase depends on the current through that phase. If the current is zero, the magnetic field is zero. If the current is a maximum, the magnetic field is a maximum. Since the currents in the three windings are 120° out of phase, the magnetic fields generated will also be 120° out of phase.

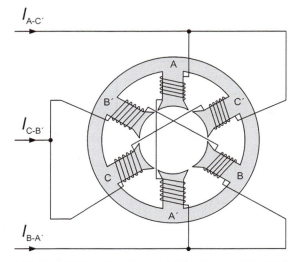

Figure 5.199 Arrangement of the field windings of a three-phase AC motor.

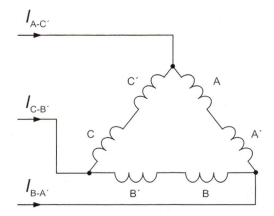

Figure 5.200 AC motor as a delta-connected three-phase load.

The three magnetic fields that exist at any instant will combine to produce one field that acts on the rotor. The magnetic fields inside the motor will combine to produce a moving magnetic field and, at the end of one complete cycle of the applied current, the magnetic field will have shifted through 360° (or one complete revolution).

Figure 5.201 shows the three current waveforms applied to the field system. These waveforms are 120° out of phase with each other. The waveforms can represent either the three

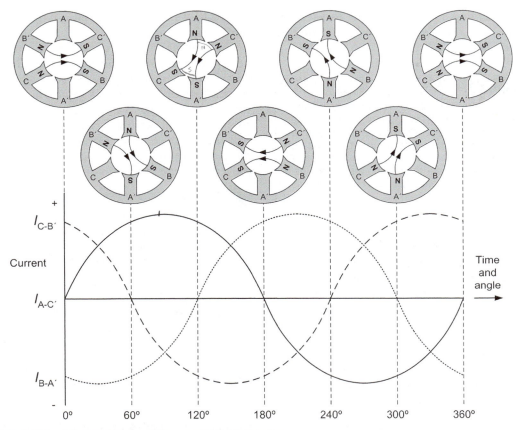

Figure 5.201 AC waveforms and magnetic field direction.

alternating magnetic fields generated by the three phases, or the currents in the phases.

We can consider the direction of the magnetic field at regular intervals over a cycle of the applied current (i.e. every 60°). To make life simple we take the times at which one of the three current waveforms passes through zero (i.e. the point at which there will be no current and therefore no field produced by one pair of field windings). For the purpose of this exercise we will use the current applied to A and C′ as our reference waveform (i.e. this will be the waveform that starts at 0° on our graph).

At 0°, waveform C-B′ is positive and waveform B-A′ is negative. This means that the current flows in opposite directions through phases B and C, and so establishes the magnetic polarity of phases B and C. The polarity is shown on the simplified diagram above. Note that B′ is a

north pole and B is a south pole, and that C is a north pole and C′ is a south pole.

Since at 0° there is no current flowing through phase A, its magnetic field is zero. The magnetic fields leaving poles B′ and C will move towards the nearest south poles C′ and B. Since the magnetic fields of B and C are equal in amplitude, the resultant magnetic field will lie between the two fields, and will have the direction shown.

At the next point, 60° later, the current waveforms to phases A and B are equal and opposite, and waveform C is zero. The resultant magnetic field has rotated through 60°. At point 120°, waveform B is zero and the resultant magnetic field has rotated through another 60°. From successive points (corresponding to one cycle of AC), you will note that the resultant magnetic field rotates through one revolution for every cycle of applied current. Hence, by applying a three-phase AC to the three windings we have been able to produce a rotating magnetic field.

5.19.3 Synchronous motors

We have already shown how a rotating magnetic field is produced when a three-phase AC is applied to the field coils of a stator arrangement. If the rotor winding is energized with DC, it will act like a bar magnet and it will rotate in sympathy with the rotating field. The speed of rotation of the magnetic field depends on the frequency of the three-phase AC supply and, provided that the supply frequency remains constant, the rotor will turn at a constant speed. Furthermore, the speed of rotation will remain constant regardless of the load applied. For many applications this is a desirable characteristic however one of the disadvantages of a synchronous motor is that it cannot be started from a standstill by simply applying three-phase AC to the stator. The reason for this is that, the instant AC is applied to the stator, a high-speed rotating field appears. This rotating field moves past the rotor poles so quickly that the rotor does not have a chance to get started. Instead, it is repelled first in one direction and then in the other.

Another way of putting this is simply that a synchronous motor (in its pure form) has no starting torque. Instead, it is usually started with the help of a small induction motor (or with windings equivalent to this incorporated in the synchronous motor). When the rotor has been brought near to synchronous speed by the starting device, the rotor is energized by connecting it to a DC voltage source. The rotor then falls into step with the rotating field. The requirement to have an external DC voltage source as well as

the AC field excitation makes this type of motor somewhat unattractive!

The amount by which the rotor lags the main field is dependent on the load. If the load is increased too much, the angle between the rotor and the field will increase to a value which causes the linkage of flux to break. At this point the rotor speed will rapidly decrease and the motor will either burn out due to excessive current or the circuit protection will operate in order to prevent damage to the motor.

5.19.4 Three-phase induction motors

The induction motor derives its name from the fact that AC currents are induced in the rotor circuit by the rotating magnetic field in the stator. The stator construction of the induction motor and of the synchronous motor are almost identical, but their rotors are completely different.

The induction motor rotor is a laminated cylinder with slots in its surface. The windings in these slots are one of two types. The most common uses so-called *squirrel cage* construction (see Figure 5.202) which is made up of heavy copper bars connected together at either end by a metal ring made of copper or brass. No insulation is required between the core and the bars because of the very low voltages generated in the rotor bars. The air gap between the rotor and stator is kept very small so as to obtain maximum field strength.

The other type of winding contains coils placed in the rotor slots. The rotor is then called

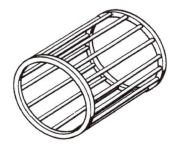

Figure 5.202 Squirrel cage rotor construction.

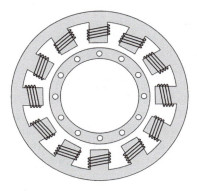

Figure 5.203 Typical stator construction.

a *wound rotor*. Just as the rotor usually has more than one conductor, the stator usually has more than one pair of poles per coil, as shown in Figure 5.203.

Key point

The induction motor is the most commonly used AC motor because of its simplicity, its robust construction and its relatively low cost. These advantages arise from the fact that the rotor of an induction motor is a self-contained component that is *not actually electrically connected to an external source of voltage*.

Regardless of whether a squirrel cage or wound rotor is used, the basic principle of operation of an induction motor is the same. The rotating magnetic field generated in the stator induces an e.m.f. in the rotor. The current in the rotor circuit caused by this induced e.m.f. sets up a magnetic field. The two fields interact, and cause the rotor to turn. Figure 5.204 shows how the rotor moves in the same direction as the rotating magnetic flux generated by the stator.

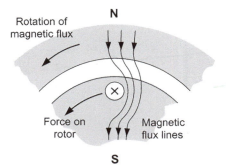

Figure 5.204 Force on the rotor of an induction motor.

From Lenz's law we know that an induced current oppose the changing field which induces it. In the case of an induction motor, the changing field is the rotating stator field and so the force exerted on the rotor (caused by the interaction between the rotor and the stator fields) attempts to cancel out the continuous motion of the stator field. Hence the rotor will move in the same direction as the stator field and will attempt to align with it. In practice, it gets as close to the moving stator field but never quite aligns perfectly with it!

Key point

The induction motor has the same stator as the synchronous motor. The rotor is different in that it does not require an external source of power. Current is induced in the rotor by the action of the rotating field cutting through the rotor conductors. This rotor current generates a magnetic field which interacts with the stator field, resulting in a torque being exerted on the rotor and causing it to rotate.

5.19.5 Slip, torque and speed

We have already said that the rotor of an induction motor is unable to turn in sympathy with the rotating field and, in practice, a small difference always exists. In fact, if the speeds were exactly the same, no relative motion would exist between the two, and so no e.m.f. would be induced in the rotor or this reason the rotor operates at a lower speed than that of the rotating magnetic field. This phenomenon is known as *slip* and it becomes more significant as the

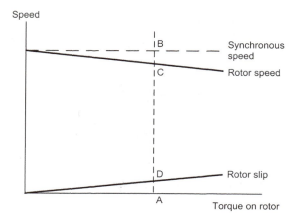

Figure 5.205 Relationship between torque and slip.

rotor develops increased torque, as shown in Figure 5.205.

From Figure 5.205, for a torque of A the rotor speed will be represented by the distance AC whilst the slip will be represented by distance AD. Now:

$$AD = AB - AC = CB$$

For values of torque within the working range of the motor (i.e. over the linear range of the graph shown in Figure 5.205), the slip is directly proportional to the torque and the *per-unit slip* is given by:

$$\text{Per-unit slip} = \frac{\text{slip}}{\text{synchronous speed}} = \frac{AD}{AB}$$

Now since $AD = AB - BC$,

$$\text{slip} = \text{synchronous speed} - \text{rotor speed}$$

thus:

$$\text{Per-unit slip} = \frac{\begin{array}{c}\text{synchronous speed}\\ - \text{rotor speed}\end{array}}{\text{synchronous speed}}$$
$$= \frac{AB - BC}{AB}$$

The percentage slip is given by:

$$\text{Percentage slip} = \frac{\begin{array}{c}\text{synchronous speed}\\ - \text{rotor speed}\end{array}}{\text{synchronous speed}} \times 100\%$$
$$= \frac{AB - BC}{AB} \times 100\%$$

The actual value of slip tends to vary from about 6% for a small motor to around 2% for a large machine. Hence, for most purposes the induction motor can be considered to provide a constant speed (determined by the frequency of the current applied to its stator) however one of its principal disadvantages is the fact that it is not easy to vary the speed of such a motor!

Note that, in general, it is not easy to control the speed of an AC motor unless a variable frequency AC supply is available. The speed of a motor with a wound rotor can be controlled by varying the current induced in the rotor but such an arrangement is not very practical as some means of making contact with the rotor windings is required. For this reason, DC motors are usually preferred in applications where the speed must be varied. However, where it is essential to be able to adjust the speed of an AC motor, the motor is invariably powered by an *inverter*. This consists of an electronic switching unit which produces a high-current three-phase pulse-width modulated (PWM) output voltage from a DC supply, as shown in Figure 5.206.

Key point

The rotor of an induction motor rotates at less than synchronous speed, in order that the rotating field can cut through the rotor conductors and induce a current flow in them. This percentage difference between the synchronous speed and the rotor speed is known as slip. Slip varies very little with normal load changes, and the induction motor is therefore considered to be a constant-speed motor.

Example 5.104

An induction motor has a synchronous speed of 3600 rpm and its actual speed of rotation is measured as 3450 rpm Determine (a) the per-unit slip and (b) the percentage slip.

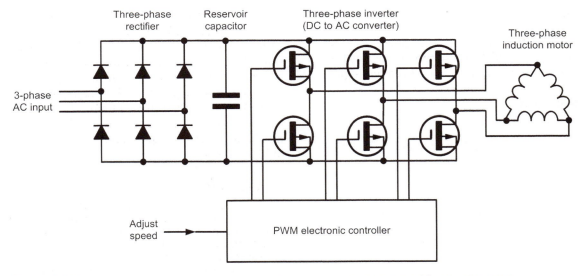

Figure 5.206 Using an inverter to produce a variable output speed from an AC induction motor.

Solution

(a) The per-unit slip is found from:

$$\text{Per-unit slip} = \frac{3600 - 3450}{3600}$$

$$= \frac{150}{3600} = 0.042$$

(b) The percentage slip is given by:

$$\text{Percentage slip} = \frac{3600 - 3450}{3600} \times 100\%$$

$$= \frac{150}{3600} \times 100\% = 4.2\%$$

Inside an induction motor, the speed of the rotating flux, N, is given by the relationship:

$$N = \frac{f}{p}$$

where N is the speed of the flux (in rev/s), f is the frequency of the applied AC (in Hz) and p is the number of pole pairs.

Now the per-unit slip, s, is given by:

$$s = \frac{AB - BC}{AB} = \frac{N - N_r}{N}$$

where N is the speed of the flux (in revolutions per second) and N_r is the rotor speed.

Now:

$$sN = N - N_r$$

from which:

$$N_r = N - sN = N(1 - s)$$

and:

$$N_r = N(1 - s) = \frac{f}{p}(1 - s)$$

where N_r is the speed of the rotor (in revolutions per second), f is the frequency of the applied AC (in Hz) and s is the per-unit slip.

Example 5.105

An induction motor has four poles and is operated from a 400 Hz AC supply. If the motor operates with a slip of 2.5% determine the speed of the output rotor.

Solution

Now:

$$N_r = \frac{f}{p}(1 - s) = \frac{400}{2}(1 - 0.025)$$

$$= 200 \times 0.975 = 195$$

Thus the rotor has a speed of 195 revolutions per second (or 11,700 rpm).

Example 5.106

An induction motor has four poles and is operated from a 60 Hz AC supply. If the rotor speed is 1700 rpm determine the percentage slip.

Solution

Now:

$$N_r = \frac{f}{p}(1 - s)$$

from which:

$$s = 1 - \frac{N_r p}{f} = 1 - \frac{\left(\frac{1700}{60}\right) \times 2}{60}$$

$$= 1 - \frac{56.7}{60} = 1 - 0.944 = 0.056$$

Expressed as a percentage, i.e. 5.6%

5.19.6 Single- and two-phase induction motors

In the case of a two-phase induction motor, two windings are placed at right angles to each other. By exciting these windings with current which is 90° out of phase, a rotating magnetic field can be created. A single-phase induction motor, on the other hand, has only one phase. This type of motor is extensively used in applications which require small low-output motors. The advantage gained by using single-phase motors is that in small sizes they are less expensive to manufacture than other types. Also they eliminate the need for a three-phase supply. Single-phase motors are used in communication equipment, fans, portable power tools, etc. Since the field due to the single-phase AC voltage applied to the stator winding is pulsating, single-phase AC induction motors develop a pulsating torque. They are therefore less efficient than three- or two-phase motors, in which the torque is more uniform.

Single-phase induction motors have only one stator winding. This winding generates a field which can be said to alternate along the axis of the single winding, rather than to rotate. Series motors, on the other hand, resemble DC machines in that they have commutators and brushes.

When the rotor is stationary, the expanding and collapsing stator field induces currents in the rotor which generate a rotor field. The opposition of these fields exerts a force on the rotor, which tries to turn it 180° from its position. However, this force is exerted through the center of the rotor and the rotor will not turn unless a force is applied in order to assist it. Hence some means of *starting* is required for all single-phase induction motors.

Key point

Induction motors are available that are designed for three-, two- and single-phase operation. The three-phase stator is exactly the same as the three-phase stator of the synchronous motor. The two-phase stator generates a rotating field by having two windings positioned at right angles to each other. If the voltages applied to the two windings are 90° out of phase, a rotating field will be generated.

Key point

A synchronous motor uses a single- or three-phase stator to generate a rotating magnetic field, and an electromagnetic rotor that is supplied with DC. The rotor acts like a magnet and is attracted by the rotating stator field. This attraction will exert a torque on the rotor and cause it to rotate with the field.

Key point

A single-phase induction motor has only one stator winding; therefore the magnetic field generated does not rotate. A single-phase induction motor with only one winding cannot start rotating by itself. Once the rotor is started rotating, however, it will continue to rotate and come up to speed. A field is set up in the rotating rotor that is 90° out of phase with the stator field. These two fields together produce a rotating field that keeps the rotor in motion.

5.19.7 Capacitor starting

In an induction motor designed for capacitor starting, the stator consists of the main winding together with a starting winding which is connected in parallel with the main winding and spaced at right angles to it. A phase difference between the current in the two windings

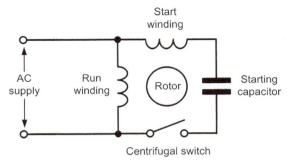

Figure 5.207 Capacitor starting arrangement.

is obtained by connecting a capacitor in series with the auxiliary winding. A switch is included solely for the purposes of applying current to the auxiliary winding in order to start the rotor (see Figure 5.207).

On starting, the switch is closed, placing the capacitor in series with the auxiliary winding. The capacitor is of such a value that the auxiliary winding is effectively a resistive–capacitive circuit in which the current leads the line voltage by approximately 45°. The main winding has enough inductance to cause the current to lag the line voltage by approximately 45°. The two field currents are therefore approximately 90° out of phase. Consequently the fields generated are also at an angle of 90°. The result is a revolving field that is sufficient to start the rotor turning.

After a brief period (when the motor is running at a speed which is close to its normal speed) the switch opens and breaks the current flowing in the auxiliary winding. At this point, the motor runs as an ordinary single-phase induction motor. However, since the two-phase induction motor is more efficient than a single-phase motor, it can be desirable to maintain the current in the auxiliary winding so that motor runs as a two-phase induction motor.

In some types of motor a more complicated arrangement is used with more than one capacitor switched into the auxiliary circuit. For example, a large value of capacitor could be used in order to ensure sufficient torque for starting a heavy load and then, once the motor has reached its operating speed, the capacitor value can be reduced in order to reduce the current in the auxiliary winding. A motor that employs such an arrangement, where two different capacitors

are used (one for *starting* and one for *running*) is often referred to as *capacitor-start, capacitor-run* induction motor. Finally, note that, since phase shift can also be produced by an inductor, it is possible to use an inductor instead of a capacitor. Capacitors tend to be less expensive and more compact than comparable inductors and therefore are more frequently used.

Since the current and voltage in an inductor are also 90° out of phase, inductor starting is also possible. Once again, a starting winding is added to the stator. If this starting winding is placed in series with an inductor across the same supply as the running winding, the current in the starting winding will be out of phase with the current in the running winding. A rotating magnetic field will therefore be generated, and the rotor will rotate.

Key point

In order to make a single-phase motor self-starting, a starting winding is added to the stator. If this starting winding is placed in series with a capacitor across the same supply as the running winding, the current in the starting winding will be out of phase with the current in the running winding. A rotating magnetic field will therefore be generated, and the rotor will rotate. Once the rotor comes up to speed, the current in the auxiliary winding can be switched-out, and the motor will continue running as a single-phase motor.

5.19.8 Shaded pole motors

A different method of starting a single-phase induction motor is based on a *shaded-pole*. In this type of motor, a moving magnetic field is produced by constructing the stator in a particular way. The motor has projecting pole pieces just like DC machines; and part of the pole surface is surrounded by a copper strap or *shading coil*.

As the magnetic field in the core builds, the field flows effortlessly through the unshaded segment. This field is coupled into the shading coil which effectively constitutes a short-circuited loop. A large current momentarily flows in this loop and an opposing field is generated as a consequence. The result is simply that the unshaded segment initially experiences a larger

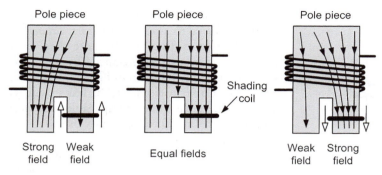

Figure 5.208 Action of a shaded pole.

(a) Field increasing   (b) Maximum field   (c) Field decreasing

magnetic field than does the shaded segment. At some time later, the fields in the two segments become equal. Later still, as the magnetic field in the unshaded segment declines, the field in the shaded segment strengthens. This is illustrated in Figure 5.208.

Key point

In the shaded pole induction motor, a section of each pole face in the stator is shorted out by a metal strap. This has the effect of moving the magnetic field back and forth across the pole face. The moving magnetic field has the same effect as a rotating field, and the motor is self-starting when switched on.

Test your understanding 5.19

1. Explain the difference between synchronous AC motors and induction motors.
2. Explain the main disadvantage of the synchronous motor.
3. Sketch the construction of a squirrel cage induction motor.
4. Explain why the induction motor is the most commonly used form of AC motor.
5. An induction motor has a synchronous speed of 7200 rpm and its actual speed of rotation is measured as 7000 rpm Determine (a) the per-unit slip and (b) the percentage slip.
6. An induction motor has four poles and is operated from a 400 Hz AC supply. If the motor operates with a slip of 1.8%, determine the speed of the output rotor.
7. An induction motor has four poles and is operated from a 60 Hz AC supply. If the rotor speed is 1675 rpm, determine the percentage slip.
8. Explain why a single-phase induction motor requires a means of starting.

9. Describe a typical capacitor starting arrangement for use with a single-phase induction motor.
10. With the aid of a diagram, explain the action of a shaded pole motor.

5.20 Multiple choice questions

The example questions set out below follow the sections of Module 3 in the Part 66 syllabus. Note that the following questions have been separated by level, where appropriate. Several of the sections (e.g. DC circuits, resistance, power, capacitance, magnetism, inductance, etc.) are not required for category A certifying mechanics. Please remember that **ALL these questions must be attempted** *without* **the use of a calculator** and that the pass mark for all JAR 66 multiple-choice examinations is 75%!

Electron theory

1. Within the nucleus of the atom, protons are:
   [A, B1, B2]
   (a) positively charged
   (b) negatively charged
   (c) neutral

2. A positive ion is an atom that has:
   [A, B1, B2]
   (a) gained an electron
   (b) lost an electron
   (c) an equal number of protons and electrons

3. Within an atom, electrons can be found:
   [A, B1, B2]
   (a) along with neutrons as part of the nucleus

(b) surrounded by protons in the center of the nucleus

(c) orbiting the nucleus in a series of shells

4. A material in which there are no free charge carrier is known as: **[A, B1, B2]**
   (a) a conductor
   (b) an insulator
   (c) a semiconductor

5. The charge carriers in a metal consist of:
                                                     [A, B1, B2]
   (a) free electrons
   (b) free atoms
   (c) free neutrons

Static electricity and conduction

6. Two charged particles are separated by a distance, d. If this distance is doubled (without affecting the charge present) the force between the particles will: **[B1, B2]**
   (a) increase
   (b) decrease
   (c) remain the same

7. A beam of electrons moves between two parallel plates, P and Q, as shown in Figure 5.209. Plate P has a positive charge whilst plate Q has a negative charge. Which one of the three paths will the electron beam follow?   **[B1, B2]**
   (a) A
   (b) B
   (c) C

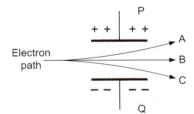

Figure 5.209

8. The force between two charged particles is proportional to the: **[B1, B2]**
   (a) product of their individual charges
   (b) sum of their individual charges
   (c) difference between the individual charges

9. Two isolated charges have dissimilar polarities. The force between them will be:
                                                     [B1, B2]
   (a) a force of attraction
   (b) a force of repulsion
   (c) zero

10. Which one of the following gives the symbol and abbreviated units for electric charge?
                                                     [A, B1, B2]
    (a) Symbol, Q; unit, C
    (b) Symbol, C; unit, F
    (c) Symbol, C; unit, V

Electrical terminology

11. Which one of the following gives the symbol and abbreviated units for resistance?
                                                     [A, B1, B2]
    (a) Symbol, R; unit, Ω
    (b) Symbol, V; unit, V
    (c) Symbol, R; unit, A

12. Current can be defined as the rate of flow of:
                                                     [A, B1, B2]
    (a) charge
    (b) resistance
    (c) voltage

13. A current of 3 A flows for a period of 2 min. The amount of charge transferred will be:
                                                     [B1, B2]
    (a) 6 C
    (b) 40 C
    (c) 360 C

14. The volt can be defined as: **[B1, B2]**
    (a) a joule per coulomb
    (b) a watt per coulomb
    (c) an ohm per watt

15. Conventional current flow is: **[A, B1, B2]**
    (a) always from negative to positive
    (b) in the same direction as electron movement
    (c) in the opposite direction to electron movement

16. Conductance is the inverse of: **[A, B1, B2]**
    (a) charge
    (b) current
    (c) resistance

Generation of electricity

17. A photocell produces electricity from:
    [A, B1, B2]
    (a) heat
    (b) light
    (c) chemical action

18. A secondary cell produces electricity from:
    [A, B1, B2]
    (a) heat
    (b) light
    (c) chemical action

19. A thermocouple produces electricity from:
    [A, B1, B2]
    (a) heat
    (b) light
    (c) chemical action

20. Which one of the following devices uses magnetism and motion to produce electricity?
    [A, B1, B2]
    (a) a transformer
    (b) an inductor
    (c) a generator

21. A small bar magnet is moved at right angles to a length of copper wire. The e.m.f. produced at the ends of the wire will depend on the:
    [B1, B2]
    (a) diameter of the copper wire and the strength of the magnet
    (b) speed at which the magnet is moved and the strength of the magnet
    (c) resistance of the copper wire and the speed at which the magnet is moved

DC sources of electricity

22. The e.m.f. produced by a fresh zinc–carbon battery is approximately: [A, B1, B2]
    (a) 1.2 V
    (b) 1.5 V
    (c) 2 V

23. The electrolyte of a fully charged lead–acid battery will have a relative density of approximately: [A, B1, B2]
    (a) 0.95
    (b) 1.15
    (c) 1.26

24. The terminal voltage of a cell falls slightly when it is connected to a load. This is because the cell: [B1, B2]
    (a) has some internal resistance
    (b) generates less current when connected to the load
    (c) produces more power without the load connected

25. The electrolyte of a conventional lead–acid cell is: [A, B1, B2]
    (a) water
    (b) dilute hydrochloric acid
    (c) dilute sulphuric acid

26. The anode of a conventional dry (Leclanché) cell is made from: [A, B1, B2]
    (a) carbon
    (b) copper
    (c) zinc

27. A junction between two dissimilar metals that produces a small voltage when a temperature difference exists between it and a reference junction is known as a: [A, B1, B2]
    (a) diode
    (b) thermistor
    (c) thermocouple

28. A photocell consists of: [A, B1, B2]
    (a) two interacting layers of a semiconductor material
    (b) two electrodes separated by an electrolyte
    (c) a junction of two dissimilar metals

29. The materials used in a typical thermocouple are: [A, B1, B2]
    (a) silicon and selenium
    (b) silicon and germanium
    (c) iron and constantan

DC circuits

30. The relationship between voltage, V, current, I, and resistance, R, for a resistor is:
    [B1, B2]
    (a) $V = IR$
    (b) $V = \frac{R}{I}$
    (c) $V = IR^2$

31. A potential difference of 7.5 V appears across a 15 Ω resistor. Which one of the

following gives the current flowing:

[B1, B2]

(a) 0.25 A
(b) 0.5 A
(c) 2 A

32. A DC supply has an internal resistance of 1 Ω and an open-circuit output voltage of 24 V. What will the output voltage be when the supply is connected to a 5 Ω load?

[B1, B2]

(a) 19 V
(b) 20 V
(c) 24 V

33. Three 9 V batteries are connected in series. If the series combination delivers 150 mA to a load, which one of the following gives the resistance of the load? [B1, B2]
(a) 60 Ω
(b) 180 Ω
(c) 600 Ω

34. The unknown current shown in Figure 5.210 will be: [B1, B2]
(a) 1 A flowing towards the junction
(b) 1 A flowing away from the junction
(c) 4 A flowing towards the junction

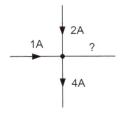

Figure 5.210

35. Which one of the following gives the output voltage produced by the circuit shown in Figure 5.211? [B1, B2]

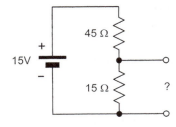

Figure 5.211

(a) 3.75 V
(b) 1.9 V
(c) 4.7 V

36. Which one of the following gives the current flowing in the 60 Ω resistor as shown in Figure 5.212? [B1, B2]
(a) 0.33 A
(b) 0.66 A
(c) 1 A

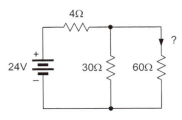

Figure 5.212

Resistance and resistors

37. A 20 m length of cable has a resistance of 0.02 Ω. If a 100 m length of the same cable carries a current of 5 A flowing in it, what voltage will be dropped across its ends?

[B1, B2]

(a) 0.02 V
(b) 0.1 V
(c) 0.5 V

38. The resistance of a wire conductor of constant cross section: [B1, B2]
(a) decreases as the length of the wire increases
(b) increases as the length of the wire increases
(c) is independent of the length of the wire

39. Three 15 Ω resistors are connected in parallel. Which one of the following gives the effective resistance of the parallel combination? [B1, B2]
(a) 5 Ω
(b) 15 Ω
(c) 45 Ω

40. Three 15 Ω resistors are connected in series. Which one of the following gives the effective resistance of the series combination?

[B1, B2]

(a) $5\,\Omega$
(b) $15\,\Omega$
(c) $45\,\Omega$

41. Which one of the following gives the effective resistance of the circuit shown Figure 5.213?   [B1, B2]
    (a) $5\,\Omega$
    (b) $6\,\Omega$
    (c) $26\,\Omega$

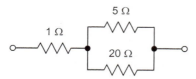

Figure 5.213

42. A $10\,\Omega$ wirewound resistor is made from 0.2 m of wire. A second wirewound resistor is made from 0.5 m of the same wire. The second resistor will have a resistance of:
                                                         [B1, B2]
    (a) $4\,\Omega$
    (b) $15\,\Omega$
    (c) $25\,\Omega$

Power

43. The relationship between power, P, current, I, and resistance, R, is: [B1, B2]
    (a) $P = I \times R$
    (b) $P = \frac{R}{I}$
    (c) $P = I^2 \times R$

44. A DC generator produces an output of 28 V at 20 A. The power supplied by the generator will be: [B1, B2]
    (a) $14\,W$
    (b) $560\,W$
    (c) $1.4\,kW$

45. A cabin reading lamp consumes 10 W from a 24 V DC supply. The current supplied will be: [B1, B2]
    (a) 0.42 A
    (b) 0.65 A
    (c) 2.4 A

46. A generator delivers 250 W of power to a $50\,\Omega$ load. The current flowing in the load will be: [B1, B2]

(a) 2.24 A
(b) 5 A
(c) 10 A

47. An aircraft cabin has 110 passenger reading lamps each rated at 10 W, 28 V. What is the maximum load current imposed by these lamps?   [B1, B2]
    (a) 25.5 A
    (b) 39.3 A
    (c) 308 A

48. An aircraft fuel heater consists of two parallel-connected heating elements each rated at 28 V, 10 A. What total power is supplied to the fuel heating system? [B1, B2]
    (a) 140 W
    (b) 280 W
    (c) 560 W

49. An aircraft battery is being charged from a bench DC supply that has an output of 28 V. If the charging current is 10 A, what energy is supplied to the battery if it is charged for 4 h?   [B1, B2]
    (a) 67 kJ
    (b) 252 kJ
    (c) 4.032 MJ

50. A portable power tool operates from a 7 V rechargeable battery. If the battery is charged for 10 h at 100 mA, what energy is supplied to it?   [B1, B2]
    (a) 25.2 kJ
    (b) 252 kJ
    (c) 420 kJ

Capacitance and capacitors

51. The high-voltage connection on a power supply is fitted with a rubber cap. The reason for this is to: [B1, B2]
    (a) provide insulation
    (b) concentrate the charge
    (c) increase the current rating

52. Which one of the following gives the symbol and abbreviated units for capacitance?
                                                         [B1, B2]
    (a) Symbol, C; unit, C
    (b) Symbol, C; unit, F
    (c) Symbol, Q; unit, C

53. A capacitor is required to store a charge of $32 \mu C$ when a voltage of 4 V is applied to it. The value of the capacitor should be:
[B1, B2]
(a) $0.125 \mu F$
(b) $0.25 \mu F$
(c) $8 \mu F$

54. An air-spaced capacitor has two plates separated by a distance, d. If the distance is doubled (without affecting the area of the plates) the capacitance will be: [B1, B2]
(a) doubled
(b) halved
(c) remain the same

55. A variable air-spaced capacitor consists of two sets of plates that can be moved. When the plates are fully meshed, the: [B1, B2]
(a) capacitance will be maximum and the working voltage will be reduced
(b) capacitance will be maximum and the working voltage will be unchanged
(c) capacitance will be minimum and the working voltage will be increased

56. A $20 \mu F$ capacitor is charged to a voltage of 50 V. The charge present will be:
[B1, B2]
(a) $0.5 \mu C$
(b) $2.5 \mu F$
(c) $1 mC$

57. A power supply filter uses five parallel-connected $2200 \mu F$ capacitors each rated at 50 V. What single capacitor could be used to replace them?   [B1, B2]
(a) $11,000 \mu F$ at 10 V
(b) $440 \mu F$ at 50 V
(c) $11,000 \mu F$ at 250 V

58. A high-voltage power supply uses four identical series-connected capacitors. If 1 kV appears across the series arrangement and the total capacitance required is $100 \mu F$, which one of the following gives a suitable rating for each individual capacitor?
[B1, B2]
(a) $100 \mu F$ at 250 V
(b) $25 \mu F$ at 1 kV
(c) $400 \mu F$ at 250 V

59. Which one of the following materials is suitable for use as a capacitor dielectric?
[B1, B2]
(a) aluminium foil
(b) polyester film
(c) carbon granules

60. The relationship between capacitance, C, charge, Q, and potential difference, V, for a capacitor is:
[B1, B2]
(a) $Q = CV$
(b) $Q = \frac{C}{V}$
(c) $Q = CV^2$

61. The material that appears between the plates of a capacitor is known as the: [B1, B2]
(a) anode
(b) cathode
(c) dielectric

Magnetism

62. Permanent magnets should be stored using
[B1, B2]
(a) anti-static bags
(b) insulating material such as polystyrene
(c) soft iron keepers

63. Lines of magnetic flux: [B1, B2]
(a) originate at the south pole and end at the north pole
(b) originate at the north pole and end at the south pole
(c) start and finish at the same pole, either south or north

64. The magnetomotive force produced by a solenoid is given by: [B1, B2]
(a) the length of the coil divided by its cross-sectional area
(b) number of turns on the coil divided by its cross-sectional area
(c) the number of turns on the coil multiplied by the current flowing in it

65. An air-cored solenoid with a fixed current flowing through it is fitted with a ferrite core. The effect of the core will be to: [B1, B2]
(a) increase the flux density produced by the solenoid

(b) decrease the flux density produced by the solenoid

(c) leave the flux density produced by the solenoid unchanged

66. The permeability of a magnetic material is given by the ratio of: [B1, B2]
    (a) magnetic flux to cross-sectional area
    (b) magnetic field intensity to magneto-motive force
    (c) magnetic flux density to magnetic field intensity

67. The relationship between permeability, μ, magnetic flux density, B, and magnetizing force, H, is: [B1, B2]
    (a) $\mu = B \times H$
    (b) $\mu = \frac{B}{H}$
    (c) $\mu = \frac{H}{B}$

68. The relationship between absolute permeability, μ, relative permeability, μ_r, and the permeability of free-space, μ_0, is given by: [B1, B2]
    (a) $\mu = \mu_0 \times \mu_r$
    (b) $\mu = \frac{\mu_0}{\mu_r}$
    (c) $\mu = \frac{\mu_r}{\mu_0}$

69. The relative permeability of steel is in the range: [B1, B2]
    (a) 1 to 10
    (b) 10 to 100
    (c) 100 to 1000

70. The feature marked X on the B–H curve shown in Figure 5.214 is: [B1, B2]
    (a) saturation
    (b) reluctance
    (c) hysteresis

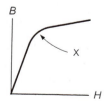

Figure 5.214

Inductance and inductors

71. Which one of the following gives the symbol and abbreviated units for inductance? [B1, B2]
    (a) Symbol, I; unit, L
    (b) Symbol, L; unit, H
    (c) Symbol, H; unit, L

72. Which one of the following materials is suitable for use as the coil winding of an inductor? [B1, B2]
    (a) brass
    (b) copper
    (c) steel

73. Which one of the following materials is suitable for use as the laminated core of an inductor? [B1, B2]
    (a) brass
    (b) copper
    (c) steel

74. Lenz's law states that: [B1, B2]
    (a) the reluctance of a magnetic circuit is zero
    (b) an induced e.m.f. will always oppose the motion that created it
    (c) the force on a current-carrying conductor is proportional to the current flowing

75. The inductance of a coil is directly proportional to the: [B1, B2]
    (a) current flowing in the coil
    (b) square of the number of turns
    (c) mean length of the magnetic path

76. The inductance of a coil can be increased by using: [B1, B2]
    (a) a low number of turns
    (b) a high permeability core
    (c) wire having a low resistance

DC motor and generator theory

77. The commutator in a DC generator is used to: [B1, B2]
    (a) provide a means of connecting an external field current supply
    (b) periodically reverse the connections to the rotating coil winding

(c) disconnect the coil winding when the induced current reaches a maximum value

78. The core of a DC motor/generator is laminated in order to: **[B1, B2]**
    (a) reduce the overall weight of the machine
    (b) reduce eddy currents induced in the core
    (c) increase the speed at which the machine rotates

79. The brushes fitted to a DC motor/generator should have: **[B1, B2]**
    (a) low coefficient of friction and low contact resistance
    (b) high coefficient of friction and low contact resistance
    (c) low coefficient of friction and high contact resistance

80. A feature of carbon brushes used in DC motors and generators is that they are: **[B1, B2]**
    (a) self-lubricating
    (b) self-annealing
    (c) self-healing

81. Self-excited generators derive their field current from: **[B1, B2]**
    (a) the current produced by the armature
    (b) a separate field current supply
    (c) an external power source

82. In a series-wound generator: **[B1, B2]**
    (a) none of the armature current flows through the field
    (b) some of the armature current flows through the field
    (c) all of the armature current flows through the field

83. In a shunt-wound generator: **[B1, B2]**
    (a) none of the armature current flows through the field
    (b) some of the armature current flows through the field
    (c) all of the armature current flows through the field

84. A compound-wound generator has: **[B1, B2]**

(a) only a series field winding
(b) only a shunt field winding
(c) both a series and a shunt field winding

AC theory

85. Figure 5.215 shows an AC waveform. The waveform is a: **[A, B1, B2]**
    (a) square wave
    (b) sine wave
    (c) triangle wave

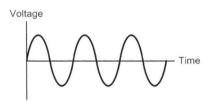

Figure 5.215

86. Figure 5.216 shows an AC waveform. The periodic time of the waveform is: **[B1, B2]**
    (a) 1 ms
    (b) 2 ms
    (c) 4 ms

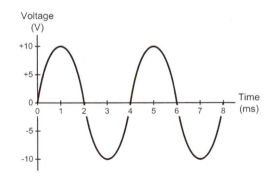

Figure 5.216

87. Figure 5.217 shows an AC waveform. The amplitude of the waveform is: **[B1, B2]**
    (a) 5 V
    (b) 10 V
    (c) 20 V

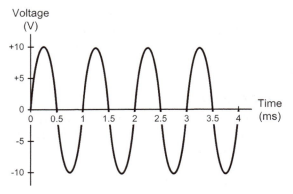

Figure 5.217

88. Figure 5.218 shows two AC waveforms. The phase angle between these waveforms is: [B1, B2]
    (a) 45°
    (b) 90°
    (c) 180°

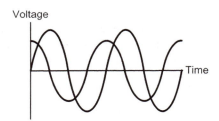

Figure 5.218

89. An AC waveform has a frequency of 400 Hz. Which one of the following gives its period? [B1, B2]

    (a) 2.5 ms
    (b) 25 ms
    (c) 400 ms

90. An AC waveform has a period of 4 ms. Which one of the following gives its frequency? [B1, B2]
    (a) 25 Hz
    (b) 250 Hz
    (c) 4 kHz

91. Which one of the following gives the angle between the successive phases of a three-phase supply? [A, B1, B2]

    (a) 60°
    (b) 90°
    (c) 120°

92. An aircraft supply has an r.m.s value of 115 V. Which one of the following gives the approximate peak value of the supply voltage? [B1, B2]
    (a) 67.5 V
    (b) 115 V
    (c) 163 V

93. The peak value of current supplied to an aircraft TRU is 28 A. Which one of the following gives the approximate value of r.m.s. current supplied? [B1, B2]
    (a) 10 A
    (b) 14 A
    (c) 20 A

Resistive, capacitive and inductive circuits

94. A circuit consisting of a pure capacitance is connected across an AC supply. Which one of the following gives the phase relationship between the voltage and current in this circuit? [B1, B2]
    (a) The voltage leads the current by 90°
    (b) The current leads the voltage by 90°
    (c) The current leads the voltage by 180°

95. An inductor has an inductive reactance of 50 Ω and a resistance of 50 Ω. Which one of the following gives the phase relationship between the voltage and current in this circuit? [B1, B2]
    (a) The current leads the voltage by 45°
    (b) The voltage leads the current by 45°
    (c) The voltage leads the current by 90°

96. A capacitor having negligible resistance is connected across a 115 V AC supply. If the current flowing in the capacitor is 0.5 A, which one of the following gives its reactance? [B1, B2]
    (a) 0 Ω
    (b) 50 Ω
    (c) 230 Ω

97. A pure capacitor having a reactance of 100 Ω is connected across a 200 V AC supply. Which one of the following gives the power dissipated in the capacitor? [B1, B2]
    (a) 0 W
    (b) 50 W
    (c) 400 W

98. The power factor in an AC circuit is defined as the: [B1, B2]
    (a) ratio of true power to apparent power
    (b) ratio of apparent power to true power
    (c) ratio of reactive power to true power

99. The power factor in an AC circuit is the same as the: [B1, B2]
    (a) sine of the phase angle
    (b) cosine of the phase angle
    (c) tangent of the phase angle

100. An AC circuit consists of a capacitor having a reactance of 40 Ω connected in series with a resistance of 30 Ω. Which one of the following gives the impedance of this circuit? [B1, B2]
    (a) 10 Ω
    (b) 50 Ω
    (c) 70 Ω

101. An AC circuit consists of a pure inductor connected in parallel with a pure capacitor. At the resonant frequency, the: [B1, B2]
    (a) impedance of the circuit will be zero
    (b) impedance of the circuit will be infinite
    (c) impedance of the circuit will be the same as at all other frequencies

Transformers

102. A transformer has 2400 primary turns and 600 secondary turns. If the primary is supplied from a 220 V AC supply, which one of the following gives the resulting secondary voltage: [B1, B2]
    (a) 55 V
    (b) 110 V
    (c) 880 V

103. Two inductive coils are placed in close proximity to one another. Minimum flux linkage will occur between the coils when the relative angle between them is: [B1, B2]
    (a) 0°
    (b) 45°
    (c) 90°

104. The primary and secondary voltage and current for an aircraft transformer is given in the table below: [B1, B2]

	Primary	Secondary
Voltage (V)	110	50
Current (A)	2	4

Which one of the following gives the approximate efficiency of the transformer? [B1, B2]
    (a) 63%
    (b) 85%
    (c) 91%

105. The "copper loss" in a transformer is a result of: [B1, B2]
    (a) the I^2R power loss in the transformer windings
    (b) the power required to magnetize the core of the transformer
    (c) eddy currents flowing in the magnetic core of the transformer

Filters

106. The frequency response shown in Figure 5.219 represents the output of a: [B1, B2]
    (a) low-pass filter
    (b) high-pass filter
    (c) band-pass filter

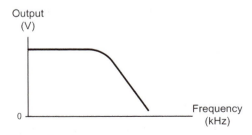

Figure 5.219

107. The frequency response shown in Figure 5.220 represents the output of a: [**B1, B2**]
    (a) low-pass filter
    (b) high-pass filter
    (c) band-pass filter

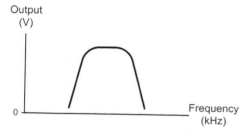

Figure 5.220

108. Signals at 10 kHz and 400 Hz are present in a cable. The 10 kHz signal can be removed by means of an appropriately designed: [**B1, B2**]
    (a) low-pass filter
    (b) high-pass filter
    (c) band-pass filter

109. Signals at 118, 125 and 132 MHz are present in the feeder to an antenna. The signals at 118 and 132 MHz can be reduced by means of: [**B1, B2**]
    (a) low-pass filter
    (b) high-pass filter
    (c) band-pass filter

110. The circuit shown in Figure 5.221 is a:
    (a) low-pass filter
    (b) high-pass filter
    (c) band-pass filter

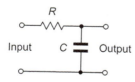

Figure 5.221

AC generators

111. The slip rings in an AC generator provide a means of: [**B1, B2**]

(a) connecting an external circuit to a rotating armature winding
(b) supporting a rotating armature without the need for bearings
(c) periodically reversing the current produced by an armature winding

112. Decreasing the field current in a generator will: [**B1, B2**]
    (a) decrease the output voltage
    (b) increase the output voltage
    (c) increase the output frequency

113. A single-phase AC generator has 12 poles and it runs at 600 rpm. Which one of the following gives the output frequency of the generator? [**B1, B2**]

    (a) 50 Hz
    (b) 60 Hz
    (c) 120 Hz

114. In a star-connected three-phase system, the line voltage is found to be 200 V. Which one of the following gives the approximate value of phase voltage? [**B1, B2**]
    (a) 67 V
    (b) 115 V
    (c) 346 V

115. In a delta-connected three-phase system, the phase current is found to be 2 A. Which one of the following gives the approximate value of line current? [**B1, B2**]
    (a) 1.2 A
    (b) 3.5 A
    (c) 6 A

116. In a balanced star-connected three-phase system the line current is 2 A and the line voltage is 110 V. If the power factor is 0.75 which one of the following gives the total power in the load? [**B1, B2**]
    (a) 165 W
    (b) 286 W
    (c) 660 W

AC motors

117. The rotor of an AC induction motor consists of a: [**B1, B2**]

(a) laminated iron core inside a "squirrel cage" made from copper or aluminium
(b) series of coil windings on a laminated iron core with connections via slip rings
(c) single copper loop which rotates inside the field created by a permanent magnet

118. The slip speed of an AC induction motor is the difference between the: [B1, B2]
    (a) synchronous speed and the rotor speed
    (b) frequency of the supply and the rotor speed
    (c) maximum speed and the minimum speed

119. When compared with three-phase induction motors, single-phase induction motors: [B1, B2]

(a) are not inherently "self starting"
(b) have more complicated stator windings
(c) are significantly more efficient

120. The use of laminations in the construction of an electrical machine is instrumental in reducing the: [B1, B2]
    (a) losses
    (b) output
    (c) weight

121. A three-phase induction motor has three pairs of poles and is operated from a 60 Hz supply. Which one of the following gives the motor's synchronous speed? [B1, B2]
    (a) 1200 rpm
    (b) 1800 rpm
    (c) 3600 rpm

Chapter 6

Electronic fundamentals

6.1 Introduction

If you have previously studied Chapter 5, you will already be aware of just how important electricity is in the context of a modern aircraft. However, whereas Chapter 5 introduced you to the fundamentals of electrical power generation, distribution and utilization, this section concentrates on developing an understanding of the electronic devices and circuits that are found in a wide variety of aircraft systems. Such devices include diodes, transistors and integrated circuits, and the systems that are used to include control instrumentation, radio and navigation aids.

We will begin this section by introducing you to some important concepts starting with an introduction to electronic systems and circuit diagrams. It is particularly important that you get to grips with these concepts if you are studying electronics for the first time!

6.1.1 Electronic circuit and systems

Electronic circuits, such as amplifiers, oscillators and power supplies, are made from arrangements of the basic electronic components (such as the resistors, capacitors, inductors and transformers that we met in Chapter 5) along with the semiconductors and integrated circuits that we shall meet for the first time in this section.

Semiconductors are essential for the operation of the circuits in which they are used, however, for them to operate correctly; there is a requirement for them to have their own supply and *bias* voltages. We will explain how this works later in this section when we introduce transistors and integrated circuits but, for the moment, it is important to understand that most electronic circuits may often appear to be somewhat more complex than they are, simply because there is a need to supply the semiconductor devices with the voltages and currents that they need in order to operate correctly.

In order to keep things simple, we often use block schematic diagrams rather than full circuit diagrams in order to help explain the operation of electronic systems. Each block usually represents a large number of electronic components and instead of showing all the electrical connections we simply show a limited number of them, sufficient to indicate the flow of signals and power between blocks. As an example, the block schematic diagram of a power supply is shown in Figure 6.1. Note that the input is taken from a 400 Hz 115 V alternating current (AC) supply, stepped down to 28 V AC, then rectified (i.e. converted to direct current (DC)) and finally regulated to provide a constant output voltage of 28 V DC.

> **Key point**
>
> Electronics is based on the application of semiconductor devices (such as diodes, transistors and integrated circuits) along with components, such as resistors, capacitors, inductors and transformers, that we met earlier in Chapter 5.

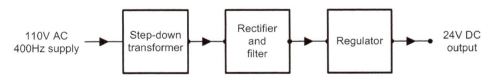

110V AC 400Hz supply → Step-down transformer → Rectifier and filter → Regulator → 24V DC output

Figure 6.1 A block schematic diagram of a power supply.

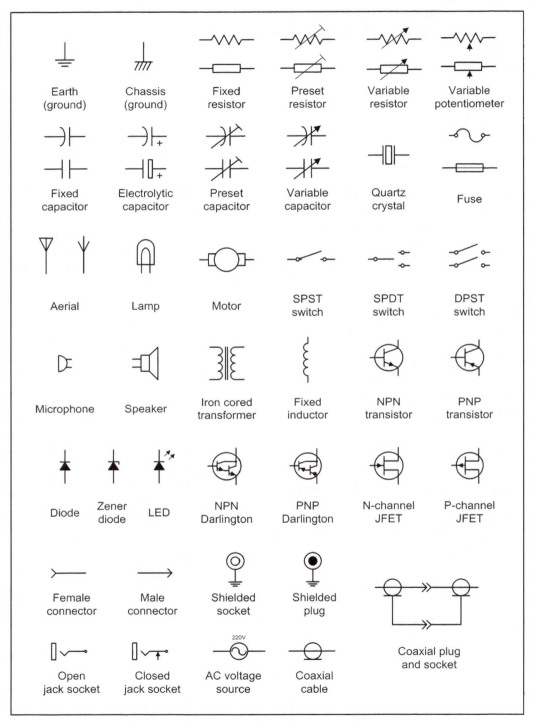

Figure 6.2 A selection of symbols used in electronic circuit schematics.

6.1.2 Reading and understanding circuit diagrams

Before you can make sense of some of the semiconductor devices and circuits that you will meet later in this section it is important to be able to read and understand a simple electronic circuit diagram. Circuit diagrams use standard conventions and symbols to represent the components and wiring used in an electronic circuit. Visually, they bear very little relationship to the physical layout of a circuit but, instead, they provide us with a "theoretical" view of the circuit.

It is important that you become familiar with reading and understanding circuit diagrams right from the start. So, a selection of some of the most commonly used symbols is shown in Figure 6.2. It is important to note that there are a few (thankfully quite small) differences between the symbols used in American and European diagrams.

As a general rule, the input should be shown on the left of the diagram and the output on the right. The supply (usually the most positive voltage) is normally shown at the top of the diagram and the common, 0 V, or ground connection is normally shown at the bottom. This rule is not always obeyed, particularly for complex diagrams where many signals and supply voltages may be present. Note also that, in order to simplify a circuit diagram (and avoid having too many lines connected to the same point), multiple connections to common, 0 V, or ground may be shown using the appropriate symbol (see the negative connection to C_1 in Figure 6.3). The same applies to supply connections that may be repeated (appropriately labelled) at various points in the diagram.

Three different types of switch are shown in Figure 6.2: single-pole single-throw (SPST), single-pole double-throw (SPDT) and double-pole single-throw (DPST). The SPST switch acts as a single-circuit *on/off switch* whilst the DPST provides the same on/off function but makes and breaks two circuits simultaneously. The SPDT switch is sometimes referred to as a *changeover switch* because it allows the selection of one circuit or another. Multi-pole switches are also available. These provide switching between many different circuits. For example, one-pole six-way (1P 6W) switch allows you to select six different circuits.

Example 6.1

The circuit of a simple intercom amplifier is shown in Figure 6.3.

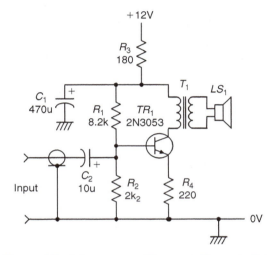

Figure 6.3   Intercom amplifier – see Example 6.1.

(a) What is the value of C_1?
(b) What is the value of R_1?
(c) Which component has a value of 220 Ω?
(d) Which component is connected directly to the positive supply?
(e) Which component is connected to the circuit via T_1?
(f) Where is coaxial cable used in this circuit?

Solution

(a) 470 μF
(b) 8.2 kΩ
(c) R_4
(d) R_3 (the top end of R_3 is marked "+12 V")
(e) LS_1 (the loudspeaker is connected via a step-down transformer, T_1)
(f) To screen the input signal (between the "live" input terminal and the negative connection on C_2).

Key point

Circuit diagrams use standard conventions and symbols to represent the components and wiring used in an electronic circuit. Circuit diagrams provide a "theoretical" view of a circuit, which is often different from the physical layout of the circuit to which they refer.

6.1.3 Characteristic graphs

The characteristics of semiconductor devices are often described in terms of the relationship between the voltage, V, applied to them and the current, I, flowing in them. With a device such as a diode (which has two terminals) this is relatively straightforward. However, with a three-terminal device (such as a transistor), a family of characteristics may be required to fully describe the behaviour of the device. This point will become a little clearer when we meet the transistor later in this section but, for the moment, it is worth considering what information can be gleaned from a simple current/voltage characteristic.

Figure 6.4(a) shows the graph of current plotted against voltage for a linear device such as a resistor whilst Figure 6.4(b) shows a similar graph plotted for a non-linear device such as a semiconductor. Since the ratio of I to V is the reciprocal of resistance, R, we can make the following inferences:

1. At all points in Figure 6.4(a) the ratio of I to V is the same showing that the resistance, R, of the device remains constant. This is exactly how we would want a resistor to perform.
2. In Figure 6.4(b) the ratio of I to V is different at different points on the graph; thus, the resistance, R, of the device does not remain constant but changes as the applied voltage and current changes. This is an important point since most semiconductor devices have distinctly non-linear characteristics!

Key point

Characteristic graphs are used to describe the behaviour of semiconductor devices. These graphs show corresponding values of current and voltage and they are used to predict the performance of a particular device when used in a circuit.

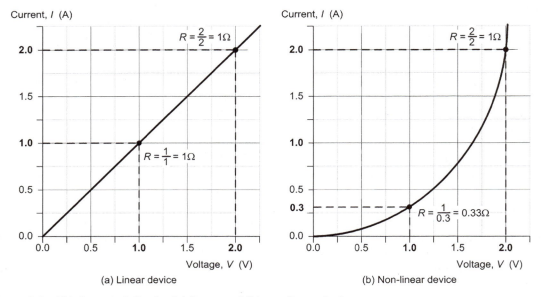

(a) Linear device   (b) Non-linear device

Figure 6.4 I/V characteristics for (a) linear and (b) non-linear device.

Example 6.2

The I/V characteristic for a non-linear electronic device is shown in Figure 6.5. Determine the resistance of the device when the applied voltage is:

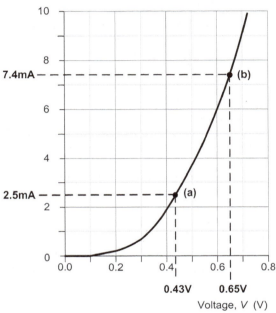

Current, I (mA)

(a) $V = 0.43V$, $R = 0.43V/2.5$ mA = 172Ω

(b) $V = 0.65V$, $R = 0.65V/7.4$ mA = 88Ω

Figure 6.5 I/V characteristic – see Example 6.2.

(a) 0.43 V

(b) 0.65 V

Solution

(a) At 0.43 V the corresponding values of I is 2.5 mA and the resistance, R, of the device will be given by:

$$R = \frac{V}{I} = \frac{0.43}{2.5} = 172\,\Omega$$

(b) At 0.65 V the corresponding values of I is 7.4 mA and the resistance, R, of the device will be given by:

$$R = \frac{V}{I} = \frac{0.65}{7.4} = 88\,\Omega$$

Test your understanding 6.1

1. Identify components shown in Figure 6.6(a)–(o).

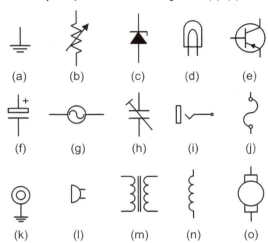

Figure 6.6 See Question 1.

2. Sketch the circuit schematic symbol for:
   (a) a PNP transistor
   (b) a variable capacitor
   (c) a chassis connection
   (d) a quartz crystal

3. Explain, with the aid of a sketch, the operation of each of the following switches:
   (a) SPST
   (b) SPDT
   (c) DPDT

Questions 4–8 refer to the motor driver circuit shown in Figure 6.7.

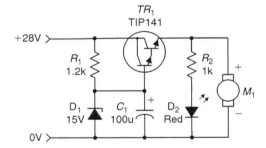

Figure 6.7

4. What type of device is: (a) D_1, (b) D_2 and (c) TR_1?
5. Which components have a connection to the 0 V rail?
6. Which two components are connected in parallel?
7. Which two components are connected in series?
8. Redraw the circuit with the following modifications:
   (a) TR_1 is to be replaced by a conventional NPN transistor,

(b) an SPST switch is to be placed in series with R_1,
(c) the value of C_1 is to be increased to $220\,\mu F$,
(d) the light emitting diode (LED) indicator and series resistor are to be removed and replaced by a single fixed capacitor of 470 nF.

9. Corresponding readings of current, I, and voltage, V, for a semiconductor device are given in the table below:

V (V)	0	0.1	0.2	0.3	0.4	0.5	0.6	0.7	0.8
I (mA)	0	0.2	0.5	1.5	3.0	5.0	8.5	13.0	20.0

Plot the I/V characteristic for the device.

10. Determine the resistance of the device in Question 9, when the applied voltage is:
(a) 0.35 V
(b) 0.75 V

6.2 Semiconductors

6.2.1 Diodes

Syllabus

Diode symbols; Diode characteristics and properties; Diodes in series and parallel; Main characteristics and use of silicon (Si) controlled rectifiers (SCRs), LED, photo-conductive diode, varistor, rectifier diodes; Functional testing of diodes.

Knowledge level key

A	B1	B2
–	2	2

Syllabus

Materials, electron configuration, electrical properties; P- and N-type materials: effects of impurities on conduction, majority and minority carriers; P–N junction in a semiconductor, development of a potential across a P–N junction in unbiased, forward- and reverse-biased conditions; Diode parameters: peak inverse voltage (PIV), maximum forward current, temperature, frequency, leakage current, power dissipation; Operation and function of diodes in the following circuits: clippers, clampers, full- and half-wave rectifiers, bridge rectifiers, voltage doublers and triplers; Detailed operation and characteristics of the following devices: SCR, LED, Schottky diode, photo-conductive

diode, varactor diode, varistor, rectifier diodes, Zener diode.

Knowledge level key

A	B1	B2
–	–	2

Semiconductor materials

This section introduces devices that are made from materials that are neither conductors nor insulators. These semiconductor materials form the basis of many important electronic components, such as diodes, SCRs, triacs, transistors and integrated circuits. We shall start with a brief introduction to the principles of semiconductors and then go on to examine the characteristics of each of the most common types that you are likely to meet.

You should recall that an atom contains both negative charge carriers (*electrons*) and positive charge carriers (*protons*). Electrons each carry a single unit of negative electric charge while protons each exhibit a single unit of positive charge. Since atoms normally contain an equal number of electrons and protons, the net charge present will be zero. For example, if an atom has 11 electrons, it will also contain 11 protons. The end result is that the negative charge of the electrons will be exactly balanced by the positive charge of the protons.

Electrons are in constant motion as they orbit around the nucleus of the atom. Electron orbits are organized into *shells*. The maximum number of electrons present in the first shell is two, in the second shell eight and in the third, fourth and fifth shells it is 18, 32 and 50, respectively. In electronics only the electron shell furthermost from the nucleus of an atom is important. It is important to note that the movement of electrons between atoms involves only those present in the outer *valence shell* (Figure 6.8).

If the valence shell contains the maximum number of electrons possible the electrons are rigidly bonded together and the material has the properties of an insulator. If, however, the valence shell does not have its full complement of electrons, the electrons can be easily detached from their orbital bonds, and the

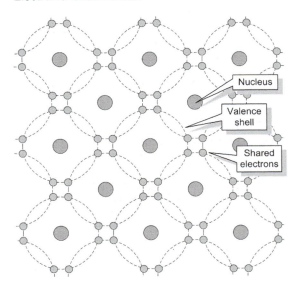

Figure 6.8 Electrons orbiting a nucleus.

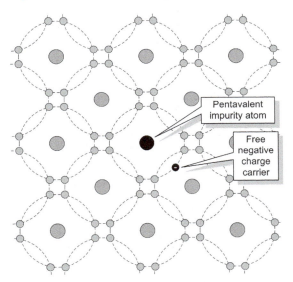

Figure 6.9 Effect of introducing a pentavalent impurity.

material has the properties associated with an electrical conductor.

In its pure state, Si is an insulator because the covalent bonding rigidly holds all of the electrons leaving no free (easily loosened) electrons to conduct current. If, however, an atom of a different element (i.e. an *impurity*) is introduced that has five electrons in its valence shell, a surplus electron will be present (see Figure 6.9).

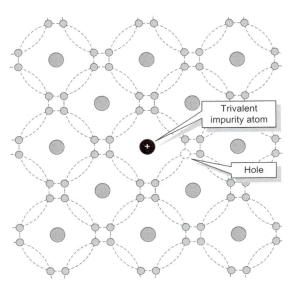

Figure 6.10 Effect of introducing a trivalent impurity.

These free electrons become available for use as charge carriers and they can be made to move through the lattice by applying an external potential difference to the material.

Similarly, if the impurity element introduced into the pure Si lattice has three electrons in its valence shell, the absence of the fourth electron needed for proper covalent bonding will produce a number of spaces into which electrons can fit (see Figure 6.10). These spaces are referred to as *holes*. Once again, current will flow when an external potential difference is applied to the material.

Regardless of whether the impurity element produces surplus electrons or holes, the material will no longer behave as an insulator, neither will it have the properties that we normally associate with a metallic conductor. Instead, we call the material a semiconductor – the term simply indicates that the substance is no longer a good insulator or a good conductor but is somewhere in between! Examples of semiconductors include germanium (Ge) and silicon (Si).

The process of introducing an atom of another (impurity) element into the lattice of an otherwise pure material is called *doping*. When the pure material is doped with an impurity with five electrons in its valence shell (i.e. a *pentavalent impurity*) it will become an N-type (i.e. negative type) material. If, however, the pure material is doped with an impurity having three electrons in its valence shell (i.e. a *trivalent impurity*) it will become P-type material (i.e. positive type). N-type semiconductor material contains an excess of negative charge carriers and P-type material contains an excess of positive charge carriers.

Key point

Circuit diagrams use standard conventions and symbols to represent the components and wiring used in an electronic circuit. Circuit diagrams provide a "theoretical" view of a circuit which is often different from the physical layout of the circuit to which they refer.

Semiconductor classification

Semiconductor devices are classified using a unique part numbering system. Several schemes are in use including the American Joint Engineering Device Engineering Council (JEDEC) system, the European Pro-Electron system and the Japanese Industrial Standard (JIS) system (which is Japanese based). In addition, some manufacturers have adopted their own coding schemes.

The JEDEC system of semiconductor classification is based on the following coding format:

leading digit, letter, serial number, suffix (optional)

The leading digit designates the number of P–N junctions used in the device. Hence, a device code starting with 1 relates to a single P–N junction (i.e. a diode) whilst a device code starting with 2 indicates a device which has two P–N junctions (usually a transistor) (Table 6.1). The letter is always N (signifying a JEDEC device) and the remaining digits are the serial number of the device. In addition, a suffix may be used in order to indicate the gain group.

The European Pro-Electron system for classifying semiconductors involves the following coding format (Table 6.2).

first letter, second letter, third letter (optional), serial number, suffix (optional)

Table 6.1

Leading digit – number of P–N junctions
1 Diode
2 Transistor
3 SCR or dual gate MOSFET
4 Optocoupler

Letter – origin
N North American JEDEC-coded device

Serial number – the serial number does not generally have any particular significance

Suffix – some transistors have an additional suffix that denotes the gain group for the device (where no suffix appears the gain group is either inapplicable or the group is undefined for the device in question)
A Low gain
B Medium gain
C High gain

Table 6.2

First letter – semiconductor material
A Ge
B Si
C Gallium arsenide, etc.
D Photodiodes, etc.

Second letter – application
A Diode, low power or signal
B Diode, variable capacitance
C Transistor, audio frequency (AF) low power
D Transistor, AF power
E Diode, tunnel
F Transistor, high frequency, low power
P Photodiode
Q LED
S Switching device
T Controlled rectifier
X Varactor diode
V Power rectifier
Z Zener diode

Third letter – if present this indicates that the device is intended for industrial or professional rather than commercial applications

Serial number – the serial number does not generally have any particular significance

Suffix – some transistors have an additional suffix that denotes the gain group for the device (where no suffix appears the gain group is either inapplicable or the group is undefined for the device in question)
A Low gain
B Medium gain
C High gain

Table 6.3

Leading digit – number of P–N junctions
1   Diode
2   Transistor
3   SCR or dual gate MOSFET
4   Optocoupler

First and second letters – application
SA   PNP high-frequency transistor
SB   PNP AF transistor
SC   NPN high frequency
SD   NPN AF transistor
SE   Diode
SF   SCR
SJ   P-channel field effect
        transistor (FET)/MOSFET
SK   N-channel FET/MOSFET
SM   Triac
SQ   LED
SR   Rectifier
SS   Signal diode
ST   Diode
SV   Varactor
SZ   Zener diode

Serial number – the serial number does not generally have any particular significance

Suffix – some devices have a suffix that denotes approval of the device for use by certain organizations

---

The JIS is based on the following coding format (Table 6.3).

leading digit, first letter, second letter, serial number, suffix (optional)

The JIS coding system is similar to the JEDEC system.

Example 6.3

Classify the following semiconductor devices:

(a) 1N4001
(b) BFY51
(c) 3N201
(d) AA119
(e) 2N3055
(f) 2SA1077

Solution

(a) Diode (JEDEC-coded)
(b) Si high-frequency low-power transistor (Pro-Electron coded)

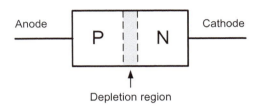

Figure 6.11   A P–N junction diode.

(c) MOSFET (JEDEC-coded)
(d) Ge low-power signal diode (Pro-Electron coded)
(e) Transistor (JEDEC-coded)
(f) PNP high-frequency transistor (JIS-coded).

The P–N junction diode

When a junction is formed between N- and P-type semiconductor materials, the resulting device is called a *diode*. This component offers an extremely low resistance to current flow in one direction and an extremely high resistance to current flow in the other. This characteristic allows diodes to be used in applications that require a circuit to behave differently according to the direction of current flowing in it. An ideal diode would pass an infinite current in one direction and no current at all in the other direction (Figure 6.11).

Connections are made to each side of the diode. The connection to the P-type material is referred to as the *anode* while that to the N-type material is called the *cathode*. With no externally applied potential, electrons from the N-type material will cross into the P-type region and fill some of the vacant holes. This action will result in the production of a region on either sides of the junction in which there are no free charge carriers. This zone is known as the *depletion region*.

If a positive voltage is applied to the anode (see Figure 6.12), the free positive charge carriers in the P-type material will be repelled and they will move away from the positive potential

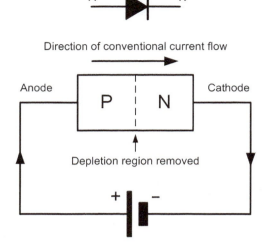

Figure 6.12 A forward-biased P–N junction diode.

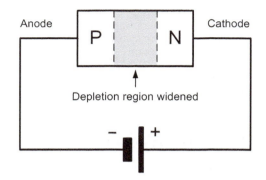

Figure 6.13 A reverse-biased P–N junction diode.

towards the junction. Likewise, the negative potential applied to the cathode will cause the free negative charge carriers in the N-type material to move away from the negative potential towards the junction.

When the positive and negative charge carriers arrive at the junction, they will attract one another and combine (recall from Chapter 5 that unlike charges attract). As each negative and positive charge carriers combine at the junction, new negative and positive charge carriers will be introduced to the semiconductor material from the voltage source. As these new charge carriers enter the semiconductor material, they will move towards the junction and combine. Thus, current flow is established and it will continue for as long as the voltage is applied. In this *forward-biased* condition, the diode freely passes current.

If a negative voltage is applied to the anode (see Figure 6.13), the free positive charge carriers in the P-type material will be attracted and they will move away from the junction. Likewise, the positive potential applied to the cathode will cause the free negative charge carriers in the N-type material to move away from the junction. The combined effect is that the depletion region becomes wider. In this *reverse-biased* condition, the diode passes a negligible amount of current.

> **Key point**
>
> In the freely conducting forward-biased state, the diode acts rather like a closed switch. In the reverse-biased state, the diode acts like an open switch.

Diode characteristics

Typical *I/V* characteristics for Ge and Si diodes are shown in Figure 6.14. It should be noted from these characteristics that the approximate *forward conduction voltage* for a Ge diode is 0.2 V whilst that for a Si diode is 0.6 V. This threshold voltage must be high enough to completely overcome the potential associated with the depletion region and force charge carriers to move across the junction.

> **Key point**
>
> The forward voltage for a Ge diode is approximately 0.2 V whilst that for a Si diode is approximately 0.6 V.

> **Example 6.4**

The characteristic of a diode is shown in Figure 6.15. Determine:

(a) the current flowing in the diode when a forward voltage of 0.4 V is applied;

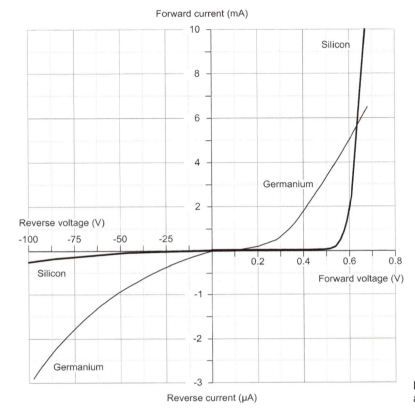

Forward current (mA)

Reverse voltage (V)

Figure 6.14 Typical *I/V* characteristics for Ge and Si diodes.

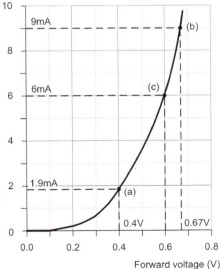

Figure 6.15 See Example 6.4.

(b) the voltage dropped across the diode when a forward current of 9 mA is flowing in it;
(c) the resistance of the diode when the forward voltage is 0.6 V;
(d) whether the diode is a Ge or Si type.

Solution

(a) When $V = 0.4$ V, $I = 1.9$ mA.
(b) When $I = 9$ mA, $V = 0.67$ V.
(c) From the graph, when $V = 0.6$ V, $I = 6$ mA. Now

$$R = \frac{V}{I} = \frac{0.6}{6 \times 10^{-3}} = 0.1 \times 10^3 = 100\,\Omega$$

(d) The onset of conduction occurs at approximately 0.2 V. This suggests that the diode is a Ge type.

Maximum ratings

It is worth noting that diodes are limited by the amount of forward current and reverse voltage

Table 6.4

Device code	Material	Maximum reverse voltage	Maximum forward current	Maximum reverse current	Application
1N4148	Si	100 V	75 mA	25 nA	General purpose
1N914	Si	100 V	75 mA	25 nA	General purpose
AA113	Ge	60 V	10 mA	200 µA	Radio frequency (RF) detector
OA47	Ge	25 V	110 mA	100 µA	Signal detector
OA91	Ge	115 V	50 mA	275 µA	General purpose
1N4001	Si	50 V	1 A	10 µA	Low voltage rectifier
1N5404	Si	400 V	3 A	10 µA	High voltage rectifier
BY127	Si	1250 V	1 A	10 µA	High voltage rectifier

they can withstand. This limit is based on the physical size and construction of the diode. In the case of a reverse-biased diode, the P-type material is negatively biased relative to the N-type material. In this case, the negative potential to the P-type material attracts the positive carriers, drawing them away from the junction. This leaves the area depleted; virtually no charge carriers exist and therefore current flow is inhibited. The reverse bias potential may be increased to the breakdown voltage for which the diode is rated. As in the case of the maximum forward current rating, the reverse voltage is specified by the manufacturer. Typical values of maximum reverse voltage or PIV range from 50 to 500 V.

The reverse breakdown voltage is usually very much higher than the forward threshold voltage. A typical general-purpose diode may be specified as having a forward threshold voltage of 0.6 V and a reverse breakdown voltage of 200 V. If the latter is exceeded, the diode may suffer irreversible damage.

Diode types and applications

Diodes are often divided into signal or rectifier types according to their principal field of application. Signal diodes require consistent forward characteristics with low forward voltage drop. Rectifier diodes need to be able to cope with high values of reverse voltage and large values of forward current; consistency of characteristics is of secondary importance in such applications. Table 6.4 summarizes the characteristics of some common semiconductor diodes.

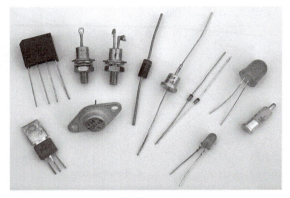

Figure 6.16 Various diodes (including signal diodes, rectifiers, Zener diodes, LEDs and SCRs).

Diodes are also available as connected in a bridge configuration for use as a rectifier in an AC power supply. Figure 6.16 shows a selection of various diode types (including those that we will meet later in this section) whilst Figure 6.17 shows the symbols used to represent them in circuit schematics.

Zener diodes

Zener diodes are heavily doped Si diodes that, unlike normal diodes, exhibit an abrupt reverse breakdown at relatively low voltages (typically <6 V). A similar effect (avalanche) occurs in less heavily doped diodes. These avalanche diodes also exhibit a rapid breakdown with negligible current flowing below the avalanche voltage and a relatively large current flowing once the avalanche voltage has been reached. For

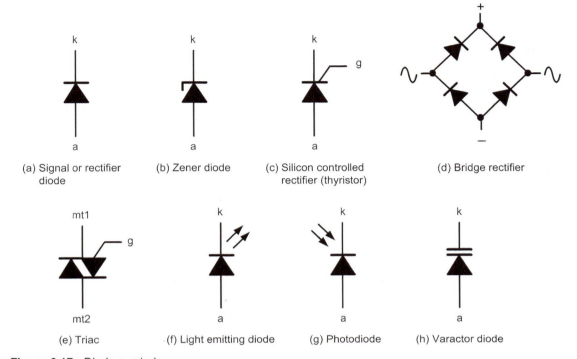

(a) Signal or rectifier diode (b) Zener diode (c) Silicon controlled rectifier (thyristor) (d) Bridge rectifier

(e) Triac (f) Light emitting diode (g) Photodiode (h) Varactor diode

Figure 6.17 Diode symbols.

avalanche diodes, this breakdown voltage usually occurs at voltages above 6 V. In practice, however, both types of diode are referred to as Zener diodes. The symbol for a Zener diode was shown earlier in Figure 6.17 whilst typical Zener diode characteristics are shown in Figure 6.18.

Though reverse breakdown is a highly undesirable effect in circuits that use conventional diodes, it can be extremely useful in the case of Zener diodes where the breakdown voltage is precisely known. When a diode is undergoing reverse breakdown and provided its maximum ratings are not exceeded, the voltage appearing across it will remain substantially constant (equal to the nominal Zener voltage) regardless of the current flowing. This property makes the Zener diode ideal for use as a voltage regulator.

Zener diodes are available in various families (according to their general characteristics, encapsulations and power ratings) with reverse breakdown (Zener) voltages in the range 2.4–91 V. Table 6.5 summarizes the characteristics of common Zener diodes.

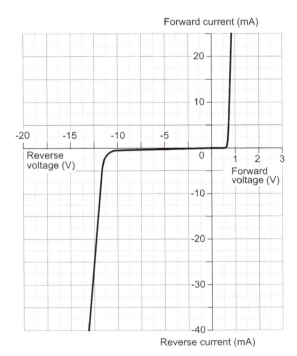

Figure 6.18 Typical Zener diode characteristic.

Table 6.5

Zener series	Description and rating
BZY88 series	Miniature glass-encapsulated diodes rated at 500 mW (at 25°C). Zener voltages range from 2.7 to 15 V (voltages are quoted for 5 mA reverse current at 25°C)
BZX61 series	Encapsulated alloy junction rated at 1.3 W (25°C ambient). Zener voltages range from 7.5 to 72 V
BZX85 series	Medium-power glass-encapsulated diodes rated at 1.3 W and offering Zener voltages in the range 5.1–62 V
BZY93 series	High-power diodes in stud mounting encapsulation. Rated at 20 W for ambient temperatures up to 75°C, Zener voltages range from 9.1 to 75 V
1N5333 series	Plastic encapsulated diodes rated at 5 W. Zener voltages range from 3.3 to 24 V

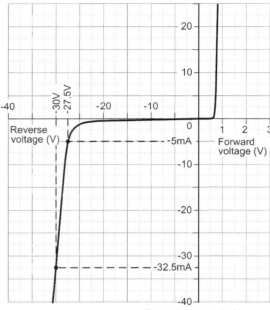

Figure 6.19

Example 6.5

The characteristic of a Zener diode is shown in Figure 6.19. Determine:

(a) the current flowing in the diode when a reverse voltage of 30 V is applied;
(b) the voltage dropped across the diode when a reverse current of 5 mA is flowing in it;
(c) the voltage rating for the Zener diode;
(d) the power dissipated in the Zener diode when a reverse voltage of 30 V appears across it.

Solution

(a) When $V = -30$ V, $I = -32.5$ mA.
(b) When $I = -5$ mA, $V = -27.5$ mA.
(c) The characteristic graph shows the onset of Zener action at 27 V. This would suggest a Zener voltage rating of 27 V.
(d) Now $P = I \times V$ from which $P = (32.5 \times 10^{-3}) \times 30 = 0.975$ W $= 975$ mW.

Key point

Zener diodes begin to conduct heavily when the applied voltage reaches a particular threshold value (known as the Zener voltage). Zener diodes can thus be used to maintain a constant voltage.

SCRs

SCRs (or *thyristors*) are three-terminal devices, which can be used for switching and for AC power control. SCRs can switch very rapidly from a conducting to a non-conducting state. In the off state, the SCR exhibits negligible leakage current, while in the on state the device exhibits very low resistance. This results in very little power loss within the SCR even when appreciable power levels are being controlled.

Once switched into the conducting state, the SCR will remain conducting (i.e. it is latched in the on state) until the forward current is removed from the device. In DC applications this necessitates the interruption (or disconnection) of the supply before the device can be reset into its non-conducting state. Where the device is used with an alternating supply, the device will automatically become reset whenever the main supply reverses. The device can then be triggered on the next half-cycle having correct polarity to permit conduction.

Like their conventional Si diode counterparts, SCRs have anode and cathode connections; control is applied by means of a gate terminal. The symbol for an SCR was shown earlier in Figure 6.17.

In normal use, an SCR is triggered into the conducting (on) state by means of the application of a current pulse to the gate terminal (see Figure 6.20). The effective triggering of an SCR requires a gate trigger pulse having a fast rise time derived from a low-resistance source. Triggering can become erratic when insufficient

gate current is available or when the gate current changes slowly.

Table 6.6 summarizes the characteristics of several common SCRs.

> **Key point**
>
> SCRs are diodes that can be triggered into conduction by applying a small current to their gate input. SCRs are able to control large voltages and currents from a relatively small (low current, low voltage) signal.

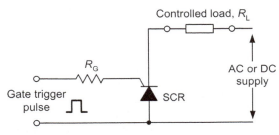

Figure 6.20 Triggering a SCR.

LEDs

LEDs can be used as general-purpose indicators and, compared with conventional filament lamps, operate from significantly smaller voltages and currents. LEDs are also very much more reliable than filament lamps. Most LEDs will provide a reasonable level of light output when a forward current of between 5 and 20 mA is applied.

LEDs are available in various formats with the round types being most popular. Round LEDs are commonly available in the 3 and 5 mm (0.2 in.) diameter plastic packages and also in a 5 mm × 2 mm rectangular format. The viewing angle for round LEDs tends to be in the region of 20°–40°, whereas for rectangular types this is increased to around 100°. The symbol for an LED was shown earlier in Figure 6.17. Table 6.7 summarizes the characteristics of several common types of LED.

In order to limit the forward current of an LED to an appropriate value, it is usually necessary to include a fixed resistor in series with

Table 6.6

Type	$I_F (A_V)$ (A)	V_{RRM} (V)	V_{GT} (V)	I_{GT} (mA)
2N4444	5.1	600	1.5	30
BT106	1.0	700	3.5	50
BT152	13.0	600	1.0	32
BTY79-400R	6.4	400	3.0	30
TIC106D	3.2	400	1.2	0.2
TIC126D	7.5	400	2.5	20

Table 6.7

Parameter	Standard LED	Standard LED	High-efficiency LED	High intensity
Diameter (mm)	3	5	5	5
Maximum forward current (mA)	40	30	30	30
Typical forward current (mA)	12	10	7	10
Typical forward voltage drop (V)	2.1	2.0	1.8	2.2
Maximum reverse voltage (V)	5	3	5	5
Maximum power dissipation (mW)	150	100	27	135
Peak wavelength (nm)	690	635	635	635

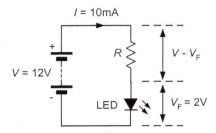

Figure 6.21   See Example 6.6.

an LED indicator as shown in Figure 6.21. The value of the resistor may be calculated from the formula:

$$R = \frac{V - V_F}{I}$$

where V_F is the forward voltage drop produced by the LED and V is the applied voltage. Note that, for most common LED, V_F is approximately 2 V.

Example 6.6

A simple LED indicator circuit is shown in Figure 6.21. Determine the value for R and the diode which is to operate with a current of 10 mA and has a forward voltage drop of 2 V.

Solution

Using the formula,

$$R = \frac{V - V_F}{I}$$

gives:

$$R = \frac{12 - 2}{10 \times 10^{-3}} = 1 \times 10^3 = 1\,k\Omega$$

Key point

LEDs produce light when a small current is applied to them. They are generally smaller and more reliable than conventional filament lamps and can be used to form larger and more complex displays.

Diodes in series and in parallel

Like other components, diodes can be connected in series and in parallel (see Figure 6.22). In the

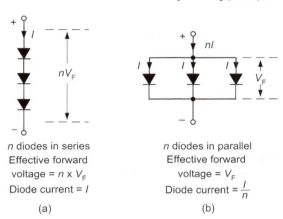

n diodes in series
Effective forward
voltage = *n* x V_F
Diode current = *I*

n diodes in parallel
Effective forward
voltage = V_F
Diode current = $\frac{I}{n}$

(a)   (b)

Figure 6.22   (a) Series connected diodes (b) parallel connected diodes.

series case shown in Figure 6.22(a), the total voltage dropped across *n* diodes connected in series will be *n* times the forward threshold voltage of a single diode. Thus, for two Si diodes connected in series the forward voltage drop will be approximately 2×0.6 or 1.2 V. It is also worth noting that the same current flows through each of the diodes. In the parallel case shown in Figure 6.22(b), the total current will be divided equally between the diodes (assuming that they are identical) but the voltage dropped across them will be the same as the forward threshold voltage of a single diode. Thus, for two Si diodes connected in parallel the forward voltage drop will be approximately 0.6 V but with the current shared between them.

Rectifiers

Semiconductor diodes are commonly used to convert AC to DC, in which case they are referred to as *rectifiers*. The simplest form of rectifier circuit makes use of a single diode and, since it operates on only either positive or negative half-cycles of the supply, it is known as a half-wave rectifier.

Figure 6.23 shows a simple half-wave rectifier circuit. The AC supply at 115 V is applied to the primary of a step-down transformer (T_1). The secondary of T_1 steps down the 115 V 400 Hz supply to 28.75 V root mean square (RMS) (the turns ratio of T_1 will thus be 115/28.75 or 4 : 1).

Diode D_1 will only allow the current to flow in the direction shown (i.e. from anode to cathode). D_1 will be forward biased during each positive half-cycle and will effectively behave like a closed switch as shown in Figure 6.24(a). When the circuit current tries to flow in the opposite direction, the voltage bias across the diode will be reversed, causing the diode to act like an open switch as shown in Figure 6.24(b).

The switching action of D_1 results in a pulsating output voltage, which is developed across the load resistor (R_L) shown in Figure 6.25. Since the supply is at 400 Hz, the pulses of voltage developed across R_L will also

be at 400 Hz even if only half the AC cycle is present. During the positive half-cycle, the diode will drop the 0.6 V forward threshold voltage normally associated with Si diodes. However,

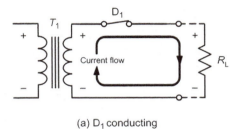

(a) D_1 conducting

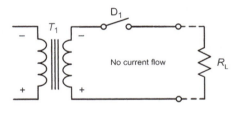

(b) D_1 non-conducting

Figure 6.24 Switching action of the diode in the half-wave rectifier (a) D_1 forward biased (b) D_1 reverse biased.

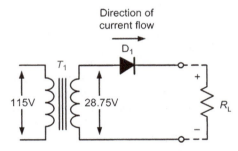

Figure 6.23 A simple half-wave rectifier circuit.

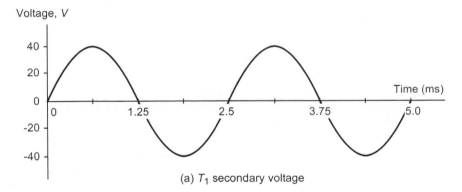

(a) T_1 secondary voltage

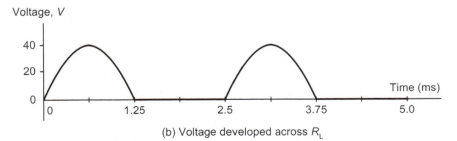

(b) Voltage developed across R_L

Figure 6.25 Waveforms of voltages in the simple half-wave power supply.

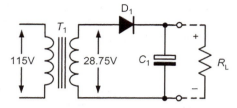

Figure 6.26 Effect of adding a reservoir capacitor to the output of the simple half-wave power supply.

during the negative half-cycle the peak AC voltage will be dropped across D_1 when it is reverse biased. This is an important consideration when selecting a diode for a particular application. Assuming that the secondary of T_1 provides 28.75 V RMS, the peak voltage output from the transformer's secondary winding will be given by:

$$V_{pk} = 1.414 \times V_{RMS} = 1.414 \times 28.75\,V$$
$$= 40.65\,V$$

The peak voltage applied to D_1 will thus be a little over 40 V. The negative half-cycles are blocked by D_1 and thus only the positive half-cycles appear across R_L. Note, however, that the actual peak voltage across R_L will be the 40.65 V positive peak being supplied from the secondary on T_1, minus the 0.6 V forward threshold voltage dropped by D_1. In other words, positive half-cycle pulses having a peak amplitude of almost exactly 40 V will appear across R_L.

Figure 6.26 shows a considerable improvement to the earlier simple rectifier. The capacitor, C_1, has been added to ensure that the output voltage remains at, or near, the peak voltage even when the diode is not conducting. When the primary voltage is first applied to T_1, the first positive half-cycle output from the secondary will charge C_1 to the peak value seen across R_L. Hence, C_1 charges to 40 V at the peak of the positive half-cycle. Because C_1 and R_L are in parallel, the voltage across R_L will be the same as that developed across C_1 (see Figure 6.25).

The time required for C_1 to charge to the maximum (peak) level is determined by the charging circuit time constant (the series resistance multiplied by the capacitance value). In this circuit, the series resistance comprises the secondary winding resistance together with the forward resistance of the diode and the (minimal) resistance of the wiring and connections. Hence, C_1 charges to 40 V at the peak of the positive half-cycle. Because C_1 and R_L are in parallel, the voltage across R_L will be the same as that across C_1.

The time required for C_1 to discharge is, in contrast, very much greater. The discharge time constant is determined by the capacitance value and the load resistance, R_L. In practice, R_L is very much larger than the resistance of the secondary circuit and hence C_1 takes an appreciable time to discharge. During this time, D_1 will be reverse biased and will thus be held in its non-conducting state. As a consequence, the only discharge path for C_1 is through R_L.

C_1 is referred to as a *reservoir capacitor*. It stores charge during the positive half-cycles of secondary voltage and releases it during the negative half-cycles. The circuit shown earlier is thus able to maintain a reasonably constant output voltage across R_L. Even so, C_1 will discharge by a small amount during the negative half-cycle periods from the transformer secondary. Figure 6.27 shows the secondary voltage waveform together with the voltage developed across R_L with and without C_1 present. This gives rise to a small variation in the DC output voltage (known as *ripple*).

Since ripple is undesirable we must take additional precautions to reduce it. One obvious method of reducing the amplitude of the ripple is that of simply increasing the discharge time constant. This can be achieved either by increasing the value of C_1 or by increasing the resistance value of R_L. In practice, however, the latter is not really an option because R_L is the effective resistance of the circuit being supplied and we do not usually have the ability to change it! Increasing the value of C_1 is a more practical alternative and very large capacitor values (often in excess of $1000\,\mu F$) are typical.

Figure 6.28 shows a further refinement of the simple power supply circuit. This circuit employs two additional components, R_1 and C_2, which act as a filter to remove the ripple. The value of C_2 is chosen so that the component exhibits a negligible reactance at the ripple frequency.

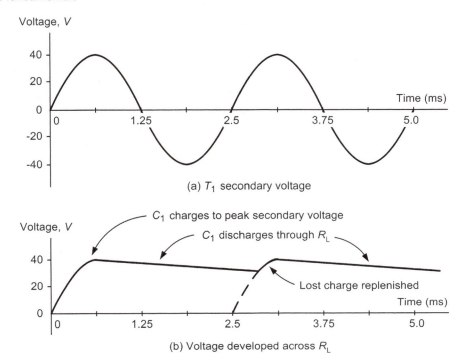

(a) T_1 secondary voltage

(b) Voltage developed across R_L

Figure 6.27 Waveforms of voltages in the half-wave power supply with reservoir capacitor.

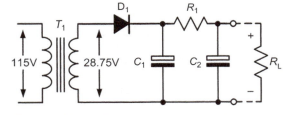

Figure 6.28 Half-wave power supply with reservoir capacitor and smoothing filter.

The half-wave rectifier circuit is relatively inefficient as conduction takes place only on alternate half-cycles. A better rectifier arrangement would make use of both positive *and* negative half-cycles. These *full-wave rectifier* circuits offer a considerable improvement over their half-wave counterparts. They are not only more efficient but are significantly less demanding in terms of the reservoir and smoothing components (Figure 6.29). There are two basic forms of full-wave rectifier: the bi-phase type and the bridge rectifier type.

Figure 6.30 shows a simple bi-phase rectifier circuit. The AC supply at 115 V is applied to the primary of a step-down transformer (T_1). This has two identical secondary windings, each providing 28.75 V RMS (the turns ratio of T_1 will still be 115/28.75 or 4 : 1 for each secondary winding).

On positive half-cycles, point A will be positive with respect to point B. Similarly, point B will be positive with respect to point C. In this condition D_1 will allow conduction (its anode will be positive with respect to its cathode) while D_2 will not allow conduction (its anode will be negative with respect to its cathode). Thus, D_1 alone conducts on positive half-cycles.

On negative half-cycles, point C will be positive with respect to point B. Similarly, point B will be positive with respect to point A. In this condition D_2 will allow conduction (its anode will be positive with respect to its cathode) while D_1 will not allow conduction (its anode will be negative with respect to its cathode). Thus, D_2 alone conducts on negative half-cycles.

Figure 6.31 shows the bi-phase rectifier circuit with the diodes replaced by switches. In Figure 6.31(a) D_1 is shown conducting on a

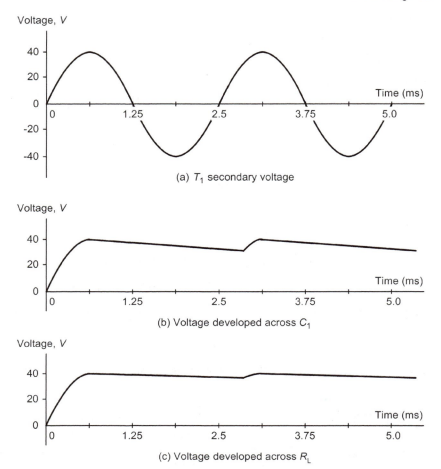

Figure 6.29 Waveforms of voltages in the half-wave power supply with reservoir capacitor and smoothing filter.

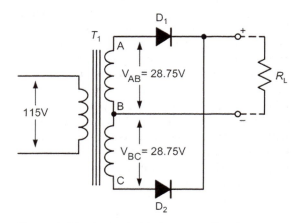

Figure 6.30 A simple bi-phase rectifier circuit.

positive half-cycle whilst in Figure 6.31(b) D_2 is shown conducting on a negative half-cycle of the input. The result is that current is routed through the load *in the same direction* on successive half-cycles. Furthermore, this current is derived alternately from the two secondary windings.

As with the half-wave rectifier, the switching action of the two diodes results in a pulsating output voltage being developed across the load resistor (R_L). However, unlike the half-wave circuit the pulses of voltage developed across R_L will occur at a frequency of 800 Hz (not 400 Hz). This doubling of the ripple frequency allows us to use smaller values of reservoir and

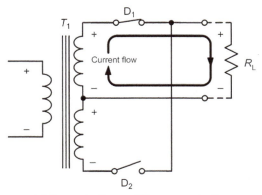

(a) D₁ conducting and D₂ non-conducting

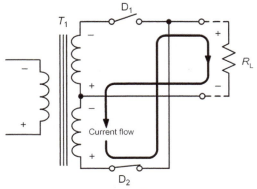

(b) D₂ conducting and D₁ non-conducting

Figure 6.31 Switching action of the diodes in the bi-phase rectifier (a) D₁ forward biased and D₂ reverse biased (b) D₁ reverse biased and D₂ forward biased.

smoothing capacitor to obtain the same degree of ripple reduction (recall that the reactance of a capacitor is reduced as frequency increases). As before, the peak voltage produced by each of the secondary windings will be approximately 17 V and the peak voltage across R_L will be about 40 V (i.e. 40.65 V less the 0.6 V forward threshold voltage dropped by the diodes).

Figure 6.32 shows how a reservoir capacitor (C_1) can be added to ensure that the output voltage remains at, or near, the peak voltage even when the diodes are not conducting. This component operates in exactly the same way as for the half-wave circuit, i.e. it charges to approximately 40 V at the peak of the positive half-cycle and holds the voltage at this level

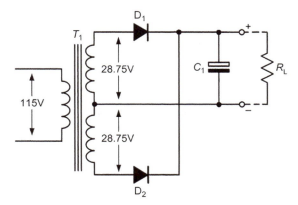

Figure 6.32 Bi-phase power supply with reservoir capacitor.

when the diodes are in their non-conducting states. The time required for C_1 to charge to the maximum (peak) level is determined by the charging circuit time constant (the series resistance multiplied by the capacitance value) (Figure 6.33).

In this circuit, the series resistance comprises the secondary winding resistance together with the forward resistance of the diode and the (minimal) resistance of the wiring and connections. Hence, C_1 charges very rapidly as soon as either D_1 or D_2 starts to conduct. The time required for C_1 to discharge is, in contrast, very much greater. The discharge time constant is determined by the capacitance value and the load resistance, R_L. In practice, R_L is very much larger than the resistance of the secondary circuit and hence C_1 takes an appreciable time to discharge. During this time, D_1 and D_2 will be reverse biased and held in a non-conducting state. As a consequence, the only discharge path for C_1 is through R_L.

An alternative to the use of the bi-phase circuit is that of using a four-diode bridge rectifier in which opposite pairs of diode conduct on alternate half-cycles (Figure 6.34). This arrangement avoids the need to have two separate secondary windings.

A full-wave bridge rectifier arrangement is shown in Figure 6.35. The 115 V AC supply at 400 Hz is applied to the primary of a step-down transformer (T_1). As before, the secondary winding provides 28.75 V RMS (approximately

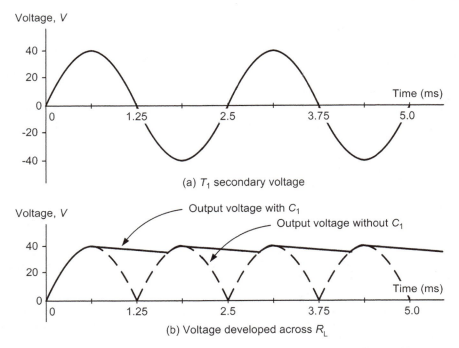

(a) T_1 secondary voltage

(b) Voltage developed across R_L

Figure 6.33 Waveforms of voltages in the bi-phase power supply with reservoir capacitor.

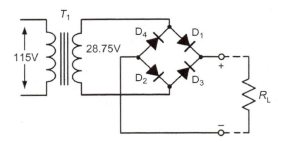

Figure 6.34 Full-power supply using a bridge rectifier.

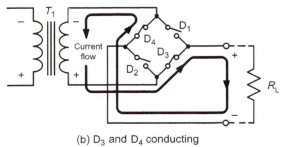

(a) D_1 and D_2 conducting

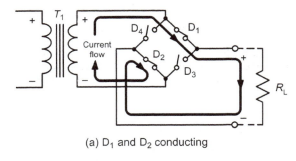

(b) D_3 and D_4 conducting

40 V peak) and has a turns ratio of 4:1. On positive half-cycles, point A will be positive with respect to point B. In this condition D_1 and D_2 will allow conduction while D_3 and D_4 will not allow conduction. Conversely, on negative half-cycles, point B will be positive with respect to point A. In this condition D_3 and D_4 will allow conduction while D_1 and D_2 will not allow conduction.

As with the bi-phase rectifier, the switching action of the two diodes results in a pulsating output voltage being developed across the

Figure 6.35 Switching action of the diodes in the full-wave bridge (a) D_1 and D_2 forward biased whilst D_3 and D_4 are reverse biased (b) D_1 and D_2 reverse biased whilst D_3 and D_4 are reverse biased.

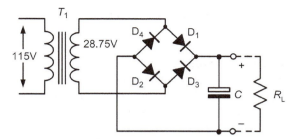

Figure 6.36 Full-wave bridge power supply with reservoir capacitor.

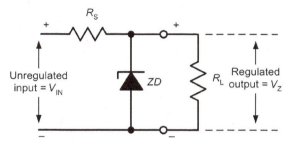

Figure 6.37 A simple Zener diode voltage regulator.

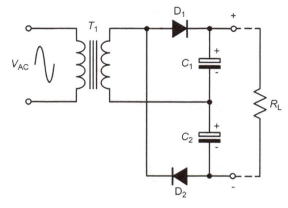

Figure 6.38 A voltage doubler.

load resistor (R_L). Once again, the peak output voltage is approximately 40 V (i.e. 40.65 V less the 2×0.6 V forward threshold voltage of the two diodes).

Figure 6.36 shows how a reservoir capacitor (C_1) can be added to the basic bridge rectifier circuit in order to ensure that the output voltage remains at, or near, the peak voltage even when the diodes are not conducting. This component operates in exactly the same way as for the bi-phase circuit, i.e. it charges to approximately 40 V at the peak of the positive half-cycle and holds the voltage at this level when the diodes are in their non-conducting states. This voltage waveforms are identical to those that we met earlier for the bi-phase rectifier.

A simple voltage regulator is shown in Figure 6.37. The series resistor, R_S, is included to limit the Zener current to a safe value when the load is disconnected. When a load (R_L) is connected, the Zener current will fall as current is diverted into the load resistance (it is usual to allow a minimum current of 2–5 mA in order to ensure that the diode regulates). The output voltage will remain at the Zener voltage (V_Z) until regulation fails at the point at which the potential divider formed by R_S and R_L produces a lower output voltage that is less than V_Z. The ratio of R_S to R_L is thus important.

Voltage doublers and voltage triplers

By adding a second diode and capacitor we can increase the output of the simple half-wave rectifier that we met earlier. A voltage doubler using this technique is shown in Figure 6.38. In this arrangement C_1 will charge to the positive peak secondary voltage whilst C_2 will charge to the negative peak secondary voltage. Since the output is taken from C_1 and C_2 connected in series the resulting output voltage is twice that produced by one diode alone.

The voltage doubler can be extended to produce higher voltages using the cascade arrangement shown in Figure 6.39. Here C_1 charges to the positive peak secondary voltage, whilst C_2 and C_3 charge to twice the positive peak secondary voltage. The result is that the output voltage is the sum of the voltages across C_1 and C_3, which is three times the voltage that would be produced by a single diode. The ladder arrangement shown in Figure 6.39 can be easily extended to provide even higher voltages but the efficiency of the circuit becomes increasingly impaired and high order *voltage multipliers* of this type are only suitable for providing relatively small currents.

Clipping and clamping circuits

Apart from their use as rectifiers, diodes are frequently used in signal-processing circuits for clipping and clamping. In clipping applications diodes are used to remove part of a waveform as shown in Figure 6.40. They do this by conducting on the positive, negative or both half-cycles of the AC waveform, effectively shunting the signal to common. On the remaining part of the cycle they are non-conducting and therefore have no effect on the shape of the waveform.

In clamping applications diodes can be used to change the DC level present on a waveform so that the waveform is all positive or all negative. Depending on which way it is connected, the diode conducts on either the negative- or positive-going half-cycles of the input AC waveform. The capacitor, C, charges to the peak value of the waveform, effectively lifting or depressing the waveform so that it all lies either above or below the zero voltage axis, as shown in Figure 6.41.

Varactor diodes

We have already shown that, when a diode is operated in the reverse-biased condition, the width of the depletion region increases as the applied voltage increases. Varying the width of the depletion region is equivalent to varying the plate separation of a very small capacitor such that the relationship between junction capacitance and applied reverse voltage will look something like that shown in Figure 6.42.

We can put this effect to good use in circuits that require capacitance to be adjusted by means

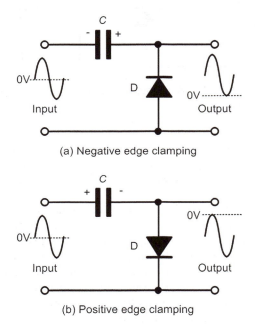

(a) Negative edge clamping

(b) Positive edge clamping

Figure 6.41 Clamping circuits (a) a positive edge clamp (b) a negative edge clamp.

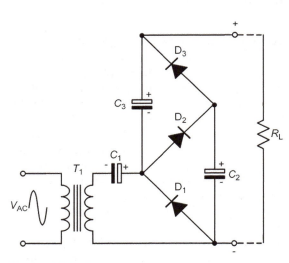

Figure 6.39 A voltage tripler.

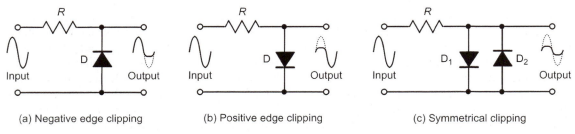

(a) Negative edge clipping   (b) Positive edge clipping   (c) Symmetrical clipping

Figure 6.40 Clipping circuits (a) a positive edge clipper (b) a negative edge clipper (c) a symmetrical clipper.

of an external voltage. This is a requirement found in many RF filter and oscillator circuits. Figure 6.43 shows a typical arrangement in which a varactor diode is used in conjunction with an L–C tuned circuit. The varactor diode

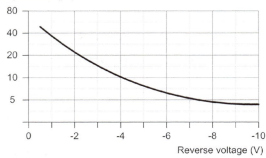

Capacitance (pF)

Figure 6.42 Capacitance plotted against reverse voltage for a typical varactor diode.

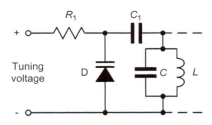

(a) Varactor tuning circuit

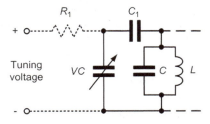

(b) Varactor shown as a variable capacitor

Figure 6.43 Using a varactor diode in conjunction with an L–C tuned circuit.

is coupled to the L–C circuit by means of a low-reactance capacitor C_1 whilst reverse bias voltage is fed to the varactor diode from the tuning voltage supply by means of a relatively high value series resistor, R_1.

The resonant frequency of the L–C circuit shown in Figure 6.43 is given by the formula:

$$f = \frac{1}{2\pi\sqrt{L\left(C + \frac{(C_1 \times VC)}{(C_1 + VC)}\right)}}$$

If $C_1 \gg VC$ then

$$f = \frac{1}{2\pi\sqrt{L(C + VC)}}$$

Table 6.8 summarizes the characteristics of several common types of varactor diode.

Example 6.7

A BB147 varactor diode is used in the circuit shown in Figure 6.43. If $L = 10\,\text{nH}$, $C = 120\,\text{pF}$ and $C_1 = 10\,\text{nF}$, determine the resonant frequency of the tuned circuit when the tuning voltage is (a) 2 V and (b) 10 V.

Solution

Since $C_1 \gg VC$ we can use the simplified formula:

$$f = \frac{1}{2\pi\sqrt{L(C + VC)}}$$

For a BB147 varactor diode when the tuning voltage is 2 V, $VC = 67\,\text{pF}$ (see Table 6.8). Hence:

$$f = \frac{1}{2\pi\sqrt{10 \times 10^{-9} \times (120 + 67) \times 10^{-12}}}$$

$$= \frac{10^{10}}{6.28 \times \sqrt{187}} = 116.28\,\text{MHz}$$

Table 6.8

Device code	Capacitance at a reverse voltage of 2 V (pF)	Capacitance at a reverse voltage of 10 V (pF)	Typical capacitance ratio (C_{max}/C_{min})	Maximum reverse voltage (V)
BB640	55	17.5	17	30
BB147	67	14	40	30
BBY31	13	4.5	8.3	30

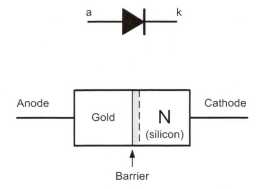

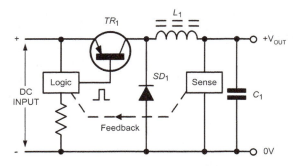

Figure 6.45 Schottky diode switching circuit.

Figure 6.44 Construction of a Schottky diode.

Similarly, when the tuning voltage is 10 V, $VC = 14\,\text{pF}$ (see Table 6.8). Hence:

$$f = \frac{1}{2\pi\sqrt{10 \times 10^{-9} \times (120 + 14) \times 10^{-12}}}$$

$$= \frac{10^{10}}{6.28 \times \sqrt{134}} = 137.6\,\text{MHz}$$

Key point

The junction capacitance of a varactor diode varies with the applied reverse voltage. Varactors are frequently used as the tuning element in an L–C circuit. The voltage applied to the varactor is varied in order to change resonant frequency of the circuit.

Schottky diodes

The conventional P–N junction diode that we met in section *The P–N junction diode* operates well as a rectifier at relatively low frequencies (i.e. 50–400 Hz) but its performance as a rectifier becomes seriously impaired at high frequencies due to the presence of stored charge carriers in the junction. These have the effect of momentarily allowing current to flow in the reverse direction when reverse voltage is applied. This problem becomes increasingly more problematic as the frequency of the AC supply is increased and the periodic time of the applied voltage becomes smaller.

To avoid these problems we use a diode that uses a metal–semiconductor contact rather than a P–N junction (see Figure 6.44). When

compared with conventional Si junction diodes, these Schottky diodes have a lower forward voltage (typically 0.35 V) and a slightly reduced maximum reverse voltage rating (typically 50–200 V). Their main advantage, however, is that they operate with high efficiency in *switched-mode power supplies* (SMPSs) at frequencies of up to several MHz. The basic arrangement of an SMPS using a Schottky diode is shown in Figure 6.45. Schottky diodes are also extensively used in the construction of integrated circuits designed for high-speed digital logic applications.

Key point

Schottky diodes are designed for use in fast switching application. Unlike conventional P–N junction diodes, negligible charge is stored in the junction of an Schottky diode.

Diode detector (demodulator)

Diodes are frequently used as detectors (or more correctly *demodulators*) which provide us with a way of recovering the modulation from an amplitude (or frequency) modulated carrier wave. Figure 6.46 shows a simple amplitude modulation (AM) demodulator. This circuit is worth studying as it not only serves as an introduction to another common diode application but should also help you to consolidate several important concepts that you have learnt in Chapter 5!

The RF AM carrier wave at the input of the demodulator is applied to a transformer, the secondary of which, *L*, forms a tuned circuit with

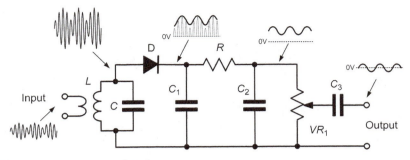

Figure 6.46 Diode detector (demodulator) circuit.

the tuning capacitor, C. This circuit provides a degree of selectivity as it is tuned to the frequency of the incoming carrier wave and rejects signals at other frequencies.

The detector diode, D, conducts on the positive half-cycles of the modulated carrier wave and charges C_1 to the peak voltage of each cycle. The waveform developed across C_1 thus follows the modulated envelope of the AM waveform. A simple low-pass C–R filter (R and C_2) removes any residual carrier-frequency components that may be present after rectification whilst C_3 removes the DC level present on the output waveform. To provide some control of the signal level at the output, a potentiometer, VR_1, acts as a simple volume control.

Test your understanding 6.2

1. Identify the types of diodes shown in Figure 6.47(a)–(d).

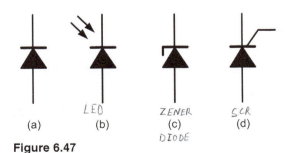

LED   ZENER   SCR
(a)   (b)   (c)   (d)
                   DIODE

Figure 6.47

2. When a diode is conducting, the most positive terminal is _ANODE_ and the most negative terminal is _CATHODE_.

3. The typical forward threshold voltage for a Si diode is _____ whilst that for a Ge diode is _____.

4. Corresponding readings of current I and voltage V for a semiconductor device are given in the table below:

V (V)	0	0.1	0.2	0.3	0.4	0.5	0.6	0.7	0.8
I (mA)	0	0	0	0	0	2.5	10	18	40

Plot the I/V characteristic for the device and identify the type of device.

5. Determine the resistance of the device in Question 4 when the applied voltage is:
   (a) 0.65 V
   (b) 0.75 V

6. Sketch the symbol for a bridge rectifier and identify the four connections to the device.

7. Sketch the circuit of a simple half-wave rectifier.

8. Sketch the circuit of a simple full-wave bi-phase rectifier.

9. Sketch a graph showing how the capacitance of a varactor diode varies with applied reverse voltage.

10. State TWO applications for Schottky diodes.

6.2.2 Transistors

Syllabus
Transistor symbols; Component description and orientation; Transistor characteristics and properties.

Knowledge level key

A	B1	B2
–	1	2

Syllabus
Construction and operation of PNP and NPN transistors; Base, collector and emitter configurations; Testing of transistors; Basic appreciation of other transistor types and their uses;

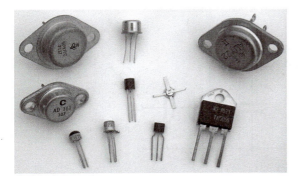

Figure 6.48 Various transistors (including low-frequency, high-frequency, high-voltage, small-signal and power types).

Application of transistors: classes of amplifier (A, B and C); Simple circuits including: bias, decoupling, feedback and stabilization; Multi-stage circuit principles: cascades, push–pull oscillators, multi-vibrators, flip–flop circuits.

Knowledge level key

A	B1	B2
–	–	2

Transistor classification

Transistors fall into two main classes (bipolar and field effect). They are also classified according to the semiconductor material employed (Si or Ge) and to their field of application (e.g. general purpose, switching, high frequency, etc.) (Figures 6.48 and 6.49). Transistors are also classified according to the application that they are designed for as shown in Table 6.9.

Bipolar junction transistors

Bipolar transistors generally comprise NPN or PNP junctions of either Si or Ge material (Figure 6.50). The junctions are, in fact, produced in a single slice of Si by diffusing impurities through a photographically reduced mask. Si transistors are superior when compared with Ge transistors in the vast majority of applications (particularly at high temperatures) and thus Ge devices are very rarely encountered in modern electronic equipment.

Note that the base–emitter junction is forward biased and the collector–base junction is reverse

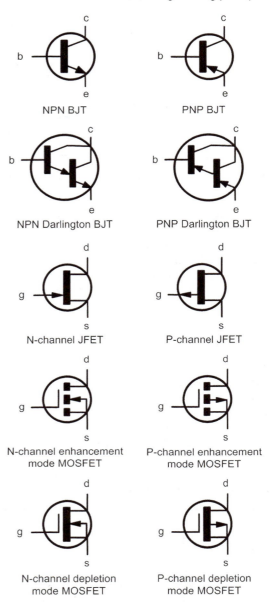

Figure 6.49 Transistor symbols.

biased. The base region is, however, made very narrow so that carriers are swept across it from emitter to collector and only a relatively small current flows in the base. To put this into context, the current flowing in the emitter circuit is typically 100 times greater than that flowing in the base. The direction of conventional current flow is from emitter to collector in the case of

Table 6.9

Low frequency	Transistors designed specifically for audio low-frequency applications (below 100 kHz)
High frequency	Transistors designed specifically for high-RF applications (100 kHz and above)
Switching	Transistors designed for switching applications
Low noise	Transistors that have low-noise characteristics and which are intended primarily for the amplification of low-amplitude signals
High voltage	Transistors designed specifically to handle high voltages
Driver	Transistors that operate at medium power and voltage levels and which are often used to precede a final (power) stage which operates at an appreciable power level
Small-signal	Transistors designed for amplifying small voltages in amplifiers and radio receivers
Power	Transistor designed to handle high currents and voltages

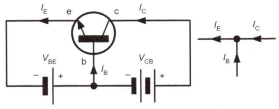

(a) NPN bipolar junction transistor (BJT)

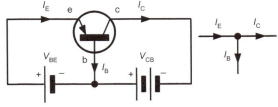

(b) PNP bipolar junction transistor (BJT)

Figure 6.51 Current flow in NPN and PNP BJTs.

a PNP transistor and collector to emitter in the case of an NPN device.

The equation that relates current flow in the collector, base and emitter currents (see Figure 6.51) is:

$$I_E = I_B + I_C$$

where I_E is the emitter current, I_B is the base current and I_C is the collector current (all expressed in the same units).

Key point

The three connections on a BJT are referred to as the base, emitter and collector. Inside a BJT there are two semiconductor junctions, the base–emitter junction and the base–collector junction. In normal operation the base–emitter junction is forward biased whilst the base–collector junction is reverse biased.

Key point

The base current of a transistor is very much smaller than either the collector or emitter currents (which are roughly the same). The direction of conventional current flow in a transistor is from emitter to collector in the case of a PNP device and collector to emitter in the case of an NPN device.

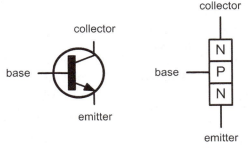

(a) NPN bipolar junction transistor (BJT)

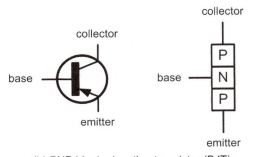

(b) PNP bipolar junction transistor (BJT)

Figure 6.50 NPN and PNP BJTs.

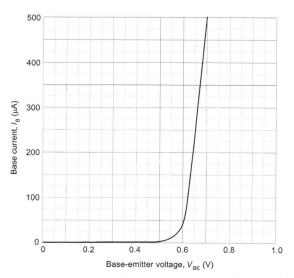

Figure 6.52 Input characteristic (I_B/V_{BE}) for an NPN BJT.

Example 6.8

A transistor operates with a collector current of 100 mA and an emitter current of 102 mA. Determine the value of base current.

Solution

Now:

$$I_E = I_B + I_C$$

thus:

$$I_B = I_E - I_C$$

Hence:

$$I_B = 102 - 100 = 2 \text{ mA}$$

Bipolar transistor characteristics

The characteristics of a BJT are usually presented in the form of a set of graphs relating voltage and current present at the transistors terminals. Figure 6.52 shows a typical *input characteristic* (I_B plotted against V_{BE}) for an NPN BJT operating in *common-emitter mode*. In this mode, the input current is applied to the base and the output current appears in the collector (the emitter is effectively common to both the input and output circuits).

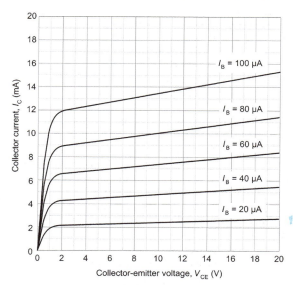

Figure 6.53 Output characteristic (I_C/V_{CE}) for an NPN BJT.

The input characteristic shows that very little base current flows until the base–emitter voltage V_{BE} exceeds 0.6 V. Thereafter, the base current increases rapidly (this characteristic bears a close resemblance to the forward part of the characteristic for a Si diode).

Figure 6.53 shows a typical set of *output (collector) characteristics* (I_C plotted against V_{CE}) for an NPN bipolar transistor. Each curve corresponds to a different value of base current. Note the "knee" in the characteristic below $V_{CE} = 2$ V. Also note that the curves are quite flat. For this reason (i.e. since the collector current does not change very much as the collector–emitter voltage changes) we often refer to this as a *constant current characteristic*.

Figure 6.54 shows a typical *transfer characteristic* for an NPN BJT. Here I_C is plotted against I_B for a small-signal general-purpose transistor. The slope of this curve (i.e. the ratio of I_C to I_B) is the common-emitter current gain of the transistor. We shall explore this further in Section *Current gain*.

Transistor parameters

The transistor characteristics that we met in the previous section provide us with some useful information that can help us to model the

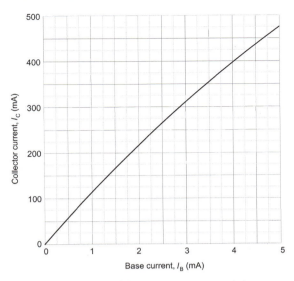

Figure 6.54 Transfer characteristic (I_C/I_B) for an NPN BJT.

behaviour of a transistor. In particular, we can use the three characteristic graphs to determine the following parameters:

Input resistance (from the input characteristic)

Static (or DC) input resistance $= \frac{V_{BE}}{I_B}$
  (from corresponding points on the graph)
Dynamic (or AC) input resistance $= \frac{\Delta V_{BE}}{\Delta I_B}$
  (from the slope of the graph)

(Note that ΔV_{BE} means "change of V_{BE}" and ΔI_B means "change of I_B")

Output resistance (from the output characteristic)

Static (or DC) output resistance $= \frac{V_{CE}}{I_C}$
  (from corresponding points on the graph)
Dynamic (or AC) output resistance $= \frac{\Delta V_{CE}}{\Delta I_C}$
  (from the slope of the graph)

(Note that ΔV_{CE} means "change of V_{CE}" and ΔI_C means "change of I_C")

Current gain (from the transfer characteristic)

Static (or DC) current gain $= \frac{I_C}{I_B}$
  (from corresponding points on the graph)

Dynamic (or AC) current gain $= \frac{\Delta I_C}{\Delta I_B}$
  (from the slope of the graph)

(Note that ΔI_C means "change of I_C" and ΔI_B means "change of I_B")
The method for obtaining these parameters from the relevant characteristic is illustrated in the three examples that follow.

Example 6.9

Figure 6.55 shows the input characteristic for an NPN Si transistor. When the base–emitter voltage is 0.65 V, determine:

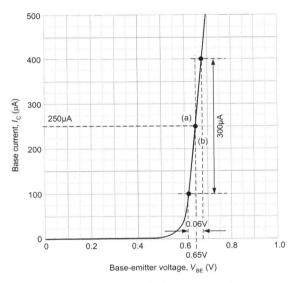

Figure 6.55 Input characteristic.

(a) the value of base current,
(b) the static value of input resistance,
(c) the dynamic value of input resistance.

Solution

(a) When $V_{BE} = 0.65$ V, $I_B = 250\,\mu$A.
(b) When $V_{BE} = 0.65$ V, $I_B = 250\,\mu$A, the static value of input resistance is given by:

$$\frac{V_{BE}}{I_B} = \frac{0.65}{250 \times 10^{-6}} = 2.6\,\text{k}\Omega$$

(c) When $V_{BE} = 0.65$ V, $I_B = 250\,\mu A$, the dynamic value of input resistance is given by:

$$\frac{\Delta V_{BE}}{\Delta I_B} = \frac{0.06}{300 \times 10^{-6}} = 200\,\Omega$$

Example 6.10

Figure 6.56 shows the output characteristic for an NPN Si transistor. When the collector voltage is 10 V and the base current is 80 μA, determine:

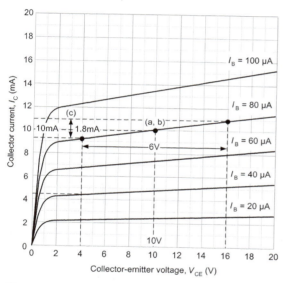

Figure 6.56 Output characteristic.

(a) the value of collector current,
(b) the static value of output resistance,
(c) the dynamic value of output resistance.

Solution

(a) When $V_{CE} = 10$ V and $I_B = 80\,\mu A$, $I_C = 10$ mA.
(b) When $V_{CE} = 10$ V and $I_B = 80\,\mu A$, the static value of output resistance is given by:

$$\frac{V_{CE}}{I_C} = \frac{10}{10 \times 10^{-3}} = 1\,k\Omega$$

(c) When $V_{CE} = 10$ V and $I_B = 80\,\mu A$, the dynamic value of output resistance is given by:

$$\frac{\Delta V_{CE}}{\Delta I_C} = \frac{6}{1.8 \times 10^{-3}} = 3.33\,k\Omega$$

Example 6.11

Figure 6.57 shows the transfer characteristic for an NPN Si transistor. When the base current is 2.5 mA, determine:

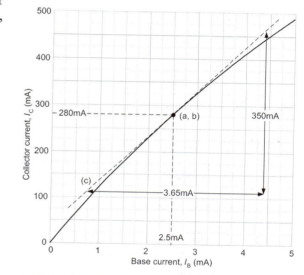

Figure 6.57 Transfer characteristic.

(a) the value of collector current,
(b) the static value of current gain,
(c) the dynamic value of current gain.

Solution

(a) When $I_B = 2.5$ mA, $I_C = 280$ mA.
(b) When $I_B = 2.5$ mA, the static value of current gain is given by:

$$\frac{I_C}{I_B} = \frac{280 \times 10^{-3}}{2.5 \times 10^{-3}} = 112$$

(c) When $I_B = 2.5$ mA, the dynamic value of current gain is given by:

$$\frac{\Delta I_C}{\Delta I_B} = \frac{350 \times 10^{-3}}{3.65 \times 10^{-3}} = 96$$

Table 6.10

Device	Type	I_C max.	V_{CE} max.	P_{TOT} max.	h_{FE} typical	Application
BC108	NPN	100 mA	20 V	300 mW	125	General-purpose small-signal amplifier
BCY70	PNP	200 mA	−40 V	360 mW	150	General-purpose small-signal amplifier
2N3904	NPN	200 mA	40 V	310 mW	150	Switching
BF180	NPN	20 mA	20 V	150 mW	100	RF amplifier
2N3053	NPN	700 mA	40 V	800 mW	150	Low-frequency amplifier/driver
2N3055	NPN	15 A	60 V	115 W	50	Low-frequency power

I_C max. is the maximum collector current, V_{CE} max. is the maximum collector–emitter voltage, P_{TOT} max. is the maximum device power dissipation and h_{FE} is the typical value of common-emitter current gain.

Bipolar transistor types and applications

The following Table 6.10 shows common examples of bipolar transistors for different applications.

Example 6.12

Which of the bipolar transistors listed in Table 6.10 would be most suitable for each of the following applications:

(a) the input stage of a radio receiver,
(b) the output stage of an audio amplifier,
(c) generating a 5 V square wave pulse.

Solution

(a) BF180 (this transistor is designed for use in RF applications).
(b) 2N3055 (this is the only device in the list that can operate at a sufficiently high-power level).
(c) 2N3904 (switching transistors are designed for use in pulse and square wave applications).

Current gain

As stated earlier, the common-emitter current gain is given as the ratio of collector current, I_C, to base current, I_B. We use the symbol h_{FE} to represent the static value of common-emitter current gain, thus:

$$h_{FE} = \frac{I_C}{I_B}$$

Similarly, we use h_{fe} to represent the dynamic value of common-emitter current gain, thus:

$$h_{fe} = \frac{\Delta I_C}{\Delta I_B}$$

As we showed earlier, values of h_{FE} and h_{fe} can be obtained from the transfer characteristic (I_C plotted against I_B). Note that h_{FE} is found from corresponding static values while h_{fe} is found by measuring the slope of the graph. Also note that, if the transfer characteristic is linear, there is little (if any) difference between h_{FE} and h_{fe}.

It is worth noting that current gain (h_{fe}) varies with collector current. For most small-signal transistors, h_{fe} is a maximum at a collector current in the range 1 and 10 mA. Current gain also falls to very low values for power transistors when operating at very high values of collector current. Furthermore, most transistor parameters (particularly common-emitter current gain, h_{fe}) are liable to wide variation from one device to the next. It is, therefore, important to design circuits on the basis of the minimum value for h_{fe} in order to ensure successful operation with a variety of different devices.

Example 6.13

A bipolar transistor has a common-emitter current gain of 125. If the transistor operates with a collector current of 50 mA, determine the value of base current.

Solution

Rearranging the formula,

$$h_{FE} = \frac{I_C}{I_B}$$

to make I_B the subject gives,

$$I_B = \frac{I_C}{h_{FE}}$$

from which:

$$I_B = \frac{50 \times 10^{-3}}{125} = 400\,\mu A$$

Key point

The current gain of a BJT is the ratio of output current to input current. In the case of common-emitter mode (where the input is connected to the base and the output is taken from the collector) the current gain is the ratio of collector current to base current.

FETs

FETs are available in two basic forms: junction gate and insulated gate. The gate–source junction of a junction gate field effect transistor (JFET) is effectively a reverse-biased P–N junction. The gate connection of an insulated gate field effect transistor (IGFET), on the other hand, is insulated from the channel and charge is capacitively coupled to the channel. To keep things simple, we will consider only JFET devices. Figure 6.58 shows the basic construction of an N-channel JFET.

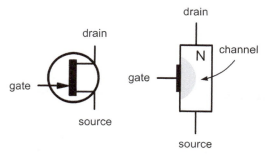

Figure 6.58 Construction of an N-channel JFET.

FETs comprise a channel of P- or N-type material surrounded by material of the opposite polarity. The ends of the channel (in which conduction takes place) form electrodes known as the source and drain. The effective width of the channel (in which conduction takes place) is controlled by a charge placed on the third (gate) electrode. The effective resistance between the source and drain is thus determined by the voltage present at the gate.

JFETs offer a very much higher input resistance when compared with bipolar transistors. For example, the input resistance of a bipolar transistor operating in common-emitter mode is usually around 2.5 kΩ. A JFET transistor operating in equivalent common-source mode would typically exhibit an input resistance of 100 MΩ! This feature makes JFET devices ideal for use in applications where a very high-input resistance is desirable.

As with bipolar transistors, the characteristics of an FET are often presented in the form of a set of graphs relating voltage and current present at the transistors terminals.

Key point

The three connections on a JFET are referred to as the gate, source and drain. Inside a JFET there is a resistive connection between the source and drain and a normally reverse-biased junction between the gate and source.

Key point

In a JFET, the effective resistance between the source and drain is determined by the voltage that appears between the gate and source.

FET characteristics

A typical *mutual characteristic* (I_D plotted against V_{GS}) for a small-signal general-purpose N-channel FET operating in common-source mode is shown in Figure 6.59. This characteristic shows that the drain current is progressively reduced as the gate–source voltage is made more negative. At a certain value of V_{GS} the drain current falls to zero and the device is said to be cut-off.

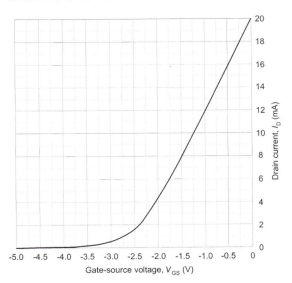

Figure 6.59 Mutual characteristic (I_D/V_{GS}) for an N-channel FET.

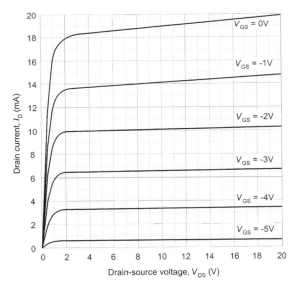

Figure 6.60 Output characteristic (I_D/V_{DS}) for an N-channel FET.

Figure 6.60 shows a typical family of output characteristics (I_D plotted against V_{DS}) for a small-signal general-purpose N-channel FET operating in common-source mode. This characteristic comprises a family of curves each relating to a different value of gate–source voltage V_{GS}. You might also like to compare this

characteristic with the output characteristic for a transistor operating in common-emitter mode that you met earlier.

As in the case of the BJT, the output characteristic curves for an N-channel FET have a "knee" that occurs at low values of V_{GS}. Also, note how the curves become flattened above this value with the drain current I_D not changing very significantly for a comparatively large change in drain–source voltage V_{DS}. These characteristics are, in fact, even flatter than those for a bipolar transistor. Because of their flatness, they are often said to represent a constant current characteristic.

The gain offered by a FET is normally expressed in terms of its *forward transconductance* (g_{fs} or Y_{fs}) in common-source mode. In this mode, the input voltage is applied to the gate and the output current appears in the drain (the source is effectively common to both the input and output circuits).

In common-source mode, the static (or DC) forward transfer conductance is given by:

$$g_{FS} = \frac{I_D}{V_{DS}} \quad \begin{array}{l}\text{(from corresponding points} \\ \text{on the graph)}\end{array}$$

whilst the dynamic (or AC) forward transfer conductance is given by:

$$g_{fs} = \frac{\Delta I_D}{\Delta V_{DS}} \quad \text{(from the slope of the graph)}$$

(Note that ΔI_D means "change of I_D" and ΔV_{DS} means "change of V_{DS}")

The method for determining these parameters from the relevant characteristic is illustrated in Example 6.14.

Forward transfer conductance (g_{fs}) varies with drain current. For most small-signal devices, g_{fs} is quoted for values of drain current between 1 and 10 mA. Most FET parameters (particularly forward transfer conductance) are liable to wide variation from one device to the next. It is, therefore, important to design circuits on the basis of the minimum value for g_{fs}, in order to ensure successful operation with a variety of different devices.

Example 6.14

Figure 6.61 shows the mutual characteristic for a JFET. When the gate–source voltage is −2.5 V, determine:

(a) the value of drain current, 5 mA
(b) the dynamic value of forward trans-conductance. 4.8 m s

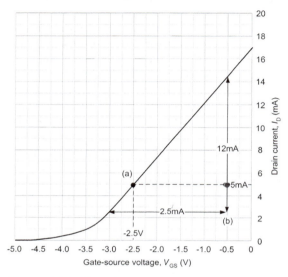

Figure 6.61 Mutual characteristic.

Solution

(a) When $V_{GS} = -2.5$ V, $I_D = 5$ mA.
(b) When $V_{GS} = -2.5$ V, the dynamic value of forward transconductance is given by:

$$\frac{\Delta I_D}{\Delta V_{GS}} = \frac{12 \times 10^{-3}}{2.5} = \text{mS}$$

Example 6.15

A FET operates with a drain current of 100 mA and a gate–source bias of −1 V. If the device has a g_{fs} of 0.25 S, determine the change in drain current if the bias voltage decreases to −1.1 V.

Solution

The change in gate–source voltage (V_{GS}) is −0.1 V and the resulting change in drain current can be determined from:

$$\Delta I_D = g_{fs} \times V_{GS} = 0.25 \times -0.1 = -0.025 \text{ A}$$
$$= -25 \text{ mA}$$

The new value of drain current will thus be $(100 - 25) = 75$ mA.

FET types and applications

The following Table 6.11 shows common examples of FETs for different applications (the list includes both depletion and enhancement types as well as junction and insulated gate types).

Example 6.16

Which of the FETs listed in the table would be most suitable for each of the following applications:

(a) the input stage of a radio receiver,
(b) the output stage of a transmitter,
(c) switching a load connected to a high-voltage supply.

Table 6.11

Device	Type	I_D max.	V_{DS} max.	P_D max.	g_{fs} typ.	Application
2N2819	N-channel	10 mA	25 V	200 mW	4.5 mS	General purpose
2N5457	N-channel	10 mA	25 V	310 mW	1.2 mS	General purpose
2N7000	N-channel	200 mA	60 V	400 mW	0.32 S	Low-power switching
BF244A	N-channel	100 mA	30 V	360 mW	3.3 mS	RF amplifier
BSS84	P-channel	−130 mA	−50 V	360 mW	0.27 S	Low-power switching
IRF830	N-channel	4.5 A	500 V	75 W	3.0 S	Power switching
MRF171A	N-channel	4.5 A	65 V	115 W	1.8 S	RF power amplifier

I_D max. is the maximum drain current, V_{DS} max. is the maximum drain–source voltage, P_D max. is the maximum drain power dissipation and g_{fs} typ. is the typical value of forward transconductance for the transistor.

Solution

(a) BF244A (this transistor is designed for use in RF applications).
(b) MRF171A (this device is designed for RF power applications).
(c) IRF830 (this device is intended for switching applications and can operate at up to 500 V).

Transistor amplifiers

Three basic circuit configurations are used for transistor amplifiers. These three circuit configurations depend upon one of the three transistor connections which is made common to both the input and the output. In the case of bipolar transistors, the configurations are known as common emitter, common collector (or emitter follower) and common base. Where FETs are used, the corresponding configurations are common source, common drain (or source follower) and common gate.

These basic circuit configurations shown in Figures 6.62 and 6.63 exhibit quite different performance characteristics as shown in Tables 6.12 and 6.13.

Classes of operation

A requirement of most amplifiers is that the output signal should be a faithful copy of the input signal or be somewhat larger in amplitude. Other types of amplifier are "non-linear", in which case their input and output waveforms will not necessarily be similar. In practice, the degree of linearity provided by an amplifier can be affected by a number of factors including the amount of bias applied and the amplitude of the input signal. It is also worth noting that a linear amplifier will become non-linear when the

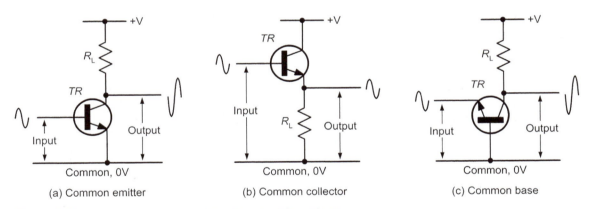

Figure 6.62   Bipolar transistor amplifier circuit configurations.

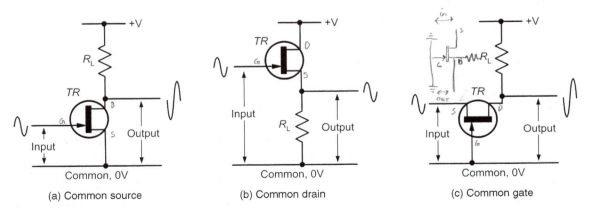

Figure 6.63   FET amplifier circuit configurations.

Table 6.12 Bipolar transistor amplifiers (see Figure 6.62)

Parameter	Common emitter	Common collector	Common base
Voltage gain	Medium/high (40)	Unity (1)	High (200)
Current gain	High (200)	High (200)	Unity (1)
Power gain	Very high (8000)	High (200)	High (200)
Input resistance	Medium (2.5 KΩ)	High (100 KΩ)	Low (200 Ω)
Output resistance	Medium/high (20 KΩ)	Low (100 Ω)	High (100 KΩ)
Phase shift	180°	0°	0°
Typical applications	General purpose, AF and RF amplifiers	Impedance matching, input and output stages	RF and VHF amplifiers

Table 6.13 FET amplifiers (see Figure 6.63)

Parameter	Common emitter	Common collector	Common base
Voltage gain	Medium/high (40)	Unity (1)	High (250)
Current gain	Very high (200,000)	Very high (200,000)	Unity (1)
Power gain	Very high (8,000,000)	Very high (200,000)	High (250)
Input resistance	Very high (1 MΩ)	Very high (1 MΩ)	Low (500 Ω)
Output resistance	Medium/high (50 KΩ)	Low (200 Ω)	High (150 KΩ)
Phase shift	180°	0°	0°
Typical applications	General purpose, AF and RF amplifiers	Impedance matching stages	RF and VHF amplifiers

applied input signal exceeds a threshold value. Beyond this value the amplifier is said to be over-driven and the output will become increasingly distorted if the input signal is further increased.

Amplifiers are usually designed to be operated with a particular value of bias supplied to the active devices (i.e. transistors). For linear operations, the active devices must be operated in the linear part of their transfer characteristics (V_{OUT} plotted against V_{IN}).

In Figure 6.64 the input and output signals for an amplifier are acting in linear mode. This form of operation is known as Class A and the bias point is adjusted to the midpoint of the linear part of the transfer characteristics. Furthermore, current will flow in the active devices used in a Class A amplifier during complete cycle of the signal waveform. At no time does the current fall to zero.

Figure 6.65 shows the effect of moving the bias point down the transfer characteristic and, at the same time, increasing the amplitude of the input signal. From this, you should notice that the extreme negative portion of the output signal has become distorted. This effect arises

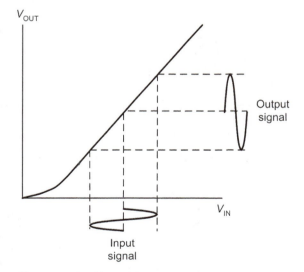

Figure 6.64 Class A operation.

from the non-linearity of the transfer characteristic that occurs near the origin (i.e. the zero point). Despite the obvious non-linearity in the output waveform, the active device(s) will conduct current during a complete cycle of the signal waveform.

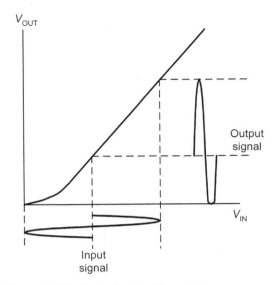

Figure 6.65   Reducing the bias point.

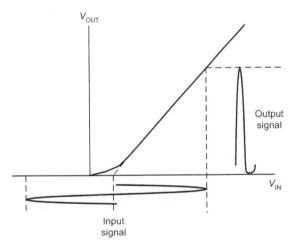

Figure 6.66   Class AB operation.

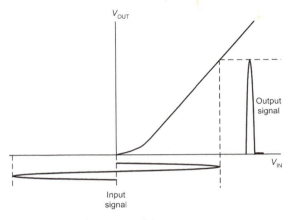

Figure 6.67   Class B operation.

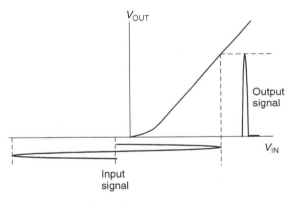

Figure 6.68   Class C operation.

Now consider the case of reducing the bias even further, while further increasing the amplitude of the input signal (see Figure 6.66). Here the bias point has been set as the projected cut-off point. The negative-going portion of the output signal becomes cut-off (or clipped) and the active device(s) will cease to conduct for this part of the cycle. This mode of operation is known as Class AB.

Now let us consider what will happen if no bias at all is applied to the amplifier (see Figure 6.67). The output signal will comprise a series of positive-going half-cycles and the active devices will be conducting only during half-cycles of the waveform (i.e. they will only be operating 50% of the time). This mode of operation is known as Class B and is commonly used in push–pull power amplifiers where the two active devices in the output stage operate on alternate half-cycles of the waveform.

Finally, there is one more class of operation to consider. The input and output waveforms for Class C operation are shown in Figure 6.68. Here the bias point is set at beyond the cut-off (zero) point and a very large input signal is applied. The output waveform will then comprise a series of quite sharp positive-going pulses. These pulses of current or voltage can be

Table 6.14

Class of operation	Bias point	Conduction angle (typical) (°)	Efficiency (typical) (%)	Application
A	Midpoint	360	5–40	Linear audio amplifiers
AB	Projected cut-off	210	20–40	Push–pull audio amplifiers
B	At cut-off	180	40–70	Push–pull audio amplifiers
C	Beyond cut-off	120	70–90	RF power amplifiers

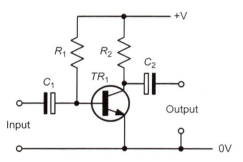

Figure 6.69 A simple Class A common-emitter amplifier.

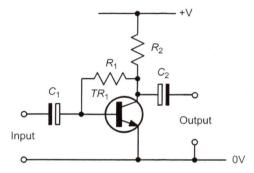

Figure 6.70 An improved Class A common-emitter amplifier.

applied to a tuned circuit load in order to recreate a sinusoidal signal. In effect, the pulses will excite the tuned circuit and its inherent flywheel action will produce a sinusoidal output waveform. This mode of operation is only used in RF power amplifiers, which must operate at high levels of efficiency.

Table 6.14 summarizes the classes of operation used in amplifiers.

We stated earlier that the optimum value of bias for Class A (linear) amplifier is that value which ensures that the active devices are operated at the midpoint of their transfer characteristics. In practice, this means that a static value of collector current will flow even when there is no signal present. Furthermore, the collector current will flow throughout the complete cycle of an input signal (i.e. conduction will take place over an angle of 360°). At no stage will the transistor be saturated nor should it be cut-off.

In order to ensure that a static value of collector current flows in a transistor, a small current must be applied to the base of the transistor. This current can be derived from the same voltage rail that supplies the collector circuit (via the load). Figure 6.69 shows a simple Class A common-emitter circuit in which the base bias

resistor, R_1, and collector load resistor, R_2, are connected to a common positive supply rail.

The signal is applied to the base terminal of the transistor via a coupling capacitor, C_1. This capacitor removes the DC component of any signal applied to the input terminals and ensures that the base bias current delivered by R_1 is unaffected by any device connected to the input. C_2 couples the signal out of the stage and also prevents DC current flowing appearing at the output terminals.

In order to stabilize the operating conditions for the stage and compensate for variations in transistor parameters, base bias current for the transistor can be derived from the voltage at the collector (see Figure 6.70). This voltage is dependent on the collector current that, in turn, depends upon the base current. A negative feedback loop thus exists in which there is a degree of self-regulation. If the collector current increases, the collector voltage will fall and the base current will be reduced. The reduction in base current will produce a corresponding reduction in collector current to offset the original change. Conversely, if the collector current falls, the collector voltage will rise and the base current will increase. This, in turn, will produce a corresponding increase in collector current to offset the original change.

Figure 6.71 shows a further improved amplifier circuit in which DC negative feedback is used to stabilize the stage and compensate for variations in transistor parameters, component values and temperature changes. R_1 and R_2 form a potential divider that determines the DC base potential, V_B. The base–emitter voltage (V_{BE}) is the difference between the potentials present at the base (V_B) and emitter (V_E). The potential at the emitter is governed by the emitter current (I_E). If this current increases, the emitter voltage (V_E) will increase, and as a consequence V_{BE} will fall. This, in turn, produces a reduction in emitter current which largely offsets the original change. Conversely, if the emitter current (V_E) decreases the emitter voltage V_{BE}

will increase (remember that V_B remains constant). The increase in bias results in an increase in emitter current compensating for the original change.

Multi-stage circuits

In many cases, a single transistor is insufficient to provide the amount of gain required in a circuit. In such an eventuality it is necessary to connect stages together so that one stage of gain follows another in what is known as a multi-stage amplifier (see Figure 6.72).

Some other common circuits involving transistors are shown in Figure 6.73(a). These include a push–pull amplifier where the two transistors work together, each amplifying a complete half-cycle of the waveform (and thus overcoming the distortion problems normally associated with a Class B amplifier). Figure 6.73(b) and (c) also shows two simple forms of oscillator. One is a ladder network oscillator and the other is an astable multi-vibrator. Both circuits use *positive feedback*, the former circuit produces a sinusoidal output whilst the latter produces a square wave output.

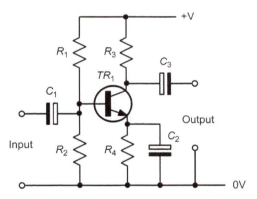

Figure 6.71 A Class A common-emitter with emitter stabilization.

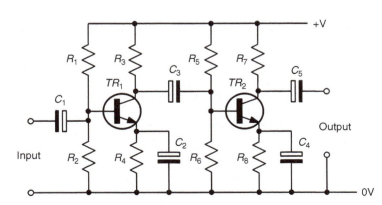

Figure 6.72 A multi-stage amplifier.

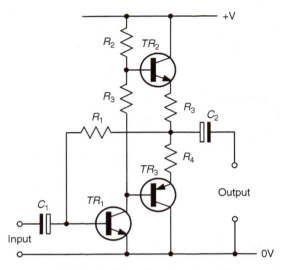

(a) A simple Class-B push-pull amplifier

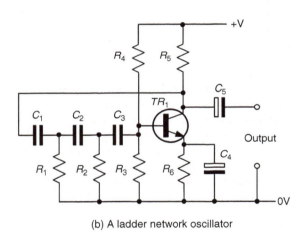

(b) A ladder network oscillator

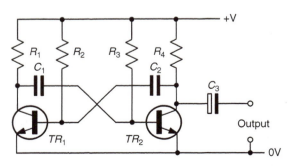

(c) An astable multivibrator

Figure 6.73 Some common circuits involving transistors.

Test your understanding 6.3

1. Identify the types of transistor shown in Figure 6.74(a)–(d).

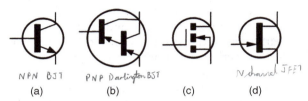

(a) NPN BJT   (b) PNP Darlington BJT   (c)   (d) N channel JFET

Figure 6.74

2. In normal operation the collector of an NPN BJT is at a more _Negitive_ potential than its emitter.

3. The three terminals of a JFET are labelled _gate_, _source_ and _drain_.

4. In normal operation the base–emitter junction of a bipolar transistor is _____ biased whilst the collector–base junction is _____ biased.

5. A BJT operates with a collector current of 1.2 A and a base current of 50 mA. What will the value of emitter current be?

6. What is the value of common-emitter current gain for the transistor in Question 4?

7. Corresponding readings of base current, I_B, and base–emitter voltage, V_{BE}, for a BJT are given in the table below:

V_{BE} (V)	0	0.1	0.2	0.3	0.4	0.5	0.6	0.7	0.8
I_B (μA)	0	0	0	0	1	3	19	57	130

Plot the I_B/V_{BE} characteristic for the device and use it to determine:
(a) the value of I_B when $V_{BE} = 0.65$ V,
(b) the static value of input resistance when $V_{BE} = 0.65$ V,
(c) the dynamic value of input resistance when $V_{BE} = 0.65$ V.

8. Corresponding readings of base current, I_B, and collector current, I_C, for a BJT are given in the table below:

I_B (μA)	0	10	20	30	40	50	60	70	80
I_C (mA)	0	1.1	2.1	3.1	4.0	4.9	5.8	6.7	7.6

Plot the I_C/I_B characteristic for the device and use it to determine the static value of common-emitter current gain when $I_B = 45$ μA.

9. Sketch a labelled circuit diagram for a simple Class A common-emitter amplifier. State the function of each component used.

10. Explain, with the aid of diagrams, the essential difference between Class A, Class B and Class C modes of operation.

6.2.3 Integrated circuits

Syllabus
Description and operation of logic circuits and linear circuits/operational amplifiers.

Knowledge level key

A	B1	B2
–	1	–

Syllabus
Description and operation of logic circuits and linear circuits; Introduction to operation and function of an operational amplifier used as: integrator, differentiator, voltage follower, comparator; Operation and amplifier stages connecting methods: resistive capacitive, inductive (transformer), inductive resistive (IR), direct; Advantages and disadvantages of positive and negative feedback.

Knowledge level key

A	B1	B2
–	–	2

Integrated circuit classification

Considerable cost savings can be made by manufacturing all of the components required for a particular circuit function on one small slice of semiconductor material (usually Si). The resulting *integrated circuit* may contain as few as 10 or more than 100,000 active devices (transistors and diodes). With the exception of a few specialized applications (such as amplification at high-power levels) integrated circuits have largely rendered conventional circuits (i.e. those based on *discrete components*) obsolete.

Integrated circuits can be divided into two general classes, linear (analogue) and digital. Typical examples of linear integrated circuits are operational amplifiers whereas typical examples of digital integrated are logic gates. A number of devices bridge the gap between the analogue and digital world. Such devices include analogue to digital converters (ADCs), digital to analogue converters (DACs) and timers. Table 6.15 and Figure 6.75 outline the main types of integrated circuit.

> **Key point**
>
> Integrated circuits combine the functions of many individual components into a single small package. Integrated circuits can be divided into three main categories: digital, linear and hybrid.

Digital integrated circuits

Digital integrated circuits have numerous applications quite apart from their obvious use in computing. Digital signals exist only in discrete steps or levels; intermediate states are disallowed. Conventional electronic logic is based on two binary states, commonly referred to as logic 0 (low) and logic 1 (high). A comparison between digital and analogue signals is shown in Figure 6.76.

The relative size of a digital integrated circuit (in terms of the number of active devices that it contains) is often referred to as its scale of integration and the following terminology is commonly used:

Scale of integration	Abbreviation	Number of logic gates*
Small	SSI	1–10
Medium	MSI	10–100
Large	LSI	100–1000
Very large	VLSI	1000–10,000
Super large	SLSI	10,000–100,000

*Or active circuitry of equivalent complexity.

Logic gates

The British Standard (BS) and American Standard (MIL/ANSI) symbols for some basic logic gates are shown, together with their truth tables in Figure 6.77. The action of each of the basic logic gates is summarized below. Note that, whilst inverters and buffers each have only one input, exclusive-OR gates have two inputs and the other basic gates (AND, OR, NAND and NOR) are commonly available with up to eight inputs.

Buffers
Buffers do not affect the logical state of a digital signal (i.e. a logic 1 input results in a logic 1 output whereas a logic 0 input results in a logic

Table 6.15

Digital	
Logic gates	Digital integrated circuits that provide logic functions, such as AND, OR, NAND and NOR.
Microprocessors	Digital integrated circuits that are capable of executing a sequence of programmed instructions. Microprocessors are able to store digital data whilst it is being processed and to carry out a variety of operations on the data, including comparison, addition and subtraction.
Memory devices	Integrated circuits that are used to store digital information.
Analogue	
Operational amplifiers	Integrated circuits that are designed primarily for linear operation and which form the fundamental building blocks of a wide variety of linear circuits, such as amplifiers, filters and oscillators.
Low-noise amplifiers	Linear integrated circuits that are designed so that they introduce very little noise which may otherwise degrade low-level signals.
Voltage regulators	Linear integrated circuits that are designed to maintain a constant output voltage in circumstances when the input voltage or the load current changes over a wide range.
Hybrid (combined digital and analogue)	
Timers	Integrated circuits that are designed primarily for generating signals that have an accurately defined time interval, such as that which could be used to provide a delay or determine the time between pulses. Timers generally comprise several operational amplifiers together with one or more bistable devices.
ADCs	Integrated circuits that are used to convert a signal in analogue form to one in digital form. A typical application would be where temperature is sensed using a thermistor to generate an analogue signal. This signal is then converted to an equivalent digital signal using an ADC and then sent to a microprocessor for processing.
DACs	Integrated circuits that are used to convert a signal in digital form to one in analogue form. A typical application would be where the speed of a DC motor is to be controlled from the output of a microprocessor. The digital signal from the microprocessor is converted to an analogue signal by means of a DAC. The output of the DAC is then further amplified before applying it to the field winding of a DC motor.

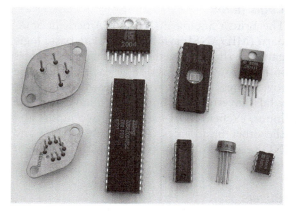

Figure 6.75 Various integrated circuits (including logic gates, operational amplifiers, memories and operational amplifiers).

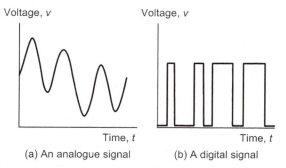

(a) An analogue signal   (b) A digital signal

Figure 6.76 Digital and analogue signals.

0 output). Buffers are normally used to provide extra current drive at the output but can also be used to regularize the logic levels present at an interface.

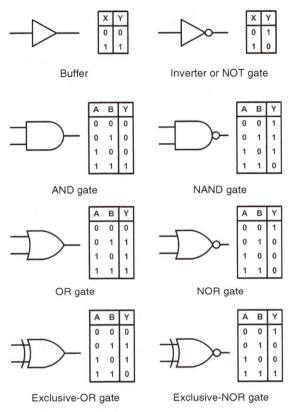

Figure 6.77 Logic gate symbols and truth tables.

Inverters

Inverters are used to complement the logical state (i.e. a logic 1 input results in a logic 0 output and vice versa). Inverters also provide extra current drive and, like buffers, are used in interfacing applications where they provide a means of regularizing logic levels present at the input or output of a digital system.

AND gates

AND gates will only produce a logic 1 output when all inputs are simultaneously at logic 1. Any other input combination results in a logic 0 output.

OR gates

OR gates will produce a logic 1 output whenever any one or more inputs are at logic 1. Putting this in another way, an OR gate will only produce

a logic 0 output whenever all of its inputs are simultaneously at logic 0.

NAND gates

NAND gates will only produce a logic 0 output when all inputs are simultaneously at logic 1. Any other input combination will produce a logic 1 output. A NAND gate, therefore, is nothing more than an AND gate with its output inverted! The circle shown at the output denotes this inversion.

NOR gates

NOR gates will only produce a logic 1 output when all inputs are simultaneously at logic 0. Any other input combination will produce a logic 0 output. A NOR gate, therefore, is simply an OR gate with its output inverted. A circle is again used to indicate inversion.

Exclusive-OR gates

Exclusive-OR gates will produce a logic 1 output whenever either one of the inputs is at logic 1 and the other is at logic 0. Exclusive-OR gates produce a logic 0 output whenever both inputs have the same logical state (i.e. when both are at logic 0 or at logic 1).

Monostables

A logic device which has only one stable output state is known as a monostable. The output of such a device is initially at logic 0 (low) until an appropriate level change occurs at its trigger input. This level change can be from 0 to 1 (positive edge trigger) or 1 to 0 (negative edge trigger) depending upon the particular monostable device or configuration. Upon receipt of a valid trigger pulse the output of the monostable changes state to logic 1. Then, after a time interval determined by external C–R timing components, the output reverts to logic 0. The device then awaits the arrival of the next trigger. A typical application for a monostable device is in stretching a pulse of very short duration.

Bistables

The output of a bistable has two stable states (logic 0 or 1) and once set, the output of the

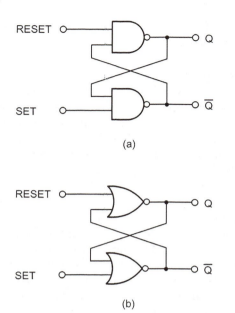

Figure 6.78 R–S bistables can be built from cross-coupled NAND and NOR gates.

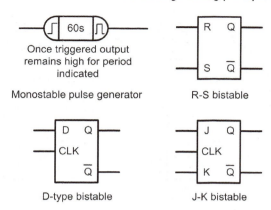

Figure 6.79 Symbols for D-type and J–K bistables.

device will remain at a particular logic level for an indefinite period until reset. A bistable thus constitutes a simple form of *memory cell* as it will remain in its latched state (whether set or reset) until commanded to change its state (or until the supply is disconnected). Various forms of bistable are available including R–S, D-type and J–K types.

R–S bistables

The simplest form of bistable is the R–S bistable. This device has two inputs SET and RESET and complementary outputs $\overline{Q}$ and Q. A logic 1 applied to the SET input will cause the $\overline{Q}$ output to become (or remain at) logic 1 whilst a logic 1 applied to the RESET input will cause the Q output to become (or remain at) logic 0. In either case, the bistable will remain in its SET or RESET state until an input is applied in such a sense as to change the state.

R–S bistables can be easily implemented using cross-coupled NAND or NOR gates as shown in Figure 6.78(a) and (b). These arrangements are, however, unreliable as the output state is indeterminate when S and R are simultaneously at logic 1.

D-type bistables

The D-type bistable has two principal inputs: D (standing variously for data or delay) and CLOCK (CLK). The data input (logic 0 or 1) is clocked into the bistable such that the output state only changes when the clock changes state. Operation is thus said to be synchronous. Additional subsidiary inputs (which are invariably active low) are provided which can be used to directly set or reset the bistable. These are usually called PRESET (PR) and CLEAR (CLR). D-type bistables are used both as latches (a simple form of memory) and as binary dividers.

J–K bistables

J–K bistables are the most sophisticated and flexible of the bistable types and they can be configured in various ways including binary dividers, shift registers and latches. J–K bistables have two clocked inputs (J and K), two direct inputs (PRESET and CLEAR), a CLOCK (CLK) input and outputs (Q and $\overline{Q}$) (Figure 6.79). As with R–S bistables, the two outputs are complementary (i.e. when one is 0 the other is 1, and vice versa). Similarly, the PRESET and CLEAR inputs are invariably both active low (i.e. a 0 on the PRESET input will set the Q output to 1 whereas a 0 on the CLEAR input will set the $\overline{Q}$ output to 0).

Logic families

Digital integrated circuit devices are often classified according to the semiconductor technology

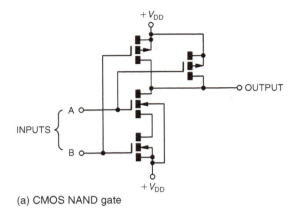

(a) CMOS NAND gate

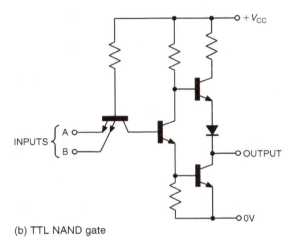

(b) TTL NAND gate

Figure 6.80 Representative circuit for a two-input AND gate using (a) CMOs and (b) TTL technology.

Infix	Meaning
None	Standard TTL device
ALS	Advanced low-power Schottky
C	CMOS version of a TTL device
F	"Fast" – a high-speed version of the device
H	High-speed version
S	Schottky input configuration (improved speed and noise immunity)
HC	High-speed CMOS version (CMOS compatible inputs)
HCT	High-speed CMOS version (TTL compatible inputs)
LS	Low-power Schottky

The most common family of CMOS devices is known as the 4000 series. Sub-families are identified by suffix letters as follows:

Suffix	Meaning
None	Standard CMOS device
A	Standard (unbuffered) CMOS device
B, BE	Improved (buffered) CMOS device
UB, UBE	Improved (unbuffered) CMOS device

used in their manufacture; the logic family to which a device belongs being largely instrumental in determining its operational characteristics (such as power consumption, speed and immunity to noise).

The two basic logic families are complementary metal oxide semiconductor (CMOS) and transistor–transistor logic (TTL). Each of these families is then further subdivided. Representative circuits of a two-input AND gate in both technologies is shown in Figure 6.80. The most common family of TTL logic devices is known as the 74 series. Devices from this family are coded with the prefix number 74. Sub-families are identified by letters that follow the initial 74 prefix as follows:

Example 6.17

Identify each of the following integrated circuits:

(i) 4001UBE
(ii) 74LS14

Solution

Integrated circuit (i) is an improved (unbuffered) version of the CMOS 4001 device.

Integrated circuit (ii) is a low-power Schottky version of the TTL 7414 device.

Key point

Logic gates are digital integrated circuits that can be used to perform logical operations, such as AND, OR, NAND and NOR.

Logic circuit characteristics

Logic levels are simply the range of voltages used to represent the logic states 0 and 1. The logic levels for CMOS differ markedly from those associated with TTL. In particular, CMOS logic levels are relative to the supply voltage used whilst the logic levels associated with TTL devices tend to be absolute. The following table usually applies:

	CMOS	TTL
Logic 1	$> \frac{2}{3} V_{DD}$	$> 2\,V$
Logic 0	$< \frac{1}{3} V_{DD}$	$< 0.8\,V$
Indeterminate	between $\frac{1}{3} V_{DD}$ and $\frac{2}{3} V_{DD}$	between 0.8 and 2 V

Note: V_{DD} is the positive supply associated with CMOS devices.

The noise margin is an important feature of any logic device. Noise margin is a measure of the ability of the device to reject noise; the larger the noise margin, the better is its ability to perform in an environment in which noise is present. Noise margin is defined as the difference between the minimum values of high-state output and high-state input voltage and the maximum values of low-state output and low-state input voltage. Hence:

$$\text{noise margin} = V_{OH(MIN)} - V_{IH(MIN)}$$

$$\text{or} \quad \text{noise margin} = V_{OL(MAX)} - V_{IL(MAX)}$$

where $V_{OH(MIN)}$ is the minimum value of high-state (logic 1) output voltage, $V_{IH(MIN)}$ is the minimum value of high-state (logic 1) input voltage, $V_{OL(MAX)}$ is the maximum value of low-state (logic 0) output voltage and $V_{IL(MIN)}$ is the minimum value of low-state (logic 0) input voltage. The noise margin for standard 7400-series TTL is typically 400 mV whilst that for CMOS is $\frac{1}{3} V_{DD}$ as shown in Figure 6.81.

The following Table 6.16 compares the more important characteristics of various members of the TTL family with buffered CMOS logic.

Operational amplifiers

Operational amplifiers are analogue integrated circuits designed for linear amplification that offer near-ideal characteristics (virtually infinite voltage gain and input resistance coupled with low-output resistance and wide bandwidth).

Operational amplifiers can be thought of as universal "gain blocks" to which external components are added in order to define their function within a circuit. By adding two resistors,

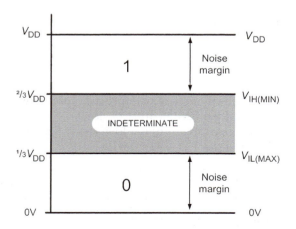
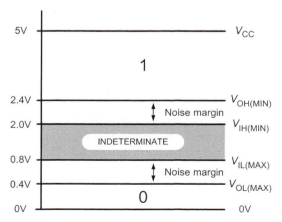

Figure 6.81   Comparison of logic levels for 7400-series TTL and 4000-series CMOS devices.

we can produce an amplifier having a precisely defined gain. Alternatively, with three resistors and two capacitors we can realize a low-pass filter. From this you might begin to suspect that operational amplifiers are really easy to use. The good news is that they are!

The symbol for an operational amplifier is shown in Figure 6.82. There are a few things to note about this. The device has two inputs and one output and no common connection. Furthermore, we often do not show the supply connections – it is often clearer to leave them out of the circuit altogether!

In Figure 6.82, one of the inputs is marked "−" and the other is marked "+". These polarity markings have nothing to do with the supply connections – they indicate the overall phase shift between each input and the output. The "+" sign indicates zero phase shift whilst the "−" sign indicates 180° phase shift. Since 180° phase shift produces an inverted (i.e. turned upside down) waveform, the "−" input is often referred to as the "inverting input". Similarly, the "+" input is known as the "non-inverting" input.

Most (but not all) operational amplifiers require a symmetrical supply (of typically ±6 to ±15 V). This allows the output voltage to swing both positive (above 0 V) and negative (below 0 V). Figure 6.83 shows how the supply

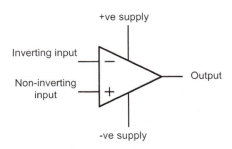

Figure 6.82 Symbol for an operational amplifier.

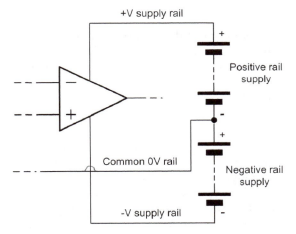

Figure 6.83 Supply rails for an operational amplifier.

Table 6.16

Characteristic	Logic family			
	74	74 LS	74 HC	40 BE
Maximum supply voltage	5.25 V	5.25 V	5.5 V	18 V
Minimum supply voltage	4.75 V	4.75 V	4.5 V	3 V
Static power dissipation (mW per gate at 100 kHz)	10	2	Negligible	Negligible
Dynamic power dissipation (mW per gate at 100 kHz)	10	2	0.2	0.1
Typical propagation delay (ns)	10	10	10	105
Maximum clock frequency (MHz)	35	40	40	12
Speed–power product (pJ at 100 kHz)	100	20	1.2	11
Minimum output current (mA at $V_{OUT} = 0.4$ V)	16	8	4	1.6
Fan-out (LS loads)	40	20	10	4
Maximum input current (mA at $V_{IN} = 0.4$ V)	−1.6	−0.4	0.001	−0.001

connections would appear if we decided to include them. Note that we usually have two separate supplies: a positive supply and an equal, but opposite, negative supply. The common connection to these two supplies (i.e. the 0 V rail) acts as the common rail in our circuit. The input and output voltages are usually measured relative to this rail.

> ### Key point
>
> Operational amplifiers are linear integrated circuits that can be used as versatile "gain blocks" within a wide variety of linear circuits.

Operational amplifier parameters

Before we take a look at some of the characteristics of "ideal" and "real" operational amplifiers, it is important to define some of the terms and parameters that we apply to these devices.

Open-loop voltage gain

The *open-loop voltage gain* of an operational amplifier is defined as the ratio of output voltage to input voltage measured with no feedback applied. In practice, this value is exceptionally high (typically >100,000) but is liable to considerable variation from one device to another.

Open-loop voltage gain may thus be thought of as the "internal" voltage gain of the device:

$$A_{\text{VOL}} = \frac{V_{\text{OUT}}}{V_{\text{IN}}}$$

where A_{VOL} is the open-loop voltage gain, V_{OUT} and V_{IN} are the output and input voltages, respectively, under open-loop conditions.

In linear voltage amplifying applications, a large amount of negative feedback will normally be applied and the open-loop voltage gain can be thought of as the internal voltage gain provided by the device.

The open-loop voltage gain is often expressed in decibels (dB) rather than as a ratio. In this case:

$$A_{\text{VOL}} = 20 \log_{10} \frac{V_{\text{OUT}}}{V_{\text{IN}}}$$

Most operational amplifiers have open-loop voltage gains of 90 dB, or more.

Closed-loop voltage gain

The *closed-loop voltage gain* of an operational amplifier is defined as the ratio of output voltage to input voltage measured with a small proportion of the output fed back to the input (i.e. with feedback applied). The effect of providing negative feedback is to reduce the loop voltage gain to a value that is both predictable and manageable. Practical closed-loop voltage gains range from 1 to several thousand but note that high values of voltage gain may make unacceptable restrictions on bandwidth, seen later.

Closed-loop voltage gain is the ratio of output voltage to input voltage when negative feedback is applied, hence:

$$A_{\text{VCL}} = \frac{V_{\text{OUT}}}{V_{\text{IN}}}$$

where A_{VCL} is the closed-loop voltage gain, V_{OUT} and V_{IN} are the output and input voltages, respectively, under closed-loop conditions. The closed-loop voltage gain is normally very much less than the open-loop voltage gain.

Example 6.18

An operational amplifier operating with negative feedback produces an output voltage of 2 V when supplied with an input of $400\,\mu$V. Determine the value of closed-loop voltage gain.

Solution

Now:

$$A_{\text{VCL}} = \frac{V_{\text{OUT}}}{V_{\text{IN}}}$$

thus:

$$A_{\text{VCL}} = \frac{2}{400 \times 10^{-6}} = \frac{2 \times 10^6}{400} = 5000$$

Input resistance

The *input resistance* of an operational amplifier is defined as the ratio of input voltage to input current expressed in ohms. It is often expedient to assume that the input of an operational amplifier is purely resistive though this is not the case at high frequencies where shunt capacitive reactance may become significant. The input resistance of operational amplifiers is very much dependent on the semiconductor technology employed. In practice, values range from about 2 MΩ for common bipolar types to over 10^{12} Ω for FET and CMOS devices.

Input resistance is the ratio of input voltage to input current:

$$R_{IN} = \frac{V_{IN}}{I_{IN}}$$

where R_{IN} is the input resistance (in ohms), V_{IN} is the input voltage (in volts) and I_{IN} is the input current (in amperes). Note that we usually assume that the input of an operational amplifier is purely resistive though this may not be the case at high frequencies where shunt capacitive reactance may become significant.

The input resistance of operational amplifiers is very much dependent on the semiconductor technology employed. In practice, values range from about 2 MΩ for bipolar operational amplifiers to over 10^{12} Ω for CMOS devices.

Example 6.19

An operational amplifier has an input resistance of 2 MΩ. Determine the input current when an input voltage of 5 mV is present.

Solution

Now:

$$R_{IN} = \frac{V_{IN}}{I_{IN}}$$

thus:

$$I_{IN} = \frac{V_{IN}}{R_{IN}} = \frac{5 \times 10^{-3}}{2 \times 10^6}$$
$$= 2.5 \times 10^{-9} \text{ A} = 2.5 \text{ nA}$$

Output resistance

The *output resistance* of an operational amplifier is defined as the ratio of open-circuit output voltage to short-circuit output current expressed in ohms. Typical values of output resistance range from <10 Ω to around 100 Ω depending upon the configuration and amount of feedback employed.

Output resistance is the ratio of open-circuit output voltage to short-circuit output current, hence:

$$R_{OUT} = \frac{V_{OUT(OC)}}{I_{OUT(SC)}}$$

where R_{OUT} is the output resistance (in ohms), $V_{OUT(OC)}$ is the open-circuit output voltage (in volts) and $I_{OUT(SC)}$ is the short-circuit output current (in amperes).

Input offset voltage

An ideal operational amplifier would provide zero output voltage when 0 V difference is applied to its inputs. In practice, due to imperfect internal balance, there may be some small voltage present at the output. The voltage that must be applied differentially to the operational amplifier input in order to make the output voltage exactly zero is known as the *input offset voltage*.

Input offset voltage may be minimized by applying relatively large amounts of negative feedback or by using the offset null facility provided by a number of operational amplifier devices. Typical values of input offset voltage range from 1 to 15 mV. Where AC rather than DC coupling is employed, offset voltage is not normally a problem and can be happily ignored.

Full-power bandwidth

The *full-power bandwidth* for an operational amplifier is equivalent to the frequency at which the maximum undistorted peak output voltage swing falls to 0.707 of its low-frequency (DC) value (the sinusoidal input voltage remaining constant). Typical full-power bandwidths range from 10 kHz to over 1 MHz for some high-speed devices.

Slew rate

Slew rate is the rate of change of output voltage with time, when a rectangular step input voltage

Inverting input voltage, V_{IN}

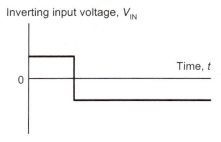

Output voltage, V_{OUT}

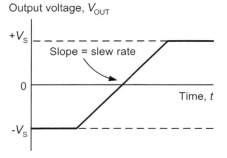

Figure 6.84 Slew rate for an operational amplifier.

is applied (as shown in Figure 6.84). The slew rate of an operational amplifier is the rate of change of output voltage with time in response to a perfect step-function input. Hence:

$$\text{Slew rate} = \frac{\Delta V_{OUT}}{\Delta t}$$

where ΔV_{OUT} is the change in output voltage (in volts) and Δt is the corresponding interval of time (in seconds).

Slew rate is measured in V/s (or V/μs) and typical values range from 0.2 V/μs to over 20 V/μs. Slew rate imposes a limitation on circuits in which large amplitude pulses rather than small amplitude sinusoidal signals are likely to be encountered.

Having now defined the parameters that we use to describe operational amplifiers we shall now consider the desirable characteristics for an "ideal" operational amplifier. These are as follows:

(a) The open-loop voltage gain should be very high (ideally infinite).
(b) The input resistance should be very high (ideally infinite).

(c) The output resistance should be very low (ideally zero).
(d) Full-power bandwidth should be as wide as possible.
(e) Slew rate should be as large as possible.
(f) Input offset should be as small as possible.

The characteristics of most modern integrated circuit operational amplifiers (i.e. "real" operational amplifiers) come very close to those of an "ideal" operational amplifier, as witnessed in the table below:

Parameter	Ideal	Real
Voltage gain	Infinite	100,000
Input resistance	Infinite	100 MΩ
Output resistance	Zero	20 Ω
Bandwidth	Infinite	2 MHz
Slew rate	Infinite	10 V/μs
Input offset	Zero	<5 mV

Operational amplifier types and applications

Some common examples of operational amplifiers for different applications are given in Table 6.17.

Example 6.20

Which of the operational amplifiers in Table 6.17 would be most suitable for each of the following applications:

(a) Amplifying the low-level output from a piezoelectric vibration sensor.
(b) A high-gain amplifier that can be faithfully used to amplify very small signals.
(c) A low-frequency amplifier for audio signals.

Solution

(a) AD548 (this operational amplifier is designed for use in instrumentation applications and it offers a very low input offset current which is important when the input is derived from a piezoelectric transducer).
(b) CA3140 (this is a low-noise operational amplifier that also offers high gain and fast slew rate).

Table 6.17

Device	Type	Open-loop voltage gain (dB)	Input bias current	Slew rate (V/μs)	Application
AD548	Bipolar	100 min.	0.01 nA	1.8	Instrumentation amplifier
AD711	FET	100	25 pA	20	Wideband amplifier
CA3140	CMOS	100	5 pA	9	Low-noise wideband amplifier
LF347	FET	110	50 pA	13	Wideband amplifier
LM301	Bipolar	88	70 nA	0.4	General-purpose operational amplifier
LM348	Bipolar	96	30 nA	0.6	General-purpose operational amplifier
TL071	FET	106	30 pA	13	Wideband amplifier
741	Bipolar	106	80 nA	0.5	General-purpose operational amplifier

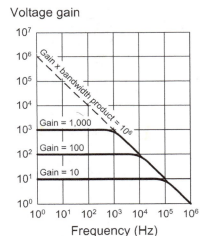

Figure 6.85 Frequency response curves for an operational amplifier.

(c) LM348 or LM741 (both are general-purpose operational amplifiers and are ideal for non-critical applications such as audio amplifiers).

Gain and bandwidth

It is important to note that the product of gain and bandwidth is a constant for any particular operational amplifier. Hence, an increase in gain can only be achieved at the expense of bandwidth, and vice versa.

Figure 6.85 shows the relationship between voltage gain and bandwidth for a typical operational amplifier (note that the axes use logarithmic, rather than linear scales). The open-loop voltage gain (i.e. that obtained with no feedback applied) is 100,000 (or 100 dB) and the bandwidth obtained in this condition is a mere 10 Hz. The effect of applying increasing amounts of negative feedback (and consequently reducing the gain to a more manageable amount) is that the bandwidth increases in direct proportion.

The frequency response curves in Figure 6.85 show the effect on the bandwidth of making the closed-loop gains equal to 10,000, 1000, 100 and 10. The following table summarizes these results. You should also note that the (gain × bandwidth) product for this amplifier is 1×10^6 Hz (i.e. 1 MHz).

We can determine the bandwidth of the amplifier when the closed-loop voltage gain is set to 46 dB by constructing a line and noting the intercept point on the response curve. This shows that the bandwidth will be 10 kHz (note that, for this operational amplifier, the (gain × bandwidth) product is 2×10^6 Hz (or 2 MHz).

Voltage gain (A_V)	Bandwidth
1	DC to 1 MHz
10	DC to 100 kHz
1000	DC to 10 kHz
10,000	DC to 1 kHz
100,000	DC to 100 Hz
1,000,000	DC to 10 Hz

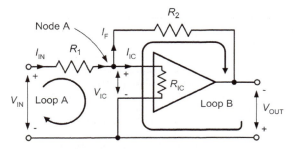

Figure 6.86 Operational amplifier with negative feedback applied.

Key point

The product of gain and bandwidth for an operational amplifier is a constant. Thus, an increase in gain can only be achieved at the expense of bandwidth and vice versa.

Inverting amplifier with feedback

Figure 6.86 shows the circuit of an inverting amplifier with negative feedback applied. For the sake of our explanation we will assume that the operational amplifier is "ideal". Now consider what happens when a small positive input voltage is applied. This voltage (V_{IN}) produces a current (I_{IN}) flowing in the input resistor R_1.

Since the operational amplifier is "ideal", we will assume that:

(a) the input resistance (i.e. the resistance that appears between the inverting and non-inverting input terminals, R_{IC}) is infinite;
(b) the open-loop voltage gain (i.e. the ratio of V_{OUT} to V_{IN} with no feedback applied) is infinite.

As a consequence of (a) and (b):

(i) the voltage appearing between the inverting and non-inverting inputs (V_{IC}) will be zero;
(ii) the current flowing into the chip (I_{IC}) will be zero (recall that $I_{IC} = V_{IC}/R_{IC}$ and R_{IC} is infinite).

Applying Kirchhoff's current law at node A gives:

$$I_{IN} = I_{IC} + I_F \quad \text{but} \quad I_{IC} = 0 \quad \text{thus} \quad I_{IN} = I_F \tag{1}$$

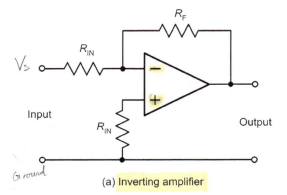

(a) Inverting amplifier

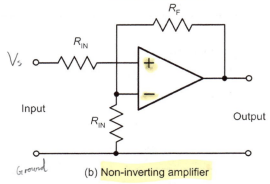

(b) Non-inverting amplifier

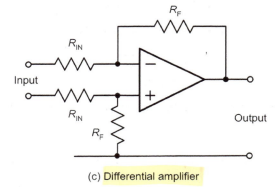

(c) Differential amplifier

Figure 6.87 The three basic configurations for operational voltage amplifiers.

(this shows that the current in the feedback resistor, R_2, is the same as the input current, I_{IN}).

Applying Kirchhoff's voltage law to loop A gives:

$$V_{IN} = (I_{IN} \times R_1) + V_{IC} \quad \text{but} \quad V_{IC} = 0$$
$$\text{thus} \quad V_{IN} = I_{IN} \times R_1 \tag{2}$$

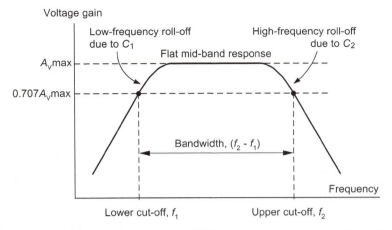

Figure 6.88 Effect of placing a capacitor in series with the input of an operational amplifier.

Applying Kirchhoff's voltage law to loop B gives:

$$V_{OUT} = -V_{IC} + (I_F \times R_2) \quad \text{but} \quad V_{IC} = 0$$

$$\text{thus} \quad V_{OUT} = I_F \times R_2 \tag{3}$$

Combining (1) and (3) gives:

$$V_{OUT} = I_{IN} \times R_2 \tag{4}$$

The voltage gain of the stage is given by:

$$A_V = \frac{V_{OUT}}{V_{IN}}$$

Combining (4) and (2) with (5) gives:

$$A_V = \frac{I_{IN} \times R_2}{I_{IN} \times R_1} = \frac{R_2}{R_1}$$

To preserve symmetry and minimize offset voltage, a third resistor is often included in series with the non-inverting input. The value of this resistor should be equivalent to the parallel combination of R_1 and R_2. Hence:

$$R_3 = \frac{R_1 \times R_2}{R_1 + R_2}$$

Operational amplifier configurations

The three basic configurations for operational voltage amplifiers, together with the expressions for their voltage gain are shown in Figure 6.87. Supply rails have been omitted from these diagrams for clarity but are assumed to be symmetrical about 0 V.

All of the amplifier circuits described previously have used direct coupling and thus have frequency response characteristics that extend to DC. This, of course, is undesirable for many applications, particularly where a wanted AC signal may be superimposed on an unwanted DC voltage level. In such cases a capacitor of appropriate value may be inserted in series with the input as shown below. The value of this capacitor should be chosen so that its reactance is very much smaller than the input resistance at the lower applied input frequency. The effect of the capacitor on an amplifier's frequency response is shown in Figure 6.88.

We can also use a capacitor to restrict the upper frequency response of an amplifier. This time, the capacitor is connected as part of the feedback path. Indeed, by selecting appropriate values of capacitor, the frequency response of an inverting operational voltage amplifier may be very easily tailored to suit individual requirements (see Figure 6.89).

The lower cut-off frequency is determined by the value of the input capacitance, C_1, and input resistance, R_1. The lower cut-off frequency is given by:

$$f_1 = \frac{1}{2\pi C_1 R_1} = \frac{0.159}{C_1 R_1}$$

where C_1 is in farads and R_1 is in ohms.

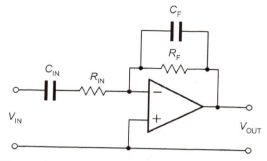

Figure 6.89 An inverting amplifier with capacitors to limit both the low and the high-frequency response.

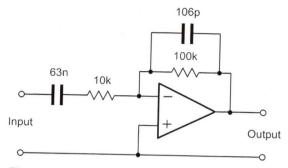

Figure 6.90 The complete amplifier circuit showing component values.

Provided the upper frequency response is not limited by the gain × bandwidth product, the upper cut-off frequency will be determined by the feedback capacitance, C_2, and feedback resistance, R_2, such that:

$$f_1 = \frac{1}{2\pi C_2 R_2} = \frac{0.159}{C_2 R_2}$$

where C_2 is in farads and R_2 is in ohms.

Example 6.21

An inverting operational amplifier is to operate according to the following specification:

Voltage gain $= 100$
Input resistance (at mid-band) $= 10\,k\Omega$
Lower cut-off frequency $= 250\,Hz$
Upper cut-off frequency $= 15\,kHz$

Devise a circuit to satisfy the above specification using an operational amplifier.

Solution

To make things a little easier, we can break the problem down into manageable parts. We shall base our circuit on a single operational amplifier configured as an inverting amplifier with capacitors to define the upper and lower cut-off frequencies as shown in Figure 6.89.

The nominal input resistance is the same as the value for R_1. Thus:

$$R_1 = 10\,k\Omega$$

To determine the value of R_2 we can make use of the formula for mid-band voltage gain:

$$A_V = R_2/R_1$$

thus:

$$R_2 = A_V \times R_1 = 100 \times 10\,k\Omega = 100\,k\Omega$$

To determine the value of C_1 we will use the formula for the low-frequency cut-off:

$$f_1 = \frac{0.159}{C_1 R_1}$$

from which:

$$C_1 = \frac{0.159}{f_1 R_1} = \frac{0.159}{250 \times 10 \times 10^3} = \frac{0.159}{2.5 \times 10^6}$$

$$= 63 \times 10^{-9}\,F = 63\,nF$$

Finally, to determine the value of C_2 we will use the formula for high-frequency cut-off:

$$f_2 = \frac{0.159}{C_2 R_2}$$

$$C_2 = \frac{0.159}{f_2 R_2} = \frac{0.159}{15 \times 10^3 \times 100 \times 10^3}$$

$$= \frac{0.159}{1.5 \times 10^9} = 0.106 \times 10^{-9}\,F = 106\,pF$$

The circuit of the amplifier is shown in Figure 6.90.

Operational amplifier circuits

In addition to their application as a general-purpose amplifying device, operational amplifiers have a number of other uses, including

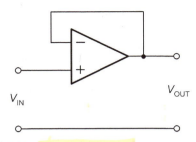

Figure 6.91   A voltage follower.

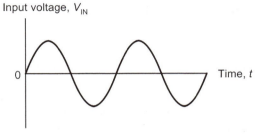

Figure 6.92   Typical input and output waveforms for a voltage follower.

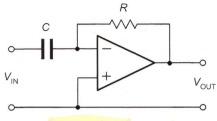

Figure 6.93   A differentiator.

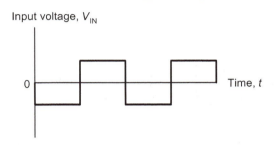

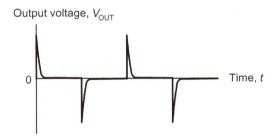

Figure 6.94   Typical input and output waveforms for a differentiator.

voltage followers, differentiators, integrators, comparators and summing amplifiers. We shall conclude this section by taking a brief look at each of these applications.

Voltage followers

A voltage follower using an operational amplifier is shown in Figure 6.91. This circuit is essentially an inverting amplifier in which 100% of the output is fed back to the input. The result is an amplifier that has a voltage gain of 1 (i.e. unity), a very high-input resistance and a very high-output resistance. This stage is often referred to as a buffer and is used for matching a high-impedance circuit to a low-impedance circuit. Typical input and output waveforms for a voltage follower are shown in Figure 6.92.

Notice how the input and output waveforms are both in-phase (they rise and fall together) and that they are identical in amplitude.

Differentiators

A differentiator using an operational amplifier is shown in Figure 6.93. A differentiator produces an output voltage that is equivalent to the rate of change of its input. This may sound a little complex but it simply means that, if the input voltage remains constant (i.e. if it is not changing) the output also remains constant. The faster the input voltage changes, the greater will the output be. In mathematics this is equivalent to the differential function. Typical input and output waveforms for a differentiator are shown in Figure 6.94. Note how the square wave input is

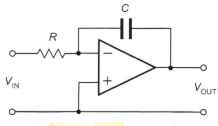

Figure 6.95 An integrator.

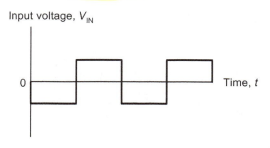

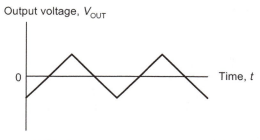

Figure 6.96 Typical input and output waveforms for an integrator.

converted to a train of short duration pulses at the output. Note also that the output waveform is inverted because the signal has been applied to the inverting input of the operational amplifier.

An integrator using an operational amplifier is shown in Figure 6.95. This circuit provides the opposite function to that of a differentiator in that its output is equivalent to the area under the graph of the input function rather than its rate of change. If the input voltage remains constant (and is other than 0 V) the output voltage will ramp up or down according to the polarity of the input. The longer the input voltage remains at a particular value, the larger the value of output voltage (of either polarity) produced. Typical input and output waveforms for an integrator are shown in Figure 6.96. Note how the

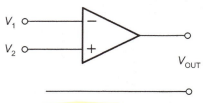

Figure 6.97 A comparator.

square wave input is converted to a wave that has a triangular shape. Once again, note that the output waveform is inverted.

Comparators
A comparator using an operational amplifier is shown in Figure 6.97. Since no negative feedback has been applied, this circuit uses the maximum gain of the operational amplifier. The output voltage produced by the operational amplifier will thus rise to the maximum possible value (equal to the positive supply rail voltage) whenever the voltage present at the non-inverting input exceeds that present at the inverting input. Conversely, the output voltage produced by the operational amplifier will fall to the minimum possible value (equal to the negative supply rail voltage) whenever the voltage present at the inverting input exceeds that present at the non-inverting input. Typical input and output waveforms for a comparator are shown in Figure 6.98. Note how the output is either +15 or −15 V depending on the relative polarity of the two inputs.

Summing amplifiers
A summing amplifier using an operational amplifier is shown in Figure 6.99. This circuit produces an output that is the sum of its two input voltages. However, since the operational amplifier is connected in inverting mode, the output voltage is given by:

$$V_{OUT} = -(V_1 + V_2)$$

where V_1 and V_2 are the input voltages (note that all of the resistors used in the circuit have the same value). Typical input and output waveforms for a summing amplifier are shown in Figure 6.100.

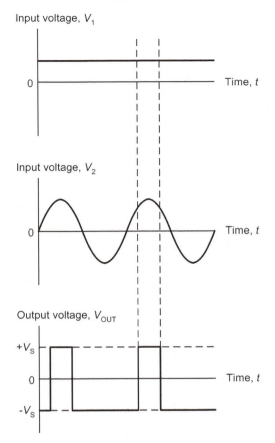

Figure 6.98 Typical input and output waveforms for a comparator.

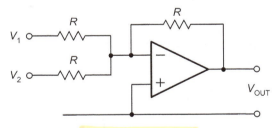

Figure 6.99 A summing amplifier.

Multi-stage amplifiers

In many cases, a single transistor or integrated circuit may be insufficient to provide the amount of gain required in a circuit. In such an eventuality it is necessary to connect stages together so that one stage of gain follows another in what is known as a multi-stage amplifier (see Figure 6.101).

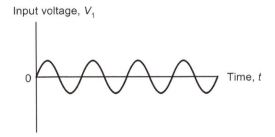

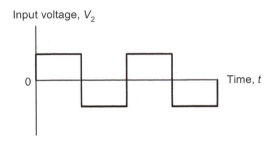

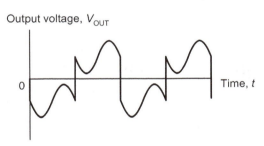

Figure 6.100 Typical input and output waveforms for a summing amplifier.

Various connecting methods are used in order to connect stages together. These coupling circuits allow the signal to be passed from one stage to another without affecting the internal bias currents and voltages required for each stage. Coupling methods include the following:

- resistor–capacitor (R–C) coupling;
- inductor–capacitor (L–C) coupling;
- transformer coupling;
- direct coupling.

Figure 6.102 illustrates these coupling methods.

Positive versus negative feedback

We have already shown how negative feedback can be applied to an operational amplifier in order to produce an exact value of gain.

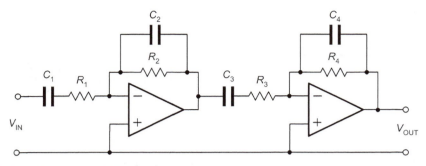

Figure 6.101   A multi-stage amplifier.

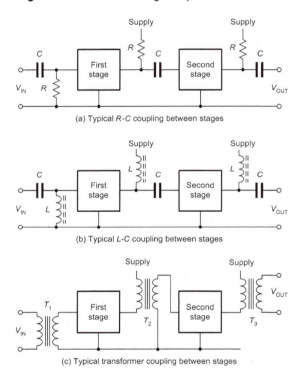

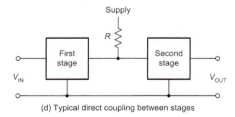

Figure 6.102   Various coupling methods.

Negative feedback is frequently used in order to stabilize the gain of an amplifier and also to increase the frequency response (recall that, for an amplifier the product of gain and bandwidth is a constant). Positive feedback, on the other hand, results in an increase in gain and a reduction in bandwidth. Furthermore, the usual result of applying positive feedback is that an amplifier becomes unstable and oscillates (i.e. it generates an output without an input being present!). For this reason, positive feedback is only used in amplifiers when the voltage gain is less than unity.

> **Key point**
>
> When negative feedback is applied to an amplifier, the overall gain is reduced and the bandwidth is increased (note that the gain × bandwidth product remains constant). When positive feedback is applied to an amplifier, the overall gain is increased and the bandwidth is reduced. In most cases this will result in instability and oscillation.

Test your understanding 6.4

1. Identify logic gates (a)–(d) shown in Figure 6.103.

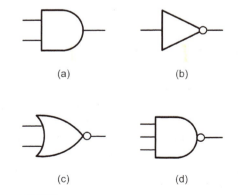

Figure 6.103

2. A two-input logic gate only produces a logic 1 output when both of its inputs are at logic 1. What type of logic gate is it?

3. Name the two basic logic families. To which family does each of the following devices belong:
   (a) 74LS04
   (b) 4001BE

4. For each of the devices listed in Question 3, state the standard range of voltages that is used to represent:
   (a) logic 0
   (b) logic 1

5. Sketch the circuit symbol for an operational amplifier. Label each of the connections.

6. List four characteristics associated with an "ideal" operational amplifier.

7. An operational amplifier with negative feedback applied produces an output of 1.5 V when an input of 7.5 mV is present. Determine the value of closed-loop voltage gain.

8. Sketch the circuit of an inverting amplifier based on an operational amplifier. Label your circuit and identify the components that determine the closed-loop voltage gain.

9. Sketch the circuit of each of the following based on the use of operational amplifiers:
   (a) a comparator
   (b) a differentiator
   (c) an integrator

10. An inverting amplifier is to be constructed having a mid-band voltage gain of 40 and a frequency response extending from 20 Hz to 20 kHz. Devise a circuit and specify all component values required.

6.3 Printed circuit boards

Syllabus
Description and use of printed circuit boards (PCBs).

Knowledge level key

A	B1	B2
–	1	2

6.3.1 PCB design considerations

PCBs comprise copper tracks bonded to an epoxy glass or synthetic resin bonded paper (SRBP) board. Once designed and tested, printed circuits are easily duplicated and the production techniques are based on automated component assembly and soldering.

A number of considerations must be taken into account when a PCB is designed including the current carrying capacity of the copper track conductors and the maximum voltage that can be safely applied between adjacent tracks.

The current rating of a PCB track depends on three factors:

(a) the width of the track,
(b) the thickness of the copper coating,
(c) the maximum permissible temperature rise.

The most common coating thickness is $35 \, \mu m$ (equivalent to 1 oz. of copper per square foot). The table below is a rough guide as to the minimum track width for various currents (assuming a temperature rise of no $>10°C$):

Current (DC or RMS, AC) in A	Minimum track width (mm)
<2.5	0.5
2.5–4	1.5
4–6	3.0
6–9	5.0

The table below is a rough guide as to the amount of track spacing required for different voltages:

Voltage between adjacent conductors (DC or peak AC) in V	Minimum track spacing (mm)
<150	1
150–300	1.5
300–600	2.5
600–900	3

Off-board connections to PCBs can be made using various techniques including:

- direct soldering to copper pads;
- soldered or crimped connections to pins inserted into the PCB which are themselves soldered into place;
- edge connectors (invariably these are gold plated to reduce contact resistance and prevent oxidation);

Figure 6.104 The upper (component) side of a typical double-sided PCB. The holes are plated-through and provide electrical links from the upper side of the board to the lower (track) side.

- indirect connector using headers soldered to a matrix of pads on the PCB.

Key point

PCBs provide us with a convenient way of mounting electronic components that also helps to simplify maintenance, since it is possible to remove and replace a complete PCB and then carry out repairs away from the aircraft using specialized test equipment.

6.3.2 Materials used for PCBs

The laminate material used to construct a PCB must have the following properties (Figures 6.104–6.107):

- very high resistivity;
- very high-flexural strength;
- ability to operate at relatively high temperatures (e.g. up to 125°C);
- high dielectric breakdown strength.

Typical materials are listed in Table 6.18.

6.3.3 PCB manufacture

Most PCBs are designed and manufactured entirely using computer-aided manufacturing (CAM) techniques. The first stage in the process involves transferring the circuit diagram data to a printed circuit layout package. The example shown in Figure 6.108 is for a single-chip

Figure 6.105 Part of the upper (component) side of a microcontroller circuit board. In order to provide some clearance above the board and to aid heat dissipation, the two 2.5 W resistors (R_{17} and R_{18}) are mounted on small ceramic spacers. Connections to other boards is made possible with multi-way connectors SK2–SK6.

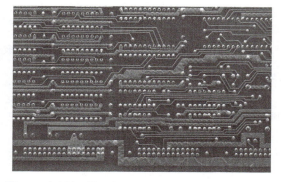

Figure 6.106 The lower (track) side of the microcontroller circuit board. The wider tracks are used for supply (+5 V) and ground (0 V).

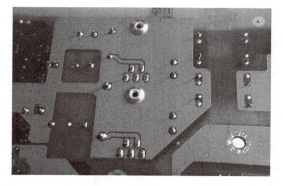

Figure 6.107 On some circuit boards the ground (0 V) track is extended over a large area. This can have a number of benefits including assisting with screening, improving high-frequency performance and helping to conduct heat away from components.

Table 6.18

Laminate type	Laminate construction
FR-2	Phenolic laminate, much cheaper but not as strong or stable. Widely used in cost conscious applications, i.e. consumer goods.
PTFE	Used in specialist high-frequency applications. This laminate material has a much lower dielectric constant and is, therefore, suitable for use in specialized high-frequency applications. Unfortunately PTFE is prohibitively expensive for most applications!
FR-4	This is the standard glass-epoxy laminate used in the industry. It is available in several variants and a range of standard thickness including 0.8, 1.0, 1.2, 1.6, 2.0, 2.4 and 3.2 mm. The most common copper thickness is 35 µm (equivalent to 1 oz. per square foot). The dielectric constant (relative permittivity) of FR-4 laminate ranges from about 4.2 to 5.0 and the maximum operating temperature is usually around 125°C.
G-10 and G-11	These are the non-flame retardant versions of the FR-4 laminate. G11 had an extended operating range of up to about 150°C.

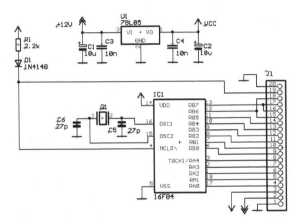

Figure 6.108 Example of a circuit diagram as it appears on the screen of a PCB design package. The circuit shown here is for a single-chip microcontroller and its regulated 5 V power supply.

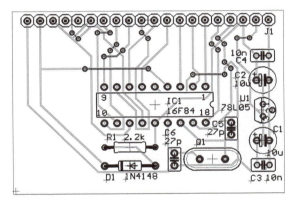

Figure 6.109 The PCB design package produces a layout diagram that has been automatically routed.

microcontroller and its regulated 5 V power supply. After drawing the circuit diagram, a *netlist* is produced which allows the circuit to be either manually or automatically routed. A database is used to hold dimensional information about each component (e.g. the distance between the component leads) and this data is combined with the netlist and basic component placement information in order to generate a fully routed PCB layout as shown in Figure 6.109.

Following etching and drilling, and the application of silk-screen component legend, the boards are coated with a solder resist before a tin–lead reflow finish is applied to the exposed

copper pads. Components are then placed (usually by machine) and the board is then passed through a flow-soldering machine to complete the soldering process.

Example 6.22

Refer to Figures 6.108 and 6.109 and answer the following questions:

(a) How many pins has J_1?
(b) How many pins has IC_1?
(c) To which pin on J_1 is pin-1 on IC_1 connected?
(d) To which pin on IC_1 is pin-9 on IC_1 connected?
(e) To which pin on J_1 is the cathode of D_1 connected?

(f) Which pins on J_1 are used for the common ground connection?

(g) Which pin on J_1 has the highest positive voltage?

(h) Between which two pins on J_1 would you expect to measure the $+5$ V supply?

(i) Which pin on IC_1 is connected directly to ground?

(j) What is the value of C_1?

(k) What is the value of R_1?

(l) What type of device is U_1?

(m) What type of device is IC_1?

Solution

(a) 20

(b) 18

(c) 6

(d) 12

(e) 19

(f) 1 and 20

(g) 2

(h) 1 ($-$) or 20 ($-$) and 3 ($+$)

(i) 5

(j) $10\,\mu$F

(k) $2.2\,k\Omega$

(l) A 78L05 three-terminal voltage regulator

(m) A 16F84 microcontroller

6.3.4 Surface mounting technology

Surface mounting technology (SMT) is now widely used in the manufacture of PCBs for avionic systems. SMT allows circuits to be assembled in a much smaller space than would be possible using components with conventional wire leads and pins that are mounted using through-hole techniques. It is also possible to mix the two technologies, i.e. some through-hole mounting of components and some surface mounted components (SMCs) present on the same circuit board. The following combinations are possible:

- SMCs on both sides of a PCB.
- SMC on one side of the board and conventional through-hole components (THCs) on the other.
- A mixture of SMC and THC on both sides of the PCB.

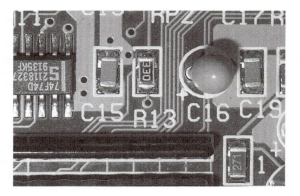

Figure 6.110 Close up of a circuit board which uses leaded components (tantalum capacitor, C_{16}), surface mounted capacitors (C_{15} and C_{19}) and surface mounted resistors (e.g. R_{13}). Note also the surface mounted integrated circuit and component legend.

SMCs are supplied in packages that are designed for mounting directly on the surface of a PCB. To provide electrical contact with the PCB, some SMC have contact pads on their surface. Other devices have contacts which extend beyond the outline of the package itself but which terminate on the surface of the PCB rather than making contact through a hole (as is the case with a conventional THC). In general, passive components (such as resistors, capacitors and inductors) are configured leadless for surface mounting, whilst active devices (such as transistors and integrated circuits) are available in both surface mountable types as well as lead and in leadless terminations suitable for making direct contact to the pads on the surface of a PCB (Figure 6.110).

Most SMCs have a flat rectangular shape rather than the cylindrical shape that we associate with conventional wire leaded components (Figure 6.111). During manufacture of a PCB, the various SMC are attached using reflow-soldering paste (and in some cases adhesives) which consists of particles of solder and flux together with binder, solvents and additives. They need to have good "tack" in order to hold the components in place and remove oxides without leaving obstinate residues.

The component attachment (i.e. soldering!) process is completed using one of several techniques including *convection ovens* in which the

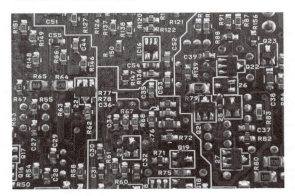

Figure 6.111 Some PCB have components mounted on both sides. This shows part of the lower (track) side of a large PCB. All of the components mounted on this side of the board are surface mounted types.

PCB is passed, using a conveyor belt, through a convection oven which has separate zones for preheating, flowing and cooling, and *infra-red reflow* in which infra-red lamps are used to provide the source of heat.

Key point

Modern SMCs take up considerably less space than conventional components that have connecting leads or pins that require fitting through holes in a PCB. SMCs need special handling techniques due to their small size and the need for soldering direct to surface pads on the PCB.

Test your understanding 6.5

1. What is the most commonly used laminate material used in the manufacture of PCBs?
2. State THREE factors that determine the current carrying capacity of the copper tracks on a PCB.
3. List FOUR important characteristics of a material used for the manufacture of a PCB.
4. State ONE advantage and ONE disadvantage of PTFE laminate when compared with FR-4 laminate.
5. The maximum operating temperature for a PCB is usually quoted as 175°C. Is this statement true or false? Explain your answer. *False   125°C*
6. SMCs must be individually soldered in place. Is this statement true or false? Explain your answer. *False*
7. Explain why FR-4 laminate is preferred to G-10 laminate in an aircraft PCB application.
8. Explain the purpose of the silk-screened legend that appears on the component side of a PCB.

9. State the typical range of coating thicknesses for the copper surface coating on a PCB.
10. Some of the stages that are involved in the production of a PCB are listed below. Organize this list into the correct sequence:
    - drilling,
    - etching,
    - screen printing of component legend,
    - application of tin–lead reflow coating,
    - application of solder-resist coating.

6.4 Servomechanisms

Syllabus
Understanding of the following terms: Open- and closed-loop systems, feedback, follow up, analogue transducers; Principles of operation and use of the following synchro-system components/features: resolvers, differential, control and torque, transformers, inductance and capacitance transmitters.

Knowledge level key

A	B1	B2
–	1	–

Syllabus
Understanding of the following terms: open and closed loop, follow up, servomechanism, analogue, transducer, null, damping, feedback and deadband; Construction operation and use of the following synchro-system components: resolvers, differential, control and torque, E and I transformers, inductance transmitters, capacitance transmitters, synchronous transmitters; Servomechanism defects, reversal of synchro-leads, hunting.

Knowledge level key

A	B1	B2
–	–	2

6.4.1 Control systems

Control systems are used in aircraft, cars and many other complex machines. A specific input,

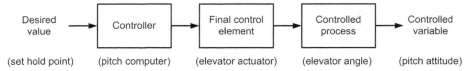

Figure 6.112 A simple control system.

Figure 6.113 The pitch attitude controller shown in block schematic form.

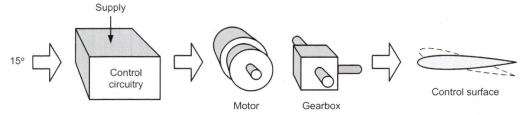

Figure 6.114 The pitch attitude controller shown in diagrammatic form.

such as moving a lever or joystick, causes a specific output, such as feeding current to an electric motor that in turn operates a hydraulic actuator that moves, e.g. the elevator of the aircraft. At the same time, the position of the elevator is detected and fed back to the pitch attitude controller, so that small adjustments can continually be made to maintain the desired attitude and altitude.

Control systems invariably comprise a number of elements, components or sub-systems that are connected together in a particular way. The individual elements of a control system interact together to satisfy a particular functional requirement, such as modifying the position of an aircraft's control surfaces.

A simple control system is shown in Figure 6.112. This system has a single input, the desired value (or *set point*) and a single output (the *controlled variable*). In the case of a pitch attitude control system, the desired value would be the hold point (set by the pilot) whereas the controlled variable would be the pitch attitude of the aircraft.

The pitch attitude control system uses three basic components:

- a controller (the pitch computer),
- a final control element (the elevator actuator),
- the controlled process (the adjustment of elevator angle).

Figures 6.113 and 6.114, respectively, show the pitch attitude controller represented in block schematic and diagrammatic forms.

6.4.2 Servomechanisms

Control systems on aircraft are frequently referred to as *servomechanisms* or *servo systems*. An important feature of a servo system is that operation is *automatic* and once set; they are usually capable of operating with minimal human intervention. Furthermore, the input (command) signal used by a servo system is generally very small whereas the output may involve the control or regulation of a very considerable amount of power. For example, the physical power required to operate the control surfaces of a large aircraft greatly exceeds the unaided physical capability of the pilot! Later in this section we shall look at the components used in some typical aircraft servo systems.

6.4.3 Control methods

System control involves maintaining the desired output from the system (e.g. aircraft turn rate) at the desired value regardless of any disturbances that may affect it (e.g. wind speed and direction). Controlling a system involves taking into account:

(a) the desired value of output from the system,
(b) the level of demand (or *loading*) on the output,
(c) any unwanted variations in the performance of the components of the system.

Different control methods are appropriate to different types of system. The overall control strategy can be based on analogue or digital techniques (or a mixture of the two). At this point, it is worth explaining what we mean by these two methods.

Analogue control

Analogue control involves the use of signals and quantities that are continuously variable. Within analogue control systems, signals are represented by voltages and currents that can take any value between two set limits. Figure 6.115 shows how a typical analogue signal varies with time.

Digital control

Digital control involves the use of signals and quantities that vary in discrete steps. Values that fall between two adjacent steps must take one or other value as intermediate values are disallowed! Figure 6.116 shows how a typical digital signal varies with time.

Digital control systems are usually based on digital logic devices or microprocessor-based controllers. Values represented within a digital system are expressed in binary coded form

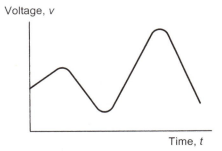

Figure 6.115   A typical analogue signal.

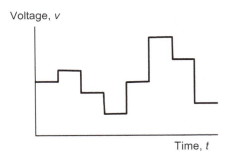

Figure 6.116   A typical digital signal.

using a number of signal lines. The voltage on each line can be either *high* (representing logic 1) or *low* (representing logic 0). The more signal lines the greater the resolution of the system. For example, with just two signal lines it is only possible to represent a number using two binary digits (or *bits*). Since each bit can be either 0 or 1 it is only possible to represent four different values (00, 01, 10 and 11) using this system. With three signal lines we can represent numbers using three bits and eight different values are possible (000, 001, 010, 011, 100, 101, 110 and 111).

The relationship between the number of bits, n, and the number of different values possible, m, is given by $m = 2^n$. So, in an 8-bit system the number of different discrete states is given by $m = 2^8 = 256$.

Example 6.23

A digital system uses a 10-bit code to represent values encoded in digital form. How many different states are possible?

Solution

The total number of different states will be given by:

$$m = 2^n$$

In this case $n = 10$, therefore, the total number of different states will be:

$$m = 2^{10} = 1024$$

Key point

The resolution of a digital system is determined by the number of bits used in the digital codes. The more bits used, the greater the resolution will be.

Example 6.24

A digital control system is required to represent values to a resolution of at least 1%. With how many bits should it operate?

Solution

Let us assume that we might use a 6-bit code. This would provide us with $2^6 = 64$ possible values. Clearly this is not enough because we will need 100 values in order to achieve a 1% resolution. If we use a 7-bit code we would have $2^7 = 128$ possible values. This will allow us to have a resolution of 1/128 or 0.78%, which is slightly better than the minimum 1% that we are aiming for. Hence, a 7-bit code should be used.

6.4.4 Transducers

Transducers are devices that convert energy in the form of sound, light, heat, etc. into an equivalent electrical signal or vice versa. Before we go further, let us consider a couple of examples that you will already be familiar with. A *loudspeaker* is a device that converts low-frequency electric current into sound. A *thermocouple*, on the other hand, is a device that converts temperature into voltage. Both of these act as a transducer.

Transducers may be used both as system inputs and system outputs. From the two previous examples, it should be obvious that a loudspeaker is an *output transducer* designed for use in conjunction with an audio system, whereas a thermocouple is an *input transducer* which can be used in a temperature control system.

Table 6.19 provides examples of transducers that can be used to input and output three physical quantities: sound, temperature and angular position.

Example 6.25

Classify the following transducers as either input transducers or output transducers:

(a) a photocell
(b) an electric motor
(c) a thermocouple

Solution

(a) A photocell produces electric current when exposed to light and, therefore, it is an input transducer.
(b) An electric motor produces motion when supplied with electric current and, therefore, it is an output transducer.
(c) A thermocouple produces electric current when exposed to heat and, therefore, it is an input transducer.

6.4.5 Sensors

A *sensor* is simply a transducer, that is used to generate an input signal to a control or measurement system. The signal produced by a sensor is an *electrical analogy* of a physical quantity, such as angular position, distance, velocity, acceleration, temperature, pressure, light level, etc. The signals returned from a sensor, together with control inputs from the operator (where appropriate), will subsequently be used to determine the output from the system. The choice of sensor is governed by a number of factors including accuracy, resolution, cost, electrical specification and physical size (Table 6.20).

Sensors can be categorized as either *active* or *passive*. An active sensor *generates* a current or voltage output. A passive transducer requires a source of current or voltage and it modifies this in some way (e.g. by virtue of a change in the sensor's resistance). The result may still be a voltage

Table 6.19

Physical quantity	Transducer	Notes
Input transducers		
Sound (pressure change)	Dynamic microphone	Diaphragm attached to a coil is suspended in a magnetic field. Movement of the diaphragm causes current to be induced in the coil.
Temperature	Thermocouple	Small e.m.f. generated at the junction between two dissimilar metals (e.g. copper and constantan). Requires reference junction and compensated cables for accurate measurement.
Angular position	Rotary potentiometer	Fine wire resistive element is wound around a circular former. Slider attached to the control shaft makes contact with the resistive element. A stable DC voltage source is connected across the ends of the potentiometer. Voltage appearing at the slider will then be proportional to angular position.
Output transducers		
Sound (pressure change)	Loudspeaker	Diaphragm attached to a coil is suspended in a magnetic field. Current in the coil causes movement of the diaphragm that alternately compresses and rarefies the air mass in front of it.
Temperature	Resistive heating element	Metallic conductor is wound onto a ceramic or mica former. Current flowing in the conductor produces heat.
Angular position	Stepper motor	Multi-phase motor provides precise rotation in discrete steps of 15° (24 steps per revolution), 7.5° (48 steps per revolution) and 1.8° (200 steps per revolution).

or current *but it is not generated by the sensor on its own.*

Sensors can also be classified as either *digital* or *analogue.* The output of a digital sensor can exist in only two discrete states, either "on" or "off", "low" or "high", "logic 1" or "logic 0", etc. The output of an analogue sensor can take any one of an infinite number of voltage or current levels. It is thus said to be *continuously variable.*

Example 6.26

Classify the following sensors as either active or passive sensors:

(a) a photocell
(b) a photodiode
(c) an LDR

Solution

(a) A photocell produces electric current when exposed to light and, therefore, is an active sensor.

(b) A photodiode cannot generate an electric current on its own and, therefore, is a passive sensor.
(c) An LDR cannot generate electric current on its own and, therefore, is a passive sensor.

Example 6.27

Classify the following sensors as either digital or analogue sensors:

(a) a reed switch
(b) a photodiode
(c) an LDR

Solution

(a) A reed switch is either on or off and, therefore, is a digital sensor.
(b) The current flowing through a thermistor varies continuously with changes in temperature and, therefore, a thermistor is an analogue sensor.

Table 6.20

Physical parameters	Type of sensors	Notes
Angular position	Resistive rotary position sensor	Rotary track potentiometer with linear law produces analogue voltage proportional to angular position.
	Optical shaft encoder	Encoded disk interposed between optical transmitter and receiver (infra-red LED and photodiode or phototransistor).
	Differential transformer	Transformer with fixed E-laminations and pivoted I-laminations acting as a moving armature.
Angular velocity	Tachogenerator	Small DC generator with linear output characteristic. Analogue output voltage proportional to shaft speed.
	Toothed rotor tachometer	Magnetic pick-up responds to the movement of a toothed ferrous disk. The pulse repetition frequency of the output is proportional to the angular velocity.
Flow	Rotating vane flow sensor	Turbine rotor driven by fluid. Turbine interrupts infra-red beam. Pulse repetition frequency of output is proportional to flow rate.
Linear position	Resistive linear position sensor	Linear track potentiometer with linear law produces analogue voltage proportional to linear position. Limited linear range.
	Linear variable differential transformer (LVDT)	Miniature transformer with split secondary windings and moving core attached to a plunger. Requires AC excitation and phase-sensitive detector.
	Magnetic linear position sensor	Magnetic pick-up responds to movement of a toothed ferrous track. Pulses are counted as the sensor moves along the track.
Light level	Photocell	Voltage-generating device. The analogue output voltage produced is proportional to light level.
	Light-dependent resistor (LDR)	An analogue output voltage results from a change of resistance within a cadmium sulphide (CdS) sensing element. Usually connected as part of a potential divider or bridge.
	Photodiode	Two-terminal device connected as a current source. An analogue output voltage is developed across a series resistor of appropriate value.
	Phototransistor	Three-terminal device connected as a current source. An analogue output voltage is developed across a series resistor of appropriate value.
Liquid level	Float switch	Simple switch element that operates when a particular level is detected.
	Capacitive proximity switch	Switching device that operates when a particular level is detected. Ineffective with some liquids.
	Diffuse scan proximity switch	Switching device that operates when a particular level is detected. Ineffective with some liquids.
Pressure	Microswitch pressure sensor	Microswitch fitted with actuator mechanism and range setting springs. Suitable for high-pressure applications.

(*continued*)

Table 6.20 (*continued*)

Physical parameter	Type of sensors	Notes
	Differential pressure vacuum switch	Microswitch with actuator driven by a diaphragm. May be used to sense differential pressure. Alternatively, one chamber may be evacuated and the sensed pressure applied to a second input.
	Piezo-resistive pressure sensor	Pressure exerted on diaphragm causes changes of resistance in attached piezo-resistive transducers. Transducers are usually arranged in the form of a four active element bridge, which produces an analogue output voltage.
Proximity	Reed switch	Reed switch and permanent magnet actuator. Only effective over short distances.
	Inductive proximity switch	Target object modifies magnetic field generated by the sensor. Only suitable for metals (non-ferrous metals with reduced sensitivity).
	Capacitive proximity switch	Target object modifies electric field generated by the sensor. Suitable for metals, plastics, wood, and some liquids and powders.
	Optical proximity switch	Available in diffuse and through scan types. Diffuse scan types require reflective targets. Both types employ optical transmitters and receivers (usually infra-red emitting LEDs and photodiodes or phototransistors). Digital input port required.
Strain	Resistive strain gauge	Foil type resistive element with polyester backing for attachment to body under stress. Normally connected in full bridge configuration with temperature-compensating gauges to provide an analogue output voltage.
	Semiconductor strain gauge	Piezo-resistive elements provide greater outputs than comparable resistive foil types. More prone to temperature changes and also inherently non-linear.
Temperature	Thermocouple	Small e.m.f. generated by a junction between two dissimilar metals. For accurate measurement, requires compensated connecting cables and specialized interface.
	Thermistor	Usually connected as part of a potential divider or bridge. An analogue output voltage results from resistance changes within the sensing element.
	Semiconductor temperature sensor	Two-terminal device connected as a current source. An analogue output voltage is developed across a series resistor of appropriate value.
Weight	Load cell	Usually comprises four strain gauges attached to a metal frame. This assembly is then loaded and the analogue output voltage produced is proportional to the weight of the load.
Vibration	Electromagnetic vibration sensor	Permanent magnet seismic mass suspended by springs within a cylindrical coil. The frequency and amplitude of the analogue output voltage are, respectively, proportional to the frequency and amplitude of vibration.

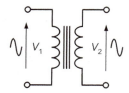

Figure 6.117 A transformer with primary and secondary windings.

(c) The current flowing through a tachogenerator varies continuously with shaft speed and, therefore, a tachogenerator is an analogue sensor.

Key point

Sensors are transducers that are used to provide the required inputs to instrumentation and control systems. Sensors can be either active (generating voltage or current) or passive (requiring a source of voltage or current in order to operate). Sensors can also be classified as digital or analogue.

6.4.6 Transformers

At this point, and before we start to explain the principle of the synchros and servos, it is worth revising what we know about transformers. A simple transformer with a single primary and a single secondary winding is shown in Figure 6.117. The primary voltage, V_1, and secondary voltage, V_2, rise and fall together and they are thus said to be *in-phase* with one another.

Now imagine that the secondary is not wound on top of the primary winding (as it is with a normal transformer) but is wound on a core which is aligned at an angle, θ, to the core of the primary winding (as shown in Figure 6.118).

The amount of magnetic flux that links the two windings will depend on the value of angle, θ. When θ is 0° (or 180°) maximum flux linkage will occur and, as a result, the secondary voltage, V_2, will have a maximum value. When θ is 90° (or 270°) minimum flux linkage will occur and, as a result, the secondary voltage, V_2, will have a minimum (zero) value. The relationship between RMS secondary voltage, V_2, and angle, θ, is shown in Figure 6.119.

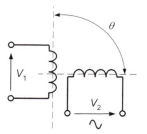

Figure 6.118 A transformer with an angle between the primary and secondary windings.

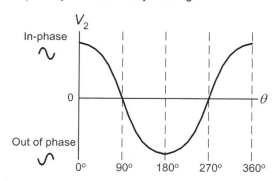

Figure 6.119 Relationship between RMS secondary voltage and angle.

In order to understand how the phase angle changes as angle, take a look at Figure 6.120. This diagram shows how the *amplitude* and *phase angle* of the secondary voltage, V_2, changes as angle, θ, changes. The important thing to note from all of this is that, when the angle between the two windings changes, two things change:

- the amplitude of the secondary voltage (V_2);
- the phase angle of the secondary voltage (V_2) relative to the primary voltage (V_1).

Key point

When the angle between transformer windings is 0° all of the lines of flux generated by the primary winding will cut through the secondary winding and maximum flux linkage will occur. When the angle between transformer windings is 90° none of the lines of flux generated by the primary winding will cut through the secondary winding and minimum flux linkage will occur. Hence, the RMS output voltage produced by the transformer will depend on the angle between the primary and secondary windings.

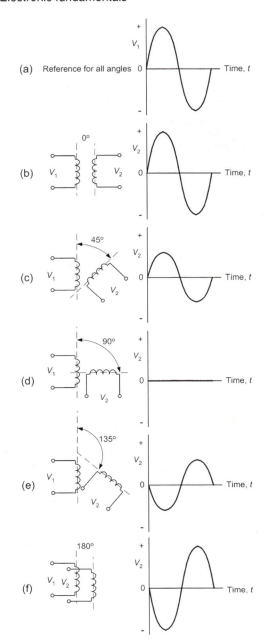

Figure 6.120 Waveform and phase angle of secondary voltage as angle θ varies.

6.4.7 The E and I transformer

You will probably recall from Chapter 5 that the most common type of transformer uses steel E- and I-laminations like those shown in

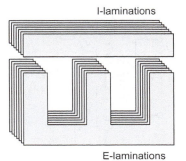

Figure 6.121 A transformer with E- and I-laminations.

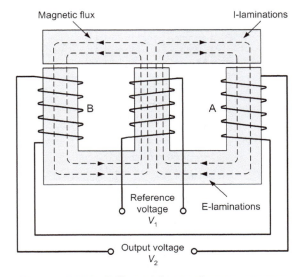

Figure 6.122 Differential transformer arrangement.

Figure 6.121. By having two separate secondary windings, one on each of the outer limbs of the E-section lamination, we can produce a useful control system component, the *differential transformer* as shown in Figure 6.122.

The circuit diagram of the differential transformer is shown in Figure 6.123. Note that the two secondary windings produce out-of-phase voltages and these subtract from each other making the output voltage, V_2, zero when $V_A = V_B$.

A simple application of the differential transformer is shown in Figure 6.124 in which the I-lamination is pivoted. Displacement in one direction will cause the lines of magnetic flux

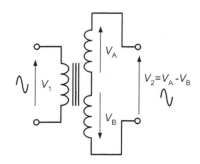

Figure 6.123 Circuit of the differential transformer.

Figure 6.125 Typical arrangement of synchros in an aircraft instrument.

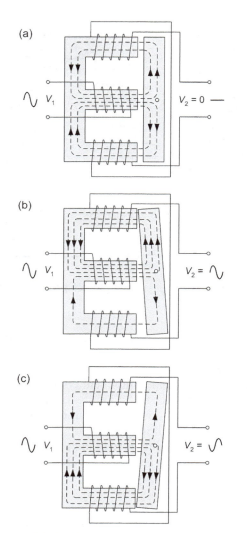

Figure 6.124 Differential transformer used to sense angular displacement.

to be strengthened in one limb whilst they are weakened in the other limb. This causes one secondary voltage to exceed the other and thus an output voltage is produced when they are combined at the output. The direction of motion will determine the phase of the output voltage (i.e. whether it is in-phase or out-of-phase) whilst the size of the displacement will determine the magnitude of the output voltage.

6.4.8 Synchros

Synchro is a generic term for a family of electromechanical devices (including *resolvers*) that are sometimes also referred to as *variable transformers*. Synchros can be used as transmitters or receivers according to whether they are providing an input or an output to a position control system. The two devices are, in fact, very similar in construction, the main difference being that the receiver has low friction bearings to follow the movement of the transmitter accurately and some form of damping mechanism designed to prevent oscillation (Figure 6.125).

The schematic of a synchro is shown in Figure 6.126. Note that the synchro has three-stator coils (S_1, S_2 and S_3) spaced by 120° and rotating rotor coils (R_1 and R_2). The three-stator coils have an internal common connection.

Synchros are designed to transmit the shaft position (i.e. the position of the primary rotor winding) to another synchro used as a receiver. A *resolver* is similar to a normal three-wire synchro but has multiple rotor and stator windings

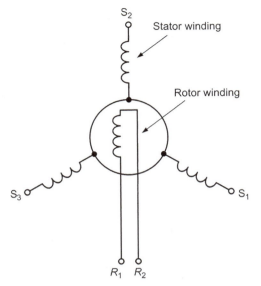

Figure 6.126 Circuit schematic symbol for a synchro.

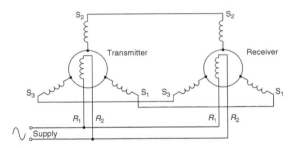

Figure 6.127 A typical synchro-transmitter and receiver arrangement.

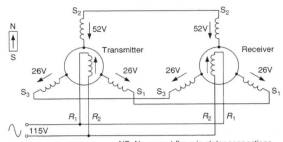

(a) System in correspondence

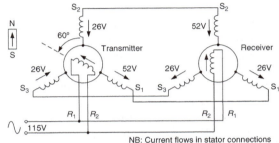

(b) System out of correspondence

(c) System again in correspondence

Figure 6.128 A synchro-based position control system.

spaced by 90° rather than the 120° found on the three-stator synchro.

A typical synchro-transmitter and receiver arrangement is shown in Figure 6.127. Note that the rotor is fed from an AC supply of typically 115 V at 400 Hz. When the rotor winding is energized, voltages are induced in the stator windings, S_1, S_2 and S_3. The magnitude and phase of the induced voltages will depend on the relative rotor and stator positions.

In order to explain the action of the synchro-based position control system you need to understand how the magnitude and phase angle of voltages induced in the stator coils varies according to the relative positions of the two

rotors. Take a look at Figure 6.128 which shows a synchro system that starts with the two rotors aligned (we say they are *in correspondence*) but which then becomes misaligned (or *out of correspondence*).

The stators of both synchros have their leads connected S_1 to S_1, S_2 to S_2 and S_3 to S_3, so the voltage in each of the transmitter-stator coils opposes the voltage in the corresponding coils of the receiver. Arrows indicate the voltage directions at a particular instant of time.

In Figure 6.128(a), the transmitter and receiver are shown in correspondence. In this condition, the rotor of the receiver induces voltages in its stator coils ($S_2 = 52$ V; S_1 and $S_3 = 26$ V) that are equal and opposite to the voltages induced into the transmitter-stator coils ($S_2 = 52$ V; S_1 and $S_3 = 26$ V). This causes the voltages to cancel and reduces the stator currents to zero. With zero current through the coils, the receiver torque is zero and the system remains in correspondence.

Now assume that the transmitter rotor is mechanically rotated through an angle of 60° as shown in Figure 6.128(b). When the transmitter rotor is turned, the rotor field follows and the magnetic coupling between the rotor and stator windings changes. This results in the transmitter S_2 coil voltage decreasing to 26 volts, the S_3 coil voltage reversing direction and the S_1 coil voltage increasing to 52 V. This imbalance in voltages, between the transmitter and receiver, causes current to flow in the stator coils in the direction of the stronger voltages.

The current flow in the receiver produces a resultant magnetic field in the receiver stator in the same direction as the rotor field in the transmitter. A force (torque) is now exerted on the receiver rotor by the interaction between its resultant-stator field and the magnetic field around its rotor. This force causes the rotor to turn through the same angle as the rotor of the transmitter. As the receiver approaches correspondence, the stator voltages of the transmitter and receiver approach equality. This action decreases the stator currents and produces a decreasing torque on the receiver. When the receiver and the transmitter are again in correspondence as shown in Figure 6.128(c), the stator voltages between the two synchros are equal and opposite ($S_1 = 52$ V; S_2 and $S_3 = 26$ V), the rotor torque is zero, and the rotors are displaced from zero by the same angle (60°). This sequence of events causes the transmitter and receiver to stay in correspondence.

The receiver's direction of rotation may be reversed by simply reversing the S_1 and S_3 connections so that S_1 of the transmitter is connected to S_3 of the receiver and vice versa as shown in Figure 6.129.

Even when the S_1 and S_3 connections are reversed, the system at 0° acts in the same way

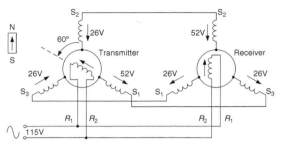

(a) S_1 and S_3 reversed, transmitter turned 60° anti-clockwise

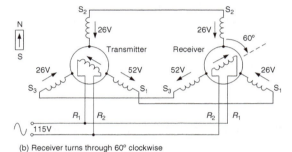

(b) Receiver turns through 60° clockwise

Figure 6.129 Effect of reversing stator connections.

as the system that we previously described at 0°. This is because the voltages induced in the S_1 and S_3 stator windings are still equal and oppose each other. This causes a cancelling effect, which results in zero-stator current and no torque. Without the torque required to move the receiver rotor, the system remains in correspondence and the reversing of the stator connections has no noticeable effect on the system at 0°.

Now suppose the transmitter rotor is turned counter-clockwise 60° as shown in Figure 6.129(a). The transmitter rotor is now aligned with S_1. This results in maximum magnetic coupling between the transmitter rotor and the S_1 winding. This maximum coupling induces maximum voltage in S_1. Because S_1 is connected to S_3 of the receiver, a voltage imbalance occurs between them. As a result of this voltage imbalance, maximum current flows through the S_3 winding of the receiver causing it to have the strongest magnetic field. Because the other two fields around S_2 and S_1 decrease proportionately, the S_3 field has the greatest effect on the resultant receiver's stator field. The strong S_3 stator field forces the rotor to turn 60° clockwise into alignment with itself as shown in Figure 6.129(b). At this point, the rotor of the receiver induces cancelling voltages in its own stator coils and causes the rotor to stop. The system is now in correspondence. Note that by reversing S_1 and S_3, both synchro rotors turn the same amount, but in opposite directions.

It is worth mentioning that the only stator leads ever interchanged, for the purpose of reversing receiver rotation, are S_1 and S_3. S_2 cannot be reversed with any other lead, since it represents the electrical zero position of the synchro. Furthermore, since the stator leads in a synchro are 120° apart, any change in the S_2 connection will result in a 120° error in the synchro-system together with a reversal of the direction of rotation. Another potential problem is the accidental reversal of the R_1 and R_2 leads on either the transmitter or receiver. This will result in a 180° error between the two synchros whilst the direction of rotation remains the same.

6.4.9 Open- and closed-loop control

In a system that employs open-loop control, the value of the input variable is set to a given value in the expectation that the output will reach the desired value. In such a system there is no automatic comparison of the actual output value with the desired output value in order to compensate for any differences. A simple open-loop system is shown in Figure 6.130.

A simple example of an open-loop control method is the manual adjustment of the regulator that controls the flow of gas to a burner on the hob of a gas cooker. This adjustment is carried out in the expectation that food will be raised to the correct temperature in a given time and without burning. Other than the occasional watchful eye of the chef, there is no means of automatically regulating the gas flow in response to the actual temperature of the food.

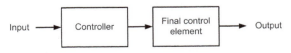

Figure 6.130 A simple open-loop system.

Clearly, open-loop control has some significant disadvantages. What is required is some means of closing the loop in order to make a continuous automatic comparison of the actual value of the output compared with the setting of the input control variable.

In the previous example, the chef actually closes the loop on an intermittent basis. In effect, the gas cooker relies on human intervention in order to ensure consistency of the food produced. If our cooking only requires boiling water, this intervention can be kept to a minimum, however, for "haute cuisine" we require the constant supervision of a skilled human operator!

Within the context of a modern passenger aircraft it is simply impossible for the flight crew to manually operate all of the systems! Hence, we need to ensure that the aircraft's control systems work with only a minimum of human intervention.

All modern aircraft systems make use of *closed-loop control*. In some cases, the loop might be closed by the pilot who determines the deviation between the desired and actual output. In most cases, however, the action of the system is made fully automatic and no human intervention is necessary other than initially setting the desired value of the output. The principle of a closed-loop control system is shown in Figure 6.131.

In general, the advantages of closed-loop systems can be summarized as follows:

1. Some systems use a very large number of input variables and it may be difficult or impossible for a human operator to keep track of them.
2. Some processes are extremely complex and there may be significant interaction between the input variables.
3. Some systems may have to respond very quickly to changes in variables (human reaction at times may not be just fast enough).
4. Some systems require a very high degree of precision (human operators may be unable to work to a sufficiently high degree of accuracy).

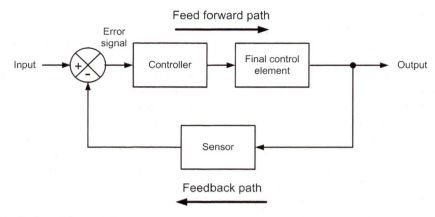

Figure 6.131 A closed-loop system.

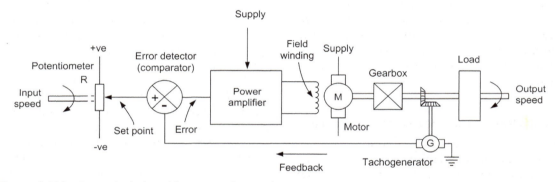

Figure 6.132 A practical closed-loop speed control system.

A practical closed-loop speed control system is shown in Figure 6.132. A power amplifier is used to provide the field current for the DC motor (M). The actual speed of the motor's output shaft is sensed by means of a small DC tachogenerator (G) coupled to the output shaft by means of suitable gearing. The voltage produced by the tachogenerator is compared with that produced at the slider of a potentiometer (R) which is used to set the desired speed.

The comparison of the two voltages (i.e. that of the tachogenerator with that corresponding to the set point) is performed by an operational amplifier connected as a comparator. The output of the comparator stage is applied to a power amplifier that supplies current to the DC motor. Energy is derived from a DC power supply

comprising transformer, rectifier and smoothing circuits.

6.4.10 Control system response

In a perfect system, the output value will respond instantaneously to a change in the input value (set point). There will be no delay when changing from one value to another and no time required for the output to settle to its final value. In practice, real-world systems take time to reach their final state. Indeed, a very sudden change in output may, in some cases, be undesirable. Furthermore, friction and inertia are present in many systems.

Consider the case of the speed control system that we met earlier. The mass of the load will effectively limit the acceleration of the

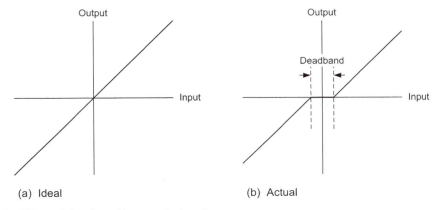

(a) Ideal   (b) Actual

Figure 6.133 Effect of deadband in a control system.

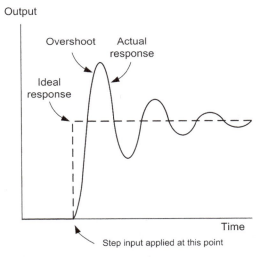

Figure 6.134 Variation of output with time for a control system.

motor speed when the set point is increased. Furthermore, as the output speed reaches the desired value, the inertia present will keep the speed increasing despite the reduction in voltage applied to the motor. Thus, the output shaft speed will overshoot the desired value before eventually falling back to the required value.

Increasing the gain present in the system will have the effect of increasing the acceleration but this, in turn, will also produce a correspondingly greater value of overshoot. Conversely, decreasing the gain will reduce the overshoot but at the expense of slowing down the response.

Finally, the term *deadband* refers to the inability of a control system to respond to a small change in the input (in other words, the input changes but the output does not). Deadband is illustrated by the ideal and actual system response shown in Figure 6.133(a) and (b). Deadband can be reduced by increasing the gain present within the system but, as we said earlier, this may have other undesirable effects!

A comparison between the ideal response and the actual response of a control system with time is shown in Figure 6.134. The ideal response consists of a step function (a sudden change) whilst the actual response builds up slowly and shows a certain amount of *overshoot*.

The response of a control system generally has two basic components: an *exponential growth curve* and a *damped oscillation* (see Figure 6.135). In some extreme cases the oscillation which occurs when the output value cycles continuously above and below the required value may be continuous. This is referred to as *hunting* (see Figure 6.136). The oscillatory component can be reduced (or eliminated) by artificially slowing down the response of the system. This is known as *damping*. The optimum value of damping is that which *just* prevents overshoot. When a system is *underdamped*, some overshoot is still present. Conversely, an *overdamped* system may take a significantly greater time to respond to a sudden change in input. The response of underdamped and overdamped control systems is shown in Figure 6.137.

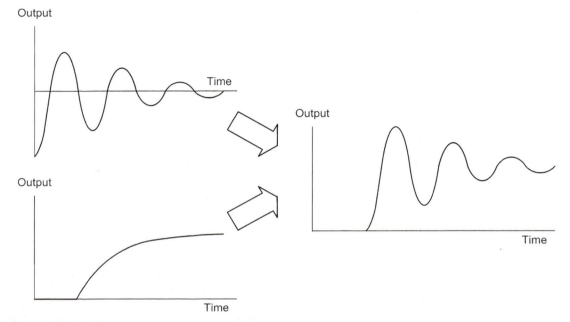

Figure 6.135   The two components present in the output of a typical control system.

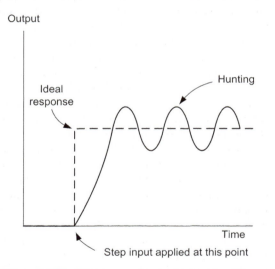

Figure 6.136   Response of a control system showing hunting.

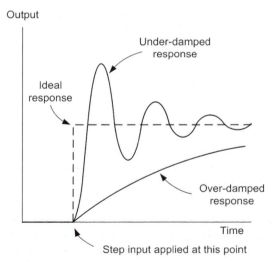

Figure 6.137   Effect of underdamping and overdamping on the response of a control system.

Test your understanding 6.6

1. Describe the operation of a servomechanism.
2. Explain the essential differences between analogue and digital control systems.
3. Identify a transducer for use in the following applications:

(a) producing a voltage that is proportional to the angle of a shaft;
(b) producing a current that depends on the incident light intensity;
(c) interrupting a current when the level of liquid in a tank exceeds a certain value;
(d) generating heat when an electric current is applied.

4. Classify the transducers in Question 3 as either input or output transducers.

5. Classify the transducers in Question 3 as either digital or analogue transducers.

6. Briefly explain the principle of the differential transformer.

7. Sketch the circuit symbol for a synchro-transmitter. Label each of the connections.

8. Which two leads of a synchro-receiver need to be reversed in order to reverse the direction of rotation?

9. Explain why (a) deadband and (b) hunting is undesirable in a control system.

10. Explain the effect of (a) underdamping and (b) overdamping in relation to a control system. Illustrate your answer with appropriate graphs.

6.5 Multiple choice questions

The set out example questions follow the sections of Module 4 in the Part 66 syllabus. Note that the following questions have been separated by level, where appropriate. Please remember that *ALL these questions must be attempted without the use of a calculator* and that the pass mark for all JAR 66 multiple choice examinations is 75%!

Semiconductors

1. Which of the following materials are semiconductors: [B2]
   (a) aluminium and copper
   (b) germanium and silicon
   (c) aluminium and zinc

2. The connections on a diode are labelled: [B1, B2]
   (a) anode and cathode
   (b) collector and emitter
   (c) source and drain

3. The stripe on a plastic encapsulated diode usually indicates the: [B1, B2]
   (a) anode connection
   (b) cathode connection
   (c) earth or ground connection

4. The direction of conventional current flow in a diode is from: [B1, B2]
   (a) anode to cathode
   (b) cathode to anode
   (c) emitter to collector

5. A diode will conduct when the: [B1, B2]
   (a) anode is more positive than the cathode
   (b) cathode is more positive than the anode
   (c) collector is more positive than the emitter

6. An ideal diode would have: [B1, B2]
   (a) zero forward resistance and infinite reverse resistance
   (b) infinite forward resistance and zero reverse resistance
   (c) zero forward resistance and zero reverse resistance

7. The region inside a diode where no free charge carriers exist is known as the: [B2]
   (a) conduction layer
   (b) depletion layer
   (c) insulation layer

8. When a diode is forward biased it exhibits: [B1, B2]
   (a) zero resistance
   (b) a very low resistance
   (c) a very high resistance

9. When a diode is reverse biased it exhibits: [B1, B2]
   (a) zero resistance
   (b) a very low resistance
   (c) a very high resistance

10. Which one of the following gives the approximate forward voltage drop for a silicon diode? [B2]
    (a) 0.2 V
    (b) 0.6 V
    (c) 1.2 V

11. Which one of the following gives the approximate forward voltage drop for a germanium diode? [B2]
    (a) 0.2 V
    (b) 0.6 V
    (c) 1.2 V

12. Which one of the following gives a typical value of forward current for a small-signal silicon signal diode? [B2]
    (a) 10 mA
    (b) 1 A
    (c) 10 A

13. Which one of the following is a typical maximum reverse voltage rating for a small-signal silicon signal diode? **[B2]**
   (a) 0.6 V
   (b) 5 V
   [(c)] 50 V

14. Which one of the following is a typical maximum forward current rating for a silicon rectifier? **[B2]**
   (a) 10 mA
   (b) 100 mA
   [(c)] 3 A

15. A diode has the following specifications:
   • Forward voltage: 0.2 V at 1 mA forward current
   • Maximum forward current: 25 mA
   • Maximum reverse voltage: 50 V

   A typical application for this diode is a: **[B2]**
   (a) rectifier in a power supply
   (b) voltage regulator reference
   [(c)] signal detector in a radio receiver

16. A diode has the following specifications:
   • Forward voltage: 0.7 V at 1 A forward current
   • Maximum forward current: 5 A
   • Maximum reverse voltage: 200 V

   A typical application for this diode is a: **[B2]**
   [(a)] rectifier in a power supply
   (b) voltage regulator reference
   (c) signal detector in a radio receiver

17. The device shown in Figure 6.138 is a: **[B1, B2]**

Figure 6.138

   (a) rectifier diode
   (b) light emitting diode
   [(c)] silicon controlled rectifier

18. The device shown in Figure 6.139 is a: **[B2]**

Figure 6.139

   (a) signal diode
   (b) light emitting diode
   [(c)] varactor diode

19. The device shown in Figure 6.140 is a: **[B2]**

Figure 6.140

   (a) signal diode
   (b) light emitting diode
   [(c)] Zener diode

20. The alternating current input to the bridge rectifier shown in Figure 6.141 should be connected at terminals: **[B2]**

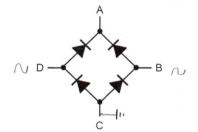

Figure 6.141

   (a) A and B
   (b) A and C
   [(c)] B and D

21. The connections to a silicon controller rectifier are labelled: [B1, B2]
    (a) emitter, base and collector
    (b) anode, cathode and gate
    (c) source, drain and gate

22. A typical application for a rectifier diode is: [B1, B2]
    (a) detecting signals in a radio receiver
    (b) converting alternating current to direct current in a power supply
    (c) switching current in an alternating current power controller

23. A typical application for a silicon controlled rectifier is: [B1, B2]
    (a) detecting signals in a radio receiver
    (b) converting alternating current to direct current in a power supply
    (c) switching current in an alternating current power controller

24. A typical application for a varactor diode is: [B2]
    (a) detecting signals in a radio receiver
    (b) converting alternating current to direct current in a power supply
    (c) varying the frequency of a tuned circuit

25. A typical application for a Zener diode is: [B2]
    (a) regulating a voltage supply
    (b) controlling the current in a load
    (c) acting as a variable capacitance in a tuned circuit

26. The connections to a Zener diode are labelled: [B1, B2]
    (a) source and drain
    (b) anode and cathode
    (c) collector and emitter

27. When a reverse voltage of 4.7 V is applied to a Zener diode, a current of 25 mA flows through it. When 4.8 V is applied to the diode, the current flowing is likely to: [B2]
    (a) fall slightly
    (b) remain at 25 mA
    (c) increase slightly

28. When a reverse voltage of 6.2 V is applied to a Zener diode, a current of 25 mA flows through it. When a reverse voltage of 3.1 V is applied to the diode, the current flowing is likely to: [B2]
    (a) be negligible
    (b) increase slightly
    (c) fall to about 12.5 mA

29. Which one of the following types of diode emits visible light when current flows through it?   [B1, B2]
    (a) A light emitting diode
    (b) A photodiode
    (c) A zener diode

30. Which one of the following gives the typical forward current for a light emitting diode?   [B2]
    (a) 2 mA
    (b) 20 mA
    (c) 200 mA

31. The forward voltage drop for a light emitting diode is approximately: [B2]
    (a) 0.2 V
    (b) 0.6 V
    (c) 2 V

32. The output of the circuit shown in Figure 6.142 will consist of a: [B2]

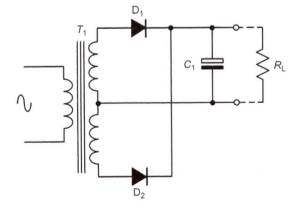

Figure 6.142

    (a) sine-wave voltage
    (b) square-wave voltage
    (c) steady direct current voltage

33. The function of C_1 in the circuit shown in Figure 6.143 is to: **[B2]**

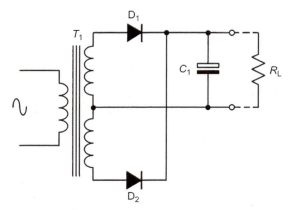

Figure 6.143

(a) form a load with R_1
(b) act as a reservoir
(c) block direct current at the output

34. The circuit provided in Figure 6.144 shows a: **[B2]**

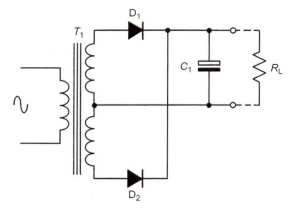

Figure 6.144

(a) full-wave power supply
(b) half-wave power supply
(c) regulated power supply

35. The device shown in Figure 6.145 is: **[B1, B2]**
(a) an NPN bipolar junction transistor
(b) a PNP bipolar junction transistor
(c) a junction gate field effect transistor

Figure 6.145

36. The connections to a JFET are labelled: **[B1, B2]**
(a) collector, base and emitter
(b) anode, cathode and gate
(c) source, gate and drain

37. In normal operation of a bipolar junction transistor the:
(a) base–emitter junction is forward biased and the collector–base junction is reverse biased
(b) base–emitter junction is reverse biased and the collector–base junction is forward biased
(c) both junctions are forward biased

38. Which one of the following statements is true?
(a) The base current for a transistor is very much smaller than the collector current
(b) The base current for a transistor is just slightly less than the emitter current
(c) The base current for a transistor is just slightly greater than the emitter current

39. Corresponding base and collector currents for a transistor are $I_B = 1\,\text{mA}$ and $I_C = 50\,\text{mA}$. Which one of the following gives the value of common-emitter current gain? **[B2]**
(a) 0.02
(b) 49
(c) 50

40. The corresponding base and collector currents for a transistor are $I_B = 1\,\text{mA}$ and $I_C = 50\,\text{mA}$. Which one of the following gives the value of emitter current? **[B2]**
(a) 49 mA
(b) 50 mA
(c) 51 mA

41. If the emitter current of a transistor is 0.5 A
and the collector current is 0.45 A, which
one of the following gives the base current?
[B2]
    (a) 0.05 A
    (b) 0.4 A
    (c) 0.95 A

42. The common-emitter current gain of a transistor is found from the ratio of: [B2]
    (a) collector current to base current
    (b) collector current to emitter current
    (c) emitter current to base current

43. The input resistance of a transistor in
common-emitter mode is found from the
ratio of: [B2]
    (a) collector–base voltage to base current
    (b) base–emitter voltage to base current
    (c) collector–emitter voltage to emitter
        current

44. The voltage gain produced by the circuit
shown in Figure 6.146 will depend on the:
[B2]

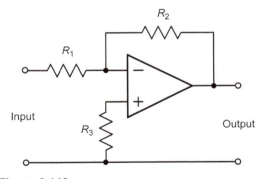

Figure 6.146

    (a) ratio of R_1 to R_2
    (b) ratio of R_1 to R_3
    (c) ratio of R_2 to R_1

45. The terminal marked 'X' in Figure 6.147
is the: [B1, B2]
    (a) inverting input
    (b) non-inverting input
    (c) positive supply connection

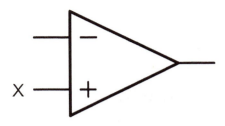

Figure 6.147

Printed circuits

46. The tracks on a printed circuit board are
made from: [B1, B2]
    (a) aluminium
    (b) copper
    (c) steel

47. The width of the track on a printed circuit
board determines the: [B1, B2]
    (a) voltage that can be carried
    (b) current that can be carried
    (c) speed at which information can be
        carried

48. Printed circuit board edge connectors are
frequently gold plated. This is because it:
[B1, B2]
    (a) increases the contact resistance
    (b) improves contact reliability
    (c) reduces contact friction

49. Which one of the following is NOT a suitable material for manufacturing a printed
circuit board: [B1, B2]
    (a) glass fibre
    (b) synthetic resin bonded paper
    (c) polystyrene

50. A surface mounted device is attached to a
printed circuit board using: [B1, B2]
    (a) solder pads
    (b) connecting pins
    (c) a printed circuit board connector

51. A cable is attached to a printed circuit board
using an indirect printed circuit board connector. This connecting arrangement uses a:
[B1, B2]
    (a) header to terminate the cable

(b) series of soldered connections at the end of the cable

(c) number of individual crimped connections

Servomechanisms

52. A servo-based position control system is an example of: [B1, B2]
    (a) an open-loop system
    (b) an automatic closed-loop system
    (c) a system that exploits positive feedback

53. The output from a control system is usually referred to as the: [B1, B2]
    (a) set point
    (b) error signal
    (c) controlled variable

54. The input to a control system is usually referred to as the: [B1, B2]
    (a) set point
    (b) error signal
    (c) controlled variable

55. In a control system the difference between the desired value and the actual value of the output is referred to as the: [B1, B2]
    (a) error signal
    (b) demand signal
    (c) feedback signal

56. A signal that varies continuously from one level to another is called: [B1, B2]
    (a) an error signal
    (b) a digital signal
    (c) an analogue signal

57. Digital signals vary: [B2]
    (a) in discrete steps
    (b) continuously between set levels
    (c) slowly from one level to another level

58. In a digital control system values are represented by an 8-bit code. How many different values are possible in this system? [B2]
    (a) 8
    (b) 80
    (c) 256

59. A digital control system is required to have a resolution of 2%. With how many bits should it operate? [B2]
    (a) 4
    (b) 5
    (c) 6

60. Which one of the following is an input transducer? [B1, B2]
    (a) An actuator
    (b) A motor
    (c) A potentiometer

61. Which one of the following is an output transducer? [B1, B2]
    (a) A heater
    (b) A photodiode
    (c) A potentiometer

62. A target modifies the electric field generated by a sensor. This is the principle of the: [B1, B2]
    (a) optical proximity sensor
    (b) inductive proximity sensor
    (c) capacitive proximity sensor

63. A foil element with polyester backing is resin bonded to mechanical component. This is the principle of the: [B1, B2]
    (a) resistive strain gauge
    (b) inductive strain gauge
    (c) capacitive strain gauge

64. Which one of the following sensor produces a digital output? [B2]
    (a) Magnetic reed switch
    (b) Piezoelectric strain gauge
    (c) Light-dependent resistor

65. Which one of the following is an active sensor? [B1, B2]
    (a) Tachogenerator
    (b) Resistive strain gauge
    (c) Light-dependent resistor

66. Minimum flux linkage will occur when two coils are aligned at a relative angle of: [B1, B2]
    (a) $0°$
    (b) $45°$
    (c) $90°$

67. A differential transformer is made from: [B2]
    (a) U- and I-laminations
    (b) E- and I-laminations
    (c) E- and H-laminations

68. The reference voltage in a differential transformer is applied to a winding on: [B2]
    (a) the center limb of the E-lamination
    (b) all three limbs of the E-laminations
    (c) one of the outer limbs of the E-lamination

69. The laminations of a differential transformer are usually made from: [B2]
    (a) ceramic material
    (b) low permeability steel
    (c) high permeability steel

70. Identical root mean square voltages of 26 V are measured across each of the secondary windings of a differential transformer. Which one of the following gives the value of output voltage produced by the transformer? [B2]
    (a) 0 V
    (b) 26 V
    (c) 52 V

71. A synchro-transmitter is sometimes also referred to as a: [B1, B2]
    (a) linear transformer
    (b) variable transformer
    (c) differential transformer

72. How many stator connections are there in a synchro-transmitter? [B1, B2]
    (a) 1
    (b) 2
    (c) 3

73. The alternating current supply to a synchro-transmitter is connected to the connections marked: [B1, B2]
    (a) R_1 and R_2
    (b) S_1 and S_2
    (c) S_1, S_2 and S_3

74. A synchro-resolver has multiple rotor and stator windings spaced by: [B1, B2]

(a) 45°
(b) 90°
(c) 120°

75. A conventional synchro-transmitter has stator windings spaced by: [B1, B2]
    (a) 45°
    (b) 90°
    (c) 120°

76. When a synchro-transmitter and -receiver system are in correspondence the current in the stator windings will: [B1, B2]
    (a) be zero
    (b) take a maximum value
    (c) be equal to the supply current

77. Which two synchro-leads should be reversed in order to reverse the direction of a synchro-receiver relative to a synchro-transmitter? [B2]
    (a) R_1 and R_2
    (b) S_1 and S_2
    (c) S_1 and S_3

78. An open-loop control system is one in which: [B1, B2]
    (a) no feedback is applied
    (b) positive feedback is applied
    (c) negative feedback is applied

79. A closed-loop control system is one in which: [B1, B2]
    (a) no feedback is applied
    (b) positive feedback is applied
    (c) negative feedback is applied

80. In a closed-loop system the error signal can be produced by: [B1, B2]
    (a) an amplifier
    (b) a comparator
    (c) a tachogenerator

81. The range of outputs close to the zero point that a control system is unable to respond to is referred to as: [B2]
    (a) hunting
    (b) deadband
    (c) overshoot

82. Overshoot in a control system can be reduced by: [B2]
    (a) increasing the gain
    (b) reducing the damping
    (c) increasing the damping

83. The output of a control system cycles continuously above and below the required value. This characteristic is known as: [B2]
    (a) hunting
    (b) deadband
    (c) overshoot

Fundamentals of aerodynamics

Chapter 7 Basic aerodynamics

7.1 Introduction

This chapter serves as an introduction to the study of aerodynamics. It covers in full all requisite knowledge needed for the successful study and completion of *Basic aerodynamics* as laid down in *Module 8* of the *Joint Aviation Requirements* (*JAR*) *66 syllabus*. The study of elementary flight theory in this module forms an essential part of the knowledge needed for all potential practicing aircraft engineers, no matter what their trade specialization. In particular, there is a need for engineers to understand how aircraft produce lift and how they are controlled and stabilized for flight. This knowledge will then assist engineers with their future understanding of aircraft control systems and the importance of the design features that are needed to stabilize aircraft during all phases of flight.

In addition, *recognition of the effects of careless actions on aircraft aerodynamic performance and the need to care for surface finish and streamlining features* will also be appreciated through the knowledge gained from your study of this module.

Our study of aerodynamics starts with a brief reminder of the important topics covered previously when you considered the atmosphere and atmospheric physics. The nature and purpose of the International Standard Atmosphere (ISA) will be looked at again. We will also be drawing on your knowledge of fluids in motion for the *effects of airflow over aircraft* and in order to demonstrate the underlying physical principles that account for the creation of *aircraft lift and drag*. We look in detail at the lift generated by *aerofoil sections* and the methods adopted to maximize the lift created by such lift producing surfaces. Aircraft flight forces (lift, drag, thrust and weight) are then studied in some detail including their interrelationship during steady state flight and during manoeuvres. The effects and limitations of flight loads on the aircraft structure will also be considered, as well as the effects of aerofoil contamination, by such things as ice accretion.

Once the concepts of aircraft flight forces and performance have been covered, we will consider the ways in which aircraft are controlled and stabilized and the relationship between aircraft control and stability. In addition to the primary flight controls, we will briefly look at secondary devices, including the methods used to augment lift and dump lift when necessary. Only manual controls will be considered at this stage, although the need for powered flight control devices and systems will be mentioned.

Note: the study of flight control and basic control devices *does not form a part of Module 8*, but is covered here for the sake of completeness. In particular, it will be found especially helpful when we consider *flight stability*.

Full coverage of aircraft flight control, control devices and high speed flight theory will be found in the third book in the series *Aircraft Aerodynamics, Structural Maintenance and Repair*, which is primarily aimed at those pursuing a *mechanical pathway*. For those individuals pursuing an *avionic pathway*, the *appropriate aircraft aerodynamics and control* will be covered in fifth book in the series on *Avionic Systems*, which will essentially be designed to cover Part 66 Module 13.

7.2 A review of atmospheric physics

You should have already covered in some detail the nature of the atmosphere in your study of Physics. In particular, you should look back (pages 247–251) and remind yourself of the gas laws, the temperature, pressure and density relationships, the variation of the air with altitude and the need for the ISA. For your convenience, a few importance definitions and key facts are summarized below.

7.2.1 Temperature measurement

You will know that the temperature of a body is the measurement of the internal energy of a body. Thus, if heat energy is added to a body the subsequent increase in the molecular vibration of the molecules that goes to make up the substance increases, and usually causes an expansion. It is this measure of internal molecular energy that we call *temperature*.

Then, the practical measurement of temperature is the comparison of temperature differences. To assist us in measuring temperature we need to identify at least two fixed points from which comparisons can be made. We use the boiling point (100°C) and freezing point (0°C) of water for this purpose. These two fixed points are referred to as *the lower fixed point of ice and the upper fixed point of steam*.

The two most common temperature scales, in general use are *degree celsius* (°C) and *degree fahrenheit* (°F). Remembering also that when measuring thermodynamic temperatures, we use the *kelvin* (K) scale.

7.2.2 Pressure measurement

Atmospheric pressure can be measured using a mercury barometer, where:

Pressure due to height = $\rho g h$ N/m^2

Density of mercury (Hg) = 13,600 kg/m^3

Thus changes in atmospheric pressure will register as changes in the height of the column of mercury. In the ISA, *standard atmospheric pressure*, i.e. the pressure of one standard atmosphere may be represented by:

1 standard atmosphere

= 760 mmHg (millimeter of mercury)

= 101,325 Pa (pascal)

= 101,325 N/m^2 = 1013.25 mb(millibar) or

= 29.92 in. Hg (inches of mercury)

= 14.69 psi (pounds per square inch)

Remembering that:

$$1 \, \text{bar} = 100,000 \, \text{Pa} = 100,000 \, \text{N/m}^2$$

$$1 \, \text{mbar} = 100 \, \text{Pa} = 100 \, \text{N/m}^2$$

$$= 1 \, \text{hPa (hecto-pascal), and}$$

$$1 \, \text{Pa} = 1 \, \text{N/m}^2$$

With respect to the measurement of pressure you should also note that pressure measured above atmospheric pressure and which use atmospheric pressure as its zero datum is referred to as *gauge pressure*, whereas pressure measured using an absolute vacuum as its zero datum is known as *absolute pressure*. Then:

$$\frac{Absolute}{pressure} = \frac{gauge}{pressure} + \frac{atmospheric}{pressure}$$

7.2.3 Humidity measurement

Humidity is the amount of water vapour that is present in the particular sample of air being measured. *Relative humidity* refers to the *degree of saturation* of the air, and may be defined as:

The ratio of the amount of water vapour present in the sample of air, to the amount that is needed to saturate it at that temperature.

The humidity of the air may be measured using a psychrometer or wet and dry bulb hygrometer. In their simplest form a psychrometer consists of two similar thermometers placed side by side, one being kept wet by use of a wick, leading from it into a container of distilled water. If the air surrounding the psychrometer is saturated, no evaporation can take place, and the two thermometers indicate the same temperature reading. Thus the drier the air, the greater the temperature difference between the wet and dry bulbs. Then using graphs with the wet and dry temperatures known, the dew point and relative humidity can be determined.

The *dew point* temperature is that temperature when condensation first starts to appear from a sample of gas. Thus, another way of measuring the amount of water vapour in a gas is to pass the gas over a surface, where the temperature is gradually lowered, until the moisture from the gas starts to condense on the surface. The "dew point" temperature is then read off. The relative humidity is then obtained by comparing the dew point temperature of the gas sample with its saturation temperature.

7.2.4 Density ratio and airspeed

The relationship between the density at sea level in the ISA and the density at altitude, the *density ratio* σ has a special significance. It is used in the calibration of pitot–static instruments to convert *equivalent airspeed* (EAS) to *true airspeed* (TAS) or vice versa, and is defined as:

$$\sigma = \frac{\text{density at altitude } (\rho_a)}{\text{density at sea level } (\rho_{sl}) \text{ in the ISA}}$$

Before we look at the relationship connecting EAS and TAS, it is first, worth defining three important airspeeds:

- *Indicated airspeed* (IAS) is that speed shown by a simple airspeed indicator (ASI).
- *Equivalent airspeed* (EAS) is that speed that would be shown by an *error-free* ASI.
- *True airspeed* (TAS) is the actual speed of an aircraft relative to the air.

Much more will be said about aircraft pitot–static instruments, in a later volume in the series, for now we will define the simple relationship connecting EAS and TAS:

$$\textbf{EAS} = \textbf{TAS}\sqrt{\sigma} \quad \text{or}$$

$$V_E = V_T\sqrt{\sigma} \quad \text{or} \quad V_E = V_T\sqrt{\frac{\rho_a}{\rho_{sl}}}$$

Example 7.1

Given that an error-free ASI registers an EAS of 220 m/s at an altitude where the density is 0.885 kg/m³, determine the TAS of the aircraft at this altitude assuming that the aircraft flies in still air.

Remembering that sea level density in the ISA is 1.2256 kg/m³, then the density ratio:

$$\sigma = \frac{0.885}{1.2256} = 0.7221$$

Therefore from the relationship given above:

$$V_T = \frac{V_E}{\sqrt{\sigma}}$$

we find that

$$V_T = \frac{220}{\sqrt{0.7221}} = \frac{220}{0.8498} = 258.9\,\text{m/s}$$

You should also note that the density ratio $\sigma = 1.0$ at sea level, therefore, from the above relationship $V_E = V_T$.

7.2.5 Changes in the air with increasing altitude

- Static pressure decreases with altitude, but not in a linear manner.
- Air density which is proportional to pressure also decreases, but at a different rate, to that of pressure.
- Temperature also decreases with altitude, the rate of decrease is linear, in the troposphere and may be found from the relationship: $T_h = T_0 - Lh$, in the ISA.

7.2.6 The International Civil Aviation Organization ISA

As you already know, the International Civil Aviation Organization (ICAO) standard atmosphere as defined in ICAO document 7488/2 lays down an arbitrary set of conditions, accepted by the international community, as a basis for comparison of aircraft and engine performance parameters and for the calibration of aircraft instruments.

The conditions adopted have been based on those observed in a temperate climate at a latitude of 40° North up to an altitude of 105,000 ft.

The principle conditions assumed in the ISA are summarized below for your convenience.

- Temperature   288.15 K or 15.15°C
- Pressure   1013.25 mb or 101,325 N/m²
- Density   1.2256 kg/m²
- Speed of sound   340.3 m/s
- Gravitational acceleration   9.80665 m/s²
- Dynamic viscosity 1.789×10^{-5} N s/m²

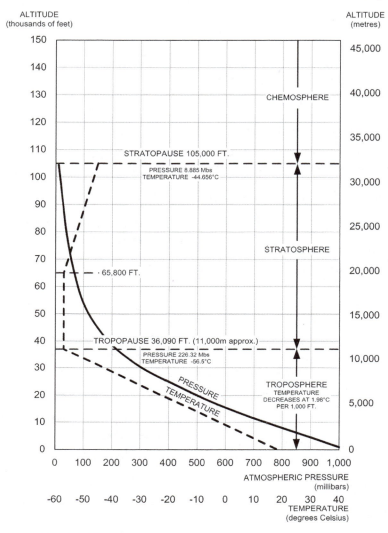

Figure 7.1 The ICAO standard atmosphere.

- Temperature lapse rate   6.5 K/km or 6.5°C/km
- Tropopause   11,000 m, 56.5°C or 216.5 K

Note the following *Imperial equivalents*, which are often quoted:

- Pressure   14.69 lb/in.²
- Speed of sound   1120 ft/s
- Temperature lapse rate   1.98°C per 1000 ft
- Tropopause   36,090 ft
- Stratopause   105,000 ft.

The changes in ICAO standard atmosphere with altitude are illustrated in Figure 7.1. Note that the temperature in the upper

stratosphere starts to rise again after 65,000 ft at a rate of 0.303°C per 1000 ft or 0.994°C per 1000 m. At a height of 105,000 ft or approximately 32,000 m, the chemosphere is deemed to begin. The *chemosphere* is the collective name for *mesosphere*, *thermosphere* and *exosphere*, which was previously identified, in atmospheric physics.

Example 7.2

Find the temperature at an altitude of 6400 m and 18,000 ft in the ISA.

Then, using the formula $T_h = T_0 - Lh$ with the first height in SI units, we get:

$$T_{6400} = 288.15 - (6.5)(6.4)$$
$$= 288.15 - 41.6 = 246.55\,\text{K}$$

Note that in the SI system the temperature lapse rate is 6.5 K/km. Also note that the 0.15 is often dropped when performing temperature calculations in Kelvin.

Similarly, for the height in preferred English Engineering units we get:

$$T_{18,000} = 15 - (1.98)(18,000)$$
$$= 15 - 35,640$$
$$= -35.6°\text{C}$$
$$\text{or} \quad \approx 237.5\,\text{K}$$

If you try to convert these values to the SI system first and then carry out the calculation, you will find a small discrepancy between the temperatures. This is due to rounding errors in the conversion process.

7.2.7 Speed of sound and Mach number

Although high speed flight is not a part of the basic aerodynamics module. The speed of sound and Mach number occur regularly in conversation about aircraft flight, therefore it is felt that they are worthy of mention at this stage.

The speed at which sound waves travel in a medium is dependent on the temperature and the bulk modulus (K) of the medium concerned i.e. the temperature and density of the material concerned. The denser the material the faster is the speed of the sound waves. For air treated as a perfect gas, *the speed of sound* (a) is given by:

$$a = \sqrt{\frac{\gamma p}{\rho}} = \sqrt{\gamma RT} = \sqrt{\frac{K}{\rho}}$$

You should recognize γ as the ratio of the specific heats from your work on thermodynamics for air at standard temperature and pressure $\gamma = 1.4$. Also remembering that R is the characteristic gas constant, which for air is 287 J/kgK.

Then as an approximation, the speed of sound may be expressed as:

$$a = \sqrt{\gamma RT} = \sqrt{(287)(1.4)T} = 20.05\sqrt{T}$$

For an aircraft in flight, the *Mach number* (M) named after the Austrian physicist Ernst Mach, may be defined as:

$$M = \frac{\text{the aircraft flight speed}}{\left(\begin{array}{c}\text{the local speed of sound}\\\text{in the surrounding atmosphere}\end{array}\right)}$$

Note that the flight speed must be the aircraft's TAS.

Example 7.3

An aircraft is flying at a TAS of 240 m/s at an altitude where the temperature is 230 K. What is the speed of sound at this altitude and what is the aircraft Mach number?

The local speed of sound is given by: $a = \sqrt{\gamma RT}$ or $a = 20.05\sqrt{230}$, then using square root tables and multiplying, $a = (20.05)(15.17) = 304$ m/s. Noting from the ISA values that the speed of sound at sea level is 340.3 m/s. You can see that the speed of sound *decreases* with *increase* in altitude.

Now using our relationship for Mach number, then

$$M = \frac{240}{304} = 0.789M$$

We next look at the effects of airflow over varying bodies and, in particular, we will consider airflow over *aerofoil* sections. Before you study the next section, you are strongly advised to review the *Fluids in motion* section of the Chapter 4, Physics (starting on page 251). You should make sure you understand Bernoulli's equations and the Venturi principle, you learnt earlier.

7.3 Elementary aerodynamics

7.3.1 Static and dynamic pressure

You have already met static and dynamic pressure in your study of fluids in the previous chapter. Look back now at the energy and pressure versions of the Bernoulli's equation in page 253.

I hope you can see that a fluid in steady motion has both static pressure energy and dynamic pressure energy (kinetic energy) due to the motion. All Bernoulli's equation showed was that for an ideal fluid.

The total energy in a steady streamline flow remains constant.

Static pressure energy	+	dynamic (kinetic) energy	=	constant total energy

or in symbols:

$$p + \tfrac{1}{2}\rho v^2 = C$$

In particular, you should note that with respect to aerodynamics, the *dynamic pressure* is dependent on the *density* of the air (treated as an ideal fluid) and the *velocity* of the air. Thus, with increase in altitude, there is a drop in density and the dynamic pressure acting on the aircraft as a result of the airflow, will also drop with increase in altitude. The *static pressure* of the air also drops with increase in altitude.

You have already met an application of the Bernoulli's equation when we considered the Venturi tube in our earlier studies of fluids in motion; where from the above Bernoulli's equation, use was made of the fact that to maintain equality:

An increase in velocity will mean a decrease in static pressure, or alternatively, *a decrease in velocity will mean an increase in static pressure.*

The above fact is key to the way aircraft wing sections create lift, as you will see later.

7.3.2 Subsonic airflow

Flow over a flat plate

When a body is moved through the air, or any fluid that has viscosity, such as water, there is a resistance produced which tends to oppose the body. For example, if you are driving in an open top car, there is a resistance from the air acting in the opposite direction to the motion of the car. This *air resistance* can be felt on your face or hands as you travel. In the aeronautical world, this air resistance is known as *drag*. It is undesirable for obvious reasons. For example, aircraft engine power is required to overcome this air resistance and unwanted heat is generated by friction as the air flows over the aircraft hull during flight.

We consider the effect of air resistance by studying the behaviour of airflow over a flat plate. If a flat plate is placed edge on to the relative airflow (Figure 7.2), then there is little or no alteration to the smooth passage of air over it. On the other hand, if the plate is offered into the airflow at some angle of inclination to it *angle of attack* (AOA), it will experience a *reaction* that tends to both lift it and drag it back. This is the same effect that you can feel on your hand when placed into the airflow as you are travelling, e.g. in the open topped car mentioned earlier. The amount of reaction depends upon the speed and AOA between the flat plate and relative airflow.

As can be seen in Figure 7.2, when the flat plate is inclined at some AOA to the relative airflow, the streamlines are disturbed. An *upwash* is created at the front edge of the plate causing the air to flow through a more constricted area, in a similar manner to flow through the throat of a Venturi meter. The net result is that as the airflows through this restricted area, it speeds up. This in turn causes a drop in static pressure above the plate (Bernoulli) when compared with the static pressure beneath it resulting in a net *upward reaction*. After passing the plate, there is a resulting *downwash* of the air stream.

The total reaction on the plate caused by it disturbing the relative airflow has two vector components as shown in Figure 7.3. One at right angles to the relative airflow known as *lift* and the other parallel to the relative airflow, opposing the motion, known as *drag*.

The above drag force is the same as that mentioned earlier, which caused a resistance to the flow of the air stream, over your hand.

Streamline flow, laminar flow and turbulent flow

Streamline flow, sometimes referred to as *viscous flow,* is flow in which the particles of the

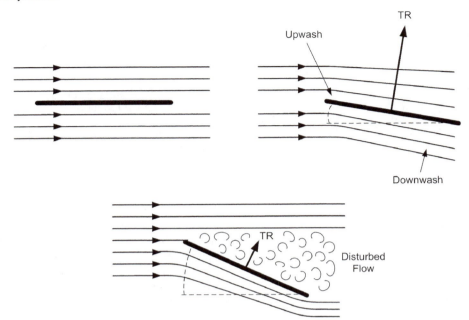

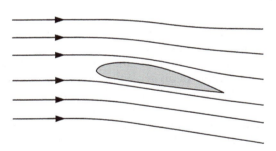

Figure 7.4 Streamline flow.

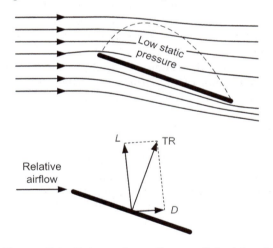

Figure 7.2 Airflow over a flat plate.

Figure 7.3 Nature of reaction on flat plate, to relative airflow.

fluid move in an orderly manner and retain the same relative positions in successive cross sections. In other words, a flow that maintains the shape of the body over which it is flowing. This type of flow is illustrated in Figure 7.4, where it can be seen that the successive cross sections are represented by lines that run parallel to one another hugging the shape of the body around which the fluid is flowing.

Laminar flow may be described as the smooth parallel layers of air flowing over the surface of a body in motion, i.e. streamline flow.

Turbulent flow is flow in which the particles of fluid move in a disorderly manner occupying different relative positions in successive cross sections (Figure 7.5).

This motion results in the airflow thickening considerably and breaking-up.

7.3.3 The aerofoil

In its simplest sense an *aerofoil section* may be defined as *that profile designed to obtain a desirable reaction from the air through which it moves.* In other words, an aerofoil is able to

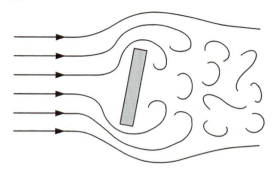

Figure 7.5 Turbulent flow.

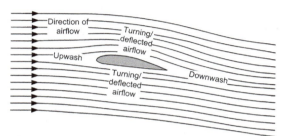

Figure 7.6 Airflow over an aerofoil section.

convert air resistance into a useful force that produces lift for flight.

The cross-section of an aircraft wing is a good example of an aerofoil section, where the top surface usually (but not always) has greater curvature than the bottom surface.

The airflow travelling over the top and bottom surfaces of the wing is split at the leading edge of the aerofoil and is deflected or **turned** to follow the aerofoil contours, due to the inner molecules of the air adhering to the surfaces of the aerofoil. This turning (change in direction) causes a change in (magnitude and/or direction) the local velocity of the air with respect to time. On the top surface of the aerofoil the turning causes the air to accelerate (increase in velocity with time), then from $F = ma$, this acceleration will produce a *force*, that acts perpendicular to the surface of the aerofoil this force is known as *lift*. A lift force is also generated on the underside of the aerofoil section, but the resulting change in direction of the airflow, is different from that created by the top surface (dependent on its shape) hence generating a force differential that acts upwards on the wing. Thus, *lift is a force generated by turning the moving airflow*

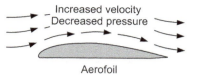

Figure 7.7 Top surface of aerofoil acting as bottom-half of a Venturi tube.

(changing its direction) over a body that is immersed in it.

From the above argument any part of a solid body (in this case an aerofoil) can deflect or turn the airflow. The air upstream of the aerofoil when it impacts with the leading edge produces a deflection upwards the *upwash* (Figure 7.6), while at the trailing edge the air is deflected downwards, *downwash*, both these deflections play their part in the generation of lift from the aerofoil and the trailing edge downwash cannot be ignored, when considering the magnitude and direction of the lift force.

A result of the changing direction of the airflow, over the top surface of an aerofoil can be demonstrated using Bernoulli's equation and the Venturi apparatus. If we consider the top surface of the aerofoil to be the bottom surface of a Venturi tube (Figure 7.7), then the *turning or deflection* of the airflow over the cambered surface causes an increase in local velocity with respect to time and produces a lift force acting upwards perpendicular to the aerofoil surface, from Bernoulli and knowing that pressure is equal to force divided by area, then an increase in velocity must be accompanied by a decrease in pressure over the surface, that may be demonstrated experimentally using the Venturi tube apparatus.

Aerofoil terminology

We have started to talk about such terms as: camber, trailing edge and AOA without defining them fully. Set out below are a few useful terms and definitions about *airflow* and *aerofoil* sections that will be used frequently throughout the remainder of this chapter.

Camber is the term used for the *upper and lower curved surfaces of the aerofoil* section. Where the **mean camber line** is that *line drawn halfway between the upper and lower cambers*.

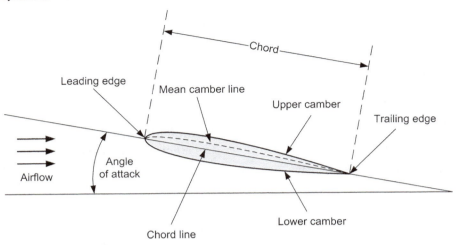

Figure 7.8   Aerofoil terminology.

Chord line is the *line joining the centers of curvatures of the leading and trailing edges.* Note that this line may fall outside the aerofoil section dependent on the amount of camber of the aerofoil being considered.

Leading and trailing edge are those *points on the center of curvature of the leading and trailing part of the aerofoil section that intersect with the chord line* as shown in Figure 7.8.

Angle of incidence (**AOI**) is *the angle between the relative airflow and the longitudinal axis of the aircraft.* It is a built-in feature of the aircraft and is a fixed "rigging angle". On conventional aircraft, the AOI is designed to minimize drag during cruise thus maximizing fuel consumption!

Angle of attack (**AOA**) is *the angle between the chord line and the relative airflow.* This will vary, dependent on the longitudinal attitude of the aircraft, with respect to the relative airflow as you will see later.

Thickness/chord ratio (**t/c**) is simply *the ratio of the maximum thickness of the aerofoil section to its chord length* normally expressed as a percentage. It is sometimes referred to as the *fineness ratio* and is a measure of the aerodynamic thickness of the aerofoil.

The aerofoil shape is also defined in terms of its t/c ratio. The aircraft designer chooses that shape which best fits the aerodynamic requirements of the aircraft.

Light aircraft and other aircraft that may fly at low velocity are likely to have a highly cambered thick aerofoil section; where the air flowing over the upper camber is forced to travel a significantly longer distance than the airflow travelling over the lower camber. This results in a large acceleration of the upper airflow significantly increasing speed and correspondingly reducing the pressure over the upper surface.

These high lift aerofoil sections may have a t/c ratio of around 15%, although the point of maximum thickness for these high lift aerofoils can be as high as 25–30%. The design will depend on whether forward speeds are of more importance compared to maximum lift. Since, it must be remembered that accompanying the large increase in lift that thick aerofoil sections bring, there is also a significant increase drag. However, thick aerofoil sections allow the use of deep spars and have other advantages, such as more room for fuel storage and for the stowage of the undercarriage assemblies.

Thin aerofoil sections are preferred on high speed aircraft that spend time flying at transonic and supersonic speeds. The reason for choosing slim wings is to reduce the time spent flying in the transonic range, where at these speeds the build up of shockwaves create stability and control problems. We need not concern ourselves here with the details of high speed flight, this will be addressed comprehensively in later books.

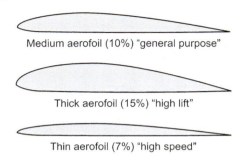

Medium aerofoil (10%) "general purpose"

Thick aerofoil (15%) "high lift"

Thin aerofoil (7%) "high speed"

Figure 7.9 t/c ratio for some common aerofoil sections.

Figure 7.10 The exceptional thin delta wing of Concorde.

However, it is worth knowing that the thinner the aerofoil section then the nearer to sonic speed an aircraft can fly before the effects of shockwave formation take effect. What limits the fineness ratio of aerofoil sections is their structural strength and rigidity, as well as providing sufficient room for fuel and the stowage of the undercarriage. A selection of aerofoil sections is shown in Figure 7.9.

Concorde has an exceptional fineness ratio (3–4%) because of its very long chord length resulting from its delta wings. It can therefore alleviate the problems of flying in the transonic range as well as providing sufficient room for fuel and the stowage of its undercarriage assemblies. In general, fineness ratios (t/c ratios) of less than 7% are unusual (Figure 7.10).

With regard to the under surface alterations in the camber have less effect. A slightly concave camber will tend to increase lift, but convex cambers give the necessary thickness to allow for the fitment of deeper and lighter spars. The convex sections are also noted for limiting the

movement of the *center of pressure* (CP). This limitation is most marked where the lower camber is identical to that of the upper camber giving a symmetrical section. Such sections have been adopted for medium and high speed main aerofoil sections and for some tail plane sections.

Aerofoil efficiency

The *efficiency* of an aerofoil is measured using the lift to drag (L/D) ratio. As you will see when we study lift and drag, this ratio varies with changes in the AOA reaching a maximum at one particular AOA. For conventional aircraft using wings as their main source of lift maximum L/D is found to be around 3° or 4°. Thus, if we set the wings at an *incidence angle* of 3° or 4°, then when the aircraft is flying straight and level in cruise, this AOI will equal the AOA at which we achieve maximum lift with minimum drag, i.e. at the *maximum efficiency* of the aerofoil. A typical lift/drag curve is shown in Figure 7.11, where the AOA for normal flight will vary from 0° to around 15° or 16° at which point the aerofoil will *stall*.

Research has shown that the most efficient aerofoil sections for general use have their maximum thickness occurring around one-third back from the leading edge of the wing. It is thus the *shape* of the aerofoil section that determines the AOA at which the wing is most efficient and the degree of this efficiency. High lift devices, such as slats, leading edge flaps and trailing edge flaps alter the shape of the aerofoil section in such a way as to increase lift. However, the penalty for this increase in lift is an increase in drag, which has the overall effect of reducing the L/D ratio.

7.3.4 Effects on airflow with changing AOA

The point on the chord line through which the resultant lift force acts is known as the **center of pressure** (CP). This was first illustrated in Figure 7.8 and for clarity, the CP through which the *resultant force* or *total reaction* acts is shown in Figure 7.12.

As the aerofoil changes its AOA, the way in which the pressure changes around the surface

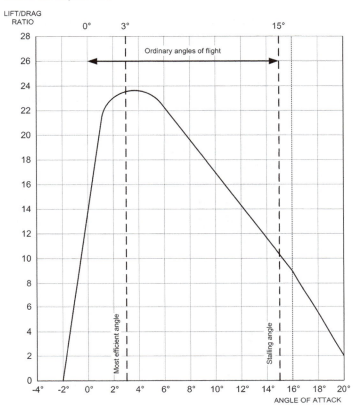

Figure 7.11 A typical lift/drag ratio curve.

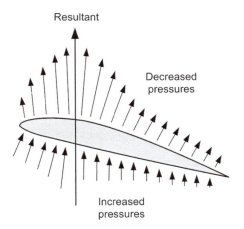

Figure 7.12 Center of pressure.

also alters. This means that the CP will move along the chord line (Figure 7.13).

For all positive AOA the CP moves forward as the aircraft attitude or pitching angle is increased, until the stall angle is reached when it suddenly moves backwards along the chord line. Note that the aircraft pitch angle should not be confused with the AOA. This is because the relative airflow will change direction in flight in relation to the pitch angle of the aircraft. This important point of recognition between pitching angle and AOA, which often causes confusion, is illustrated in Figure 7.14.

Aerofoil stall

When the AOA of the aerofoil section is increased gradually towards a positive angle, the lift component increases rapidly up to a certain point and then suddenly begins to drop off. When the AOA increases to a point of maximum lift, the *stall point* is reached, this is known as the *critical angle* or stall angle.

When the stall angle is reached, the air ceases to flow smoothly over the top surface of the aerofoil and it begins to break away (Figure 7.15) creating turbulence.

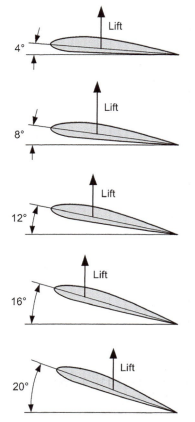

Figure 7.13   Changes in CP with changing AOA.

In fact, at the critical angle the pressure gradient is large enough to actually push a flow up the wing against the normal flow direction. This has the effect of causing a reverse flow region below the normal *boundary layer*, which is said to separate from the aerofoil surface. When the aerofoil stalls, there is a dramatic drop in lift.

7.3.5 Viscosity and the boundary layer

Viscosity

The ease with which a fluid flows is an indication of its viscosity. Cold heavy oils such as those used to lubricate large gearboxes have a

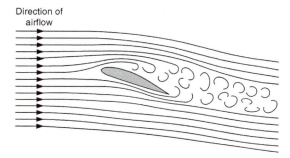

Figure 7.15   Effects on airflow over aerofoil when stall angle is reached.

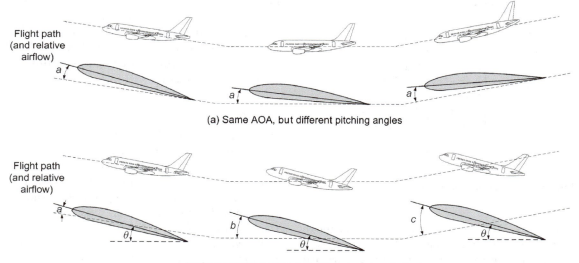

(a) Same AOA, but different pitching angles

(b) Same pitching angle, but different angles of attack

Figure 7.14   AOA and aircraft pitching angles.

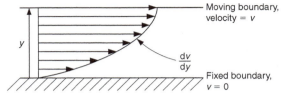

Figure 7.16 Illustration of the velocity change in the boundary layer.

high viscosity and flow very slowly, whereas petroleum spirit is extremely light and volatile and flows very easily and so has low viscosity. Air is a viscous fluid and as such offers resistance to its flow. We thus define *viscosity as the property of a fluid that offers resistance to the relative motion of the fluid molecules*. The energy losses due to friction in a fluid are dependent on the viscosity of the fluid and the distance travelled over a body by a fluid.

The *dynamic viscosity* (μ) of a fluid is in fact the *constant of proportionality* used in the relationship:

$$\tau = \mu \frac{\Delta v}{\Delta y}$$

where $\frac{\Delta v}{\Delta y}$ = the velocity gradient or shear rate of the fluid, i.e. the rate of change of velocity v with respect to linear distance y (Figure 7.16) and τ = the shear stress that takes place between successive molecular layers of the fluid (see boundary layer). The exact relationship need not concern us here. It is given to verify that the *units of dynamic viscosity are* $N\,s/m^2$ or in terms of mass they are kg/ms. When the dynamic viscosity is divided by the density of the fluid it is known as *kinematic viscosity* (v) = μ/ρ and therefore has units m^2/s.

The boundary layer

When a body such as an aircraft wing is immersed in a fluid, which is flowing past it, the fluid molecules in contact with the wings surface tend to be brought to rest by friction and stick to it. The next molecular layer of the fluid tends to bind to the first layer by molecular attraction but tends to shear slightly creating movement with respect to the first stationary layer. This process

continues as successive layers shear slightly relative to the layer underneath them. This produces a gradual increase in velocity of each successive layer of the fluid (say air) until the free stream relative velocity is reached some distance away from the body immersed in the fluid.

In Figure 7.16 the fixed boundary represents the skin of an aircraft wing, where the initial layer of air molecules has come to rest on its surface. The moving boundary is the point where the air has regained its free stream velocity relative to the wing. The region between the fixed boundary and moving boundary, where this shearing takes place is known as the *boundary layer*.

For an aircraft, subject to laminar flow over its wing section, the thickness of the boundary layer is seldom more than 1 mm. The thinner the boundary layer, the less the drag and the greater the efficiency of the lift producing surface. Since friction reduces the energy of the air flowing over an aircraft wing, it is important to keep wing surfaces and other lift producing devices as clean and as smooth as possible. This will ensure that energy losses in the air close to the boundary are minimized and efficient laminar flow is maintained for as long as possible.

Boundary layer separation and control

Now, irrespective of the smoothness and condition of the lift producing surface, as the airflow continues back from the leading edge, friction forces in the boundary layer continue to use up the energy of the air stream gradually slowing it down. This results in an increase in thickness of the laminar boundary layer with increase in distance from the leading edge (Figure 7.17). Some distance back from the leading edge, the laminar flow begins an oscillatory disturbance which is unstable. A waviness or eddying starts to occur, which grows larger and more severe until the smooth laminar flow is destroyed. Thus, a transition takes place in which laminar flow decays into turbulent boundary layer flow.

Boundary layer control devices provide an additional means of increasing the lift produced across an aerofoil section. In effect all these devices are designed to increase the energy of the air flowing in the boundary layer, thus reducing

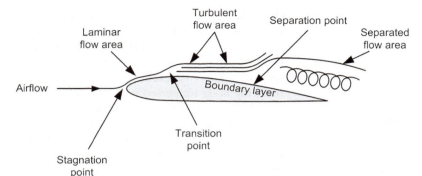

Figure 7.17 Boundary layer separation.

the rate of boundary layer separation, from the upper surface of the aerofoil. At high AOA the propensity for boundary layer separation increases (Figure 7.15), as the airflow over the upper surface tends to separate and stagnate.

Boundary layer control devices include *leading edge slats, trailing and leading edge flaps* for high lift applications as mentioned previously. With slats the higher pressure air from beneath the aerofoil section is sucked over the aerofoil upper surface, through the slot created by deploying the slat. This high velocity air re-energizes the stagnant boundary layer air moving the transition point further back increasing lift.

One fixed device that is used for boundary layer control is the *vortex generator*. These generators are literally small metal plates that are fixed obliquely to the upper surface of the wing or other lift producing surface and effectively create a row of convergent ducts close to the surface. These accelerate the airflow and provide higher velocity air to re-energize the boundary layer.

Many other devices exist or are being developed to control the boundary layer, which include blown air, suction devices and use of smart devices. Again, more will be said on this subject, when Aircraft Aerodynamics is covered in a later book in the series.

7.3.6 Lift, drag and pitching moment of an aerofoil

Both lift and drag have already been mentioned. Here, we will consider the nature of the lift and drag forces and the methods used to estimate their magnitude.

We have already discovered that the lift generated by an aerofoil surface is dependent on the *shape* of the aerofoil and its *AOA* to the relative airflow. Thus, *the magnitude of the negative pressure distributed over the top surface of the wing is dependent on the wing camber and the wing AOA.*

The shape of the aerofoil may be represented by a shape coefficient, which alters with AOA, this is known as the *lift coefficient* (C_L), which may be found experimentally for differing aerofoil sections. A corresponding *drag coefficient* (C_D) may also be determined experimentally or analytically, if the lift coefficient is known.

In addition to the lift and drag coefficients, the *pitching moment* (C_M), or the tendency of the aerofoil to revolve about its center of gravity (CG), can also be determined experimentally.

Figure 7.18(a) shows a photograph of the setup for a 1 in 48 scale model of a Boeing 727 mounted upside down in a closed section wind tunnel. It is mounted in this manner so that the lift, drag and pitching moment apparatus is above the wind-tunnel working section. This allows the operator to work at normal eye-level. If the complex apparatus (Figure 7.18(b)) was placed below the working section it would be at a height where observations would need to be made standing on a ladder or platform! The digital read out for lift, drag and pitching moment values are shown in Figure 7.18(c). These readings need to be evaluated using formulae in order to obtain the actual lift and drag

(a)

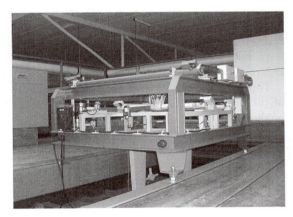

(b)

(c)

Figure 7.18 Photographs of wind-tunnel set-up using a Boeing 727 model.

values in Newton (N) and the pitching moment in Newton-metres (Nm).

Now, since pressure = force/area then the lift force, i.e. the total reaction of a wing surface is given by pressure × area = lift force and so, the magnitude of the *lift force* also depends on the *plan area* of the lift producing surface. It is the plan area because the lift component of the total reaction acts at right angles to the direction of motion of the aircraft and to the lift producing surface, whereas the *drag force* acts parallel and opposite to the direction of motion. Normally when measuring drag we would consider the frontal area of the body concerned, whereas for the drag on aerofoil sections, we take the *plan area*. This is because the vast majority of the drag produced is lift producing drag that acts over the wing plan area.

In addition to the above factors the results of experiments show that within certain limitations the lift and drag produced by an aerofoil is also dependent on the dynamic pressure of the relative airflow, where:

$$\text{Dynamic pressure } (q) = \tfrac{1}{2}\rho v^2$$

as discussed earlier.

Thus the results of experiments and modern computational methods show that the lift, drag and pitching moment of an aerofoil depend on:

- the shape of an aerofoil,
- the plan area of an aerofoil,
- the dynamic pressure.

Thus lift, drag and pitching moment may be expressed mathematically as:

$$\text{Lift} = C_L \tfrac{1}{2}\rho v^2 S$$

$$\text{Drag} = C_D \tfrac{1}{2}\rho v^2 S$$

$$\text{Pitching moment} = C_M \tfrac{1}{2}\rho v^2 S c$$

where $c =$ the mean chord length of the aerofoil section.

Finally remember that lift is the component of the *total reaction* that is at right angles or perpendicular to the relative airflow, and drag is that component of the total reaction that acts parallel to the relative airflow, in such a way as to oppose the motion of the aircraft.

Example 7.4

Determine the lift and drag of an aircraft flying in straight and level flight, with a constant velocity of 190 m/s at an altitude where the air density is 0.82 kg/m^3. Given that the aircraft has a wing area of 90 m^2 and for straight and level flight $C_L = 0.56$ and C_D is related to C_L by the drag equation,

$$C_D = 0.025 + 0.05C_L^2$$

The lift may be found straight away from the relationship, $L = C_L \frac{1}{2}\rho v^2 S$ then substituting for the given values, we get:

$$L = (0.56)(0.5)(0.82)(190)^2(90)$$

so that

$$L = 745,970.4\,\text{N} \quad \text{or} \quad L = 745.97\,\text{kN}$$

Now in order to find the drag, we must first calculate the drag coefficient from the drag equation.

Then:

$$C_D = 0.0025 + (0.05)(0.56)^2 = 0.0407$$

and the drag is given by:

$$D = C_D \frac{1}{2}\rho v^2 S = (0.0407)(0.5)(0.82)(190)^2(90)$$
$$= 54,216\,\text{N} \quad \text{or} \quad 54.216\,\text{kN}$$

Note that the expression $\frac{1}{2}\rho v^2 S$ is the same at any given altitude when an aircraft is flying at constant velocity. The expression is the product of the dynamic pressure (q) and the wing plan form area (S). Thus knowing the C_L and C_D, the total lift or drag can be determined when the dynamic pressure at altitude and the wing area are known.

7.3.7 Total aircraft drag and its components

To complete our understanding of lift and drag, we need to define the different types of drag that affect the performance of the whole aircraft. Mention of one or two different types of drag was made earlier. For clarity we will discuss all the types of drag, that go to make up the total drag acting on an aircraft, as shown in Figure 7.19.

Total drag is the total resistance to the motion of the aircraft as it passes through the air that is the sum total of the various drag forces acting on the aircraft. These drag forces may be divided into subsonic drag and supersonic drag. Although supersonic drag is shown in Figure 7.19 for completeness, it will not be studied now but later when you study supersonic

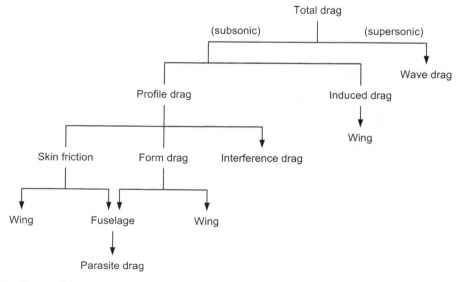

Figure 7.19 Types of drag.

flight as part of Module 11. The subsonic flight drag may be divided into two major categories: these are *profile drag* and *induced drag*. Profile drag is further subdivided into skin friction drag, form drag and interference drag. The total drag of an aircraft may be divided in another way. Whereby, the drag of the lift producing surfaces, *lift dependent drag*, is separated from those parts of the aircraft hull that do not produce lift. This non-lift dependent drag is often known as *parasite drag* and is shown in Figure 7.19, as that drag which results from the wing to fuselage shape and frictional resistance.

Skin friction drag

Skin friction drag results from the frictional forces that exist between a body and the air through which it is moving. The magnitude of the skin friction drag depends on:

- Surface area of the aircraft, since the whole surface area of the aircraft experiences skin friction drag, as it moves through the air.
- Surface roughness, the rougher the surface the greater the skin friction drag. Hence, as mentioned earlier, the need to keep surfaces polished or with a good paint finish, in order to maintain a smooth surface finish.
- The state of boundary layer airflow, i.e. whether laminar or turbulent.

Form drag

Form drag is that part of the air resistance, that is created by virtue of the shape of the body subject to the airflow. Those shapes which encourage the airflow to separate from their surface create eddies and the streamline flow is disturbed. The turbulent wake that is formed increases drag. Form drag can be reduced by streamlining the aircraft in such a way as to reduce the drag resistance to a minimum. A definite relationship exists between the length and thickness of a streamlined body, this is known as the *fineness ratio*, which you met earlier, when we looked at streamlined aerofoil sections.

The act of streamlining shapes reduces their form drag by decreasing the curvature of surfaces and avoiding sudden changes of cross-sectional area and shape. Apart from the streamlining of aerofoil sections, where we look for a finer t/c other parts of the airframe may also be streamlined, by adding fairings. Figure 7.20 shows how streamlining helps to substantially reduce form drag.

The photographs shown in Figure 7.21, again, illustrate how a streamlined shape maintains the laminar flow while the cylinder produces eddies and a greater turbulent wake.

Interference drag

The total drag acting on an aircraft is greater than the sum of the component drag. This is

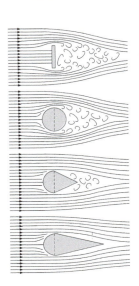
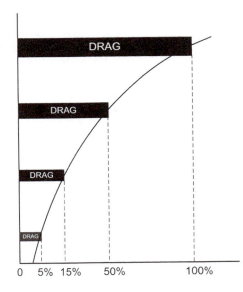

Figure 7.20 Streamlining and relative reduction of form drag.

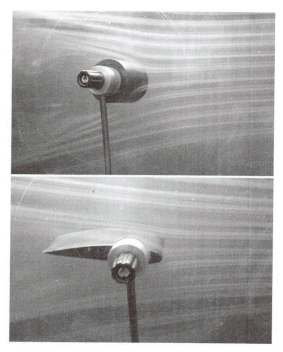

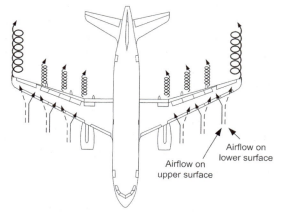

Figure 7.21 Smoke generator traces, showing differences in flow behind a cylinder and an aerofoil section. With subsequent reduction in form drag, as indicated in Figure 7.20.

Figure 7.22 Production of lift induced drag, resulting from the creation of wingtip vortices.

because, due to flow, interference occurs at the various junctions of the surfaces. These include the wing/fuselage junctions, wing/engine pylon junctions and those between tail plane, fin and fuselage. This flow interference results in additional drag, that we call *interference drag*. As this type of drag is not directly associated with lift, it is another form of parasite drag.

When the airflows from the various aircraft surfaces meet, they form a wake behind the aircraft. The additional turbulence that occurs in the wake causes a greater pressure difference between the front and rear surfaces of the aircraft and therefore increases drag. As mentioned earlier, interference drag can be minimized by using suitable fillets, fairings and stream lined shapes.

Induced drag

Induced drag results from the production of lift. It is created by differential pressures acting on the top and bottom surfaces of the wing. The pressure above the wing is slightly below atmospheric, while the pressure beneath the wing is at or slightly above atmospheric. This results in the migration of the airflow at the wing tips from the high pressure side to the low pressure side. Since this flow of air is spanwise, it results in an overflow at the wing tip, that sets up a whirlpool action (Figure 7.22). This whirling of the air at the wing tip is known as a *vortex*. Also, the air on the upper surface of the wing tends to move towards the fuselage and off the trailing edge. This air current forms a similar vortex at the inner portion of the trailing edge of the wing. These vortices increase drag due to the turbulence produced and this type of drag is known as **induced drag**.

In the same way as lift increases with increase in AOA, so too does induced drag. This results from the greater pressure difference produced with increased AOA, creating even more violent vortices, greater turbulence and greater downwash of air behind the trailing edge. These vortices can be seen on cool moist days when

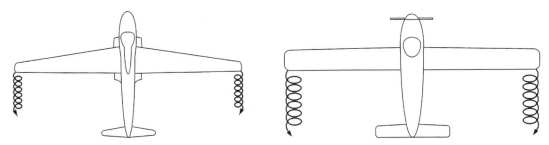

Figure 7.23 Reduction of induced drag by use of high aspect ratio tapered wings.

condensation takes place in the twisting vortices and they can be seen from the ground as vortex spirals.

If the speed of the aircraft is increased, then lift will be increased. Thus to maintain straight and level flight the AOA of the aircraft must also be reduced. We have just seen that increase in AOA increases induced drag. Therefore, *by reducing the AOA and increasing speed, we reduce the ferocity of the wingtip vortices and so reduce the induced drag.* This is the direct opposite to form drag, which clearly increases with increase in velocity. In fact it can be shown that induced drag, reduces in proportion to the square of the airspeed, while profile drag increases proportionally with the square of the airspeed.

Wing tip stall

A situation can occur when an aircraft is flying at high AOA, say on the approach to landing, where due to losses incurred by strong wingtip vortices in this situation, one wingtip may stall while the remainder of the main plane is still lifting. This will result in more lift being produced by one wing, than the other, resulting in a roll motion towards the stalled wingtip. Obviously if this happens at low altitude, the aircraft might sideslip into the ground. Thus wingtip stall is most undesirable under any conditions and methods have been adopted to reduce losses at the wingtip.

Three of the most common methods for reducing induced drag and so, wingtip stall, are to use *washout*, introduce *fixed leading edge spoilers* or use *long narrow tapered wings*.

If the AOI of the wing is decreased towards the wingtip, there will be less tendency for wingtip vortices to form at high AOA, due to the fact that the wingtip is at a lower AOA than the remaining part of the wing. This design method is known as *washout* and the opposite i.e., an increase in the AOI towards the wingtip is known as *wash-in*. Some aircraft are fitted with *fixed spoilers* to their inboard leading edge. These have the effect of disturbing the airflow and inducing the stall over the inboard section of the wing, before it occurs at the wingtip, thus removing the possibility of sudden wingtip stall.

Another method of reducing induced drag is to have *long narrow tapered wings*, i.e. wings with a *high aspect ratio*. Unfortunately from a structural point of view a long narrow tapered wing is quite difficult to build, and this is often the limiting factor in developing high aspect ration wings. The result of this type of design is to create smaller vortices that are a long way apart and therefore will not readily interact. Figure 7.23 shows how the taper at the end of a long thin wing, such as those fitted to a glider, helps reduce the strength of the wing tip vortices and so induced drag.

Aspect ratio may be calculated using any one of the following three formulae, dependent on the information available:

$$\text{Aspect ratio} = \frac{\text{span}}{\text{mean chord}} \quad \text{or}$$

$$= \frac{(\text{span})^2}{\text{area}} \quad \text{or}$$

$$= \frac{\text{wing area}}{(\text{mean chord})^2}$$

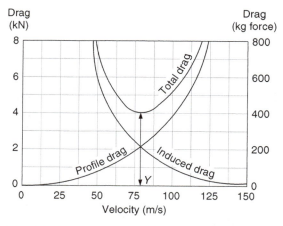

Figure 7.24   Minimum total drag and airspeed.

Total drag

It is of utmost importance that aircraft designers know the circumstances under which the total drag of an aircraft is at a minimum. Because then it is possible to fly a particular sortie pattern, that keeps drag to a minimum, reduces fuel burn and improves aircraft performance and operating costs. We know that profile drag increases with the square of the airspeed and that induced drag decreases with the square of the airspeed. Therefore there must be an occasion when at a particular airspeed and AOA, drag is at a minimum.

The drag curves for induced drag and profile drag (Figure 7.24) show when their combination, i.e. total drag is at a minimum.

7.3.8 Aerodynamic effects of ice accretion

When aircraft operate at altitude or in climates where ground temperatures are at or below freezing point, ice may form on the leading edge of wings, as well as over flying controls and other areas of the airframe.

The build-up of ice can have a severe detrimental effect on aircraft performance, in terms of extra weight and drag, loss of lift, and also the freezing or unbalancing of control surfaces. Any of these events, if severe enough, can and have caused fatal accidents.

It is not the intention here, to cover in any detail the methods used to detect and remove

snow and ice. The subject of ice and rain protection will be covered in detail in a later book in the series. However, the nature of snow and ice build-up, together with its effect on aircraft aerodynamic performance, must be understood and for that reason, is looked at next.

Ice is caused by coldness acting on the water in the atmosphere and depending on the type of water present and its temperature, will depend on the type of ice formed. Ice accretion is generally classified under three main types; *hoar frost*, *rime ice* and *glaze ice*.

Hoar frost is likely to occur on a surface that is at a temperature below which frost is formed in the adjacent moist air. Thus water in contact with this surface is converted into a white semi-crystalline coating, normally feathery in appearance, that we know as frost.

If hoar frost is not removed from an aircraft which is on the ground, it may interfere with the laminar airflow over the leading edge and lift producing surfaces, causing loss of lift during take-off. Free movement of the control surfaces may also be affected.

Hoar frost on an aircraft in flight commences with a thin layer of glaze ice on leading edges followed by the formation of frost which will spread over the whole surface area. If this frost is not removed prior to landing, then some changes in the handling characteristics of the aircraft can be expected.

Rime ice is a light, porous, opaque, rough deposit which at ground level forms in freezing fog from individual water droplet particles, with little or no spreading.

When an aircraft at a temperature below freezing, flies through a cloud of small water droplets, an ice build-up is formed on the wing leading edge. This ice formation has no great weight, but does interfere with the airflow over the wing.

Glaze ice forms in flight, when the aircraft encounters clouds or freezing rain, where the air temperature and that of the airframe are both below freezing point. Glaze ice may form as either a transparent deposit or opaque deposit with a glassy surface. This results from water in the form of a liquid, flowing over the airframe surface prior to freezing. Glaze ice is heavy, dense and tough. It adheres firmly to the airframe surface and is not easy to remove. However, when it does break away, it does so in large lumps!

The main dangers associated with glaze ice accretion include, aerodynamic instability, unequal wing loading that may affect aircraft trim and for propellers there is an associated loss of efficiency accompanied by an excessive amount of vibration. *Glaze ice is the most dangerous ice to be present on aircraft.*

From what has been said, it is evident that if ice continues to form on aircraft one or more of the following undesirable events is (are) likely to occur:

- decrease in lift due to changes in wing section;
- increase in drag due to increase in friction from rough surface;
- loss or restriction of control surface movement;
- increase in wing loading due to extra weight, resulting in possible loss of aircraft altitude;
- aerodynamic instability due to displacement of CG;
- decrease in propeller efficiency due to altered blade profile and subsequent increase in vibration.

If aircraft are properly de-iced on the ground, prior to flight, when icing conditions prevail and aircraft are fitted with appropriate ice detection, anti-icing and de-icing equipment, then the detrimental effects of ice accretion on aerodynamic performance can be minimized or eliminated.

Test your understanding 7.1

1. Explain the changes that take place to the air in the atmosphere with increase in altitude, up to the outer edge of the stratosphere.
2. Explain the reasons for setting up the ICAO standard atmosphere.
3. Given that an aircraft is flying with an EAS of 180 m/s in still air at an altitude where the density is 0.93 kg/m^3, determine the TAS of the aircraft assuming that ICAO standard atmospheric conditions apply.
4. Without consulting tables, determine the temperature of the air at an altitude of 7000 m, in the ICAO standard atmosphere.

5. Determine the local speed of sound, at a height where the temperature is $-20°C$
6. What will be the Mach number of an aircraft flying at a TAS of 319 m/s, where the local air temperature is $-20°C$.
7. Define, "streamline flow".
8. Explain the difference between AOA and AOI.
9. Define "aerofoil efficiency".
10. What are the symbols and units of (i) dynamic viscosity (ii) kinematic viscosity?
11. Explain the nature and importance of the "boundary layer" with respect to airflows.
12. What is the dynamic pressure created by the airflow in a wind-tunnel travelling with a velocity of 45 m/s. You may assume standard atmospheric conditions prevail.
13. Determine the lift and drag of an aircraft flying straight and level, with a constant velocity of 160 m/s, at an altitude where the *relative density* $\sigma = 0.75$. Given that the aircraft has a wing of 100 m^2 and for straight and level flight $C_L = 0.65$ and C_D is related to C_L by the *drag polar*, $C_D = 0.03 + 0.04\,C_L^2$.
14. What aircraft design features can be adopted to reduce the effects of wingtip stall?
15. Define "aspect ratio".
16. Explain how *glaze ice* is formed and the effects its formation may have on aircraft aerodynamic performance.

7.4 Flight forces and aircraft loading

In this very short section, we take a brief look at the nature of the forces that act on an aircraft when in straight and level flight and also during steady correctly applied manoeuvres, such as the climb, dive and turn. We will also consider the use and nature of the aircraft *flight envelope* as defined in JAR 25.

7.4.1 The four forces acting on the aircraft

We have already dealt with *lift* and *drag* when we considered *aerofoil* sections, we now look at these two forces and two others, *thrust* and *weight*, in particular with respect to their effect on the aircraft as a whole.

For the aircraft to maintain constant height then the lift force created by the aerofoil sections must be balanced by the weight of the aircraft (Figure 7.25). Similarly for an aircraft to fly with constant velocity, or zero acceleration, the thrust force must be equal to the drag force that opposes it.

Figure 7.25 shows the four flight forces acting at right angles to one another with their appropriate lines of action:

- *Lift* of the main planes acts perpendicular to the relative airflow through the CP of the main aerofoil sections.
- *Weight* acts vertically downwards through the aircraft's CG.
- *Thrust* of the engines works along the engine axis approximately parallel to the direction of flight.
- *Drag* is the component acting rearwards parallel to the direction of the relative airflow and is the resultant of two components: induced drag and profile drag. For convenience the *total drag is said to act at a point known as the center of drag.*

Then as mentioned above in un-accelerated straight and level flight:

$$\text{Lift} = \text{weight} \quad \text{and} \quad \text{Thrust} = \text{drag}$$

In practice, for normal flight modes, changes in AOA will causes changes in the CP, thus the lift component which acts through the CP will change as the AOA changes.

The weight which acts through the CG depends on every individual part of the aircraft and will vary depending on the distribution of passengers, crew, freight and fuel consumption.

The line of action of the thrust is set in the basic design and is totally dependent on the position of the propeller shaft or the center line of the exhaust jet.

The drag may be found by calculating its component parts separately or by experiment with models in a wind tunnel (Figure 7.18). The four forces do not, therefore, necessarily act at the same point so that equilibrium can only be maintained providing that the moments produced by the forces are in balance. In practice, the lift and weight forces may be so designed as to provide a nose-down couple (Figure 7.26(a)), so that in the event of engine failure a nose-down gliding attitude is produced. For straight and level flight the thrust and drag must provide an equal and opposite nose-up couple.

However, the design of an aircraft will not always allow a high drag and low thrust line, so that some other method of balancing the flight forces must be found. This involves the use of the *tail plane* or *horizontal stabilizer*. One reason for fitting a tail plane is to counter the out-of-balance pitching moments that arise as a result of inequalities with the two main couples. The tail plane is altogether a lot smaller than the wings, however because it is positioned some distance behind the CG, it can exert considerable leverage from the moment produced (Figure 7.26(b)).

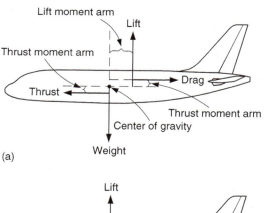

(a)

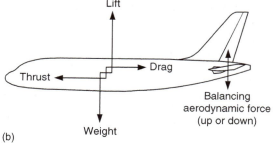

(b)

Figure 7.26 Force couples for straight and level flight.

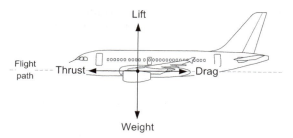

Figure 7.25 The four flight forces.

At high speed the AOA of the main plane will be small. This causes the CP to move rearwards creating a nose-down *pitching moment*. To counteract, this the tail plane will have a downward force acting on it to re-balance the aircraft. Quite clearly, following the same argument, for high AOA at slow speeds, the CP moves forward creating a nose-up pitching moment. Thus, tail planes may need to be designed to carry loads in either direction. A suitable design for this purpose is the symmetrical cambered tail plane, which at zero AOA will allow the chord line of the section to be the neutral line.

Most tail planes have been designed to act at a specified AOA for normal flight modes. However, due to variables (such as speed) changing AOA with changing load distribution and other external factors, there are times when the tail plane will need to act with a different AOI, to allow for this some tail planes are moveable in flight and are known as the *all-moving tail plane*.

Example 7.5

The system of forces that act on an aircraft at a particular time during horizontal flight are shown in Figure 7.27; where the lift acts 0.6 m behind the weight and the drag acts 0.5 m above the thrust line equidistantly spaced about the CG. The CP of the tail plane is 14 m behind the CG. For the system of forces shown determine the magnitude and direction of the load that needs to act on the tail in order to maintain balance.

Now in order to solve this problem all we need to do is apply the principle of moments that you learnt earlier!

Then for balance, the sum of the clockwise moments must equal the sum of the anticlockwise moments. Our only problem is that because we do not know at this time whether the load on the tail acts downwards or upwards, we do not know the direction of the moment. Let us make the *assumption that the load acts downwards, so creating a clockwise moment.*

Knowing this, all that is required to proceed is to choose a point about which to take moments. We will take moments about the CG, which will eliminate the unknown weight force from the calculation. Also noting that the lines of action for the thrust/drag couple are equidistantly spaced about the CG as shown, then:

Sum of the CWM = sum of the ACWM

$$14F_T + (0.25)(4000) + (0.25)(16,000)$$
$$= (0.6)(50,000)\,\text{Nm}$$
$$14F_T = 30,000\,\text{Nm} - 5000\,\text{Nm}$$
$$= 25,000\,\text{Nm}$$

and so, $F_T = 1786\,\text{N}$

this is positive and therefore acts in the assumed direction, i.e. *downwards*.

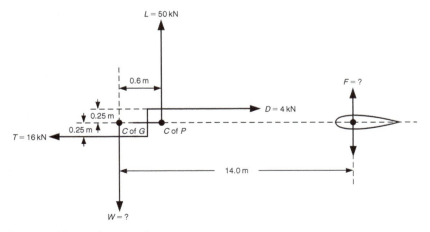

Figure 7.27 System of forces for aircraft.

Note that weight W acts at zero distance from the CG, when we take moments about this point, therefore it produces zero moment and is eliminated from the above calculation.

7.4.2 Flight forces in steady manoeuvres

We now consider the forces that act on the aircraft when gliding, diving, climbing and moving in a horizontal banked turn.

Gliding flight

Aircraft with zero thrust cannot maintain height indefinitely. Gliders or aircraft with total engine failure usually descend in a shallow flight path at a steady speed. The forces that act on an aircraft during gliding flight are shown in Figure 7.28(a).

If the aircraft is descending at steady speed, we may assume that it is in equilibrium and a vector force triangle may be drawn as shown in Figure 7.28(b), where:

D = drag, W = weight, L = lift and γ = the glide angle. Then from the vector triangle:

$$\sin \gamma = \frac{\text{drag}}{\text{weight}} \quad \text{and} \quad \cos \gamma = \frac{\text{lift}}{\text{weight}}$$

also because

$$\frac{\sin \gamma}{\cos \gamma} = \tan \gamma \quad \text{then,}$$

$$\tan \gamma = \frac{\frac{\text{drag}}{\text{weight}}}{\frac{\text{lift}}{\text{weight}}} = \frac{\text{drag } (D)}{\text{lift } (L)}$$

Example 7.6

An aircraft weighing 30,000 N descends with engines off at a glide angle of 3°. Find the drag and lift components that act during the glide.

Then from,

$$\sin \gamma = \frac{\text{drag}}{\text{weight}}$$

$$\text{drag} = W \sin \gamma = (30,000)(0.0523)$$

$$= 1570 \, \text{N}$$

Similarly from,

$$\cos \gamma = \frac{\text{lift}}{\text{weight}}$$

$$\text{lift} = W \cos \gamma = (30,000)(0.9986)$$

$$= 29,959 \, \text{N}$$

There is another parameter that we can find for gliding flight that is the *range*. This is the horizontal distance an aircraft can glide before reaching the ground. Figure 7.29 shows diagrammatically the relationship between the range, vertical height and aircraft flight path.

Then from the triangle,

$$\tan \gamma = \frac{\text{height}}{\text{range}}$$

and from above

$$\tan \gamma = \frac{D}{L}$$

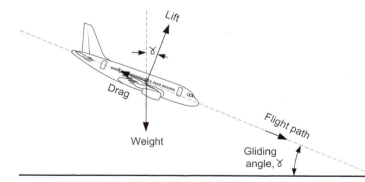
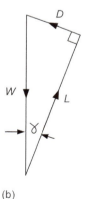

(a)   (b)

Figure 7.28 Gliding flight.

so that:

$$\frac{\text{height}}{\text{range}} = \frac{D}{L}$$

therefore,

$$\frac{\text{range}}{\text{height}} = \frac{L}{D}$$

and so,

$$\text{range} = (\text{height})\left(\frac{L}{D}\right)$$

So considering again Example 7.6, if the aircraft starts the glide from a height of 10 km then from above:

$$\text{range} = (\text{height})\left(\frac{L}{D}\right)$$

so,

$$\text{range} = (10\,\text{km})\left(\frac{29{,}959}{1570}\right)$$

$$= (10)(19.082) = 190.82\,\text{km}.$$

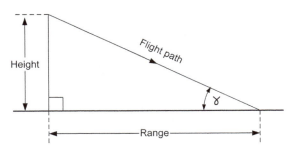

Figure 7.29 The range vector triangle for gliding flight.

To achieve the greatest range the L/D should be as large as possible. An AOA of about 3° to 4° gives the best L/D ratio.

Diving flight

If an aircraft suffers a loss of power and has less thrust than drag, then it can only maintain constant speed by diving.

Figure 7.30(a) shows the forces acting in this situation together with their vector triangle (as shown in Figure 7.30(b)) where from the triangle:

$$\sin \gamma = \frac{D-T}{W} \quad \text{and} \quad \cos \gamma = \frac{L}{W}$$

Example 7.7

An aircraft weighing 20 kN has a thrust, $T = 900\,\text{N}$ and a drag, $D = 2200\,\text{N}$, when in a constant speed dive. What is the aircraft dive angle?

This is a very simple application, where we have all the unknowns, therefore from:

$$\sin \gamma = \frac{D-T}{W} \quad \text{we see that,}$$

$$\sin \gamma = \frac{2200 - 900}{20{,}000} = 0.065$$

and the dive angle $\qquad \gamma = 3.73°$

Climbing flight

In a constant speed climb, the thrust produced by the engines must be greater than the drag

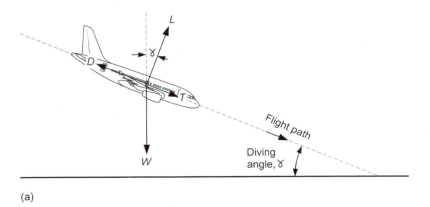

(a)

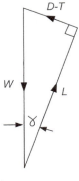

(b)

Figure 7.30 Forces on aircraft in diving flight.

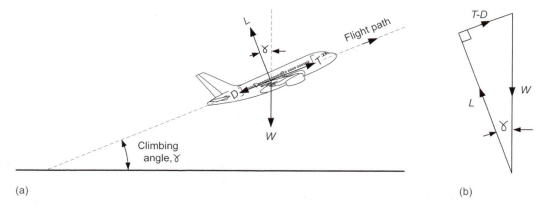

(a)   (b)

Figure 7.31 Forces acting on aircraft in a steady speed climb.

to maintain a steady speed. The steady speed climb is illustrated in Figure 7.31(a), where again the vector triangle of the forces is given in Figure 7.31(b).

From the vector triangle of forces, we find that:

$$\sin \gamma = \frac{T - D}{W} \quad \text{and} \quad \cos \gamma = \frac{L}{W}$$

For example, if an aircraft weighing 50,000 N is climbing with a steady velocity, where $\gamma = 12°$ and $D = 2500$ N and we need to find the required thrust.

Then, $\sin \gamma = 0.2079$ so,

$$0.2079 = \frac{T - 2500}{50,000} \quad \text{and}$$

$$T = (0.2079)(50,000) + 2500$$

from which the required thrust $T = 12,895$ N.

If an aircraft is in a **vertical climb** at constant speed, the aircraft must have more thrust than weight in order to overcome the drag, i.e.:

$$\text{Thrust} = W + D$$

(for steady vertical climb where the lift is zero).

Turning flight

When gliding, diving and climbing, the aircraft has been in equilibrium, where its speed and direction were fixed. If the aircraft manoeuvres

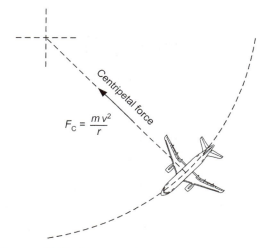

Figure 7.32 Centripetal force acting towards center of turn.

by changing speed or direction, an acceleration takes place and equilibrium is lost. When an aircraft turns, centripetal force (F_C) is required to act towards the center of the turn, in order to hold the aircraft in the turn (Figure 7.32).

This centripetal force must be balanced by the lift component in order to maintain a constant radius (steady) turn. This is achieved by banking the aircraft. In a correctly banked turn, the forces are as shown in Figure 7.33.

The horizontal component of lift is equal to the centrifugal force holding the aircraft in

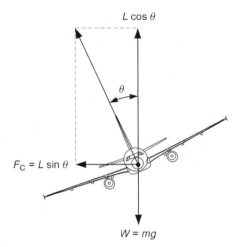

Figure 7.33 Forces acting in a correctly banked steady turn.

the turn. Then, resolving forces horizontally we get:

$$L \sin \theta = \frac{mv^2}{r}$$

where θ = radius of turn in m, m = mass in kg and v = velocity in m/s.

Also from Figure 7.33, resolving vertically, we get: $L \cos \theta = W = mg$, where g = acceleration due to gravity in m/s^2.

Now remembering from your trigonometry the identity $\dfrac{\sin \theta}{\cos \theta} = \tan \theta$, then:

$$\frac{L \sin \theta}{L \cos \theta} = \frac{\frac{mv^2}{r}}{mg} = \frac{v^2}{gr} \quad \text{therefore}$$

$$\tan \theta = \frac{v^2}{gr}$$

Example 7.8

An aircraft enters a correctly banked turn of radius 1800 m at a velocity of 200 m/s. If the aircraft has a mass of 80,000 kg. Determine:
1. the centripetal force acting towards the center of the turn,
2. the angle of bank.

1. $F_C = \dfrac{mv^2}{r} = \dfrac{(80,000)(200)^2}{1800} = 1.777\,\text{MN}$

2. Since we do not have the lift, we can only use the relationship, $\tan \theta = \dfrac{v^2}{gr}$ so that:

$$\tan \theta = \frac{200^2}{(1800)(9.81)} = 2.26$$

giving an angle of bank $\boldsymbol{\theta = 66.2°}$

Load factor

The high forces created in tight turns stress both aircraft and flight crew. With respect to the airframe, the degree of stress is called the *load factor*, which is the relationship between lift and weight given as:

$$\text{Load factor} = \frac{\text{lift}}{\text{weight}}$$

For example, if $L = 90,000\,\text{N}$ and $W = 30,000\,\text{N}$, then the load factor $= 3$. In more common language there is a loading of 3 g. High load factors may also occur when levelling out from a dive. The lift force has to balance the weight and provide centripetal force to maintain the aircraft in the manoeuvre.

7.4.3 Aircraft loading and the flight envelope

As mentioned previously, subjecting an aircraft to high load factors may easily damage the airframe structure. Apart from the aerodynamic loads that affect aircraft as a result of manoeuvres, other loads stress the airframe that occur during taxying, take-off, climb, descent, go-round and landing. These loads may be static or dynamic, e.g. fatigue loads subject the airframe to repeated fluctuating stresses. They are dynamic loads that may cause fatigue damage at stress levels very much below a materials yield stress.

In order to ensure that aircraft hulls are able to withstand a degree of excessive loading, whether that be static or dynamic loading, the manufacturer must show that the airframe is able to meet certain strength standards, as laid down in JAR 25. These strength requirements are specified in terms of *limit loads* (the maximum loads to be expected during service) and **ultimate loads** (the limit loads multiplied by prescribed factors of safety normally 1.5 unless otherwise specified). Strength and deformation criteria are also specified in JAR 25. In that, structure must be able to support limit loads without detrimental permanent deformation. The structure must also be capable of supporting ultimate loads for at least 3 s. The ultimate loads that an aircraft structure can withstand may be ascertained by conducting static tests, which must include ultimate deformation and ultimate deflections of the structure, when subject to the loading.

Flight load factors have already been defined as the particular relationship between aerodynamic lift and aircraft weight. A positive load factor is one in which the aerodynamic force acts upward with respect to the aircraft.

Flight loads, need to meet the criteria laid down in JAR 25. To do this loads need to be calculated:

- for each critical altitude;
- at each weight from design minimum to the design maximum weight, appropriate to flight conditions;
- for any practical distribution of disposal load, within the operating limitations.

Also, the analysis of symmetrical flight loads must include:

- manoeuvring balanced conditions, assuming the aircraft to be in equilibrium;
- manoeuvring pitching conditions;
- gust conditions.

The above loading analysis to which each aircraft design must comply is summarized in the aircraft *flight envelope*.

The flight envelope

The *flight operating strength limitations of an aircraft* are presented at varying combinations of airspeed and load factor (*g*-loading), on and within the boundaries of manoeuvre envelopes and gust envelopes. An illustration of a typical manoeuvring envelope and a typical gust envelope is shown in Figure 7.34.

In both diagrams the load factor (*n*) is plotted against the EAS. This is the reason for referring

to these plots as **V–n diagrams**. The load factor, sometimes known as the inertia loading, is the same load factor that we defined earlier as:

$$\text{Load factor } (n) = \frac{\text{lift}}{\text{weight}}$$

Each aircraft type has its own particular *V–n* diagram with specific velocities and load factors applicable to that aircraft type. Each flight envelope (illustration of aircraft strength) is dependent on four factors being known:

- the aircraft gross weight;
- the configuration of the aircraft (clean, external stores, flaps and landing gear position, etc.);
- symmetry of loading (non-symmetrical manoeuvres, such as a rolling pull-out, can reduce the structural limits);
- the applicable altitude.

A change in any one of these four factors can cause important changes in operating limits. The limit airspeed is a design reference point for the aircraft and an aircraft in flight above this speed may encounter a variety of adverse effects, such as destructive flutter, aileron reversal, wing divergence, etc.

Note
We will not concern ourselves here with the exact nature of these effects nor with the methods used to construct and interpret the flight envelopes. This is covered in the third Book in the series, when aircraft loading is looked at in rather more detail.

As potential maintenance technicians, you should be aware of the nature and magnitude of the loads that your aircraft may be subjected to, by being able to interpret the aircraft flight envelopes. This is the primary reason for introducing this topic here.

Test your understanding 7.2

1. Produce a sketch showing the four forces that act on an aircraft during straight and level constant velocity flight and explain why in practice the lift and weight forces may be designed to provide a nose down couple.

2. If the four flight forces that act on an aircraft do not produce balance in pitch, what method is used to balance the aircraft longitudinally?

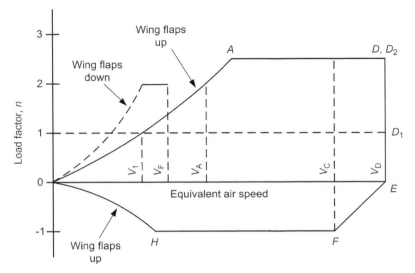

(a) Manouvering envelope

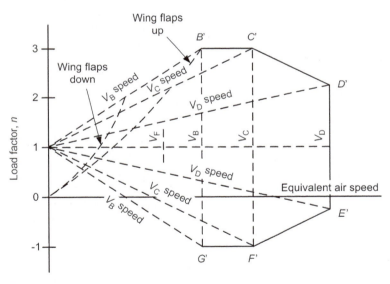

(b) Gust envelope

V_A = Design manoeuvring speed
V_B = Design speed for maximum gust intensity
V_C = Design cruising speed
V_D = Design diving speed
V_F = Design wing flap speed
V_{S1} = Stalling speed with wing flaps retracted
C_N = Coefficient of normal force

A = Positive high angle of attack position
D = Positive low angle of attack position
E = Negative low angle of attack position
F = Negative intermediate angle of attack position
H = Negative high angle of attack position
B' to J = gust conditions that require investigation

Figure 7.34 Typical manoeuvring and gust envelopes.

3. A light aircraft of mass 3500 kg descends with engines off at a glide angle of 4° from an altitude of 5 km. Find:
   (a) the drag and lift components that act during the glide,
   (b) the range covered from the start of the glide to touch down.

4. An aircraft in a steady 10° climb requires 15,000 N of thrust to overcome 3000 N of drag. What is the weight of the aircraft?

5. An aircraft, at sea level, enters a steady turn and is required to bank at an angle of 50°. If the radius of the turn is 2000 m, determine the velocity of the aircraft in the turn.

6. An aircraft weighing 40,000 N is in a manoeuvre where the load factor is 3.5. What is the lift required by the aircraft to remain in the manoeuvre?

7. With respect to aircraft loading, JAR 25 specifies the criteria that aircraft must meet. Upon what criteria are flight loads calculated *and* how is this information displayed?

8. With respect to aircraft loading, why is it important that passenger baggage and other stores and equipment carried by aircraft is loaded in a manner specified in the aircraft weight and balance documentation?

7.5 Flight stability and dynamics

7.5.1 The nature of stability

The stability of an aircraft is a measure of its tendency to return to its original flight path after a displacement. This displacement caused by a disturbance can take place in any of three planes of reference, these are the *pitching rolling* and *yawing* planes (Figure 7.35).

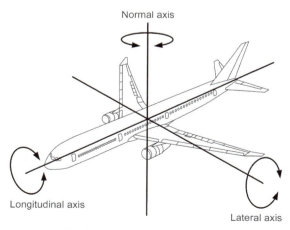

Figure 7.35 Aircraft axes and planes of reference.

The planes are not constant to the earth but are constant relative to the three axes of the aircraft. Thus, the disturbance may cause the aircraft to rotate about one or more of these axes. These axes are imaginary lines passing through the CG of the aircraft, which are mutually perpendicular to one another, i.e. they are at right angles to one another. All the complex dynamics concerned with aircraft use these axes to model and mathematically define stability and control parameters.

Any object which is in equilibrium when displaced by a disturbing force will react in one of three ways once the disturbing force is removed, thus:

- When the force is removed and the object returns to the equilibrium position, it is said to be *stable*.
- When the force is removed and the object continues to move in the direction of the force and never returns to the equilibrium position, it is said to be *unstable*.
- When the force is removed and the object stops in the position to which it has been moved to, neither returning or continuing, it is said to be *neutrally stable*.

These reactions are illustrated in Figure 7.36 where a ball bearing is displaced and then released. In Figure 7.36(a), it can be clearly seen that the ball bearing once released in the bowl, will gradually settle back to the equilibrium point after disturbance. In Figure 7.36(b), it can be clearly seen that due to the effects of gravity the ball bearing will never return to its original equilibrium position on top of the cone. While in Figure 7.36(c), after disturbance, the ball bearing eventually settles back into a new equilibrium position somewhere remote from its original resting place.

There are in fact two types of stability that we need to consider, static stability and dynamic stability.

*An object such as an aircraft is said to have **static stability** if, once the disturbing force ceases, it starts to return to the equilibrium position.*

Now with respect to dynamic stability, consider again the situation with the ball bearing in the bowl (Figure 7.36(a)), where it is statically stable and starts to return to the equilibrium position. In returning, the ball bearing oscillates backwards and forwards before it settles. This oscillation is damped out and grows smaller until the ball bearing finally returns to the equilibrium position. *An object is said to be **dynamically stable** if it returns to the equilibrium position, after a disturbance, with decreasing oscillations.* If an increasing oscillation occurs, then the object may be *statically stable* but *dynamically unstable*. This is a very dangerous situation which can happen to moving objects, if the force balance is incorrect. An example of dynamic instability is helicopter rotor vibration, if the blades are not properly balanced.

7.5.2 Aircraft stability dynamics

Aircraft response to a disturbance

The static and dynamic responses of an aircraft, after it has been disturbed by a small force, are represented by the series of diagrams shown in Figure 7.37.

Figure 7.37(a) shows the situation for deadbeat static stability, where the aircraft returns to the equilibrium position, without any dynamic oscillation, caused by the velocities of motion. This of course, is very unlikely to occur. *Figure 7.37(b)* shows the situation for an aircraft that is both statically and dynamically stable (the ball bearing in the bowl). Under these circumstances, the aircraft will return to its equilibrium position after a few diminishing oscillations. *Figure 7.37(c)* illustrates the undesirable

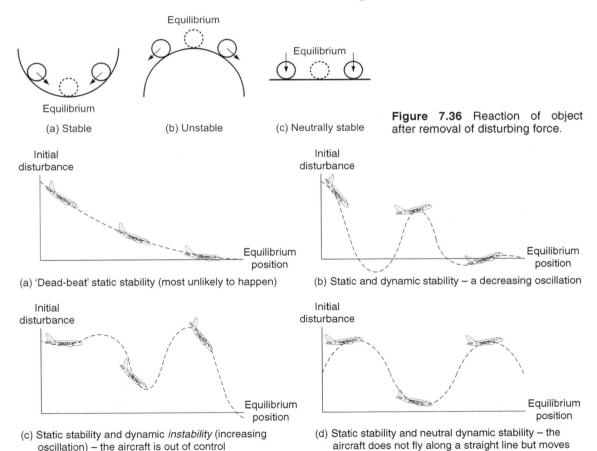

(a) Stable   (b) Unstable   (c) Neutrally stable

Figure 7.36 Reaction of object after removal of disturbing force.

(a) 'Dead-beat' static stability (most unlikely to happen)

(b) Static and dynamic stability – a decreasing oscillation

(c) Static stability and dynamic *instability* (increasing oscillation) – the aircraft is out of control

(d) Static stability and neutral dynamic stability – the aircraft does not fly along a straight line but moves slowly up and down. A *phugoid* oscillation.

Figure 7.37 Static and dynamic response of an aircraft after an initial disturbance.

situation where the aircraft may be statically stable but is dynamically unstable, in other words, the aircraft is out of control. This situation is similar to that of a suspension bridge that oscillates at its resonant frequency, the oscillations getting larger and larger until the bridge fails. It is worth noting here, that it is not possible for an aircraft to be statically unstable and dynamically stable but the reverse situation (Figure 7.37(c)) is possible. *Figure 7.37(d)* illustrates the situation for an aircraft that has static stability and neutral dynamic stability. Under these circumstances the aircraft does not fly in a straight line but is subject to very large low frequency oscillations known as *phugoid* oscillations.

Types of stability

When considering stability, we assume that the CG of the aircraft continues to move in a straight line and that the disturbances to be overcome cause rotational movements about the CG. These movements can be:

- *rolling* movements about (around) the longitudinal axis – *lateral stability*;
- *yawing* movements about the normal axis – *directional stability*;
- *pitching* movements about the lateral axis – *longitudinal stability*.

Figure 7.38 illustrates the type of movement that must be damped, if the aircraft is to be considered stable. Thus lateral stability is the inherent ability of an aircraft to recover from a disturbance around the longitudinal plane (axis), i.e. rolling movements. Similarly, longitudinal stability is the inherent (built-in) ability of the aircraft to recover from disturbances around the lateral axis i.e. pitching movements. Finally, directional stability is the inherent ability of the aircraft to recover from disturbances around the normal axis.

There are many aircraft features specifically designed to either aid stability or reduce the amount of inherent stability an aircraft possesses, depending on aircraft configuration and function. We look at some of these design features next when we consider lateral,

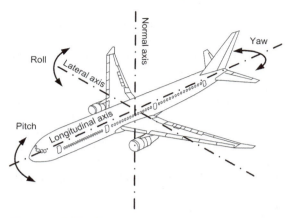

Figure 7.38 Rolling, yawing and pitching movements.

longitudinal and directional stability in a little more detail.

Lateral stability

From what has been said above an aircraft has lateral stability if, following a roll displacement, a restoring moment is produced which opposes the roll and returns the aircraft to a wings level position. In that, aerodynamic coupling produces rolling moments that can set up side slip or yawing motion. It is therefore necessary to consider these interactions when designing an aircraft to be *inherently statically stable* in roll. The main contributors to lateral static stability are:

- wing dihedral,
- sweepback,
- high wing position,
- keel surface.

A design feature that has the opposite effect to those given above, i.e. that reduces stability is *anhedral*. The need to reduce lateral stability may seem strange, but combat aircraft and many high speed automatically controlled aircraft, use anhedral to provide more manoeuvrability.

Wing dihedral and lateral stability

Dihedral angle is defined as the upward inclination of the wings from the horizontal. The amount of dihedral angle being dependent on aircraft type and wing configuration i.e. whether the wings are positioned high or low

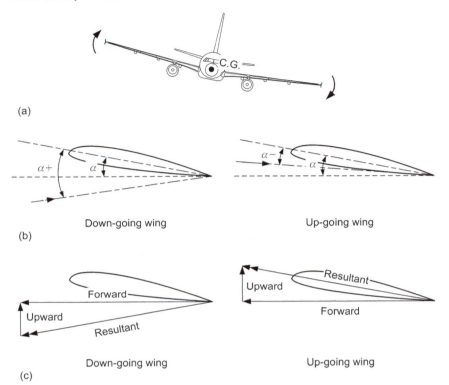

(a)

(b)

Down-going wing   Up-going wing

(c)

Down-going wing   Up-going wing

Figure 7.39 Stopping the rolling motion.

with respect to the fuselage and whether or not they are straight or swept back.

The righting effect from a roll using wing dihedral angle may be considered as a two stage process; where the rolling motion is first stopped and then the down-going wing is returned to the horizontal position.

So we first stop the roll. In Figure 7.39(a) we see that for an aircraft in a roll, one wing will move down and the other will move up, as a result of the rolling motion. The vector diagrams (Figure 7.39(b)) show the velocity resultants for the up-going and down-going wings. The direction of the free stream airflow approaching the wing is changed and the AOA on the down-going wing is increased, while the AOA on the up-going wing is decreased (Figure 7.39(c)). This causes a larger C_L and lift force to be produced on the lower wing and a smaller lift forces on the upper wing, so the roll is stopped. When the roll stops the lift forces equalize again and the restoring effect is lost.

In order to return the aircraft to the equilibrium position, dihedral angle is necessary. A natural consequence of banking the aircraft is to produce a component of lift which acts in such a way as to cause the aircraft to sideslip (Figure 7.40).

In Figure 7.40(a) the component of lift resulting from the angle of bank can clearly be seen. It is this force that is responsible for sideslip. Now if the wings were straight the aircraft would continue to sideslip, but if dihedral angle is built-in the sideways air stream will create a greater lift force on the down-going wing (Figure 7.40(b)). This difference in lift force will restore the aircraft until it is no longer banked over and side-slipping stops.

If anhedral is used, lateral stability is decreased. *Anhedral* is the downward inclination of the wings (Figure 7.41(a)). In this case, as the aircraft sideslips, the lower wing, due to its anhedral will meet the relative airflow at a reduced AOA (Figure 7.41(b)) so reducing lift, while the upper wing will meet the relative

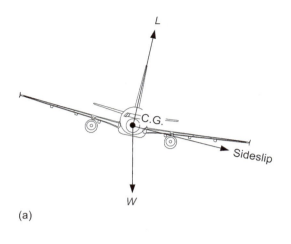

(a)

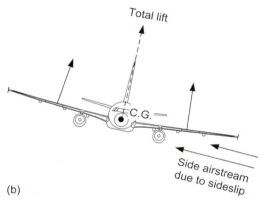

(b)

Figure 7.40 Returning the aircraft to the equilibrium position using wing dihedral.

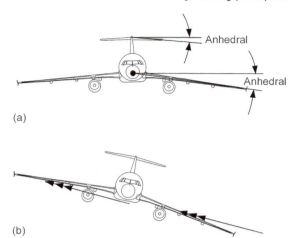

Figure 7.41 Reducing lateral stability by use of anhedral.

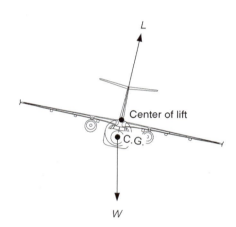

Figure 7.42 Roll correction using high CP and low CG.

airflow at a higher AOA and will produce even more lift. The net effect will be to increase the roll and thus reduce lateral stability.

High wing and keel surface

When an aircraft is fitted with a *high wing* (Figure 7.42), the CG lies in a low position within the aircraft hull, which can create a pendulum effect in a sideslip. The wing and body drag, resulting from the relative airflow in the sideslip and the forward motion of the aircraft, produces forces that act parallel to the longitudinal axis and at right angles to it in the direction of the raised wing. These forces produce a turning moment about the CG, which together with a certain loss of lift on the upper main plane

(caused by turbulence over the fuselage) and the pendulum effect tends to lift the aircraft.

Again when an aircraft is in a sideslip as a result of a roll, air loads will act on the side of the fuselage and on the vertical stabilizer (fin/rudder assembly) which together form the *keel surface*, i.e. the cross-sectional area of the aircraft when viewed from the side. These loads produces a rolling moment, which has a stabilizing effect. The magnitude of this moment is mainly dependent on the size of the fin and its distance from the aircraft CG (Figure 7.43).

Sweepback and lateral stability

Wings with sweepback can also enhance lateral stability. As the aircraft sideslips following a disturbance in roll, the lower sweptback wing generates more lift than the upper wing. This results from the fact that in the sideslip the lower wing presents more of its span to the airflow

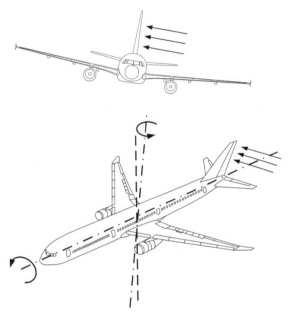

Figure 7.43 Restoring moment created by relative airflow acting on the fin.

than the upper wing (Figure 7.44(a)), therefore the lower wing generates more lift and tends to restore the aircraft to a wings level position.

Figure 7.44(b) shows how the component of the velocity perpendicular to the leading edge is increased on the down-going mainplane. It is this component of velocity that produces the increased lift and together with the increase in effective wing span restores the aircraft to a wings level position.

In addition, the surface of the down-going mainplane will be more steeply cambered to the relative airflow than that of the up-going mainplane. This will result in the down-going mainplane having a higher lift coefficient compared with the up-going mainplane during sideslip, thus the aircraft tends to be restored to its original attitude.

Lateral dynamic stability

The relative effect of combined rolling, yawing and sideslip motions, resulting from *aerodynamic coupling* (see directional stability), determine the lateral dynamic stability of an aircraft. If the aircraft stability characteristics are not sufficient, the complex motion interactions produce three possible types of instability, these are:

- directional divergence
- spiral divergence
- dutch roll.

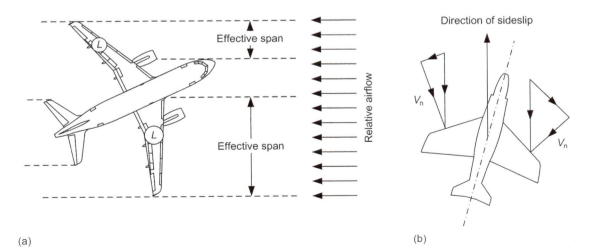

(a)   (b)

Figure 7.44 Sweepback enhances lateral stability.

If an aircraft is directionally unstable a divergence in yaw may result from an unwanted yaw disturbance. In addition a side force will act on the aircraft, while in the yawed position, it will curve away from its original flight path. If under these circumstances the aircraft has lateral static stability, *directional divergence* will occur without any significant degree of bank angle and the aircraft will still fly in a curved path with a large amount of sideslip.

Spiral divergence exits when directional static stability is very large when compared with lateral stability. This may occur on aircraft with a significant amount of anhedral coupled with a large fin, such as the old military Lightning fighter aircraft. If an aircraft is subject to a yaw displacement, then because of the greater directional stability the yaw would be quickly eliminated by the stabilizing yawing moment set up by the fin. However, a rolling moment would also be set up in the same direction as the yaw and if this rolling moment were strong enough to overcome the restoring moment due to static stability, the angle of bank would increase and cause the aircraft nose to drop into the direction of the yaw. The aircraft then begins a nose spiral which may develop into a spiral dive.

Dutch roll is an oscillatory mode of instability which may occur if an aircraft has positive directional static stability, but not so much in relation to static lateral stability, as to lead to spiral divergence. Thus Dutch roll is a form of lateral dynamic instability that does not quite have the inherent dangers associated with spiral divergence. Dutch roll may occur where there is a combination of high wing loading, sweepback, high altitude and where the weight is distributed towards the wing tips. If an aircraft is again subject to a yaw disturbance, it will roll in the same direction as the yaw. Directional stability will then begin to reduce the yaw and due to inertia forces, the aircraft will over-correct and start to yaw in the opposite direction. Now each of these continuing oscillations in yaw act in such a manner as to cause further displacements in roll, the resulting motion being a combination of roll and yaw oscillations which have the same frequency, but are out of phase with each other. *The development of Dutch roll is prevented by fitting aircraft with yaw damping systems*, which will be looked at in more detail when aircraft control is studied in third book in the series.

Longitudinal stability

As mentioned previously, an aircraft is longitudinally statically stable if it has the tendency to return to a trimmed AOA position following a pitching disturbance. Consider an aircraft, initially without a tailplane or horizontal stabilizer (Figure 7.45), which suffers a disturbance causing the nose to *pitch-up*. The CG will continue to move in a straight line, so the effects will be:

- an increase in the AOA;
- the CP will move forward;
- a clockwise moment about the CG provided by the lift force.

This causes the nose to keep rising so that it will not return to the equilibrium position. The aircraft is thus unstable.

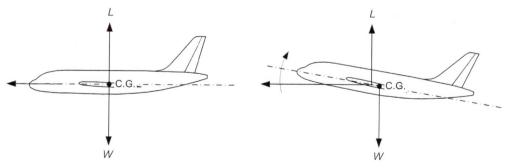

Figure 7.45 Unchecked nose-up pitching moment for an aircraft without tailplane, after a disturbance.

If the pitching disturbance causes a *nose-down* attitude, the CP moves to the rear and the aircraft is again unstable (Figure 7.46).

For an aircraft to be *longitudinal statically stable*, it must meet two criteria:

- A nose-down pitching disturbance must produce aerodynamic forces to give a nose-up restoring moment.
- This restoring moment must be large enough to return the aircraft to the trimmed AOA position after the disturbance.

Thus, the requirements for longitudinal stability are met by the tailplane (horizontal stabilizer). Consider now, the effects of a nose-up pitching moment on an aircraft with tailplane (Figure 7.47). The CG of the aircraft will still continue to move around a vertical straight line. The effects will now be:

- an increase in AOA for both wing and tailplane;
- the CP will move forward and a lift force will be produced by the tailplane;

- the tailplane will provide an anticlockwise restoring moment ($L_T y$), i.e. greater than the clockwise moment ($L_W x$), produced by the wing lift force as the CP moves forward.

A similar restoring moment is produced for a nose-down disturbance except that the tailplane lift force acts downwards and the direction of the moments are reversed.

From the above argument, it can be seen that the *restoring moment* depends on:

- the size of the tailplane (or horizontal stabilizer);
- the distance of the tailplane behind the CG;
- the amount of elevator movement (or complete tailplane movement, in the case of aircraft with all-moving slab tailplanes) which can be used to increase tailplane lift force.

All of the above factors are limited, therefore as a consequence, there will be a limit to the restoring moment that can be applied. It is therefore necessary to ensure that the disturbing moment produced by the wing lift moment

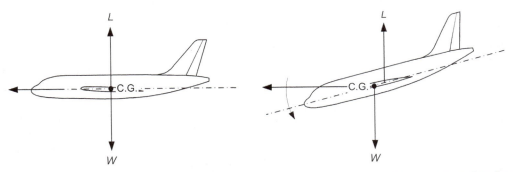

Figure 7.46 Unchecked nose-down pitching moment for an aircraft without tailplane, after a disturbance.

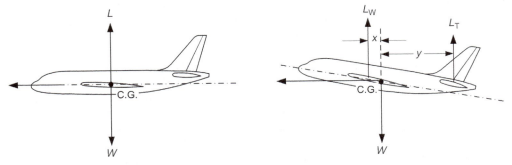

Figure 7.47 Nose-up pitching moment being counteracted by tailplane restoring moment.

about the CG is also limited. This moment is affected by movements of the CG due to differing loads and load distributions, in addition to fuel load distribution.

It is therefore vitally important that *the aircraft is always loaded so that the CG stays within the limits specified in the aircraft weight and balance documentation* detailed by the manufacturer. The result of failure to observe these limits may result in the aircraft becoming unstable with subsequent loss of control or worse!

Dynamic longitudinal stability

Longitudinal dynamic stability consists of two basic modes, one of which you have already met, phugoid (Figure 7.37). *Phugoid* motion consists of long period oscillations that involve noticeable changes in pitch attitude, aircraft altitude and airspeed. The pitching rate is low and because only very small changes in AOA occur, damping is weak and sometimes negative.

The second mode involves short period motion of relatively high frequency that involves negligible changes in aircraft velocity. During this type of motion, static longitudinal stability restores the aircraft to equilibrium and the amplitude of the oscillation is reduced by the pitch damping contributed by the tailplane (horizontal stabilizer). If instability was to exist in this mode of oscillation, *porpoising* of the aircraft would occur and because of the relative high frequency of oscillation, the amplitude could reach dangerously high proportions with severe flight loads being imposed on the structure.

Directional stability

As you already know, directional stability of an aircraft is its inherent (built-in) ability to recover from a disturbance in the yawing plane, i.e. about the normal axis. However, unlike longitudinal stability, it is not independent in its influence on aircraft behaviour because as a result of what is known as *aerodynamic coupling*, yaw displacement moments also produce roll displacement moments about the longitudinal axis. As a consequence of this aerodynamic coupling aircraft directional motions have an

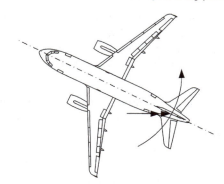

Figure 7.48 Restoring moment created by the fin after a yawing disturbance.

effect on lateral motions and vice versa. The nature of these motions are yawing, rolling and sideslip, or any combination of the three.

With respect to yawing motion only, the primary influence on directional stability is provided by the fin (or vertical stabilizer). As the aircraft is disturbed from its straight and level path by the nose or tail being pushed sideways (yawed), then due to its inertia the aircraft will continue to move in the direction created by the disturbance. This will expose the keel surface to the on-coming airflow. Now the fin, acting as a vertical aerofoil, will generate a sideways lift force which tends to swing the fin back towards its original position, straightening the nose as it does so.

It is thus the powerful turning moment created by the vertical fin, due to its large area and distance from the aircraft CG, which restores the aircraft nose back to its original position (Figure 7.48). The greater the keel surface area (which includes the area of the fin) behind the CG, and the greater the moment arm, then the greater will be the directional stability of the aircraft. Knowing this, it can be seen that a forward CG is preferable to an aft CG, since it provides a longer moment arm for the fin.

We finish our study of basic aerodynamics by looking briefly at the way in which aircraft are controlled. This introduction to the subject is provided here for the sake of completeness and in order to better understand the interactions between stability and control. It does not form part of Module 8, rather, it is part of the aerodynamics covered in Modules 11 and 13, which as mentioned before will be covered in detail in later books in the series.

7.6 Control and controllability

7.6.1 Introduction

To ensure that an aircraft fulfils its intended operational role it must have, in addition to varying degrees of stability, the ability to respond to the requirements for manoeuvring and trimming about its three axes. Thus the aircraft must have the capacity for the pilot to control it in roll, pitch and yaw, so that all desired flying attitudes may be achieved throughout all phases of flight.

Controllability is a different problem from stability in that it requires aerodynamic forces and moments to be produced about the three axes, where these forces *always oppose the inherent stability restoring moments*, when causing the aircraft to deviate from its equilibrium position. Thus if an aircraft is highly stable, the forces required to deviate the aircraft from its current position will need to be greater than those required to act against an aircraft that is less inherently stable. This is one of the reasons why aircraft that are required to be highly manoeuvrable and respond quickly to pilot or autopilot demands, are often designed with an element of instability built-in.

Different control surfaces are used to control the aircraft about each of the three axes. Movement of these control surfaces changes the airflow over the aircraft's surface. This, in turn, produces changes in the balance of the forces that keep the aircraft flying straight and level, thus creating the desired change necessary to manoeuvre the aircraft. No matter how unusual the aircraft configuration may be, conventional flying control surfaces are always disposed so that each gives control about the aircraft axes (Figure 7.49). Thus:

- *Ailerons* provide roll control about the longitudinal axis.
- *Elevators* on the tailplane provide longitudinal control in pitch about the lateral axis.

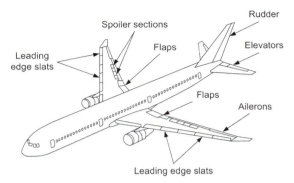

Figure 7.49 The Airbus A320 showing the ailerons elevators and rudder control surfaces. The spoiler sections, leading edge slats and trailing edge flaps can also be seen in the diagram.

Figure 7.50 Concorde, with the elevon control surfaces clearly visible.

- The *rudder* on the fin gives control in yawing about the normal axis.

On some aircraft a control surface group is provided, one group known as *elevons*, provide the combined control functions of the elevators and ailerons. This type of control is often fitted to delta-wing aircraft such as Concorde (Figure 7.50).

When these two control surfaces are lowered or raised together they act as elevators, when operated differentially they act as ailerons.

Another common control grouping is the *taileron*, which has been designed to provide the combined control functions of the tailplane and ailerons. Here the two sides of the slab tail will

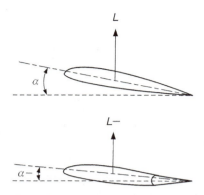

Figure 7.51 Down-going wing created by an up-going aileron.

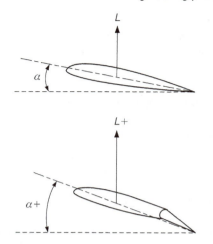

Figure 7.52 Up-going wing created by a down-going aileron.

act collectively to provide tailplane control in pitch or differentially to provide aileron control.

7.6.2 Ailerons

As you are already aware, aileron movement causes roll by producing a difference in the lift forces over the two wings. One aileron moves up and the other simultaneously moves down. The aileron that is deflected upwards causes an effective decrease in the angle of attack (Figure 7.51) of the wing with a subsequent reduction in C_L and lift force, so we have a down-going wing.

Similarly an aileron deflected downwards causes an effective increase in the AOA of the wing, increasing C_L and lift force, so we have an up-going wing.

While the ailerons remain deflected the aircraft will continue to roll. To maintain a steady angle of bank, the ailerons must be returned to the neutral position after the required angle has been reached.

Aileron drag

Ailerons cause more complicated aerodynamic problems than other control surfaces. One of these problems is *aileron drag* or *adverse yaw*. Ailerons not only produce a difference in lift forces between the wings, but also produce a difference in drag force. The drag force on the up-going aileron (down-going wing) becomes greater due to air loads and turbulence than the drag force on the down-going wing

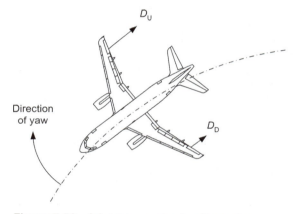

Figure 7.53 Adverse yaw due to aileron drag.

(Figure 7.52). The effect of this is to produce an adverse yaw away from the direction of turn (Figure 7.53).

There are two common methods of reducing aileron drag. The first involves the use of *differential ailerons*, where the aileron that is deflected downwards moves through a smaller angle than the aileron that is deflected upwards (Figure 7.54). This tends to equalize the drag on the two wings.

The second method, only found on older low speed aircraft, uses *frise ailerons*. A frise aileron has a *beak*, which projects downwards into the airflow (Figure 7.55), when the aileron is

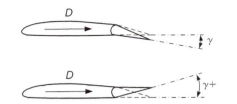

Figure 7.54 Differential ailerons used to minimize aileron drag.

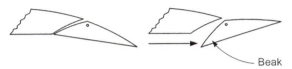

Figure 7.55 The frise type aileron used to reduce aileron drag.

Figure 7.56 High speed aileron reversal.

deflected upwards, but does not project when it is deflected downwards.

The beak causes an increase in drag on the down-going wing, helping to equalise the drag between the wings.

Aileron reversal

At *low speeds*, an aircraft has a relatively high AOA that is close to the stall angle. If the ailerons are operated while the wings are at this high angle of attack, the increase in the effective angle of attack may cause the wing with the aileron deflected downwards to have a *lower* C_L than the other, instead of the normal higher C_L. This will cause the wing to drop instead of rise and the aircraft is said to have suffered *low speed aileron reversal*.

When ailerons are deflected at high speeds the aerodynamic forces set-up may be sufficiently large to twist the outer end of the wing (Figure 7.56).

This can cause the position of the chord line to alter so that the result is the opposite of what would be expected. That is, a downward deflection of the aileron causes the wing to drop and an upward deflection causes the wing to rise,

under these circumstances we say that the aircraft has suffered a *high speed aileron reversal*. On modern large transport aircraft, that fly at relatively high speed, and high speed military aircraft this becomes a serious problem.

Solutions to this problem include:

- Building sufficiently stiff wings that can resist torsional divergence beyond the maximum speed of the aircraft.
- Use of two sets of ailerons, one outboard pair that operate at low speeds and one inboard pair that operate at high speeds, where the twisting moment will be less than when the ailerons are positioned outboard.
- Use of spoilers (Figure 7.49), either independently or in conjunction with ailerons, where their use reduces the lift on the down-going wing by interrupting the airflow over the top surface. Spoilers do not cause the same torsional divergence of the wing and have the additional advantage of providing increased drag on the down-going wing, thus helping the adverse yaw problem, created by aileron drag.

7.6.3 Rudder and elevators

The rudder

Movement of the rudder to port gives a lift force to starboard which yaws the aircraft nose to port. Although this will cause the aircraft to turn, eventually, it is much more effective to use the ailerons to bank the aircraft, with minimal use of the rudder. The main functions of the rudder are:

- Application during take-off and landing to keep the aircraft straight while on the runway.
- To provide limited assistance during the turn by helping the aircraft to yaw correctly into the turn.
- Application during spin to reduce the roll rate and aid recovery from the spin
- Application at low speeds and high angles of attack to help raise a dropping wing that has suffered aileron reversal.
- Application on multi-engine aircraft to correct yawing when asymmetric power conditions exist.

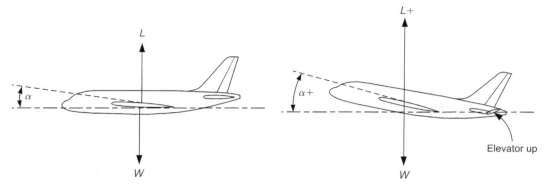

Figure 7.57 Upward deflection of elevator causes a downward deflection of tailplane and a subsequent increase in the AOA.

Elevators

When the elevator is deflected upwards it causes a downward lift force on the tailplane (Figure 7.57).

Thus the upward movement of the elevators causes an increase in the AOA and the forward movement of the CP. This means that the aircraft will rise if the speed is maintained or increased because the C_L rises as the AOA rises and a larger lift force is created. To maintain an aircraft in a steady climb the elevators are returned to the neutral position. If elevator deflection is maintained, the aircraft would continue up into a loop.

If the speed is reduced as the elevators are raised the aircraft continues to fly level because the increase in C_L due to the increased AOA, is balanced by the decrease in velocity and so from, $L = C_L \frac{1}{2}\rho v^2 S$ the total lift force will remain the same. Under these circumstances, the elevator deflection must be maintained to keep the nose high unless the speed is increased again.

If we wish to pull an aircraft out of a glide or dive, then upward elevator movement must be used to increase the total lift force, necessary for this manoeuvre.

7.6.4 Lift augmentation devices

Lift augmentation devices fall into two major categories, trailing and leading edge flaps and slats and slots. We finish our introduction to control by looking briefly at these two categories of lift augmentation device.

Figure 7.58 Plain flap.

If an aircraft is to take-off and land in a relatively short distance, its wings must produce sufficient lift at a much slower speed than in normal cruising flight. It is also necessary during landing, to have some means of slowing the aircraft down. Both these requirements can be meet by the use of flaps and slats or a combination of both.

Flaps are essentially moving wing sections which increase wing camber and therefore angle of attack. In addition, in some cases, the effective wing area is also increased. Dependent on type and complexity flap systems are capable of increasing C_{Lmax} by up to approximately 90% of the clean wing value.

Flaps also greatly increase the drag on the wings thus slowing the aircraft down. Thus on take-off, flaps are partially deployed and the increase in drag is overcome with more thrust, while on landing they are fully deployed for maximum effect.

Trailing edge flaps

There are many types of trailing edge flaps and a few of the more common types are described below.

The *plain flap* (Figure 7.58) is normally retracted to form a complete section of trailing edge, and hinged downward in use.

Figure 7.59 Split flap.

Figure 7.60 Slotted flap.

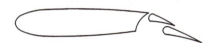

Figure 7.61 Double slotted Fowler flap.

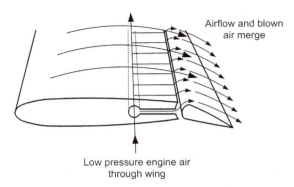

Airflow and blown air merge

Low pressure engine air through wing

Figure 7.62 Blown flap.

Figure 7.63 Trailing edge multi-slotted Fowler flap system on Boeing 747-400.

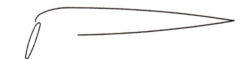

Figure 7.64 Leading edge Kruger flap.

The *split flap* (Figure 7.59) is formed by the hinged lower part of the trailing edge only.

When the flap is lowered the top surface is unchanged thus eliminating airflow breakaway which occurs over the top surface of the plain flap at large angles of depression.

During the operation of the *slotted flap*, a gap or slot is formed between the wing and flap (Figure 7.60).

Air flows through the gap from the lower surface and over the top surface of the flap. This increases lift by speeding up the airflow. This more energetic laminar flow remains in contact with the top surface of the flap for longer delaying boundary layer separation and maintaining a high degree of lift.

The *Fowler flap* (Figure 7.61) is similar to the split flap but this type of flap moves rearwards as well as downwards on tracks, creating slots, if more than one fowler is connected as part of the system. Thus, both wing camber and wing area are increased.

In the *blown flap* (Figure 7.62), air bled from the engines is ducted over the top surface of the flap to mix with and re-energize the existing airflow.

Figure 7.63 shows the trailing edge flap system of a Boeing 747-400, in the deployed position. This system is a *multi-slotted Fowler combination*, which combines and enhances the individual attributes of the slotted flap and fowler flap, greatly enhancing lift.

Leading edge flaps

As mentioned earlier *leading edge flaps* are used to augment low speed lift especially on swept wing aircraft. Leading edge flaps further increase the camber and are normally coupled to operate together with trailing edge flaps. They also prevent leading edge separation that takes place on thin sharp-edged wings at high angles of attack. This type of flap is often known as a *Kruger flap* (Figure 7.64).

Slats and slots

Slats are small, high cambered aerofoils (Figure 7.65) fitted to the wing leading edges.

When open, *slats* form a slot between themselves and the wing through which air from the higher pressure lower surface accelerates and flows over the wing top surface to maintain lift and increase the stalling angle of the wing. Slats may be fixed, controlled or automatic.

A *slot* is a suitably shaped aperture built into the wing structure near the leading edge (Figure 7.66).

Slots guide and accelerate air from below the wing and discharge it over the upper surface to re-energize the existing airflow. Slots may be fixed, controlled, automatic or blown.

In Figure 7.67 a typical leading and trailing edge lift enhancement system is illustrated.

This system consists of a triple-slotted Fowler flap at the trailing edge, with a slat and Kruger flap at the leading edge. This combination will significantly increase the lift capability of the aircraft.

Figure 7.65 The leading edge slat.

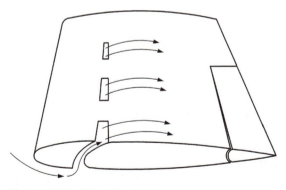

Figure 7.66 Wing-tip slots.

Figure 7.67 Modern aircraft lift enhancement system.

7.6.5 Aerodynamic balance, mass balance and control surface tabs

Aerodynamic balance

On all but the smallest of low speed *aircraft*, the size of the control force will produce hinge moments that produce control column forces that are too high for easy control operation. Non-sophisticated light *aircraft* do not necessarily have the advantage of powered controls and as such , they are usually designed with some form of inherent *aerodynamic balance* that assists the pilot during their operation. There are several methods of providing aerodynamic balance, three such methods are given below.

Inset hinges are set back so that the airflow striking the surface in front of the hinge to assist with control movement (Figure 7.68). A rule of thumb with this particular design is to limit the amount of control surface forward of the hinge line to about 20% of the overall control surface area. Following this rule, helps prevent excessive snatch and over-balance.

Another way of achieving lower hinge moments and so assisting the pilot to move the controls is to use a horn balance (Figure 7.69).

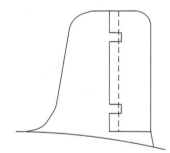

Figure 7.68 Inset hinge control surface.

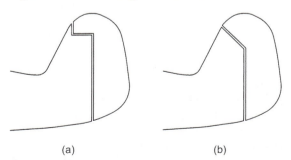

(a)   (b)

Figure 7.69 The horn balance.

The principle of operation is the same as for the insert hinge. These balances can be fitted to any of the primary flying control surfaces. Figure 7.69(a) shows the **standard horn balance**, this device is sometimes prone to snatch; a **graduated horn balance** (Figure 7.69(b)) overcomes this problem by introducing a progressively increasing amount of control surface area into the airflow forward of the hinge line, rather than a sudden change in area that may occur with the standard horn balance.

Figure 7.70 illustrates the *internal balance*, where the balancing area is inside the wing.

For example, downward movement of the control surface creates a decrease in pressure above the wing and a relative increase below the wing. Since the gap between the low and high pressures areas is sealed (using a flexible strip) the pressure acting on the strip, creates a force that acts upwards, which in turn produces the balancing moment that assists the pilot to move the controls further. This situation would obviously work in reverse when the control surface is moved up.

Mass balance

If the CG of a control surface is some distance behind the hinge line, then the inertia of the conrol surface may cause it to oscillate about the hinge, as the structure distorts during flight.

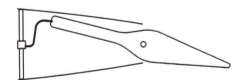

Figure 7.70 The internal balance.

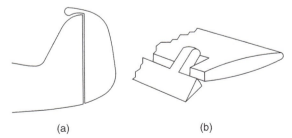

(a)   (b)

Figure 7.71 Mass balance helps prevent control surface flutter.

This undesirable situation is referred to as **control surface flutter** and in certain circumstances these flutter oscillations can be so severe as to cause damage or possible failure of the structure.

Flutter, may be prevented by adding a carefully determined mass to the control surface in order to bring the CG closer to the hinge line. This procedure, known as *mass balance*, helps reduce the inertia moments and so prevents flutter developing. Figure 7.71 shows a couple of examples of mass balance, where the mass is adjusted forward of the hinge line, as necessary.

Control surfaces that have been re-sprayed or repaired must be check weighed and the CG re-calculated to ensure that it remains within laid down limits.

Tabs

A tab is a small hinged surface forming part of the trailing edge of a primary control surface. Tabs may be used for:

- control balancing, to assist the pilot to move the control;
- servo operation of the control;
- trimming.

The balance tab (Figure 7.72) is used to reduce the hinge moment produced by the control and is therefore a form of aerodynamic balance, which reduces the effort the pilot needs to apply to move the control.

The tab arrangement described above may be reversed to form an *anti-balance tab* (Figure 7.73).

The anti-balance tab is connected in such a way, as to move in the same direction as the

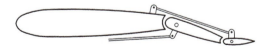

Figure 7.72 The balance tab.

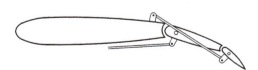

Figure 7.73 The anti-balance tab.

control surface so increasing the control column loads. This tab arrangement is used to give the pilot *feel*, so that the aircraft will not be overstressed as a result of excessive movement of the control surface by the pilot.

The spring tab arrangement is such that tab movement is proportional to the load on the

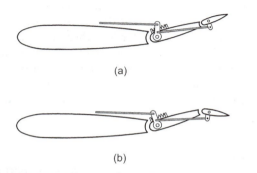

(a)

(b)

Figure 7.74   Spring tab operation.

Figure 7.75   Servo-tab arrangement.

control rather than the control surface deflection angle (Figure 7.74). Spring tabs are used mainly to reduce control loads at high speeds.

The spring is arranged so that below a certain speed it is ineffective (Figure 7.74(a)). The aerodynamic loads are such that they are not sufficient to overcome the spring force and the tab remains in line with the primary control surface. As speed is increased the aerodynamic load acting on the tab is increased sufficiently to overcome the spring force and the tab moves in the opposite direction to the primary control to provide assistance (Figure 7.74(b)).

Servo tab
The servo tab is designed to operate the primary control surface (Figure 7.75). Any deflection of the tab produces an opposite movement of the free-floating primary control surface, thereby reducing the effort the pilot has to apply to fly the aircraft.

Trim tab
The trim tab is used to relieve the pilot from any sustained control force that may be needed to

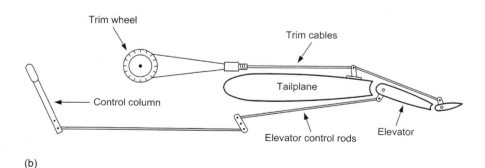

(a)

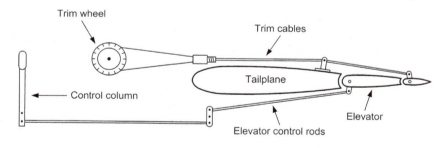

(b)

Figure 7.76   A typical manual elevator trim system.

maintain level flight. These out of balance forces may occur as a result of fuel use, variable freight and passenger loadings or out of balance thrust production from the aircraft engines. A typical trim system is illustrated in Figure 7.76(a).

On a manual trim system, the pilot operates a trim wheel, which provides instinctive movement. So, for example, if the pilot pushes the elevator trim wheel forward the nose of the aircraft would drop, the control column would also be trimmed forward of the neutral position, as shown in Figure 7.76(b). Under these circumstances the trim tab moves up, the elevator is moved down by the action of the trim tab, thus the tail of the aircraft will rise and the nose will pitch down. At the same time the elevator control rods move in such a way as to pivot the control column forward. A similar set up may be used for aileron trim, where in this case the trim wheel would be mounted parallel to the aircraft lateral axis and rotation of the wheel clockwise would drop the starboard wing, again the movement being instinctive.

The subject matter contained within the sections on stability and control, lends itself to multiple-choice answer type questions. For this reason there are no test your understanding questions to end these sections. Instead, a typical example of multiple-choice objective question paper that covers *all* of the subject matter contained in *Module 8* is given next.

7.7 Multiple choice questions

The example questions set out below follow the sections of Module 8 in the Part-66 syllabus. In addition there are questions set on aircraft manual flying controls. It is felt that an introduction to control and controllability must accompany the knowledge required on aircraft stability, which forms a core element of this module.

Please note that for this module, there are only a few questions that are considered inappropriate for those strictly following the Category A pathway and these have been annotated as **B1**, **B2**. However, *it is felt by the authors* that it would be to the advantage of potential Category

A Mechanics, if they studied this module at the Category B level. All questions have thus been designed to test the knowledge required to the highest Category B certifying technician level, in line with the chapter content.

Please remember, as always, that *ALL questions must be attempted* without *the use of a calculator* and that the pass mark for all JAR 66 multiple-choice examinations is 75%.

1. The composition by volume of the gases in the atmosphere is:
   (a) 78% oxygen, 21% nitrogen, 1% other gases
   (b) 78% nitrogen, 21% oxygen, 1% other gases
   (c) 81% oxygen 18% nitrogen, 1% other gases

2. The layer of the atmosphere next to the surface of the earth is called:
   (a) Ionosphere
   (b) Stratosphere
   (c) Troposphere

3. Approximately 75% of the mass of the gases in the atmosphere is contained in the layer known as the:
   (a) Chemosphere
   (b) Stratosphere
   (c) Troposphere

4. If the temperature of the air in the atmosphere increases but the pressure remains constant, then the density will:
   (a) decrease
   (b) increase
   (c) remain the same

5. In the International Civil Aviation Organization standard atmosphere, the stratosphere commences at an altitude of:
   (a) 11 km
   (b) 30 km
   (c) 11,000 ft

6. In the International Standard Atmosphere, the mean sea-level temperature is set at:
   (a) 15 K
   (b) 288°C
   (c) 288 K

7. In the International Standard Atmosphere, the mean sea-level density is set at:
   (a) 1.2256 kg/m³
   (b) 1.01325 kg/m³
   (c) 14.7 kg/m³

8. With increase in altitude with respect to the pressure and density in the atmosphere:
   (a) pressure increases, density decreases
   (b) pressure decreases, density increases
   (c) both decrease

9. The temperature at the tropopause in the International standard atmosphere is approximately:
   (a) 15 K
   (b) −56°C
   (c) −6 K

10. The transition level between the troposphere and stratosphere is known as:
    (a) Tropopause
    (b) Stratopause
    (c) Chemopause

11. The rate of decrease of temperature in the first layer of the International Standard Atmosphere, is assumed linear and is given by the formula: [B1, B2]
    (a) $T_0 = T_h - Lh$
    (b) $L_h = T_0 - T_h$
    (c) $T_h = T_0 - Lh$

12. With increase in altitude, the speed of sound will:
    (a) increase
    (b) decrease
    (c) remain the same

13. If the density ratio at an altitude in the International Standard Atmosphere is 0.5, then the density at that altitude will be approximately:
    (a) 2.4512 kg/m³
    (b) 1.2256 kg/m³
    (c) 0.6128 kg/m³

14. Given that the speed of sound at altitude may be estimated from the relationship $a = 20.05\sqrt{T}$. Then the speed of sound at the tropopause in the International Standard

Atmosphere will be approximately:
                                        [B1, B2]
   (a) 400 m/s
   (b) 340 m/s
   (c) 295 m/s

15. The dynamic viscosity of the air in the International Standard atmosphere is given a value of:
    (a) 1.789×10^{-5} N s/m²
    (b) 6.5 K/km
    (c) 9.80665 m/s²

16. Bernoulli's theorem may be represented by the equation:
    (a) $p + v = $ constant
    (b) $pT + \rho v^2 = $ constant
    (c) $p + \frac{1}{2}\rho v^2 = $ constant

17. The component of the total reaction that acts parallel to the relative airflow is known as:
    (a) lift
    (b) drag
    (c) thrust

18. Flow in which the particles of the fluid move in an orderly manner and maintain the same relative positions in successive cross sections, is known as:
    (a) turbulent flow
    (b) streamline flow
    (c) downwash flow

19. The dimension from port wing tip to starboard wing tip is known as:
    (a) wing span
    (b) wing chord
    (c) aspect ratio

20. The mean camber line is defined as:
    (a) the line drawn halfway between the upper and lower curved surfaces of an aerofoil
    (b) the line joining the center of curvature of the trailing and leading edge of an aerofoil
    (c) the straight line running from wing root to wing tip

21. The angle of attack is defined as:
    (a) the angle between the relative airflow and the longitudinal axis of the aircraft

(b) the angle between the chord line and the relative airflow

(c) the angle between the maximum camber line and the relative airflow

22. Which of the three graphs (Figure 7.77) correctly shows the relationship between C_L and angle of attack for a symmetrical aerofoil?

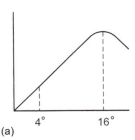

(a)

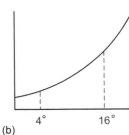

(b)

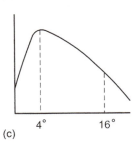

(c)

Figure 7.77

23. The angle of incidence:
    (a) is a fixed rigging angle on conventional layout aircraft
    (b) varies with aircraft attitude
    (c) is altered using the tailplane

24. A primary reason for having thin aerofoil sections on high speed aircraft is to:
    (a) increase the speed of the relative airflow over the top surface of the aerofoil section

(b) help reduce the time spent in the transonic range

(c) increase the fuel dispersion throughout the wing and improve handling quality

25. The lift of an aerofoil is created by:
    (a) decrease in pressure on both the upper and lower surfaces
    (b) increase in pressure on both the upper and lower surfaces
    (c) decrease in pressure on upper surface and increase in pressure on the lower surface

26. The efficiency of an aerofoil is measured using:
    (a) W/L ratio
    (b) L/D ratio
    (c) T/L ratio

27. If the angle of attack of an aerofoil is increased the center of pressure will:
    (a) move forward
    (b) move backward
    (c) stay the same

28. Once the stalling angle of an aerofoil has been reached the center of pressure will:
    (a) move rapidly backwards to about the mid-chord position
    (b) move rapidly forwards towards the leading edge
    (c) oscillate rapidly around the center of gravity

29. Boundary layer separation may be delayed using:
    (a) all moving tailplane
    (b) elevons
    (c) vortex generator

30. At the transition point the boundary layer becomes:
    (a) thicker with turbulent flow
    (b) thinner with turbulent flow
    (c) thinner with laminar flow

31. An equation for calculating the lift produced by an aerofoil is:
    (a) $\frac{1}{2}\rho v S C_L^2$
    (b) $\frac{1}{2}\rho v^2 S C_L$
    (c) $\frac{1}{2}\rho v S C_D$

32. The components of zero lift profile drag are:
    (a) skin friction drag, form drag and interference drag
    (b) induced drag, form drag and interference drag
    (c) skin friction drag, vortex drag and induced drag

33. Interference drag may be reduced by:
    (a) highly polished surface finish
    (b) high aspect ratio wings
    (c) fairings at junctions between fuselage wing

34. Form drag may be reduced by:
    (a) streamlining
    (b) highly polished surface finish
    (c) increased use of high lift devices

35. The term "wash-out" is defined as:
    (a) decrease of incidence towards the wing tip
    (b) increase of incidence towards the wing tip
    (c) a chord wise decrease in incidence angle

36. If lift increases, vortex drag:
    (a) increases
    (b) decreases
    (c) remains the same

37. The aspect ratio may be defined as:
    (a) span squared/chord
    (b) span squared/area
    (c) chord/span

38. Profile drag:
    (a) is not affected by airspeed
    (b) increases with the square of the airspeed
    (c) decreases with the square of the airspeed

39. Tapered wings will produce:
    (a) less vortex drag than a non-tapered wing
    (b) more vortex drag than a non-tapered wing
    (c) the same vortex drag as a non-tapered wing

40. Glaze ice:
    (a) forms on the surface of a wing at a temperature below which frost is formed in the adjacent air
    (b) forms in freezing fog from individual water droplet particles
    (c) forms in freezing rain, where the air temperature and that of the airframe are both below freezing point

41. With respect to ice accretion and aircraft performance select the *one correct* statement:
    (a) Increases in lift and drag will occur as a result of changes to the wing section
    (b) A decrease in drag and increase in lift will occur due to decrease in friction over wing surface
    (c) aerodynamic instability may occur

42. The ratio of the length of a streamlined body to its maximum diameter is the:
    (a) aspect ratio
    (b) thickness ratio
    (c) finess ratio

43. Which figure (Figure 7.78) correctly shows the relationship between the forces acting on an aircraft in a steady climb?

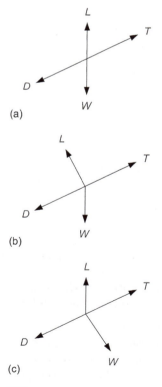

Figure 7.78

44. To fly an aircraft close to the stalling speed, the aircraft must be flown with wings:
    (a) at a high angle of attack
    (b) at zero angle of incidence
    (c) at or near the angle of incidence

45. Stalling speed increases with increase in altitude because: [B1,B2]
    (a) temperature decreases
    (b) air density decreases
    (c) the lift coefficient is increased

46. In a climb at steady speed the:
    (a) thrust is greater than the drag
    (b) thrust is equal to the drag
    (c) thrust is less than the drag

47. The glide angle of an aircraft may be found from the relationship:
    (a) height/range
    (b) range/height
    (c) lift/weight

48. The angle of bank for an aircraft in a steady turn may be calculated from the formula:
                                                                                         [B1,B2]
    (a) $\tan \theta = \dfrac{v^2}{gr}$
    (b) $\sin \theta = \dfrac{v^2}{gr}$
    (c) $\cos \theta = \dfrac{v^2}{gr}$

49. Aircraft load factor is found from the relationship:
    (a) lift/drag
    (b) lift/weight
    (c) weight/drag

50. The flight maneouvring envelope is a means of displaying:
    (a) gust conditions requiring no further investigation
    (b) discharge coefficients
    (c) flight operating strength limitations

51. The taper ratio is the ratio of the wing:
    (a) tip chord to root chord
    (b) root thickness to tip thickness
    (c) root thickness to mean chord

52. If a disturbing force is removed from a body and the body immediately tends to return towards the equilibrium, then it is:

(a) statically stable
(b) dynamically stable
(c) dynamically unstable

53. The function of the tailplane is to assist:
    (a) lateral stability about the longitudinal axis
    (b) longitudinal stability about the lateral axis
    (c) directional stability about the normal axis

54. If a disturbing force is removed from a body and the body settles in a position away from its previous equilibrium position, it is said to be:
(a) statically stable
(b) dynamically stable
(c) neutrally stable

55. The function of the horizontal stabilizer is to assist:
    (a) lateral stability about the longitudinal axis
    (b) longitudinal stability about the lateral axis
    (c) lateral stability about the lateral axis

56. Spiral divergence is a form of: [B1,B2]
    (a) lateral dynamic instability
    (b) longitudinal dynamic instability
    (c) lateral static stability

57. The effect on an aircraft subject to a nose-up pitching moment is:
    (a) to cause the center of pressure to move backwards
    (b) an increase in the angle of attack
    (c) to cause a nose-down pitching moment

58. Phugoid motion is a form of: [B1,B2]
    (a) longitudinal dynamic stability
    (b) directional dynamic instability
    (c) lateral dynamic instability

59. When an aircraft starts to roll, the effective angle of attack is:
    (a) increased on the up-going wing and decreased on the down-going wing
    (b) decreased on the up-going wing and increased on the down-going wing
    (c) increased on both wings

60. To ensure longitudinal stability in flight, the position of the center of gravity should:
    (a) be aft of the neutral point
    (b) coincide with the neutral point
    (c) be forward of the neutral point

61. Anhedral is defined as:
    (a) the upward and outward inclination of the wings
    (b) the downward and outward inclination of the aircraft wings
    (c) the forward sloping canard stabilizer

62. The fin of an aircraft helps to provide a restoring moment when an aircraft:
    (a) dives
    (b) pitches
    (c) yaws

63. The dihedral angle of a wing provides a restoring moment when an aircraft:
    (a) climbs
    (b) pitches
    (c) rolls

64. On a swept wing aircraft, that enters a sideslip, the air velocity normal to the leading edge increases: [B1, B2]
    (a) on both wings
    (b) on the up-going wing
    (c) on the down-going wing

65. Control of yaw is mainly influenced by:
    (a) the fin
    (b) the rudder
    (c) the tailplane

66. Movement of an aircraft about its normal axis is called
    (a) yawing
    (b) rolling
    (c) pitching

67. For a symmetrical aerofoil, downward deflection of a control surface:
    (a) increases both lift and drag
    (b) increases lift, decreases drag
    (c) decreases lift, increases drag

68. Different drag force between up-going and down-going ailerons is counteracted by:
    (a) aerodynamic balance control

(b) static balance
(c) differential aileron movement

69. The drag produced by aileron movement is:
    (a) greater on the down-going aileron
    (b) less on the down-going aileron
    (c) equal on both ailerons

70. At low speeds and at high angles of attack the wing with the down-going aileron may:
    (a) bend
    (b) stall
    (c) twist

71. At high speeds, the wing with the down-going aileron may:
    (a) turn at wing tip
    (b) yaw at the wing tip
    (c) twist at the wing tip

72. The frise type aileron is used to:
    (a) increase directional control
    (b) reduce high speed aileron reversal
    (c) reduce aileron drag

73. The lift augmentation device shown in Figure 7.79, is a:

Figure 7.79

(a) plain flap
(b) Kruger flap
(c) split flap

74. A Fowler flap:
    (a) increases the wing camber and the angle of attack
    (b) increases the wing camber and reduces the effective wing area
    (c) decrease the lift coefficient and stalling angle

75. The device shown at the leading edge of the Figure 7.80 is:
    (a) flap
    (b) slat
    (c) slot

Figure 7.80

76. All types of trailing edge flaps
    (a) decrease C_{Lmax} and increase C_D
    (b) increase C_{Lmax} and decrease C_D
    (c) increase both C_{Lmax} and C_D

77. If air is blown over the top surface of an aerofoil from within, the effect is to:
    (a) reduce surface friction drag
    (b) increase the boundary layer and so reduce form drag
    (c) re-energize the boundary layer and delay separation

78. The device used to produce steady flight conditions and reduce control column forces to zero is called:
    (a) a servo-tab
    (b) a trim tab
    (c) a balance tab

79. The device used to assist the pilot to move the controls is called:
    (a) a servo-tab
    (b) a trim tab
    (c) a balance tab

80. A servo-tab is deflected:
    (a) in the same direction as the control surface
    (b) parallel to the direction of the control surface
    (c) in the opposite direction to the control surface

81. An anti-balance tab is used to:
    (a) reduce pilot control column forces to zero
    (b) assist the pilot to move the control
    (c) provide more feel to the control column

82. When the control column of a manual control system is pushed forward, a balance tab on the elevator would:
    (a) move to the neutral position
    (b) move down
    (c) move up.

Final note: This ends our study of the Basic aerodynamics contained in JAR66 Module 8. As mentioned earlier, aircraft flying controls both manual and powered, together with a comprehensive study of high speed flight, will be covered in detail by the Mechanical Maintenance specialists, in Book 3 in the series (Aircraft Aerodynamics, Structural Maintenance and Repair). Individuals pursuing the Avionics pathway, will find the further required Aircraft Aerodynamic knowledge in the fifth book (avionic systems).

Appendices

Engineering licensing examinations

Individuals wishing to take *multiple-choice examinations* for specific modules will find a list of Part 147 approved training organizations in Appendix B. These organizations may also be approved Civil Aviation Authority (CAA) examination centres for multiple-choice individual module examinations that may also be open to external candidates. The appropriate organization should be contacted to ascertain the range and times for such Part 66 individual module examinations.

Set out below are the main UK approved personnel licensing centres, who run Engineer Licensing *written examinations* throughout the year. Their contact details and further information on dates and examination sitting availability may be obtained by contacting the CAA web site: *www.caa.co.uk/srg/licensing/el/documents.asp*.

Oral examinations are conducted in the UK at CAA National and Regional Offices, some of which are detailed below:

Aberdeen	Civil Aviation Authority, SRG 1st Floor, North Norfolk House Annexe Pitmedden Road Dyce, Aberdeen AB2 OBP Scotland, UK
East Midlands	Civil Aviation Authority, SRG Building 65 Ambassador Road East Midlands, Derbyshire DE74 2SA England, UK
Gatwick	Civil Aviation Authority, SRG Ground Floor, Consort House Consort Way Horley, Surrey RH6 7AF England, UK
Heathrow	Civil Aviation Authority, SRG Sipson House 595 Sipson Road Sipson, West Drayton Middlesex UB7 0JD England, UK
Irvine	Civil Aviation Authority, SRG Galt House Bank Street Irvine, Ayrshire KA12 0LL Scotland, UK
Luton	Civil Aviation Authority, SRG 1st Floor, Barrett House 668 Hitchen Road Stopsley Luton, Beds LU2 7HX England, UK
Manchester	Suite 12 Manchester International Of"ce Centre Styal Road Wythenshawe, Manchester M22 5WB England, UK
Stansted	Civil Aviation Authority, SRG Walden Court, Parsonage Lane Bishops Stortford Herts CM23 5DB England, UK
Weston-Super-Mare	Civil Aviation Authority, SRG Unit 101, Parkway Worle Weston-Super-Mare BS22 0WB England, UK

Given below are some *European EASA, National Aviation Authorities (NAA)*, with their office contact addresses as on July 2003.

Athens	Ministry of Transport and Communications Civil Aviation Authority PO Box 73751 GR16604 Helliniko Greece
Berlin	Luftfahrt-Bundesamt Aussenstelle Berlin Postfach 11 D-12508 Berlin Germany
Berne	Federal Office for Civil Aviation Maulbeerstrasse 9 Ch-3003 Berne Switzerland
Brussels	Administration de l'Aeronautique Central Communication Nord 4th Floor, Rue du Progres 80 B01210 Brussels Belgium
Copenhagen	Civil Aviation Administration Department of Aviation Inspection PO Box 744 DK-2450 Copenhagen SV Denmark
Czech Republic	Civil Aviation Authority Czech Republic Ruzyne Airport 160 08 Praha 6 Czech Republic
Dublin	Irish Aviation Authority Airworthiness Standards Department 3rd Floor, Aviation House Hawkins Street Dublin 2 Ireland
Frankfurt	Luftfahrt-Bundesamt Langer Kornweg 19-23 D-65451 Kelsterbach Germany
Helsinki	Civil Aviation Administration Box 50 SF-01531 Helsinki-Vantaa Finland
Hoofddorp	Civil Aviation Authority The Netherlands PO Box 575 2130 An Hoofddorp The Netherlands
Lisbon	Direccao-Geral da Aviacao Civil Direccao Do Material Aeronautico Rua B – Edificio No. 6 Aeroporto de Lisboa 1700 Lisboa Portugal
Luxembourg	Ministry of Transport 19–21 Boulevard Royal L-2938 Luxembourg Luxembourg
Madrid	Ministerio de Fomento Subdireccion General de Control del Transporte Aereo Pza. San Juan de la Cruz, s/n 28071 Madrid Spain
Malta	The Director General of Aviation Malta Department of Civil Aviation Malta International Airport Luga LQA05 Malta
Monaco	Civil Aviation Authority Heliport de Monaco MC 98000 Monaco Monaco
Oslo	Civil Aviation Administration Head Office Aeronautical Inspection Department PO Box 8124 Department N-0032 Oslo Norway

Paris   GS de L'Aviation Civile
             72-78 Grande Rue
             92310 Sevres, Paris
             France

Reykjavik   Civil Aviation
             Administration (FMS)
             PO Box 50
             Reykjavik Airport
             212 Reykjavik
             Iceland

Rome HQ   Registro Aeronautico Italiano
             Via di Villa Ricotti 42
             00161 Roma
             Italy

Stockholm   Luftfartsverket
             Flight Safety Department
             Surveyance Section
             Bergkallavagen 32
             Box 304
             192 30 Sollentuna
             Sweden

Vienna   Federal Ministry of Science,
             Transport and the Arts
             Department of Civil Aviation
             Radetzkystrasse 2
             A-1030 Vienna
             Austria

Warszawa   General Inspectorate of
             Civil Aviation
             Grojecka 17
             02-021 Warszawa
             Poland

Note: A *current full list* of all JAR-NAA offices will be found in the Joint Aviation Authority (JAA) publication catalogue under its administrative and guidance material in the JAR Directory. General information on all related JAR publications may be found on the JAA web site at: www.jaa.nl/catalogue/pubcat.html. The above list is correct at the time of going to the press. Since the arrival of EASA the above publications may still be found at the above web address or on the CAA web site.

Organizations offering aircraft maintenance engineering training and education

In the tables below, there will be found details on a selection of UK and European organizations known to the authors, who offer a variety of aircraft maintenance engineering training and educational programmes. **Table B.1** provides details on those EASA Part 147 approved organizations within the UK, who offer *ab initio* and other forms of training for categories A and B certifying staff. **Table B.2** provides similar detail to that given in Table B.1 for a selection of Part 147 approved *European* organizations. A *complete current unabridged list of Part 147 approved organizations* may be found in the Joint Aviation Authority (JAA) publication

Table B.1

Name and contact address of UK organization

ATC Lasham Ltd
Lasham Airfield
Lasham, Hampshire,
GU34 5SP, UK

Air Service Training (Engineering) Ltd
Perth Airport
Perth, PH2 6NP, UK

Airline Maintenance Training Inc
Building D1
Fairoaks Airport
Chobham, Surrey
GU24 8HX, UK

Britannia Airways
Technical Training School
Luton Airport, Luton
Bedfordshire, LU2 9ND, UK

City of Bristol College
Ashley Down, Bristol
BS7 9BU, UK

KLM UK Engineering Ltd
27 Hurricane Way
Norwich Airport
Norfolk, NR6 6HE, UK

FLS Aerospace (UK) Ltd
Long Border Road
Stanstead Airport
Essex, CM24 1RE, UK

Marshall of Cambridge Aerospace Ltd
Technical Training Department
The Airport, Cambridge, CB5 8RX, UK

Table B.2

Name and contact address of European organization

Agusta S.P.A.
Via G. Agusta 520
Cascina Costa di Samarate (VA), Italy

Airbus
1 Rond point Maurice Bellonte
31707 Blagnac Cedex, France

Austrian Airlines
Osterreichische Luftverkehrs
Aktiengesellschaft
A-1300 Vienna Airport, Austria

Dornier Luftfahrt GMBH
Airport Oberpfaffenhofen
82234 Wessling, Germany

EADS Airbus GMBH
Kreetslag 10
21129 Hamburg, Germany

FLS Aerospace (IRL) Ltd
Dublin Airport
County Dublin, Ireland

Lufthansa Technical Training
Weg beim Jager 193
22335 Hamburg, Germany

Scandinavian Airlines System
Maintenance Training Organization
Arlanda Airport, S-19587 Stockholm
Sweden

Shannon Aerospace Ltd
Shannon, County Clare, Ireland

Swiss Aviation Training Ltd
CH-4002 Basel, Switzerland

Table B.3

UK training and/or educational organization	Contact address	Type of course
Barry College	Colcot Road Barry, Vale of Glamorgan CF62 8YJ, UK www.barry.ac.uk	JAR 66 theory, to practising aircraft engineering personnel. Other BTEC qualifications, using their good practical facilities.
Brooklands College	Heath Road Weybridge, Surrey KT13 8TT, UK www.brooklands.ac.uk www.jar66now.com	On-line learning programme that will eventually cover the associated theory for all JAR66 Modules.
University of Bristol	IGDS Office Faculty of Engineering University of Bristol University Gate Park Row Bristol BS1 5UB, UK www.fen.bris.ac.uk/igds/aero	In conjunction with the University of the West of England, the Integrated Graduate Development Scheme (IDGS) is offered. Primarily concerned with aerospace design, manufacture and management; some maintenance technology biased modules offered.
Coventry University	School of Engineering Coventry University Prior Street Coventry CV1 5FB, UK www.coventry.ac.uk	Offers B.Eng. and I.Eng. degrees in aerospace systems technology. Novel delivery method with strong aircraft technology focus.
Kingston University	School of Engineering Kingston University Roehampton Vale Friars Avenue London SW15 3DW, UK www.kingston.ac.uk	Offers a unique foundation degree (FD), with B.Eng. (Hons) top-up, covering JAR 66 skills and knowledge that may lead to JAA categories B and C licenses. Delivered in conjunction with KLM, UK and City of Bristol College at various venues. Particular routes are JAR 147 approved.
Macclesfield College	Macclesfield College Park Lane, Macclesfield Cheshire SK1 18L, UK www.macclesfield.ac.uk	Offers JAR 66 modular programmes in mechanical and avionic specialisms. In addition, mainstream BTEC qualifications are offered in aerospace engineering.
Newcastle College	Newcastle College Rye Hill Campus Scotswood Road Newcastle-upon-Tyne Tyne and Wear NE4 7SA, UK	Offers a new 2 year full-time BTEC national diploma in aerospace engineering (mechanical). This course acts as a suitable feeder for the JAR 66 program offered by Kingston University at Newcastle Airport.
Northbrook College	Northbrook College Shoreham Airport West Sussex BN43 5FJ, UK www.northbrook.ac.uk	Offers JAR 66 modular training programmes, with access to training aircraft and workshop facilities. In addition, mainstream BTEC qualifications in aerospace engineering are also offered.
Oxford Aviation Training	Engineer Training Oxford Aviation Training Oxford Airport, Kidlington Oxford OX5 1RA, UK www.oxfordaviation.co.uk	Offers modular JAR 66 tailor made courses, covering all JAR 66 modules. Also offers conversion courses particularly for conversion of military personnel to civil aviation.

under *maintenance*, where at the time of going to press the required publication is called *147 list: JAR 147 approved organizations*. **Table B.3** provides details on UK educational and training establishments, who offer programmes, ***with an aircraft engineering maintenance bias***, which may or may not operate under the auspices of a Part 147 organization.

The role of the European Aviation Safety Agency

With an ever-expanding European Community (EC), where individual nations have a variety of ways of administering and controlling their own aviation industries, with varying degrees of safety, there has developed the need to provide a unified regulatory framework for the European aviation industry at large that is totally dedicated to aviation safety.

For these reasons, the EC and other entities involved in the sector have set up the European Aviation Safety Agency (EASA), which will have real authority for aviation safety, in a similar manner to that of the Federal Aviation Administration (FAA) in the US.

Thus the primary aims of EASA, in accordance with EC Regulation 1592/2002, are to:

- draw-up common standards to ensure the highest level of aviation safety,
- oversee their uniform application across Europe,
- promote these standards worldwide.

An additional aim for *international standards of aircraft environmental compatibility, for noise and emissions* is already ensconced into European Union (EU) Regulations. However, with the formation of EASA the unification and harmonization of the technical rules and procedures will be much more easily administered and regulated.

So under EC Regulation 1592, the primary objectives of EASA are to:

- establish and maintain a high uniform level of civil aviation safety and environmental protection in Europe;
- facilitate the free movement of goods, persons and services;
- promote cost efficiency in the regulatory and certification process;
- assist member states in fulfilling their ICAO obligations on a common basis;

- promote worldwide community views regarding civil aviation safety standards.

In addition, EASA will develop its knowledge in all fields of aviation safety in order to assist Community legislators in the development of common rules for:

- the certification of aeronautical products, parts and appliances;
- *the approval of organizations and personnel engaged in the maintenance of these products*;
- the approval of air operations;
- the licensing of air crew;
- the safety oversight of airports and air traffic services operators.

Reference source: www.europa.eu.int/comm/transport/air/safety/agency_en.htm.

The establishment of the EASA will inevitably cause changes to the JAA system and organisation. Since July 2002 the JAA has actively participated in the transition from the JAA system to EASA by developing, in consultation with the European Commission, a transition plan that focuses on regulatory aspects.

One of the key issues is the interaction of EASA rules with the already existing European and National legislation. A *core group* of experts have been drafted to look into this issue and produce a report on *Regulatory Interactions*. The above may have far reaching implications for the future role of the NAA, in particular in the case of the UK, for the role of the CAA safety regulation group (SRG).

Since this book was first published EASA was brought into being, as planned, on the 28th of September 2003. EASA's maintenance related activities are now executed in accordance with *Regulation (EC) 2042/2003* and its associated annexes. On an organisational level, certification of maintenance is now handled by the Continuing

Airworthiness team of the Organisation Approvals unit within the certification directorate of EASA.

Lists of approved maintenance organisations (EASA PART-145) may be found at: http://www.easa.eu.int/home/maint_en.html by clicking on the Organisation Approvals-Continuing Airworthiness Organisations, page.

Regulation EC 2042/2003, dated 20 November 2003, defines the relevant legislation for continuing airworthiness of aircraft and aeronautical products, parts and appliances, and also the legislation for the approval of organisations and personnel involved in:

Maintaining aircraft (Annex II – Part 145 – Maintaining Organisational Approvals)

Maintenance Training Organisations (Annex IV – Part 147)

Certifying Staff (Annex III – Part 66)

The *Essential Requirements for Engineering Licensing (Part – 66)*, have not, as suspected, brought about any significant changes to the existing regulations for Engineer Licensing, as contained in old JAR publications. However, those intending to enter the aircraft maintenance world should always ensure that they read the new legislation for approval of personnel as contained in the Part – 66, identified above.

Finally in the remit for the EASA, it is intended that they will look beyond EC borders, with a view to associating as many European partners as possible with their ever-evolving Safety System.

For the latest information on the activities of EASA, you should consult their website: http://www.easa.eu.int/

Mathematical tables

Table D.1 Logarithms

N	0	1	2	3	4	5	6	7	8	9		Mean Differences							
											1	2	3	4	5	6	7	8	9
10	0000	0043	0086	0128	0170	0212	0253	0294	0334	0374	4	8	12	17	21	25	29	33	37
11	0414	0453	0492	0531	0569	0607	0645	0682	0719	0755	4	8	11	15	19	23	26	30	34
12	0792	0828	0864	0899	0934	0969	1004	1038	1072	1106	3	7	10	14	17	21	24	28	31
13	1139	1173	1206	1239	1271	1303	1335	1367	1399	1430	3	6	10	13	16	19	23	26	29
14	1461	1492	1523	1553	1584	1614	1644	1673	1703	1732	3	6	9	12	15	18	21	24	27
15	1761	1790	1818	1847	1875	1903	1931	1959	1987	2014	3	6	8	11	14	17	20	22	25
16	2041	2068	2095	2122	2148	2175	2201	2227	2253	2279	3	5	8	11	13	16	18	21	24
17	2304	2330	2355	2380	2405	2430	2455	2480	2504	2529	2	5	7	10	12	15	17	20	22
18	2553	2577	2601	2625	2648	2672	2695	2718	2742	2765	2	5	7	9	12	14	16	19	21
19	2788	2810	2833	2856	2878	2900	2923	2945	2967	2989	2	4	7	9	11	13	16	18	20
20	3010	3032	3054	3075	3096	3118	3139	3160	3181	3201	2	4	6	8	11	13	15	17	19
21	3222	3243	3263	3284	3304	3324	3345	3365	3385	3404	2	4	6	8	10	12	14	16	18
22	3424	3444	3464	3483	3502	3522	3541	3560	3579	3598	2	4	6	8	10	12	14	15	17
23	3617	3636	3655	3674	3692	3711	3729	3747	3766	3784	2	4	6	7	9	11	13	15	17
24	3802	3820	3838	3856	3874	3892	3909	3927	3945	3962	2	4	5	7	9	11	12	14	16
25	3979	3997	4014	4031	4048	4065	4082	4099	4116	4133	2	3	5	7	9	10	12	14	15
26	4150	4166	4183	4200	4216	4232	4249	4265	4281	4298	2	3	5	7	8	10	11	13	15
27	4314	4330	4346	4362	4378	4393	4409	4425	4440	4456	2	3	5	6	8	9	11	13	14
28	4472	4487	4502	4518	4533	4548	4564	4579	4594	4609	2	3	5	6	8	9	11	12	14
29	4624	4639	4654	4669	4683	4698	4713	4728	4742	4757	1	3	4	6	7	9	10	12	13
30	4771	4786	4800	4814	4829	4843	4857	4871	4886	4900	1	3	4	6	7	9	10	11	13
31	4914	4928	4942	4955	4969	4983	4997	5011	5024	5038	1	3	4	6	7	8	10	11	12
32	5051	5065	5079	5092	5105	5119	5132	5145	5159	5172	1	3	4	5	7	8	9	11	12
33	5185	5198	5211	5224	5237	5250	5263	5276	5289	5302	1	3	4	5	6	8	9	10	12
34	5315	5328	5340	5353	5366	5378	5391	5403	5416	5428	1	3	4	5	6	8	9	10	11
35	5441	5453	5465	5478	5490	5502	5514	5527	5539	5551	1	2	4	5	6	7	9	10	11
36	5563	5575	5587	5599	5611	5623	5635	5647	5658	5670	1	2	4	5	6	7	8	10	11
37	5682	5694	5705	5717	5729	5740	5752	5763	5775	5786	1	2	3	5	6	7	8	9	10
38	5798	5809	5821	5832	5843	5855	5866	5877	5888	5899	1	2	3	5	6	7	8	9	10
39	5911	5922	5933	5944	5955	5966	5977	5988	5999	6010	1	2	3	4	5	7	8	9	10
40	6021	6031	6042	6053	6064	6075	6085	6096	6107	6117	1	2	3	4	5	6	8	9	10
41	6128	6138	6149	6160	6170	6180	6191	6201	6212	6222	1	2	3	4	5	6	7	8	9
42	6232	6243	6253	6263	6274	6284	6294	6304	6314	6325	1	2	3	4	5	6	7	8	9
43	6335	6345	6355	6365	6375	6385	6395	6405	6415	6425	1	2	3	4	5	6	7	8	9
44	6435	6444	6454	6464	6474	6484	6493	6503	6513	6522	1	2	3	4	5	6	7	8	9
45	6532	6542	6551	6561	6571	6580	6590	6599	6609	6618	1	2	3	4	5	6	7	8	9
46	6628	6637	6646	6656	6665	6675	6684	6693	6702	6712	1	2	3	4	5	6	7	7	8
47	6721	6730	6739	6749	6758	6767	6776	6785	6794	6803	1	2	3	4	5	5	6	7	8
48	6812	6821	6830	6839	6848	6857	6866	6875	6884	6893	1	2	3	4	4	5	6	7	8
49	6902	6911	6920	6928	6937	6946	6955	6964	6972	6981	1	2	3	4	4	5	6	7	8
50	6990	6998	7007	7016	7024	7033	7042	7050	7059	7067	1	2	3	3	4	5	6	7	8
51	7076	7084	7093	7101	7110	7118	7126	7135	7143	7152	1	2	3	3	4	5	6	7	8
52	7160	7168	7177	7185	7193	7202	7210	7218	7226	7235	1	2	2	3	4	5	6	7	7
53	7243	7251	7259	7267	7275	7284	7292	7300	7308	7316	1	2	2	3	4	5	6	6	7
54	7324	7332	7340	7348	7356	7364	7372	7380	7388	7396	1	2	2	3	4	5	6	6	7

N	0	1	2	3	4	5	6	7	8	9		Mean Differences							
											1	2	3	4	5	6	7	8	9
55	7404	7412	7419	7427	7435	7443	7451	7459	7466	7474	1	2	2	3	4	5	5	6	7
56	7482	7490	7497	7505	7513	7520	7528	7536	7543	7551	1	2	2	3	4	5	5	6	7
57	7559	7566	7574	7582	7589	7597	7604	7612	7619	7627	1	2	2	3	4	5	5	6	7
58	7634	7642	7649	7657	7664	7672	7679	7686	7694	7701	1	1	2	3	4	4	5	6	7
59	7709	7716	7723	7731	7738	7745	7752	7760	7767	7774	1	1	2	3	4	4	5	6	7
60	7782	7789	7796	7803	7810	7818	7825	7832	7839	7846	1	1	2	3	4	4	5	6	6
61	7853	7860	7868	7875	7882	7889	7896	7903	7910	7917	1	1	2	3	4	4	5	6	6
62	7924	7931	7938	7945	7952	7959	7966	7973	7980	7987	1	1	2	3	3	4	5	6	6
63	7993	8000	8007	8014	8021	8028	8035	8041	8048	8055	1	1	2	3	3	4	5	5	6
64	8062	8069	8075	8082	8089	8096	8102	8109	8116	8122	1	1	2	3	3	4	5	5	6
65	8129	8136	8142	8149	8156	8162	8169	8176	8182	8189	1	1	2	3	3	4	5	5	6
66	8195	8202	8209	8215	8222	8228	8235	8241	8248	8254	1	1	2	3	3	4	5	5	6
67	8261	8267	8274	8280	8287	8293	8299	8306	8312	8319	1	1	2	3	3	4	5	5	6
68	8325	8331	8338	8344	8351	8357	8363	8370	8376	8382	1	1	2	3	3	4	4	5	6
69	8388	8395	8401	8407	8414	8420	8426	8432	8439	8445	1	1	2	2	3	4	4	5	6
70	8451	8457	8463	8470	8476	8482	8488	8494	8500	8506	1	1	2	2	3	4	4	5	6
71	8513	8519	8525	8531	8537	8543	8549	8555	8561	8567	1	1	2	2	3	4	4	5	5
72	8573	8579	8585	8591	8597	8603	8609	8615	8621	8627	1	1	2	2	3	4	4	5	5
73	8633	8639	8645	8651	8657	8663	8669	8675	8681	8686	1	1	2	2	3	4	4	5	5
74	8692	8698	8704	8710	8716	8722	8727	8733	8739	8745	1	1	2	2	3	4	4	5	5
75	8751	8756	8762	8768	8774	8779	8785	8791	8797	8802	1	1	2	2	3	3	4	5	5
76	8808	8814	8820	8825	8831	8837	8842	8848	8854	8859	1	1	2	2	3	3	4	5	5
77	8865	8871	8876	8882	8887	8893	8899	8904	8910	8915	1	1	2	2	3	3	4	4	5
78	8921	8927	8932	8938	8943	8949	8954	8960	8965	8971	1	1	2	2	3	3	4	4	5
79	8976	8982	8987	8993	8998	9004	9009	9015	9020	9025	1	1	2	2	3	3	4	4	5
80	9031	9036	9042	9047	9053	9058	9063	9069	9074	9079	1	1	2	2	3	3	4	4	5
81	9085	9090	9096	9101	9106	9112	9117	9122	9128	9133	1	1	2	2	3	3	4	4	5
82	9138	9143	9149	9154	9159	9165	9170	9175	9180	9186	1	1	2	2	3	3	4	4	5
83	9191	9196	9201	9206	9212	9217	9222	9227	9232	9238	1	1	2	2	3	3	4	4	5
84	9243	9248	9253	9258	9263	9269	9274	9279	9284	9289	1	1	2	2	3	3	4	4	5
85	9294	9299	9304	9309	9315	9320	9325	9330	9335	9340	1	1	2	2	3	3	4	4	5
86	9345	9350	9355	9360	9365	9370	9375	9380	9385	9390	1	1	2	2	3	3	4	4	5
87	9395	9400	9405	9410	9415	9420	9425	9430	9435	9440	0	1	1	2	2	3	3	4	4
88	9445	9450	9455	9460	9465	9469	9474	9479	9484	9489	0	1	1	2	2	3	3	4	4
89	9494	9499	9504	9509	9513	9518	9523	9528	9533	9538	0	1	1	2	2	3	3	4	4
90	9542	9547	9552	9557	9562	9566	9571	9576	9581	9586	0	1	1	2	2	3	3	4	4
91	9590	9595	9600	9605	9609	9614	9619	9624	9628	9633	0	1	1	2	2	3	3	4	4
92	9638	9643	9647	9652	9657	9661	9666	9671	9675	9680	0	1	1	2	2	3	3	4	4
93	9685	9689	9694	9699	9703	9708	9713	9717	9722	9727	0	1	1	2	2	3	3	4	4
94	9731	9736	9741	9745	9750	9754	9759	9763	9768	9773	0	1	1	2	2	3	3	4	4
95	9777	9782	9786	9791	9795	9800	9805	9809	9814	9818	0	1	1	2	2	3	3	4	4
96	9823	9827	9832	9836	9841	9845	9850	9854	9859	9863	0	1	1	2	2	3	3	4	4
97	9868	9872	9877	9881	9886	9890	9894	9899	9903	9908	0	1	1	2	2	3	3	4	4
98	9912	9917	9921	9926	9930	9934	9939	9943	9948	9952	0	1	1	2	2	3	3	4	4
99	9956	9961	9965	9969	9974	9978	9983	9987	9991	9996	0	1	1	2	2	3	3	3	4

Table D.2 Antilogarithms

Continuation (.50–.99) with Mean Differences:

	0	1	2	3	4	5	6	7	8	9	1	2	3	4	5	6	7	8	9
.50	3162	3170	3177	3184	3192	3199	3206	3214	3221	3228	1	2	2	3	4	5	5	6	7
.51	3236	3243	3251	3258	3266	3273	3281	3289	3296	3304	1	2	2	3	4	5	6	6	7
.52	3311	3319	3327	3334	3342	3350	3357	3365	3373	3381	1	2	2	3	4	5	5	6	7
.53	3388	3396	3404	3412	3420	3428	3436	3443	3451	3459	1	2	2	3	4	5	6	6	7
.54	3467	3475	3483	3491	3499	3508	3516	3524	3532	3540	1	2	2	3	4	5	6	6	7
.55	3548	3556	3565	3573	3581	3589	3597	3606	3614	3622	1	2	2	3	4	5	6	6	7
.56	3631	3639	3648	3656	3664	3673	3681	3690	3698	3707	1	2	3	3	4	5	6	7	8
.57	3715	3724	3733	3741	3750	3758	3767	3776	3784	3793	1	2	3	3	4	5	6	7	8
.58	3802	3811	3819	3828	3837	3846	3855	3864	3873	3882	1	2	3	4	4	5	6	7	8
.59	3890	3899	3908	3917	3926	3936	3945	3954	3963	3972	1	2	3	4	5	5	6	7	8
.60	3981	3990	3999	4009	4018	4027	4036	4046	4055	4064	1	2	3	4	5	5	6	7	8
.61	4074	4083	4093	4102	4111	4121	4130	4140	4150	4159	1	2	3	4	5	6	7	8	9
.62	4169	4178	4188	4198	4207	4217	4227	4236	4246	4256	1	2	3	4	5	6	7	8	9
.63	4266	4276	4285	4295	4305	4315	4325	4335	4345	4355	1	2	3	4	5	6	7	8	9
.64	4365	4375	4385	4395	4406	4416	4426	4436	4446	4457	1	2	3	4	5	6	7	8	9
.65	4467	4477	4487	4498	4508	4519	4529	4539	4550	4560	1	2	3	4	5	6	7	8	9
.66	4571	4581	4592	4603	4613	4624	4634	4645	4656	4667	1	2	3	4	5	6	7	9	10
.67	4677	4688	4699	4710	4721	4732	4742	4753	4764	4775	1	2	3	4	5	7	8	9	10
.68	4786	4797	4808	4819	4831	4842	4853	4864	4875	4887	1	2	3	4	5	7	8	9	10
.69	4898	4909	4920	4932	4943	4955	4966	4977	4989	5000	1	2	4	5	6	7	8	9	10
.70	5012	5023	5035	5047	5058	5070	5082	5093	5105	5117	1	2	4	5	6	7	8	9	11
.71	5129	5140	5152	5164	5176	5188	5200	5212	5224	5236	1	2	4	5	6	7	8	10	11
.72	5248	5260	5272	5284	5297	5309	5321	5333	5346	5358	1	2	4	5	6	7	9	10	11
.73	5370	5383	5395	5408	5420	5433	5445	5458	5470	5483	1	3	4	5	6	8	9	10	11
.74	5495	5508	5521	5534	5546	5559	5572	5585	5598	5610	1	3	4	5	6	8	9	10	12
.75	5623	5636	5649	5662	5675	5689	5702	5715	5728	5741	1	3	4	5	7	8	9	11	12
.76	5754	5768	5781	5794	5808	5821	5834	5848	5861	5875	1	3	4	5	7	8	9	11	12
.77	5888	5902	5916	5929	5943	5957	5970	5984	5998	6012	1	3	4	6	7	8	10	11	12
.78	6026	6039	6053	6067	6081	6095	6109	6124	6138	6152	1	3	4	6	7	8	10	11	13
.79	6166	6180	6194	6209	6223	6237	6252	6266	6281	6295	1	3	4	6	7	9	10	11	13
.80	6310	6324	6339	6353	6368	6383	6397	6412	6427	6442	1	3	4	6	7	9	10	12	13
.81	6457	6471	6486	6501	6516	6531	6546	6561	6577	6592	2	3	4	6	8	9	11	12	14
.82	6607	6622	6637	6653	6668	6683	6699	6714	6730	6745	2	3	5	6	8	9	11	12	14
.83	6761	6776	6792	6808	6823	6839	6855	6871	6887	6902	2	3	5	6	8	9	11	13	14
.84	6918	6934	6950	6966	6982	6998	7015	7031	7047	7063	2	3	5	6	8	10	11	13	15
.85	7079	7096	7112	7129	7145	7161	7178	7194	7211	7228	2	3	5	7	8	10	11	13	15
.86	7244	7261	7278	7295	7311	7328	7345	7362	7379	7396	2	3	5	7	8	10	12	13	15
.87	7413	7430	7447	7464	7482	7499	7516	7534	7551	7568	2	3	5	7	9	10	12	14	16
.88	7586	7603	7621	7638	7656	7674	7691	7709	7727	7745	2	4	5	7	9	11	12	14	16
.89	7762	7780	7798	7816	7834	7852	7870	7889	7907	7925	2	4	5	7	9	11	12	14	16
.90	7943	7962	7980	7998	8017	8035	8054	8072	8091	8110	2	4	5	7	9	11	13	14	16
.91	8128	8147	8166	8185	8204	8222	8241	8260	8279	8299	2	4	6	8	9	11	13	15	17
.92	8318	8337	8356	8375	8395	8414	8433	8453	8472	8492	2	4	6	8	10	11	13	15	17
.93	8511	8531	8551	8570	8590	8610	8630	8650	8670	8690	2	4	6	8	10	12	14	16	18
.94	8710	8730	8750	8770	8790	8810	8831	8851	8872	8892	2	4	6	8	10	12	14	16	18
.95	8913	8933	8954	8974	8995	9016	9036	9057	9078	9099	2	4	6	8	10	12	15	17	19
.96	9120	9141	9162	9183	9204	9226	9247	9268	9290	9311	2	4	6	8	11	13	15	17	19
.97	9333	9354	9376	9397	9419	9441	9462	9484	9506	9528	2	4	7	9	11	13	15	17	20
.98	9550	9572	9594	9616	9638	9661	9683	9705	9727	9750	2	4	7	9	11	13	16	18	20
.99	9772	9795	9817	9840	9863	9886	9908	9931	9954	9977	2	5	7	9	11	14	16	18	20

First section (.00–.49) with Mean Differences:

	0	1	2	3	4	5	6	7	8	9	1	2	3	4	5	6	7	8	9
.00	1000	1002	1005	1007	1009	1012	1014	1016	1019	1021	0	0	1	1	1	1	2	2	2
.01	1023	1026	1028	1030	1033	1035	1038	1040	1042	1045	0	1	1	1	1	2	2	2	2
.02	1047	1050	1052	1054	1057	1059	1062	1064	1067	1069	0	1	1	1	1	2	2	2	2
.03	1072	1074	1076	1079	1081	1084	1086	1089	1091	1094	0	1	1	1	1	2	2	2	2
.04	1096	1099	1102	1104	1107	1109	1112	1114	1117	1119	0	1	1	1	1	2	2	2	2
.05	1122	1125	1127	1130	1132	1135	1138	1140	1143	1146	0	1	1	1	1	2	2	2	2
.06	1148	1151	1153	1156	1159	1161	1164	1167	1169	1172	0	1	1	1	1	2	2	2	2
.07	1175	1178	1180	1183	1186	1189	1191	1194	1197	1199	0	1	1	1	1	2	2	2	3
.08	1202	1205	1208	1211	1213	1216	1219	1222	1225	1227	0	1	1	1	1	2	2	2	3
.09	1230	1233	1236	1239	1242	1245	1247	1250	1253	1256	0	1	1	1	1	2	2	2	3
.10	1259	1262	1265	1268	1271	1274	1276	1279	1282	1285	0	1	1	1	1	2	2	2	3
.11	1288	1291	1294	1297	1300	1303	1306	1309	1312	1315	0	1	1	1	2	2	2	2	3
.12	1318	1321	1324	1327	1330	1334	1337	1340	1343	1346	0	1	1	1	2	2	2	3	3
.13	1349	1352	1355	1358	1361	1365	1368	1371	1374	1377	0	1	1	1	2	2	2	3	3
.14	1380	1384	1387	1390	1393	1396	1400	1403	1406	1409	0	1	1	1	2	2	2	3	3
.15	1413	1416	1419	1422	1426	1429	1432	1435	1439	1442	0	1	1	1	2	2	2	3	3
.16	1445	1449	1452	1455	1459	1462	1466	1469	1472	1476	0	1	1	1	2	2	2	3	3
.17	1479	1483	1486	1489	1493	1496	1500	1503	1507	1510	0	1	1	2	2	2	3	3	3
.18	1514	1517	1521	1524	1528	1531	1535	1538	1542	1545	0	1	1	2	2	2	3	3	3
.19	1549	1552	1556	1560	1563	1567	1570	1574	1578	1581	0	1	1	2	2	2	3	3	3
.20	1585	1589	1592	1596	1600	1603	1607	1611	1614	1618	0	1	1	2	2	2	3	3	4
.21	1622	1626	1629	1633	1637	1641	1644	1648	1652	1656	0	1	1	2	2	2	3	3	4
.22	1660	1663	1667	1671	1675	1679	1683	1687	1690	1694	0	1	1	2	2	2	3	3	4
.23	1698	1702	1706	1710	1714	1718	1722	1726	1730	1734	0	1	1	2	2	2	3	3	4
.24	1738	1742	1746	1750	1754	1758	1762	1766	1770	1774	0	1	1	2	2	2	3	3	4
.25	1778	1782	1786	1791	1795	1799	1803	1807	1811	1816	0	1	1	2	2	3	3	3	4
.26	1820	1824	1828	1832	1837	1841	1845	1849	1854	1858	0	1	1	2	2	3	3	3	4
.27	1862	1866	1871	1875	1879	1884	1888	1892	1897	1901	0	1	1	2	2	3	3	4	4
.28	1905	1910	1914	1919	1923	1928	1932	1936	1941	1945	0	1	1	2	2	3	3	4	4
.29	1950	1954	1959	1963	1968	1972	1977	1982	1986	1991	0	1	1	2	2	3	3	4	4
.30	1995	2000	2004	2009	2014	2018	2023	2028	2032	2037	0	1	1	2	2	3	3	4	4
.31	2042	2046	2051	2056	2061	2065	2070	2075	2080	2084	0	1	1	2	2	3	3	4	4
.32	2089	2094	2099	2104	2109	2113	2118	2123	2128	2133	0	1	1	2	2	3	3	4	4
.33	2138	2143	2148	2153	2158	2163	2168	2173	2178	2183	0	1	1	2	2	3	3	4	5
.34	2188	2193	2198	2203	2208	2213	2218	2223	2228	2234	0	1	1	2	3	3	4	4	5
.35	2239	2244	2249	2254	2259	2265	2270	2275	2280	2286	1	1	2	2	3	3	4	4	5
.36	2291	2296	2301	2307	2312	2317	2323	2328	2333	2339	1	1	2	2	3	3	4	4	5
.37	2344	2350	2355	2360	2366	2371	2377	2382	2388	2393	1	1	2	2	3	3	4	4	5
.38	2399	2404	2410	2415	2421	2427	2432	2438	2443	2449	1	1	2	2	3	3	4	5	5
.39	2455	2460	2466	2472	2477	2483	2489	2495	2500	2506	1	1	2	2	3	4	4	5	5
.40	2512	2518	2523	2529	2535	2541	2547	2553	2559	2564	1	1	2	2	3	4	4	5	5
.41	2570	2576	2582	2588	2594	2600	2606	2612	2618	2624	1	1	2	2	3	4	4	5	5
.42	2630	2636	2642	2649	2655	2661	2667	2673	2679	2685	1	1	2	2	3	4	4	5	6
.43	2692	2698	2704	2710	2716	2723	2729	2735	2742	2748	1	1	2	3	3	4	4	5	6
.44	2754	2761	2767	2773	2780	2786	2793	2799	2805	2812	1	1	2	3	3	4	5	5	6
.45	2818	2825	2831	2838	2844	2851	2858	2864	2871	2877	1	1	2	3	3	4	5	5	6
.46	2884	2891	2897	2904	2911	2917	2924	2931	2938	2944	1	1	2	3	3	4	5	5	6
.47	2951	2958	2965	2972	2979	2985	2992	2999	3006	3013	1	1	2	3	3	4	5	6	6
.48	3020	3027	3034	3041	3048	3055	3062	3069	3076	3083	1	1	2	3	4	4	5	6	6
.49	3090	3097	3105	3112	3119	3126	3133	3141	3148	3155	1	2	2	3	4	4	5	6	6

Table D.3 Natural sines

	0′ 0.0°	6′ 0.1°	12′ 0.2°	18′ 0.3°	24′ 0.4°	30′ 0.5°	36′ 0.6°	42′ 0.7°	48′ 0.8°	54′ 0.9°	1′	2′	3′	4′	5′
0°	0.0000	0017	0035	0052	0070	0087	0105	0122	0140	0157	3	6	9	12	15
1	0.0175	0192	0209	0227	0244	0262	0279	0297	0314	0332	3	6	9	12	15
2	0.0349	0366	0384	0401	0419	0436	0454	0471	0488	0506	3	6	9	12	15
3	0.0523	0541	0558	0576	0593	0610	0628	0645	0663	0680	3	6	9	12	15
4	0.0698	0715	0732	0750	0767	0785	0802	0819	0837	0854	3	6	9	12	15
5	0.0872	0889	0906	0924	0941	0958	0976	0993	1011	1028	3	6	9	12	15
6	0.1045	1063	1080	1097	1115	1132	1149	1167	1184	1201	3	6	9	12	14
7	0.1219	1236	1253	1271	1288	1305	1323	1340	1357	1374	3	6	9	12	14
8	0.1392	1409	1426	1444	1461	1478	1495	1513	1530	1547	3	6	9	12	14
9	0.1564	1582	1599	1616	1633	1650	1668	1685	1702	1719	3	6	9	11	14
10°	0.1736	1754	1771	1788	1805	1822	1840	1857	1874	1891	3	6	9	11	14
11	0.1908	1925	1942	1959	1977	1994	2011	2028	2045	2062	3	6	9	11	14
12	0.2079	2096	2113	2130	2147	2164	2181	2198	2215	2233	3	6	9	11	14
13	0.2250	2267	2284	2300	2317	2334	2351	2368	2385	2402	3	6	8	11	14
14	0.2419	2436	2453	2470	2487	2504	2521	2538	2554	2571	3	6	8	11	14
15	0.2588	2605	2622	2639	2656	2672	2689	2706	2723	2740	3	6	8	11	14
16	0.2756	2773	2790	2807	2823	2840	2857	2874	2890	2907	3	6	8	11	14
17	0.2924	2940	2957	2974	2990	3007	3024	3040	3057	3074	3	5	8	11	14
18	0.3090	3107	3123	3140	3156	3173	3190	3206	3223	3239	3	5	8	11	14
19	0.3256	3272	3289	3305	3322	3338	3355	3371	3387	3404	3	5	8	11	14
20°	0.3420	3437	3453	3469	3486	3502	3518	3535	3551	3567	3	5	8	11	14
21	0.3584	3600	3616	3633	3649	3665	3681	3697	3714	3730	3	5	8	11	14
22	0.3746	3762	3778	3795	3811	3827	3843	3859	3875	3891	3	5	8	11	13
23	0.3907	3923	3939	3955	3971	3987	4003	4019	4035	4051	3	5	8	11	13
24	0.4067	4083	4099	4115	4131	4147	4163	4179	4195	4210	3	5	8	11	13
25	0.4226	4242	4258	4274	4289	4305	4321	4337	4352	4368	3	5	8	11	13
26	0.4384	4399	4415	4431	4446	4462	4478	4493	4509	4524	3	5	8	10	13
27	0.4540	4555	4571	4586	4602	4617	4633	4648	4664	4679	3	5	8	10	13
28	0.4695	4710	4726	4741	4756	4772	4787	4802	4818	4833	3	5	8	10	13
29	0.4848	4863	4879	4894	4909	4924	4939	4955	4970	4985	3	5	8	10	13
30°	0.5000	5015	5030	5045	5060	5075	5090	5105	5120	5135	3	5	8	10	13
31	0.5150	5165	5180	5195	5210	5225	5240	5255	5270	5284	2	5	7	10	12
32	0.5299	5314	5329	5344	5358	5373	5388	5402	5417	5432	2	5	7	10	12
33	0.5446	5461	5476	5490	5505	5519	5534	5548	5563	5577	2	5	7	10	12
34	0.5592	5606	5621	5635	5650	5664	5678	5693	5707	5721	2	5	7	10	12
35	0.5736	5750	5764	5779	5793	5807	5821	5835	5850	5864	2	5	7	9	12
36	0.5878	5892	5906	5920	5934	5948	5962	5976	5990	6004	2	5	7	9	12
37	0.6018	6032	6046	6060	6074	6088	6101	6115	6129	6143	2	5	7	9	11
38	0.6157	6170	6184	6198	6211	6225	6239	6252	6266	6280	2	5	7	9	11
39	0.6293	6307	6320	6334	6347	6361	6374	6388	6401	6414	2	4	7	9	11
40°	0.6428	6441	6455	6468	6481	6494	6508	6521	6534	6547	2	4	7	9	11
41	0.6561	6574	6587	6600	6613	6626	6639	6652	6665	6678	2	4	7	9	11
42	0.6691	6704	6717	6730	6743	6756	6769	6782	6794	6807	2	4	6	9	11
43	0.6820	6833	6845	6858	6871	6884	6896	6909	6921	6934	2	4	6	8	11
44	0.6947	6959	6972	6984	6997	7009	7022	7034	7046	7059	2	4	6	8	10

	0′ 0.0°	6′ 0.1°	12′ 0.2°	18′ 0.3°	24′ 0.4°	30′ 0.5°	36′ 0.6°	42′ 0.7°	48′ 0.8°	54′ 0.9°	1′	2′	3′	4′	5′
45°	0.7071	7083	7096	7108	7120	7133	7145	7157	7169	7181	2	4	6	8	10
46	0.7193	7206	7218	7230	7242	7254	7266	7278	7290	7302	2	4	6	8	10
47	0.7314	7325	7337	7349	7361	7373	7385	7396	7408	7420	2	4	6	8	10
48	0.7431	7443	7455	7466	7478	7490	7501	7513	7524	7536	2	4	6	8	10
49	0.7547	7559	7570	7581	7593	7604	7615	7627	7638	7649	2	4	6	8	9
50°	0.7660	7672	7683	7694	7705	7716	7727	7738	7749	7760	2	4	6	7	9
51	0.7771	7782	7793	7804	7815	7826	7837	7848	7859	7869	2	4	5	7	9
52	0.7880	7891	7902	7912	7923	7934	7944	7955	7965	7976	2	4	5	7	9
53	0.7986	7997	8007	8018	8028	8039	8049	8059	8070	8080	2	4	5	7	9
54	0.8090	8100	8111	8121	8131	8141	8151	8161	8171	8181	2	3	5	7	8
55	0.8192	8202	8211	8221	8231	8241	8251	8261	8271	8281	2	3	5	7	8
56	0.8290	8300	8310	8320	8329	8339	8348	8358	8368	8377	2	3	5	6	8
57	0.8387	8396	8406	8415	8425	8434	8443	8453	8462	8471	2	3	5	6	8
58	0.8480	8490	8499	8508	8517	8526	8536	8545	8554	8563	2	3	5	6	8
59	0.8572	8581	8590	8599	8607	8616	8625	8634	8643	8652	1	3	4	6	7
60°	0.8660	8669	8678	8686	8695	8704	8712	8721	8729	8738	1	3	4	6	7
61	0.8746	8755	8763	8771	8780	8788	8796	8805	8813	8821	1	3	4	6	7
62	0.8829	8838	8846	8854	8862	8870	8878	8886	8894	8902	1	3	4	5	7
63	0.8910	8918	8926	8934	8942	8949	8957	8965	8973	8980	1	3	4	5	6
64	0.8988	8996	9003	9011	9018	9026	9033	9041	9048	9056	1	3	4	5	6
65	0.9063	9070	9078	9085	9092	9100	9107	9114	9121	9128	1	2	4	5	6
66	0.9135	9143	9150	9157	9164	9171	9178	9184	9191	9198	1	2	4	5	6
67	0.9205	9212	9219	9225	9232	9239	9245	9252	9259	9265	1	2	3	5	5
68	0.9272	9278	9285	9291	9298	9304	9311	9317	9323	9330	1	2	3	4	5
69	0.9336	9342	9348	9354	9361	9367	9373	9379	9385	9391	1	2	3	4	5
70°	0.9397	9403	9409	9415	9421	9426	9432	9438	9444	9449	1	2	3	4	5
71	0.9455	9461	9466	9472	9478	9483	9489	9494	9500	9505	1	2	3	4	5
72	0.9511	9516	9521	9527	9532	9537	9542	9548	9553	9558	1	2	3	4	4
73	0.9563	9568	9573	9578	9583	9588	9593	9598	9603	9608	1	2	3	3	4
74	0.9613	9617	9622	9627	9632	9636	9641	9646	9650	9655	1	2	2	3	4
75	0.9659	9664	9668	9673	9677	9681	9686	9690	9694	9699	1	2	2	3	4
76	0.9703	9707	9711	9715	9720	9724	9728	9732	9736	9740	1	1	2	3	4
77	0.9744	9748	9751	9755	9759	9763	9767	9770	9774	9778	1	1	2	3	3
78	0.9781	9785	9789	9792	9796	9799	9803	9806	9810	9813	1	1	2	3	3
79	0.9816	9820	9823	9826	9829	9833	9836	9839	9842	9845	1	1	2	2	3
80°	0.9848	9851	9854	9857	9860	9863	9866	9869	9871	9874	1	1	2	2	3
81	0.9877	9880	9882	9885	9888	9890	9893	9895	9898	9900	1	1	2	2	2
82	0.9903	9905	9907	9910	9912	9914	9917	9919	9921	9923	0	1	1	2	2
83	0.9925	9928	9930	9932	9934	9936	9938	9940	9942	9943	0	1	1	2	2
84	0.9945	9947	9949	9951	9952	9954	9956	9957	9959	9960	0	1	1	1	2
85	0.9962	9963	9965	9966	9968	9969	9971	9972	9973	9974	0	1	1	1	1
86	0.9976	9977	9978	9979	9980	9981	9982	9983	9984	9985	0	1	1	1	1
87	0.9986	9987	9988	9989	9990	9990	9991	9992	9993	9993	0	0	1	1	1
88	0.9994	9995	9995	9996	9996	9997	9997	9997	9998	9998	0	0	0	1	1
89°	0.9998	9999	9999	9999	9999	1.000	1.000	1.000	1.000	1.000	0	0	0	0	0

Table D.4 Natural cosines

Top half (SUBTRACT Mean Differences)

	0' 0.0°	6' 0.1°	12' 0.2°	18' 0.3°	24' 0.4°	30' 0.5°	36' 0.6°	42' 0.7°	48' 0.8°	54' 0.9°	1'	2'	3'	4'	5'
45°	0.7071	7059	7046	7034	7022	7009	6997	6984	6972	6959	2	4	6	8	10
46	0.6947	6934	6921	6909	6896	6884	6871	6858	6845	6833	2	4	6	8	11
47	0.6820	6807	6794	6782	6769	6756	6743	6730	6717	6704	2	4	6	9	11
48	0.6691	6678	6665	6652	6639	6626	6613	6600	6587	6574	2	4	7	9	11
49	0.6561	6547	6534	6521	6508	6494	6481	6468	6455	6441	2	4	7	9	11
50°	0.6428	6414	6401	6388	6374	6361	6347	6334	6320	6307	2	4	7	9	11
51	0.6293	6280	6266	6252	6239	6225	6211	6198	6184	6170	2	5	7	9	11
52	0.6157	6143	6129	6115	6101	6088	6074	6060	6046	6032	2	5	7	9	12
53	0.6018	6004	5990	5976	5962	5948	5934	5920	5906	5892	2	5	7	9	12
54	0.5878	5864	5850	5835	5821	5807	5793	5779	5764	5750	2	5	7	10	12
55	0.5736	5721	5707	5693	5678	5664	5650	5635	5621	5606	2	5	7	10	12
56	0.5592	5577	5563	5548	5534	5519	5505	5490	5476	5461	2	5	7	10	12
57	0.5446	5432	5417	5402	5388	5373	5358	5344	5329	5314	2	5	7	10	12
58	0.5299	5284	5270	5255	5240	5225	5210	5195	5180	5165	3	5	8	10	13
59	0.5150	5135	5120	5105	5090	5075	5060	5045	5030	5015	3	5	8	10	13
60°	0.5000	4985	4970	4955	4939	4924	4909	4894	4879	4863	3	5	8	10	13
61	0.4848	4833	4818	4802	4787	4772	4756	4741	4726	4710	3	5	8	10	13
62	0.4695	4679	4664	4648	4633	4617	4602	4586	4571	4555	3	5	8	11	13
63	0.4540	4524	4509	4493	4478	4462	4446	4431	4415	4399	3	5	8	11	13
64	0.4384	4368	4352	4337	4321	4305	4289	4274	4258	4242	3	5	8	11	13
65	0.4226	4210	4195	4179	4163	4147	4131	4115	4099	4083	3	5	8	11	13
66	0.4067	4051	4035	4019	4003	3987	3971	3955	3939	3923	3	5	8	11	14
67	0.3907	3891	3875	3859	3843	3827	3811	3795	3778	3762	3	5	8	11	14
68	0.3746	3730	3714	3697	3681	3665	3649	3633	3616	3600	3	5	8	11	14
69	0.3584	3567	3551	3535	3518	3502	3486	3469	3453	3437	3	5	8	11	14
70°	0.3420	3404	3387	3371	3355	3338	3322	3305	3289	3272	3	5	8	11	14
71	0.3256	3239	3223	3206	3190	3173	3156	3140	3123	3107	3	6	8	11	14
72	0.3090	3074	3057	3040	3024	3007	2990	2974	2957	2940	3	6	8	11	14
73	0.2924	2907	2890	2874	2857	2840	2823	2807	2790	2773	3	6	8	11	14
74	0.2756	2740	2723	2706	2689	2672	2656	2639	2622	2605	3	6	8	11	14
75	0.2588	2571	2554	2538	2521	2504	2487	2470	2453	2436	3	6	8	11	14
76	0.2419	2402	2385	2368	2351	2334	2317	2300	2284	2267	3	6	8	11	14
77	0.2250	2233	2215	2198	2181	2164	2147	2130	2113	2096	3	6	8	11	14
78	0.2079	2062	2045	2028	2011	1994	1977	1959	1942	1925	3	6	8	11	14
79	0.1908	1891	1874	1857	1840	1822	1805	1788	1771	1754	3	6	8	11	14
80°	0.1736	1719	1702	1685	1668	1650	1633	1616	1599	1582	3	6	9	12	14
81	0.1564	1547	1530	1513	1495	1478	1461	1444	1426	1409	3	6	9	12	14
82	0.1392	1374	1357	1340	1323	1305	1288	1271	1253	1236	3	6	9	12	14
83	0.1219	1201	1184	1167	1149	1132	1115	1097	1080	1063	3	6	9	12	14
84	0.1045	1028	1011	0993	0976	0958	0941	0924	0906	0889	3	6	9	12	14
85	0.0872	0854	0837	0819	0802	0785	0767	0750	0732	0715	3	6	9	12	14
86	0.0698	0680	0663	0645	0628	0610	0593	0576	0558	0541	3	6	9	12	15
87	0.0523	0506	0488	0471	0454	0436	0419	0401	0384	0366	3	6	9	12	15
88	0.0349	0332	0314	0297	0279	0262	0244	0227	0209	0192	3	6	9	12	15
89	0.0175	0157	0140	0122	0105	0087	0070	0052	0035	0017	3	6	9	12	15

Bottom half (SUBTRACT Mean Differences)

	0' 0.0°	6' 0.1°	12' 0.2°	18' 0.3°	24' 0.4°	30' 0.5°	36' 0.6°	42' 0.7°	48' 0.8°	54' 0.9°	1'	2'	3'	4'	5'
0°	1.000	1.000	1.000	1.000	1.000	1.000	.9999	9999	9999	9999	0	0	0	0	0
1	0.9998	9998	9998	9997	9997	9997	9996	9996	9995	9995	0	0	0	0	0
2	0.9994	9993	9993	9992	9991	9990	9990	9989	9988	9987	0	0	0	0	0
3	0.9986	9985	9984	9983	9982	9981	9980	9979	9978	9977	0	0	0	1	1
4	0.9976	9974	9973	9972	9971	9969	9968	9966	9965	9963	0	0	1	1	1
5	0.9962	9960	9959	9957	9956	9954	9952	9951	9949	9947	0	1	1	1	2
6	0.9945	9943	9942	9940	9938	9936	9934	9932	9930	9928	0	1	1	1	2
7	0.9925	9923	9921	9919	9917	9914	9912	9910	9907	9905	0	1	1	2	2
8	0.9903	9900	9898	9895	9893	9890	9888	9885	9882	9880	0	1	1	2	2
9	0.9877	9874	9871	9869	9866	9863	9860	9857	9854	9851	0	1	1	2	2
10°	0.9848	9845	9842	9839	9836	9833	9829	9826	9823	9820	1	1	2	2	3
11	0.9816	9813	9810	9806	9803	9799	9796	9792	9789	9785	1	1	2	2	3
12	0.9781	9778	9774	9770	9767	9763	9759	9755	9751	9748	1	1	2	3	3
13	0.9744	9740	9736	9732	9728	9724	9720	9715	9711	9707	1	1	2	3	3
14	0.9703	9699	9694	9690	9686	9681	9677	9673	9668	9664	1	1	2	3	4
15	0.9659	9655	9650	9646	9641	9636	9632	9627	9622	9617	1	2	2	3	4
16	0.9613	9608	9603	9598	9593	9588	9583	9578	9573	9568	1	2	2	3	4
17	0.9563	9558	9553	9548	9542	9537	9532	9527	9521	9516	1	2	3	3	4
18	0.9511	9505	9500	9494	9489	9483	9478	9472	9466	9461	1	2	3	4	5
19	0.9455	9449	9444	9438	9432	9426	9421	9415	9409	9403	1	2	3	4	5
20°	0.9397	9391	9385	9379	9373	9367	9361	9354	9348	9342	1	2	3	4	5
21	0.9336	9330	9323	9317	9311	9304	9298	9291	9285	9278	1	2	3	4	5
22	0.9272	9265	9259	9252	9245	9239	9232	9225	9219	9212	1	2	3	4	5
23	0.9205	9198	9191	9184	9178	9171	9164	9157	9150	9143	1	2	3	5	6
24	0.9135	9128	9121	9114	9107	9100	9092	9085	9078	9070	1	2	4	5	6
25	0.9063	9056	9048	9041	9033	9026	9018	9011	9003	8996	1	3	4	5	6
26	0.8988	8980	8973	8965	8957	8949	8942	8934	8926	8918	1	3	4	5	6
27	0.8910	8902	8894	8886	8878	8870	8862	8854	8846	8838	1	3	4	5	7
28	0.8829	8821	8813	8805	8796	8788	8780	8771	8763	8755	1	3	4	6	7
29	0.8746	8738	8729	8721	8712	8704	8695	8686	8678	8669	1	3	4	6	7
30°	0.8660	8652	8643	8634	8625	8616	8607	8599	8590	8581	2	3	4	6	7
31	0.8572	8563	8554	8545	8536	8526	8517	8508	8499	8490	2	3	5	6	8
32	0.8480	8471	8462	8453	8443	8434	8425	8415	8406	8396	2	3	5	6	8
33	0.8387	8377	8368	8358	8348	8339	8329	8320	8310	8300	2	3	5	6	8
34	0.8290	8281	8271	8261	8251	8241	8231	8221	8211	8202	2	3	5	7	8
35	0.8192	8181	8171	8161	8151	8141	8131	8121	8111	8100	2	3	5	7	8
36	0.8090	8080	8070	8059	8049	8039	8028	8018	8007	7997	2	3	5	7	8
37	0.7986	7976	7965	7955	7944	7934	7923	7912	7902	7891	2	4	5	7	9
38	0.7880	7869	7859	7848	7837	7826	7815	7804	7793	7782	2	4	5	7	9
39	0.7771	7760	7749	7738	7727	7716	7705	7694	7683	7672	2	4	6	7	9
40°	0.7660	7649	7638	7627	7615	7604	7593	7581	7570	7559	2	4	6	7	9
41	0.7547	7536	7524	7513	7501	7490	7478	7466	7455	7443	2	4	6	8	10
42	0.7431	7420	7408	7396	7385	7373	7361	7349	7337	7325	2	4	6	8	10
43	0.7314	7302	7290	7278	7266	7254	7242	7230	7218	7206	2	4	6	8	10
44	0.7193	7181	7169	7157	7145	7133	7120	7108	7096	7083	2	4	6	8	10

Table D.5 Natural tangents

	0' 0.0°	6' 0.1°	12' 0.2°	18' 0.3°	24' 0.4°	30' 0.5°	36' 0.6°	42' 0.7°	48' 0.8°	54' 0.9°	Mean Differences 1'	2'	3'	4'	5'
0°	0.0000	0017	0035	0052	0070	0087	0105	0122	0140	0157	3	6	9	12	15
1	0.0175	0192	0209	0227	0244	0262	0279	0297	0314	0332	3	6	9	12	15
2	0.0349	0367	0384	0402	0419	0437	0454	0472	0489	0507	3	6	9	12	15
3	0.0524	0542	0559	0577	0594	0612	0629	0647	0664	0682	3	6	9	12	15
4	0.0699	0717	0734	0752	0769	0787	0805	0822	0840	0857	3	6	9	12	15
5	0.0875	0892	0910	0928	0945	0963	0981	0998	1016	1033	3	6	9	12	15
6	0.1051	1069	1086	1104	1122	1139	1157	1175	1192	1210	3	6	9	12	15
7	0.1228	1246	1263	1281	1299	1317	1334	1352	1370	1388	3	6	9	12	15
8	0.1405	1423	1441	1459	1477	1495	1512	1530	1548	1566	3	6	9	12	15
9	0.1584	1602	1620	1638	1655	1673	1691	1709	1727	1745	3	6	9	12	15
10°	0.1763	1781	1799	1817	1835	1853	1871	1890	1908	1926	3	6	9	12	15
11	0.1944	1962	1980	1998	2016	2035	2053	2071	2089	2107	3	6	9	12	15
12	0.2126	2144	2162	2180	2199	2217	2235	2254	2272	2290	3	6	9	12	15
13	0.2309	2327	2345	2364	2382	2401	2419	2438	2456	2475	3	6	9	12	16
14	0.2493	2512	2530	2549	2568	2586	2605	2623	2642	2661	3	6	9	12	16
15	0.2679	2698	2717	2736	2754	2773	2792	2811	2830	2849	3	6	9	13	16
16	0.2867	2886	2905	2924	2943	2962	2981	3000	3019	3038	3	6	9	13	16
17	0.3057	3076	3096	3115	3134	3153	3172	3191	3211	3230	3	6	10	13	16
18	0.3249	3269	3288	3307	3327	3346	3365	3385	3404	3424	3	6	10	13	16
19	0.3443	3463	3482	3502	3522	3541	3561	3581	3600	3620	3	7	10	13	16
20°	0.3640	3659	3679	3699	3719	3739	3759	3779	3799	3819	3	7	10	13	17
21	0.3839	3859	3879	3899	3919	3939	3959	3979	4000	4020	3	7	10	13	17
22	0.4040	4061	4081	4101	4122	4142	4163	4183	4204	4224	3	7	10	14	17
23	0.4245	4265	4286	4307	4327	4348	4369	4390	4411	4431	3	7	10	14	17
24	0.4452	4473	4494	4515	4536	4557	4578	4599	4621	4642	4	7	11	14	18
25	0.4663	4684	4706	4727	4748	4770	4791	4813	4834	4856	4	7	11	14	18
26	0.4877	4899	4921	4942	4964	4986	5008	5029	5051	5073	4	7	11	15	18
27	0.5095	5117	5139	5161	5184	5206	5228	5250	5272	5295	4	7	11	15	18
28	0.5317	5340	5362	5384	5407	5430	5452	5475	5498	5520	4	8	11	15	19
29	0.5543	5566	5589	5612	5635	5658	5681	5704	5727	5750	4	8	12	15	19
30°	0.5774	5797	5820	5844	5867	5890	5914	5938	5961	5985	4	8	12	16	20
31	0.6009	6032	6056	6080	6104	6128	6152	6176	6200	6224	4	8	12	16	20
32	0.6249	6273	6297	6322	6346	6371	6395	6420	6445	6469	4	8	12	16	20
33	0.6494	6519	6544	6569	6594	6619	6644	6669	6694	6720	4	8	13	17	21
34	0.6745	6771	6796	6822	6847	6873	6899	6924	6950	6976	4	9	13	17	21
35	0.7002	7028	7054	7080	7107	7133	7159	7186	7212	7239	4	9	13	18	22
36	0.7265	7292	7319	7346	7373	7400	7427	7454	7481	7508	5	9	14	18	23
37	0.7536	7563	7590	7618	7646	7673	7701	7729	7757	7785	5	9	14	18	23
38	0.7813	7841	7869	7898	7926	7954	7983	8012	8040	8069	5	9	14	19	24
39	0.8098	8127	8156	8185	8214	8243	8273	8302	8332	8361	5	10	15	19	24
40°	0.8391	8421	8451	8481	8511	8541	8571	8601	8632	8662	5	10	15	20	25
41	0.8693	8724	8754	8785	8816	8847	8878	8910	8941	8972	5	10	16	21	26
42	0.9004	9036	9067	9099	9131	9163	9195	9228	9260	9293	5	11	16	21	27
43	0.9325	9358	9391	9424	9457	9490	9523	9556	9590	9623	6	11	17	22	28
44	0.9657	9691	9725	9759	9793	9827	9861	9896	9930	9965	6	11	17	23	29

	0' 0.0°	6' 0.1°	12' 0.2°	18' 0.3°	24' 0.4°	30' 0.5°	36' 0.6°	42' 0.7°	48' 0.8°	54' 0.9°	Mean Differences 1'	2'	3'	4'	5'
45°	1.0000	0035	0070	0105	0141	0176	0212	0247	0283	0319	6	12	18	24	30
46	1.0355	0392	0428	0464	0501	0538	0575	0612	0649	0686	6	12	18	25	31
47	1.0724	0761	0799	0837	0875	0913	0951	0990	1028	1067	6	13	19	25	32
48	1.1106	1145	1184	1224	1263	1303	1343	1383	1423	1463	7	13	20	27	33
49	1.1504	1544	1585	1626	1667	1708	1750	1792	1833	1875	7	14	21	28	34
50°	1.1918	1960	2002	2045	2088	2131	2174	2218	2261	2305	7	14	22	29	36
51	1.2349	2393	2437	2484	2527	2572	2617	2662	2708	2753	8	15	23	30	38
52	1.2799	2846	2892	2938	2985	3032	3079	3127	3175	3222	8	16	24	31	39
53	1.3270	3319	3367	3416	3465	3514	3564	3613	3663	3713	8	16	25	33	41
54	1.3764	3814	3865	3916	3968	4019	4071	4124	4176	4229	9	17	26	34	43
55	1.4281	4335	4388	4442	4496	4550	4605	4659	4715	4770	9	18	27	36	45
56	1.4826	4882	4938	4994	5051	5108	5166	5224	5282	5340	10	19	29	38	48
57	1.5399	5458	5517	5577	5637	5697	5757	5818	5880	5941	10	20	30	40	50
58	1.6003	6066	6128	6191	6255	6319	6383	6447	6512	6577	11	21	32	43	53
59	1.6643	6709	6775	6842	6909	6977	7045	7113	7182	7251	11	23	34	45	56
60°	1.7321	7391	7461	7532	7603	7675	7747	7820	7893	7966	12	24	36	48	60
61	1.8040	8115	8190	8265	8341	8418	8495	8572	8650	8728	13	26	38	51	64
62	1.8807	8887	8967	9047	9128	9210	9292	9375	9458	9542	14	27	41	55	68
63	1.9626	9711	9797	9883	9970	0057	0145	0233	0323	0413	15	29	44	58	73
64	2.0503	0594	0686	0778	0872	0965	1060	1155	1251	1348	16	31	47	63	78
65	2.1445	1543	1642	1742	1842	1943	2045	2148	2251	2355	17	34	51	68	85
66	2.2460	2566	2673	2781	2889	2998	3109	3220	3332	3445	18	37	55	73	92
67	2.3559	3673	3789	3906	4023	4142	4262	4383	4504	4627	20	40	60	79	99
68	2.4751	4876	5002	5129	5257	5386	5517	5649	5782	5916	22	43	65	87	108
69	2.6051	6187	6325	6464	6605	6746	6889	7034	7179	7326	24	47	71	95	119
70°	2.7475	7625	7776	7929	8083	8239	8397	8556	8716	8878	26	52	78	104	131
71	2.9042	9208	9375	9544	9714	9887	0061	0237	0415	0595	29	58	87	116	145
72	3.0777	0961	1146	1334	1524	1716	1910	2106	2305	2506	32	64	96	129	161
73	3.2709	2914	3122	3332	3544	3759	3977	4197	4420	4646	36	72	108	144	180
74	3.4874	5105	5339	5576	5816	6059	6305	6554	6806	7062	41	81	122	163	204
75	3.7321	7583	7848	8118	8391	8667	8947	9232	9520	9812	46	93	139	186	232
76	4.0108	0408	0713	1022	1335	1653	1976	2303	2635	2972					
77	4.3315	3662	4015	4374	4737	5107	5483	5864	6252	6646					
78	4.7046	7453	7867	8288	8716	9152	9594	0045	0504	0970					
79	5.1446	1929	2422	2924	3435	3955	4486	5026	5578	6140					
80°	5.6713	7297	7894	8502	9124	9758	0405	1066	1742	2432					
81	6.3138	3859	4596	5350	6122	6912	7720	8548	9395	0264					
82	7.1154	2066	3002	3962	4947	5958	6996	8062	9158	0285					
83	8.1443	2636	3863	5126	6427	7769	9152	0579	2052	3572					
84	9.514	9.677	9.845	10.02	10.20	10.39	10.58	10.78	10.99	11.20					
85	11.43	11.66	11.91	12.16	12.43	12.71	13.00	13.30	13.62	13.95					
86	14.30	14.67	15.06	15.46	15.89	16.35	16.83	17.34	17.89	18.46					
87	19.08	19.74	20.45	21.20	22.02	22.90	23.86	24.90	26.03	27.27					
88	28.64	30.14	31.82	33.69	35.80	38.19	40.92	44.07	47.74	52.08					
89	57.29	63.66	71.62	81.85	95.49	114.6	143.2	191.0	286.5	573.0					

Mean differences no longer sufficiently accurate. (rows 76°–89°)

Natural Cotangents:
Cot x° = tan (90−x)° and use above table.

Table D.6(a) Square roots: from 1 to 9.9

	0	1	2	3	4	5	6	7	8	9	1	2	3	4	5	6	7	8	9
											\<Mean Differences\>								
1.0	1.000	1.005	1.010	1.015	1.020	1.025	1.030	1.034	1.039	1.044	0	1	1	2	2	3	3	3	4
1.1	1.049	1.054	1.058	1.063	1.068	1.072	1.077	1.082	1.086	1.091	0	1	1	2	2	3	3	4	4
1.2	1.095	1.100	1.105	1.109	1.114	1.118	1.123	1.127	1.131	1.136	0	1	1	2	2	3	3	4	4
1.3	1.140	1.145	1.149	1.153	1.158	1.162	1.166	1.171	1.175	1.179	0	1	1	2	2	3	3	3	4
1.4	1.183	1.187	1.192	1.196	1.200	1.204	1.208	1.212	1.217	1.221	0	1	1	2	2	2	3	3	4
1.5	1.225	1.229	1.233	1.237	1.241	1.245	1.249	1.253	1.257	1.261	0	1	1	2	2	2	3	3	4
1.6	1.265	1.269	1.273	1.277	1.281	1.285	1.288	1.292	1.296	1.300	0	1	1	2	2	2	3	3	4
1.7	1.304	1.308	1.312	1.315	1.319	1.323	1.327	1.330	1.334	1.338	0	1	1	2	2	2	3	3	3
1.8	1.342	1.345	1.349	1.353	1.357	1.360	1.364	1.368	1.371	1.375	0	1	1	1	2	2	3	3	3
1.9	1.378	1.382	1.386	1.389	1.393	1.396	1.400	1.404	1.407	1.411	0	1	1	1	2	2	3	3	3
2.0	1.414	1.418	1.421	1.425	1.428	1.432	1.435	1.439	1.442	1.446	0	1	1	1	2	2	2	3	3
2.1	1.449	1.453	1.456	1.460	1.463	1.466	1.470	1.473	1.477	1.480	0	1	1	1	2	2	2	3	3
2.2	1.483	1.487	1.490	1.493	1.497	1.500	1.503	1.507	1.510	1.513	0	1	1	1	2	2	2	3	3
2.3	1.517	1.520	1.523	1.526	1.530	1.533	1.536	1.539	1.543	1.546	0	1	1	1	2	2	2	3	3
2.4	1.549	1.552	1.556	1.559	1.562	1.565	1.568	1.572	1.575	1.578	0	1	1	1	2	2	2	2	3
2.5	1.581	1.584	1.587	1.591	1.594	1.597	1.600	1.603	1.606	1.609	0	1	1	1	2	2	2	2	3
2.6	1.612	1.616	1.619	1.622	1.625	1.628	1.631	1.634	1.637	1.640	0	1	1	1	2	2	2	2	3
2.7	1.643	1.646	1.649	1.652	1.655	1.658	1.661	1.664	1.667	1.670	0	1	1	1	1	2	2	2	3
2.8	1.673	1.676	1.679	1.682	1.685	1.688	1.691	1.694	1.697	1.700	0	1	1	1	1	2	2	2	3
2.9	1.703	1.706	1.709	1.712	1.715	1.718	1.720	1.723	1.726	1.729	0	1	1	1	1	2	2	2	3
3.0	1.732	1.735	1.738	1.741	1.744	1.746	1.749	1.752	1.755	1.758	0	1	1	1	1	2	2	2	3
3.1	1.761	1.764	1.766	1.769	1.772	1.775	1.778	1.780	1.783	1.786	0	1	1	1	1	2	2	2	2
3.2	1.789	1.792	1.794	1.797	1.800	1.803	1.806	1.808	1.811	1.814	0	1	1	1	1	2	2	2	2
3.3	1.817	1.819	1.822	1.825	1.828	1.830	1.833	1.836	1.839	1.841	0	1	1	1	1	2	2	2	2
3.4	1.844	1.847	1.849	1.852	1.855	1.857	1.860	1.863	1.866	1.868	0	1	1	1	1	2	2	2	2
3.5	1.871	1.874	1.876	1.879	1.882	1.884	1.887	1.889	1.892	1.895	0	1	1	1	1	2	2	2	2
3.6	1.897	1.900	1.903	1.905	1.908	1.911	1.913	1.916	1.918	1.921	0	1	1	1	1	2	2	2	2
3.7	1.924	1.926	1.929	1.931	1.934	1.937	1.939	1.942	1.944	1.947	0	0	1	1	1	1	2	2	2
3.8	1.949	1.952	1.955	1.957	1.960	1.962	1.965	1.967	1.970	1.972	0	0	1	1	1	1	2	2	2
3.9	1.975	1.977	1.980	1.982	1.985	1.988	1.990	1.993	1.995	1.998	0	0	1	1	1	1	2	2	2
4.0	2.000	2.002	2.005	2.007	2.010	2.012	2.015	2.017	2.020	2.022	0	0	1	1	1	1	2	2	2
4.1	2.025	2.027	2.030	2.032	2.035	2.037	2.040	2.042	2.045	2.047	0	0	1	1	1	1	2	2	2
4.2	2.049	2.052	2.054	2.057	2.059	2.062	2.064	2.066	2.069	2.071	0	0	1	1	1	1	2	2	2
4.3	2.074	2.076	2.078	2.081	2.083	2.086	2.088	2.091	2.093	2.095	0	0	1	1	1	1	2	2	2
4.4	2.098	2.100	2.102	2.105	2.107	2.110	2.112	2.114	2.117	2.119	0	0	1	1	1	1	2	2	2
4.5	2.121	2.124	2.126	2.128	2.131	2.133	2.135	2.138	2.140	2.142	0	0	1	1	1	1	2	2	2
4.6	2.145	2.147	2.149	2.152	2.154	2.156	2.159	2.161	2.163	2.166	0	0	1	1	1	1	2	2	2
4.7	2.168	2.170	2.173	2.175	2.177	2.179	2.182	2.184	2.186	2.189	0	0	1	1	1	1	2	2	2
4.8	2.191	2.193	2.195	2.198	2.200	2.202	2.205	2.207	2.209	2.211	0	0	1	1	1	1	2	2	2
4.9	2.214	2.216	2.218	2.220	2.223	2.225	2.227	2.229	2.232	2.234	0	0	0	1	1	1	2	2	2
5.0	2.236	2.238	2.241	2.243	2.245	2.247	2.249	2.252	2.254	2.256	0	0	1	1	1	1	1	2	2
5.1	2.258	2.261	2.263	2.265	2.267	2.269	2.272	2.274	2.276	2.278	0	0	1	1	1	1	1	2	2
5.2	2.280	2.283	2.285	2.287	2.289	2.291	2.294	2.296	2.298	2.300	0	0	1	1	1	1	1	2	2
5.3	2.302	2.304	2.307	2.309	2.311	2.313	2.315	2.317	2.320	2.322	0	0	1	1	1	1	1	2	2
5.4	2.324	2.326	2.328	2.330	2.332	2.335	2.337	2.339	2.341	2.343	0	0	1	1	1	1	1	2	2

	0	1	2	3	4	5	6	7	8	9	1	2	3	4	5	6	7	8	9
											\<Mean Differences\>								
5.5	2.345	2.347	2.350	2.352	2.354	2.356	2.358	2.360	2.362	2.364	0	0	1	1	1	1	1	2	2
5.6	2.366	2.369	2.371	2.373	2.375	2.377	2.379	2.381	2.383	2.385	0	0	1	1	1	1	1	2	2
5.7	2.388	2.390	2.392	2.394	2.396	2.398	2.400	2.402	2.404	2.406	0	0	1	1	1	1	1	2	2
5.8	2.408	2.410	2.412	2.415	2.417	2.419	2.421	2.423	2.425	2.427	0	0	1	1	1	1	1	2	2
5.9	2.429	2.431	2.433	2.435	2.437	2.439	2.441	2.443	2.445	2.447	0	0	1	1	1	1	1	2	2
6.0	2.449	2.452	2.454	2.456	2.458	2.460	2.462	2.464	2.466	2.468	0	0	1	1	1	1	1	2	2
6.1	2.470	2.472	2.474	2.476	2.478	2.480	2.482	2.484	2.486	2.488	0	0	1	1	1	1	1	2	2
6.2	2.490	2.492	2.494	2.496	2.498	2.500	2.502	2.504	2.506	2.508	0	0	1	1	1	1	1	2	2
6.3	2.510	2.512	2.514	2.516	2.518	2.520	2.522	2.524	2.526	2.528	0	0	1	1	1	1	1	2	2
6.4	2.530	2.532	2.534	2.536	2.538	2.540	2.542	2.544	2.546	2.548	0	0	1	1	1	1	1	2	2
6.5	2.550	2.551	2.553	2.555	2.557	2.559	2.561	2.563	2.565	2.567	0	0	1	1	1	1	1	2	2
6.6	2.569	2.571	2.573	2.575	2.577	2.579	2.581	2.583	2.585	2.587	0	0	1	1	1	1	1	2	2
6.7	2.588	2.590	2.592	2.594	2.596	2.598	2.600	2.602	2.604	2.606	0	0	1	1	1	1	1	2	2
6.8	2.608	2.610	2.612	2.613	2.615	2.617	2.619	2.621	2.623	2.625	0	0	1	1	1	1	1	2	2
6.9	2.627	2.629	2.631	2.632	2.634	2.636	2.638	2.640	2.642	2.644	0	0	1	1	1	1	1	2	2
7.0	2.646	2.648	2.650	2.651	2.653	2.655	2.657	2.659	2.661	2.663	0	0	1	1	1	1	1	2	2
7.1	2.665	2.667	2.668	2.670	2.672	2.674	2.676	2.678	2.680	2.681	0	0	1	1	1	1	1	2	2
7.2	2.683	2.685	2.687	2.689	2.691	2.693	2.694	2.696	2.698	2.700	0	0	1	1	1	1	1	2	2
7.3	2.702	2.704	2.706	2.707	2.709	2.711	2.713	2.715	2.717	2.719	0	0	1	1	1	1	1	2	2
7.4	2.720	2.722	2.724	2.726	2.728	2.729	2.731	2.733	2.735	2.737	0	0	1	1	1	1	1	2	2
7.5	2.739	2.740	2.742	2.744	2.746	2.748	2.750	2.751	2.753	2.755	0	0	1	1	1	1	1	2	2
7.6	2.757	2.759	2.760	2.762	2.764	2.766	2.768	2.769	2.771	2.773	0	0	1	1	1	1	1	2	2
7.7	2.775	2.777	2.778	2.780	2.782	2.784	2.786	2.787	2.789	2.791	0	0	1	1	1	1	1	2	2
7.8	2.793	2.795	2.796	2.798	2.800	2.802	2.804	2.805	2.807	2.809	0	0	1	1	1	1	1	2	2
7.9	2.811	2.812	2.814	2.816	2.818	2.820	2.821	2.823	2.825	2.827	0	0	1	1	1	1	1	2	2
8.0	2.828	2.830	2.832	2.834	2.835	2.837	2.839	2.841	2.843	2.844	0	0	1	1	1	1	1	1	2
8.1	2.846	2.848	2.850	2.851	2.853	2.855	2.857	2.858	2.860	2.862	0	0	1	1	1	1	1	1	2
8.2	2.864	2.865	2.867	2.869	2.871	2.872	2.874	2.876	2.877	2.879	0	0	1	1	1	1	1	1	2
8.3	2.881	2.883	2.884	2.886	2.888	2.890	2.891	2.893	2.895	2.897	0	0	1	1	1	1	1	1	2
8.4	2.898	2.900	2.902	2.903	2.905	2.907	2.909	2.910	2.912	2.914	0	0	1	1	1	1	1	1	2
8.5	2.915	2.917	2.919	2.921	2.922	2.924	2.926	2.927	2.929	2.931	0	0	1	1	1	1	1	1	2
8.6	2.933	2.934	2.936	2.938	2.939	2.941	2.943	2.944	2.946	2.948	0	0	1	1	1	1	1	1	2
8.7	2.950	2.951	2.953	2.955	2.956	2.958	2.960	2.961	2.963	2.965	0	0	1	1	1	1	1	1	2
8.8	2.966	2.968	2.970	2.972	2.973	2.975	2.977	2.978	2.980	2.982	0	0	1	1	1	1	1	1	2
8.9	2.983	2.985	2.987	2.988	2.990	2.992	2.993	2.995	2.997	2.998	0	0	1	1	1	1	1	1	2
9.0	3.000	3.002	3.003	3.005	3.007	3.008	3.010	3.012	3.013	3.015	0	0	0	1	1	1	1	1	1
9.1	3.017	3.018	3.020	3.022	3.023	3.025	3.027	3.028	3.030	3.032	0	0	0	1	1	1	1	1	1
9.2	3.033	3.035	3.036	3.038	3.040	3.041	3.043	3.045	3.046	3.048	0	0	0	1	1	1	1	1	1
9.3	3.050	3.051	3.053	3.055	3.056	3.058	3.059	3.061	3.063	3.064	0	0	0	1	1	1	1	1	1
9.4	3.066	3.068	3.069	3.071	3.072	3.074	3.076	3.077	3.079	3.081	0	0	0	1	1	1	1	1	1
9.5	3.082	3.084	3.085	3.087	3.089	3.090	3.092	3.094	3.095	3.097	0	0	0	1	1	1	1	1	1
9.6	3.098	3.100	3.102	3.103	3.105	3.106	3.108	3.110	3.111	3.113	0	0	0	1	1	1	1	1	1
9.7	3.115	3.116	3.118	3.119	3.121	3.123	3.124	3.126	3.127	3.129	0	0	0	1	1	1	1	1	1
9.8	3.131	3.132	3.134	3.135	3.137	3.139	3.140	3.142	3.143	3.145	0	0	0	1	1	1	1	1	1
9.9	3.146	3.148	3.150	3.151	3.153	3.154	3.156	3.158	3.159	3.161	0	0	0	1	1	1	1	1	1

Table D.6(b) Square roots: from 10 to 99

	0	1	2	3	4	5	6	7	8	9	1	2	3	4	5	6	7	8	9
											Mean Differences									
10	3.162	3.178	3.194	3.209	3.225	3.240	3.256	3.271	3.286	3.302	2	3	5	6	8	9	11	12	14
11	3.317	3.332	3.347	3.362	3.376	3.391	3.406	3.421	3.435	3.450	1	3	4	6	7	9	10	12	13
12	3.464	3.479	3.493	3.507	3.521	3.536	3.550	3.564	3.578	3.592	1	3	4	6	7	9	10	11	13
13	3.606	3.619	3.633	3.647	3.661	3.674	3.688	3.701	3.715	3.728	1	3	4	5	7	8	10	11	12
14	3.742	3.755	3.768	3.782	3.795	3.808	3.821	3.834	3.847	3.860	1	3	4	5	7	8	9	11	12
15	3.873	3.886	3.899	3.912	3.924	3.937	3.950	3.962	3.975	3.988	1	3	4	5	6	8	9	10	11
16	4.000	4.012	4.025	4.037	4.050	4.062	4.074	4.087	4.099	4.111	1	2	4	5	6	7	9	10	11
17	4.123	4.135	4.147	4.159	4.171	4.183	4.195	4.207	4.219	4.231	1	2	4	5	6	7	8	10	11
18	4.243	4.254	4.266	4.278	4.290	4.301	4.313	4.324	4.336	4.347	1	2	3	5	6	7	8	9	10
19	4.359	4.370	4.382	4.393	4.405	4.416	4.427	4.438	4.450	4.461	1	2	3	5	6	7	8	9	10
20	4.472	4.483	4.494	4.506	4.517	4.528	4.539	4.550	4.561	4.572	1	2	3	4	6	7	8	9	10
21	4.583	4.594	4.604	4.615	4.626	4.637	4.648	4.658	4.669	4.680	1	2	3	4	5	7	8	9	10
22	4.690	4.701	4.712	4.722	4.733	4.743	4.754	4.765	4.775	4.785	1	2	3	4	5	6	8	9	10
23	4.796	4.806	4.817	4.827	4.837	4.848	4.858	4.868	4.879	4.889	1	2	3	4	5	6	7	8	9
24	4.899	4.909	4.919	4.930	4.940	4.950	4.960	4.970	4.980	4.990	1	2	3	4	5	6	7	8	9
25	5.000	5.010	5.020	5.030	5.040	5.050	5.060	5.070	5.079	5.089	1	2	3	4	5	6	7	8	9
26	5.099	5.109	5.119	5.128	5.138	5.148	5.158	5.167	5.177	5.187	1	2	3	4	5	6	7	8	9
27	5.196	5.206	5.215	5.225	5.235	5.244	5.254	5.263	5.273	5.282	1	2	3	4	5	6	7	8	9
28	5.292	5.301	5.310	5.320	5.329	5.339	5.348	5.357	5.367	5.376	1	2	3	4	5	6	7	7	8
29	5.385	5.394	5.404	5.413	5.422	5.431	5.441	5.450	5.459	5.468	1	2	3	4	5	6	6	7	8
30	5.477	5.486	5.495	5.505	5.514	5.523	5.532	5.541	5.550	5.559	1	2	3	4	5	5	6	7	8
31	5.568	5.577	5.586	5.595	5.604	5.612	5.621	5.630	5.639	5.648	1	2	3	4	4	5	6	7	8
32	5.657	5.666	5.675	5.683	5.692	5.701	5.710	5.718	5.727	5.736	1	2	3	4	4	5	6	7	8
33	5.745	5.753	5.762	5.771	5.779	5.788	5.797	5.805	5.814	5.822	1	2	3	3	4	5	6	7	8
34	5.831	5.840	5.848	5.857	5.865	5.874	5.882	5.891	5.899	5.908	1	2	3	3	4	5	6	7	8
35	5.916	5.925	5.933	5.941	5.950	5.958	5.967	5.975	5.983	5.992	1	2	2	3	4	5	6	7	8
36	6.000	6.008	6.017	6.025	6.033	6.042	6.050	6.058	6.066	6.075	1	2	2	3	4	5	6	7	7
37	6.083	6.091	6.099	6.107	6.116	6.124	6.132	6.140	6.148	6.156	1	2	2	3	4	5	6	6	7
38	6.164	6.173	6.181	6.189	6.197	6.205	6.213	6.221	6.229	6.237	1	2	2	3	4	5	6	6	7
39	6.245	6.253	6.261	6.269	6.277	6.285	6.293	6.301	6.309	6.317	1	2	2	3	4	5	6	6	7
40	6.325	6.332	6.340	6.348	6.356	6.364	6.372	6.380	6.387	6.395	1	2	2	3	4	5	5	6	7
41	6.403	6.411	6.419	6.427	6.434	6.442	6.450	6.458	6.465	6.473	1	2	2	3	4	5	5	6	7
42	6.481	6.488	6.496	6.504	6.512	6.519	6.527	6.535	6.542	6.550	1	2	2	3	4	5	5	6	7
43	6.557	6.565	6.573	6.580	6.588	6.595	6.603	6.611	6.618	6.626	1	2	2	3	4	5	5	6	7
44	6.633	6.641	6.648	6.656	6.663	6.671	6.678	6.686	6.693	6.701	1	1	2	3	4	4	5	6	7
45	6.708	6.716	6.723	6.731	6.738	6.745	6.753	6.760	6.768	6.775	1	1	2	3	4	4	5	6	7
46	6.782	6.790	6.797	6.804	6.812	6.819	6.826	6.834	6.841	6.848	1	1	2	3	4	4	5	6	7
47	6.856	6.863	6.870	6.878	6.885	6.892	6.899	6.907	6.914	6.921	1	1	2	3	4	4	5	6	7
48	6.928	6.935	6.943	6.950	6.957	6.964	6.971	6.979	6.986	6.993	1	1	2	3	4	4	5	6	7
49	7.000	7.007	7.014	7.021	7.029	7.036	7.043	7.050	7.057	7.064	1	1	2	3	4	4	5	6	6
50	7.071	7.078	7.085	7.092	7.099	7.106	7.113	7.120	7.127	7.134	1	1	2	3	4	4	5	6	6
51	7.141	7.148	7.155	7.162	7.169	7.176	7.183	7.190	7.197	7.204	1	1	2	3	3	4	5	6	6
52	7.211	7.218	7.225	7.232	7.239	7.246	7.253	7.259	7.266	7.273	1	1	2	3	3	4	5	5	6
53	7.280	7.287	7.294	7.301	7.308	7.314	7.321	7.328	7.335	7.342	1	1	2	3	3	4	5	5	6
54	7.349	7.355	7.362	7.369	7.376	7.382	7.389	7.396	7.403	7.410	1	1	2	3	3	4	5	5	6

	0	1	2	3	4	5	6	7	8	9	1	2	3	4	5	6	7	8	9
											Mean Differences									
55	7.416	7.423	7.430	7.436	7.443	7.450	7.457	7.463	7.470	7.477	1	1	2	3	3	4	5	5	6
56	7.483	7.490	7.497	7.503	7.510	7.517	7.523	7.530	7.537	7.543	1	1	2	3	3	4	5	5	6
57	7.550	7.556	7.563	7.570	7.576	7.583	7.589	7.596	7.603	7.609	1	1	2	3	3	4	5	5	6
58	7.616	7.622	7.629	7.635	7.642	7.649	7.655	7.662	7.668	7.675	1	1	2	3	3	4	5	5	6
59	7.681	7.688	7.694	7.701	7.707	7.714	7.720	7.727	7.733	7.740	1	1	2	3	3	4	5	5	6
60	7.746	7.752	7.759	7.765	7.772	7.778	7.785	7.791	7.797	7.804	1	1	2	3	3	4	4	5	6
61	7.810	7.817	7.823	7.829	7.836	7.842	7.849	7.855	7.861	7.868	1	1	2	3	3	4	4	5	6
62	7.874	7.880	7.887	7.893	7.899	7.906	7.912	7.918	7.925	7.931	1	1	2	3	3	4	4	5	6
63	7.937	7.944	7.950	7.956	7.962	7.969	7.975	7.981	7.987	7.994	1	1	2	2	3	4	4	5	6
64	8.000	8.006	8.012	8.019	8.025	8.031	8.037	8.044	8.050	8.056	1	1	2	2	3	4	4	5	6
65	8.062	8.068	8.075	8.081	8.087	8.093	8.099	8.106	8.112	8.118	1	1	2	2	3	4	4	5	5
66	8.124	8.130	8.136	8.142	8.149	8.155	8.161	8.167	8.173	8.179	1	1	2	2	3	4	4	5	5
67	8.185	8.191	8.198	8.204	8.210	8.216	8.222	8.228	8.234	8.240	1	1	2	2	3	4	4	5	5
68	8.246	8.252	8.258	8.264	8.270	8.276	8.283	8.289	8.295	8.301	1	1	2	2	3	4	4	5	5
69	8.307	8.313	8.319	8.325	8.331	8.337	8.343	8.349	8.355	8.361	1	1	2	2	3	4	4	5	5
70	8.367	8.373	8.379	8.385	8.390	8.396	8.402	8.408	8.414	8.420	1	1	2	2	3	4	4	5	5
71	8.426	8.432	8.438	8.444	8.450	8.456	8.462	8.468	8.473	8.479	1	1	2	2	3	4	4	5	5
72	8.485	8.491	8.497	8.503	8.509	8.515	8.521	8.526	8.532	8.538	1	1	2	2	3	4	4	5	5
73	8.544	8.550	8.556	8.562	8.567	8.573	8.579	8.585	8.591	8.597	1	1	2	2	3	4	4	5	5
74	8.602	8.608	8.614	8.620	8.626	8.631	8.637	8.643	8.649	8.654	1	1	2	2	3	3	4	5	5
75	8.660	8.666	8.672	8.678	8.683	8.689	8.695	8.701	8.706	8.712	1	1	2	2	3	3	4	5	5
76	8.718	8.724	8.729	8.735	8.741	8.746	8.752	8.758	8.764	8.769	1	1	2	2	3	3	4	5	5
77	8.775	8.781	8.786	8.792	8.798	8.803	8.809	8.815	8.820	8.826	1	1	2	2	3	3	4	4	5
78	8.832	8.837	8.843	8.849	8.854	8.860	8.866	8.871	8.877	8.883	1	1	2	2	3	3	4	4	5
79	8.888	8.894	8.899	8.905	8.911	8.916	8.922	8.927	8.933	8.939	1	1	2	2	3	3	4	4	5
80	8.944	8.950	8.955	8.961	8.967	8.972	8.978	8.983	8.989	8.994	1	1	2	2	3	3	4	4	5
81	9.000	9.006	9.011	9.017	9.022	9.028	9.033	9.039	9.044	9.050	1	1	2	2	3	3	4	4	5
82	9.055	9.061	9.066	9.072	9.077	9.083	9.088	9.094	9.099	9.105	1	1	2	2	3	3	4	4	5
83	9.110	9.116	9.121	9.127	9.132	9.138	9.143	9.149	9.154	9.160	1	1	2	2	3	3	4	4	5
84	9.165	9.171	9.176	9.182	9.187	9.192	9.198	9.203	9.209	9.214	1	1	2	2	3	3	4	4	5
85	9.220	9.225	9.230	9.236	9.241	9.247	9.252	9.257	9.263	9.268	1	1	2	2	3	3	4	4	5
86	9.274	9.279	9.284	9.290	9.295	9.301	9.306	9.311	9.317	9.322	1	1	2	2	3	3	4	4	5
87	9.327	9.333	9.338	9.343	9.349	9.354	9.359	9.365	9.370	9.375	1	1	2	2	3	3	4	4	5
88	9.381	9.386	9.391	9.397	9.402	9.407	9.413	9.418	9.423	9.429	1	1	2	2	3	3	4	4	5
89	9.434	9.439	9.445	9.450	9.455	9.460	9.466	9.471	9.476	9.482	1	1	2	2	3	3	4	4	5
90	9.487	9.492	9.497	9.503	9.508	9.513	9.518	9.524	9.529	9.534	1	1	2	2	3	3	4	4	5
91	9.539	9.545	9.550	9.555	9.560	9.566	9.571	9.576	9.581	9.586	1	1	2	2	3	3	4	4	5
92	9.592	9.597	9.602	9.607	9.613	9.618	9.623	9.628	9.633	9.638	1	1	2	2	3	3	4	4	5
93	9.644	9.649	9.654	9.659	9.664	9.670	9.675	9.680	9.685	9.690	1	1	2	2	3	3	4	4	5
94	9.695	9.701	9.706	9.711	9.716	9.721	9.726	9.731	9.737	9.742	1	1	2	2	3	3	4	4	5
95	9.747	9.752	9.757	9.762	9.767	9.772	9.778	9.783	9.788	9.793	1	1	2	2	3	3	4	4	5
96	9.798	9.803	9.808	9.813	9.818	9.823	9.829	9.834	9.839	9.844	1	1	2	2	3	3	3	4	4
97	9.849	9.854	9.859	9.864	9.869	9.874	9.879	9.884	9.889	9.894	1	1	2	2	3	3	3	4	4
98	9.899	9.905	9.910	9.915	9.920	9.925	9.930	9.935	9.940	9.945	1	1	2	2	2	3	3	4	4
99	9.950	9.955	9.960	9.965	9.970	9.975	9.980	9.985	9.990	9.995	0	1	1	2	2	3	3	4	4

Table D.7 Reciprocals of four-figure numbers

x	0	1	2	3	4	5	6	7	8	9	1	2	3	4	5	6	7	8	9
											SUBTRACT Mean Differences								
1.0	1.000	9901	9804	9709	9615	9524	9434	9346	9259	9174	9	18	28	37	46	55	64	74	83
1.1	0.9091	9009	8929	8850	8772	8696	8621	8547	8475	8403	8	15	23	31	38	46	53	61	69
1.2	0.8333	8264	8197	8130	8065	8000	7937	7874	7813	7752	7	13	20	26	33	39	46	52	59
1.3	0.7692	7634	7576	7519	7463	7407	7353	7299	7246	7194	6	11	17	22	28	33	39	44	50
1.4	0.7143	7092	7042	6993	6944	6897	6849	6803	6757	6711	5	10	14	19	24	29	33	38	43
1.5	0.6667	6623	6579	6536	6494	6452	6410	6369	6329	6289	4	8	13	17	21	25	29	33	38
1.6	0.6250	6211	6173	6135	6098	6061	6024	5988	5952	5917	4	7	11	15	18	22	26	29	33
1.7	0.5882	5848	5814	5780	5747	5714	5682	5650	5618	5587	3	7	10	13	16	20	23	26	29
1.8	0.5556	5525	5495	5464	5435	5405	5376	5348	5319	5291	3	6	9	12	15	18	21	23	26
1.9	0.5263	5236	5208	5181	5155	5128	5102	5076	5051	5025	3	5	8	11	13	16	18	21	24
2.0	0.5000	4975	4950	4926	4902	4878	4854	4831	4808	4785	2	5	7	10	12	14	17	19	22
2.1	0.4762	4739	4717	4695	4673	4651	4630	4608	4587	4566	2	4	7	9	11	13	15	17	19
2.2	0.4545	4525	4505	4484	4464	4444	4425	4405	4386	4367	2	4	6	8	10	12	14	16	18
2.3	0.4348	4329	4310	4292	4274	4255	4237	4219	4202	4184	2	4	5	7	9	11	13	14	16
2.4	0.4167	4149	4132	4115	4098	4082	4065	4049	4032	4016	2	3	5	7	8	10	11	13	15
2.5	0.4000	3984	3968	3953	3937	3922	3906	3891	3876	3861	2	3	5	6	8	9	11	12	14
2.6	0.3846	3831	3817	3802	3788	3774	3759	3745	3731	3717	1	3	4	6	7	9	10	11	13
2.7	0.3704	3690	3676	3663	3650	3636	3623	3610	3597	3584	1	3	4	5	7	8	9	11	12
2.8	0.3571	3559	3546	3534	3521	3509	3497	3484	3472	3460	1	2	4	5	6	7	9	10	11
2.9	0.3448	3436	3425	3413	3401	3390	3378	3367	3356	3344	1	2	3	5	6	7	8	9	10
3.0	0.3333	3322	3311	3300	3289	3279	3268	3257	3247	3236	1	2	3	4	5	6	7	9	10
3.1	0.3226	3215	3205	3195	3185	3175	3165	3155	3145	3135	1	2	3	4	5	6	7	8	9
3.2	0.3125	3115	3106	3096	3086	3077	3067	3058	3049	3040	1	2	3	4	5	6	7	8	9
3.3	0.3030	3021	3012	3003	2994	2985	2976	2967	2959	2950	1	2	3	4	4	5	6	7	8
3.4	0.2941	2933	2924	2915	2907	2899	2890	2882	2874	2865	1	2	3	3	4	5	6	7	8
3.5	0.2857	2849	2841	2833	2825	2817	2809	2801	2793	2786	1	2	2	3	4	5	6	6	7
3.6	0.2778	2770	2762	2755	2747	2740	2732	2725	2717	2710	1	2	2	3	4	5	5	6	7
3.7	0.2703	2695	2688	2681	2674	2667	2660	2653	2646	2639	1	1	2	3	4	4	5	6	6
3.8	0.2632	2625	2618	2611	2604	2597	2591	2584	2577	2571	1	1	2	3	3	4	5	5	6
3.9	0.2564	2558	2551	2545	2538	2532	2525	2519	2513	2506	1	1	2	3	3	4	4	5	6
4.0	0.2500	2494	2488	2481	2475	2469	2463	2457	2451	2445	1	1	2	2	3	4	4	5	5
4.1	0.2439	2433	2427	2421	2415	2410	2404	2398	2392	2387	1	1	2	2	3	3	4	4	5
4.2	0.2381	2375	2370	2364	2358	2353	2347	2342	2336	2331	1	1	2	2	3	3	4	4	5
4.3	0.2326	2320	2315	2309	2304	2299	2294	2288	2283	2278	1	1	2	2	3	3	4	4	5
4.4	0.2273	2268	2262	2257	2252	2247	2242	2237	2232	2227	0	1	1	2	2	3	3	4	4
4.5	0.2222	2217	2212	2208	2203	2198	2193	2188	2183	2179	0	1	1	2	2	3	3	4	4
4.6	0.2174	2169	2165	2160	2155	2151	2146	2141	2137	2132	0	1	1	2	2	3	3	4	4
4.7	0.2128	2123	2119	2114	2110	2105	2101	2096	2092	2088	0	1	1	2	2	3	3	4	4
4.8	0.2083	2079	2075	2070	2066	2062	2058	2053	2049	2045	0	1	1	2	2	2	3	3	4
4.9	0.2041	2037	2033	2028	2024	2020	2016	2012	2008	2004	0	1	1	2	2	2	3	3	4
5.0	0.2000	1996	1992	1988	1984	1980	1976	1972	1969	1965	0	1	1	2	2	2	3	3	4
5.1	0.1961	1957	1953	1949	1946	1942	1938	1934	1931	1927	0	1	1	1	2	2	3	3	3
5.2	0.1923	1919	1916	1912	1908	1905	1901	1898	1894	1890	0	1	1	1	2	2	3	3	3
5.3	0.1887	1883	1880	1876	1873	1869	1866	1862	1859	1855	0	1	1	1	2	2	2	3	3
5.4	0.1852	1848	1845	1842	1838	1835	1832	1828	1825	1821	0	1	1	1	2	2	2	3	3

x	0	1	2	3	4	5	6	7	8	9	1	2	3	4	5	6	7	8	9
											SUBTRACT Mean Differences								
5.5	0.1818	1815	1812	1808	1805	1802	1799	1795	1792	1789	0	1	1	1	2	2	2	2	3
5.6	0.1786	1783	1779	1776	1773	1770	1767	1764	1761	1757	0	1	1	1	2	2	2	3	3
5.7	0.1754	1751	1748	1745	1742	1739	1736	1733	1730	1727	0	1	1	1	1	2	2	2	3
5.8	0.1724	1721	1718	1715	1712	1709	1706	1704	1701	1698	0	1	1	1	1	2	2	2	3
5.9	0.1695	1692	1689	1686	1684	1681	1678	1675	1672	1669	0	1	1	1	1	2	2	2	3
6.0	0.1667	1664	1661	1658	1656	1653	1650	1647	1645	1642	0	1	1	1	1	2	2	2	2
6.1	0.1639	1637	1634	1631	1629	1626	1623	1621	1618	1616	0	1	1	1	1	2	2	2	2
6.2	0.1613	1610	1608	1605	1603	1600	1597	1595	1592	1590	0	0	1	1	1	1	2	2	2
6.3	0.1587	1585	1582	1580	1577	1575	1572	1570	1567	1565	0	0	1	1	1	1	2	2	2
6.4	0.1562	1560	1558	1555	1553	1550	1548	1546	1543	1541	0	0	1	1	1	1	2	2	2
6.5	0.1538	1536	1534	1531	1529	1527	1524	1522	1520	1517	0	0	1	1	1	1	2	2	2
6.6	0.1515	1513	1511	1508	1506	1504	1502	1499	1497	1495	0	0	1	1	1	1	1	2	2
6.7	0.1493	1490	1488	1486	1484	1481	1479	1477	1475	1473	0	0	1	1	1	1	1	2	2
6.8	0.1471	1468	1466	1464	1462	1460	1458	1456	1453	1451	0	0	1	1	1	1	1	2	2
6.9	0.1449	1447	1445	1443	1441	1439	1437	1435	1433	1431	0	0	1	1	1	1	1	2	2
7.0	0.1429	1427	1425	1422	1420	1418	1416	1414	1412	1410	0	0	1	1	1	1	1	2	2
7.1	0.1408	1406	1404	1403	1401	1399	1397	1395	1393	1391	0	0	1	1	1	1	1	1	2
7.2	0.1389	1387	1385	1383	1381	1379	1377	1376	1374	1372	0	0	1	1	1	1	1	1	2
7.3	0.1370	1368	1366	1364	1362	1361	1359	1357	1355	1353	0	0	1	1	1	1	1	1	2
7.4	0.1351	1350	1348	1346	1344	1342	1340	1339	1337	1335	0	0	0	1	1	1	1	1	1
7.5	0.1333	1332	1330	1328	1326	1325	1323	1321	1319	1318	0	0	0	1	1	1	1	1	1
7.6	0.1316	1314	1312	1311	1309	1307	1305	1304	1302	1300	0	0	0	1	1	1	1	1	1
7.7	0.1299	1297	1295	1294	1292	1290	1289	1287	1285	1284	0	0	0	1	1	1	1	1	1
7.8	0.1282	1280	1279	1277	1276	1274	1272	1271	1269	1267	0	0	0	1	1	1	1	1	1
7.9	0.1266	1264	1263	1261	1259	1258	1256	1255	1253	1252	0	0	0	0	1	1	1	1	1
8.0	0.1250	1248	1247	1245	1244	1242	1241	1239	1238	1236	0	0	0	0	1	1	1	1	1
8.1	0.1235	1233	1232	1230	1229	1227	1225	1224	1222	1221	0	0	0	0	1	1	1	1	1
8.2	0.1220	1218	1217	1215	1214	1212	1211	1209	1208	1206	0	0	0	0	1	1	1	1	1
8.3	0.1205	1203	1202	1200	1199	1198	1196	1195	1193	1192	0	0	0	0	1	1	1	1	1
8.4	0.1190	1189	1188	1186	1185	1183	1182	1181	1179	1178	0	0	0	0	0	1	1	1	1
8.5	0.1176	1175	1174	1172	1171	1170	1168	1167	1166	1164	0	0	0	0	0	1	1	1	1
8.6	0.1163	1161	1160	1159	1157	1156	1155	1153	1152	1151	0	0	0	0	0	1	1	1	1
8.7	0.1149	1148	1147	1145	1144	1143	1142	1140	1139	1138	0	0	0	0	0	1	1	1	1
8.8	0.1136	1135	1134	1133	1131	1130	1129	1127	1126	1125	0	0	0	0	0	1	1	1	1
8.9	0.1124	1122	1121	1120	1119	1117	1116	1115	1114	1112	0	0	0	0	0	1	1	1	1
9.0	0.1111	1110	1109	1107	1106	1105	1104	1103	1101	1100	0	0	0	0	0	1	1	1	1
9.1	0.1099	1098	1096	1095	1094	1093	1092	1090	1089	1088	0	0	0	0	0	0	1	1	1
9.2	0.1087	1086	1085	1083	1082	1081	1080	1079	1078	1076	0	0	0	0	0	0	1	1	1
9.3	0.1075	1074	1073	1072	1071	1070	1068	1067	1066	1065	0	0	0	0	0	0	1	1	1
9.4	0.1064	1063	1062	1060	1059	1058	1057	1056	1055	1054	0	0	0	0	0	0	1	1	1
9.5	0.1053	1052	1050	1049	1048	1047	1046	1045	1044	1043	0	0	0	0	0	0	1	1	1
9.6	0.1042	1041	1039	1038	1037	1036	1035	1034	1033	1032	0	0	0	0	0	0	0	1	1
9.7	0.1031	1030	1029	1028	1027	1026	1025	1024	1022	1021	0	0	0	0	0	0	0	1	1
9.8	0.1020	1019	1018	1017	1016	1015	1014	1013	1012	1011	0	0	0	0	0	0	0	1	1
9.9	0.1010	1009	1008	1007	1006	1005	1004	1003	1002	1001	0	0	0	0	0	0	0	1	1

e.g. $\dfrac{1}{3.7} = 0.2703$, $\dfrac{1}{3.74} = 0.2674$, $\dfrac{1}{3.748} = 0.2668$, $\dfrac{1}{374.8} = 0.002668$, $\dfrac{1}{0.0003748} = 2668$.

System international and imperial units

E.1 Introduction

As mentioned in the introduction familiarity with both the System International (SI) and Imperial system of units is important not only because accurate conversion is important, but also mistakes in such conversions can jeopardize safety. Therefore if you are unfamiliar with the SI system or the English Engineering (Imperial) system or both, *then this appendix should be treated as essential reading*!

In this appendix, you will find the fundamental units for the SI d'unites together with the important units for the English Engineering (Imperial) system.

In addition to tables of units, you will also find examples of commonly used conversions that are particularly applicable to aircraft maintenance engineering.

We start by introducing the SI system and the Imperial system in the form of tables of fundamental units and the more commonly derived units, together with the multiples and submultiples that often prefix many of these units.

The complete set of units, given below, are not all applicable to your study of Physics (Module 2, Chapter 4 of this book). However, they will also act as a source of reference for your study of Modules 3 and 4 (Chapters 5 and 6 of this book) that cover the electrical and electronic fundamentals, respectively, required for your license.

E.2 SI units

E.2.1 SI base units and their definitions

What follows are the **true and accurate** definitions of the SI base units (Table E.1); at

Table E.1 Base units

Basic quantity	SI unit name	SI unit symbol
Mass	kilogram	kg
Length	metre	m
Time	second	s
Electric current	ampere	A
Temperature	kelvin	K
Amount of substance	mole	mol
Luminous intensity	candela	cd

first these definitions may seem quite strange. They are detailed below for reference; you would have come across most of them during your study of Chapter 4 and also in Chapter 5.

Kilogramme

The kilogramme or kilogram is the unit of mass; it is equal to the mass of the international prototype of the kilogram, as defined by the *General Conference on Weights and Measures*, which has initials (CGPM).

Metre

The metre is the length of the path travelled by light in vacuum during the time interval of $1/299{,}792{,}458$ s.

Second

The second is the duration of $9{,}192{,}631{,}770$ periods of radiation corresponding to the transition between the two hyperfine levels of the ground state of the caesium 133 atom.

Ampere

The ampere is that constant current which if maintained in two straight parallel conductors of infinite length, of negligible circular cross-section, and placed 1 m apart in a vacuum would produce between these conductors a force equal to 2×10^{-7} N/m length.

Kelvin

The kelvin, unit of thermodynamic temperature, is the fraction 1/273.16 of the thermodynamic temperature of the triple point of water.

Mole

The mole is the amount of substance of a system, which contains as many elementary particles as there are atoms in 0.012 kg of carbon 12. When the mole is used, the elementary entities must be specified and may be atoms, molecules, ions, electrons or other particles, or specified groups of such particles.

Candela

The candela is the luminous intensity, in a given direction, of a source that emits monochromatic radiation of frequency 540×10^{12} Hz and that has a radiant intensity in that direction of 1/683 W/srad (see Table E.2).

E.2.2 SI supplementary and derived units

In addition to the seven base units given above, as mentioned before, there are two supplementary units (Table E.2): the radian for plane angles (which you may have already come across when you studied Chapter 2) and the steradian for solid three-dimensional angles. Both of these relationships are ratios and *ratios have no units*, e.g. metres/metres = 1. Again, do not worry too much at this stage, it may become clearer later, when we look at radian measure in our study of dynamics.

Table E.2 SI supplementary units

Supplementary unit	SI unit name	SI unit symbol
Plane angle	radian	rad
Solid angle	steradian	srad or sr

The SI derived units are defined by simple equations relating two or more base units. The names and symbols of some of the derived units may be substituted by special names and symbols. Some of the derived units, which you may be familiar with, are listed in Table E.3 with their special names as appropriate.

Table E.3 SI derived units

SI unit name	SI unit symbol	Quantity
Coulomb	C	Quantity of electricity, electric charge
Farad	F	Electric capacitance
Henry	H	Electrical inductance
Hertz	Hz	Frequency
Joule	J	Energy, work, heat
Lux	lx	Luminance
Newton	N	Force, weight
Ohm	Ω	Electrical resistance
Pascal	Pa	Pressure, stress
Sieman	S	Electrical conductance
Tesla	T	Induction field, magnetic flux density
Volt	V	Electric potential, electromotive force
Watt	W	Power, radiant flux
Weber	Wb	Induction, magnetic flux

E.2.3 SI prefixes

The SI prefixes are provided in Table E.4.

For example, 1 millimetre = 1 mm = 10^{-3} m, $1 \, cm^3 = (10^{-2} \, m)^3 = 10^{-6} \, m^3$ and $1 \, mm = 10^{-6}$ km.

Note the way in which powers of ten are used. The above examples show us the correct way for representing multiples and sub-multiples of units.

E.2.4 Some acceptable non-SI units

Some of the more commonly used, legally accepted, *non-SI units* are given in detail in Table E.5.

E.3 English Engineering system (Imperial) base units

The English systems (Imperial) base units are given in Table E.6.

Table E.4   SI prefixes

Prefix	Symbol	Multiply by
Yotta	Y	10^{24}
Zetta	Z	10^{21}
Exa	E	10^{18}
Peta	P	10^{15}
Tera	T	10^{12}
Giga	G	10^{9}
Mega	M	10^{6}
Kilo	k	10^{3}
Hecto	h	10^{2}
Deca	da	10^{1}
Deci	d	10^{-1}
Centi	c	10^{-2}
Milli	m	10^{-3}
Micro	μ	10^{-6}
Nano	n	10^{-9}
Pico	p	10^{-12}
Femto	f	10^{-15}
Atto	a	10^{-18}
Zepto	z	10^{-21}
Yocto	y	10^{-24}

E.4 Table of conversions

Throughout Table E.7 *to convert SI units to Imperial and other units* of measurement, *multiply* the unit given by the *conversion factor*, i.e. in the direction of the arrow. To *reverse the process*, i.e. to convert from non-SI units to SI units, divide by the conversion factor.

For example, from Table E.7:

$$14\,kg = (14)(2.20462) = 30.865\,lb$$

and 70 bar

$$= \frac{70}{0.01} = 7000\,kPa \text{ or } 7.0\,MPa$$

E.5 Examples using the SI system

1. **How many cubic centimetres (cc) are there in a cubic metre?**
   When converting cubic measure, mistakes are often made. You need to remember

Table E.5   Non-SI units

Name	Symbol	Physical quantity	Equivalent in SI base units
Ampere-hour	Ah	Electric charge	$1\,Ah = 3600\,C$
Day	d	Time, period	$1\,d = 86,400\,s$
Degree	°	Plane angle	$1° = (\pi/180)\,rad$
Electronvolt	eV	Electric potential	$1\,eV = (e/C)\,J$
Kilometre per hour	kph, km/h	Velocity	$1\,kph = (1/3.6)\,m/s$
Hour	H	Time, period	$1\,h = 3600\,s$
Litre	L, l	Capacity, volume	$1\,L = 10^{-3}\,m^3$
Minute	min	Time, period	$1\,min = 60\,s$
Metric tonne	T	Mass	$1\,t = 10^3\,kg$

Table E.6   English systems (Imperial) base units

Basic quantity	English engineering name	English engineering symbol	Other recognized units
Mass	Pound	lb	ton, hundredweight (cwt)
Length	Foot	ft	inch, yard, mile
Time	Second	s	min, hour, day
Electric current	Ampere	A	mA
Temperature	Rankin	R	°F (Fahrenheit)
Luminous intensity	Foot candle	lm/ft^2	lux, cd/ft^2

Table E.7 Conversion factors

Quantity	SI unit	Conversion factor →	Imperial/other unit
Acceleration	metre/second^2 (m/s^2)	3.28084	feet/second^2 (ft/s^2)
Angular measure	radian (rad)	57.296	degrees (°)
	radian/second (rad/s)	9.5493	revolutions per minute (rpm)
Area	metre^2 (m^2)	10.7639	feet^2 (ft^2)
	metre^2 (m^2)	6.4516×10^4	inch^2 (in.^{2})
Density	kilogram/metre^3 (kg/m^3)	0.062428	pound/foot^3 (lb/ft^3)
	kilogram/metre^3 (kg/m^3)	3.6127×10^{-5}	pound/inch^3 (lb/in.^{3})
	kilogram/metre^3 (kg/m^3)	0.010022	pound/gallon (UK)
Energy, work, heat	joule (J)	0.7376	foot pound-force (ft lbf)
	joule (J)	9.4783×10^{-4}	British thermal unit (btu)
	joule (J)	0.2388	calorie (cal)
Flow rate	m^3/s (Q)	35.315	ft^3/s
	m^3/s (Q)	13,200	gal/min (UK)
Force	newton (N)	0.2248	pound-force (lbf)
	newton (N)	7.233	poundal
	kilo-newton	0.1004	ton-force (UK)
Heat transfer	watt (W)	3.412	btu/h
	watt (W)	0.8598	kcal/h
	watt/metre^2 kelvin (W/m^{2}K)	0.1761	btu/h ft^2 °F
Illumination	lux (lx)	0.0929	foot candle
	lux (lx)	0.0929	lumen/foot^2 (lm/ft^2)
	candela/metre^2 (cd/m^2)	0.0929	candela/ft^2 (cd/ft^2)
Length	metre (m)	1×1010	angstrom
	metre (m)	39.37008	inch (in.)
	metre (m)	3.28084	feet (ft)
	metre (m)	1.09361	yard (yd)
	kilometre (km)	0.621371	mile
	kilometre (km)	0.54	nautical miles
Mass	kilogram (kg)	2.20462	pound (lb)
	kilogram (kg)	35.27392	ounce (oz)
	kilogram (kg)	0.0685218	slug
	tonne (t)	0.984207	ton (UK)
	tonne (t)	1.10231	ton (US)
Moment, torque	newton-metre (Nm)	0.73756	foot pound-force (ft lbf)
	newton-metre (Nm)	8.8507	inch pound-force (in.lbf)
Moment of inertia (mass)	kilogram-metre squared (kg m^2)	0.7376	slug-foot squared (slug ft^2)
Second moment of area	millimetres to the fourth (mm^4)	2.4×10^{-6}	inch to the fourth (in.^{4})
Power	watt (W)	3.4121	British thermal unit/hour (btu/h)
	watt (W)	0.73756	foot pound-force/second (ft lbf/s)
	kilowatt (kW)	1.341	horsepower
	horsepower (hp)	550	foot pound-force/second (ft lbf/s)
Pressure, stress	kilopascal (kPa)	0.009869	atmosphere (atm)
	kilopascal (kPa)	0.145	pound-force/inch^2 (psi)
	kilopascal (kPa)	0.01	bar
	kilopascal (kPa)	0.2953	inches of mercury
	pascal	1.0	newton/metre^2 (N/m^2)
	megapascal (MPa)	145.0	pound-force/inch^2 (psi)

(continued)

Table E.7 (*continued*)

Quantity	SI unit	Conversion factor →	Imperial/other unit
Temperature	kelvin (K)	1.0	celsius (°C)
	kelvin (K)	1.8	rankin (R)
	kelvin (K)	1.8	fahrenheit (°F)
	kelvin (K)		°C + 273.16
	kelvin (K)		(°F + 459.67)/1.8
	celsius (°C)		(°F − 32)/1.8
Velocity	metre/second (m/s)	3.28084	feet/second (ft/s)
	metre/second (m/s)	196.85	feet/minute (ft/min)
	metre/second (m/s)	2.23694	miles/hour (mph)
	kilometre/hour (kph)	0.621371	miles/hour (mph)
	kilometre/hour (kph)	0.5400	knot (international)
Viscosity (kinematic)	square metre/second (m^2/s)	1×10^6	centi-stoke
	square metre/second (m^2/s)	1×10^4	stoke
	square metre/second (m^2/s)	10.764	square feet/second (ft^2/s)
Viscosity (dynamic)	pascal second (Pa s)	1000	centipoises (cP)
	centipoise (cP)	2.419	pound/feet hour (lb/ft h)
Volume	cubic metre (m^3)	35.315	cubic feet (ft^3)
	cubic metre (m^3)	1.308	cubic yard (yd^3)
	cubic metre (m^3)	1000	litre (l)
	litre (l)	1.76	pint (pt) UK
	litre (l)	0.22	gallon (gal) UK

that there are "three linear" dimensions in any one cubic dimension. So that we know there are 100 cm in 1 m or 10^2 cm in 1 m. Therefore, there are $10^2 \times 10^2$ cm^2 in 1 m^2 and finally there are $100 \times 100 \times 100$ cm^3 in 1 m^3 or $10^2 \times 10^2 \times 10^2 = 10^6$ cm^3 in 1 m^3 or 1,000,000 cm^3 = 1 m^3.

2. **Convert 20°C into Kelvin.**
   Then from Table E.7, noting that there is no multiplying factor we simply add 273.16, i.e. 20°C + 273.16 = 293.16 K. Note that when expressing temperature in absolute units (kelvin) we drop the degree sign. Also there is no plural, it is kelvin not kelvins. In practice the 0.16 is also dropped, unless this particular degree of accuracy is required. For thermodynamics section (Physics module), you should always convert temperature to kelvin. Thermodynamic temperature is represented by upper case "T", while temperature in celsius may be represented using lower case "t".

You should memorize the factor (**273**) for converting degree celsius to kelvin.

3. **Add 300 megawatts (MW) to 300 gigawatts (GW).**
   The only problem here is that we are dealing with different sized units. In index form (powers of ten) we have 300×10^6 W plus 300×10^9 W. So all we need to do is to express these quantities in the same units, e.g. where $200 \times 10^6 = 0.2 \times 10^9$ W, so that:

$$300 \times 10^6 + 300 \times 10^9 \text{ W}$$
$$= 0.3 \times 10^9 + 300 \times 10^9 \text{ W}$$
$$= 300.3 \text{ GW}.$$

4. **Convert 60,000 kg into the SI unit of weight.**
   The SI unit of weight is the newton (N) and unless told differently to convert a mass into a weight we multiply the mass by the accepted value of the acceleration due to gravity, which is **9.81 m/s^2** or ms^{-2}. This

is another conversion factor that you should commit to memory.

Then $\quad 60,000\,kg = 60,000 \times 9.81$

$$= 588,600\,N$$

(by long multiplication)

5. **How many pascal are there in 270 bar?**
This requires us to divide by the conversion factor (0.01) in Table E.7 to convert from bar to kilopascal and then convert kilopascal to pascal.

Then $\quad 270/0.01 = 27,000\,kPa$ or

$$= 27,000 \times 1000$$

$$= 27 \times 10^6\,Pa \text{ or } 27\,MPa$$

The following conversions will be particularly useful when you study stress and pressure and should be committed to memory:

- $1\,N/m^2 = 1\,Pa$
- $100,000\,N/m^2 = 100,000\,Pa = 1 \times 10^5\,Pa = 1\,bar$
- $1\,MPa = 1\,MN/m^2 = 1\,N/mm^2$

So in the above example:

$$270\,bar = 270 \times 10^5\,Pa = 27 \times 10^6\,Pa$$

$$= 27\,MPa$$

6. **How many newton-metres (Nm) are there in 600 MJ?**
This example introduces another important relationship, i.e.:

$$1\,Nm = 1\,J$$

The Nm is sometimes used as the unit of work, rather than the joule, which is often reserved for energy (the capacity to do work).
Thus $600\,MJ = 600$ mega-newton-metres, or $600 \times 10^6\,Nm$, as required.

7. **What is 36 kJ/h in watts?**
You will learn during your study of Chapter 4 that the rate of doing work (Nm/s) or the rate of the transfer of energy (J/s) is in fact power, where $1\,W = 1\,J/s$ (joule per second). When we refer to "unit time", we are talking

about per second. Thus the transfer of energy per unit time (the rate of transfer) has units of J/s.
Now, for the above, we are saying that 360 kJ (360,000 J) of energy is transferred per hour. Knowing that there are 3600 s in 1 h (Table E.5), then we have transferred:

$$\frac{360,000\,J}{3600\,s} = 100\,J/s = 100\,W \text{ (from above)}$$

8. **How many litres are there in 40 m³?**
This question is easily answered by consulting Table E.7, where $1\,m^3 = 1000\,L$, so that:

$$40\,m^3 = 40,000\,L \text{ or } 4 \times 10^4\,L$$

This is another useful conversion factor to commit to memory, $1\,m^3 = 1000\,L$.
Another useful conversion factor to memorize is that $1000\,cc = 1\,L$. Thus, e.g. a 1000 cc engine is a 1 L engine.

E.6 Examples of useful conversions

1. **Convert 10 kN into pound-force (lbf).**
This is simply done by multiplying the force in newton by 0.2248 (Table E.7). Then, $10,000 \times 0.2248 = 2248\,lbf$. It may be easier to remember the rule of thumb approximation that $1\,N = 0.225\,lb$, then approximately $4.45\,N = 1\,lb$.
2. **Convert 20,000 kg to pound-mass.**
Then we need to remember that, as a rule of thumb, roughly, $2.2\,lb = 1\,kg$. So that $20,000\,kg = 44,000\,lb$. In Table E.7 the more exact conversion factor is given as 2.20462.
3. **You are refuelling an aircraft, where the refuelling vehicle is calibrating the fuel in Imperial gallons. You require 60,000 L of fuel to fill the aircraft. How many Imperial gallons must go in?**
This is, again, where you should remember the conversion factor and make sure you get your conversion right!

$$1\,L = 0.22 \text{ (UK) gallon}$$

thus $\quad 60,000\,L = 60,000 \times 0.22$

$$= 13,200 \text{ gallons}$$

The inverse factor for converting gallons to litres is **1 (UK) gallon = 4.545 L.**

4. **An aircraft has a wingspan of 160 ft. The door of your hangar opens up to a maximum of 49 m; can you get the aircraft in the hangar?**

From Table E.7, you will note that to convert feet to metres we *divide* by 3.28084, then:

$$\frac{160 \, \text{ft}}{3.28084} = 48.77 \, \text{m}$$

This means that you have 13 cm clearance; I most certainly would not like to be the tug driver! As a rule of thumb, you can use the approximation that **1 m = 3.28 ft.** This approximation would give us 48.8 m, for the above calculation, still too close for comfort!

5. **Standard atmospheric pressure in the Imperial system is 14.7 lb/in.^{2}. What is its value in pascal?**

From Table E.7, we divide by the factor 0.145 to obtain a value in kilopascal.

So:

$$\frac{14.7 \, \text{lb/in}^2}{0.145} = 101.379 \, \text{kPa} = 101379 \, \text{Pa}$$

In fact the value in the table is already an approximation, a more accurate value is 0.145037738, which gives an answer of 101,350 Pa, which is a little nearer the value of atmospheric pressure used in the SI version of the International Standard Atmosphere, i.e. 101,325 Pa.

6. **An aircraft engine produces 200 kN of thrust. What is the equivalent in pounds of thrust?**

Thrust is a force; therefore, we require to convert newton (N) to pound-force (lbf). Using our approximate conversion factor 0.225, then:

$$200 \, \text{kN} = 200,000 \, \text{N} = 200,000 \times 0.225$$
$$= 45,000 \, \text{lbf of thrust}$$

Final Note: You should commit to memory all the base units and derived units in the SI system, as well as become familiar with the units of distance, mass, length and time in the English Engineering (Imperial) systems. Try also to memorize the important conversion factors given in the above examples.

Answers to "Test your understanding"

TYU 2.1

1. Natural numbers and positive integers
2. Rational numbers
3. $\frac{30}{6}, \frac{78}{6}, \frac{96}{6}$
4. $-\frac{16}{4}, -\frac{28}{4}, -\frac{48}{4}$
5. positive integer, 4
6. rational, real
7. (a) 0.333333 (b) 0.142857 (c) 1.999999
8. (a) 9 (b) 66 (c) 39
9. $-31 = 31$
10. 11
11. 14
12. (a) -5 (b) -18 (c) 7
13. (a) 96 (b) 90
14. 80
15. (a) 191.88 (b) 4304.64 (c) 1.05 (d) 2.1672
    (e) 28000 (f) 0.1386

TYU 2.2

1. (a) 0.43 (b) 5080
2. (a) 3.1862×10^2 (b) 4.702×10^{-5}
   (c) 5.1292×10^{10} (d) -4.1045×10^{-4}
3. (a) 2.71 (b) 0.000127 (c) 5.44×10^4
4. (a) -5×10^4 (b) 8.2×10^{-5}

TYU 2.3

1. (a) $\frac{1}{10}$ (b) $\frac{25}{3}$ (c) $\frac{9}{10}$
2. (a) $\frac{11}{9}$ (b) $3\frac{3}{10}$ (c) 2
3. $\frac{3}{32}$
4. 1.0
5. $\frac{16}{15}$
6. $\frac{38}{45}$

TYU 2.4

1. 7.5
2. 1.215 million
3. 720 km/h
4. 60 km/h

5. 33.17 litres
6. 20 men
7. £23.44
8. $y = 35$
9. $h = \frac{kV}{r^2}$

TYU 2.5

1. (a) 10001 (b) 10111 (c) 101000
2. (a) 11 (b) 31 (c) 85
3. (a) 1702_{16} (b) $41FC_{16}$
4. (a) 110 (b) 61208

TYU 2.6

1. (a) (2, 8), (4, 4), (2, 2, 2, 2) (b) (n, n)
   (c) $(wx, yz), (wxy, z), (xyz, w), (wyz, x)$
2. $ab^2 c$
3. (a) 32 (b) $\frac{8}{27}$ (c) b^2
4. (a) 70 (b) $\frac{10}{9}$

TYU 2.7

1. (a) $a^5 b^{-1} c^3 d$ (b) $4(6x^3 y^2 - xy^2)$
2. (a) $\frac{3}{4} a^{-6} b^2$ (b) d
3. (a) $6a^2 + 4a - 2$ (b) $4 - x^4$ (c) $3a^3 b + a^2 b^2 - 2ab^3$ (d) $s^3 - t^3$
4. (a) $(x + 3)(x - 1)$ (b) $(a + 3)(a - 6)$
   (c) $(2p + 3)(2p + 4)$ (d) $(3z + 4)(3z - 6)$
5. (a) $3x(x+7)(x+2)$, (b) $3xy(3xy+2)(3xy-1)$
6. (a) 0 (b) 0.279

TYU 2.8

1. (a) $xy + xyz + 2xz - 2x + 8y$ (b) $5ab + abc$
2. $-13p^2 s + 2pqr - 8s$
3. (a) $(u - 2)(u - 3)$ (b) $6abc(ab^2 + 2c - 5)$
   (c) $(3x - 5)(4x + 2)$ (d) $(2a + 2b)(a^2 - ab + b^2)$
4. $(a - b)(a^2 - ab + b^2) = a^3 - b^3$ so $(a - b)$ is a factor
5. $(x^2 - 1)(2x + 1) = 2x^3 + x^2 - 2x - 1$ so $(2x + 1)$ is the quotient

TYU 2.9

1. $\sqrt{70,000}$ and from square root tables
   $= 264.6$

2. $r = \sqrt{\dfrac{v}{\pi h}}$

3. $x = \left(\dfrac{y}{8} + 2\right)^2$

4. $2.25\,\Omega$

5. $x = \left(\dfrac{5}{y - 20}\right)^{\frac{3}{4}}$

6. $t = \sqrt{\dfrac{s + 4}{18a - 6}}$

7. $a = \dfrac{s}{n} - \dfrac{1}{2}(n - 1)d$

8. $x = \dfrac{bc + ac + b^2}{b + c}$

9. $C = 4.834 \times 10^{-6}$

10. (a) 137.4 (b) 0.0152 (c) 21925 (d) 229.8
    (e) 9.956 (f) 85.66

TYU 2.10

1. $157.14\,\text{cm}^2$
2. $47.143\,\text{cm}^2$
3. $10\,\text{mm}$
4. $12.73\,\text{cm}$
5. $4.33 \times 10^{-3}\,\text{m}^3$

TYU 2.11

1. $V = 45$ when $I = 3$
2. (a) 1 (b) 4 (c) $-\frac{1}{3}$
3. (a) 2.5, 1 (b) 2, 3 (c) $\frac{3}{5}, -\frac{1}{5}$ (d) 2, 3
4. 13
5. 7, 3
6. (a) $x = 3$ or $x = -1$ (b) $x = -8$ or
   $x = -2$ (c) 2.62, 0.38 (d) ± 1.58
7. $x = 2.12$ or $x = -0.79$
8. $u = -1$ or $u = +2$

TYU 2.12

1. (a) 0.839 (b) 0.1392 (c) 0.2309 (d) 0.2187
   (e) 0.8892 (f) 1.3238
2. (a) 0.5 (b) 0.94
3. (a) 48.6° (b) 22.62°
4. Angles are: 45.2°, 67.4°, 67.4° and
   height $= 6\,\text{cm}$

5. $9.22\angle 49.4$
6. (a) (4.33, 2.5) (b) $(-6.93, 4)$
8. $9.6\,\text{m}$
9. $53.3\,\text{m}$
10. $h = 3.46\,\text{cm}, x = 1.04\,\text{cm}$

TYU 3.1

1. $Q = 0.017$
2. $C = 4.83 \times 10^{-6}$
3. $\dfrac{1}{2(x^2 - 1)}$
4. $p_2 = 374.28$
5. 14
6. $\mu = 0.4$
7. $I = 0.0017R^2$

TYU 3.2

1. (a) $(-1 - 7i)$ (b) $36 + 26i$ (c) $(-\frac{1}{5} + \frac{2}{5}i)$
2. $\sqrt{72}\angle 45$ (b) $\sqrt{25}\angle 36.9$ (c) $\sqrt{1681}\angle -12.7$
3. (a) $2.74 + 4.74i$ (b) $\dfrac{\sqrt{13}}{\sqrt{2}} + \dfrac{\sqrt{13}}{\sqrt{2}}i$
4. (a) 651.92 (b) $21250 - 7250j$ (c) $\dfrac{710}{37} - \dfrac{550}{37}j$
   (d) $\dfrac{87}{185} - \dfrac{181}{185}j$

TYU 3.3

1. $10.82\,\text{cm}$
2. $6.93\,\text{cm}$
3. $48.1\,\text{cm}$
4. (a) $27.05\,\text{m}$ (b) $58\,\text{m}$
6. $\angle A = 45.3, \angle B = 37, \angle C = 97.7, a = 37.2,$
   $b = 31.6, c = 52$
7. $\angle A = 94.78, \angle B = 56.14, \angle C = 29.08,$
   Area $= 29.9\,\text{cm}^2$
9. (a) $2.56\,\text{cm}$ (b) $10.88\,\text{cm}^2$
10. $10,088\,\text{m}^2$

TYU 3.4

1. $(0 < \mu < 0.67)$
3. (a) $\sin 6\theta$ (b) $\cos 11t$

TYU 3.5

1. Business and administration $= 29.23\%$,
   Humanities and social science $= 42.28\%$,

Physical and life sciences $= 15.74\%$, Technology $= 12.75\%$

2.

x	35	36	37	38	39	40	41	42	43	44	45
f	1	2	4	5	7	5	4	7	2	2	1

3. Percentage height of column relates to average for class interval

Class interval	62	67	72	77	82	87
Percentage height	6.67	18.33	30	26.67	11.67	6.66

TYU 3.6

1. mean $(\bar{x}) = 127$
2. mean $= 20$, median $= 8.5$, mode $= 9$
3. $\bar{x} = 38.6$ cm, mean deviation $= 1.44$ cm
4. $\bar{x} = 169.075$ mm, mean deviation $= 0.152$ mm
5. $\bar{x} = 8.5, \sigma = 34.73$
6. $\bar{x} = 3.42, \sigma = 0.116$

TYU 3.7

1. $\dfrac{dy}{dx} = nax^{n-1}$
2. $f(3) = 51, f(-2) = 76$
3. (i) $\dfrac{dy}{dx} = 12x - 3$ (ii) $\dfrac{ds}{dt} = 6t + 6t^{-2} - \dfrac{t^{-4}}{4}$
   (iii) $\dfrac{dp}{dr} = 4r^3 - 3r^2 + 12$ (iv) $\dfrac{dy}{dx} = \dfrac{27}{2}x^{7/2} - \dfrac{15}{2}\sqrt{x} + \dfrac{1}{2\sqrt{x}}$
4. Gradient $= -1.307$ to 3 decimal places
5. $x = 4, y = 9$
6. At $x = -2$, rate of change $= -56$
7. 40.7 to 3 significant figures
8. (a) $\dfrac{1}{x}$ (b) $\dfrac{3}{x}$ (c) $\dfrac{1}{x}$ then it can be seen that $\dfrac{dy}{dx}$ of In $ax = \dfrac{dy}{dx}$ of In x
9. -0.423
10. 866.67 Cs^{-1}, 684.2 Cs^{-1}

TYU 3.8

1. (a) $\dfrac{4x^3}{3} - x - 2 + c$ (b) $\dfrac{x^{\frac{3}{2}}}{3} - \dfrac{2x^{\frac{3}{2}}}{3} + \dfrac{2x^{\frac{5}{2}}}{5} + c$
   (c) $+\dfrac{3}{2}\cos 2x + c$ (d) $+\dfrac{2}{3}\sin x + c$
   (e) $-0.75e^{3\theta} + c$ (f) $\dfrac{-3}{x} + c$

2. (a) $10\frac{5}{12}$ (b) $-\frac{1}{15}$ (c) -3 (d) -287.5
3. $v = \dfrac{3t^2}{2} + 4t + 8$, $s = \dfrac{t^3}{2} + 2t^2 + 8t$ and $s = 1762.5$
4. $12\frac{2}{3}$ sq. units
5. $\frac{4}{3}$ sq. units

TYU 4.1

1.

Base quantity	SI unit name	SI unit symbol
Mass	kilogram	kg
Length	metre	m
Time	second	s
Electric current	ampere	A
Temperature	kelvin	K
Amount of substance	mole	mol
Luminous intensity	candela	cd

2. Radian
3. Centimetre-gram-second
4. (a) 1219.5 kg (b) 1.784 m^3 (c) 1.4×10^{-3} m^2/s (d) 1.00575 hp
5. 217.5 psi
6. 20 m^2

TYU 4.2

1. It is decreased in proportion to $\frac{1}{d^2}$
2. The Newton (N) which is equal to 1 kg m/s^2
3. 9.81 m/s^2
4. 3636.36 litres
5. 4281.3 kg
6. (a) The *poundal* is $\frac{1}{32.17\text{th}}$ of a pound-force (lbf) (b) the *pound-force* (lbf) is that force required to accelerate 1 lb mass at 1 ft/s^2

TYU 4.3

1. kg/m^3
2. (a) 2700 kg/m (b) 168.5 lb/ft^3
3. Density is likely to decrease (i.e. become less dense)
4. It is a ratio
5. Approximately 1000 kg/m^3

TYU 4.4

1. Magnitude, direction and point of application
2. (a) Scalar quantities have magnitude only, e.g. speed (b) vector quantities have both magnitude and direction, e.g. velocity
3. Force = mass × acceleration $(F = ma)$. For weight force the acceleration is that due to gravity, i.e. $W = mg$
4. Strut is member in *compression* and a tie is a member in *tension*
5. Pressure $= \frac{\text{Force}}{\text{Area}}$. Units: Pascal (Pa) or Nm^{-2}, etc.
6. (a) 193103 Pa (b) 101592 Pa

TYU 4.5

1. (a) 372.8 mph (b) 313.17 mph (c) 82 ft/s² (d) 24.4 m/s (e) 241.4 m/s (f) 123.5 m/s
2. 4.4 m/s²
3. Inertia is the force resisting change in momentum, i.e. resisting acceleration. Therefore it has units in the SI system of the Newton (N).
4. *Force* = rate of change of momentum of a body
5. The "degree of hotness" of a body

TYU 4.6

1. An *ion* is an atom with either more or less electrons than protons. When there is an *excess of electrons,* we have a *negative ion,* when there are *less electrons* we have a *positive ion*
2. A noble gas configuration is when all the outer electron shells of the molecule are full. Atoms and molecules in combination try to achieve this state because then they sit in their lowest energy level
3. Rows indicate the number of shells in the atom, the columns indicate the number of valence electrons, the atom has in its outer *p* and/or *s* shells
4. Two outer valence electrons are available for chemical combination (i.e. bonding)
5. Simply, *Stage one* involves lose or gain of an electron/s to form a positive or negative ion. *Stage two* involves the oppositely charged ions, electro-statically bonding, that is forming the ionic bond

6. A covalent bond, because it is difficult to shed all 4 valence electrons, or gain another 4, to form the noble gas configuration. Therefore, electron sharing is more likely

TYU 4.7

1. At the atomic level the inter-atomic spacing of solid and liquid molecules are very similar. However, in the liquid the molecules spend less time under the influence of the inter-atomic bonding forces of their neighbors. This is due to the higher speeds the molecules attain in the liquid, which is generally associated with higher molecular energy due to the increase in temperature
2. Within the range of zero to one or two atomic diameters
3. It is defined as the energy available, due to molecular vibration

TYU 4.8

1. Coplanar forces are deemed to act in the same two-dimensional space, such as the face of this paper
2. (a) See key point page 188 (b) See key point page 187
3. The algebraic sum of the forces, acting at a point on a body equal zero.
4. See Figure 4.17 on page 189
5. 12.048 tonne

TYU 4.9

1. Moment $M =$ Force × perpendicular distance from axis of reference
2. Turning effect = Force × distance, when distance is zero then $F \times 0 = 0$ and so no effect.
3. Use simple trigonometric ratio's to determine perpendicular components
4. (a) Point or axis about which rotation takes place (b) perpendicular distance to the line of action of the force to the fulcrum (c) difference between total clockwise moment and total anti-clockwise moment
5. Upward forces = downward force and sum of CWM = sum of ACWM
6. (a) A couple occurs when two equal forces acting in opposite directions have their line

of action parallel (b) one of the equal forces multiplied by the perpendicular distance between them

7. 108.5 Nm

TYU 4.10

1. Then $\bar{x}$ is the sum of the moments of the masses, divided by, the total mass.
2. If any single mass is altered, this will alter the total mass and total mass moment of the aircraft
3. In all cases stress = force/area. (a) Tensile stress set up by forces tending to pull material apart (b) Shear, results from forces tending to cut through the material (c) Compressive, set up by forces tending to crush the material.
4. See page 199, for definition. The elastic modulus is given by the slope of the Hooke's law plot; that is, stress/strain
5. Spring stiffness k = force/deflection so units of N/m
6. See definitions on pages 199 and 200
7. (a) 240,000 N/m² (b) 2.28 × 10⁸ N/m² (c) 6 × 10⁸ N/m² (d) 3300 N/m² (e) 1.0 × 10¹⁰ N/m²
8. (a) strut takes compressive loads (b) a tie take tensile loads

TYU 4.11

1. By measuring the yield stress or proof stress. See pages 203 and 204 for a full definition
2. To provide a margin of safety and to allow for "a factor of ignorance" in design, manufacture and integrity of materials
3. See definitions on page 204
4. (a) upper limit of validity of Hooke's law (b) the ultimate tensile strength, that is the maximum load divided by the original cross-sectional area (c) start of plastic phase (d) material is permanently deformed
5. (a) The axis about which the shaft rotates (b) a measure of the way the area or mass is distributed in rotating solids, that is the shaft resistance to bending (see formulae on page 203) (c) torque is simply the applied twisting moment, created by the load, that sets up shear stresses in the shaft
6. See full explanation given in Section 4.7.11 on page 203

TYU 4.12

1. *acceleration*
2. *distance*
3. *area under graph* by *time interval*
4. *zero* and the distance travelled is equal to *vt*
5. $\frac{1}{2}vt$
6. *uniformly retarded motion.* (1) $s = ut + \frac{1}{2}at^2$; (2) *variable*
7. Graph (d) represents uniformly accelerated motion having initial velocity *u*, final velocity *v* and acceleration *a*. So distanced traveled is equal to *a*.
8. Graph (e) represents *a* acceleration.
9. (a) Inertia force is equal and opposite to the accelerating force that produced it (b) momentum of a body is equal to its mass multiplied by its velocity
10. speed has *magnitude* only, velocity has *magnitude and direction*
11. Weight force is dependent on acceleration due to gravity, which varies with distance from the Earth. Mass is the amount of matter in a body which remains unchanged
12. Momentum is the mass of a body multiplied by its velocity. The rate of change in momentum is given by $mv - mu/t$ or $m(v - u)/t$. The latter of these two expressions is simply *mass × acceleration* and we know that Force = mass × the acceleration producing it. So that $F = ma$
13. (a) V_{je} = velocity of slipstream (b) V_{je} = velocity of exhaust gas stream
14. Thrust is a maximum, when engine is stationary, then $V_a = 0$

TYU 4.13

1. (a) This is the angular distance moved in radian, divided by the time taken with units of rad/s or rad s⁻¹ (b) Angular acceleration is the change in angular velocity divided by the time taken, with units of rad/s² or rad s⁻²
2. 142.9 rad/s
3. (a) 26.18 rad/s (b) 21.8 rad/s (c) 1100 rad/s
4. (a) Torque = force × radius = Fr, units (Nm) (b) The point mass multiplied by the radius squared, or $1 = mk^2$, with units kgm²
5. The moment of inertia I, is used instead of the mass because it provides a more accurate

picture of the distribution of the mass from the center of rotation, since the centripetal acceleration of the mass is proportional to the radius squared. Thus mass positioned furthest from the radius has the greatest effect on the inertia of the rotating body

6. (a) In rotary motion, an acceleration acting towards the center of rotation, given by $a = \omega^2 r$ (b) centripetal acceleration acting on a mass produces force $Fc = m\omega^2 r$.

7. The *weight* of the aircraft will act vertically down, the *lift force* from the wings will act normal to the angle of bank and the *centripetal force* will act towards the center of the turn, *holding* the aircraft into the turn. This will be opposed by an equal and opposite force the *centrifugal force*, trying to throw the aircraft out of the turn

8. (a) momentum of a body is the product of its mass and velocity units are kg/s (b) the force that resists acceleration of a body is its inertia, units are Newton (N)

9. Rigidity is the resistance of a body to change its motion. The greater the momentum of the body the greater is this resistance to change. Therefore it depends on the mass of the rotor, the distribution of this mass and the angular velocity of the rotor

10. Precession is simply the reaction to a force applied to the axis of rotation of the gyro assembly. The nature of this phenomenon is described in italics under Sperry's rule on page 215

TYU 4.14

1. *Free vibration* occurs in an elastic system after an initial disturbance, where it is allowed to oscillate unhindered. *Forced vibration* refers to a vibration that is excited by an external force applied at regular intervals

2. See definitions on page 217

3. Resonance occurs where the natural frequency of the system coincides with the frequency of the driving oscillation. Example of undesirable resonance include all large structures such as bridges, pylons, etc. A radio tuner is an example of desirable resonance being used

4. This is the periodic motion of a body where the acceleration is always towards a fixed point in its path and is proportional to its displacement from that point e.g. the motion of a pendulum bob

5. For SHM (a) velocity is a maximum at the equilibrium position of the motion (b) acceleration is a maximum at the extremities of the motion

6. See explanation on pages 219 and 220

7. Spring stiffness is the force per unit change in length measured in N/m

8. In this formula, s is the arc length, r is the radius of the body from the center of oscillation and θ is the angle of swing in radian

TYU 4.15

1. Mechanical work may be defined as, the force required to overcome the resistance (N) multiplied by the distance moved against the resistance (m)

2. The equation for work done is $W = Fd$, giving units of Newton-metres (Nm) or Joules

3. As given on page 225. The principle of the conservation of energy states that: *energy may neither be created nor destroyed; only changed from one form into another*

4. (a) mechanical energy into electrical energy (b) chemical and heat energy into kinetic energy (c) chemical energy into electrical energy (d) electrical energy into sound energy

5. The spring constant, units are N/m

6. Linear or translational $KE = \frac{1}{2}mv^2$, where m is the mass of the body (kg) and v^2 is its velocity squared (m²/s²) with KE in Joules (J). Rotational $KE = \frac{1}{2}I\omega^2$, where $I = mk^2$, the total mass of the rotating object multiplied by the square of the radius of gyration (m²) and ω is the angular velocity in radian/second (rad/s), again with KE in Joules (J)

7. Power is the rate of doing work (Nm/s) and since energy is the capacity to do work then power is the rate of consumption of energy (J/s). Therefore machine A produces 1500 W, while machine B produces 1548 W. So machine B is more powerful

TYU 4.16

1. Nature of surfaces in contact
2. False – not necessarily true for very low speeds and in some cases very high speeds
3. (a) The angle of friction is that angle between the frictional force and the resultant of the frictional force and normal force (b) the coefficient of friction is the ratio of the frictional force divided by the normal force. They are related by $\mu = \tan\theta$
4. See Figure 4.62 on page 228
5. Angle between resultant and vertical component of weight, up slope $= \phi + \theta$ and down slope $\phi = \theta$
6. See Figure 4.66 on page 231
7. See Figures 4.66, 4.67 and 4.68 plus the explanation given on pages 231 and 232

TYU 4.17

1. A machine may be defined as: the combination of components that transmit or modify the action of a force or torque to do useful work
2. (a) $\text{VR} = \dfrac{\text{Distance moved by effort}}{\text{Distance moved by load}}$

   (b) $\text{MA} = \dfrac{\text{Load}}{\text{Effort}}$
3. $\text{MA} = 183.75$
4. $E = aW + b$ where $E = $ effort, $W = $ load, $a = $ slope $= 1/\text{MA}$, $b = $ the effort intercept
5. Count the cable sections supporting the load
6. (a) $\text{VR} = 98.17$ (b) $\text{MA} = 66.7$ (c) 67.9%
7. $\text{VR} = 0.075$ step-up

TYU 4.18

1. 1.97
2. (a) 29,000 psi (b) 11.6 psi (c) 1044 psi
3. See laws on page 241
4. 18 m
5. (a) Gauge pressure $= \rho g h$ (b) absolute pressure $=$ gauge pressure $+$ atmospheric pressure
6. See explanation for Buoyancy on page 244
7. See measurement of pressure on page 244
8. (a) Gauge pressure $= 6.148$ psi (b) absolute pressure $= 20.84$ psi

9. See explanation of fluid viscosity on pages 246 and 247
10. $v = \dfrac{\mu}{\rho} = \dfrac{Nsm^{-2}}{kgm^{-3}} = \dfrac{kgm^1 s^1 s^{-2} m^{-2}}{kgm^{-3}} = \dfrac{m^{-1}s}{m^{-3}} = m^2 s^{-1}$

TYU 4.19

1. A gas that is seen to obey a gas law
2. (a) 553 K (b) 103 K
3. Temperature
4. 11 km or 36,000 ft
5. Provide a standard for comparison of aircraft performance and the calibration of aircraft instruments
6. In the troposphere; temperature, density and humidity all fall. In the stratosphere; pressure, density and humidity fall, while the temperature remains constant at 216.7 K
7. (a) 761 mph (b) 661 knots (c) 1117 ft/s
8. 225.7 K

TYU 4.20

1. (a) Flow in which the fluid particles move in an orderly manner and retain the same relative positions in successive cross-sections (b) flow in which the density does not vary from point-to-point
2. $\dot{Q} = A_1 v_1 = A_1 v_2$ and $\dot{m}\rho_1 A_1 v_1 = \rho_2 A_2 v_2$. Incompressible flow is assumed
3. See detail on page 253
4. Shape at throat causes an increase in velocity so that dynamic pressure increases. Then, from Bernoulli, the static pressure at the throat must decrease
5. See explanation on page 255
6. When air flow velocities exceed 130–150 m/s, where compressibility errors exceed 4% to 5%. Small errors do occur at lower speeds

TYU 4.21

1. (a) 253 K (b) 48.9 °C (c) 227.6 K
2. Although a resistance thermometer could be used, a thermocouple is more suitable. It is robust and easily capable of measuring temperatures up to 1200°C, in fact it has a maximum measuring capacity of up to 1600°C.

There physical composition also makes thermocouples eminently suitable for this kind of harsh environment, where hot exhaust gas temperatures are being measured

3. α = the amount of a material will expand per °C or Kelvin: Surface expansion and volumetric expansion can be approximated using 2α and 3α, respectively, as expansion coefficients. See also page 258

4. Heat energy Q is the transient energy brought about by the interaction of bodies by virtue of their temperature difference, when in contact. Internal energy of a material is the energy of vibration of the molecules, which is directly proportional to the temperature of the material

5. See explanation on pages 259–261

6. Since, for a constant pressure process, volume change must take place, then pressure-volume work is done. So heat energy is required for both this work and for any increase in the internal energy (U), while for constant volume only internal energy is increased

7. $Q = mc\Delta T$ but when calculating latent heat no temperature change takes place, so we use $Q = mL$, where L = latent heat of evaporation or condensation, as required

8. $c_p = 940\,J/kgK$

9. Either increase temperature, reduce pressure, or increase the surface area

10. (a) The evaporator allows refrigerant to absorb heat from the medium being cooled (b) the condenser allows heat to be dissipated to an external medium, so reducing the temperature of the refrigerant

TYU 4.22

1. (a) A system where particular amounts of a thermodynamic substance, normally compressible fluids, such as vapours and gases, are surrounded by an identifiable boundary (b) Heat Q, is energy in transit brought about by the interaction of bodies by virtue of their temperature difference when they communicate

2. When it has a moveable boundary

3. (a) **NFEE:** $Q - W = \Delta U$, where Q is the heat energy entering or leaving the system, W is work done by the system or on the system, ΔU is the change in internal energy of the working fluid. All terms have units of the Joule (J) (b) **SFEE:** $Q - W = (U_2 - U_1) + (p_2 V_2 - p_1 V_1) + (mgz_2 - mgz_1) + (\frac{1}{2}mv_2^2 - \frac{1}{2}mv_1^2)$ where Q, W, and $U_2 - U_1$ or ΔU, have the same meaning as above. $(p_2 V_2 - p_1 V_1)$ = the change in pressure/volume energy, within the working fluid, $(mgz_2 - mgz_1)$ = the change in PE of the working fluid and $(\frac{1}{2}mv_2^2 - \frac{1}{2}mv_1^2)$ = the change in KE of the working fluid. See also pages 269 and 270

4. In a closed system no working fluid crosses the system boundary, as it does in an open system

5. Internal energy involves only the vibration of the fluid molecules. Whereas, enthalpy is the combined energies both internal and pressure/volume of the fluid. Enthalpy is a very useful characteristic of the working fluid used in open systems

6. According to the second law of thermodynamic and substantiated in practice, no thermodynamic system can produce more work energy than the heat energy supplied. In other words Q_{in} is always greater than W_{out}. With practical working systems, energy is dissipated as sound, heat etc and cannot be reversed back to its original form

7. (a) Isothermal – temperature remains constant (b) polytropic – both heat and work may be transferred (c) reversible adiabatic – no heat energy is transferred to or from the working fluid

8. Heat source, engine, sink

9. Net work done is always less than the heat supplied

10. See pages 275 and 276

TYU 4.23

1. Approximately 671 million miles

2. The angle of incidence is equal to the angle of reflection. The incident ray, the reflected ray and the normal all lie within the same plane

3. 60 cm

4. Rays close to the principal axis and therefore, where the mirror aperture may be represented by a straight line

5. See Figure 4.119 on page 279

6. The angle of the refracted ray increases as the light ray enters the material having the lower refractive index

7. The greater the refractive index of a medium then the lower the speed of light as it passes through it

8. Total internal reflection

9. To reduce energy losses due to dirt at the boundary and due to the Fresnel effect and impurities in the glass

10. The principle focus for convex lenses is the point at which all paraxial rays converge. In the case of concave lenses these same rays after refraction, appear to diverge

11. The plane that is at right angles to the principal axis is the focal plane

12. By using the relationship:

$$\text{Image height} = \frac{\dfrac{\text{image}}{\text{distance}} \times \dfrac{\text{object}}{\text{height}}}{\dfrac{\text{object}}{\text{distance}}}$$

TYU 4.24

1. Waves in the electromagnetic spectrum have vastly different frequencies. Their energy is in direct proportion to their frequency and inversely proportional to their wavelength. Thus for the above reason, ultraviolet waves with the higher frequency, will have more energy than infrared waves

2. Simply that the oscillatory motion of the wave is at right angles to the direction of travel of the wave front

3. Diffraction of transverse waves takes place, where they spread out to produce circular wave fronts

4. When two wave sets are in-phase, they reinforce one another as they meet, creating constructive interference. If they are out of phase peaks and troughs meet and cancel one another, causing destructive interference

5. See characteristics detailed on pages 289 and 290

6. VHF waves have very short wavelengths and are not reflected by the ionosphere and therefore, it is not practical to transmit them as sky waves

7. Around $\frac{1}{50}$ to $\frac{1}{100}$ of a metre

8. To reduce the possibility of static interference, which is not a problem with VHF or UHF communication

9. (a) Skip distance, is the first point from the transmitter at which the first sky wave can be reached (b) Dead space, is the area that cannot receive either ground waves or the first sky wave and is known as the silent zone

10. Carrier waves and satellite communication – see explanation on pages 289–291

11. This phenomena is due to a change in frequency brought about by the relative motion, known as the Doppler effect

TYU 4.25

1. Sound waves are caused by a source of vibration, creating pressure pulses

2. Sound waves are longitudinal mechanical waves that require a medium through which to be transmitted and received. Electromagnetic waves travel at the speed of light through a vacuum

3. The speed of sound depends on the temperature and density of the material through which the sound passes. The denser the material the faster the speed of sound

4. 5.55 ms

5. (a) Intensity is a measure of the energy of the sound passing through unit area every second and is measured in W/m^2 (see page 295) (b) pitch is dependent on the frequency of the sound being generated. The higher the frequency the higher the pitch (c) amplitude, is the maximum displacement of a particle from its rest position. The greater the amplitude, the louder the sound

TYU 5.1

1. Ampere
2. Hertz
3. C
4. G

5. 7.5 ms
6. 0.44 kV
7. 15,620 kHz
8. 0.57 mA
9. 220 nF
10. 0.47 MΩ

TYU 5.2

1. protons, electrons
2. positively, positive ion
3. negative, negative ion
4. free electrons
5. insulators
6. copper, silver (and other metals), carbon (any two)
7. plastics, rubber, and ceramic materials
8. silicon, germanium, selenium, gallium (any two)
9. See page 315
10. See page 315

TYU 5.3

1. positive
2. repel
3. See page 316
4. one quarter of the original force
5. 2 kV/m
6. 8 V
7. 1.264 mC
8. ions
9. negatively, electrons
10. See page 319

TYU 5.4

1. charge, Ampere
2. positive, negative
3. negative, positive
4. Ohm, Ω
5. (a) silver (b) aluminium
6. 900 C
7. 1 V
8. power, time
9. See page 320
10. See page 320

TYU 5.5

1. positive, negative
2. static discharger
3. zinc, carbon
4. sulphuric acid
5. e.m.f., induced
6. photovoltaic
7. negatively, positively
8. smoke detector
9. thermocouple
10. piezoelectric

TYU 5.6

1. voltage, chemical reaction
2. Primary
3. cathode
4. Carbon
5. sulphuric acid
6. 2.2 V
7. 1.2 V
8. 1.26
9. 1.15
10. See page 331

TYU 5.7

1. algebraic sum, zero
2. 1.4 A flowing away from the junction
3. 11 A flowing towards the junction
4. algebraic sum, zero
5. 6 V
6. R_4 and R_5
7. R_2 and R_5
8. 12 V
9. internal, falls
10. 0.36 Ω

TYU 5.8

1. 0.138 Ω
2. 504 Ω to 616 Ω
3. 1.5 kΩ, ±5%
4. (a) 47 Ω (b) 4.71 Ω
5. See pages 334 and 336
6. 14.1 V
7. 20 mA
8. Wheatstone Bridge, balanced
9. 0.965 Ω
10. increases

TYU 5.9

1. rate, work
2. energy, Joule, second
3. heat, resistor; light, lamp; sound, loud-speaker
4. 5 W
5. 3 kJ
6. 388.8 kJ
7. 224 W
8. 0.424 A
9. 3.136 Ω
10. 75 kJ

TYU 5.10

1. charge, voltage
2. 44 mC
3. 0.05 V
4. charge stored
5. electric field
6. 2 mJ
7. vacuum
8. (a) 1.33 μF (b) 6 μF
9. 39.8 nF
10. (a) 19.7 V (b) 43.3 V

TYU 5.11

1. current, magnetic field
2. Weber, Wb
3. Tesla, T
4. 0.768 N
5. $B = \Phi/A$
6. 0.8 mWb
7. directly, inversely
8. See page 377
9. 2,270
10. See page 379

TYU 5.12

1. magnetic, conductor, induced, conductor
2. See page 379
3. See page 381
4. See page 381
5. See page 381
6. 1.326 V
7. −120 V

8. See page 382
9. 88.5 mH
10. 1.15 A

TYU 5.13

1. conductor, e.m.f., induced
2. 6 V
3. slip rings, brushes
4. 180°, commutator
5. 90°
6. 200 mV
7. 0.012 N
8. See pages 394 and 395
9. See page 394
10. See page 395

TYU 5.14

1. zero
2. 0.636
3. 1.414
4. 0.707
5. 25 Hz
6. 2 ms
7. effective
8. peak
9. heat
10. See page 398

TYU 5.15

1. current, voltage, 90°
2. (a) 1.81 kΩ (b) 36.1 Ω
3. (a) 7.54 Ω (b) 1.51 kΩ
4. 0.138 A
5. 200 Ω, 1 A
6. 0.108 A, 10.8 V (resistor), 21.465 V (capacitor)
7. 9.26 A
8. 66.5°, 0.398
9. 330 W
10. 534 VA, 0.82

TYU 5.16

1. See page 414
2. laminated, eddy current

3. See page 417
4. 27.5 V
5. 550 turns
6. 1.023 A
7. See pages 414 and 415
8. See page 417
9. 0.0545 (or 5.45%)
10. 96.1%

TYU 5.17

1. See page 419
2. See page 419
3. 2.12 kHz
4. 115 kHz, 185 kHz
5. π-section low-pass filter
6. output, 0.707, input
7. 1.414 V
8. 5.11 mH
9. See pages 420 and 421
10. 462 Ω

TYU 5.18

1. See page 423
2. 1,200 r.p.m.
3. See pages 425 and 426
4. See page 425
5. 381.04 V
6. 69.3 V
7. 6.93 A
8. 4.056 kW
9. 2.286 kW
10. See pages 428 and 429

TYU 5.19

1. See page 430
2. See page 432
3. See page 432
4. See page 433
5. (a) 0.0277 (b) 2.77%
6. 11,784 r.p.m.
7. 6.9%
8. See page 436
9. See page 436
10. See pages 437 and 438

TYU 6.1

1. (a) earth (b) variable resistor (c) zener diode (d) lamp (e) PNP transistor (f) electrolytic (polarised) capacitor (g) AC generator (or signal source) (h) pre-set capacitor (i) jack socket (female connector) (j) fise (k) coaxial connector (l) microphone (m) transformer (n) inductor (o) motor
2. See page 452
3. See page 453
4. (a) zener diode (b) light emitting diode (LED) (c) NPN Darlington transistor
5. D_1, C_1, D_2, M_1
6. D_1 and C_1
7. R_2 and D_2
8. See page 452
9. See page 454 and Example 6.2
10. (a) 159 Ω (b) 46.9 Ω

TYU 6.2

1. (a) zener diode (b) light emitting diode (c) variable capacitance diode (d) light sensitive diode (photodiode)
2. anode, cathode
3. 0.6 V, 0.2 V
4. See page 461, silicon diode
5. (a) 46.4 Ω (b) 27.8 Ω, see page 461 and Example 6.4
6. See page 463
7. See page 467
8. See page 470
9. See page 475
10. fast switching, switched-mode power supplies, high-speed digital logic (any two)

TYU 6.3

1. (a) NPN bipolar transistor (b) PNP Darlington transistor (c) N-channel enhancement mode MOSFET (d) N-channel JFET
2. positive
3. gate, source, drain
4. forward, reverse
5. 1.25 A
6. 24
7. (a) 34 μA (b) 19.1 kΩ (c) 2.7 kΩ, see Example 6.9

8. 98.9, see Example 6.11
9. See page 490
10. See pages 487, 488 and 489

TYU 6.4

1. (a) two-input AND gate (b) inverter or
   NOT gate (c) two-input NOR gate
   (d) three-input NAND gate
2. AND gate
3. (a) low-power Schottky TTL (b) buffered
   CMOS
4. (a) TTL: 0 V to 0.8 V, CMOS: 0 to $\frac{1}{3}V_{DD}$
   (b) TTL: 2 V to 5 V, CMOS: 2/3 V_{DD} to V_{DD}
5. See page 498
6. See page 499
7. 200
8. See pages 504, 505 and 506
9. See pages 507 and 508
10. See page 506 and Example 6.21

TYU 6.5

1. FR-4
2. width of copper track, thickness of copper
   track, permissible temperature rise
3. See page 512
4. advantage: low dielectric constant (therefore excellent at high frequencies), disadvantage: very expensive
5. false, see Table 6.18 on page 513
6. false, see page 514
7. See Table 6.18 on page 513
8. See pages 513 and 514
9. See Table 6.18 on page 513

10. etching; drilling; screen printing of com-
    ponent legend; application of solder resist
    coating; application of tin-lead reflow
    coating

TYU 6.6

1. See page 516
2. See page 517
3. rotary potentiometer; light dependent resistor (LDR); float switch; resistive heating
   element
4. input; input; input; input; output
5. analogue; analogue; digital; analogue
6. See pages 523 and 524
7. See page 524
8. S_1 and S_3
9. See page 529
10. See pages 529 and 530

TYU 7.1

3) TAS = 209.2 m/s
4) Temperature at altitude = 242.5 K
5) Local speed of sound = 319 m/s
6) Then from question 5, Mach no. = 1.0
10) (i) μ, Ns/m^2 (ii) v, m^2/s
12) $q = 1240.92$ N/m^2

TYU 7.2

3) (a) Drag = 2395 N and Lift = 34251 N
   (b) Range = 71.505 km
4) $W = 69105$ N
5) $V = 152.9$ m/s
6) Lift = 140 kN

Index